Social Problems

Readings with Four Questions

Social Problems

Readings with Four Questions

FOURTH EDITION

JOEL M. CHARON
Minnesota State University Moorhead, Emeritus

LEE GARTH VIGILANT
Minnesota State University Moorhead

Australia • Brazil • Japan • Korea • Mexico • Singapore • Spain • United Kingdom • United States

WADSWORTH
CENGAGE Learning™

Social Problems: Readings with Four Questions, Fourth Edition

Joel M. Charon and Lee Garth Vigilant

Editor: Linda Schreiber-Ganster

Acquisitions Editor: Erin Mitchell

Assistant Editor: John Chell

Editorial Assistant: Mallory Ortberg

Media Editor: Melanie Cregger

Marketing Manager: Andrew Keay

Marketing Assistant: Sean Foy

Marketing Communications Manager: Tami Strang

Content Project Management: PreMediaGlobal

Design Director: Rob Hugel

Senior Art Director: Caryl Gorska

Print Buyer: Rebecca Cross

Rights Acquisitions Specialist: Roberta Broyer

Production Service: PreMediaGlobal

Copy Editor: PreMediaGlobal

Cover Designer: Riezebos Holzbaur/ Tim Heraldo

Cover Image: ©Michael Turek/Getty Images

Compositor: PreMediaGlobal

Library of Congress Control Number: 2010942661

ISBN 13: 978-1-133-31824-8

ISBN 10: 1-133-31824-X

Wadsworth
20 Davis Drive
Belmont, CA 94002-3098
USA

Cengage Learning is a leading provider of customized learning solutions with office locations around the globe, including Singapore, the United Kingdom, Australia, Mexico, Brazil, and Japan. Locate your local office at **www.cengage.com/global.**

Cengage Learning products are represented in Canada by Nelson Education, Ltd.

To learn more about Wadsworth, visit **www.cengage.com/wadsworth**

Purchase any of our products at your local college store or at our preferred online store **www.cengagebrain.com.**

Instructors: Please visit **login.cengage.com** and log in to access instructor-specific resources.

Printed in the United States of America
1 2 3 4 5 6 7 15 14 13 12 11

We would like to dedicate this book to our families,
Susan Charon
Andrew, Bridget, Quincy, and Corbin Charon
Daniel Charon and Natalie Desch

Ayuko Vigilant
Jonah and Aya Vigilant
Mark and Madelaine Vigilant

They contributed to this work in a multitude of ways. Thank you.

Contents

Preface

Like many sociologists, we, the authors, Joel Charon and Lee Vigilant were drawn to sociology because we were disturbed by the social problems we saw around us. We believed that the science of sociology would help us understand why these problems existed and perhaps show us how they could be solved. We wanted to help bring about a more just and fair society. We came to realize that no matter what we thought the world should become, we must try to be as objective as possible in our understanding of its problems. We also came to realize that we knew very little about society and its social problems, and that it would take patience and a critical attitude to overcome our own ignorance before gaining even a partial understanding.

We believe that social problems should be the concern of everyone in society, not social scientists and politicians. Sociology should give students the vocabulary, the theory, and the critical thinking skills to understand and explain not only social problems in general but also their own life situations in particular. This is a reflexive approach because we believe the sociological study of social problems should reflect a concern for how people are affected by social problems. The true gift of sociology is that it helps us to see the "small picture" by assisting in our understanding of how social problems have affected our life experiences thus far and our future life chances.

PURPOSE OF THIS BOOK

The purpose of this reader is to examine society and its problems *critically* and *sociologically*. By *critically*, we mean that these readings should encourage us to recognize the biases that we have developed as individuals, as groups, and as a society, and to question our most basic beliefs about society and its problems. We are all victims to some extent: victims of our cultural bias. If that cultural bias is not faced and understood, it will inevitably get in the way of our understanding and dealing with social problems intelligently. By *sociologically*, we mean that the authors of the readings regard society itself to be the source of these problems. The advantage of a collection of readings over a standard textbook is that it can display a wider range of problems and diverse approaches; students gain an opportunity to examine a number of perspectives rather than a single individual's views. Social problems such as discrimination, oppression, and exploitation both fascinate and disturb us. As sociologists, we want to understand them; as people who seek a better society, we want to deal with them in a cogent manner. As teachers, we believe that learning is accomplished in part through organizing what is to be learned. Students are often told simply to read a collection of unrelated articles and to figure out what is important. We have sought a useful structure for understanding what the authors are saying about society, one that might help students read, understand, listen to, or express an intelligent opinion about any social problem they might encounter.

CHANGES TO THE FOURTH EDITION

The fourth edition of *Social Problems: Readings with Four Questions* comes with substantial changes over the previous one. The work for the fourth edition began by reading the comments, criticisms, and constructive suggestions from professors who were using the third edition. Of particular interest were the articles they found intellectually engaging for their students. Our next concern was how to make the fourth edition a *significant* improvement over the third while ensuring that the structure of the text would be similar to what instructors have come to expect over the previous editions. With this mind, the fourth edition of *Social Problems: Readings with Four Questions* comes with **25 new articles**, each with a renewed emphasis on inequalities of race, social class, and gender as well as other intractable and topical social problems such as consumerism, the immigration conundrum, terrorism and its political responses, our current health care dilemma, and natural disasters as a social problem.

The fourth edition includes two entirely new sections on the social problems of American culture and with immigration. The section on American culture takes on the current economic crisis by linking it to the American Dream ideology, to the cultural goal of profligate consumerism, and to market-dominant values that have seeped into all arenas of our life-worlds. The section on immigration introduces students to the major controversies and theories that leading immigration scholars are debating: discourses that will broaden their understanding of this important part of American history and contemporary society.

The fourth edition bolsters our emphasis on violence in all of its forms: from the symbolic violence of chronic joblessness, poverty, and entrenched deprivation, to the brutality of "natural disasters" that always impact the poor and vulnerable members of society more intensely than other groups; to the gender-based physical violence of rape and sexual harassment, and the global violence of "wars without exits" and genocide.

The authors of this reader were keen to ensure that the organization of the articles and their chapters continue to match leading social problems textbooks so that professors can bundle this book with any of the popular social problems textbooks on the market today. Finally, the questions that accompany each selection provide a cogent guide to assist students in grasping the focal point or central argument of each reading.

FEATURES AND ORGANIZATION OF THE READER

This reader's structure is sociological: all problems have their origin in society. It is also critical: social patterns are analyzed to ascertain how society actually works. The reader is organized around four central questions:

What is a problem?

What makes a problem a *social* problem?

What *causes* a social problem?

What can be done about a social problem?

These four questions should help students determine the central issues in each article.

This text is divided into 13 parts. These parts are prefaced by a brief summary of the articles to follow, and each article is introduced by a list of topics covered and the four basic questions: What exactly is the problem? How is it a *social* problem? What does the author tell us is the cause of the problem? Does the author suggest what to do about this problem? Finally, the selections are followed by discussion questions that allow students to respond to the central idea of the reading and to formulate their own opinions and critical thoughts on the readings.

SUPPLEMENTS

Print Resources:

- Instructor's Resource Manual:

This instructor's manual provides Internet, video and print resources related to social problems to stimulate student interest and understanding. This resource manual provides additional information on social problems, social policy, current social movements, and ways to become involved.

Online Resources:

- The website for all Wadsworth sociology texts is located at http://academic.cengage.com/sociology.
- InfoTrac College Edition with InfoMarks!:

It is possible to bundle free with this text a four-month access to InfoTrac College Edition's online database of more than 18 million reliable, full-length articles from 5,000 academic journals and periodicals (including *The New York Times, Science, Forbes*, and *USA Today*). This includes access to InfoMarks—stable URLs that can be linked to articles, journals, and searches. InfoMarks allows you to use a simple copy-and-paste technique to create instant and continually updated online readers, content services, bibliographies, electronic reserve readings, and current topic sites. Consult your Wadsworth Cengage sales representative for details.

ACKNOWLEDGMENTS

We wish to thank the following instructors and colleagues who generously provided reviews for the preparation of the fourth edition: Anissa Breaux-Schropp, Ph.D., University of North Texas; Buffy Smith, Ph.D., University of St. Thomas; Wilbrod Madzura, Normandale Community College; Tony S. Jugé, Ph.D., Pasadena City College; Bennett M. Judkins, Emory and Henry College; Josh Packard, Ph.D., Midwestern State University.

PART I

An Introduction to Social Problems

For many students, one of the attractions of sociology is a passion to understand social problems in order to make the world a better place. This inspiration is what sociologist Peter L. Berger describes as "a Boy Scout view" of sociology. Sociology, of course, is much more than this once people study it, but this "social problems" inspiration remains part of what we sociologists are and part of what the discipline has become.

It goes without saying that every society has social problems and that social problems will never cease to exist. We should also recognize that social problems are not usually obvious to people, that they are not easily solved, that they are caused by many complex social conditions, and that their origins are almost always in society itself. The topic of social problems is a difficult one in part because it is hard to understand exactly what social problems are, because people are often not familiar with the sociological approach to understanding social problems, and because sociologists differ from one another as to which problems are significant.

Part I comprises two selections. Each one attempts to introduce and explore the four questions asked throughout this book: What is a problem? What makes a problem a *social* problem? What causes a social problem? What can be done about a social problem?

Joel Charon, one of the editors of this reader, wrote the first selection to highlight these questions and lay out systematically some of the issues involved in trying to answer them. He also introduces three perspectives that sociologists take toward studying social problems: the *conflict perspective*, the *consensus perspective*, and the *interactionist perspective*. Throughout this selection he emphasizes the importance of understanding all social problems in the context of the larger society—its social patterns, its social forces, and its social conditions.

Joel Best has one basic goal in the second selection: to criticize the use of war metaphors in relation to social problems. For example, he examines the so-called "war on drugs" and "war on poverty." Since 9/11, our society now has a "war on terrorism." What is misleading in this metaphor is the idea that social problems should be and can be "solved," that we can ultimately "defeat" them and be "victorious." Through criticism of this metaphor, Best explores this book's four questions, describing what problems are, what social problems are, what causes social problems and, most important, what can be realistically done about social problems.

These two selections raise many issues, and they do not necessarily agree on the answers to the four questions that structure this book. Together, however, they introduce us to some of the possible approaches to these questions:

What is a problem?
What makes a problem a *social* problem?
What causes a social problem?
What can be done about social problems?

1

An Introduction to the Study of Social Problems

JOEL M. CHARON

The Four Questions

1. What is a problem?
2. What is a *social* problem?
3. What is a social cause?
4. How can we solve social problems?

Topics Covered

Social problems
Values and goals
Objective and subjective problems
Social cause
Social condition
Social issue
Conflict perspective
Consensus perspective
Interactionist perspective
Social pattern
Solving problems

WHY BOTHER?

In 1759, a man named Voltaire published a book in France. It was called *Candide*, and it had a great and lasting effect on European intellectuals. It is still read in universities today all over the world, being one of the most outstanding satires in the history of thought.

Candide is the name of a naive young man. To him, as bad as the world was, it still was "the best it can be." His teacher taught him a philosophy of optimism: to accept the world as it is, to believe that God meant the world to be as it is, to always look on the bright side. Candide saw instances of evil, including murder, rape, robbery, and poverty, yet he always declared "the world is the very best it can be." Terrible situations were made into acceptable and even good ones. Like Candide, Europe was also naive, Voltaire believed; Voltaire would say that many of us today are naive.

This optimistic philosophy might be tempting for many of us to accept, but according to Voltaire and other philosophers who wrote during his time, humans should not accept evil in the world, and they need to address and solve social problems. Progress is possible, and we should work toward it. This philosophy became the basis for democratic thought, which led to the modern search for a democratic society. This philosophy also had a lot to do with the development of sociology in the late 18th and early 19th centuries. It is the basic inspiration for the study of social problems as well as the efforts of many people in the world today to work to shape a better world for all.

For many people, the study of social problems has little use, either because, like Candide, they do not recognize the problems, or because, like Candide, they believe that nothing can be done about them. Others might understand the existence of problems and even the possibility for improvement, yet they have other priorities in life.

I would like to make a simple case for studying social problems, a case that I have tried to make explicit through my teaching and writing.

- From a *moral standpoint,* social problems need to be identified, understood, and dealt with because a large number of people are being seriously harmed.
- From a *democratic standpoint,* social problems need to be identified, understood, and dealt with because democracy's most basic concern should be a commitment to bettering the human condition.
- From a *societal standpoint,* social problems need to be identified, understood, and dealt with because they may seriously threaten the continuation of society as we know it.

This book is meant to be an introduction to social problems. It is one sociologist's attempt to identify what he considers to be serious social problems according to his personal understanding and value system. I know that many readers will, and should, disagree with my choices partly because their values or understandings are different from mine. But I believe that these articles illustrate well a sociological view of social problems; they are also insightful and well written and will inspire thinking and discussion.

Throughout the book readers must continue to ask themselves: Why should I care? What difference does it make anyway? Is understanding all this a waste of my time, or is there something important here that I should know and think about?

In order to help you understand social problems and to press you to consider their importance, I encourage you to examine each selection from the standpoint of four questions. Sometimes the author will answer these four questions directly; sometimes the four questions, although not addressed directly will be implied by what the author writes. You should therefore try to infer how the author might address these questions.

Please remember: These selections are usually from previously published books and you do not have the entire book in front of you, so please do not blame the author for not answering the four questions directly. Of course, this should not let the author off the hook either; be prepared to criticize the author if his or her arguments are not clear or convincing or if the questions are poorly answered.

Here are the four questions that address each selection in one way or another. These questions should help you understand what this book is all about and, I hope, inspire you to understand and think about social problems.

1. **What is a problem?** What makes something into a problem? Specifically, what is the problem in this selection that the author is identifying, and do I agree that it is a problem?
2. **What makes a problem a *social* problem?** What distinguishes a *social* problem from other types of problems? Specifically, is the problem described in this selection really a *social* problem?
3. **What causes a social problem?** What conditions or forces in society create social problems? Specifically, what is the author identifying as the cause or causes of the problem, and do I agree that these are the causes?
4. **What can be done about a social problem?** Can the problem be solved? Can we—as individuals, groups, or society—successfully do something about it? Specifically, does the author suggest a way of dealing with the problem, or, at least, does the author imply a strategy for making it less serious?

QUESTION 1: WHAT IS A PROBLEM?

As you will see throughout this book, it is not easy to define what constitutes a problem for an individual, for a group, or for a society.

Values and Problems

A problem depends, first of all, on people's *values.* What are values and how do they relate to problems?

Values are the cherished beliefs of what is good that people are committed to. They are broad, abstract guides to what people do. They have the force of "right," "good," "valuable," or "desirable," and underlie an individual's belief as to what life "should be." Values shape people's actions. If people believe in goodness, they try to act morally. If they value materialism, they try to accumulate material goods. If people value freedom, they might fight for the right for all people to think and act as they choose.

Groups develop values, as do societies. Individuals are socialized to accept these values, and most do although some do not. Sometimes we are punished because we do not accept the values of our society; too often, we do not critically evaluate its values. In a complex society such as the United States, it might seem as though we agree on few values, but even here, there tends to be general, although not perfect, agreement about some values. In my own analysis of American society, I believe that we tend to value materialism, the present or the immediate future, physical beauty, individualism, the free market, and the family. Some would say we value equality, education, creativity, and freedom of thought, but I am less certain that these are important to many of us, especially when we try to define what these values are. Some groups and societies value war, oppression, slavery, violence, and power. Others value God and salvation above all else, or friendship, order, or the search for truth. Values are what people consider to be worthwhile; values give meaning to people's lives and direction to their thoughts and actions.

Values are important to defining problems. *A problem is an existing condition that is inconsistent with or threatening to our most important values.* People can never totally agree on what constitutes a problem because they have different values. A problem for some is inevitably seen by others as a positive quality.

A problem is not simply something that objectively exists "out there" that reveals itself to people; instead, it is a condition that is "wrong" because it violates or threatens people's values. To those who value a liberal arts education, the growing emphasis on education aimed at career placement is probably a problem. They do not believe that this is what education should be. To those who value order, extremist attacks on the courts and law might be a problem. They do not believe that people should act this way if we are going to have order and cooperation in society. Because what is and is not a problem depends on violations of values, there is always a subjective aspect to what a problem is—for the individual and for the group or society. For many people, abortion is a problem; for others, threats to abortion rights is a problem. For many people, violation of the law is a serious problem; for others, unjust laws are the problem.

Goals and Problems

Goals are ends that people work toward. People establish goals and tend to organize their actions to achieve these goals. In most cases, goals are not achieved completely; sometimes a serious condition gets in the way and makes a goal difficult to achieve. This condition becomes a problem to the extent that it interrupts the actor's achievement. In society, a condition that prevents the achievement of a goal that many people hold important is a problem.

Goals, like values, are guides to our actions. They are specific and practical and are often guided by values, which are more general, more morally based, more cherished, and more long range. Goals can be achieved, whereas values are continuous guides. Values guide our actions and our goals; goals organize our actions. We might have a strong commitment to education (a value), so we might try to choose classes in which we are going to learn something that will be useful for our understanding the world (a goal). On the other hand, we might value being successful in the world of computer science, so we might try to seek classes that are relevant to our career and necessary for our graduating on time. We might value contributing to the lives of others and decide to act toward a degree in social work.

Goals and values influence what we do and determine the problems that exist for us. Problems are either threats to or violations of values that the individual, group, or society believes in, or they are conditions that stand in the way of people achieving their goals. The individual holds values and pursues goals; problems arise in relation to these. A group or society also holds values and pursues goals; social problems arise in relation to these.

Any problem is therefore relative, relative to the particular values and goals held by an individual, group, or society. People will ultimately disagree as to what constitutes a problem in large part because each has a different set of values and goals.

Throughout this book, the authors of the selections identify what they consider to be a problem; always basic to their judgments are their own values and goals. Each one is trying to persuade you that these are important problems to solve. You might or might not share the author's values and goals. Of course, every article in every magazine that describes a "problem," every book, every sermon in church, and every lunchroom discussion that describes a "problem" assumes a value orientation or a goal to be achieved. This is one of the central points of this book.

Difficulties in Identifying Problems

It is often very difficult to pin down the problem exactly. We know something is "wrong," but we do not understand what it is, or we miss the "real problem." Something violates our values or prevents us from achieving our goal, but we aren't really sure what it is. For example, if we believe in justice, then we must try to determine if the problem is racism, segregation, prejudice, or poverty. Perhaps something like laissez-faire capitalism is really the problem or maybe it is poverty. The first step in understanding social problems is to identify what the problem is and why we think it is a real problem. Even if we believe in justice, we still might miss the problem, the majority of people might miss it, academics might miss it, and government officials might miss it.

It is important to recognize that problems exist because of a complex mixture of subjective values and goals on one hand and careful study and understanding of the situation as it actually is on the other. We must realize that our values and goals are different from those that other people hold, which makes the problem subjective; we also must realize that the so-called problems we identify might not be problems at all or might be irrelevant to our values or goals.

The first question—what is the problem?—must be identified if we wish to understand social problems. Only then can we really go to the second question.

QUESTION 2: WHAT MAKES A PROBLEM A *SOCIAL* PROBLEM?

Some problems are social, and some are individual or natural, such as earthquakes or brain genetics. A social problem:

1. Must be social in origin, caused by social factors. The more its causes are social, the more it becomes a social problem.
2. Must be harmful to many individuals. The greater the harm, and the greater number of individuals, the more it becomes a social problem.
3. Harms society in the sense that it threatens what we believe society should be. The more threat to our view of society, the more it becomes a social problem.

Some problems are caused by factors besides social ones. For example, earthquakes, genetics, and a physical illness can cause our problems. An automobile accident might be caused by daydreaming or falling asleep at the wheel. I may fail a test because I did not study. I may act out of impulse because of a mental disorder that arises in my genes. To the extent it is not caused by social causes, the problem is not social. Only as we begin to see the role of society, social patterns, culture, and social inequality, for example, we begin to see a social problem.

Some problems affect only a small number of people, or the harm to people is relatively minor. At one time we believed that smoking hurt only some people, and usually not seriously. As we studied the problem, we realized that large numbers of people were being affected and that the harm was deadly for many. To the extent that numbers of people are harmed, and to the extent that the harm is serious, we begin to identify a social problem.

Some problems do not harm society. The death of a child may be serious, family and friends mourn, but it does not necessarily harm society. However, the deaths of children through parental abuse affects society itself, and people recognize that allowing this abuse to continue may affect society as we wish it to be. We might believe that society is going in the direction of disorder, or immorality, or inequality, or oppression, for example. To the extent the problem we identify harms society, to that extent we begin to identify a social problem.

Normally but not always, these three points are interrelated: *The more the origin/cause is social, the more people are affected and society itself is affected.* If crime, for instance, is caused by social conditions, then large numbers of people will be affected, and society will be threatened in some important way. Furthermore, *the more people who are affected, the more likely the cause is social and society is affected.* Thus, if large numbers of people are experiencing unemployment, poverty, crime, or alienation, the cause is likely to be social and there will be a negative effect on society. And finally, *the more society is affected, the more likely the cause is social and many people are affected.* Thus, if society is negatively affected because people are not able to read, then the origin of the problem is undoubtedly social, and large numbers of people are hurt by that fact.

A problem is a social problem, then, to the degree that its causes are social, to the extent that large numbers of people are affected, and to the extent that conditions threaten society as we know and accept it. Crime is a serious social problem if its causes are overwhelmingly social, if large numbers of people are harmed one way or another, and if the social order is undermined and institutions are threatened.

There is another matter to consider here. Even though a social problem is objectively something that exists because it is harmful to people, threatening to society, and caused by social forces, most sociologists consider one more point in its definition. *A full-blown social problem also relies on an agreed-upon recognition by a large number of people that something is wrong and needs to be corrected.*

This might be clearer if we distinguish between a social *condition,* a social issue, and a *social problem.* A social condition exists as a neutral quality identified in society: for example, globalization, a market economy, marriage, divorce, class distinctions. It evolves into a social issue when people begin to debate whether or not the condition is a problem. The social issue evolves into a social problem when those who consider it a problem are able to persuade enough people that it is, and a consensus develops among many people that something needs to be done about it. There is a *contest of meaning,* a contest among people to identify a social problem. Always, social power is important in this contest, and numbers of people, the media, wealth, people in positions in authority, charismatic leaders, well-organized groups willing to work for their views, elections, and many other resources can be used to persuade people to commit societal resources to the problem identified.

A social problem is said to exist to the extent that

- the cause of the condition is social.
- a social condition seriously harms large numbers of people in society.
- a social condition harms society as we believe it should be.
- people agree that the condition should be changed.

Sociologists, like others, disagree on whether a certain condition is social. In part it is because they use different perspectives in studying social problems. The following three key sociological perspectives—*conflict, consensus,* and *interactionist*—each view society and social problems differently.

1. **The Conflict Perspective:** This perspective emphasizes social conditions that cause harm to people, especially societal conditions that create poverty and inequality of class and power.

Society is defined as a conflict of various interests, and whereas some people are able to meet their needs and desires in society, others are systematically excluded and harmed. Sexism, racism, and harmful economic conditions are especially singled out as problems and causes of other problems.

2. **The Consensus Perspective:** This perspective tends to emphasize those social conditions that threaten the continuation of society as it is. Its major concerns are too much disorder in society, too little consensus, and too few institutions that work well to uphold society as we know it. The questions here are "What is going to happen to us as we become less and less committed to society and as institutions fail to do what they have traditionally done?"
3. **The Interactionist Perspective:** Also known as *constructionism,* the interactionist perspective highlights how social conditions become social problems through communication (interaction) and definition. The interactionist emphasizes the idea that social problems exist because certain conditions are identified in society as unacceptable. The question for the interactionist interested in social problems is "How do people successfully influence others to accept what they regard as a social problem?"

QUESTION 3: WHAT CAUSES A SOCIAL PROBLEM?

What do we mean when we say that a problem has a social cause? A good deal of the sociological approach to social problems focuses on cause. This is a complex issue. Cause will become a critical subject in every selection in this book and will usually become the most controversial topic in any discussion of social problems.

In the above discussion we included social cause as one of the central points to defining a social problem. Let us examine this issue more carefully.

Do most people believe that social forces cause social problems? There is a great deal of research to support the idea that Americans shy away from believing that something in society causes social problems. They hesitate to "blame it on society." Instead, they try to explain social problems by focusing on three other causes:

1. Many people believe that the victims of social problems cause the problems. They think that poor personal choices cause most of the problems that exist in our society. People *choose* poverty, people *choose* crime, people *choose* to have unsafe sex, people *choose* to turn away from God, people *choose* to abuse alcohol or drugs. Such an approach ultimately means that society or social conditions do not cause social problems; rather, individuals making free choices in their lives cause them. This leads many people to believe that somehow people "make their own beds," that they create their own problems, and that society does not need to be changed to correct what individuals bring on themselves.
2. Some people attribute social problems to evil people. One argument is that *all* people are naturally evil, and thus wars, cruelty, selfishness, and so on are the natural results of the evil in everyone. Little can be done because people always will be bad. Another argument is that *some* people are evil, and these evil people create societal problems. "Crazy" people create problems, or selfish people, or stupid people, or immigrants, or poor people, or Jewish people, or African Americans, or the rich and powerful.
3. Some people will attribute social problems to biological, physical, or psychological causes. Thus, societal problems have their source in individuals who are brain damaged, who have a chemical imbalance, or who have serious personality disorders. These people argue that to change society we need to cure or rehabilitate such individuals.

To the sociologist, social problems are not primarily caused by personal choices made by individuals or evil people, by biology, or by the physical world or psychology; social problems arise from the nature of society itself. They are caused by the operation of society, the limitations of its workings, and its qualities that act on many individuals. Émile Durkheim, an important sociologist who wrote in the early part of the 20th century, named these social origins of behavior *social facts,* what we might today call social forces or patterns that exist in reality and act on individuals. How people act in a society results from the nature of that society; the social problems created in a society also result from the nature of that society. Durkheim was adamant: It is a mistake to try to understand social problems as resulting from individuals. Explaining social problems requires an understanding of social facts.

To argue that people make free choices and cause their own problems is to ignore cause completely. Of course, our individual choices are important, and to some extent they might even be free, but always various factors, some social and some not social, enter into every "free" choice.

There is, of course, nothing wrong with arguing that evil people might cause problems, but then the questions become, Why are some people evil and others not? Or are all people equally evil? What influences the existence of evil people? Where do they come from? Evil people exist in the world, at least in part, because of their specific life circumstances, their upbringing, their biology, their social life. Many people are tempted to blame social problems on other people who are different from themselves, calling them evil.

Psychologists, biologists, and physical scientists are correct when they maintain that human problems arise because of individual characteristics, but these deal with *individual* problems, not social problems. If large numbers of people share a problem, making the problem no longer simply individual but widespread in society, how can we explain this without examining social foundations? People may have miserable lives because of personal qualities that they do not simply choose to develop, but how do we explain why so many people may share these qualities at a particular time and place? We must begin to see that such qualities are linked to a larger context—a social one.

It is too simple to stop with the individual and refuse to understand this larger social context. This allows us to ignore cause entirely, or it turns our attention to seeking the cause of an individual behaving as he or she does rather than asking what causes *so many* people to behave in a particular way. The more people who share a problem, the less likely we can understand cause without looking at social conditions.

As soon as we start seeing the larger social context of cause, two dramatic changes occur. First, there is a strong tendency for our understanding of cause to become broader and broader until we come to see society itself as cause. We might first see cause in the immediate social surroundings of the individual—for instance, in parents, friends, or school. But then if we are careful and honest, we encounter broader issues: family life in society, the role of friendship in people's lives, the success and failure of the American system of education. The fact is that Jack's or Jill's troubles at high school are not simply a result of their particular isolated families or neighborhoods. Their personal troubles are linked to broader social problems, or social issues. Many Jacks and Jills in the United States today share a number of troubles. What are they? What in our society contributes to an epidemic of a certain problem? We might feel threatened by allowing the social context into our causal analysis: It will ultimately reach society itself.

The second change is that by examining the social we begin to recognize the complexity of cause and the interdependence of a number of problems in society. If you want simplicity, do not start down this road of social context, for you will find everything interlinked. You will find that each identified social problem causes other social problems and that each identified social problem results from other social problems.

This is a complicated issue; let us look at an example to illustrate. If someone asks me what the most serious problem facing American society is today, I might say poverty. They ask, "What is the cause of poverty?" I might answer, "A class system that assures that some people remain poor." "And what is the cause of this class system?" I might say, "A set of institutions that protects the wealthy and does not respond to the poor." "And what causes these institutions to exist?" I might say, "These institutions have over time become biased toward rewarding those who are successful rather than those who are struggling to survive and improve their position." What exactly *is* the social problem here, and what is its cause? One might argue successfully that poverty is a social problem, but so is the class system and so are the institutions that contribute to poverty. And each of these, we can argue, causes other problems as well as being the outcome of others. It is naive to identify a problem and a cause without looking at them in the larger context of other social problems. Ultimately, the cause of all social problems becomes some aspect of society. This is the essence of the sociological approach to social problems.

Karl Marx believed that the real problem and the cause of every other problem is capitalism. Of course, Marx is too simple here in his analysis, but his point should not be lost. He identifies one huge problem—capitalism—and argues that this problem is the cause of all other problems. We might ask, Why does capitalism exist? What are the social processes that created it? And, as we answer these questions, we will begin to uncover more causes, which can also be seen as social problems. Joe R. Feagin (1990: xiv–xv), who has contributed much to the sociological approach to understanding social problems, addresses this same point:

> What is often discussed in this country as separate problems are, in fact, closely interrelated. For example, the drive for ever renewed profit in capitalistic enterprises—profit whose use and reinvestment are ultimately in the hands of small capitalist elites—links together such apparently diverse problems as environmental pollution, persisting unemployment, and corporate price-fixing.... Many problems appear at first to be isolated, but on closer examination they are linked together because of their roots in the structure and process of our advanced capitalistic system.

Before we leave this issue of social cause, let us look at one more point. For sociologists, *society* usually has to do with more than simply a bunch of individuals who act together. The world of the social means more than the influence of one individual on another. Society has certain qualities; over time, it develops certain ways, certain patterns. It has a structure, a culture, and social institutions. It has a system of inequality, a set of values, ideas, and rules that govern it, and a set of ways that seem to work to keep it functioning over time. When sociologists claim that the cause of social problems is something that exists in the social world, they generally mean *social patterns* that have developed over time. Society and its ways become, in truth, the cause of society's own problems. This means two things to the sociologist:

1. Many of the social patterns that have developed work for some people, but they are unable to work well for everyone. Some people are left out; they fall to the bottom of society or through the cracks; they do not achieve in society as it has developed. Even the most successful society will develop social problems because nothing can meet everyone's needs. So, for example, although capitalism is an important social pattern in the United States and might work well for many people, one could fairly argue that it has also produced many of the social problems we face as a society. Charles Dickens in *A Tale of Two Cities* declared that for the people of 18th-century London and Paris, "It was the best of times, it was the worst of times." Why? It was prosperous for some and created horrible poverty for others. It is important to understand that although society might benefit many people, it does not benefit everyone and may actually hurt many others.
2. Many of the social patterns that have developed do not work well anymore. For instance, marriage, public schools, private medicine, huge bureaucratic corporations, the family farm, the free market, welfare payments to the poor, or the "war" on drugs may have once

been successful but may no longer work, given changes that have occurred in society over time.

Therefore, when sociologists argue that social problems are caused by society, they do not simply mean that problems are caused by other people, or by the "environment," or by "bad parents." They mean, instead, that we must search the essence of society itself to understand why such problems exist. How does society work? How do the problems we face as a society tie into the workings of society? Of course, what we are left with is what the Enlightenment taught long ago: If we are going to deal with social problems and do something about them, we need to change society. That is difficult to do, and many people do not want to change something if they believe that it works well for them.

That is why people are attracted to ideas that explain cause in terms of something other than society, and that is why it is difficult for many students to carefully consider sociology as a perspective that regards society itself to cause social problems.

As you read the selections in this book, you will find this one basic point of agreement: Society, and its various patterns that have developed over time, is ultimately the cause of social problems. What patterns are right for some people may actually hurt others. What patterns used to work may no longer work.

QUESTION 4: HOW CAN WE SOLVE SOCIAL PROBLEMS?

Believing that social problems can be solved is really a myth, and the myth is sometimes harmful. The problem of crime will never go away, nor will poverty, nor will inequality, nor will scapegoating, poor schools, racism, or destructive families.

There are three reasons for this. First, whenever we try to solve a serious problem, our definition of the problem will change as we learn more about it and are successful in confronting it. For instance, as we have dealt with the problem of child abuse, we have expanded our definition of "child abuse" to include actions that were not previously seen as abusive, such as psychological abuse, or even spanking. Another example could be pollution. Recognizing how gas-powered vehicles polluted the air, we began building "cleaner" engines; yet, even as we were successful in doing that, we turned our attention to eliminating gasoline engines altogether. Over time, we simply come to understand more about the social problem at hand and redefine it.

Second, problems are too complex to be solved. All the causes of any given problem are too complicated and interrelated to be changed successfully. Third, problems are embedded in the nature of society. Finally, solving one would mean that society would have to change dramatically. This is too much to expect and probably simpleminded to believe.

Dealing with solving social problems is like dealing with the issue of freedom. It is not an either/or thing; it is a constant effort. We are never fully successful in our fight for freedom; there are always new controls that have to be dealt with, new threats to freedom, new understandings of what it means to be free. At any time, we might begin to lose whatever freedom we have, or we might take freedom for granted and think we will always have it—only to find ourselves suddenly without it.

Trying to deal with social problems means that as a society we need to understand what is wrong and to work at making it *less wrong*. If we leave problems alone, they might get better on their own, but they will more likely get worse and might eventually cause even more serious problems or destroy whatever works well in society.

Democracy is a commitment to dealing intelligently with social problems, because democracy above all else means that all people are important and to some extent we are responsible for one another. Society's problems can be confronted and made less serious, and covering them up or running from them can bring about destructive conflict and disorder, one of the most serious social problems of all. If we demand to *solve* a problem, we will discourage efforts to deal rationally with it.

This book is not about solutions; it is about doing something to correct our most serious problems so that fewer people are hurt and a better society can be built. In some selections the author may not seem to be offering a recommendation for improvement, but if you carefully read the author's description of the problem and its cause, you will find that action is implied. In some selections changing government policies seems sensible, or relying on concerned people organizing and exerting their collective power to influence corporations or schools. In some selections the author criticizes approaches we have already tried and suggests alternatives. Almost all believe that improvement is possible, and all seem to argure that simply ignoring these problems will not make them go away.

THE FOUR QUESTIONS: A SUMMARY

The point of this book is to provide a way of approaching social problems so that we can better understand them, and it is hoped, better understand ways to deal with them. Each selection identifies a social problem. As you read, identify first what the author has to say, and then ask yourself what you believe. Try to identify exactly what the problem is, asking yourself the four questions that we have explored: Is this really a problem? Is this really a *social* problem? What does the author say causes this problem? Is there an intelligent way to make this problem less serious?

The goals are to encourage you to critically evaluate society and its problems as well as to help you form a useful approach to understanding society and its problems.

REFERENCE

Feagin, Joe R. 1990. *Social Problems: A Critical Power-Conflict Perspective*. 3rd ed. Upper Saddle River, N.J.: Prentice-Hall.

DISCUSSION QUESTIONS

1. Charon tries to make a case for why we should be concerned about social problems. Is it a good case? Is he leaving something out?
2. What values would you list as important to Americans? Can you identify some conditions in society that violate these values? Are these, in your mind, social problems?
3. How does Charon define a social problem? Evaluate his definition.
4. Social problems are caused by other social problems and ultimately by society. Explain what this means. Is this useful for understanding?
5. What is wrong with believing that social problems can be solved? Consider the problem of terrorism as you understand it. Will it ever be solved?

2

What's Wrong with Declaring War on Social Problems?

JOEL BEST

The Four Questions

1. Why is "declaring war on social problems" a problem, according to Best?
2. Why is this kind of approach to problems a *social* problem?
3. Why are "wars" against social problems occurring? What causes this kind of approach to defining problems?
4. Best suggests that we change the way we think about "solving problems." How should we change our way? Does he suggest any ways that we might change our culture so that those of us living in American society might understand social policies more realistically?

Topics Covered

Social policies

The war metaphor

Drugs

Poverty

Just as we create new crimes, new victims, and other social problems, we create social policies to deal with them. Of course, we rarely talk about "creating" social problems; we act as though our problems exist independent of our actions, that social problems are simply objective conditions. Instead of acknowledging that stalking or sexual harassment are categories created and brought to public attention through advocates' claims, we speak of "discovering" or "recognizing" these problems, as though they have always existed and we have simply ignored them up until now.

In contrast, we recognize that social policies are our creations, the products of our choices.[1] We understand that there are many ways in which society might respond to different problems, and we debate these choices: should we emphasize punishment or rehabilitation for criminals, promote curative or preventive medicine, try to eradicate or regulate vice? Most social problems inspire debates along these lines. Whatever we choose, the policies must be articulated and justified, and this means that language is an inherent feature of social policy. How we talk about policy shapes what we do, what we think we are doing, why we think we are doing it, and so on. Policy talk has consequences—often unintended, unexamined consequences. This [selection] explores the consequences of one popular form of policy talk: "declaring war" on social problems.

THE WAR METAPHOR AND SOCIAL POLICY

President Ronald Reagan's 1986 declaration of a "war on drugs" was preceded by several earlier presidential proclamations of drug wars, including one by Richard Nixon and an earlier declaration by Reagan himself in 1982. As a statement of drug policy, then, declaring war on drugs is somewhat trite. Nor is the metaphor of declaring war limited to drug policy or Republican rhetoric. Other twentieth-century presidents from both parties declared war on numerous social problems. The most familiar example is Lyndon Johnson's War on Poverty, but we may forget that Johnson also spoke of "the war against crime," or that Dwight Eisenhower called for "a new kind of war … upon the brute forces of poverty and need." Nixon declared wars on crime and cancer, Jimmy Carter called the energy crisis "the moral equivalent of war," Gerald Ford declared "an all-out war against inflation," Franklin Roosevelt's New Deal declared wars on the Depression, farm surpluses, rural poverty, and crime; and so on (Powers 1983; Sherry 1995). [And, of course, President George W. Bush declared the "war on terrorism."]

Nor is the rhetorical power to declare war on social problems restricted to presidents. Throughout the twentieth century, other federal, state, and local officials have declared war on various social problems; academics often use the metaphor to characterize social policies; activists and the press regularly call for—or make their own—declarations of war (Bricklin 1993; Gorelick 1989; Sherry 1995); and even corporations present themselves as waging war against their customers' problems (Russell 1996). The columnist George Will (1992) deflates policymaking enthusiasms with his sarcastic acronym for the "moral equivalent of war"—MEOW—while policy critics invert the term for their own ironic purposes (e.g., at least three recent books have titles that are variants of "the war on the poor" [Albelda et al. 1996; Gans 1995; Sidel 1996], just as drug-war critics denounce "the war on our children" [Trebach and Ehlers 1996]). In short, declaring war on social problems is a familiar, commonplace way of talking about social policy.

Declaring war is simply one instance of a broader tendency to use militarized language to describe social problems and social policy. References to battles, campaigns, attacks, mobilizations, enlistments, recruits, arsenals, weapons, targets, enemies, fronts, strategies, tactics, and similar military terms regularly appear in the rhetoric of policymakers, advocates who hope to influence policy, and academics who analyze policymaking (Blain 1994; Sherry 1995). This fondness for military language reveals that declarations of war are part of a more general tendency to use war as a metaphor. Critics argue that the United States is a militaristic society—that militaristic values are deeply embedded in our culture, shaping the way we understand our world (cf. Englehardt 1995; Gibson 1994):

> [After 1945] war defined much of the American imagination, as the fear of war penetrated it and the achievements of war anchored it, to the point that Americans routinely declared "war" on all sorts of things that did not involve physical combat at all.… Militarization reshaped every realm of American life—politics and foreign policy, economics and technology, culture and social relations—making America a profoundly different nation. To varying degrees, almost all groups were invested in it and attracted to it—rich and poor, whites and nonwhites, conservatives and liberals. (Sherry 1995: x)[2]

In their most extreme form, social policies can be conducted in an overtly military manner; for instance, police paramilitary units, wearing military-style uniforms and equipped with assault weapons and other military gear, play an increasingly active role in policing some cities (Kraska and Kappeler 1997), while military units conduct missions against drug smugglers. Policymakers don't just declare war; they sometimes carry out military operations against social problems.

Although there are many analyses and critiques of the current war on drugs, the War on Poverty, and other specific "wars" against social problems, the frequency with which social policymakers adopt the metaphor of warfare and the implications of that metaphor have not received as much critical attention. This [selection] begins by exploring why Americans are fond of declaring war on social problems, then examines the disadvantages of using this metaphor. It concludes by considering why we need to pay more attention to policy rhetoric.

THE ATTRACTIONS OF DECLARING WAR

All metaphors, including "declaring war," carry implications:

> Metaphors are important devices for strategic representation in policy analysis. On the surface, they simply draw a comparison between one thing and another, but in a more subtle way they usually imply a whole narrative story and a prescription for action.... Buried in every policy metaphor is an assumption that "if *a* is like *b*, then the way to solve *a* is to do what you would do with *b*." Because policy metaphors imply prescription, they are a form of advocacy. (Stone 1988:118; see also Edelman 1971:65–72)

What, then, are the implications of declaring war? What similarities between warfare and social policy do those adopting this metaphor mean to evoke? First, of course, warfare involves open conflict with an enemy, so the metaphor defines the social problem—drugs or poverty, cancer or crime—as an enemy to be fought:

> The "war on poverty" suggests massive mobilization against a universally hated enemy, and thereby helps win political support. It gives people the gratification of seeing themselves support a crusade against evil. (Edelman 1971:71)[3]

Wartime propaganda routinely transforms human enemies into evil, inhuman abstractions (Keen 1986). Because the enemy is evil, war is justified; this is what the sociologist James A. Aho (1994:11) calls the first paradox of the enemy: "My violation of you grows from my yearning to rectify the wrong I sense you have done me. Violence emerges from my quest for good and my experience of you as the opponent of good." Of course, this transformation requires little effort when the enemy is a social problem; in such cases, the enemy is already an inhuman abstraction, generally agreed to be evil. Declarations of war against social problems, then, invoke a sense of righteousness. Society can demand the support and sacrifice of its members only in pursuit of a just cause, so declarations of war must justify conflict in terms of the greater evil of allowing the problem to remain.

Of course, it can prove difficult to wage a prolonged war against an abstraction. Over time, a more specific enemies list often emerges. Official rhetoric in the current drug war, for instance, variously blames foreign farmers and domestic growers (for growing the crops from which drugs are manufactured), drug cartels and dealers, the media (for glamorizing drug use), drug users (particularly blacks living in urban ghettos), hopelessness, and "everyone who looks the other way" (President George Bush, quoted in Elwood 1994:34). Once one is committed to the war metaphor, it becomes easy to characterize all obstacles and opponents as enemies.

The opposition to and fight against these enemies is public; wars are not covert operations. Declaring war is dramatic, an uncommon step, a last resort, taken when nothing less will do. Wars therefore demand unity and commitment: "If our government is at war, be it against poverty or fraud or crime, then we are traitors if we do not support the cause. The symbol of war is an obvious tactic used by leaders to create support for their policies" (Stone 1988:121). Leaders invoke the war metaphor quite intentionally. Lyndon Johnson (1971:74) recalled that his advisers debated what to call their antipoverty policy, and he deliberately chose the "War on Poverty": "I wanted to rally the nation, to sound a call to arms which would stir people in the government, in private industry, and on the campuses to lend their talents to a massive effort to eliminate the evil." The language of unity and commitment runs throughout calls for wars against social problems: "Our country needs to swiftly launch a comprehensive war on breast cancer that attacks it on all fronts—basic research, education of the public and more innovative treatments" (Bricklin 1993:39).

Warfare presumes that fighting the enemy is a common cause of the entire society; individuals should set aside their doubts and reservations and join in the larger struggle. Society stands committed to the war effort, willing to make the sacrifices necessary to defeat the enemy: "War making is one of the few activities that people are not supposed to view 'realistically'; that is, with an eye to expense and practical outcome. In all-out war, expenditure is all-out, unprudent—war being defined as an emergency in which no sacrifice is excessive" (Sontag 1989:11). And uniting to make war strengthens a society; as Aho's second paradox (1994:15) notes: "There can be no harmony without chaos, no peace without war. While groups ostensibly fight only to secure their own short-term interests at the expense of others, the latent 'function' or unintended end of such fights is social solidarity."

Undoubtedly, these are the key elements policymakers mean to evoke when they declare war on social problems: the problem is an enemy of society at large; society is justified in fighting this enemy; society's members should rally to this cause; and they should be willing to make sacrifices for the war effort. Declaring war, then, is a call for a united, committed campaign against a social problem.

WHY WARS AGAINST SOCIAL PROBLEMS FAIL

But the war metaphor has other, more troubling implications. We need to begin by examining Americans' historical experience with war and their expectations for warfare. Beginning with the American Revolution, the United States has fought ten major wars (the Revolution, the War of 1812, the Mexican War, the Civil War, the Spanish-American War, World Wars I and II, and the Korean, Vietnam, and Gulf Wars).[4] In general, U.S. wars have been brief; aside from the Revolution and the Vietnam conflict, no U.S. war lasted much more than four years; our history does not feature "thirty years'" or "hundred years'" wars. And, in general, the United States has won—or at least not lost—its wars. Prior to Vietnam, it was common to hear that "we've never lost a war." Although Vietnam might seem to be an exception—a long conflict that ended in failure—it is important to recognize that at least a vocal segment of the public continues to insist that, had the government only been committed to the cause, willing to adopt different strategies and tactics, that war, too, might have been won in a short period of time. The Gulf War—a quick, decisive campaign with relatively few American casualties—is far more representative of the U.S. experience with war than is Vietnam.

In other words, declaring war on social problems enlists the U.S. public in a particular sort of enterprise. The metaphor invokes a complex set of meanings: a righteous cause, unity of purpose, commitment to the war's goals, and a clear-cut enemy, but also a concentrated yet relatively brief struggle, ending in ultimate victory. This is what war means to Americans; it is what we expect of war. Policymakers declare war on social problems precisely because they hope to invoke some of these expectations—commitment, unity against a common enemy. The problem is that social policies rarely live up to all that the metaphor promises.

In practice, declaring war against social problems is very different from waging international war. The war metaphor may be useful as a dramatic device for inaugurating policies: it highlights the new social policy, makes it a priority, and indicates official commitment to the cause. However, war rhetoric seems ill suited to the actual conduct of social policy.

The Problem of Complexity

Wars—at least in Americans' collective memory—usually are recalled as simple narratives: treachery by the enemy leads to a period of struggle as the nation mobilizes; then Americans begin a steady march toward victory. Of course, this simple story ignores a great deal: debates over strategies, setbacks and difficulties, complicated relationships with allies and enemies, and so on. But our underlying vision of warfare—at least from a distance—is fairly straightforward.

In contrast, social problems are not simple. Our *names* for our problems are simple—"poverty," "drugs"—but the problems those names describe are not. Social problems tend to be multifaceted. "Poverty" encompasses great diversity—poor people living in rural Appalachia and in urban ghettos, the elderly poor, the homeless, and so on. And social problems ordinarily have multiple, often hotly debated causes. Is there a culture of poverty? Is poverty an inevitable product of capitalism? Is it caused by structural shifts in the economy? How it is linked to racism and sexism? There are many competing explanations for most social problems, and probably most of them contain at least a grain of truth. The war metaphor depicts the social problem as a single enemy, but even a moment's thought reveals this view to be unrealistically simplistic.

Nixon's 1971 decision to wage war on cancer is a case in point. "Cancer" is a wishful oversimplification. Cancer is not one disease, but a broad category that includes many distinct diseases; although we speak of "breast cancer" or "stomach cancer," even these names are crude classifications encompassing varieties of malignancies affecting particular organs. In all likelihood, these diseases must be "fought" one by one as medical scientists seek to identify their causes and then search for particular therapies that are effective against particular diseases (usually at particular stages in their development). Although there might be progress in addressing particular diseases, there was, at least in 1971, no hope that a war on cancer could be won—that "cancer" could be cured or eliminated anytime in the foreseeable future. Declaring war—"the House of Representatives passed a bill mandating a cure for cancer by 1976, a present for the country's 200th birthday" (Culliton 1976:60)—created unrealistic expectations, and invited critics to quickly declare the war on cancer a failure (Beardsley 1994; Patterson 1987).

The Problem of Victory

Social policy, however warlike, is unlikely to result in clear-cut victory. In international wars, opponents suffer defeat: they sue for peace, accept terms, or surrender, publicly acknowledging their defeat and the other party's victory. Wars end—if only because the enemy can call a halt.

In contrast, social problems cannot surrender, and they rarely disappear. This raises an important question: how can we know when a war against a social problem has been won? Does victory mean that the problem has been reduced (and if so, how much of a reduction is involved? and of what sort? relative to projected increases? in absolute terms?)? Or must the problem be eradicated (what exactly does this mean? and how might it be demonstrated?)? The answers to such questions are not at all obvious.

Consider LBJ's War on Poverty. Perhaps the single most publicized war against a social problem, the War on Poverty became a target for a broad range of critics who insisted that the policy failed. Initially claims about the policy's defeat came from the left; liberal activists and social scientists were quick to announce the poverty war's failure in books such as *How We Lost the War on Poverty* (Pilisuk and Pilisuk 1973) and *Poverty: A History of the Richest Nation's Unwon War* (Lens 1969). Marc and Phyllis Pilisuk (1973:8) complained: "The war on poverty—never a real war at all—has been lost." Liberals' references to a lost war continue to appear:

> The War on Poverty didn't fail. It was called off.... In effect, [in the late 1970s] state legislators called a truce, if not a full-scale retreat. Then, in 1980 [*sic*], President Reagan's inaugural speech rallied conservatives around a war on welfare. (Albelda et al. 1996:10)

Notice how the war metaphor contains its own vocabulary of failure, easily adopted by a policy's critics: wars are lost, are called off, or end in retreat.

The policy's more recent critics have been conservatives, who insist that the failure of the War on Poverty demonstrates the limitations of government-led social policies. Charles Murray's *Losing Ground* (1984) concludes that antipoverty programs actually made things worse:

> Social programs in a democratic society tend to produce net harm in dealing with the most difficult problems. They will

> inherently tend to have enough of an inducement to produce bad behavior and not enough of a solution to stimulate good behavior; and the more difficult the problem is, the more likely that this relationship will prevail. (Murray 1984:218)

Or, as President Reagan liked to quip: "In the sixties we waged a war on poverty, and poverty won" (Lemann 1988:37).

It is possible to argue that the War on Poverty was a success. For example, Christopher Jencks (1992) insists that War on Poverty programs caused the material condition of the poor to improve and the proportion of the population living in poverty to fall. In other words, the policy worked to the degree that it reduced poverty; it made things better. But—and of course this is the critics' point—poverty did not disappear. Poor people remain: therefore, the War on Poverty failed; it ended in defeat.

Critics of the current war on drugs adopt much the same argument. They argue that the social organization of illicit markets makes it impossible to eradicate drug use, so drug wars are doomed from the start. Moreover, traffic in and use of illicit drugs continues—proof that the war on drugs has failed (see, e.g., McCoy and Block 1992; Johns 1992). Again, the rhetoric of warfare opens the door to partisan critiques. Democratic Senator Joseph Biden attacked George Bush's policy: "The President says he wants to wage war on drugs, but if that's true, what we need is another 'D Day' not another Vietnam, not another limited war fought on the cheap and destined for stalemate and human tragedy" (quoted in Inciardi 1992:271), just as Republicans in the 1996 presidential campaign charged that President Clinton had been "AWOL in the war on drugs."

Thus drug warriors find themselves sandwiched between pessimists who insist that no punitive policy can succeed and antidrug enthusiasts arguing for escalation because the existing forces, tactics, and so on aren't enough to win. Even the drug war's defenders, such as the sociologist James Inciardi, find it necessary to redefine success in narrow, ambiguous ways:

> Is the war on drugs being won? Can the war on drugs be won? Should the war on drugs be fought? To these three questions, one could answer, respectively, "Yes, to some extent, at least in the middle class"; "Perhaps, to some degree at any rate"; and "Yes, why not?" (Inciardi 1992:267)

While drug warriors can point to reductions in drug use, the policy's critics charge that these declines are not evidence of success, that drug use began falling before the current drug war started. And, more important, decline is not enough; the critics define victory as the eradication of drug use, not merely its reduction. By this standard, the war on drugs has been lost.

Note that the wars on poverty and drugs show a neat reversal in their critics' ideological postures. As a policy promoted by a liberal president, the War on Poverty was called an inevitable failure by its conservative critics, while liberals insisted that the policy could have been successful if only the government had tried harder. In contrast, when a conservative president declared war on drugs, most of the critics predicting certain failure were liberals, while many conservatives claimed that greater commitment could produce victory. Clearly, partisan politics, not philosophical principles, determines critics' rhetorical postures in these debates.

In short, the war metaphor—at least in U.S. society—carries the implication that the enemy can be defeated. When the metaphor is applied to social problems, there is an implicit expectation that victory will be total, that the problem will be eliminated. But this is unrealistic. Most social problems are products of complex cultural and structural arrangements and are unlikely to be completely eradicated, regardless of which social policies are chosen. Therefore, declaring war on social problems creates an unrealistic, unachievable expectation of total victory. And, when total victory fails to materialize—as it most surely will—the policy stands vulnerable to attack as a failure.

The Problem of Duration

A third issue concerns speed. Social change tends to occur over decades. The most effective social policies cannot produce overnight changes in social arrangements; it can take generations for policies' effects to permeate a social system. In contrast, because wars—at least in U.S. history—have tended to be concentrated, relatively brief experiences, the war metaphor invites expectations that the conflict will soon end, that the problem can be solved quickly.

Of course, enthusiasts can point to terrible problems that seem to have had quick solutions; the dramatic success of the polio vaccine became a touchstone for post-World War II believers in scientific progress. But to do so ignores the years of research required to develop the vaccine, as well as the people already affected by polio when the vaccine was released. Polio did not suddenly vanish.[5] And vaccines are preventive measures—they eradicate diseases by keeping them from spreading. The analogy doesn't fit most social problems; because social problems have complex causes, they are rarely amenable to "magic bullet" solutions, technological fixes. Most successful social policies produce modest improvements over time.

The expectation that wars don't take too long is related to the expectation that they end in victory. It is noteworthy how quickly critics assess wars on social problems as failures. The war on cancer, inaugurated in late 1971, was being attacked as a failure by 1975; criticisms of the War on Poverty and the current war on drugs came at least as quickly. It is as though the fact that U.S. involvement in major wars (e.g., the Civil War, World War II) tends to last about four years means that a social-policy "war" should produce total victory within that time. It is one thing to direct social policy toward a long-range goal of reducing or eliminating some socially problematic condition; it is quite another matter to expect cancer or poverty or drug use to disappear within five or ten years. Yet the war metaphor seems to encourage these short-range expectations.

The Problem of Unity

Declarations of war on social problems are dramatic events: they call for society to rally behind a single policy, against a common foe. Typically, the initial pronouncements receive favorable attention in the mass media; the press details the nature of the problem and outlines the efforts designed to wage war against it. Usually, the enemy—cancer or poverty—has no one speaking on its behalf. There is the sense that society is united behind the war effort. Declaring war seizes the moral high ground; at least at first, the policy seems above criticism.

But, as the war continues (for all the reasons noted above), this sense of moral purpose and unity is easily lost and the policy comes under attack. One set of critics affirms the war's goal—the eradication of the social problem—but argues that policymakers have adopted the wrong means, so that the war effort is half-hearted or wrong-headed. Officials ought to be doing more, or doing something different, if they really want to end the problem.

Other critics denounce the lack of unity in prosecuting the war. In practice, not everyone unites behind the war effort, and this dismays the policy's proponents. Because the war metaphor implies a united effort, the failure to achieve unity can be bitter. Thus, a drug-war official complains about defense lawyers accused of obstructing justice in drug cases: "I look upon these attorneys, if they're guilty, as traitors. Hell, are we in a war or not? Why aren't these guys being charged with treason?" (Dannen 1995:31–32).

Still other critics oppose the policy but not the metaphor; they argue that the war targets the wrong enemy. Some of the same liberals who criticized Reagan for pursuing a war on drugs were outraged by his failure to declare war on homelessness or AIDS: "Reagan recognized that war rhetoric sanctioned federal action on AIDS—by refusing to use it.... His choices about when to use and not use the war metaphor clearly reflected his priorities" (Sherry 1995:457).

Finally, critics may denounce war itself. This has been the stance adopted by critics of the current drug war, who argue that drug education coupled with legalization or decriminalization are more sensible goals for drug policy than eradication. Thus, the Drug Policy Foundation sells baseball caps with the slogan WAR IS NOT A DOMESTIC POLICY on the front. Although wars against social problems rarely inspire large antiwar movements, unity tends to dissolve over time.

Again, the expectation implicit in the war metaphor—that wars can be won quickly—gives the policy's critics the ammunition they need:

> The search for war's moral equivalent had invited government to pursue quick solutions to chronic problems resistant to speedy solutions. When the inevitable failure, waste, and repression surfaced, many Americans bewailed government's ineffectiveness, finding confirmation of the very weaknesses that had initially inspired their rhetoric. (Sherry 1995:460)

The Consequences of Choosing the War Metaphor

Choosing to declare war on social problems is consequential. It has obvious advantages. The metaphor is dramatic; it attracts favorable press attention, encourages everyone to move the issue higher on the policy agenda, and—at least temporarily—seems to silence critics. Undoubtedly, it is these obvious advantages that lead policymakers to continue to declare war on social problems.

At the same time, the very drama and visibility created by declarations of war make these social policies especially vulnerable to criticism:

> In the short run, the Johnson administration must be judged amazingly successful at its rhetorical task. Over a period of but a few months, strong support was obtained for measures for which there had been no public clamor or demand. Tragically, however, the very rhetorical choices which were so useful in gaining initial support proved to be dysfunctional in the long run. (Zarefsky 1986:xiv)

The war metaphor carries implications—of societal unity and speedy victory—that few social policies can fulfill. Within four or five years of declaring war on a social problem, we can expect to find, regardless of whether the policy has achieved some modest successes, the core problematic condition enduring and critics denouncing the policy's failure. With each passing year, the sense that the policy has failed is likely to become more widespread.

At the same time, the melodrama of the war metaphor, its insistence that social problems can be understood as a straightforward struggle between good and evil, constrains discussion of alternative policies. If society is at war with drugs, if drugs are our common enemy, it is easy to respond to antidrug policy failures by escalating the war, by cracking down even harder. It is much more difficult to shift paradigms, to redefine drugs as, say, a public health problem that ought to be addressed with a very different set of policies (Bertram et al. 1996). The popularity of the war metaphor, the ease with which it portrays policy as melodrama, makes it harder to adopt alternative imagery. It is much easier for the policy's critics to turn the metaphor to their own uses (e.g., criticizing defeatism) than to reframe the issue in completely different terms, to think of drugs as a health problem rather than an enemy. Public health policies produce slow changes over the long term: they seem to admit compromise and moral ambiguity. No wonder waging war offers more popular imagery.

We live in a pessimistic age. As this century's end approaches, we are surrounded by evidence of social progress: longer life expectancy, a higher standard of living, a more educated population, greater social equality, and so on. One might write the history of the twentieth-century United States in terms of this social progress. But the idea of progress has fallen into disrepute. There is a widespread sense that our society is deteriorating. In particular, there is considerable doubt that social policies make things better.

In part, this doubt reflects a sense that recent social policies have failed. Although this assessment is not limited to the policies called wars against social problems, those policies, because they were dramatic efforts, highly publicized and designed to create high expectations for social progress, have been especially visible and easily defined as policy failures. The lesson seems clear: even when society dedicates itself to solving a particular social problem, it cannot succeed. Pessimism, then, is justified.

SOCIAL POLICY AS MELODRAMA

Declaring war on social problems is dramatic. It is a headline-grabbing move that tries to enlist everyone in a collective melodrama, uniting as good guys in a just, all-out struggle against evil. A similarly melodramatic vision runs through much contemporary talk about social problems. [Previously], we examined how melodramatic imagery shapes our understanding of wilding, stalking, and other new crimes of random violence, and how advocates use melodrama to typify the new victims as vulnerable innocents. Defining social problems in terms of villains and victims makes it easy to understand these issues: propelled by evil motives in pursuit of evil goals, criminals become the equivalent of the dark-cloaked, mustache-twirling villain of nineteenth-century melodrama, while the exploited new victims are like the fair-haired maiden tied to the railroad track. Similarly, in declaring war on some social problem, the state/society/everyone assumes the hero's role, dedicating ourselves to protecting the innocent, righting wrongs, and stamping out evil. No wonder we are so fond of melodramatic imagery; it simplifies complex issues, resolving the blurry grays into stark black and white.

In contrast, effective social policy tends to be boring. It seems mundane, if only because we take what works for granted. But no high-tech medical miracle has saved anything like the number of lives that have been spared since people recognized the need and devised the means to keep sewage separate from drinking water. The mundane social changes that caused nineteenth-century crime rates to fall included industry's growing demand for workers, compulsory schooling, the spread of street lights, and the presence of police on patrol (Tobias 1967). It is the seemingly minor, taken-for-granted reforms, such as requiring vaccinations and seat belts and smoke detectors, or establishing building codes and water treatment and Social Security, that eventually change the social landscape.

We aren't much interested in these mundane policies, and we usually don't hear much about them. The mass media thrive on novelty, drama, and excitement. All television formats for news and quasi news, from talk shows and made-for-TV movies to regular news broadcasts, depend on being interesting enough, dramatic enough, to hold the audience's attention. Newspapers are more likely to report on gradual progress, but stories that lack drama don't make the front page. Politicians, depending on media coverage for visibility, learn to couch their actions in new, dramatic terms. Even bureaucrats, hoping to enlist politicians' support, discover the need to make bureaucracy seem exciting. Thus, policymakers come to favor dramatic gestures, such as declaring war. Social policy, like social problems, attracts attention when it seems new and dramatic; otherwise it gets ignored.

No wonder those trying to define social issues favor melodramatic imagery. Both advocates promoting new social problems and policymakers introducing new policies recognize melodrama's advantages. By defining social issues as straightforward struggles between good and evil, melodrama compels our attention and enlists our emotions. It simplifies complexity, and reduces opposition to whatever is being proposed. Who, after all, opposes fighting against evil, protecting the vulnerable, and preserving innocence?

Melodrama's simplicity—and its costs—become apparent in hotly contested social issues. Consider the vivid rhetoric injected into the debate over social policies on guns. Opponents of gun controls envision jack-booted government thugs, intent on ravaging the Constitution, rendering citizens defenseless, and preparing the way for an authoritarian world government, while gun-control advocates portray guns, the irrationality of "gun nuts," and the evil influence of the National Rifle Association as being at the root of American violence and crime. Both visions are fundamentally melodramatic; both seek to enlist right-thinking people in a struggle against evil. Similarly, the rhetoric of both those attacking and those defending abortion can be melodramatic, as advocates mount crusades on behalf of murdered babies or oppressed women. By characterizing one's opponents as evil villains and one's cause as the protection of innocents, melodramatic rhetoric makes such debates that much more difficult to resolve.

Melodrama stands opposed to subtlety. In *Body Count,* their analysis of "how to win America's war against crime and drugs," former U.S. "drug czar" William Bennett and his colleagues blame "moral poverty" for "America's violent crime plague," warning that "thickening ranks of juvenile 'super-predators'" mean that "America is a ticking crime bomb" (Bennett et al. 1996:13, 21, 27). Like other claims about callous criminals and deviant conspiracies, this portrayal stands opposed to the bulk of criminological knowledge that finds crime's motivations in relatively mundane, situationally specific thrills and fears (Katz 1988). Similarly, advocates present new victims as faultless, as pawns in the hands of their evil victimizers; in melodrama, victimization is the product of straightforward exploitation. And, of course, declarations of war identify social problems as enemies preying on society rather than the result of existing social arrangements. Melodrama's simplicity is appealing precisely because it makes complex issues so easy to understand.

But, of course, melodrama comes at a cost. Simplifying complexity inevitably distorts what we know, and melodrama lends itself to over-simplification and, therefore, serious distortion. In particular, by equating social problems with evil, melodramatic rhetoric discourages us from understanding their mundane roots. By exaggerating threats and arousing our fears, melodrama also raises the emotional stakes, making it harder to calmly consider policy options. Whenever we respond to social problems as emergencies, we lose sight of alternatives and sacrifice our ability to weigh costs, set priorities, and make reasoned choices (Lipsky and Smith 1989).

In part, such short-circuited reasoning is precisely why policymakers and other advocates favor melodramatic claims. In the competition for attention in the social-problems marketplace, dramatic, compelling claims have an advantage; unassuming, boring claims get overlooked. Therefore, advocates package their claims to attract attention, making new crimes seem especially threatening, new victims especially sympathetic, and proposed policies especially bright with promise. This need not be cynical or dishonest; caught up in their enthusiasm for the cause, advocates often believe their own rhetoric. But claims that define issues in melodramatic terms contain the seeds of their own destruction. This is apparent in declaring war on social problems. Declarations of war seem to offer—*do* offer—a means of rallying people around a common cause. But the war metaphor also encourages other expectations of quick, decisive victories, expectations that inevitably lead to disappointment and disillusionment. Similarly, denouncing random violence compels our attention, but it proves to be a singularly unmanageable, almost useless way to think about crime.

Although melodramatic rhetoric characterizes social issues in exaggerated, oversimplified ways, some distortion is inevitable in all claims about social problems. All claims carve off a slice of reality and give it a name, specify its causes, identify its consequences, and so on. Claims focus our attention on a particular social condition and encourage us to define it as a problem of a particular sort. In drawing our attention to some things, claims lead us to ignore others; inevitably, we see less than the whole. Every time we focus on some new social problem, we risk losing sight of its larger context.

ENDNOTES

1. Like social problems, social policies are socially constructed. Whereas political scientists have developed constructionist interpretations of policy-making (cf. Edelman 1988; Rochefort and Cobb 1994; Schneider and Ingram 1993; Stone 1988), sociologists have paid less attention to policy claims.

2. Other countries favor other rhetoric and are far less likely to declare war on social problems. Jun Ayukawa informs me that, in Japan, militarized metaphors would not be considered appropriate descriptions for social policy. Japanese media sometimes speak of "eradicating" or "exterminating"

crime, but usually restrict such extreme language to discussions of crime in other countries or to Japanese crimes involving firearms or drugs.

3. As Edelman notes elsewhere (1977:32–35), political rhetoric usually characterizes groups of people, rather than social conditions, as enemies. Ibarra and Kitsuse (1993:47–48) call for the study of "motifs" in social-problems rhetoric. They seem to equate motifs with metaphors and include declaring war on their list of examples.
4. This list is based on common perceptions; it includes those major conflicts that concentrated government and public attention and consumed substantial resources against powerful opposition—that is, the conflicts generally perceived to have been wars, regardless of whether there was a formal declaration of war. Obviously, this list ignores a variety of lesser armed conflicts, including military campaigns against various Native American peoples and assorted minor foreign invasions/police actions/occupations. (For a longer list of American wars, see Aho 1994:95). I have excluded these other conflicts precisely because they rarely became the focus for either extensive federal government action or public opinion. However, expanding my list of wars would not alter my argument; these other campaigns also tended to be short and victorious.
5. It is interesting that this successful campaign used less dramatic, albeit militarized, rhetoric to mobilize support: the "March of Dimes" suggests long-term, steady progress rather than a brief, all-out push to victory.

REFERENCES

Aho, James A. 1994. *This Thing of Darkness: A Sociology of the Enemy*. Seattle: University of Washington Press.

Albelda, Randy, Nancy Folbre, and the Center for Popular Economics. 1996. *The War on the Poor: A Defense Manual*. New York: New Press.

Beardsley, Tim. 1994. "A War Not Won." *Scientific American* 270 (January): 130–38.

Bennett, William J., John J. DiIulio, Jr., and John P. Walters. 1996. *Body Count: Moral Poverty … and How to Win America's War against Crime and Drugs*. New York: Simon and Schuster.

Bertram, Eva, Morris Blachman, Kenneth Sharpe, and Peter Andreas. 1996. *Drug War Politics*. Berkeley: University of California Press.

Blain, Michael. 1994. "Power, War, and Melodrama in the Discourses of Political Movements." *Theory and Society* 23:805–37.

Bricklin, Mark. 1993. "Let's Declare War on Breast Cancer!" (editorial). *Prevention* 145 (September): 39–40, 42.

Culliton, Barbara J. 1976. "Mrs. Lasker's War." *Harper's* 252 (June): 60.

Danner, Frederic. 1995. "The Thin White Line." *New Yorker* 71(July 31): 30–34.

Edelman, Murray. 1971. *Politics as Symbolic Action*. New York: Academic Press.

———. 1977. *Political Language: Words That Succeed and Policies That Fail*. Orlando, Fla.: Academic Press.

———. 1988. *Constructing the Political Spectacle*. Chicago: University of Chicago Press.

Elwood, William N. 1994. *Rhetoric in the War on Drugs*. New York: Praeger.

Englehardt, Tom. 1995. *The End of Victory Culture*. New York: Basic Books.

Gans, Herbert J. 1995. *The War against the Poor: The Underclass and Antipoverty Policy*. New York: Basic Books.

Gibson, James William. 1994. *Warrior Dreams: Paramilitary Culture in Post-Vietnam America*. New York: Hill and Wang.

Gorelick, Steven M. 1989. "'Join Our War': The Construction of Ideology in a Newspaper Crime-fighting Campaign." *Crime and Delinquency* 35: 421–36.

Ibarra, Peter R., and John I. Kitsuse. 1993. "Vernacular Constituents of Moral Discourse." In *Reconsidering Social Constructionism*, edited by James A. Holstein and Gale Miller, 25–58. Hawthorne, N.Y.: Aldine de Gruyter.

Inciardi, James A. 1992. *The War on Drugs II*. Mountain View, Calif.: Mayfield.

Jencks, Christopher. 1992. *Rethinking Social Policy*. New York: HarperCollins.

Johns, Christina Jacqueline. 1992. *Power, Ideology, and the War on Drugs*. New York: Praeger.

Johnson, Lyndon Baines. 1971. *The Vantage Point: Perspectives on the Presidency, 1963–1969*. New York: Holt, Rinehart and Winston.

Katz, Jack. 1988. *Seductions of Crime*. New York: Basic Books.

Keen, Sam. 1986. *Faces of the Enemy: Reflections of the Hostile Imagination*. San Francisco: Harper & Row.

Kraska, Peter B., and Victor E. Kappeler. 1997. "Militarizing American Police: The Rise and Normalization of Paramilitary Units." *Social Problems* 44: 1–18.

Lemann, Nicholas. 1988. "The Unfinished War." *Atlantic Monthly* 262 (December): 37–49, 52–56.

Lens, Sidney. 1969. *Poverty: America's Enduring Paradox; A History of the Richest Nation's Unwon War*. New York: Crowell.

Lipsky, Michael, and Steven Smith. 1989. "When Social Problems Are Treated as Emergencies." *Social Service Review* 63:5–25.

McCoy, Alfred W., and Alan A. Block, eds. 1992. *War on Drugs: Studies in the Failure of U.S. Narcotics Policy*. Boulder, Colo.: Westview.

Murray, Charles. 1984. *Losing Ground: American Social Policy, 1950–1980*. New York: Basic Books.

Patterson, James T. 1987. *The Dread Disease: Cancer and Modern American Culture*. Cambridge: Harvard University Press.

Pilisuk, Marc, and Phyllis Pilisuk, eds. 1973. *How We Lost the War on Poverty*. New Brunswick, N.J.: Transaction.

Powers, Richard Gid. 1983. *G-Men: Hoover's FBI in American Popular Culture*. Carbondale: Southern Illinois University Press.

Rochefort, David A., and Roger W. Cobb, eds. 1994. *The Politics of Problem Definition*. Lawrence: University of Kansas Press.

Russell, Edmund P., III. 1996. "'Speaking of Annihilation': Mobilizing for War against Human and Insect Enemies, 1914–1945." *Journal of American History* 82:1505–29.

Schneider, Ann, and Helen Ingram. 1993. "Social Construction of Target Populations." *American Political Science Review* 87:334–47.

Sherry, Michael S. 1995. *In the Shadow of War: The United States since the 1930s*. New Haven: Yale University Press.

Sidel, Ruth. 1996. *Keeping Women and Children Last: America's War on the Poor*. New York: Penguin.

Sontag, Susan. 1989. *AIDS and Its Metaphors*. New York: Farrar, Straus and Giroux.

Stone, Deborah A. 1988. *Policy Paradox and Political Reason*. New York: HarperCollins.

Tobias, J. J. 1967. *Crime and Industrial Society in the 19th Century*. New York: Schocken.

Trebach, Arnold S., and Scott Ehlers. 1996. "The War on Our Children." *Drug Policy Letter* 30:13–17.

Will, George. 1992. "From MEOW to Meow." *Newsweek* 119 (February 10): 82.

Zarefsky, David. 1986. *President Johnson's War on Poverty: Rhetoric and History*. Tuscaloosa: University of Alabama Press.

DISCUSSION QUESTIONS

1. What do you think is Best's central point concerning the wars we make against social problems?
2. Of course, Best's point is a weak one if in fact people can ultimately be victorious against a social problem. Do you believe there is a serious social problem that can be solved? Explain your answer.
3. Why is the metaphor of war to characterize social problems used as much as it is?
4. Best makes the understanding of and policies toward social problems highly complex. Society is not simple; social problems are not simple; solutions are not simple. Why is this? Do you agree with him?
5. Consider crime or poverty as a social problem. What would you consider "victory"? Is this possible?

PART II

Social Problems and American Culture

The readings in this section deal with American culture and key social problems that stem from society's customs and worldviews. Many sociologists point to consumerism, the unending desire to buy goods and services at excessive levels, as a cultural value that is playing a significant role in many of society's social problems. The overconsumption of resources to attain an insatiable level of material comfort is its main feature. We are now fully in the throes of what economists call *The Great Recession*, a moniker for the worst economic crisis since The Great Depression, and many observers point to consumerism as playing a significant role in this predicament.

The American sociologist and economist Thorstein Veblen, in his book *The Theory of the Leisure Class* published 1899, introduced the concept *conspicuous consumption* to describe the sumptuous and hedonistic buying behavior of the nouveau riche for the purpose of displaying their new wealth in pursuit of material comfort. Indeed, this pattern of lavish consumption is not only found among the middle and upper classes alone; conspicuous and unending buying is omnipresent: Everyone is "Keeping up Appearances" now. How does the American Dream's emphasis on material comfort and success contribute to this situation? Moreover, how are American cultural values and beliefs implicated in the social problems that stem from consumerism and overconsumption?

In the first selection Jennifer Hochschild asks us to examine the "American dream." What is the American dream? What beliefs does it teach us to accept? More importantly, is the dream real, and how is it possible that the dream might itself be a problem and the cause of other problems? This selection tries to show us that basic cultural beliefs may help create and perpetuate the social problems we face.

In a selection that was published originally in *The New Republic*, the American sociologist Amitai Etzioni reflects on the social problem of consumerism, namely the use of excessive buying to satisfy those higher needs for esteem and self-actualization that the psychologist Abraham Maslow described so eloquently. He argues passionately for the need to move away from the "social disease" of consumerism which he believes played a major role in the current national crises.

The final selection is from Barry Schwartz's book *The Costs of Living: How Market Freedom Erodes the Best Things in Life*. Schwartz moves away from the problem of inequality and asks readers to question the nature of education today. He believes that because we have made the school a part of the "market culture," extrinsic incentives have become more important than education itself. This philosophy influences teachers, students, parents, and government, leading to a loss of the love of learning.

It is not easy to read articles that are critical of the cultural practices and commonly accepted worldviews that are taken-for-granted by all of us. Each of these selections will emphasize an aspect of American culture that is a social problem. A recommended companion to this selection of readings on social problems and American culture is a publicly available,online video by the American environmental activist Annie Leonard entitled "The Story of Stuff" (www.storyofstuff.com).

3

What's Wrong with the American Dream?

JENNIFER L. HOCHSCHILD

The Four Questions

1. What is the problem Hochschild identifies? What values are applied that make this a problem?
2. Does she consider the problem a *social* problem? Does it seriously hurt anyone?
3. What does she describe as the cause of this problem? Is the cause identified primarily as a social cause? Does she claim that the problem adversely affects the future of society?
4. Does Hochschild tell us what can be done about the problem? If not, what do you think she might recommend to resolve it?

Topics Covered

Success

American dream

Ideology

Class

Economic inequality

Racial inequality

Work

"In the beginning," wrote John Locke, "all the world was *America*."[1] Locke was referring specifically to the absence of a cash nexus in primitive society. But the sentence evokes the unsullied newness, infinite possibility, limitless resources that are commonly understood to be the essence of the "American dream." The idea of the American dream has been attached to everything from religious freedom to a home in the suburbs, and it has inspired emotions ranging from deep satisfaction to disillusioned fury. Nevertheless, the phrase elicits for most Americans some variant of Locke's fantasy—a new world where anything can happen and good things might.

Millions of immigrants and internal migrants have moved to America, and around within it, to fulfill their version of the American dream. By objective measures and their own accounts, many have achieved success. Probably just as many have been defeated and disillusioned. Millions of other immigrants—predominantly but not exclusively from Africa—were moved to America despite their preferences and have been forced to come to terms with a dream that was not originally theirs. How they have done so, and how their experiences compare with those who came to America to seek their dream, is the chief subject of this [selection].

But one cannot address that subject, nor eventually move beyond it to evaluate the future of the American dream and its society, without knowing what the dream is and how it operates. That knowledge is the goal of this [selection].

THE MEANING OF SUCCESS

The American dream consists of tenets about achieving success. Let us first explore the meaning of "success" and then consider the rules for achieving it.

SOURCE: Jennifer Hochschild, *Facing Up To the American Dream,* © 1995 Princeton University Press, 1996 paperback edition. Reprinted by permission of Princeton University Press.

People most often define success as the attainment of a high income, a prestigious job, economic security. My treatment is no exception. But, *pace* President Reagan, material well-being is only one form of accomplishment. People seek success from the pulpit to the stage of the Metropolitan Opera, from membership in the newest dance club to membership in the Senate. Success can be as amorphous and encompassing as "a right to say what they wanta say, do what they wanta do, and fashion a world into something that can be great for everyone."[2]

Different kinds of success need not, but often do, conflict. A classic plot of American family sagas is the children's rejection of the parents' hard-won wealth and social standing in favor of some "deeper," more meaningful form of accomplishment.[3] The rejection may be reversed, as Cotton Mather sadly reported:

> There have been very fine settlements in the north-east regions; but what is become of them? … One of our ministers once preaching to a congregation there, urged them to approve themselves a religious people from this consideration, "that otherwise they would contradict the main end of planting this wilderness"; whereupon a well-known person, then in the assembly, cryed out, "Sir, you are mistaken: you think you are preaching to the people at the [Plymouth] Bay; our main end was to catch fish."[4]

Mather "wished that something more excellent had been the main end of the settlements in that brave country," but the ideology of the American dream itself remains agnostic as to the meaning of "something more excellent."[5]…

TENETS OF SUCCESS

> The American dream that we were all raised on is a simple but powerful one—if you work hard and play by the rules you should be given a chance to go as far as your God-given ability will take you.
>
> —PRESIDENT BILL CLINTION, SPEECH TO DEMOCRATIC LEADERSHIP COUNCIL, 1993

In one sentence, President Clinton has captured the bundle of shared, even unconsciously presumed, tenets about achieving success that make up the ideology of the American dream. Those tenets answer the questions: *Who* may pursue the American dream? In *what* does the pursuit consist? *How* does one successfully pursue the dream? *Why* is the pursuit worthy of our deepest commitment?

The answer to "who" in the standard ideology is "everyone, regardless of ascriptive traits, family background, or personal history." The answer to "what" is "the reasonable anticipation, though not the promise, of success, however it is defined." The answer to "how" is "through actions and traits under one's own control." The answer to "why" is "true success is associated with virtue." Let us consider each rule in turn.

Who May Pursue Success?

The first tenet, that everyone may always pursue their dream, is the most direct connotation of Locke's "in the beginning …" But the idea extends beyond the image of a pristine state of nature waiting for whoever "discovers" it. [Even in the distinctly nonpristine, nonnatural world of Harlem or Harlan County, anyone can pursue a dream.] A century ago, one moved to the frontier to hide a spotted past and begin afresh; Montana frontierswomen "never ask[ed] women where they come from or what they did before they came to live in our neck of the woods. If they wore a wedding band and were good wives, mothers, and neighbors that was enough for us to know."[6] Today one finds new frontiers….

What Does One Pursue?

The second tenet, that one may reasonably anticipate success, is less straightforward. "Reasonable anticipation" is far from a guarantee, as all children on the morning of their birthday know. But "reasonable anticipation" is also much more than simply longing; most children are fairly sure of getting at least some of what they wish for on their birthday. On a larger scale, from its inception America has been seen by many as an extravagant birthday party:

> Seagull: A whole countrie of English is there, man, ... and ... the Indians are so in love with 'hem that all the treasure they have they lay at their feete.... Golde is more plentiful there than copper is with us ... Why, man, all their dripping pans and their chamberpots are pure golde; and all the chaines with which they chaine up their streets are massie golde; all the prisoners they take are fettered in golde; and for rubies and diamonds they goe forthe on holy dayes and gather 'hem by the sea shore to hang on their childrens coats.[7]

Presumably few Britons even in 1605 took this message literally, but the hope of abundant riches—whether material, spiritual, or otherwise—persists.

Thus Americans are exhorted to "go for it" in their advertisements as well as their commencement addresses. And they do; three-quarters of Americans, compared with only one-third of Britons, West Germans, and Hungarians (and fewer Dutch), agree that they have a good chance of improving their standard of living. Twice as many Americans as Canadians or Japanese think future generations of their nationality will live better than the present generation.[8]

How Does One Pursue Success?

The third premise, for those who do not take Seagull literally, explains how one is to achieve the success that one anticipates. Ralph Waldo Emerson is uncharacteristically succinct on the point: "There is always a reason, *in the man,* for his good or bad fortune, and so in making money."[9] Other nineteenth-century orators exhorted young men to

> Behold him [a statue of Benjamin Franklin], ... holding out to you an example of diligence, economy and virtue, and personifying the triumphant success which may await those who follow it! Behold him, ye that are humblest and poorest...—lift up your heads and look at the image of a man who rose from nothing, who owed nothing to parentage or patronage, who enjoyed no advantages of early education, which are not open,—a hundredfold open,—to yourselves, who performed the most menial services in the business in which his early life was employed, but who lived to stand before Kings, and died to leave a name which the world will never forget.[10]

Lest we smile at the quaint optimism (or crude propaganda) of our ancestors, consider [a] recent advertisement from Citicorp Bank.... [that shows a] carefully balanced group of shining faces—young and old, male and female, black, Latino, Nordic, and Asian—all gazing starry-eyed at the middle distance over the words "THE WILL TO SUCCEED IS PART OF THE AMERICAN SPIRIT" [and] conveys the message of the third tenet in no uncertain terms.

This advertisement is well aimed; surveys unanimously show Americans' strong support for rewarding people in the marketplace according to their talents and accomplishments rather than their needs, efforts, or simple existence.[11] And Americans mostly believe that people are in fact rewarded for their acts. In 1952 fully 88 percent of Americans agreed that "there is plenty of opportunity and anyone who works hard can go as far as he wants"; in 1980, 70 percent concurred.[12]

Comparisons across space yield the same results as comparisons across time. In a 1973 survey of youth in ten nations, only Swedes and British disagreed more than did Americans that a man's [*sic*] future is "virtually determined" by his family background. A decade later only 31 percent of Americans agreed that in their nation "what you achieve in life depends largely on your family background," compared with over 50 percent of Austrians and Britons, and over 60 percent of Italians.[13] Most pointedly, half of American adolescents compared with one-fourth of British adolescents agreed in 1972 that "people get to be poor...[because] they don't work hard enough."[14]

Americans also believe more than do Europeans that people ought not to be buffered from the consequences of their actions, so long as they have a fair start in life. Thus up to four times as many more Americans think college opportunities should be increased, but roughly half as many think the government should reduce the income disparity between high- and low-income citizens, or provide jobs or income support for the poor.[15]

Why Is Success Worth Pursuing?

Implicit in the flows of oratory and survey responses is the fourth tenet of the American dream, that the pursuit of success warrants so much fervor because it is associated with virtue. "Associated with" means at least four things: virtue leads to success, success makes a person virtuous, success indicates virtue, or apparent success is not real success unless one is also virtuous.

That quintessential American, Benjamin Franklin, illustrates three of these associations: the *Autobiography* instructs us that "no Qualities were so likely to make a poor Man's Fortune as those of Probity & Integrity." Conversely, "Proverbial Sentences, chiefly such as inculcated Industry and Frugality," are included in *Poor Richard's Almanack* as "the Means of procuring Wealth and thereby securing Virtue, it being more difficult for a Man in Want to act always honestly, as … *it is hard for an empty Sack to stand upright.*" Finally, mere wealth may actually impede true success, the attainment of which requires a long list of virtues: "Fond *Pride of Dress,* is sure a very Curse;/ E'er *Fancy* you consult, consult your Purse"; "A Ploughman on his Legs is higher than a Gentleman on his Knees"; and "Pride that dines on Vanity sups on Contempt."[16]

Americans have learned Franklin's lessons well: they distinguish between the worthy and unworthy rich, as well as the deserving and undeserving poor. For example, most Americans characterize "yuppies" as people who "play fashionable games" and "eat in trendy restaurants," and on the whole they enjoy watching such forms of conspicuous consumption. But they also characterize yuppies as selfish, greedy, inclined to flaunt their wealth, and imbued with a false sense of superiority. These traits they mostly find unacceptable. Overall, Americans overwhelmingly deplore the 1980s sentiment of "making it fast while you can regardless of what happened to others."[17] This is not simply a reaction against the Reagan years. In surveys throughout the 1970s, four in ten Americans deemed honesty to be the most important quality for a child to learn, compared with 2 percent proclaiming that a child should try hard to succeed. Virtually all Americans require that their friends be "honest" and "responsible"—core components of the third and fourth tenets.[18]

Americans also focus more on virtue than do citizens of other nations, at least in their self-descriptions. A survey of youth in ten nations found that more Americans than people in any other country described their chief goal in life as "sincerity and love between myself and others," and in only one other nation (the Philippines) did more youth seek "salvation through faith." Conversely, only in Sweden did fewer youths seek "money and position," and only in three other countries did fewer seek "freedom from restrictions." More Americans than Europeans gain strength from religion, report prayer to be an important part of their daily life, and agree that there are universally applicable "clear guidelines about what is good or evil."[19] In short, "this country succeeds in living a very sinful life without being deeply cynical. That is the difference between Europe and America, and it signifies that ethics means something here."[20]

The American Dream as Fantasy

… I … use the four tenets just described to analyze Americans' changing beliefs about the American dream. But we must beware reducing the dream to its components; as a whole it has an evocative resonance greater than the sum of its parts. The theme of most Walt Disney movies boils down to the lyric in *Pinocchio:* "When you wish upon a star, makes no difference who you are, your dreams come true." It is no coincidence that

Disney movies are so durable; they simply update Locke's fantasy. And the global, amorphous vision of establishing a city upon the hill, killing the great white whale, striking a vein of gold, making the world safe for democracy—or simply living a life of decency and dignity—underlies all analyses of what success means or what practices will attain it.

VIRTUES OF THE AMERICAN DREAM

Combining the amorphous fantasy or the more precise tenets of the American dream with the various meanings of success shows the full richness—and seductiveness—of the ideology. If one measures success absolutely and accepts a wide array of indicators of success, the ideology portrays America as a land of plenty, and Americans as "people of plenty."[21] This is the great theme of one of the most powerful children's sagas ever written in America, the *Little House in the Big Woods* series. Decades (and nine volumes) of grasshopper plagues, ferocious blizzards, cheating and cowardly railroad bosses, even hostile Indians cannot prevent Pa and his girls from eventually "winning their bet with Uncle Sam" and becoming prosperous homesteaders. In the words of one of Pa's songs:

> I am sure in this world there are plenty of good things enough for us all.... It's cowards alone that are crying And foolishly saying, "I can't." It is only by plodding and striving And laboring up the steep hill Of life, that you'll ever be thriving, Which you'll do if you've only the will.[22]

If success is measured competitively and defined narrowly, however, the ideology portrays a different America. Hard work and virtue combined with scarce resources produce a few spectacular winners and many dismissible losers. This is the theme of John Rockefeller's turn-of-the-century Sunday school address:

> The growth of a large business is merely a survival of the fittest ...The American Beauty rose can be produced in the splendor and fragrance which bring cheer to its beholder only by sacrificing the early buds which grow up around it. This is not an evil tendency in business. It is merely the working out of a law of nature and a law of God.[23]

The *Little House* series has sold well over four million copies; Americans prefer the self-image of universal achievement to that of a few stalwarts triumphing over weaker contenders.[24] What matters most, however, is not any single image but rather the elasticity and range of the ideology of the American dream. People can encourage themselves with soft versions, congratulate themselves with harder ones, and exult with the hardest, as their circumstances and characters warrant.

Thus the American dream is an impressive ideology. It has for centuries lured people to America and moved them around within it, and it has kept them striving in horrible conditions against impossible odds. Most Americans celebrate it unthinkingly, along with apple pie and motherhood; criticism typically is limited to imperfections in its application. But like apple pie and motherhood, the American dream turns out upon closer examination to be less than perfect. Let us look, then, at flaws intrinsic to the dream.

FLAWS IN THE TENETS OF THE AMERICAN DREAM

The First Tenet

The first tenet, that everyone can participate equally and can always start over, is troubling to the degree that it is not true. It is, of course, never true in the strongest sense; people cannot shed their existing selves as snakes do their skin. So the myth of the individual mini-state of nature is just that—a fantasy to be sought but never achieved.

Fantasies are fine so long as people understand that that is what they are. For that reason, a weaker formulation of the first tenet—people start the pursuit of success with varying advantages, but no one is barred from the pursuit—is more troubling because the gap between ideological claim and actual fact is harder to recognize. As a factual claim, the first tenet is largely false; for most of American history, women of any race and men who were Native American, Asian, black, or poor were barred from all but a narrow range of "electable futures."[25] Ascriptive constraints have arguably been weakened over time,[26] but until recently no more than about a third of the population was able to take seriously the first premise of the American dream.

This flaw has implications beyond the evident ones of racism and sexism. The emotional potency of the American dream has made people who *were* able to identify with it the norm for everyone else. White men, especially European immigrants able to ride the wave of the Industrial Revolution (and to benefit from the absence of competition from the rest of the population) to comfort or even prosperity, are the epitomizing demonstration of America as the bountiful state of nature. Those who do not fit the model disappear from the collective self-portrait. Thus the irony is doubled: not only has the ideal of universal participation been denied to most Americans, but also the very fact of its denial has itself been denied in our national self-image.

This double irony creates deep misunderstandings and correspondingly deep political tensions. Much of this book [of which this selection is a part] examines one such tension—the fact that whites increasingly believe that racial discrimination is slight and declining, and blacks increasingly believe the opposite. But this form of racial conflict is not unique. For example, surveys show that more women than men believe that women are discriminated against in employment and wages, in "being able to combine family and work," and in their overall chance to pursue their dreams. Similarly, regardless of when the survey was conducted, more men than women believe that women are better off now than a decade earlier with regard to these issues. Not surprisingly, bitter disagreements about the need for affirmative action, policies to stem sexual harassment, family leave policies, and the like ensue.[27]

The Second Tenet

The flaws of the second tenet of the American dream, the reasonable anticipation of success, stem from the close link between anticipation and expectation. That link presents little problem so long as there are enough resources and opportunities that everyone has a reasonable chance of having some expectations met. Indeed, panegyrics to the American dream always expound on the bounty and openness of the American continent. Governor James Glen typified eighteenth-century entrepreneurs of colonization by promising that

> Adventurers will be pleased to find a Change from Poverty and Distress to Ease and Plenty; they are invited to a Country not yet half settled, where the Rivers are crouded with Fish, and the Forests with Game; and no Game-Act to restrain them from enjoying those Bounties of Providence, no heavy Taxes to impoverish them, nor oppressive Landlords to snatch the hard-earned Morsel from the Mouth of Indigence, and where Industry will certainly inrich them.[28]

Three centuries later, the message was unchanged:

> All my life I am thinking to come to this country. For what I read in the magazines, and the movies.... I would have a beautiful castle in the United States. I will have a thousand servants. I will have five Rolls-Royces in my door.... We thinking everybody has this kind of life.... I have this kind of dream.[29]

These fantasies are innocuous so long as resources roughly balance dreams for enough people enough of the time. But if they do not—worse yet, if they used to but do no longer—then the dream rapidly loses its appeal. The circumstances that cause resources no longer to balance dreams vary, from an economic downturn to a rapid increase in the number of dreamers to a narrowing of the grounds on which success is publicly recognized. The general point, however, always holds: no one promises that dreams will be fulfilled, but the distinction between the right to dream and the right to succeed is psychologically hard to maintain and politically always blurred. It is especially hard to maintain because the dream sustains Americans against daily nightmares only if they believe that they have a significant likelihood, not just a formal chance, of reaching their goals.

In short, the right to aspire to success works as an ideological substitute for a guarantee of success only if it begins to approach it. When people recognize that chances for success are slim or getting slimmer, the whole tenor of the American dream changes dramatically for the worse.

The general problem of scarcity varies depending on how people measure success and how broadly they define possible goals. It is most obvious and acute for those focused on competitive success in only a few arenas; by definition, resources and opportunities are insufficient to satisfy all dreamers in such a case. But it may be more problematic for those who measure success relatively or who admit a wide array of outcomes into their picture of success. After all, there are more such people and they have no a priori reason to assume that many will fail.

The problem of scarcity may be most devastating, however, for people anticipating absolute success or for people willing to see success almost anywhere. They, after all, have the least reason to expect failure. Losers of this type have an unmatched poignancy: "I don't dream any more like I used to. I believed that in this country, we would have all we needed for the decent life. I don't see that any more."[30]

Conversely, the availability of resources and opportunities may shape the kind of success that Americans dream of. If resources are profoundly scarce (as in a famine) or inherently limited (as in election to the presidency), people almost certainly envision competitive success in that arena. If resources are moderately scarce, people will be concerned about their position relative to that of others, but will not necessarily see another's gain as their loss. When resources and opportunities seem wide open and broadly defined—anyone can achieve salvation, get an "A" on the exam, claim 160 acres of western prairie—people are most free to pursue their idiosyncratic dreams and to measure their achievement by their own absolute standard.

This logic suggests a dynamic: as resources become tighter or success is more narrowly defined, Americans are likely to shift their understanding of success from absolute to relative to competitive. Before the 1980s, claims one journalist, "there was always enough to go around, plenty of places in the sun. It didn't even matter much about the rich—so long as everyone was living better, it seemed the rich couldn't be denied their chance to get richer." But "today [in 1988] that wave [of prosperity] has crested.... Now when the rich get richer, the middle class stagnates—and the poor get decidedly poorer. If left unchecked, a polarization of income...is likely to provoke consequences that will affect America's politics and power, to say nothing of its psyche."[31]

The risks of anticipating success do not stop with anticipation. Attaining one's dreams can be surprisingly problematic. From William Shakespeare to William Faulkner, writers have limned the loneliness of being at the top, the spiritual costs of cutthroat competition, the shallowness of a society that rewards achievement above all else. Alexis de Tocqueville characteristically provides one of the most eloquent of such admonitions:

> Every American is eaten up with longing to rise.... In America I have seen the freest and best educated of men in circumstances the happiest in the world; yet it seemed to me that a cloud habitually hung on their brow, and they seemed serious and almost sad even in their pleasures. The chief reason for this is that...[they] never stop thinking of the good things they have not got.... They clutch everything but hold

> nothing fast, and so lose grip as they hurry after some new delight.[32]

The obsession with ever more material success threatens the body politic as well as the individual soul:

> When the taste for physical pleasures has grown more rapidly than either education or experience of free institutions, the time comes when men are carried away and lose control of themselves at sight of the new good things they are ready to snatch.… There is no need to drag their rights away from citizens of this type; they themselves voluntarily let them go.… The role of government is left unfilled. If, at this critical moment, an able and ambitious man once gets power, he finds the way open for usurpations of every sort.[33]

Not only nineteenth-century romantics cautioned against the failures of success. Today psychotherapists specialize in helping "troubled winners" or the "working wounded," for whom "a life too much devoted to pursuing money, power, position, and control over others ends up being emotionally impoverished."[34] In short, material—and perhaps other forms of—success is not all it's cracked up to be, even (or especially) in a nation where it is the centerpiece of the pervasive ideology.

The problems of success, however, pale beside the problems of failure. Because success is so central to Americans' self-image,[35] and because they expect as well as hope to achieve, Americans are not gracious about failure. Others' failure reminds them that the dream may be just that—a dream, to be distinguished from waking reality. Their own failure confirms that fear. As Zora Neale Hurston puts it, "there is something about poverty that smells like death."[36]

Furthermore, the better the dream works for other people, the more devastating is failure for the smaller and smaller proportion of people left behind. In World War II, members of military units with a high probability of promotion were less satisfied with advancement opportunities than members of units with a much lower probability of promotion, because failure to be promoted in the former case was both more salient and more demonstrably a personal rather than a systemic flaw. The "tunnel effect" is a more nuanced depiction of this phenomenon of relative deprivation. The first stage is one of relative gratification, in which others' success enhances one's own well-being. After all, drivers in a traffic jam in a tunnel are initially pleased when cars in the adjacent lane begin to move "because advances of others supply information about a more benign external environment; receipt of this information produces gratification; and this gratification overcomes, or at least suspends, *envy*." At some point, however, those left behind come to believe that their heightened expectations will not be met; not only are their hopes now dashed, but they are also worse off than when the upward mobility began. "Nonrealization of the expectation ['that my turn to move will soon come'] will at some point result in my 'becoming furious.'"[37] And one is still stuck in the tunnel. In short, the ideology of the American dream includes no provision for failure; a failed dream denies the loser not only success but even a safe harbor within which to hide the loss.

The Third Tenet

Failure is made more harsh by the third premise of the American dream—the belief that success results from actions and traits under one's own control. Logically, it does not follow that if success results from individual volition, then failure results from lack of volition. All one needs in order to see the logical flaw here is the distinction between necessary and sufficient. But that distinction is not obvious or intuitive, and in any case the psychologic of the American dream differs from strict logic. In the psychologic, if one may claim responsibility for success, one must accept responsibility for failure.

Americans who do everything they can and still fail may come to understand that effort and talent alone do not guarantee success. But they have a hard time persuading others. After all, they are losers—why listen to them? Will we not benefit more by listening to winners (who seldom challenge the premise that effort and talent breed success)?

The Fourth Tenet

Failure, then, is unseemly for two reasons: it challenges the blurring between anticipation and promise that is the emotional heart of the American dream, and people who fail are presumed to lack talent or will. The coup de grace comes from the fourth tenet of the dream, the association of success with virtue. By the psychologic just described, if success implies virtue, failure implies sin.

American history and popular culture are replete with demonstrations of the connection between failure and sin. In the 1600s, indentured servants—kidnapped children, convicts, and struggling families alike—were described by earlier immigrants as "strong and idle beggars, vagabonds, egyptians, common and notorious whoores, theeves, and other dissolute and lousy persons."[38] Nineteenth-century reformers concurred: fallen women are typically "the daughters of the ignorant, depraved, and vicious part of our population, trained up without culture of any kind, amidst the contagion of evil example, and enter upon a life of prostitution for the [g]ratification of their unbridled passions, and become harlots altogether by choice."[39]

Small wonder that in the late twentieth century even the poor blame the poor for their condition. Despite her vivid awareness of exploitation by the rich, an aging cleaning woman insists that many people are poor because they "make the money and drink it all up. They don't care about the kids or the clothes. Just have a bottle on that table all the time." Losers even blame themselves: an unemployed factory worker, handicapped by a childhood accident, "wish[es] to hell I could do it [save money for his children]. I always said for years, 'I wanna get rich, I wanna get rich.' But then, phew! My mind doesn't have the strong will. I say, 'Well, I'm *gonna* do it.' Only the next day's different." These people are typical. In 1985, over 60 percent of poor people but only 45 percent of the nonpoor agreed that "poor young women often have babies so they can collect welfare." Seven years later, the same proportions of poor and well-off agreed that welfare recipients "are taking advantage of the system."[40]

The equation of failure with evil and success with virtue cannot be attributed to poor education or low status. College students "who learned that a fellow student had been awarded a cash prize as a result of a random drawing were likely to conclude that he had in fact worked especially hard." In another experiment, subjects rated a presumed victim of electric shocks who was randomly selected to receive compensation for her pain more favorably than a victim who would not be compensated. "The sight of an innocent person suffering without possibility of reward or compensation motivated people to devalue the attractiveness of the victim in order to bring about a more appropriate fit between her fate and her character."[41] Devaluing losers allows people to maintain their belief that the world is fundamentally just, even when it patently is not.

Losers are obviously harmed by the association of success with virtue. But the association creates equally important, if less obvious, problems for winners. Fitzwilliam Darcy, in Jane Austen's *Pride and Prejudice,* epitomizes the defect of pride: if I believe that virtue produced my success, or that success has made me even more virtuous, I am likely to become insufferably smug. That may not bother me much, but the fact that people around me feel the same way will.[42] In addition, this equation raises the stakes very high for further rounds of endeavor. If I continue to win, all is well; if I falter, I lose my *amour propre* as well as my wealth or power. Alternatively, if I recognize that I partly owe my success to lying to a few clients, evading a few taxes, cheating a few employees, then I am likely to feel considerable guilt. This guilt might induce reform and recompense, but it may instead induce drinking to assuage the unease, persecuting other nonvirtuous winners, striving to show that losers are even more sinful, or simple hypocrisy.[43]

These problems intensify when patterns of group success rather than the idiosyncrasies of individual success are at issue. When members of one group seem disproportionately successful, that group acquires a halo of ascribed virtue....

This effect of the fourth tenet can be taken a further, and most dangerous, step. For some Americans always, and for many Americans in some periods of our history, virtuous success has been defined as the dominance of some groups over others. This phenomenon extends the idea of competitive success from individual victories to collective hierarchies. If women are weak and emotional, it is *right* for men to control their bodies and wealth; if blacks are childlike pagans, it is *right* for whites to ensure their physical and spiritual survival through enslavement and conversion; if citizens of other nations refuse to recognize the value of capitalism and free elections, it is *right* for Americans to install a more enlightened government in their capitol. I find it hard to present these sentiments with a straight face, but they have arguably done almost as much as the American dream to shape Americans' beliefs, practices, and institutions....[44]

The Ideology as Deception

I have argued that the American dream need not be individualistic in the narrow sense, given that one can under its rubric pursue success for one's family or community as well as for oneself. But it is highly *individual,* in that it leads one to focus on people's behaviors rather than on economic processes, environmental constraints, or political structures as the causal explanation for social orderings. That focus is not itself a flaw; it is simply an epistemological choice with methodological implications for the study of American politics. But to the degree that the focus carries a moral message, it points to a weakness at the very heart of the dream.

The idea of the blank slate in the first tenet, the almost-promise of success of the second, the reliance on personal attributes in the third, the association of failure with sin in the fourth—all these elements of the dream make it extremely difficult for Americans to see that everyone cannot simultaneously attain more than absolute success. Capitalist markets require some firms to fail; elections require some candidates and policy preferences to lose; status hierarchies must have a bottom in order to have a top. But the optimistic language of and methodological individualism built into the American dream *necessarily* deceive people about these societal operations. We need not invoke hypocrites out of Mark Twain or "blue-eyed white devils" in order to understand why some people never attain success; hypocrisy or bias only enter the picture in determining *who* fails. But our basic institutions are designed to ensure that some fail, at least relatively, and the dream does nothing to help Americans cope with or even to recognize that fact....

ENDNOTES

1. Locke (1980: sec. 49, p. 29).
2. Ed Sadlowski, in Terkel (1980: 236). The proportion of first-year college students seeking to "be very well off financially" grew from 44% to 74% between 1967 and 1994; the proportion seeking to "develop a meaningful philosophy of life" sank from 83% to 43% over the same period (Cooperative Institutional Research Program 1987: 97; 1994: 26; see also People for the American Way 1989: 152; Warden 1994).
3. Hochschild (1986); Chandler and Chandler (1987).
4. Mather (1970: 27).
5. Mather's dilemma continues: college freshmen who seek financial success overlap little with those who seek a meaningful philosophy of life or to improve the nation's social and political life (Easterlin and Crimmins 1991: 505).
6. Jeffrey (1979: 141). The lyrics of a Burl Ives song from the nineteenth-century western frontier include, "Did you murder your wife and fly for your life? Say, what was your name in the States?"
7. Ben Jonson, George Chapman, and John Marston, *Eastward Ho!* (1605), quoted in Beeman (1971: 618–19).

8. Tom Smith (1988: 14); "Public Opinion and Demographic Report" (1993b: 89); see also "Public Opinion and Demographic Report" (1993a: 85, 87).
9. Emerson (1863: 86).
10. Robert Winthrop, "Oration at the Inauguration of the Statue of Benjamin Franklin," Boston, 1856, quoted in Wyllie (1954: 14–15). Franklin himself was more succinct: "In America,...people do not inquire concerning a Stranger, What is he?, but, What can he do?" (Franklin 1987: 175). See Shklar (1991: 19–22, 63–101) on the importance of working to earn one's own way in defining American citizenship.
11. Miller (1992: 564–70); Ladd (1994: 55–58).
12. Kluegel and Smith (1986: 44). See also Lynd and Lynd (1930: 65); Huber and Form (1973); Caplow and Bahr (1979: table 1); Ladd (1994: 53–56, 68–69). In 1993 fully 94% of Americans agreed that hard work was crucial to success; the next most popular choice was God's will, with 53% agreement (Marsden and Swingle 1994: 277).
13. *Gallup Opinion Index* (1973: 28); Tom Smith (1987a: 411). See also Tom Smith (1987b); Miller (1992: 586–88); Ladd (1994: 79). Working-class respondents in all western countries are slightly less committed to the third tenet than are members of the middle class. But the most striking feature of comparisons by class is the degree to which the poor, especially in the United States, support norms that benefit the rich more than themselves (Miller 1992: 582–86; see also Ladd 1994: 80; Hochschild 1981).
14. Stern and Searing (1976: 198). Perhaps the equation of poverty with laziness makes fewer American youths than youths of ten other countries (except for India) agree that "it is important...to take it easy and not to work too hard" (*Gallup Opinion Index* 1973: 36).
15. Ladd (1994: 75, 79, 80).
16. Franklin (1987: 1298–1302, 1386, 1391, 1392, 1397). Diggins (1984) gives probably the best analysis of the role of virtue in American political culture.
17. Louis Harris (1986; 1990: 2).
18. Davis and Smith (1982: vars. 127–29); Marsden and Swingle (1994: 279). Other desirable features of friends, such as being fun-loving or intelligent, received considerably less than unanimous support.
19. *Gallup Opinion Index* (1973: 34); Ladd (1994: 72–73).
20. Gunnar Myrdal, in Baldwin et al. (1964: 33). Lamont (1992; 1995) analyzes the greater weight placed by Americans than by the French on including morality in their definition of success. Nackenoff (1994) shows the interactions of virtue and material success in modern American history.
21. Potter (1954).
22. Wilder (1940: 334).
23. Quoted in Ghent (1902: 29).
24. My thanks to Walter Lippincott and Hugh Van Dusen for providing me with this figure.
25. The phrase is from Rae (1988).
26. Although not without backsliding, as a survivor of Japanese internment camps points out: "The American Dream? I think: for whites only. I didn't feel that way before World War Two" (Terkel 1980: 161–71; more generally, see Rogers Smith 1993).
27. Louis Harris (1978: 59–67); Kluegel and Smith (1986: 62–72, 222–35); Verba and Orren (1985: 83–88); Simon and Landis (1989); "Public Opinion and Demographic Report" (1993c: 88–95).
28. Message from Governor James Glen of South Carolina in 1749, quoted in Warren Smith (1961: 51).
29. Miguel Cortéz, in Terkel (1980: 131).
30. Florence Scala, in Terkel (1980: 116).
31. Goldstein (1988: 77).
32. de Tocqueville (1969: 627, 536). Once again, Studs Terkel's respondents parallel learned discourse. To a wealthy professional, "the American Dream always has a greater force when you don't already have it. People who grew up without it are told if you can only work long enough and hard enough, you can get that pot of gold at the end of the rainbow. When you already have the pot of gold, the dream loses its force." A struggling ex-convict is more rueful: "It was always competition. I went from competing in sports to competing in crime.... I always wanted to be at the top of something. So I became the first dope fiend in the neighborhood" (Leon Duncan and Ken Jackson, in Terkel 1980: 123, 218).
33. de Tocqueville (1969: 540).
34. Quotations from Douglas LaBier, in Skrzycki (1989: H1, H4). See also LaBier (1986) and Berglas (1986).

35. Consider the effects of the 1971 draft lottery on self-esteem. In one experiment, young men completed paper-and-pencil measures of self-esteem, then listened to the lottery, then retook the self-esteem index. "Subjects whose numbers put them in the fortunate half of their group tended to experience increased self-esteem, while those whose numbers put them in the unfortunate half of their group tended to experience decreased self-esteem" (Rubin and Peplau 1973: 81).
36. Hurston (1942: 116).
37. Stouffer et al. (1949: 250–57); Hirschman (1981: 41, 47). In the Latin American context that he was studying, Hirschman equated "becoming furious" with "turning into an enemy of the established order." That sometimes happens in the United States.... But "becoming furious" may also result in spouse-battering, a lawsuit, or "the embrace of victimhood" (Sykes 1992; Taylor 1991; Hughes 1992).
38. From a 1669 warrant of the Scottish Privy Council to local authorities, quoted in Nash (1982: 217).
39. Readers are, however, assured that such prostitutes "have a short career, generally dying of the effects of intemperance and pollution soon after entering upon this road to ruin." *Magdalen Report: First Annual Report of the Executive Committee of the N.Y. Magdalen Society* (New York, 1831), quoted in Hugins (1972: 42).
40. Hochschild (1981: 113, 116); Lewis and Schneider (1985: 7); "Public Opinion and Demographic Report" (1993c: 86).
41. Rubin and Peplau (1975: 67, 68); Lerner (1980); see also Rubin and Peplau (1973); Lerner and Lerner (1981).
42. Young (1958) gives the classic depiction of the costs to society as a whole of smugness among the successful.
43. Huntington (1981: 30–41, 61–70), Shklar (1984: 67–78), and McWilliams (1990: 177) examine hypocrisy in American liberal democracy. Mark Twain remains, however, its best analyst.
44. Rogers Smith (1993).

REFERENCES

Baldwin, James, et al. (1964) "Liberalism and the Negro: A Round-Table Discussion," *Commentary* 37, 3: 25–42.

Beeman, Richard (1971) "Labor Forces and Race Relations: A Comparative View of the Colonization of Brazil and Virginia," *Political Science Q.* 86, 4: 609–36.

Berglas, Steven (1986) *The Success Syndrome: Hitting Bottom When You Reach the Top* (Plenum).

Caplow, Theodore, and Howard Bahr (1979) "Haija Century of Change in Adolescent Attitudes: Replications of a Middletown Survey by the Lynds," *Public Opinion Q.* 43, 1: 1–17.

Chandler, David, and Mary Chandler (1987) *The Binghams of Louisville* (Crown).

Cooperative Institutional Research Program (1987) *The American Freshman: Twenty Year Trends* (UCLA, Graduate School of Education, Higher Education Research Institute).

Davis, James, and Tom Smith (1982) *General Social Surveys, 1972–1982: Cumulative Codebook* (U. of Chicago, National Opinion Research Center).

de Tocqueville, Alexis (1969 [1848]) *Democracy in America,* trans. George Lawrence; ed. J. P. Mayer (Doubleday).

Diggins, John (1984) *The Lost Soul of American Politics* (U. of Chicago Press).

Easterlin, Richard, and Eileen Crimmins (1991) "Private Materialism, Personal Self-fulfillment, Family Life, and Public Interest," *Public Opinion Q.* 55, 4: 499–533.

Emerson, Ralph Waldo (1863) "Wealth," in *The Conduct of Life* (Ticknor and Fields), 71–110.

Franklin, Benjamin (1987) *Writings,* ed. J. A. Lemay (Library of America).

Gallup Opinion Index (1973) Report no. 100, Oct.

Ghent, W. J. (1902) *Our Benevolent Feudalism* (Macmilkn).

Goldstein, Mark (1988) "The End of the American Dream?" *Industry Week,* April 4: 77–80.

Harris, Louis (1978) *A Study of Attitudes toward Racial and Religious Minorities and toward Women,* for National Conference of Christians and Jews, Nov.

———(1986) "Yuppie Lifestyle Felt to Be Unattractive to Americans" (N.Y.: Harris Survey), Feb. 3.

Hirschman, Albert (1981) *Essays in Trespassing* (Cambridge U. Press).

Hochschild, Adam (1986) *Half the Way Home: A Memoir of Father and Son* (Viking).

Hochschild, Jennifer (1981) *What's Fair? American Beliefs about Distributive Justice* (Harvard U. Press).

Huber, Joan, and William Form (1973) *Income and Ideology* (Free Press).

Hughes, Robert (1992) *The Culture of Complaint* (Oxford U. Press).

Hugins, Walter, ed. (1972) *The Reform Impulse, 1825–1850* (U. of South Carolina Press).

Huntington, Samuel (1981) *American Politics: The Promise of Disharmony* (Harvard U. Press).

Hurston, Zora Neale (1942) *Dust Tracks on a Road*, 2 ed. (U. of Illinois Press).

Jeffrey, Julie (1979) *Frontier Women: The Trans-Mississippi West 1840–1880* (Hill and Wang).

Kluegel, James, and Eliot Smith (1986) *Beliefs About Inequality* (Aldine de Gruyter).

LaBier, Douglas (1986) *Modern Madness: The Emotional Fallout of Success* (Addison-Wesley).

Ladd, Everett (1994) *The American Ideology* (Storrs, CT: Roper Center for Public Opinion Research).

Lamont, Michéle (1992) *Money, Morals, and Manners* (U. of Chicago Press).

———(1995) "National Identity and National Boundary Patterns in France and the United States," *French Historical Studies,* 19, 2: 349–65.

Lerner, Melvin (1980) *Belief in a Just World* (Plenum Press).

Lerner, Melvin, and Sally Lerner, eds. (1981) *The Justice Motive in Social Behavior* (Plenum Press).

Lewis, I. A., and William Schneider (1985) "Hard Times: The Public on Poverty," *Public Opinion* 8, 3: 2–7, 59–60.

Locke, John (1980 [1690]) *Second Treatise of Government*, ed. C. B. MacPherson (Hackett).

Lynd, Robert, and Helen Lynd (1930) *Middletown* (Harcourt, Brace, and Co.).

McWilliams, Wilson Carey (1990) "*Pudd'nhead Wilson on Democratic Governance,*" in Susan Gillman and Forrest Robinson, eds., *Mark Twain's Pudd'nhead Wilson* (Duke U. Press), 177–89.

Marsden, Peter, and Joseph Swingle (1994) "Conceptualizing and Measuring Culture in Surveys," *Poetics* 22: 269–89.

Mather, Cotton (1970 [1702]) *Magnolia Christi Americana,* ed. and abr. Raymond Cunningham (Frederick Ungar).

Miller, David (1992) "Distributive Justice: What the People Think," *Ethics* 102, 3: 555–93.

Nackenoff, Carol. 1994. *The Fictional Republic: Horatio Alger and American Political Discourse* (New York: Oxford U. Press).

Nash, Gary (1982) *Red, White, and Black: The Peoples of Early America* (Prentice-Hall).

People for the American Way (1989) *Democracy's Next Generation: American Youth Attitudes on Citizenship, Government and Politics* (Washington, D.C.: People for the American Way).

Potter, David (1954) *People of Plenty: Economic Abundance and the American Character* (U. of Chicago Press).

"Public Opinion and Demographic Report," *Public Perspective:*
(1993a) 4, 4: 82–104;
(1993b) 4, 5: 82–104;
(1993c) 4, 6: 82–104

Rae, Douglas (1988) "Knowing Power," in Ian Shapiro and Grant Reeher, eds., *Power, Inequality, and Democratic Politics* (Westview Press), 17–49.

Rubin, Zick, and Letitia Peplau (1973) "Belief in a Just World and Reactions to Another's Lot," *J. of Social Issues* 29, 4: 73–93.

———(1975) "Who Believes in a Just World?" J. *of Social Issues,* 31, 3: 65–89.

Shklar, Judith (1984) *Ordinary Vices* (Harvard U. Press).

———(1991) *American Citizenship* (Harvard U. Press).

Simon, Rita, and Jean Landis (1989) "Women and Men's Attitudes about a Woman's Place and Role," *Public Opinion* Q. 53, 2: 265–76.

Skrzycki, Cindy (1989) "Healing the Wounds of Success," *Washington Post,* July 23, H1 H4.

Smith, Rogers (1993) "Beyond Tocqueville, Myrdal, and Hartz," *APSR* 87, 3: 549–66.

Smith, Tom (1987a) "The Welfare State in Cross-national Perspective," *Public Opinion* Q. 51, 3: 404–21.

———(1987b) "Public Opinion and the Welfare State: A Crossnational Perspective," paper at the annual meeting of the ASA, Chicago.

———(1988) "Social Inequality in Cross-National Perspective" (U. of Chicago, National Opinion Research Center).

Smith, Warren (1961) *White Servitude in Colonial South Carolina* (U. of South Carolina Press).

Stern, Alan, and Donald Searing (1976) "The Stratification Beliefs of English and American Adolescents," *British J. of Political Science* 6, part 2: 177–201.

Stouffer, Samuel, et al. (1949) *The American Soldier: Adjustment during Army Life* (Princeton U. Press).

Sykes, Charles (1992) *A Nation of Victims: The Decay of the American Character* (St. Martin's Press).

Taylor, John (1991) "Don't Blame Me!: The New Culture of Victimization," *New York,* June 3: 26–34.

Terkel, Studs (1980) *American Dreams, Lost and Found* (Pantheon).

Verba, Sidney, and Gary Orren (1985) *Equality in America: The View From the Top* (Harvard U. Press).

Warden, Sharon (1994) "What's Happened to Youth Attitudes since Woodstock?" *Public Perspective* 5, 4: 19–24.

Wilder, Laura Ingalls (1940) *The Long Winter* (Harper & Row).

Wyllie, Irvin (1954) *The Self-Made Man in America* (Free Press).

Young, Michael (1958) *The Rise of the Meritocracy, 1870–2033* (Penguin Books).

DISCUSSION QUESTIONS

1. Do you believe in the American dream as stated, or do you think, like Hochschild, that it is a fantasy?
2. So what if the American dream is not fully true? Is that really harmful to anyone?
3. "The American dream" is an attempt to ignore the importance of social structure. Explain.
4. Who do you think "makes it" in American society? What are the qualities that matter?
5. Would the poor be better off without the American dream?

4

Consumerism and Americans

AMITAI ETZIONI

The Four Questions

1. What is the problem Etzioni describes?
2. Is it a *social* problem? Who does this problem harm?
3. What is the cause of this problem?
4. What can be done about this problem?

Topics Covered

Consumerism
Economic crisis
Capitalism
Credit-card debt
Over-consumption
Communitarianism
Happiness
Contentment
Social justice
American dream

Much of the debate over how to address the economic crisis has focused on a single word: regulation. And it's easy to understand why. Bad behavior by a variety of businesses landed us in this mess—so it seems rather obvious that the way to avoid future economic meltdowns is to create, and vigorously enforce, new rules proscribing such behavior. But the truth is quite a bit more complicated. The world economy consists of billions of transactions every day. There can never be enough inspectors, accountants, customs officers, and police to ensure that all or even most of these transactions are properly carried out. Moreover, those charged with enforcing regulations are themselves not immune to corruption, and, hence, they too must be supervised and held accountable to others—who also have to be somehow regulated. The upshot is that regulation cannot be the linchpin of attempts to reform our economy. What is needed instead is something far more sweeping: for people to internalize a different sense of how one ought to behave, and act on it because they believe it is right.

That may sound far-fetched. It is commonly believed that people conduct themselves in a moral manner mainly because they fear the punishment that will be meted out if they engage in anti-social behavior. But this position does not stand up to close inspection. Most areas of behavior are extralegal; we frequently do what is expected because we care or love. This is evident in the ways we attend to our children (beyond a very low requirement set by law), treat our spouses, do volunteer work, and participate in public life. What's more, in many of those areas that *are* covered by law, the likelihood of being caught is actually quite low, and the penalties are often surprisingly mild. For instance, only about one in 100 tax returns gets audited, and most cheaters are merely asked to pay back what they "missed," plus some interest. Nevertheless, most Americans pay the taxes due. Alan Lewis's classic study *The Psychology of Taxation* concluded that people don't just pay taxes because they fear the government; they do it because they consider the

SOURCE: From Amitai Etzioni, "Spent: America After Consumerism." In *The New Republic*, June 17, 2009, pp. 20–23.

burden fairly shared and the monies legitimately spent. In short, the normative values of a culture matter. Regulation is needed when culture fails, but it cannot alone serve as the mainstay of good conduct.

So what kind of transformation in our normative culture is called for? What needs to be eradicated, or at least greatly tempered, is consumerism: the obsession with acquisition that has become the organizing principle of American life. This is not the same thing as capitalism, nor is it the same thing as consumption. To explain the difference, it is useful to draw on Abraham Maslow's hierarchy of human needs. At the bottom of this hierarchy are basic creature comforts; once these are sated, more satisfaction is drawn from affection, self-esteem, and, finally, self-actualization. As long as consumption is focused on satisfying basic human needs—safety, shelter, food, clothing, health care, education—it is not consumerism. But, when the acquisition of goods and services is used to satisfy the higher needs, consumption turns into consumerism—and consumerism becomes a social disease.

The link to the economic crisis should be obvious. A culture in which the urge to consume dominates the psychology of citizens is a culture in which people will do most anything to acquire the means to consume—working slavish hours, behaving rapaciously in their business pursuits, and even bending the rules in order to maximize their earnings. They will also buy homes beyond their means and think nothing of running up credit-card debt. It therefore seems safe to say that consumerism is, as much as anything else, responsible for the current economic mess. But it is not enough to establish that which people ought not to do, to end the obsession with making and consuming evermore than the next person. Consumerism will not just magically disappear from its central place in our culture. It needs to be supplanted by *something*.

A shift away from consumerism, and toward this something else, would obviously be a dramatic change for American society. But such grand cultural changes are far from unprecedented. Profound transformations in the definition of "the good life" have occurred throughout human history. Before the spirit of capitalism swept across much of the world, neither work nor commerce were highly valued pursuits—indeed, they were often delegated to scorned minorities such as Jews. For centuries in aristocratic Europe and Japan, making war was a highly admired profession. In China, philosophy, poetry, and brush painting were respected during the heyday of the literati. Religion was once the dominant source of normative culture; then, following the Enlightenment, secular humanism was viewed in some parts of the world as the foundation of society. In recent years, there has been a significant increase in the influence of religious values in places like Russia and, of course, the Middle East. (Details can be found in John Micklethwait and Adrian Wooldridge's new book, *God is Back*—although, for many, he never left.) It is true that not all these changes have elevated the human condition. The point is merely that such change, especially during times of crisis, is possible.

To accomplish this kind of radical change, it is neither necessary nor desirable to imitate devotees of the 1960s counterculture, early socialists, or followers of ascetic religious orders, all of whom have resisted consumerism by rejecting the whole capitalist project. On the contrary, capitalism should be allowed to thrive, albeit within clear and well-enforced limits. This position does not call for a life of sackcloth and ashes, nor of altruism. And it does not call on poor people or poor nations to be content with their fate and learn to love their misery; clearly, the capitalist economy must be strong enough to provide for the basic creature comforts of all people. But it does call for a new balance between consumption and other human pursuits.

There is strong evidence that when consumption is used to try to address higher needs—that is, needs beyond basic creature comforts—it is ultimately Sisyphean. Several studies have shown that, across many nations with annual incomes above $20,000, there is no correlation between increased income and increased happiness. In the United States since World War II, per capita income has tripled, but levels of life satisfaction remain about the same, while the people of Japan, despite experiencing a sixfold increase in income since

1958, have seen their levels of contentment stay largely stagnant. Studies also indicate that many members of capitalist societies feel unsatisfied, if not outright deprived, however much they earn and consume, because others make and spend even more: Relative rather than absolute deprivation is what counts. This is a problem since, by definition, most people cannot consume more than most others. True, it is sometimes hard to tell a basic good from a status good, and a status good can turn into a basic one (air conditioning, for instance). However, it is not a matter of cultural, snobbery to note that no one *needs* inflatable Santas or plastic flamingos on their front lawn or, for that matter, lawns that are strikingly green even in the scorching heat of summer. No one needs a flat-screen television, not to mention diamonds as a token of love or a master's painting as a source of self-esteem.

Consumerism, it must be noted, afflicts not merely the upper class in affluent societies but also the middle class and many in the working class. Large numbers of people across society believe that they work merely to make ends meet, but an examination of their shopping lists and closets reveals that they spend good parts of their income on status goods such as brand-name clothing, the "right" kind of car, and other assorted items that they don't really need.

This mentality may seem so integral to American culture that resisting it is doomed to futility. But the current economic downturn may provide an opening of sorts. The crisis has caused people to spend less on luxury goods, such as diamonds and flashy cars; scale back on lavish celebrations for holidays, birthdays, weddings, and bar mitzvahs; and agree to caps on executive compensation. Some workers have accepted fewer hours, lower salaries, and unpaid furloughs.

So far, much of this scaling-back has been involuntary, the result of economic necessity. What is needed next is to help people realize that limiting consumption is not a reflection of failure. Rather, it represents liberation from an obsession—a chance to abandon consumerism and focus on … well, what exactly? What should replace the worship of consumer goods?

The kind of culture that would best serve a Maslowian hierarchy of needs is hardly one that would kill the goose that lays the golden eggs—the economy that can provide the goods needed for basic creature comforts. Nor one that merely mocks the use of consumer goods to respond to higher needs. It must be a culture that extols sources of human flourishing besides acquisition. The two most obvious candidates to fill this role are communitarian pursuits and transcendental ones.

Communitarianism refers to investing time and energy in relations with the other, including family, friends, and members of one's community. The term also encompasses service to the common good, such as volunteering, national service, and politics. Communitarian life is not centered around altruism but around mutuality, in the sense that deeper and thicker involvement with the other is rewarding to both the recipient and the giver. Indeed, numerous studies show that communitarian pursuits breed deep contentment. A study of 50-year-old men shows that those with friendships are far less likely to experience heart disease. Another shows that life satisfaction in older adults is higher for those who participate in community service.

Transcendental pursuits refer to spiritual activities broadly understood, including religious, contemplative, and artistic ones. The lifestyle of the Chinese literati, centered around poetry, philosophy, and brush painting, was a case in point, but a limited one because this lifestyle was practiced by an elite social stratum and based in part on exploitation of other groups. In modern society, transcendental pursuits have often been emphasized by bohemians, beginning artists, and others involved in lifelong learning who consume modestly. Here again, however, these people make up only a small fraction of society. Clearly, for a culture to buy out of consumerism and move to satisfying higher human needs with transcendental projects, the option to participate in these pursuits must be available on a wider scale.

All this may seem abstract, not to mention utopian. But one can see a precedent of sorts for a society that emphasizes communitarian and transcendental pursuits among retired people, who spend the final decades of their lives painting not for a market or galleries but as a form of self-expression, socializing with each other, volunteering, and, in some cases, taking classes. Of course, these citizens already put in the work that enables them to lead this kind of life. For other ages to participate before retirement, they will have to shorten their workweek and workday, refuse to take work home, turn off their BlackBerrys, and otherwise downgrade the centrality of labor to their lives. This is, in effect, what the French, with their 35-hour workweeks, tried to do, as did other countries in "old" Europe. Mainstream American economists—who argue that a modern economy cannot survive unless people consume evermore and hence produce and work evermore—have long scoffed at these societies and urged them to modernize. To some extent, they did, especially the Brits. Now it seems that maybe these countries were onto something after all.

A society that downplayed consumerism in favor of other organizing principles would not just limit the threat of economic meltdown and feature a generally happier populace; it would have other advantages as well. Such a society would, for example, use fewer material resources and, therefore, be much more compatible with protecting the environment. It would also exhibit higher levels of social justice.

Social justice entails redistribution of wealth, taking from those disproportionally endowed and giving to those who are underprivileged through no fault of their own—for reasons ranging from past injustices and their lingering contemporary effects to technological changes to globalization to genetic differences. The reason these redistributions have been surprisingly limited in free societies is that those who command the "extra" assets tend also to be those who are politically powerful. Promoting social justice by organizing those with less and forcing those in power to yield has had limited success in democratic countries and led to massive bloodshed in others. So the question arises: Are there other ways to reduce the resistance of elites to redistribution?

The answer is found when elites derive their main source of contentment not from acquiring more goods and services, but from activities that are neither labor nor capital intensive and, hence, do not require great amounts of money. Communitarian activities require social skills and communication skills as well as time and personal energy—but, as a rule, minimal material or financial outlays. The same holds for transcendental activities such as prayer, meditation, music, art, sports, adult education, and so on. True, consumerism has turned many of these pursuits into expensive endeavors. But one can break out of this mentality and find that it is possible to engage in most transcendental activities quite profoundly using minimal goods and services. One does not need designer clothes to enjoy the sunset or shoes with fancy labels to benefit from a hike. Chess played with plastic pieces is the same game as the one played with carved mahogany or marble pieces. And I'm quite sure that the Lord does not listen better to prayers read from a leather-bound Bible than those read from a plain one, printed on recycled paper. (Among several books that depict how this kind of culture can flourish is *Seven Pleasures* by Willard Spiegelman.) In short, those who embrace this lifestyle will find that they can achieve a high level of contentment even if they give up a considerable segment of the surplus wealth they command.

As for actually putting this vision into practice: The main way societies will determine whether the current crisis will serve as an event that leads to cultural transformation or merely constitute an interlude in the consumerism project is through a process I call "moral megalogues." Societies are constantly engaged in mass dialogues over what is right and wrong. Typically, only one or two topics dominate these megalogues at any given time. Key recent issues have included the legitimacy of the 2003 invasion of Iraq and whether gay couples should be allowed to marry. In earlier decades, women's rights and minority rights were topics of such discussions. Megalogues involve millions of members of a society exchanging views with one another at workplaces, during family gatherings, in the media, and at public events. They are often contentious and passionate, and, while they have no clear beginning or endpoint, they tend to lead to changes in a society's culture and its members' behavior.

The megalogue about the relationship between consumerism and human flourishing is now flickering but has yet to become a leading topic—like regulation. Public intellectuals, pundits, and politicians are those best-positioned to focus a megalogue on this subject and, above all, to set the proper scope for the discussion. The main challenge is not to pass some laws, but, rather, to ask people to reconsider what a good life entails.

Having a national conversation about this admittedly abstract question is merely a start, though. If a new shared understanding surrounding consumption is to evolve, education will have a crucial role to play. Schools, which often claim to focus solely on academics, are actually major avenues through which changes in societal values are fostered. For instance, many schools deeply impress on young children that they ought to respect the environment, not discriminate on racial or ethnic grounds, and resolve differences in a peaceful manner. There is no reason these schools cannot push back against consumerism while promoting communitarian and transcendental values as well. School uniforms (to counter conspicuous consumption) and an emphasis on community service are just two ways to work these ideas into the culture of public education.

For adults, changes in the workplace could go a long way toward promoting these values. Limits on overtime, except under special conditions (such as natural disasters); shorter workweeks; more part- and flex-time jobs; increased freedom to work from home; allowing employees to dress down and thereby avoid squandering money on suits and other expensive clothes—all these relatively small initiatives would encourage Americans to spend more time on things besides work.

Finally, legislation has a role to play. Taxes can discourage the purchase of ever-larger houses, cause people to favor public transportation over cars, and encourage the use of commercial aviation rather than private jets. Government could also strike a blow against consumerism by instituting caps on executive pay.

Is all this an idle, abstract hypothesis? Not necessarily. Plenty of religious Americans have already embraced versions of these values to some extent or other. And those whose secular beliefs lead them to community service are in the same boat. One such idealist named Barack Obama chose to be a community organizer in Chicago rather than pursue a more lucrative career.

I certainly do not expect that most people will move away from a consumerist mindset overnight. Some may keep one foot in the old value system even as they test the waters of the new one, just like those who wear a blazer with jeans. Still others may merely cut back on conspicuous consumption without guilt or fear of social censure. Societies shift direction gradually. All that is needed is for more and more people to turn the current economic crisis into a liberation from the obsession with consumer goods and the uberwork it requires—and, bit by bit, begin to rethink their definition of what it means to live a good life.

DISCUSSION QUESTIONS

1. Amitai Etzioni makes the argument that the incessant drive and uncontrollable urge to consume is at the heart of the nation's economic crisis. How exactly is consumerism a social problem? Moreover, do you agree with his claim that it is the central problem behind our economic crisis? Why or why not?

2. After the terrorist attack of Sept. 11, 2001, President Bush said that the best way to deal with this national shock is to "go shopping." And we did. In light of how indelible shopping is to the American economy, what is the likelihood of having a critical "national conversation" about the "social problem" of consumerism? Who will start this conversation,

and where, in your opinion, should these conversations take place?

3. To move away from a consumerist mindset would require a broad cultural shift. How long will it take to shift from a consumerist mindset to one that is more aligned to Amitai Etzioni's vision for the United States?
4. What role can the government play in encouraging the shift away from the consumerist mindset? What role can individuals play in this process? Finally, is there a constructive role for American corporations in the national discussion on consumerism? If so, what is it?
5. Go to the following address: www.storyofstuff.com. Watch the short video entitled "The Story of Stuff." How do Amitai Etzioni's ideas about the social problem of consumerism connect to this documentary?

5

The Debasing of Education

BARRY SCHWARTZ

The Four Questions

1. What is the problem, according to Schwartz?
2. What makes it a *social* problem?
3. What is its cause?
4. Can anything be done about the problem?

Topics Covered

Market system

Education

Teachers

Student motivation

The college

Culture

When one of my daughters was about three years old, our family set off on a long drive to a house we had rented for our summer vacation. While my wife and I chatted in the front seat of the car, our daughter was amusing herself in the backseat singing songs. We recognized them from the "Sesame Street" show, but we weren't really listening. Before long, though, the singing drifted into something that was unfamiliar, and my wife and I stopped talking and listened. Our daughter was making up a song of her own. When she finished, we turned to her, beaming, and told her how wonderful her song was. She smiled happily and sang it again. We enthused again. She sang it again. We enthused again. She sang it again. We enthused again. Then she politely asked us to go back to talking to each other so she could have fun. So we went back to ignoring her, and she went back to making up songs.

Most of the residents of my community teach their kids from a very early age to value education and take it seriously. Many of the kids are reading early, even before kindergarten, and those who

SOURCE: From Barry Schwartz, *The Costs of Living: How Market Freedom Erodes the Best Things in Life* (Philadelphia, PA: Xlibris, 2000); originally published in 1994 by W. W. Norton and Co. Reprinted by permission of the author.

aren't reading by then are being read to constantly. Whatever problems the elementary school teachers may have in running their classes, motivating the kids to read and to learn is not one of them. One year, a long-term substitute teacher came to the school to teach fourth-graders. Her previous experience had been at a school in which developing discipline and willingness to do schoolwork had been a major challenge. She introduced to our school a technique that had been quite effective in her previous one. Every time one of her students read a book, he or she would get a point. By accumulating a certain number of points, the kids could win prizes. And whichever kid accumulated the most points would win a grand prize. The results of this procedure were amazing. The kids were reading like demons. Some were reading more than a book a day. I learned this from a neighbor, whose daughter was in the class. When I told her how impressed I was, she told me that things were not as impressive as they seemed. Her daughter, who had already been a pretty voracious reader, was now choosing books to read on the basis of two criteria: how long they were—the shorter, the better—and how big the print was—the bigger, the better. And she seemed unable to remember what was in the books almost immediately after she finished them. She was certainly reading an enormous amount, but the only point to her reading seemed to be to get to the end of one book so that she could start another.[1]

Several years ago, a team of psychologists conducted an experiment with nursery school children. They gave the children an opportunity to draw with special felt-tipped drawing pens, an activity about which the kids were extremely enthusiastic. After a period of observation, in which the psychologists measured the amount of time the children spent drawing with the pens, the children were taken into a separate room, given the pens, and asked to draw pictures. Some of the children were told they might receive "Good Player" awards for their drawings; others were not. A week later, back in the regular nursery school setting, the drawing pens were again made available, this time with no promise of any award. The children who had received awards previously were *less* likely than the others to draw with the pens at all. And if they did draw, they spent less time at it than other children and drew pictures that were judged to be less complex, interesting, and creative. Without the prospect of further awards, their interest in drawing was now only perfunctory.[2]

Three stories. Two anecdotes and an experiment. They converge on the same point. Young children are intensely interested, active, and inventive. They derive enormous pleasure from learning and doing. They love to make up little songs, to draw pictures, to read. They love to experiment with their world and to develop mastery over it. They are curious about everything. Learning and doing are their own reward. No external incentives are required.

And yet, when external incentives are introduced—whether attention and praise from parents, prizes from teachers, or "Good Player" awards from experimenters—these external incentives exert a substantial influence on the activities that produce them. Instead of singing songs for the sake of the songs themselves, kids may come to sing songs for the praise it gets them. Or they read books not to find out about the world but to win prizes. Kids will produce for rewards, but the quality of their activity and their interest in it will be dramatically altered. Spontaneity, exploration, and inventiveness will be replaced by mechanical, perfunctory efficiency. And if the external incentives are removed, the activities may cease all together. The introduction of external incentives may succeed in turning play into work—into the most limited kind of work.

This [selection] is about how, with the help of pressure from the market, we have turned play—the excitement of education—into work. It's about the instrumentalization of education. It's about how five-year-old kids who enter school full of enthusiasm and alive to everything get turned into bored adolescents who would rather be anywhere than where they are. It's about how various efforts to prop up motivation to learn by introducing external incentives—rewards and punishments of various kinds—actually make the motivation problem worse. It's about how the dependence on external incentives, established early in the grade school years, carries all the way through the educational process. It's about how kids who start out their education seeking knowledge move from seeking knowledge to seeking approval and, from there, to seeking good grades, to seeking admission to good colleges, and to seeking good jobs. And finally, it's about how the instrumentalization of education affects not only the *process* of education but its *product,* making many of the best things about education increasingly difficult to achieve....

If we read enough popular books, magazine articles, and reports from government commissions and private foundations, we cannot help seeing that every aspect of the educational system comes in for substantial criticism. Incompetent or uninterested teachers, unmotivated students, uninvolved parents, undemanding curricula, unnecessary and constraining administrators—all are viewed as part of the problem. Most of the various criticisms of contemporary education that focus on one or another of these factors contain a piece of the truth. But in criticizing uncommitted teachers or undisciplined students, they tend to miss a crucial causal factor that underlies many of the problems they do discuss—the penetration of contemporary education by the ideology of the market. It is on this factor—the instrumentalization of education—that I will focus. It is illustrated by the examples with which this [selection] began. My reason for this focus is that, paradoxically, many of the remedies that are now being proposed for problems of teacher competence and motivation, student motivation, absence of standards, and the like promise to make education even more instrumental than it already is. And by doing so, by turning play into work, they will create problems that are just as severe as the ones they are trying to solve.[3]

TEACHERS AND THE MARKET

Let's begin with teachers. The claim of many critics of education is that children don't learn what they should in school, because teachers are either untalented or unmotivated, or both. Standards are lax because it is easier to let children move through the system than to struggle to make sure that they learn what they should at every step of the process. Suppose this claim is true. How do you go about getting more talented people to go into teaching, keeping them interested in their work, and holding their performance accountable to reasonable standards? Most of the measures that have been proposed are predicated on the idea that teaching is a job, or maybe a career, but certainly not a calling. That is, the presumption is that people teach in order to get the salary and benefits that teaching provides. To get better teachers, we need to offer better salary and benefits. To keep teachers motivated, we need a set of incentives—bonuses of various sorts, or opportunities for advancement—that are contingent on the performance of their pupils. And to make sure that reasonably stringent standards are being maintained, we need close and careful monitoring, largely through standardized tests, of what the children are actually learning. In other words, to get good people into teaching, make it pay to be a teacher. To keep good people involved, make it pay to be a *good* teacher. And to prevent standards from slipping, make it cost if student performance slips.

The logic of these kinds of recommendations is straightforward. It's the same logic that guides our approach to transactions on the market. Right now, too many talented people are becoming lawyers and not enough are becoming teachers. Why? Because lawyers earn several times what teachers earn. If we reduce the earnings discrepancy between the two occupations, some potential lawyers will become teachers instead. And it isn't just the money; it's also the prestige that comes along with the money. Though we claim to value education enormously as a society, we don't think much of the people who provide it. "Those that can, do; those that can't, teach," the saying goes. We ask ourselves why anyone who is really good would settle for the salary and status that comes with teaching instead of something else? And this suspicion about the talents of those who enter teaching is reinforced by patterns of career choice among college students. It is rare to find students at the top of the class preparing for careers in education. Much more commonly, it's the C students who do it.

It's probably true that teaching is not attracting the best people, because of the relatively low salaries that teachers earn. But the explanation is more complex than a market orientation would suggest. Teachers have never been paid high salaries. Indeed, they probably do better now in comparison with other, similarly trained people than they ever have before. Nevertheless, a few generations ago, teachers were deeply respected by society despite their low salaries. Two things have happened in recent years to make the combination of low salary and high status virtually disappear from the social landscape. The first is the growth of economic imperialism. Values and standards that were internal to certain social practices and institutions and unique to them have slowly been replaced by external economic standards that are meant to be common to all practices. Increasingly, we have come to use the amount of money society is willing to pay people as the primary indication of the value society places on what those people do. Well, there goes the value of teaching.

Ironically, a second factor is the growth of modern feminism. For much of our history, the majority of teachers were women. Teaching was one of the few professions that was open to women. Teaching also afforded a short school day and a summer vacation, both of which left time for raising children. Women could be paid less than they were worth because they would willingly sacrifice some salary for the other benefits that teaching afforded, and because women were paid less than they were worth in all domains anyway. But they were talented and enthusiastic, and society respected and admired them. Their low salary was not equated with their social worth. Modern feminism has successfully opened up many other professions for women, where they receive both respect and higher salaries than those of teachers. So education has lost many of its best and most dedicated teachers.

Many suggest that we try to get better teachers by offering better salaries, and a handful of communities have done this. For most, though, there simply isn't enough money. While token increases in salary might be possible, substantial increases—of the kind that might make the financial returns for teaching comparable to the alternatives—are just not in the cards. Perhaps this is a lucky break. Perhaps paying teachers more is not the way to get better teachers. We have to pay teachers enough that they and their families will not be deprived of access to the most important things in life. But salary increases beyond this might only succeed in attracting those people who are interested primarily in money. It does not guarantee getting any more people who regard teaching as a calling—who believe they are performing a service of enormous social value and who derive their principal reward from performing that service well. To get, and to keep, people like this requires not increases in salary but improvements in the conditions under which teachers work so that they can freely exercise their intelligence, creativity, and judgment in the service of the goals—all the goals—of education.

And in this domain, education is moving in precisely the wrong direction. It is subjecting itself to the rules of the market. In the name of maintaining standards and holding teachers accountable for the success or failure of their students, the last few years have seen public education in America embark on a frenzied program of standardized testing. Virtually every recent effort for educational reform has had a substantial testing program at its core. The people who pay the bills, and the legislators they elect, are demanding a good "product." And the quality of the product is usually assessed by measuring how local schools stack up against the national average in this or that subject, as measured by this or that standardized test. It is estimated that more than fifty million standardized tests of basic competency are administered to students each year in the United States. More than half the states now require students to pass standardized tests to graduate from high school, a number that is double what it was only five years ago. And in some areas the fate of teachers is in the hands of students as they take the tests. Salary increases, and even continued employment, depend on how well the students do.

It is certainly important to have a means of assessing how effective an educational program is. And it is certainly reasonable to expect individual teachers to take some responsibility for the performance of the students in their classes. The trouble is that assessments of student performance are helpful only in the hands of dedicated teachers. They don't succeed in *making* teachers dedicated; and in the hands of the wrong kinds of teachers, standardized assessments can be a disaster.[4]

Let's look at an example. For years, New York has depended on statewide exams, given in all the major high school subjects, to ensure a certain degree of commonality of curriculum across the state. Students can get a high school diploma without passing all the exams given in each subject at the end of the year, but they get a special diploma if they pass the exams. Passing the exams and getting the special diploma are testimony that a student has met the state's rather high standards of competency.

I had a high school chemistry course that was a joke. I mean this quite literally; the teacher came into class each day and provided us with an extended, often riotous monologue, which only occasionally had anything to do with chemistry. Neither I nor any of my classmates were troubled by this. The class was fun, and the exams were tailored to cover the very little bit of chemistry that we had actually learned. So we were quite satisfied. But then, as spring came, we began to realize that we would soon be facing a state exam. And the people who made up the exam were unlikely to know or care about our chemistry teacher's jokes. Exam day drew nearer, and still we were learning almost no chemistry. About a week before the exam, the teacher started filling all the blackboards in the room with notes. Memorize this stuff, he was telling us, and you'll pass the exam. He had seen state exams many times before, and each year's edition was more or less like the previous year's. So I did a week's worth of memorizing, learned no chemistry, got a B+ on the exam, and went on to my next year of science.

What my chemistry teacher did was, of course, appalling. Standardized tests, given over a three-hour period, can't possibly measure everything a student has learned in a full year. Instead, the tests ask about a representative sample of the material that should have been covered. The assumption that underlies such a test is that since the test is representative of the class material as a whole, the more students have learned of the material, the better they will do on the tests. The tests are meant to be not an exhaustive measure of what was learned but an *index* of what was learned.

My chemistry teacher subverted the rationale and the legitimacy of the test by teaching us what he thought would be on it, and nothing more. In the case of my class, our performance on the test was not an *index* of what we knew—it was *all* we knew. Any reasonably clever teacher who worries that his or her future salary and employment depend on how well students do on these tests can tailor what goes on in the classroom to fit what is usually on the tests. Student performance on the tests may keep getting better and better, while students are actually learning less and less. If we thought that, in general, the answers to standardized test questions were becoming all that students learned in the course of a school year, we would abandon them altogether as a yardstick for assessing the performance of teachers and their students. But if a teacher regards his work as a job, what will stop him from doing precisely what my chemistry teacher did? In the case of teachers who take their work seriously, we don't have to worry about subverting tests. But in the case of these kinds of committed teachers, we probably don't have to worry too much about how well their students are doing in the first place....

Accountability in education is important, and standardized tests can provide useful information. But the intentions and goals of teachers matter. The effectiveness of standardized tests depends upon a dedicated teaching force that will use tests appropriately to guide what it does in the classroom. Testing will not root out bad and unprincipled teachers. They will find ways to use the tests to their own advantage. So testing programs may have the perverse effect of making the bad teachers look good and the good ones look bad. If untalented or unmotivated teachers are indeed a significant source of the problem with contemporary education, testing does not appear to be the way to solve it.

Teaching kids how to take tests is not the only way to improve test scores. We can also do it by teaching kids the skills that the tests are designed to measure, that is, teaching them what they are supposed to be taught. But this strategy involves something of a risk. It is ever so much more difficult and time consuming to teach real skills. It takes patience on the part of teachers and commitment on the part of students. It may even take some cooperation from parents and communities. So teaching the test is a kind of insurance policy. It is insurance against lack of dedication and resolve on the part of all the relevant parties. Of course, in the long term, such an insurance policy is a sham. Scoring well on tests doesn't serve the kids, or anybody else, once school is behind them. It protects educators from criticism or dismissal, and it protects communities from having to commit significant additional resources to education. But it serves no one else.

STUDENTS, MOTIVATION, AND THE MARKET

While some critics of modern education focus on the teachers, others focus on what they see as unmotivated students. But do students really lack motivation? There are two ways to tackle the problem of student motivation. One is to look for things that are *intrinsic* to the tasks at hand that will generate and sustain interest, enthusiasm, and commitment—that is, to make the work children do interesting and important to them without the prodding of the teacher. The other way is to look for things that are *extrinsic* to the tasks at hand that will keep kids attentive and involved. A system of rewards for good work and punishments for bad work is the most obvious example of this sort of approach. The incentives may be honors and awards, attention and approval from teachers, good and bad grades, special privileges, and even material things like money, or toys, or candy. Because these rewards and punishments are extrinsic, that is, because they bear no special relation to the tasks that children are actually being asked to perform, they can be almost anything.[5]

The examples that began this [selection] were about the difference between extrinsic and intrinsic incentives. My daughter was simply having fun learning to "compose" new songs. Our rewarding comments were a distraction; they put a damper on her fun. The fourth-grade kids read more when they competed for points, but they remembered little of what they read, and read less when not rewarded. Reading became work for hire. These examples suggest that while learning isn't always work, and doesn't have to be work, a system of extrinsic incentives can turn learning into work. If this suggestion is correct, relying on a system of external incentives to increase student motivation may be a mistake. Let us, then, examine this suggestion more carefully.

It's hard to find a very young child for whom learning is work. Little kids are extremely curious. They try to do things that they've never done before, things that the people around them are doing. They invent new ways to play with old toys. They invent ways to turn pots and pans and large cardboard boxes into an afternoon of fun. They learn an enormous amount every day, and each new bit of knowledge, each new skill, each new sign of mastery of the world, is a source of intense delight. Most kindergarten and first-grade teachers will tell you that motivating little kids to participate in school activities is not a problem. The problem, if anything, is to restrain their enthusiasm, so that large groups of kids can learn the same things together.

As schooling proceeds, much of this early energy and enthusiasm disappears. It's possible that the waning of enthusiasm has nothing to do with what happens in the schools themselves, but instead is the result of developmental changes that all kids go through. It's possible, in other words, that "natural" curiosity and enthusiasm for learning just diminishes as kids get older. Possible, but I think unlikely. Children who come to regard school as a boring, irrelevant chore somehow become alert, inventive, and intellectually alive in the streets after school. I believe that what goes on in school makes a substantial contribution to the motivation problem. Children are taught in school that learning is work, not play, and they are further taught that the kind of learning that does go on in school—book learning—is reserved exclusively *for* school. Kids learn that there's school and there's life, and what goes on in one place has nothing to do with what goes on in the other. So if you learn from books in school, that's the only place you learn from books. If I'm right, then at the same time our schools are teaching kids the basic skills and facts that will be the building blocks of all future learning, they are also teaching them that learning stops, and fun begins, when the school doors close at three….[6]

When extrinsic incentives are used in the classroom, they may have [two] adverse effects…. They may undermine motivation to learn, by teaching children that what they thought was play is actually work. And they may interfere with imaginative problem solving, by teaching children to be safe and repeat only solutions that have worked in the past. And much that goes on during the typical school day contributes to both of these effects. From about the third grade on, the school day is very much driven by the rhythm of tests and grades. Weekly spelling tests, math tests, science tests, social studies tests; interim reports sent home to parents, official report cards, honor rolls—before long, students learn that you work to do well on tests and get good grades. You study for tests, and only for tests. You study only the material that will be on tests. You try to "psych out" the teacher, guessing at what he usually asks, and learn just that material. And you try to get by doing as little work as possible.

Several years ago, to examine how malleable this motivational effect is, I did a little experiment in a class I was teaching. At the first class meeting, I read off the names of the students and then held up a final grade sheet, the thing we normally turn in at the end of a course. Next to each name was a B. I told them that if they chose to stay in this class, all of them would get a B, no matter what they did. There would be regular reading assignments, exams, and even term papers, and the exams and papers would be evaluated, but none of the evaluations would affect their final grade in the course. They could walk out of the class right then, never appear again, and still get B's in the course. A few students protested: "Why a B and not an A?" One or two dropped the course. One stood up, asked me if I was serious, and when I said yes, he thanked me very much for the B and left, never to be seen by me again. But most were enthusiastic. For the moment, they appreciated being treated as serious, committed students.

For the first several weeks of the course, all went well. Everyone came to class—prepared. But as the students experienced the demands of other classes (midterm exams and papers), their work in my class dropped off. More and more students skipped class, and those who came were less well prepared. As the course drew to a close, only a handful of the original forty students were still participating. Only two of them turned in their required final paper. The experiment was clearly a failure. These students and I discovered that they had become so dependent on exams and grades to provide them with motivation to do their work that when these external incentives were removed, they were left with a motivational vacuum. And these were students who were as dedicated as any college students in the country….

I can imagine someone reading this discussion and responding that a willingness to experiment, to take risks, to fail, is all well and good—but only under a set of very limited circumstances. You don't want people experimenting with the bread they bake and learning from their failures if their family has to eat those failures for dinner. You don't want engineers building bridges in experimental new ways and learning from their failures if people actually have to drive over those failures. You don't want subsistence farmers experimenting with their agricultural techniques and learning from their failures if their failures mean that their families will have to go hungry. Much of life demands results, demands successes. When the consequences of failure are severe, you want people to know *a* way to do things successfully even if it isn't the best way.

This observation is true and important. There are many situations in life where people can't afford to fail and thus can't afford to experiment. But school isn't—or shouldn't be—one of them. If the only consequence of doing something wrong is that you learn from your mistake, you can risk doing things wrong and thus learn from mistakes. But if the consequence is bad grades, disapproval, failure to get into a good college, and failure to get a good job, then experimentation is risky—perhaps too risky.

Perhaps the main reason that modern science offers the most powerful techniques in human history for finding things out about the world is that science invented the laboratory. In the laboratory, you can risk failure. In the laboratory, you can try new ways to build bridges or to plant crops without worrying that people will fall into rivers or starve to death if the new ways are unsuccessful. The laboratory, like the classroom, is a place where people can afford to learn from failure and can afford to take risks. But if the scientist's entire career—including promotion, tenure, salary, research grants—depends on success, it becomes harder to take risks, because the costs of failure have escalated. The result is less informative, less imaginative science. And if the student's entire future depends on classroom success, it becomes harder to take risks, for the same reason. By instrumentalizing education—by relying on a system of extrinsic incentives—we raise the stakes for what goes on in the classroom. We engender a performance orientation in students and thus undermine their inclination to learn. By doing this, by making the consequences of failure in the classroom just as steep as the consequences of failure in life, we undermine the classroom as a kind of "laboratory for life."

COLLEGE AND THE "MARKETPLACE OF IDEAS": FROM METAPHOR TO FACT

... For the last half of this century, the United States has had the luxury of regarding university education as an institution for the creation not just of skilled workers and professionals but of informed, responsible citizens. The modern educational revolution in this country has been the extension of the ideal of general education from the wealthy classes to the population as a whole. But this conception of university education is now changing. Growing competition among members of society for good jobs has been altering the character of the university. With college education so closely tied to job prospects, it is increasingly being thought of as an "investment" in the student's future. The money spent on school is expected to be returned, with interest, later on.

Universities have contributed to the problem in the last twenty years by increasing fees at a rate that was often twice the rate of inflation.[7] There are many reasons for this dramatic rise in college costs, and the market has had its hand in several of them. As universities compete in the "market" for good students, they are pressured to offer increasingly elaborate support and social services (athletic facilities, telephones in rooms, good food, career and psychological counseling, and a wide variety of entertainments) that are very expensive. And as they compete in the "market" for star professors and research support from private industry, they are pressured to provide high-tech laboratories and equipment, low teaching loads, high salaries, and plentiful research assistants. As a result of financial pressures like these, the cost of college has loomed as a larger and larger bite out of the family's resources. It is hard not to think about what college is going to "get you" in terms of a job and an income when your family is faced with paying $25,000 a year. It is hard not to think about what college is going to get you, in terms of a job and an income, when you may be faced with thousands of dollars worth of student loans that have to be paid off. The more expensive a college education becomes, the more people are pushed to think about college as a means to financial success.

And thinking of a college education as an economic investment can affect what people want out of education, and thus how they evaluate what they get. It encourages people to take an instrumental view of education, a performance orientation instead of a mastery orientation. An instrumental orientation emphasizes the kinds of careers that a college degree makes possible. It emphasizes curricular and extracurricular choices that will enhance the student's resume. Courses are taken on the basis of how impressive they sound, and of how likely they are to yield good grades, which will "cash in" as impressive résumés. At prices like these, pursuing one's intellectual interests is a luxury that few can afford.[8]

As people begin to assess their education in these economic terms, what actually goes on in the college classroom changes. Because colleges and universities have to be sensitive to market demand, they provide what students want. And, increasingly, what students (and their parents) want is an education that will pay them back, in kind, for their huge financial investment (in the decade between the early 1970s and the early 1980s, the number of business majors in college doubled while the number of literature majors halved). The goal of education shifts from creating well-informed, responsible, and enlightened citizens to creating highly paid workers (in 1987, 73 percent of the college students surveyed said that being well-off financially was their top goal, as compared with 39 percent in 1970). Extra salary potential becomes the yardstick for evaluating the effectiveness of an educational institution. What happens as considerations like these come to dominate the university is that institutions change what they do, so the creation of extra salary potential becomes the goal itself, instead of just a measuring stick. Once again, the index of a thing replaces the thing itself, as in the case of tests of basic competency in grade school. The "marketplace of ideas" is transformed from a metaphor that describes the free and open inquiry that characterizes the college curriculum into a literal description of that curriculum....

As educational institutions, whether by choice or by outside pressure, come increasingly to be literal marketplaces of ideas, they teach their students the *implicit* lesson that knowledge is a means to extrinsic ends, that knowledge can be sold, that credentials pay for themselves on the market. And this implicit lesson not only shapes the programs of study that students pursue but also affects the way in which they pursue them. Students find the most efficient ways to do well in school. In some cases, this means choosing easy courses. In other cases, it means cheating.

There are clear indications that cheating—at all levels of education—occurs with much higher frequency now than it did in the past. A recent survey of almost 300,000 college freshmen indicated that 30 percent had cheated at some time during their senior year in high school (and other studies come up with higher estimates—some as high as 50 percent). When a similar question was asked in 1966, the number was 20 percent. Moreover, indications are that cheating gets more and more frequent as you move up the educational ladder; more in high school than in junior high, and more in college than in high school. In addition, cheating seems to be more common among gifted students than among mediocre ones; more common in high-powered academic high schools than in trade schools, and more common in honors classes than in ordinary ones. It is the honors students who feel the most intense pressure to do well, so that they get into the best colleges or, later, the best medical or law schools....

What is it that influences students to cheat? No doubt the intense competition to succeed in school and in life is part of it. No doubt the pervasive cheating that they see around them in the adult community—both petty and serious—is part of it. And perhaps the casual neglect of instruction in ethical values by their parents and their teachers is part of it. But I think the largest and most significant influence on cheating is the instrumental orientation to education. It is only if you have this orientation that cheating even surfaces as an option....

THE VALUE OF EDUCATION AND THE EDUCATION OF VALUE

There is a very important lesson to be learned from my discussion of the instrumentalization of education. The lesson is that while students are learning all the things that are explicitly taught to them in school, they are also learning other things that are being taught implicitly. No one may intend the system of tests and grades that operates in school to teach children to view their education as instrumental. The tests may be intended only to monitor student progress and to provide feedback to the teacher on the effectiveness of her instruction. Nevertheless, the tests *will* teach the children to view learning as instrumental.

If we don't want children to think of education as instrumental—if we want them to be mastery oriented rather than performance oriented, and to continue to derive joy from meeting new challenges and learning new things, what are we to do? Must we abandon tests and grades all together? A radical move like this is, I think, as unnecessary and undesirable as it is impractical. Students and teachers alike need feedback, and tests are a good way to get it. What we *do* need to do is make sure that students and teachers alike understand what tests are really for, that they are primarily tools for learning and only secondarily, if at all, tools for individual evaluation. And what we need to do is make sure that students and teachers alike know what education is really for. Rather than taking it for granted that everyone knows that an informed citizenry is the lifeblood of our society, or that individual freedom and autonomy depend upon knowledge and understanding, these goods of education must be made explicit. Perhaps the most important explicit lesson that should be taught in schools is why schools are important, and why the things people do in them are important. In short, people must be taught about the real value of education....

Most of us place a high value on being "reasonable." We aspire to it in our own lives, admire it in others, and value it in our social institutions. We want the people we live with to do things for *reasons,* reasons that make sense. And we hope and expect that they will change what they do if we can give them good reasons for changing. Living with "reasonable" people is what makes real conversation possible. Reason allows us to influence one another with ideas and arguments. The alternative to reasoned conversation is the exercise of power. And reason denies privilege to power. Or better, it substitutes the power of evidence and argument for the power of physical force, or of position, or of money.

Reasonable people who disagree talk, argue, attempt to persuade. They engage in conversation with one another, conversation based on mutual respect. But conversation is not the only means at our disposal to get people to go along with us. Instead of reasoning with people—instead of giving them *reasons*—we can try to manipulate them. We can manipulate people by means of bribery ("Say you agree with me, and I'll give you a thousand dollars), or by means of threat ("Say you agree with me, or you won't get promoted"). We can manipulate people by intentionally misleading them, as advertisers routinely do, by giving them distorted or incomplete information. When all we care about is results, the various techniques of manipulation that are available to us look very tempting.

What keeps us from resorting to manipulation all the time is that we care not only about results but about how those results are achieved. We want people to agree with us, but we want them to agree for the right reasons—because our view is correct, and not because of who we are or who we know or how much we're willing to pay. Conversation rather than manipulation is the proper stance to take with people we respect and value. To manipulate people is to use them, and you don't use people if you respect them.

But however much we may aspire to reasonable conversation in our relations with other people, we don't always attain it. The possibilities for conversation, as opposed to manipulation, vary with the social conditions in which we live. The competitive individualism of the market, for example, undermines, or at least discourages, conversation. If the point is to get what you want as efficiently as possible, power and manipulation may well be the most effective means to given ends. If competitive individualism so pervades social life that we are interested in other people only to the extent that they can serve our individual desires, conversation loses its privileged position and becomes merely one technique among many for getting what we want. With conversation just instrumental—just a means—rather than a reflection of a fundamental moral stance toward other persons, the presumption of respect and value for others disintegrates. What disintegrates with it is the social character of intellectual life, and of the practice of education more generally.

You might suppose that our regard for other people as worthy of value and respect is so deep and basic a part of us that it—and we—will resist social forces that threaten it; that, in effect, we will correct changes in our social institutions that challenge this fundamental stance we take toward others. But I have been suggesting ... that this hopeful supposition is mistaken. As practices like education and the institutions that embody those practices lose sight of the goods that make them distinctive, as they join the march to the drumbeat of the market, not only is their own integrity threatened but so is the integrity of our moral stance toward other people. Thus, by attempting to combat the instrumentalization of education—by attempting to prevent the measurement of educational goods with an economic yardstick—we will not only be protecting the distinctive values of one of the best things in life. We will also be protecting an institution that teaches us how to interact with other people as agents worthy of respect rather than as objects worthy only of manipulation.

ENDNOTES

1. For a thorough, up-to-date discussion of the empirical evidence regarding the use of rewards in schools, see A. Kohn, *Punished by Rewards* (Boston: Houghton Mifflin, 1993). In this book, Kohn argues not only that the use of rewards has the kind of unfortunate side effects I will be describing ... but also that rewards are not even especially effective in producing the effects for which they are intended.
2. For the study with nursery school children, see M. R. Lepper, D. Greene, and R. E. Nisbett, "Undermining Children's Intrinsic Interest with Extrinsic Rewards: A Test of the 'Overjustification' Hypothesis," *Journal of Personality and Social Psychology* 28 1973: 129–37.
3. National Commission on Excellence in Education, *A Nation at Risk* (Washington, D.C: Government Printing Office, 1984).
4. E. B. Fiske, "America's Test Mania," *New York Times Educational Supplement,* September 10, 1989, pp. 16–20.
5. See Kohn, *Punished by Rewards.*
6. See E. Deci, *Intrinsic Motivation* (New York: Plenum, 1975); M. R. Lepper and D. Greene, eds. *The Hidden Costs of Reward* (Hillsdale, N.J: Erlbaum, 1978); and, for the rule-finding experiments, B. Schwartz, "Reinforcement-Induced Behavioral Stereotypy: How Not to Teach People to Discover Rules," *Journal of Experimental Psychology: General* 111 (1982): 23–59; and idem, "The Creation and Destruction of Value," *American Psychologist* 45 (1990): 7–15.
7. For an illuminating discussion and analysis of the recent explosion in college costs, see E. Negin, "Why College Tuitions Are So High," *The Atlantic,* March 1993, pp. 32–34, 43–44.
8. For a discussion of the dramatic changes occurring in the course selections and career plans of students, see B. Ehrenreich, *Fear of Falling* (New York: Harper Perennial, 1990), pp. 209–14; and L. Menand, "What Are Universities For?" *Harper's,* December 1991, pp. 47–56.

DISCUSSION QUESTIONS

1. Schwartz seems to be against giving students external rewards and punishments for their performance in school. What is his concern? Do you agree with him?
2. Why does the teaching profession have difficulty attracting good candidates? Schwartz does not think that higher salaries will attract better or more people to teaching. Why? Do you agree?
3. Are standardized tests good for education? What is Schwartz's view? What is your view?
4. Schwartz's point in this selection is that we need to recognize the difference between extrinsic and intrinsic incentives. Explain his position.
5. Do you think that it is a good thing or a bad thing to think of a college education as an economic investment? What does Schwartz think?
6. Is it necessary to emphasize competition to create enough incentive to encourage students to learn in school?

PART III

Social Problems: Economic Inequality and Poverty

For many sociologists, nothing is more important than social class. Society is patterned, organized, and structured, and one of the most important ways it is structured is by class, usually defined as a system of economic inequality. Thus, sociologists will repeatedly turn to class as the source of social problems. Class structure, or economic inequality, is not an attempt to understand human problems from the point of view of the individual actor but almost always focuses on the "macro" social world—the forces that exist in society that direct people's lives.

Many articles and books address economic inequality. The four selections in Part III try to make a case against economic inequality in American society, and ask us to see inequality on a social-structural level rather than on the level of simply blaming the individual.

In a selection from their excellent book on economic class called *The New Class Society,* Robert Perrucci and Earl Wysong make the argument that the United States is undergoing a major shift in the distribution of wealth that affects all classes in significant ways. In the globalization of the American economy, "international outsourcing" and "offshoring" are the new buzzwords for productivity, and many individuals who were once securely in the middle class are now part of a contingent workforce.

Thomas Shapiro focuses on the wealth gap between Americans arising as a result of the vast differences in inheritance assets. Inherited wealth poses a challenge to the ideal of meritocracy—namely, that class positions should result from hard work.

Harrell Rodgers, Jr. also introduces us to the problem of poverty, focusing on its causes. He discusses five theories of poverty, each identifying one set of causes and each recommending what we can do about it. The first theory focuses on the culture of the poor and the welfare system that creates that culture; Rodgers labels this a conservative approach. The second theory is liberal; he calls it structural/economic, and he emphasizes the role of employment in the inner city that has created a black underclass. The third theory is one put forward by Cornel West, who focuses on the historical and contemporary causes that include both lack of employment opportunities and racism; Rodgers calls this "culture as structure." The fourth theory identifies a number of causes in the economic forces in society. Finally, Rodgers suggests some factors that must be considered if we really want to do something about poverty.

The final selection, written by Stephen McNamee and Robert Miller, Jr., details some of the ways wealth is perpetuated within families. Wealth creates cumulative advantages, many of which are economic, but also social and educational as well. McNamee and Miller suggest that the United States is far from a society that simply rewards those who have merit; the race for success, they assert, has a starting line that is staggered.

It is easy for people who are successful in an unequal society to believe that everything in that society is just and necessary. It is easy for such people to regard poverty as the fault of the individual. All four of these selections are about social structure, and thus they share a view that social problems are indeed "social," in the sense that their causes are social.

6

The Global Economy, the Privileged Class, and the Working Class

ROBERT PERRUCCI AND EARL WYSONG

The Four Questions

1. What is the problem, according to Perrucci and Wysong?
2. What makes it a *social* problem? Are many people affected? Is the cause social? Does it affect society in a problematic way?
3. What causes this problem?
4. Do the authors state or imply what to do?

Topics Covered

Privileged class

Global economy

Multinational corporations

Downsizing

Core workers

Temporary workers

Contingent workers

In 2001, record layoffs led to the worst U.S. job market since the recession of 1990–91. In the period from January to June, 2001, U.S. companies announced 652,510 layoffs. From manufacturing to high-tech, workers lost jobs at the fastest rate in years. Although the 2001 job cuts were dramatic, they were merely the latest chapter in what has been a long story for U.S. workers. Twenty years earlier we followed 850 workers through what has since become an all-too-familiar pattern for millions of workers.

On December 1, 1982, an RCA television cabinet-making factory in Monticello, Indiana, closed its doors and shut down production. Monticello, a town of five thousand people in White County (population twenty-three thousand), had been the home of RCA since 1946. The closing displaced 850 workers who were members of Local 3154 of The United Brotherhood of Carpenters and Joiners. Officials at RCA cited the high manufacturing costs and foreign competition as key factors leading to the closing....

Whether the personal response to the closing was faith, fear, or anger, the common objective experience of the displaced workers was that they had been "dumped" from the "middle class." These displaced factory workers viewed themselves as middle class because of their wages and their lifestyles (home ownership, cars, vacations). Most had worked at RCA for two decades or more. They had good wages, health care benefits, and a pension program. They owned their homes (with mortgages), cars, recreational vehicles, boats, and all the household appliances associated with middle-class membership. All the trappings of the American Dream were threatened as their seemingly stable jobs and secure incomes disappeared. In the space of a few months these workers and their families joined the growing new working class—the

SOURCE: From Robert Perrucci and Earl Wysong, *The New Class Society: Goodbye American Dream?* 2nd ed. (New York: Rowman & Littlefield Publishers, 2003), pp. 91–108. Reprinted by permission.

80 percent of Americans without stable resources for living....

The experiences of the 850 RCA workers from Monticello, Indiana, were part of a national wave of plant closings that swept across the land two decades ago. According to a study commissioned by the U.S. Congress, between the late 1970s and mid-1980s more than 11 million workers lost jobs because of plant shutdowns, relocation of facilities to other countries, or layoffs. Most of these displaced workers were in manufacturing. Subsequent displaced worker surveys commissioned by the Bureau of Labor Statistics estimated that between 1986 and 1991 another 12 million workers were displaced, but now they were predominantly from the service sector (about 7.9 million).[1] When these displaced workers found new jobs, it was often in industry sectors where wages were significantly lower than what they had earned and jobs were often part-time and lacked health insurance and other benefits....

The rush to downsize in some of America's largest and most prestigious corporations became so widespread in the 1990s that a new occupation was needed to handle the casualties. The "outplacement professional" was created to put the best corporate face on a decision to downsize, that is, to terminate large numbers of employees—as many as ten thousand. The job of these new public relations types is to get the general public to accept downsizing as the normal way of life for corporations that have to survive in the competitive global economy. Their job is also to assist the downsized middle managers to manage their anger and to get on with their lives.

The *Human Resources Development Handbook* of *the American Management Association* provides the operating philosophy for the outplacement professional: "Unnecessary personnel must be separated from the company if the organization is to continue as a viable business entity. To do otherwise in today's globally competitive world would be totally unjustified and might well be a threat to the company's future survival."[2]

The privileged 20 percent of the population are hard at work telling the other 80 percent about the harsh realities of the changing global economy. "Lifetime employment" is out. The goal is "lifetime employability" which workers try to attain by accumulating skills and being dedicated and committed employees. Even Japan's highly touted commitment to lifetime employment (in some firms) is apparently unraveling, as reported in a prominent feature article in the *New York Times*.[3] It should be no surprise that an elite media organization like the *Times,* whose upper-level employees belong to the privileged class, should join in disseminating the myth of the global economy as the "hidden hand" behind the downsizing of America. The casualties of plant closings and downsizings are encouraged to see their plight as part of the "natural laws" of economics.

This enormous transformation of the U.S. economy over a thirty-year period has been described by political leaders and media as the inevitable and therefore normal workings of the emerging global economy. Some, like former president Reagan, even applauded the changes as a historic opportunity to revitalize the economy. In a 1985 report to Congress, he stated, "The progression of an economy such as America's from the agricultural to manufacturing to services is a natural change. The move from an industrial society toward a postindustrial service economy has been one of the greatest changes to affect the developed world since the Industrial Revolution."[4]

A contrasting view posits that the transformation of the U.S. economy is not the result of natural economic laws or the "hidden hand" of global economic markets but, rather, the result of calculated actions by multinational corporations to expand their profits and power. When corporations decide to close plants and move them overseas where they can find cheap labor and fewer government regulations, they do so to enhance profits and not simply as a response to the demands of global competition. In many cases, the U.S. multinationals themselves are the global competition that puts pressure on other U.S. workers to work harder, faster, and for lower wages and fewer benefits.

THE GLOBAL ECONOMY AND CLASS STRUCTURE

... Discussion about the new global economy by mainstream media reporters and business leaders generally focuses on three topics. First is the appearance of many new producers of quality goods in parts of the world that are normally viewed as less developed. Advances in computer-based production systems have allowed many countries in Southeast Asia and Latin America to produce goods that compete with those of more advanced industrial economies in Western Europe and North America. Second is the development of telecommunications systems that permit rapid economic transactions around the globe and the coordination of economic activities in locations separated by thousands of miles. The combination of advances in computer-based production and telecommunications makes it possible for large firms, especially multinationals, to decentralize their production and locate facilities around the globe. Third is the existence of an international division of labor that makes it possible for corporations to employ engineers, technicians, or production workers from anywhere in the world. This gives corporations great flexibility when negotiating with their domestic workforce over wages and benefits. These changes in how we produce things and who produces them have resulted in expanded imports and exports and an enlarged role for trade in the world economy. Leading this expansion has been increased foreign investments around the world by the richer nations. It is estimated that two-thirds of international financial transactions have taken place within and between Europe, the United States, and Japan.[5]

The changes just noted are often used as evidence of a "new global economy" *out there* constraining the actions of all corporations to be competitive if they hope to survive. One concrete indicator of this global economy *out there* is the rising level of international trade between the United States and other nations. In the 1960s, the United States was the dominant exporter of goods and services, while the imports of foreign products played a small part in the U.S. economy. Throughout the 1970s foreign imports claimed an increasing share, and by 1981 the United States "was importing almost 26 percent of its cars, 25 percent of its steel, 60 percent of its televisions, tape recorders, radios, and phonographs, 43 percent of its calculators, 27 percent of its metal-forming machine tools, 35 percent of its textile machinery, and 53 percent of its numerically controlled machine tools."[6] Imports from developing nations went from $3.6 billion in 1970 to $30 billion in 1980.

Throughout the 1980s, the United States became a debtor nation in terms of the balance between what we exported to the rest of the world and what we import. By 2000, the U.S. trade deficit indicated that the import of goods and services exceeded exports by $370 billion. This is the largest deficit since the previous high in 1987 of $153.4 billion. But what do these trade figures tell us? On the surface, they appear to be a function of the operation of the global economy, because the figures indicate that we have an $81.3 billion deficit with Japan, $83.8 billion with China, and $24.9 billion with Mexico.[7] It appears that Japanese, Chinese, and Mexican companies are doing a better job of producing goods than the United States and thus we import products rather than producing them ourselves. But is this the correct conclusion? The answer lies in how you count imports and exports.

Trade deficit figures are based on balance of payment statistics, which tally the dollar value of U.S. exports to other countries and the dollar value of foreign exports to the United States; if the dollar value of Chinese exports to the United States exceeds the dollar value of U.S. exports to China, the United States has a trade deficit with China. This would appear to mean that Chinese companies are producing the goods being exported to the United States. But that is not necessarily the case. According to the procedures followed in calculating trade deficits, "the U.S. balance of payments statistics are intended to capture the total amount of transactions between U.S. *residents* and *residents* of the rest of the world."[8] If "resident" simply identifies the geographical location of the source of an import, then some unknown portion of the $49.7 billion U.S. trade deficit with China could be from U.S.-owned firms that are producing goods in China and exporting them to the United States. Those U.S. firms are residents of China, and their exports are counted as Chinese exports to the United States.

Thus, the global economy that is *out there* forcing U.S. firms to keep wages low so we can be more competitive might actually be made up of U.S. firms that have located production plants in countries other than in the United States. Such actions may be of great benefit to the U.S. multinational firms that produce goods around the world and export them to the U.S. market. Such actions may also benefit U.S. consumers, who pay less for goods produced in low-wage areas. But what about the U.S. worker in a manufacturing plant whose wages have not increased in twenty years because of the need to compete with "foreign companies"? What about the worker who may never get a job in manufacturing because U.S. firms have been opening plants in other countries rather than in the United States? As the comic strip character Pogo put it: "We have met the enemy and it is us."

American multinational corporations' foreign investments have changed the emphasis in the economy from manufacturing to service. This shift has changed the occupational structure by eliminating high-wage manufacturing jobs and creating a two-tiered system of service jobs. There have been big winners and big losers in this social and economic transformation. The losers have been the three out of four Americans who work for wages—wages that have been declining since 1973; these American workers constitute the new working class... The big winners have been the privileged classes, for whom jobs and incomes have expanded at the same time that everyone else was in decline. Corporate executives, managers, scientists, engineers, doctors, corporate lawyers, accountants, computer programmers, financial consultants, health care professionals, and media professionals have all registered substantial gains in income and wealth in the last thirty years. And the changes that have produced the "big losers" and "big winners" have been facilitated by the legislative actions of the federal government and elected officials of both political parties, whose incomes, pensions, health care, and associated "perks" have also grown handsomely in the past two decades....

WHEN YOUR DOG BITES YOU

... While corporate profits from the domestic U.S. economy were declining steadily from the mid-1970s, investment by U.S. corporations abroad showed continued growth. The share of corporate profits from direct foreign investment increased through the 1970s, as did the amount of U.S. direct investment abroad. In 1970, direct investment by U.S. firms abroad was $75 billion, and it rose to $167 billion in 1978. In the 1980–85 period it remained below $400 billion, but thereafter increased gradually each year, reaching $716 billion in 1994. The 100 largest U.S. multinational corporations reported foreign revenue in 1994 that ranged from 30 to 70 percent of their total revenue: IBM had 62 percent of total revenue from foreign sources; Eastman Kodak 52 percent; Colgate-Palmolive 68 percent; and Johnson and Johnson, Coca Cola, Pepsi, and Procter and Gamble each 50 percent.[9]

American multinational corporations sought to maintain their profit margins by increasing investments in affiliates abroad. This strategy may have kept stockholders happy, and maintained the price of corporate stocks on Wall Street, but it would result in deindustrialization—the use of corporate capital for foreign investment, mergers, and acquisitions rather than for investment in domestic operations.[10] Instead of investing in the U.S. auto, steel, and textile industries, companies were closing plants at an unprecedented rate and using the capital to open production facilities in other countries. By 1994, U.S. companies employed 5.4 million people abroad, more than 4 million of whom worked in manufacturing.[11] Thus, millions of U.S. manufacturing workers who were displaced in the 1980s by plant closings saw their jobs shifted to foreign production facilities. Although most criticism of U.S. investment abroad is reserved for low-wage-countries like Mexico and Thailand, the biggest share of manufacturing investment abroad is in Germany and Japan—hardly low-wage countries. The United States has large trade deficits with Japan and Western Europe, where the hourly wages in manufacturing are 15–25 percent higher than in the United States.[12] This fact challenges the argument made by multinational corporations that if they did not shift production abroad, they would probably lose the sale of that product.

The movement of U.S. production facilities to foreign countries in the 1980s and 1990s was not simply the result of a search for another home where they could once again be productive and competitive. It appeared as if RCA closed its plant in Monticello, Indiana, because its high-wage workers made it impossible to compete with televisions being produced in Southeast Asia. Saddened by having to leave its home in Indiana of thirty-five years, RCA would have to search for another home where, it was hoped, the company could stay at least another thirty-five years, if not longer. Not likely: Plants did not close in the 1980s to find other homes; the closures were the first step in the creation of the homeless and stateless multinational corporation—an entity without ties to place, or allegiances to people, communities, or nations.

Thus, the rash of plant closings in the 1970s and 1980s began as apparent responses to economic crises of declining profits and increased global competition. As such, they appeared to be rational management decisions to protect stock-holder investments and the future of individual firms. Although things may have started in this way, it soon became apparent that what was being created was the *spatially decentered firm:* a company that could produce a product with components manufactured in a half-dozen different plants around the globe and then assembled at a single location for distribution and sale. Although spatially decentered, the new transnational firm was also centralized in its decision making, allowing it to coordinate decisions about international investment. The new firm and its global production system were made possible by significant advances in computer-assisted design and manufacturing that made it unnecessary to produce a product at a single location. They were also made possible by advances in telecommunications that enabled management at corporate headquarters to coordinate research, development, design, manufacturing, and sales decisions at various sites scattered around the world.

The homeless and stateless multinational firm is able to move its product as quickly as it can spot a competitive advantage associated with low wages, cheaper raw materials, advantageous monetary exchange rates, more sympathetic governments, or proximity to markets. This encourages foreign investment because it expands the options of corporations in their choice of where to locate, and it makes them less vulnerable to pressure from workers regarding wages and benefits.

The advantages of the multinational firm and foreign investments are also a product of the U.S. tax code. In addition to providing the largest firms with numerous ways to delay, defer, and avoid taxes, corporate profits made on overseas investments are taxed at a much lower rate than profits from domestic operations. Thus, as foreign investments by U.S. firms increased over the last two decades, the share of total taxes paid by corporations declined. In the 1960s, corporations in the United States paid about 25 percent of all federal income taxes, and in 1991 it was down to 9.2 percent. A 1993 study by the General Accounting Office reported that more than 40 percent of corporations with assets of more than $250 million either paid no income tax or paid less than $100,000.[13] Another study of 250 of the nation's largest corporations reported that in 1998, twenty-four of the corporations received tax rebates totaling $1.3 billion, despite reporting U.S. profits before taxes of $12.0 billion. A total of forty-one corporations paid less than zero federal income tax in at least one year from 1996 to 1998, despite reporting a total of $25.8 billion in pretax profits.[14] In testimony before the Committee on the Budget of the U.S. House of Representatives, Ralph Nader reported that in fiscal year 1999 corporations received $76 billion in tax exclusions, exemptions, deductions, credits, and so forth, and that the estimates for the years 2000–2004 will reach $394 billion in corporate tax subsidies.[15]

CREATING THE NEW WORKING CLASS

… When the large multinational firm closes its U.S. facilities and invests in other firms abroad or opens new facilities abroad, the major losers are the production workers who have been displaced and the communities with lower tax revenues and increased costs stemming from expanded efforts to attract new businesses. But this does not mean that the firms are losers, for they are growing and expanding operations elsewhere. This growth creates the need for new employees in finance, management, computer operations, information systems, and clerical work. The total picture is one of shrinking production plants and expanding corporate headquarters; shrinking blue-collar employee rolls and two-tiered expansion of high-wage professional-managerial and low-wage clerical positions.

Having been extraordinarily successful in closing U.S. plants, shifting investment and production abroad, and cutting both labor and labor costs (both the number of production workers and their wage-benefit packages), major corporations now turned their attention to saving money by cutting white-collar employees. In the 1990s, there were no longer headlines about "plant closings," "capital flight," or "deindustrialization." The new strategy was "downsizing," "rightsizing," "reengineering," or how to get the same amount of work done with fewer middle managers and clerical workers.

When Sears, Roebuck and Company announced that it could cut 50,000 jobs in the 1990s (while still employing 300,000 people) its stock climbed 4 percent on the New York Stock Exchange. The day Xerox announced a planned cut of 10,000 employees, its stock climbed 7 percent. Eliminating jobs was suddenly linked with cutting corporate waste and increasing profits. Hardly a month could pass without an announcement by a major corporation of its downsizing plan. Tenneco Incorporated would cut 11,000 of its 29,000 employees. Delta Airlines would eliminate 18,800 jobs, Eastman Kodak would keep pace by eliminating 16,800 employees, and AT&T announced 40,000 downsized jobs, bringing its total of job cuts since 1986 to 125,000. Not to be outdone, IBM cut 180,000 jobs between 1987 and 1994. The practice continues into the new century; as reported in the *New York Times* (July 13, 2001), Motorola, Inc., announced on July 12, 2001, that it would cut 30,000 jobs in 2001. On that same day, although it reported an operating loss in the second quarter of eleven cents per share, Motorola stock rose by 16 percent….

Job loss in the 1990s appeared to hit hardest at those who were better educated (some college or more) and better paid ($40,000 or more). Job loss aimed at production workers in the 1980s was "explained" by the pressures of global competition and the opportunities to produce in areas with lower-wage workers. The "explanation" for the 1990s downsizing was either new technology or redesign of the organization. Some middle managers and supervisors were replaced by new computer systems that provide surveillance of clerical workers and data entry jobs. These same computer systems also eliminate the need for many middle managers responsible for collecting, processing, and analyzing data used by upper-level decision makers.

Redesign of organizations was achieved by eliminating middle levels within an organization and shifting work both upward and downward. The downward shift of work is often accompanied by new corporate plans to "empower" lower-level workers with new forms of participation and opportunities for career development. All of this redesign reduced administrative costs and increased the work load for continuing employees.

Investors, who may have been tentative about the potential of profiting from the deindustrialization of the 1980s because it eroded the country's role as a manufacturing power, were apparently delighted by downsizing. During the 1990s and continuing beyond 2000 the stock market skyrocketed from below 3000 points on the Dow

Jones Industrial Average to 10,498 in mid-July 2001—an increase of almost 200 percent. The big institutional investors apparently anticipated that increasing profits would follow the broadly based actions of cutting the workforce.

Downsizing is often viewed by corporations as a rational response to the demands of competition and thereby a way to better serve their investors and ultimately their own employees. Alan Downs, in his book *Corporate Executions,* challenges four prevailing myths that justify the publicly announced layoffs of millions of workers.[16] First, downsizing firms do not necessarily wind up with a smaller workforce. Often, downsizing is followed by the hiring of new workers. Second, Downs questions the belief that downsized workers are often the least productive because their expertise is obsolete: According to his findings, increased productivity does not necessarily follow downsizing. Third, jobs lost to downsizing are not replaced with higher-skill, better-paying jobs. Fourth, the claim that companies become more profitable after downsizing, and that workers thereby benefit, is only half true—many companies that downsize do report higher corporate profits and, as discussed earlier, often achieve higher valuations of their corporate stock. But there is no evidence that these profits are being passed along to employees in the form of higher wages and benefits.

After challenging these four myths, Downs concludes that the "ugly truth" of downsizing is that it is an expression of corporate self-interest to lower wages and increase profits. This view is shared by David Gordon, who documents the growth of executive, administrative, and managerial positions and compensation during the period when "downsizing" was at its highest.[17] Gordon describes bureaucratic "bloat" as part of a corporate strategy to reduce the wages of production workers and increase and intensify the level of managerial supervision. Slow wage growth for production workers and top-heavy corporate bureaucracies reinforce each other, and the combination produces a massive shift of money out of wages and into executive compensation and profits. This "wage squeeze" occurred not only in manufacturing (because of global competition) but also in mining, construction, transportation, and retail trade.[18] Although it is to be expected that foreign competition will have an impact on wages in manufacturing, it should not affect the nontrade sector to the same extent. Thus, the "wage squeeze" since the mid-1970s that increased income and wealth inequality in the United States is probably the result of a general assault on workers' wages and benefits rather than a response to global competition.

The impact of these corporate decisions on the working class was hidden from public view by the steady growth of new jobs in the latter part of the 1990s, and by the relatively low rate of unemployment. In his second term in office, President Clinton made frequent mention of the high rate of job creation (without mentioning that they were primarily low-wage service jobs) and the historically low unemployment rate. Unfortunately, the official rate of unemployment can hide the real facts about the nation's economic health. For example, an unemployment rate of 4.2 percent in 1999 excludes part-time workers who want full-time work, and discouraged workers who have given up looking. If these workers are added to the unemployed we have an "underemployment rate" of 7.5 percent, or about 10.5 million workers. The official unemployment rate also hides the fact that unemployment for Black Americans was 8.0 percent in 1999, or that in urban areas there were pockets of unemployment that approached 25 percent.[19]

Thus, the result of more than a decade of plant closings and shifting investment abroad, and less than a decade of downsizing America's largest corporations, has been the creation of a protected privileged class and a working class with very different conditions of employment and job security. The three major segments of the working class are core workers, temporary workers, and contingent workers.

Core Workers

Core workers are employees possessing the skills, knowledge, or experience that are essential to the operation of the firm. Their income levels place them in the "comfort class."... They are essential for the firm, regardless of how well it might be doing from the standpoint of profits and growth; they are simply needed for the firm's continuity. Being in the core is not the same as being in a particular occupational group. A firm may employ many engineers and scientists, only some of whom might be considered to be in the core. Skilled blue-collar workers may also be in the core. Core employees have the greatest job security with their employing organizations; they also have skills and experiences that can be "traded" in the external labor market if their firm should experience an unforeseen financial crisis. Finally, core employees enjoy their protected positions precisely because there are other employees just like them who are considered temporary.

Temporary Workers

The employment of temporary workers is linked to the economic ups and downs that a firm faces. When sales are increasing, product demand is high, and profits match those of comparable firms, the employment of temporary workers is secure. When inventories increase, or sales decline sharply, production is cut back, and temporary employees are laid off or fired. The temporary workers' relationship to the firm is a day-to-day matter. There is no tacit commitment to these employees about job security and no sense that they "belong to the family."

A good example of the role of temporary workers is revealed in the so-called transplants—the Japanese auto firms like Toyota, Nissan, and Honda that have located assembly plants in Kentucky, Ohio, Michigan, Illinois, Indiana, and Tennessee. Each of these firms employs between two thousand and three thousand American workers in their plants, and they have made explicit no-layoff commitments to workers in return for high work expectations (also as a way to discourage unionization). However, in a typical plant employing 2,000 production workers, the no-layoff commitment was made to 1,200 hires at start-up time; the other 800 hires were classified as temporary. Thus, when there is a need to cut production because of weak sales or excessive inventory, the layoffs come from the pool of temporary workers rather than from the core workers. Sometimes these temporary workers are not even directly employed by the firm but are hired through temporary help agencies like Manpower. Employment through temporary help agencies doubled between 1982 and 1989, and doubled again between 1989 and 1999.[20] These temporary workers are actually contingent workers.

Contingent Workers

Workers in nonstandard employment arrangements (part time, temporary, independent contractors) are often described as contingent workers. Some of these workers, as noted earlier, are employees of an agency that contract with a firm for their services.... About one in four persons in the labor force is a contingent worker, that is, a temporary or part-time worker.[21] These workers can be clerk-typists, secretaries, engineers, computer specialists, lawyers, or managers. They are paid by the temp agency and do not have access to a company's benefit package of retirement or insurance programs. Many of the professionals and specialists who work for large firms via temp firms are often the same persons who were downsized by those same companies....

These three groups of workers fit into the bottom part of the double-diamond class structure ... and it is only the core workers who have even the slightest chance to make it into the privileged class. Core workers with potential to move up generally have the credentials, skills, or social capital to have long-term job security, or to start their own business, and therefore the possibility of having substantial consumption capital (a good salary) and capital for investment purposes. Let us now consider how the privileged class holds on to its advantaged position in the double-diamond class structure.

CARE AND FEEDING OF THE PRIVILEGED CLASS

Most people who are in the privileged class are born there, as the sons, daughters, and relatives of highly paid executives, professionals, and business owners. Of course, they do not view their "achievements" that way. As one wag once said of former President George Bush, "He woke up on second base and thought he'd hit a double." But some members of the privileged class have earned their places, whether by means of exceptional talent, academic distinctions, or years of hard work in transforming a small business into a major corporation. Regardless of how much effort was needed to get where they are, however, members of the privileged class work very hard to stay where they are. Holding on to their wealth, power, and privilege requires an organized effort by businessmen, doctors, lawyers, engineers, scientists, and assorted political officials. This effort is often cited to convince the nonprivileged 80 percent of Americans that the privileged are deserving of their "rewards" and that, in general, what people get out of life is in direct proportion to what they put in. This effort is also used to dominate the political process so that governmental policies, and the rules for making policy, will protect and advance the interests of the privileged class.

However, before examining the organized effort of the privileged class to protect its privilege, it is first necessary to examine how members of the privileged class convince one another that they are deserving. Even sons and daughters from the wealthiest families need to develop biographical "accounts" or "stories" indicating they are deserving. This may involve accounts of how they worked their way up the ladder in the family business, starting as a clerk but quickly revealing a grasp of the complexities of the business and obtaining recognition from others of their exceptional talent.

Even without the biographical accounts used by the privileged class to justify exceptional rewards, justification for high income is built into the structure of the organizations they join. In every organization—whether an industrial firm, bank, university, movie studio, law firm, or hospital—there are multiple and distinct "ladders" that locate one's position in the organization. New employees get on one of these ladders based on their educational credentials and work experience. There are ladders for unskilled employees, for skilled workers, and for professional and technical people with specialized knowledge. Each ladder has its own distinct "floor" and "ceiling" in terms of what can be expected regarding salary benefits, and associated perks. In every organization, there is typically only one ladder that can put you in the privileged class, and this usually involves an advanced technical or administrative career line. This career line can start at entry levels of $70,000–80,000 annual compensation, with no upper limit beyond what the traffic will bear. These are the career ladders leading to upper executive positions providing high levels of consumption capital and opportunities for investment capital....

Resistance to the Global Economy

...We have tried to provide a glimpse of the meaning of the bogeyman global economy. The term has been used to threaten workers and unions and to convince everyone that they must work harder if they want to keep their jobs. The global economy is presented as if it is *out there* and beyond the control of the corporations, which must continually change corporate strategies in order to survive in the fiercely competitive global economy. It is probably more accurate to view the current global economy as an accelerated version of what U.S. financial and industrial corporations have been doing since the end of World War II—roaming the globe in search of profits. The big change is that since the 1980s, U.S. firms have found it easier to invest overseas. They have used this new opportunity to create new international agreements like NAFTA and FTAA that attack organized labor and threaten workers to keep their wage demands to a minimum. In this view, the global economy is composed primarily of U.S. companies investing abroad and exporting their products to the United States (as the largest consumer market in the world) and other countries. These multinational corporations have an interest in creating the fiction that the global economy is some abstract social development driven by "natural laws" of economics, when it is actually the product of the deliberate actions of 100 or so major corporations.

The problem posed by the global economy is that it has increased the influence of large corporations over the daily lives of most Americans. This influence is revealed in corporate control over job growth and job loss, media control of information, and the role of big money in the world of national politics. At the same time that this growing influence is revealed on a daily basis, it has become increasingly clear that the major corporations have abandoned any sense of allegiance to, or special responsibilities toward, American workers and their communities.

This volatile mix of increasing influence and decreasing responsibility has produced the double-diamond class structure, where one in five Americans is doing very well indeed, enjoying the protection that comes with high income, wealth, and social contacts. Meanwhile, the remaining four out of five Americans are exploited and excluded.

ENDNOTES

1. Office of Technology Assessment, *Technology and Structural Unemployment* (Washington, D.C.: Congress of the United States, 1986); Thomas S. Moore, *The Disposable Work Force* (New York: Aldine de Gruyter, 1996).
2. Joel Bleifuss, "The Terminators," *In These Times,* March 4, 1996, 12–13.
3. Sheryl Wu Dunn, "When Lifetime Jobs Die Prematurely: Downsizing Comes to Japan, Fraying Old Workplace Ties," *New York Times,* June 12, 1996.
4. John Miller and Ramon Castellblanch, "Does Manufacturing Matter?" *Dollars and Sense,* October 1988.
5. Noam Chomsky, *The Common Good* (Monroe, Me.: Common Courage Press., 2000).
6. Robert B. Reich, *The Next American Frontier* (New York: Times Books, 1983).
7. U.S. Bureau of the Census, Foreign Trade Division, Washington, D.C. 20233, 2000.
8. John Pomery, "Running Deficits with the Rest of the World—Part I," *Focus on Economic Issues,* Purdue University (Fall 1987). (Emphasis added).
9. "The 100 Largest U.S. Multinationals," *Forbes,* July 17, 1995, 274–76.
10. Barry Bluestone and Bennett Harrison, *The Deindustrialization of America* (New York: Basic Books, 1982).
11. Louis Uchitelle, "U.S. Corporations Expanding Abroad at a Quicker Pace," *New York Times,* July 25, 1998.
12. David M. Gordon, *Fat and Mean: The Corporate Squeeze of Working Americans and the Myth of Managerial Downsizing* (New York: Free Press, 1996).
13. Richard J. Barnet and John Cavanagh, *Global Dreams: Imperial Corporations and the New World Order* (New York: Simon and Schuster, 1994).
14. Robert S. McIntyre, "Testimony on Corporate Welfare," U.S. House of Representatives Committee on the Budget, June 30, 1999. On the Internet at http://www.ctj.org/html/corpwelf.htm (visited June 25, 2001).
15. Ralph Nader, "Testimony on Corporate Welfare," U.S. House of Representatives Committee on the Budget, June 30, 1999. On the Internet at www.nader.org/releases/63099.html (visited June 25, 2001).
16. Alan Downs, *Corporate Executions* (New York: AMACOM, 1995).
17. See Gordon, *Fat and Mean,* chap. 2.
18. Ibid., 191.
19. Lawrence Mishel, Jared Bernstein, and John Schmitt, *The State of Working America 2000–2001* (Ithaca, NY.: Cornell University Press, 2001), 220; Marc Breslow, "Job Stats: Too Good to Be True," *Dollars and Sense,* September–October 1996, 51.
20. Mishel et al., op cit., 252.
21. Chris Tilly, *Half a Job: Bad and Good Part-Time Jobs in a Changing Labor Market* (Philadelphia: Temple University Press, 1996); Kevin D. Henson, *Just a Temp* (Philadelphia: Temple University Press, 1996).

DISCUSSION QUESTIONS

1. "Downsizing and outsourcing are the natural consequences of free and open global markets, and American workers should simply adjust to the new realities of globalization." How would you respond to this statement?
2. What or who is responsible for the declining wages that three out of four Americans have been faced with since 1973? What, if anything, should be the response of government? Of business?
3. How would you justify decisions to downsize the workforce if you were on the board of directors of a major corporation? Do you believe someone will be hurt by your actions, or, do you believe that in the long run everyone will benefit?
4. Is the problem described in this selection the fault of greedy corporate executives, shareholders, government, education, those workers not ready to adjust to new circumstances, or is it simply due to the "globalization of the world's economy"?

7

Inheritance and Privilege

THOMAS M. SHAPIRO

The Four Questions

1. What is the problem Shapiro describes?
2. What makes it a *social* problem? Who is hurt? Is society harmed? Is the cause social?
3. What causes this problem?
4. Is there anything that can be done? Why is it so difficult to lessen the problem?

Topics Covered

Baby boom generation

Class

Racial inequality

Head-start assets

Inheritance

Meritocracy perpetuation of inequality

PARENTAL WEALTH AND HEAD-START ASSETS

... The baby boom generation, born between 1946 and 1964, marked a critical transformation in the American experience because a considerable number

SOURCE: From "Inheritance: That Parent Thing," from *The Hidden Costs of Being African American* by Thomas Shapiro;

of them grew up in middle-class families that accumulated substantial wealth for the first time. Now adults with families of their own, since 1990 they have been collecting a $9 trillion bounty from their parents. And this in turn has allowed them to live in houses in neighborhoods that they simply could not have afforded without parental wealth. I do not begrudge an average family inheritance, but I am concerned about how weakened public commitments to children, families, schools, and communities encourage people to use inheritances for private advantage. How adult baby boomers use this unprecedented wealth transfer is crucial to understanding racial inequality. Moreover, most families think they have earned everything and success is entirely of their own making, and this attitude makes progress toward equality more difficult....

A core part of my argument is that family inheritances, especially financial resources, are the primary means of passing class and race advantages and disadvantages from one generation to another. Examining information about parental wealth is a good starting point to explore empirically the potential significance of legacies in perpetuating racial inequality. After all, if whites tend not to come from families with greater wealth accumulation than blacks, there are no financial advantages to pass along, and we can immediately reject a notion of racial legacy secured by family wealth. If whites do come from families with greater wealth, on the other hand, then it is important to examine the extent and effects of this disparity. This seems like a simple question but it is difficult to study because collection of data on parental wealth did not occur until 1988, when the Panel Study of Income Dynamics asked respondents how much wealth their living parents had. Further, parental wealth is not a proxy for future inheritance because we know nothing about individual parents' plans for their money, health care, or financial portfolios, or their ideas about giving, for that matter. Nonetheless, net worth is the single most important piece of information about the prospective *capacity of parents* to assist their adult children financially at critical times like buying a first home or paying for private schooling. The parental-wealth measure captures the capacity of parents to give their children a head start in life. The parental-wealth charts (Figures 7.1 and 7.2) below show the percentage of all American families whose living parents possess

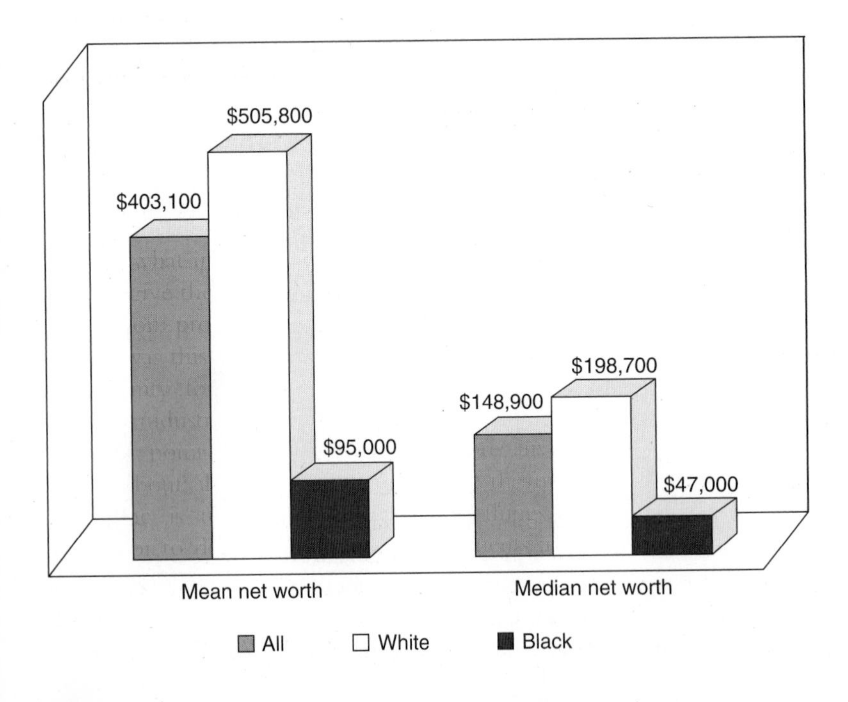

FIGURE 7.1 Parental wealth.*

*1999 dollars.

SOURCE: PSID, 1988.

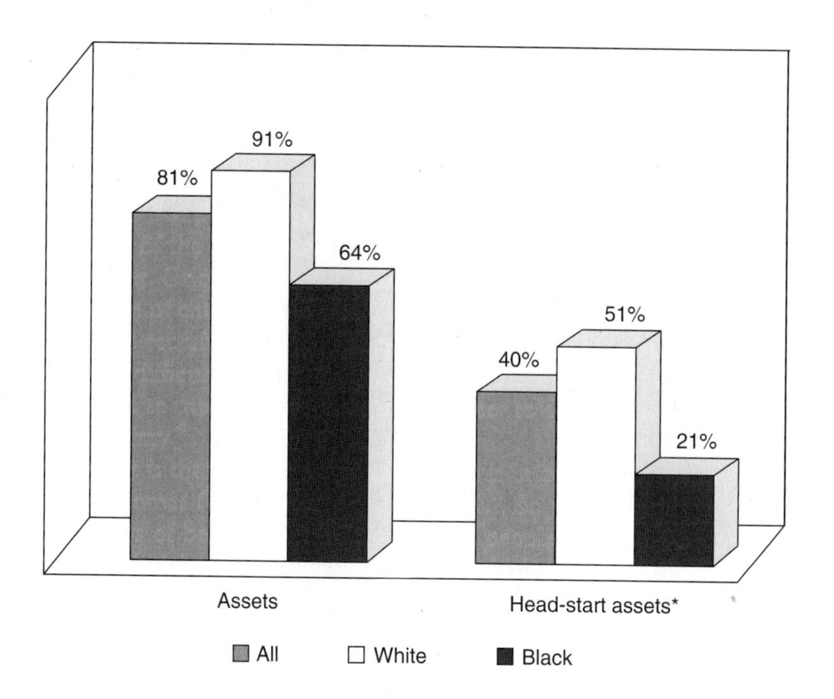

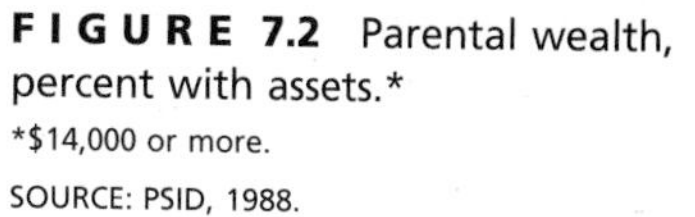
FIGURE 7.2 Parental wealth, percent with assets.*
*$14,000 or more.
SOURCE: PSID, 1988.

wealth. This includes 81 percent of all families, which seems like a healthy state of affairs but does not give any indication of how much wealth a family possesses. More than nine-tenths of white parents (91 percent) surveyed hold assets compared to fewer than two-thirds of black parents (64 percent).[1]

Of families reporting wealth, the typical family in this survey has parents with just under $150,000 in net worth. Among families who reported positive asset figures for their parents, average net worth skyrockets to over $400,000, with home equity representing the largest source of assets. The huge difference between the mean and median figures again illustrates the huge disparity between rich and poor; most of the respondents' parents have some wealth, but only a few have a great deal. Not surprisingly, among those with wealth, these data demonstrate large racial differences in parental wealth: The parental net worth for typical black families was $46,700, compared to almost $200,000 for white families. In other words, among families with positive assets, the financial capacities of the parents of white families are four times greater than those of the parents of black families. The dollar gap is the difference between parents' ability or lack thereof to buy an average house and a midsize station wagon. The average figures are even bigger, $503,800 to $93,000, or 5.4 times greater for white families than black families. These results pinpoint serious differences in wealth between previous generations of blacks and whites.

The important figures in these charts show that whites are about one and a half times more likely to come from families with assets than blacks and that an enormous racial wealth gap exists among families with these resources. In general, whites and blacks come from families with substantially different wealth capacities. What does this mean in practical terms? As a way of thinking about this question, I will describe the concept of *head-start assets* as a way of measuring what considerable financial assistance entails.

... We will look at head-start assets as the amount needed for the down payment and closing costs on the typical home in the United States. In 1999 the median-priced home in the United States sold for approximately $160,100, with typical down-payments ranging from at least 5 percent of the purchase price to about 10 percent. In addition, banks customarily charge a finance fee, usually 1 percent of the loan. At closing, then, a family buying this typical house needs between $9,600 and $17,600. We will use a figure three-quarters along this range to represent head-start assets in 1999—$14,000.

Figure 7.2 displays the percentage of families in the United States whose parents possess enough assets to help their child buy a typical house. This is a quick way of identifying families that might potentially receive large enough financial assistance to transform biographies, improve their class standing, and attain advantages for at least one child. Among all American families, only two in five have this capacity. Breaking this down by race provides a key piece of evidence for my argument. Among those with living parents, half of all whites come from families with the ability to deliver head-start assistance versus only a fifth of blacks. White families are 2.4 times more likely than blacks to have parents with substantial wealth resources. Another way to think about this finding is that monetary support among whites is most likely to go from parents to adult children, whereas elderly blacks are more likely to need help themselves from their adult children.

INHERITANCES

We commonly think of inheritance as limited to bequests at the death of a parent, and this sense of inheritance echoed throughout our interviews. When we asked people if they ever inherited money, they always responded with money received at the death of a parent. This notion of inheritance is quite restricting, because it does not include wealth given between living people, usually from parents to adult children. It omits major transfers like parental assistance for down payments for first-time homeowners or paying for college. These transfers are not inheritances in the traditional sense of the word, but they have critical short- and long-term impacts on the well-being of the receiver and his children. One study determined that 43 percent of wealth transfers occur between living relatives.[2] Therefore, to arrive at a reliable idea about the role of inheritance in passing advantages and disadvantages along from generation to generation, these kinds of wealth transfers must be included and evaluated....

In our interviews, not one white or black family even mentioned parental payment of college expenses as inheritance.[3] Almost all acknowledged the helpfulness of parental support; they just did not consider it part of an inheritance. We therefore asked who paid their college expenses and probed for an estimate of what portion came from parental assistance. I am convinced that omitting payment of college expenses from how we think about inheritances neglects a significant wealth accumulation dynamic, just as it renders comparing white and black savings rates highly suspect. Simply, college graduates without debts and college loans start their working lives with a huge head start over those who start their working lives having to pay back $50,000 or more in college loans. In this instance, the intergenerational transfer shows up not as accumulated wealth in the possession of families but rather as lack of debt. Of course, it also could be a decisive factor in thinking about going to college. In our interviews, 15 African American and 7 white families talked about still paying off student loans from college and technical schools, which they invariably called burdensome. Student loans weighed heavily on one person who told me her "student loan was going to follow me the rest of my life." Another person told me, "I have my student loans and I have bad credit. And we didn't even graduate!"

Cultural capital is yet another form of inheritance that allows families with ample assets to pass along nonmonetary benefits to their children that give them a competitive edge in school, the job market, and other areas. Cultural capital refers to an understanding of what gives a person advantages or disadvantages in school, business, and social situations; for example, knowing the work of the painter Jacob Lawrence signifies a particular knowledge of and taste in art and might add points on college entrance exams, reveal class standing in a business meeting, or provide a connection in a social setting. It also refers to other intangible preferences associated with different classes or groups that are important parts of inheritance and upbringing parents give to children.... Closing times, school sports schedules, and extracurricular activities in most upper-middle-class public and private schools typically assume that one parent who is not working will be available to drive children. A child whose family does not fit this standard sticks out. On an individual level, cultural capital may seem as silly as knowing a salad fork from a fish fork, but it often is the kind of informal knowledge that signals one's class—whether one "belongs" or not. Cultural capital is acquired through family life, formal education, informal educational experiences such as visiting museums and zoos, social connections, networks, friendships, proficiency with cultural codes and nuances, and community. Cultural capital is typically found where financial wealth is high.[4] The point here is that individuals inherit different opportunities for cultural capital, and lack of cultural capital can pose a significant impediment for advancement.

WHO INHERITS?

... The first examinations of financial inheritances among normal American families appeared in the 1990s. Studies indicate that nearly 1 in 4 white families (.244) received an inheritance after the death of a parent, averaging $144,652. In stark contrast, about 1 in 20 African American families had inherited in this way, and their average inheritance amounted to $41,985. White families were four times as likely as blacks to benefit from a significant inheritance, and whites were much more likely to inherit considerably larger amounts, by a $102,167 disadvantage.[5]

Another recent study suggests that about one-third of baby boom whites in 1989 were due to receive future inheritances worth more than $25,000 ($34,718 in 2000 dollars), versus fewer than 1 in 20 blacks. Over the lifetime, whites' inheritances are on average seven times larger than blacks' inheritances. The study estimates that the white-black gap in the value of inheritances for baby boomers will be much larger than it is for those born before 1946. Black boomers will inherit 13 cents for every dollar inherited by white boomers. The mean white baby boomers' life-time inheritance will be worth $125,000 (in 2000 dollars) at age 55, as compared to only $16,000 for black baby boomers. The preceding generation of whites inherited around $70,000. The black-white inherited gap is larger than the noninherited-wealth gap among baby boomers.[6]

We also know that about one in five families receives help from living relatives, averaging about $2,500.[7] Surprisingly, family assistance is just as common among poorer families as among wealthy families, though the amounts differ considerably. Black Americans were just as likely to receive this sort of family assistance as whites. The difference obviously was the largesse of the helping hand. The gift for the average white recipient was $2,824, compared to $805 for black families. Black families are just as willing to help their adult children, but their circumstances limit their ability to do so.[8]...

Table 7.1 compares families who inherit wealth with those who do not inherit or have not yet inherited. Inheritors can be distinguished from noninheritors in that they are older, have fewer children, are better educated, and are more likely to own their homes. One of every two families who inherited money works in upper-middle-class occupations, compared to fewer than one in three families (30 percent) who have not inherited wealth. Perhaps more fundamental, key resource differences separate families who have inherited from those who have

TABLE 7.1 Who Inherits?

	Inheritors	Noninheritors
Mean Net Financial Assets	$282,643	$94,301
Median Net Financial Assets	$57,000	$10,000
Mean Net Worth	$361,907	$129,800
Median Net Worth	$129,236	$28,700
Median Income	$49,230	$38,000
Median Number of Children in Family	0.0	1.0
Median Years of Education	13.0	12.0

SOURCE: PSID, 1984–1999.

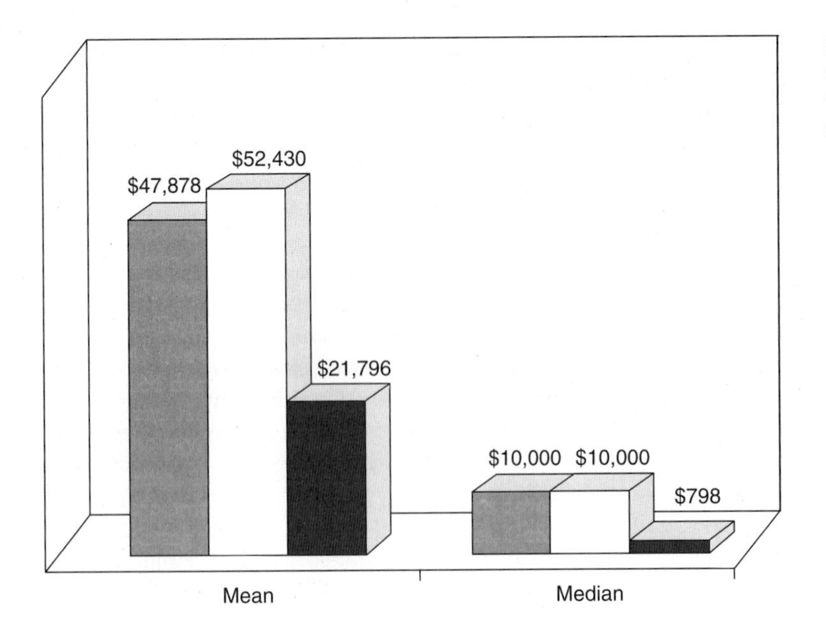

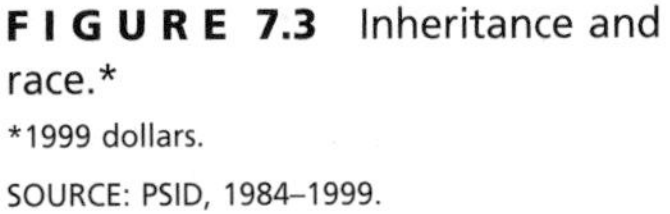

FIGURE 7.3 Inheritance and race.*

*1999 dollars.

SOURCE: PSID, 1984–1999.

not. Families with a financial inheritance report significantly higher incomes, $49,200 to $38,000. Net financial assets divulge the widest and most revealing breach between inheritors and noninheritors. Mean and median net financial assets among inheritors amount to $282,600 and $57,000, compared to $94,300 and $10,000 for families who have not inherited wealth….

The most dramatic findings concern the black-white inheritance gap. Twenty-eight percent of whites received bequests, compared to just 7.7 percent of black families. Three and a half times as many white families already have received an inheritance as black families. Even this wide disparity does not represent the full measure of inequality, because for whites the average inheritance amounted to $52,430, while for black Americans, it amounted to $21,796. Median inheritance figures registered $10,000 for white families and $798 for black families. Thus, among those fortunate enough to receive bequests, blacks received 8 cents of inheritance for every dollar inherited by whites….

Two sets of observations seem pertinent here. First, matching virtually all the other data we have presented on the racial wealth gap, whites receive inheritances at least three times larger than African Americans and up to 20 times greater (among 45–65-year-olds). Inherited money is the most obvious form of nonmerit wealth, and in this regard Figure 7.3 discloses a black-to-white inheritance ratio that is larger than the baseline racial wealth gap, 8 cents on the dollar of inherited wealth versus 10 cents on the dollar total wealth. This is prime evidence for our understanding that inheritances are reversing gains earned in schools and on jobs and making racial inequality worse.

The difference between average and typical inheritance for black Americans is striking. The typical (median) inheritance is less than $1,000, which simply means that at least one-half of blacks who inherit receive less than $1,000. By way of comparison, more than one-half of all white inheritances amount to $10,000 or more….

In our interviews, families spoke about the advantages family assistance and inheritances—both when parents pass away and at important life events like graduation, marriage, and birth of children—give them in getting a head start in life. As we already suggested, and our interviews emphasize, the single event (other than parental death) that triggers by far the largest transfer of wealth between generations is the purchase of a first home. Homeownership provides the pathway to community and schooling, and parental assistance in buying a first home is key to setting opportunities for their adult children and their families.

The families we interviewed correspond with the national studies, even as they furnish another layer of information. Only three black families out of 85 had received a head-start inheritance, at least $14,000, upon the death of a family member. In contrast, just about one in four white families already had received similar substantial inheritances. In our interviews, whites were seven times as likely to be given substantial inheritances.

Among white families who received an inheritance, the amount averages $76,000, compared to $31,000 for blacks. Among all black inheritors, half received less than $10,000; among all white inheritors, only five bequests fell below the $14,000 mark. Whites received about two and a half times as much money when they did inherit.

Beneath the numbers, the patterns and stories are revealing. The mention of grandparents in our interviews provides an interesting clue linking financial assistance to previous generations. Remember that we interviewed people who were mostly between 25 and 45 years old, when it is likely that one or both parents are still alive. Two black families mentioned receiving money from grandparents, and in both cases the amounts were quite small. More striking, however, is that 14 (out of 22) white inheritors mentioned wealth passed to them upon the death of grandparents. For the most part, the amounts are not staggering or even substantial. What the low level of inheritance from grandparents corroborates is that the legacy of grandparents of black baby boomers, who lived and toiled under harsh discrimination and glaringly different conditions, did not include financial resources. We see a glimpse of the racial reality of two generations ago continuing to impose and structure differences onto the present generation of young adults and a generation of children still coming up.

Passing wealth along to children or grandchildren apparently involves passing along class standing as well. The racial wealth gap is an important reason why black families have more difficulty passing along achieved middle-class status than white families do. A large proportion of families who were given inheritances could claim middle-class status. This is verifiable for both white and black inheritors, as only 2 of 8 black inheritors and 4 of 22 white inheritors are below middle-class standing, as measured by income, jobs, or education. Interestingly, if we think of people who inherit wealth but cannot sustain a middle-class existence as falling from middle-class grace, then those falling from grace are mostly women inheritors who subsequently experienced marital difficulties and women who were single mothers at the time we talked. The income consequences for divorced or separated women have been a cause for concern for some time.[9] Our interviews hint forebodingly at the devastating asset consequences for these women, their children, and the subsequent well-being of the family....

TRANSFORMED LIVES, DESERVED INHERITANCES

Americans believe in a strict form of meritocracy that mandates that we should only get what we earn through ability and effort. Listening to how families tell the story of inheriting wealth and the meaning they attach to this advantage in light of our meritocratic ideals provides valuable insights into understanding how families view success and reconcile conflicting values of inheritance and earned wealth.

Twenty miles from the Arch in downtown St. Louis, the Barrys and their three children moved in 1996 into a "real cute little house" on a bluff alongside the Mississippi River. Joe Barry is a computer specialist, and Briggette is a sign language interpreter who works part-time. Both earned college degrees, and their combined income is $70,000 a year. This is a solid middle-class family—educated suburban homeowners with professional jobs. They live close to Briggette's parents in a smallish white and gray vinyl-sided house, with a basketball hoop, surrounded by trees and open space. It looks like a picture-perfect place to raise kids. Their daughter attends parochial school; the two boys are still in day care. Joe and Briggette's parents enjoyed successful careers (one was a teacher, one a lawyer, another a school principal), and they accumulated considerable wealth during the post-World War II economic prosperity. This family is on track for the American Dream, but the road has not always been straight or smooth. They have relied upon parental financial assistance to maintain their success, middle-class life-styles, and identities.

Our conversation about parental helping hands starts when I ask Briggette if they had any trouble coming up with the down payment for their new house. "Not at all," she said. "My parents gave it to me. So it was a real easy matter of depositing a check." Before her parents offered this money, she explains, it was not going quite so easily at all.

> *We were trading in the kids' savings bonds. We were working two jobs each. We were working at night. We were working around the clock. We were saving every penny. We were having garage sales. And finally my mom said, "Well, this is stupid. We've got a lot of money here."*

When I ask if they encountered any unanticipated expenses with closing costs on the house or moving, Briggette remembers the difficulty they had because of two open Federal Housing Administration—insured loans. "We had to convert from an FHA loan on the new home to a conventional one," she says, "so we had to come up with an additional ten thousand dollars in closing costs within the span of seven days." For most families, coming up with an additional $10,000 within a week would throw a gigantic wrench into their plans. How did they do it? Briggette said to her mother, "'Mom, we need more than we thought.' And my mom and dad gave it to us. They gave us a chunk; we had the whole chunk of money for furniture and other things. But instead we had to take ten thousand of the money they gave us and put it into the closing costs." All together, her parents gave Briggette and Joe $30,000, and her grandmother chipped in another $2,000 to help with the move.

When I ask about other kinds of financial assistance, Briggette says she probably has "bought a total of five outfits for my children," because "every season" her mother buys "a whole wardrobe for each of my three children." I ask if her parents help with day care. Briggette says, "Not on a daily basis. They're too old for that." This ubiquitous parental assistance must leave them lots of discretionary spending money in their $70,000 earnings, which could boost their savings rate, so I wonder how they spend their money. "Oh," Briggette says, "we piss it away on our kids."

> *We go to movies. Oh, here I go. Because we are both full-time working parents, and we don't have, like, Mommy at home—I don't know about you, but when I was a kid, Mommy was home. In the summertime, every night, it's Discovery Zone or the swimming pool, or McDonald's or Raging Rivers or Six Flags. I mean, almost every night, unless we're just too tired to do anything. We spend our money on our children. We spend at Blockbuster Video. We spend a hell of a lot of money on groceries. My kids eat me out of house and home. Yeah, we spend a lot on day care and private school. But seriously, when I add it up, it's about fourteen thousand dollars a year. It's unbelievable. But it's worth it, you know? What are my choices? That's what I wanted. I wanted kids. You have to pay for them.*

Parental help started much earlier in their marriage, when Briggette was pregnant with their first child.

> *Early in our marriage, we had a lot of financial difficulty. When I was—let's see—when I was seven months' pregnant, Joe got laid off from his job. And then I had to go out on medical leave because I could no longer work. We almost went into financial disaster. We almost lost our home.*
>
> How did they save their home?
>
> *My mom and dad and his mom and dad chipped in and gave us the money.* Parents again. That parent thing [emphasis added]. *They gave us the money. And we turned all of our bills over to Consumer Credit.*

What about presents from family that help financially—or other major gifts, like a television, car or dishwasher?

> *Sure. Lots. Yes to all of it. They've given us money for the down payment on the house. His father bought us two of our cars, and my parents bought us one of our cars. My parents bought us a washer and a dryer. My parents paid the tuition for my daughter's school, because it's the Catholic school and they just wanted her to be educated for that. Let's see. Yeah,* you name it, we get it [emphasis added]. *They paid our mortgage payment for two months, before my husband found his other job. They have set up trust funds for my children, for each of my children; so that when it's time for them to go to college, they will have money to do so. This isn't money they gave us, but they helped us set up financial plans for Joe and me for retirement.*

I ask Briggette to estimate the value of the help her parents give them by buying clothes, household appliances, and other presents, taking them out to dinner, and so on—things one normally does not count. She is taken aback somewhat—"Oh, my God! Like I can really put a price tag on it"—but manages an estimate: "Oh, Jesus, I would seriously say fifty or sixty thousand dollars [since her marriage] … maybe five thousand a year."

Briggette Barry's memory seems accurate as she catalogues all sorts of parental wealthfare with matching dollar figures. At various points in our conversation, she expresses great appreciation and gratitude for this familial help, commenting, "Gotta love parents." However, as soon as the conversation turns to how she and her husband acquired assets like their home, cars, and savings account, her attitude changes dramatically. "As far as anything else [their assets], no," Briggette asserts, "we worked our butts off for what we have." How did they acquire the cars? "Worked our butts off to pay for them." How did they accumulate their $3,000 savings account? "Worked for it. All worked for it." The $1,000 retirement account? "Yes. It's taken out of my check."

She is emphasizing an important point I heard from many other families. While acknowledging a generous parental helping hand and the loving bonds between generations, the Barrys—like other families we interviewed—adamantly maintain that they deserve the unearned wealth benefits that transform their lives and opportunities. The Barrys describe themselves as self-made, conveniently forgetting that they inherited much of what they own. I do not doubt how hard they work to improve their lives, and I am sure their hard work has improved their well-being, but hard work alone has not brought them to their current level. The flawed and uncritical attribution of success to hard work precludes coming to terms with their unearned advantages. It redefines what is fair and what is unfair in a way that puts the onus for lack of wealth on those without the same advantages. Simply, what a family inherits cannot be earned. The idea of deserving unearned things is very important to the Barrys and families like them in that deservedness and worthiness substitutes for earning and merit. I emphasize this because we so often confuse advantages and connections with ability.

Had their biographies been different, they would not be able to afford their suburban middle-class home, and their daughter probably would not be attending private school. Even with a $70,000 family income, with their bad credit histories, sizable credit card debt, and no savings available for a down payment, homeownership would be highly improbable. It is entirely possible, had they not come from prosperous and generous families, that they might have made different choices in the past to avoid the financial mess that necessitated parental bailouts. The point here is not to ponder counterfactual, hypothetical scenarios so much as to underscore the taken-for-granted sense of entitlement around deservedness we found in many of our interviews with white middle-class families. Presented with the Barrys' information without all the parental assistance, a mortgage officer most likely would advise them that their financial situation exceeds commonly used industry standards and might suggest that, although mortgage underwriters use discretion in approving loans for people who have large debts in relation to their income, it would be prudent to look for ways to lower their housing sights or change plans.

Many of the families we interviewed expressed similar thoughts about inherited money. They rationalize handed-down advantages as deserved by attributing success to their own endeavors and hard work. Sociologist Heather Beth Johnson writes about a sense of entitlement among middle-class white families.[10] In her reading of these interviews, she describes an ideology of meritocracy: the idea that positions are earned through hard work and personal achievement and through no resources other than one's own. Families talking about "earning" their assets through hard work, a solid work ethic, and playing by the rules legitimize this entitlement. For many of these families, it is as if the effort of earning the reward is more important than actual accomplishments.

Glee and Barry Putnam, 36 and 38 years old, live in a suburban middle-class subdivision in South County St. Louis. Barry works as a computer network technician for a large Fortune 500 corporation, where he earns about $45,000. Glee takes care of their six children, who range in age from 3 months to 12 years. Their financial resources are typical of America's broad middle class, $56,000 in net worth and $26,300 in net financial assets. I ask them about the role of parental help in their lives.

GLEE: My family helps us, yes.... When we got married, our parents gave us some money [about $5,000]. When we came into this house, my mom gave us some money to help with the carpeting, and just things like that. My parents help out with my kids.

INT: What gifts did they give you when you were married?

GLEE: Well, my parents give us gifts all the time. I mean, they're both retired, so it's like "Let's help you out here." They'll come out here, they'll give us gifts. "Oh, let's buy dinner the entire week we're out here" or "Here's some extra money for gas since you're bringing us all around," you know. They'll do things like that.

INT: When they helped with the home repairs when you first moved here, do you have an idea of how much they may have given you as a gift?

GLEE: They didn't really help with the home repairs.... They wanted to help with the carpet, that was my mother's gift. "Here's some money, go out and buy carpets, carpet, whatever you can with this amount of money."

INT: Okay, do you remember how much that was?

BARRY: A thousand dollars.

INT: Have you received financial help from your family, like did they help pay for college or help out with the purchase of a house or home repairs or any other kinds of assistance that you can think of?

GLEE: Well, like, my parents helped me get my first house.

BARRY: Six thousand dollars.

INT: Was that a loan or a gift?

BARRY: A gift.

GLEE: It was a gift.

INT: That's fine. Any other gifts that you can recall over time?

GLEE: She gave us about $450 for the kitchen, didn't she?

BARRY: Just two years ago we redid the kitchen, and she gave us some money for that.

GLEE: My parents help us out all the time by just giving us money.

BARRY: Oh, they have flown everybody at some point or another out there [Utah] to vacation for a while.

Glee and Barry have just catalogued how their parents gave them a large cash gift as a wedding present, a $6,000 down payment to buy their home, financial help in remodeling their kitchen, $1,000 for carpet; in addition, their parents help pay for family vacations and give them money regularly. In other parts of the interview they acknowledge that they could not be living where they live and their children would not be in the schools they are without parental help. Yet when I ask how they acquired their assets, they put a different spin on it.

INT: How did you acquire the different assets that you own, like the money in the savings account?

BARRY: We worked for it.

INT: That was just, like, from every paycheck?

BARRY: Yeah.

INT: And the stocks, that was from?

BARRY: Every paycheck.

INT: Okay, and the IRA?

GLEE: I worked for it, I saved from work.

BARRY: The same thing.

INT: Same with the home equity and—

BARRY: Um-hum, everything, overtime for that.

Ambiguity and denial about inherited money provided poignant moments in many of our interviews, affecting how families told their stories as well as what they said. Loans became "gifts," as we will see when we examine down-payment monies. Wives corrected husbands about amounts given by the wife's parents, and vice versa—in every case adjusting to a higher amount, as the original answer downplayed the importance of that gift in the couple's success. Precise answers in previous sections of the interview to questions like the class of the neighborhood they grew up in sometimes gave way to hesitation. The reason for this, I believe, is that these families were responding to the strongly held American belief of meritocracy, knowing that the advantages they were describing were not wholly earned. For many of these families, the language of deservedness had replaced the fact of achievement and merit.

The concept of transformative assets challenges this understanding and instead examines the ability of inherited wealth—nonearned assets—to structure class standing beyond a family's earnings. The test is simple: Are a family's life chances bettered sufficiently by their inherited assets to move them upward in class, community, and educational environment for their children? In the mosaic of America, there are no typical families, but we can look at lots of different pieces that make up the larger pattern. From our interviews, at least, we get a solid foundation for determining the validity of the concept of transformative assets and the frequency with which such assets are found in families. The interviews also give us an appreciation of the large number of ways in which assets influence life chances and change families' lives, including the way they think about themselves and their future....

The Fergusons are accomplished professionals with a family income in the top 10 percent. But even with their $100,000 income, if it were all they had, Shauna and Shawn would not be close to the financial position and status they enjoy. Wealth accumulated over two generations assists the Ferguson family in living a good and secure life.

Nancy and Mark Hollings also are white and live with their daughter in Dorchester, an economically and racially diverse section of Boston with old but solid housing stock. The prices there seem low in the otherwise hot Boston real estate market because some sections of the community are poor and black. Nancy is an administrator in the nonprofit sector, and Mark is a teacher. Together their salaries are a very respectable $70,000. They bought their condominium just as the Boston housing market reached a low point, in 1995. In fact, they paid $55,000 for it, whereas the previous owner had paid twice that amount just a few years earlier. Most middle-class white families, and some black ones, see Dorchester as "iffy" due to its high level of class and race diversity and rapidly changing neighborhoods. The traditional valuing of location does not work well in such mixed communities, and families like the Hollingses feel that they have found good value for their money as long as they are willing to take the risks. They were able to make a down payment of half of the purchase price with monies inherited from an uncle and a grandfather and a "loan against the inheritance" from Nancy's parents.

Financially, this family seems to be doing well, as represented by their good income, low living expenses, and over $100,000 in net worth. As is the case for most middle-income American families, their largest asset pool is their home equity; for the Hollingses, it represents one-half of their net worth. When I ask how they acquired their assets, Nancy says, "From working. I'd say at this point, most of the money represents earned income." In essence, their families gave them a large down payment for their condominium, so they had to take out only a small mortgage and therefore pay a relatively small amount in fixed monthly housing costs. Because of this, they are able to put money aside from their salaries and invest it. Since 1995 these assets from savings and investments have grown to about $50,000. Truly, then, Nancy is correct when she says that most of their assets come out of earnings. In another sense, however, one needs to recognize that this family's ability to save, invest, and accumulate $50,000 is highly structured by the large down payment, inherited or given by family, that fixes their housing costs at a small amount of their monthly income and leaves considerable discretionary income for investments at the end of each month.

What if there were no family inheritance? The point is to examine how they put their good fortune together. It would cast their choices about homeownership, community, and education in an entirely different light—the same light as many of the families we talked to who do not have assets. And this, in turn, would affect where they live and the schools their children attend. Because of their large down payment and relatively small mortgage, their current fixed monthly housing costs are $250, far below typical rental housing costs in Boston. With a solid income of $72,000, well above what most American families earn, they would qualify for a mortgage of about $170,000 and pay $1,500 monthly housing costs. The Hollingses, in fact, could buy the same house they live in now without family assistance, but the family finances would look very different. The current low monthly costs would rise considerably, by $1,250 in additional mortgage and interest payments. In this light, their ability to save, invest, and accumulate over $50,000 in financial assets is made possible directly by their inheritances. In essence, family inheritances give them an additional $1,250 in discretionary resources each month....

American society is of two minds about inheritance, and we seem to want it both ways. We take pride in our accomplishments, often marking them in monetary terms, and see nothing wrong in passing on what we earned to our children. Indeed, part of the motivation for working hard and acquiring things includes bettering our family and our children for future generations. This notion, however, collides with the equally strongly held notion of meritocracy because inheritances are unearned, represent a different playing field entirely, and have precious little to do with merit, achievements, or accomplishments. We live with this duality, partly because we deny what inheritances represent, partly because we see it in individual and family terms, and partly because the current political balance heavily favors those with advantages and privileges.

We need to recognize that inherited wealth can perpetuate inequality. The sociologist Georg Simmel called money a frightful leveler, but it might be more appropriate to understand inherited money as a frightful conveyor and transmitter of inequality. Inheritance is not an achievement, and we are conflicted by it. Asking how inheritance has affected our lives can deepen our understanding of inequality....

ENDNOTES

1. PSID families were asked questions concerning parental wealth only if they had at least one living parent. Unfortunately, the collection method for this information is less than optimal. If respondents said their parents had positive assets, they were asked how much. Those who knew approximate dollar amounts specified the figure, while those who did not know or were hesitant about answering were then given wide bracket choices and dollar figures that did not require specific amounts. The available data, then, reflect this two-tiered approach: Some families specified amounts, some gave brackets, and

still others combined both methods. In our reporting for the brackets, we used figures for each bracket. For example, if a family reported that their parents' wealth fell in the $1,000 to $25,000 range, we used $12,500 as the mean. The total effect of this procedure raises the floor and lowers the ceiling; that is, amounts tend to compress to the middle.

2. See Gale and Scholz (1994).
3. In the inheritance tabulations from our interviews we did not count college expenses as a direct inheritance partially because I decided not to spend the time required in each interview for people to remember and calculate the exact dollars.
4. Bourdieu (1973); Miller and McNamee (1998); Keister (2000).
5. See Wilhelm (2001).
6. Avery and Rendall (2002).
7. These data are taken from Wilhelm (2001).
8. In some ways, these 1987 data are as disturbing as the bequest data because the $2,000 racial difference represents a single year's financial assistance. Suppose this $2,000 racial difference recurred every year?
9. See Weitzman (1985) for instance.
10. See Johnson (2001). Her insights were critical in shaping my interest in this material.

REFERENCES

Avery, Robert, and Robert Rendall. 1993. "Lifetime Inheritances of Three Generations of Whites and Blacks." *American Journal of Sociology* 107, no. 5: 1300–1346.

Bourdieu, Pierre. 1973. "Cultural Reproduction and Social Reproduction." *In Knowledge, Education, and Cultural Change*, ed. R. Brown. London: Tavistock.

Gale, William, and John Scholz. 1994. "Intergenerational Transfers and the Accumulation of Wealth." *Journal of Economic Perspective* 8, no. 4: 145–60.

Johnson, Heather. 2001. "The Ideology of Meritocracy and the Power of Wealth." Ph.D. dissertation, Northeastern University.

Keister, Lisa. 2000. *Wealth in America*. New York: Cambridge University Press.

Miller, Robert K., Jr., and Stephen J. McNamee. 1998. "The Inheritance of Wealth in America." In *Inheritance and Wealth in America*, ed. Robert K. Miller Jr. and Stephen J. McNamee. New York and London: Plenum Press.

Weitzman, Lenore. 1985. *The Divorce Revolution*. New York: Free Press/Macmillan.

Wilhelm, Mark. 2001. "The Role of Intergenerational Transfers in Spreading Asset Ownership." In *Assets for the Poor*, ed. Thomas Shapiro and Edward Wolff. New York: Russell Sage Foundation.

DISCUSSION QUESTIONS

1. Americans believe in the idea of a meritorious society and largely see personal success as the product of hard work and individual effort. Shapiro is challenging that view in this essay. Is he convincing?
2. What are the head-start assets and how do they create privilege? Is this contrary to your view of a democratic society?
3. What historical factors have contributed to the vast difference in inheritable assets between African- and European-Americans?
4. How might the inheritance gap create a problem for society?
5. What are some of the ways the inheritance gap might be lessened? Do you favor any of these ways?
6. Examine your own situation. What, if any, advantages have you had because of inheritance?

8

Why Are People Poor in America?

HARRELL R. RODGERS, JR.

The Four Questions

1. What exactly is the problem? What values and goals make it a problem?
2. Is it a *social* problem? Do all the theories assert that the causes are social?
3. What are the causes of the problem?
4. What can we do about the problem, according to the author?

Topics Covered

Poverty

Cultural/Behavioral theories

Structural/Economic theories

Culture

Underclass

Dysfunctional behavior

Welfare

Inner city employment

Out-migration

Racism

Economic trends

THEORIES OF POVERTY

Theories of poverty generally fall into one of two categories: cultural/behavioral or structural/economic. Cultural/behavioral theories make the case that the only real cause of poverty is the behavior, values, and culture of the poor. Structural/economic theories usually contain some behavioral component, but argue the precipitating cause of poverty is a lack of equal opportunities for all Americans. Structural/economic theories usually focus on unequal economic opportunities (high unemployment or sub employment), educational systems that disadvantage the poor, or discrimination in its various forms. Conservatives favor cultural/behavioral arguments, while moderates and liberals usually stress structural/economic explanations that contain a behavioral component.

Cultural/Behavioral Theories

In *The Dream and the Nightmare* (1993), Myron Magnet offers a conservative interpretation of American poverty. Magnet argues the basic cause of poverty is the culture and behavior of the poor. People become poor, Magnet argues, not because they lack social, political, and economic opportunities, but because they lack the inner resources to seize the ample opportunities that surround them. Magnet believes the poverty of the poor is mainly a destitution of the soul, a failure to develop the habits of education, reasoning, judgment, sacrifice, and hard work required to succeed in the world. To Magnet the poor are not hard-working, decent people who have fallen victim to economic problems. Rather, he says, they are people who engage in leaving school, illegitimacy, drug and alcohol abuse, welfare misuse, and often crime.

SOURCE: From Harrell R. Rodgers Jr., *American Poverty in a New Era of Reform*, pp. 63–76.

The primary job of any civilization, Magnet says, is soulcraft: the transmission of values that combine to develop mature, educated, honest, hardworking, caring, and responsible people. The poor represent society's failures, those who have not inherited the central values of mainstream culture. Magnet believes the values of today's poor reflect the revolutionized culture of the 1960s, which taught that it was legitimate to blame the system for personal failures and expect government handouts rather than seek success through education, hard work, and sacrifice. This same 1960s culture also promoted the sexual revolution, endorsing premarital sex, promiscuity, and illegitimacy, while degrading marriage, sacrifice, education, and industriousness.

Magnet's concluding point is that an individual's values and behavior determines his/her opportunities, economic or otherwise. Economic opportunity can abound, but if a person lacks the inner resources to take advantage of opportunities, he/she will fail. Giving welfare to those who fail only makes matters worse. Welfare reinforces the belief that people are not responsible for their behavior, that they cannot overcome their problems through hard work, and that welfare is something they deserve.

Moderate and liberal critics believe the image of the poor drawn by Magnet and others who accept his view is nothing but a narrowly drawn stereotype. This vision of the poor, they argue, is based on a subset of the poor, which ignores the complexity of the poverty population, and as noted above, empirical evidence about poverty spells and welfare use. Out of a population of 35 million poor people, Magnet, they argue, bases his theory of poverty on the behavior and welfare use of a few million of the poor, a group often labeled as the undeserving poor or, more recently, the underclass. Those poor people with obviously dysfunctional habits or lifestyles have long interested society. But what part of the poor do they represent?

A considerable body of research has tried to define and identify the underclass. Scholars who have attempted to identify and count the members of this class have used a number of definitions and methodologies. Some studies have tried to define the underclass as those poor who are isolated in inner-city impoverished neighborhoods, essentially separated from the workforce (Bane and Jargowsky, 1988; Gottschalk and Danziger, 1986; Hughes, 1989; Nathan, 1986; Ricketts and Sawhill, 1988). These studies identify a small underclass population ranging from under 1 million to about 6 million.

Another group of scholars has tried to identify those long-term poor people, often including those who live in neighborhoods with high concentrations of poverty, who engage in dysfunctional behavior such as chronic unemployment, drug use, out-of-wedlock births, crime, leaving school, and antisocial attitudes and behavior (Adams, Duncan, and Rodgers, 1988; Reischauer, 1989; Kasarda, 1992; O'Hare and Curry-White, 1992). These studies identify a population that ranges from fewer than 1 million to slightly more than 8 million, depending on definitions, methodologies, and number of cities studied.

While these various studies arrive at estimates that differ, they agree that most of the poor are not the highly dysfunctional people described by Magnet. The proportion of the poor population that does display serious social problems, however, is large, visible, and expensive users of social services. It is not too surprising that this group is often thought of as the poor.

Many conservative scholars, on the other hand, are not concerned about stereotyping the poor because they think Magnet's basic arguments are correct. They believe the welfare system is so perverse in design that it has the capacity to spread a contagion of pathologies among the poor. Charles Murray in his influential book *Losing Ground* (1984) argued the real culprit of the poor is welfare. Welfare, Murray argues, robs the poor of initiative, breaks up families by encouraging men either not to marry the mothers of their children or to abandon them, provides an incentive to women to have children out of wedlock, so they can get on welfare or increase their benefits, and discourages work by providing a combination of cash and noncash benefits that amounts to better compensation than could be earned through employment.

Murray's arguments have attracted strong criticism, in part, because he proposes a radical solution. Murray argues the best way to lessen poverty is simply to abolish all welfare programs (Murray, 1994). Labeled the cold turkey approach, Murray's arguments are almost the perfect foil for those sympathetic toward the poor. But even critics of Murray often agree that welfare does produce some antifamily impacts. Danziger et al. (1982), Ellwood and Bane (1985), and Bassi (1987) all found a significant link between welfare receipt and increased rates of divorce. Garfinkel and McLanahan concluded that although welfare receipt often delays remarriage, it has a modest impact on divorce and out-of-wedlock births (Garfinkel and McLanathan, 1986; 1994, 211).

Also there is little relationship between state cash relief and state out-of-wedlock birthrates (Acs, 1996; Ellwood and Bane, 1985; Duncan and Hoffman, 1988; Plotmck, 1989). The out-of-wedlock birthrate, in other words, is not higher in states that pay high cash aid benefits than the rate in states that pay low benefits. In fact, Mississippi, one of the lowest-paying states in the nation, has one of the highest out-of-wedlock birthrates. Murray's response is that the generosity of state benefits is less important than the fact that an out-of-wedlock birth makes women eligible for welfare or may increase her benefits, even modestly. Murray argues that if states gave no money to single mothers, especially teenagers, parents would put more pressure on teenagers to avoid pregnancy, and women would be more careful about having children with men who did not have the prospect or interest in being parental partners.

Other scholars take a different approach than Murray, but argue that welfare programs play an important role in making and keeping people poor. In a pair of influential books, Lawrence Mead (1986, 1992) argued the welfare system had become too permissive, placing too few obligations on recipients. By not requiring the healthy poor to work, Mead contended, the welfare system undermined both the confidence of the poor and public sympathy for them. One of Mead's theses was that passive poverty, that is an idle poverty population, reflected a welfare-nurtured defeatism among recipients rather than a real lack of economic opportunity. Mead focused, first, on proving the healthy poor could work and, second, on documenting ample economic opportunity for those with motivation. Mead argued the poor would be substantially better off economically and psychologically if they were moved into the workforce and that society would be much more positive about helping low-income working families than in aiding passive welfare families. Thus, Mead did not contend that all welfare is wrong or harmful, but rather that welfare, especially long-term, is the wrong solution to the problems of the able-bodied poor.

James L. Payne (1998) agrees with the fundamental position of conservatives that welfare is harmful to the poor. Payne makes the familiar argument that welfare destroys the integrity of the poor while promoting and prolonging poverty by asking too little in return. Payne advocates what he calls "expectant giving," basing most assistance to the poor on the understanding that they must give something in return. "What the poor need, " argues Payne, "is to be asked to give, not be given to" (p. xii). "Expectant giving" requires the poor to engage in productive efforts to help themselves. The best assistance programs, Payne argues, are those that require the greatest contribution from the poor.

Payne takes the conservative argument one step further. He contends that government can never pass or administer effective welfare reform. The reason, he argues, is that governments will always yield to a hand-out mentality, based on the belief that the poor, regardless of why they are poor, deserve help. "Expectant giving" requires judgments about why people are poor, what they need, what they should give in return, and whether they have lived up to the terms of their agreement or contract. Effectively helping the poor, Payne argues, requires that they be treated selectively or unequally, a role government is unsuited to play. With heavy caseloads, endless rigid rules, and a bias toward uniformity, government programs, he says, will always drift to handouts. The best alternative, Payne suggests, is to turn welfare over to private, voluntary efforts that can be more flexible and creative. Only then, says Payne, can we avoid the extremes of indulgence and cruelty that aggravate America's poverty problems.

The cultural/behavioral approach, in summary, places responsibility for America's poverty problem on the personal inadequacies of the poor, welfare in general, or the design flaws of welfare programs.

These conservative arguments played an important role in shaping the 1996 welfare reform. PRWORA gives states the authority to treat welfare applicants and recipients selectively (rather than uniformly), depending upon their personal situation and the state of the local economy. PRWORA presumes that the poor have an obligation to take responsibility for themselves and their families. It requires mutual obligation contracts between recipients and the state. In return for assistance, recipients agree to engage in responsible behavior. If the state decides that recipients have not lived up to their side of the agreement, their support can be terminated. PRWORA also incorporates work obligations and gives states the authority to regulate and even terminate support for teen mothers. Considerable emphasis is placed on forcing absent parents to support their children. Thus, the influence of conservatives can be clearly seen in the design and implementation of PRWORA.

Structural/Economic Theories

Moderates and liberals explain poverty in terms of limited and unfair economic opportunities, inadequate and unfair educational systems, lack of political power, biased government policies, and sometimes racial and/or sexual discrimination.

An interesting example of a structural/economic theory is offered by William J. Wilson in *The Truly Disadvantaged* (1987). Wilson examines the impact of civil rights laws on changes in economic opportunities for inner-city minorities. Wilson's theory focuses on the growth of poverty and social problems in urban black ghettoes. In the early to mid-twentieth century, blacks, Wilson notes, migrated from the South to the major cities of the Midwest and North in search of economic opportunities. Because of housing discrimination, almost all black people, regardless of education or skills, became concentrated in central city ghettos. By the 1970s, Wilson maintains, civil rights laws and changes in the nation's economy were producing important changes that would have major impacts on black ghettos. Wilson explains these dynamics with four hypotheses.

First, Wilson hypothesizes that changes in the inner-city employment market harmed low-income, low-skilled blacks, especially black men. These changes involved the general decline of manufacturing jobs, which paid good wages for low-skill work, and the migration of hundreds of thousands of these same types of manufacturing jobs from central cities to the suburbs. Replacing these manufacturing jobs in the inner city were clerical and white-collar jobs. These jobs, Wilson argues, required more post-secondary or specialized skills than many ghetto residents possessed. Wilson labels these changes slow economic growth, skills mismatch, and spatial mismatch. Slow economic growth reduced the number of manufacturing jobs, the jobs that increased in urban areas were those that many poor blacks were unqualified to fill, while those jobs they could perform were increasingly found in the suburbs.

Second, Wilson hypothesizes the resulting joblessness among inner-city men encouraged many of them to turn to idleness and hustling, including criminal activities such as the drug trade. The combination of unemployment and crime among so many men left ghetto women with fewer qualified or desirable partners, significantly reducing the marriage rate while increasing the incidence of out-of-wedlock births and the use of welfare.

Third, ironically, Wilson argues, the civil rights laws passed in the 1960s played an important role in making these problems worse. Civil rights and affirmative action laws extending equal employment opportunities and fair housing increased the income of the most educated and skilled inner-city residents and allowed them to move to the suburbs. Wilson calls this selective out-migration.

Fourth, Wilson theorizes that selective out-migration harmed the ghetto community in important ways. Those who migrated had been the communities' best role models and civic leaders. Gone were those community leaders who had championed the importance of education, quality schools, and high academic standards while resisting crime, illegitimacy, and idleness. In turn, those left behind had less education, skill, and motivation. This produced what Wilson labels a contagion effect—the degeneration of aspirations, morals, schools, and the general health of the community—leading to ever-increasing rates of poverty and other social problems.

Wilson's theory is interesting and important because it provides a tightly reasoned explanation of the growth of the black underclass in American ghettos. Wilson highlights a loss of low-skilled, decent-paying jobs as the precipitating event in increasing poverty in ghettos, but there is also a behavioral component. Wilson openly says that many poor people engage in behaviors that make and keep them poor. He believes, however, that economic problems set off the chain of events that increased poverty and other social problems.

Wilson's theory has attracted opposition on two points. First, some scholars condemn Wilson, a black scholar, because he places no emphasis on racial discrimination as a cause of ghetto poverty. Wilson, in fact, believes that racism plays a declining role in black life and black poverty (1980). Other scholars have argued that racism is the primary reason for the increasing marginalization of black males (Massey, 1990; Darity and Myers, 1983). Second, other scholars fault Wilson for not crediting welfare with making any real contribution to ghetto poverty. Scholars such as Mead, Murray, and Payne, of course, believe that welfare played a pivotal role. Mead (Wilson and Mead, 1987) criticizes Wilson's theory by pointing out that low-skilled immigrants have a history of preserving high employment levels, often doing so despite language barriers. Mead believes that immigrants sustain high employment levels because they were not raised in welfare families. Blacks, on the other hand, Mead says, often grew up in or around welfare families and are, therefore, more willing to turn to welfare rather than accept low-paying jobs that they consider unpleasant, degrading, or unrewarding or gain the education or job skills required for better employment.

Mickey Kaus in *The End of Equality* (1995) takes a different position. Kaus argues that welfare might not be the cause of poverty, but that welfare sustains it. Kaus raises the question of why inner-city blacks have often failed to follow low-skill jobs as they moved to the suburbs. Wilson's answer is that for many ghetto residents the jobs, once increased transportation and housing costs were factored in, do not pay enough to make relocation realistic. Kaus finds that answer suspect. He notes that blacks left the South in large numbers and traveled long distances in search of employment opportunities, but they often failed to follow jobs as they moved to the suburbs. Why, asks Kaus, did they travel a thousand miles to seek improved economic opportunities before the 1960s, but later failed to follow jobs that were twenty-five miles away? His answer is welfare. Blacks migrated from the South, he says, because they had no choice. The welfare system did not exist and was not an alternative. However, when urban jobs migrated to the suburbs, black citizens often had a choice. Rather than follow the jobs, they could qualify for welfare and become idle; and, that, Kaus says, is exactly what many of them did.

There is empirical evidence to support some of Wilson's hypotheses. Considerable research supports Wilson's skills-mismatch and spatial-mismatch hypotheses. Kasarda (1990, 1989, 1988) and Johnson and Oliver (1991) have documented the loss of decent-paying manufacturing jobs to the suburbs, especially in the major cities of the Northeast and North-Central regions. Holzer (1991) found that when jobs left the central cities for the suburbs, inner-city blacks lost more jobs and employment opportunities than did whites or suburban blacks. Research also documents a general decline in the wages of low-income workers (Holzer and Stoll, 2001; Hoynes, 2000). Since the mid-1970s wage increases have mainly taken place only in jobs that require higher levels of skill and education.

The empirical literature provides no support for Wilson's hypothesis that lack of economic opportunity explains the declining black marriage rate (Wilson and Neckerman, 1986). The marriage rate for blacks has been declining at least since the

1950s, and seems to be little influenced by shifting black male employment levels or the skill-level, education, or income of black males (Ellwood and Crane, 1990; Lerman, 1989; Hoffman, Duncan, and Mincy, 1991). As we will note below, Cornel West has a different explanation for the decline in black marriage rates.

Wilson's out-migration hypothesis is supported by empirical research (Gramlich and Laren, 1991). There is clear evidence that black citizens with the economic ability have fled the inner city for better neighborhoods, as have many other people. Wilson's contagion hypothesis has not been rigorously tested, but there is some evidence in support. Jencks and Mayer (1990) found that peers significantly affect academic achievement. Students perform better when they attend schools with peers who take education seriously and perform at higher levels. Hoffman, Duncan, and Mincy (1991) found the higher the proportion of single welfare mothers in a neighborhood, the higher the rate of teenage pregnancy. Anderson (1990) and Fagan (1992) found that young inner-city men turn to hustling when good jobs are unavailable.

In summary, much of Wilson's theory is supported by empirical and ethnographic research, although the role that welfare and racism do, or do not play, is seriously contested, and there is little support for the marriage hypothesis.

CULTURE AS STRUCTURE

Another prominent black scholar offers a structural theory of black poverty, which includes some of the fundamentals of Wilson's theory. Wilson believes that black poverty stems from economic problems, and he believes that economic reform would go a long way toward reducing the poverty and social problems of low-income neighborhoods and individuals. But Cornel West in his book *Race Matters* (1993) argues that while there is an economic antecedent, the problems of the black population are more complex than Wilson imagines. West agrees with Wilson that many black people, especially those who live in central city ghettos, engage in self-destructive behavior. But West argues that this dysfunctional behavior has complicated antecedents, reflecting problems not just of the economy, but also of America's culture, and, thus, cannot be reversed simply by economic reform.

West attributes the poverty and social problems of poor blacks to four causes:

- First, the economic decline that has taken place within the inner cities of America and the change in compensation for low-skill jobs. His analysis of economic problems that have harmed the black population is similar to Wilson's.
- Second, a history of racism with continuing discrimination that West believes has done great emotional harm to millions of black people.
- Third, a prevailing American culture that stresses materialism and material gain to the exclusion of other, more important values, especially intellectual, moral, and spiritual growth. Materialism, West argues, is such a dominant value in American society that it is the core of American culture. As such, says West, it should be admitted that this perverse culture is a form of structure, much like the economy and the political system (p. 12). When mainstream culture is so shallow that it reflects primarily market moralities, he argues, it misdirects and teaches low value, thus keeping people from becoming fully developed, successful humans with meaningful values and culture.
- Fourth, a lack of quality black leadership. West is critical of black leadership, arguing that most black leaders are unprepared to transcend narrow racial issues, be examples of moral and spiritual growth, and lead with real vision and courage (p. 23).

West believes the combination of flawed cultural values, racism (past and present), economic problems, and poor leadership has produced battered identities in much of black America. These battered identities express themselves in two ways. First, West contends, many black people suffer from nihilism, a disease of the soul that reveals itself by producing lives of "horrifying meaninglessness, hopelessness, and (most important) lovelessness" (p. 14). "Life without meaning, hope, and love," West concludes, "breeds a cold-hearted, mean-spirited outlook that destroys both the individual and others" (pp. 14–15). Second, is racial reasoning. Racial reasoning, West says, is not moral reasoning. "The humanity of black people," says West, "does not rest on deifying or demonizing others" (p. 28). Racial reasoning leads to closing-ranks thinking that takes the form of inchoate xenophobia (poorly thought out fears and dislikes about foreigners), systematic sexism (which West says contributes significantly to the declining black marriage rate), and homophobia.

West believes that racial reasoning must be replaced with moral reasoning, which would produce mature black identities, coalition strategies that cross racial lines, and black cultural democracy. Mature blacks could be honest about themselves and other black people, would be willing to work with other groups to achieve collective goals, and would promote a society in which black people treated everyone with respect, regardless of race, ethnicity, sex, or sexual orientation.

West's recommended solutions go beyond reducing black poverty, and they can be summarized in four parts. First, West believes that black people need a love ethic that would confront the self-destructive actions of many black people. Second, black people, he says, must also understand that the best source of help, hope, and power is themselves. Black people, argues West, must take charge of their lives and become involved in the civil soul of their communities. They must play a role in the debate, design, and passage of enlightened public policies that teach and celebrate quality values and stable, prosperous families. Third, government programs to aid low-income and poor people must be increased. Last, there needs to be a new generation of black leadership with genuine vision, courage, and commitment to ethical and religious ideas.

West's theory is provocative, controversial, and challenging. He argues that just improving economic opportunities will not solve the fundamental problems of American society. He believes that millions of black Americans suffer from battered identities, destructive reasoning, perverse cultural values, and inadequate leadership. Solving the problems of many people, he argues, requires a change in the basic values of American culture, more insight on the part of whites about racism, along with enlightened public policies and better leadership.

TOBIN'S STRUCTURAL THEORY OF POVERTY

The economist James Tobin (1994) offered another structural theory, one much more purely economic in nature. Tobin's theory rests first on his empirical research on the relationship between the yearly rate of poverty in America and the health of the economy. Tobin found that in those years in which there have been real (inflation adjusted) increases in wages and decreases in unemployment, poverty has declined (p. 148). The performance of the economy, in other words, is important. Tobin's findings are consistent with those of a considerable body of research (Bartik, 1996; Blank 1997; Fitzgerald, 1995; Harris, 1993; Hoynes, 1996; Moffit, 1992). However, as Tobin documented, between 1973 and the early 1990s the economy often performed less well, and poverty rates declined less during good economic times than in the past (p. 149).

Tobin argued there are five reasons for the decreased impact of economic performance on poverty rates:

1. An expanding proportion of the population is excluded from the more viable sectors of the market economy because they do not have the increasingly sophisticated skills required by employers (p. 161).
2. Jobs, especially good-paying low-skill jobs, are increasingly found outside the central cities (p. 160).

3. Wage rates have been held down because corporations can export jobs to low-wage markets in other nations and because corporations must spend more on fringe benefits. Corporations, Tobin notes, think about labor costs in terms of the total expense, rather than just in wages. As the cost of fringe benefits (especially health care) has increased, employers have tried to hold down wages to cover some of the increasing expense of fringes. However, fringes, says Tobin, go mostly to better-paid workers. The lowest-paid workers often do not benefit from fringes, but their wages still suffer.
4. The welfare population has changed over time, becoming increasingly comprised of female heads of households and poorly educated unattached males, with both groups less willing or able to take advantage of economic opportunities (p. 155).
5. The Federal Reserve has allowed the economy to grow too slowly (p. 164).

Tobin argued that the Federal Reserve's policies were too conservative, constituting a drag on the growth of the economy. Tobin argued that fundamental changes have taken place in the economy that allow faster growth without setting off inflation. The major changes Tobin documented are the increasing globalization and deregulation of the economy. Globalization increases available labor and product markets. The increase in labor pools and products lessens the chances of product or labor shortages that set off inflation. Since consumers have access to more markets for goods, there is less chance of consumption producing shortages and thus inflation. Global markets also hold down prices by restraining labor costs and labor demands because employers have options in other nations. Deregulation lowers business costs while allowing companies to be more creative, reducing costs and improving productivity.

Tobin argued that these changes allow the economy to grow faster than in the past, without setting off inflation. He recommended the Federal Reserve consent to the economy growing at a rate of about 3 to 3.5 percent a year, one percentage point higher than in the past. If the economy were allowed to grow faster, Tobin maintained, more jobs would be created, employers would pay their employees better because labor markets would be tighter and profits higher. Companies, Tobin argued, would be more willing to invest in their employees by giving them more training and more opportunities.

The economic history of the United States in the 1990s and early twentieth-first century bear out Tobin's arguments. The economy has grown faster than the Federal Reserve intended without setting off inflation. Also, in the second half of the Clinton administration a healthy economy resulted in a lower poverty rate, dropping the black poverty rate below 30 percent for the first time in American history. As the economy has struggled during the Bush administration, poverty rates have risen.

Tobin's theory is about as pure as structural theories ever are, but even his theory recognizes that some poor people do not take advantage of opportunities as they become available and that other poor people need considerable help to get into and stay in the job market.

Comparative studies of poverty rates in advanced capitalist democracies (for example France, Germany, Italy, Belgium, United Kingdom, United States, Switzerland, Canada, etc.) also support structural theories. These studies find the poverty rate is lowest in those nations in which a large percentage of the working population is employed in industry, where the unemployment rate is low, and where emphasis is placed on paying decent wages to workers (Moller et al., 2003). Also, the more generous the welfare state, the lower the poverty rate.

PRWORA was clearly a conservative victory, but some of the arguments made by moderates and liberals are incorporated in the law. First, PRWORA recognizes that not all poor adults can work. States can exempt up to 20 percent of their adult recipients from work requirement and time limits. Second, states cannot start the clock until support services are available to those adults required to move into the job market. Third, the food stamp and Medicaid programs remain as entitlements and there are no time limits on assistance. Fourth, states are given the latitude to adopt policies that make work pay better. Last, Congress recognized that many jobs pay poverty wages by significantly expanding the Earned Income Tax Credit (EITC).

WHAT DO THESE THEORIES TELL US ABOUT POVERTY AND WELFARE REFORM?

There are many insights gained by this review of theories of poverty. First, poverty is not a single problem; it is a series of rather complex problems. There are both cultural/behavioral and structural/economic causes of poverty, which vary by the subgroups of people found among the poor. Culture and structure are also connected. Opportunities influence culture, which influences behavior, which influences opportunity, which influences culture, ad infinitum.

Even when the same set of factors causes poverty, the problem of poverty varies because of the differential reactions of those who become poor. Some single mothers, for example, might seize the chance to gain job training, and transitional health and child care to allow them to enter the workforce, while others may be unwilling to accept responsibility for themselves. Therefore, both variations in causes and reactions to poverty make the problem complex.

Second, the economy is important. The economy must be kept healthy and growing to provide quality opportunities for low-income and poor people. Government policies, such as the EITC, can smooth out some of the deficiencies of the economy. Third, many families cannot take advantage of economic opportunities without a better education or advanced job training. Fourth, similarly, many of today's poor people, especially parents, cannot work without child care and health care aid. Fifth, some people will need extra help, encouragement, even sanctions, to make the transition to employment. These people will need not only education and skill training, but also help with interpersonal skills, self-esteem, hope, and confidence before they believe in themselves enough and interact with others well enough to benefit from assistance and become self-sufficient. Some people will need help more than once. When they fail, they will need a second or even third push to get on their feet. Some people will need support in overcoming alcohol and drug dependency and domestic violence problems before they can be successful in the job market.

Sixth, discrimination based on sex, race, or ethnicity cannot be tolerated. Assistance programs and the employment market must be free of bias. Last, poverty, especially among the elderly, would be lessened if the government sponsored more programs to help people save during their employment years.

REFERENCES

Acs, G. 1996. "The Impact of Welfare on Young Mothers' Subsequent Childbearing Decisions." *Journal of Human Resources* 31(4): 898–915.

Adams, T., Duncan, G., and Rodgers, W. L. 1988. "The Persistence of Urban Poverty." In *Quiet Riots: Race and Poverty in the United States*, eds. F. R. Harris and R. W. Wilkins, 212–41. New York: Pantheon.

Anderson, E. 1990. *Streetwise: Race, Class, and Change in an Urban Community*. Chicago: University of Chicago Press.

Bane, M. J., and Jargowsky, P. A. 1988. "Urban Poverty Areas: Basic Questions Concerning Prevalence, Growth, and Dynamics." Center for Health and Human Resources Policy Discussion Paper Series, Harvard University.

Bartik, T. J. 1996. *Who Benefits from State and Local Economic Development Policies?* Kalamazoo, MI: W.E. Upjohn Institute for Employment Research.

Bassi, L. 1987. "Family Structure and Poverty among Women and Children: What Accounts for Change?" Mimeo. Washington, DC: Georgetown University. June.

Blank, R. 1997. "Why Has Economic Growth Been Such an Ineffective Tool Against Poverty in Recent Years?" In *Poverty and Inequality: The Political Economy of Redistribution*, ed. J. Neil, 188–210. Kalamazoo, MI: W.E. Upjohn Institute for Employment Research.

Danziger, Sheldon, et al. 1982. "Work and Welfare as Determinants of Female Poverty and Household Headship." *Quarterly Journal of Economics* 98(2): 519–34.

Duncan, G. J., and Hoffman, S. D. 1988. "Welfare Dependence Within and Across Generations." *Science* 239: 467–71.

Ellwood, D., and Bane, M. J. 1985. "The Impact of AFDC on Family Structure and Living Arrangements." *Research in Labor Economics* 7(2): 137–49.

Fagan, J. 1992. "Drug Selling and Illicit Income in Distressed Neighborhoods: The Economic Lives of Street-Level Drug Users and Sellers." In *Drugs, Crime, and Social Isolation: Barriers to Urban Opportunity*, eds. A. V. Harrell and G. E. Peterson, 292–312. Washington, DC: Urban Institute Press.

Ellwood, D. T., and Crane, J., 1990. "Family Change among Black Americans." *Journal of Economic Perspectives* 4(4): 65–84.

Fitzgerald, J. 1995. "Local Labor Markets and the Local Area Effects on Welfare Duration." *Journal of Policy Analysis and Management* 14(1): 43–67.

Garfinkel, I., and McLanahan, S. S. 1986. *Single Mothers and Their Children: A New American Dilemma*. New York: Russell Sage Foundation.

Gottschalk, P., McLanahan, S., and Sandefur, G. 1994. "The Dynamics and Intergenerational Transmission of Poverty and Welfare Participation." In *Confronting Poverty: Prescriptions for Change*, eds. S. Danziger, G. Sandefur, and D. Weinberg, 85–108. Cambridge, MA: Harvard University Press.

Gramlich, E., and Laren, D. 1991. "Geographical Mobility and Persistent Poverty." Paper presented at the Conference on Urban Labor Markets and Labor Mobility, Airlie House, VA, March 7–8.

Harris, K. M. 1993. "Work and Welfare among Single Mothers in Poverty." *American Journal of Sociology* 99(3):317–52.

Hoffman, S. D., Duncan, G. J., and Mincy, R. B. 1991. "Marriage and Welfare Use among Young Women: Do Labor Market, Welfare and Neighborhood Factors Account for Declining Rates of Marriage among Black and White Women?" Paper presented at the annual meetings of the American Economic Association, New Orleans, December.

Holzer, H. J. 1991. "The Spatial Mismatch Hypothesis: What Has the Evidence Shown?" *Urban Studies* 28 (4): 104–22.

Holzer, H. J., Stoll, M. A., and Wissoker, D. 2001. "Job Performance and Retention among Welfare Recipients." Discussion paper no. 1237–01, Institute Research on Poverty, University of Wisconsin, Madison.

Hoynes, H. W. 2000. "The Employment, Earnings and Income of Less Skilled Workers over the Business Cycle." In *Finding Jobs: Work and Welfare Reform*, eds. R. Blank and D. Card, 23–71. New York: Russell Sage Foundation.

——— 1996. "Local Labor Markets and Welfare Spells: Do Demand Conditions Matter?" Institute for Research on Poverty, Discussion paper no. 1104–96. Madison, WI.

Hughes, M. A. 1989. "Misspeaking Truth to Power: A Geographical Perspective on the Urban Fallacy." *Economic Geography* 65(1): 185–207.

Jencks, C., and Mayer, S. 1990. "Residential Segregation, Job Proximity, and Black Job Opportunities." In *Inner-City Poverty in the United States*, eds. L.E. Lynn, Jr. and M.G. McGeary, 319–37. Washington, DC: National Academy Press.

Kasarda, J. D. 1992. "The Severely Distressed in Economically Transforming Cities." In *Drugs, Crime, and Social Isolation: Barriers to Urban Opportunity*, eds. A. V. Harrell and G. E. Peterson, 61–84. Washington, DC: Urban Institute Press.

——— 1990. "Structural Factors Affecting the Location and Timing of Urban Underclass Growth." *Urban Geography* 11(1): 234–64.

——— 1989. "Urban Industrial Transition and the Underclass." *Annals of the American Academy of Political and Social Science* 501 (January): 26–47.

——— 1988, "Jobs, Migration, and Emerging Urban Mismatches." In *Urban Change and Poverty*, eds. M. G. McGeary and L. L. Lynn, Jr., 98–109. Washington, DC:

Kaus, M. 1995. *The End of Equality*. New York: Basic Books.

Lerman, R. I. 1989. "Employment Opportunities of Young Men and Family Formation." *American Economic Review* 79 (May): 62–66.

Magnet, M. 1993. *The Dream and the Nightmare: The Sixties' Legacy to the Underclass*. New York: William Morrow.

Massey, D. S. 1990. "American Apartheid: Segregation and the Making of the Underclass." *American Journal of Sociology* 96(2): 329–357.

Mead, L. M. 1992. *The New Politics of Poverty: The Nonworking Poor in America*. New York: Basic Books.

——— 1986. *Beyond Entitlement: The Social Obligations of Citizenship*. New York: Free Press.

Moffitt, Robert A. 1992. "Incentive Effects of the U.S. Welfare System: A Review." *Journal of Economic Literature* 30(1): 1–61.

Moller, S., et al. 2003. "Determinants of Relative Poverty in Advanced Capitalist Democracies." *American Sociological Review* 68: 22–51.

Murray, C. 1994. "Does Welfare Bring More Babies?" *Public Interest* 115: 17–30.

——— 1984. *Losing Ground: American Social Policy 1950–1980*. New York: Basic Books.

Nathan, R. P. 1986. "The Underclass: Will It Always Be With Us?" Paper presented to a symposium at the New School for Social Research, New York, November.

O'Hare, W. P., and Curry-White, B. 1992. "The Rural Underclass: Examination of Multiple-Problem Populations in Urban and Rural Settings." Mimeo. Washington, DC: Population Reference Bureau. January.

Payne, J. L. 1998. *Overcoming Welfare: Expecting More From the Poor and Ourselves*. New York: Basic Books.

Plotnick, R. D., and Winters, R. F. 1985. "A Politico-Economic Theory of Income Redistribution." *American Political Science Review* 79: 458–73.

Reischauer, R. D. 1989. "The Size and Characteristics of the Underclass." Paper presented at the Annual Meeting of the Association for Public Policy Analysis and Management, Bethesda, MD, October.

Ricketts, E. R., and Sawhill, I. V. 1988. "Defining and Measuring the Underclass." *Journal of Policy Analysis and Management* 7(2): 316–25.

Tobin, J. 1994. "Poverty in Relation to Macroeconomic Trends, Cycles, and Policies." In *Confronting Poverty: Prescriptions for Change*, eds. S. H. Danziger, G. D. Sandefur, and D. H. Weinberg, 147–67. Cambridge, MA: Harvard University Press.

West, C. 1993. *Race Matters*. Boston: Beacon Press.

Wilson, W. J. 1980. *The Declining Significance of Race: Blacks and Changing American Institutions*. 2nd. ed. Chicago: University of Chicago Press.

Wilson, W. J., and Mead, L. M. 1987. "The Obligation to Work and the Availability of Jobs: A Dialogue Between L. M. Mead and W. J. Wilson." *Focus* 10(2): 11–19.

Wilson, W. J., and Neckerman, K. M. 1986. "Poverty and Family Structure: The Widening Gap Between Evidence and Public Policy Issues." In *Fighting Poverty: What Works and What Doesn't*, eds. S. H. Danziger and D. H. Weinberg, 232–83. Cambridge, MA: Harvard University Press.

DISCUSSION QUESTIONS

1. What exactly is a "cultural/behavioral theory"? What are its central ideas? Do you agree with them?
2. What is the philosophy of "welfare"? Why is it thought to be necessary? Does it create poverty? Does it help the poor?
3. What is William Julius Wilson's position on the cause of the black poverty? In what sense does he emphasize economic and social causes rather than racism per se? Does he bring racism into his theory?
4. Cornel West's theory is described as "culture as structure." Explain the causes of black poverty according to West.
5. In the conclusion, the author describes the problem of poverty as "complex." What is his point?

9

The Silver Spoon: Inheritance and the Staggered Start

STEPHEN J. MCNAMEE AND ROBERT K. MILLER, JR.

The Four Questions

1. What exactly is the problem according to the authors?
2. Is it a *social* problem? Does it hurt anyone? Does it hurt society? Are its causes social?
3. What are the causes of the problem?
4. Do the authors suggest ways to deal with the problem?

Topics Covered

Income
Wealth
Economic inequality
Inheritance
Cumulative advantages
Mobility
Meritocracy
Wealth and happiness

INCOME AND WEALTH INEQUALITY

In considering how parents pass on advantages to children in the race to get ahead, researchers have usually looked at occupational mobility; that is, how the occupations of parents affect the occupations of children. The results of this research show that parental occupation has strong effects on children's occupational prospects. Some of this effect is mediated through education; that is, the prestige of parental occupation increases the educational attainment of children, which, in turn, increases the prestige of the occupations they attain. Looking at occupational prestige alone, however, underestimates the full extent of inequality in society and overestimates the amount of movement within the system. A fuller appreciation of what is at stake requires examination of the kind and extent of economic inequality within the system—who gets how much of what there is to get. Economic inequality includes inequalities of both income and wealth. Income is typically defined as the total flow of financial resources from all sources (e.g., wages and salaries, interest on savings and dividends, pensions, and government transfer payments such as social security, welfare payments, or other government payments) in a given time period, usually annually. Wealth refers not to what people "earn" but to what they "own." Wealth is usually measured as net worth that includes the total value of all assets owned (such as real estate, trusts, stocks, bonds, business equity, homes, automobiles, banking deposits, insurance policies, and the like) minus the total value of all liabilities (e.g., loans, mortgage, credit card, and other forms of debt). For purposes

SOURCE: From Stephen J. McNamee and Robert K. Miller, Jr., *The Silver Spoon: Inheritance and the Staggered Start*, Rowman & Littlefield Publishers, Inc., 2009, pp. 51–62. Reprinted by permission.

of illustration, income and wealth inequalities are usually represented by dividing the population into quintiles and showing how much of what there is to get goes to each fifth, from the richest fifth of the population down to the poorest fifth. These proportions are illustrated in Table 9.1.

In terms of income, in 2001 the richest 20% of households received a 47.7% share of all income, compared to only 4.2% received by the bottom 20%. As income increases, so does the level of concentration. The top 15% alone accounts for 26.7% of the total, and the top 5% alone accounts for 21% of the total. Moreover, income has great staying power over time. That is, the same households that are at the top income group now were very likely to have been at the top of the income group in previous years. Using longitudinal data and comparing income quintiles between 1969 and 1994, Mishel et al. (2003) note that most households in 1994 had remained at or near where they were in 1969. Forty-one percent of those in the lowest income group in 1969 remained in the lowest income fifth twenty-five years later, while 39% of those who were in the top income fifth remained there twenty-five years later. With regard to movement over the twenty-five year period, Mishel and his associates note that large intragenerational transitions are uncommon:

> Only 5.8% of those who began the period in the first fifth ended up in the top fifth, while only 9.5% fell from the top fifth to the lowest fifth. Those transitions that do occur are most likely to be a move up or down to the neighboring fifth. For example among the middle three-fifths, slightly less than two thirds of the transitions were to the neighboring fifths. (2003, 76–77)

TABLE 9.1 Share of Total Available Household Income and Total Net Worth

Income Group	Share of Income	Share of Net Worth
Top Fifth	47.7%	83.4%
Fourth Fifth	22.9%	11.9%
Third Fifth	15.5%	4.5%
Second Fifth	9.7%	.8%
Bottom Fifth	4.2%	.6%
Total	100.0%	100.0%

SOURCE: Reprinted from Lawrence Mishel, Jared Bernstein, and Heather Bouchey, *The State of Working America 2002/2003*. Copyright © 2003 by Cornell University. Used by permission of the publisher, Cornell University Press.

When wealth is considered, the disparities are much greater. In 1998, the richest 20% of American households accounted for 83.4% of total net household wealth. The bottom 40% combined, by contrast, held *less than one-half of 1%* of all available net worth. At the bottom end of the wealth scale, 18% of households had zero or negative net worth (Mishel et al. 2003, 285). In other words, a significant number of Americans either own nothing or owe more than they own.

As MIT economist Lester Thurow has observed, "Even among the wealthy, wealth is very unequally distributed" (1999, 200). In 1998, for instance, the top 1% of all wealth holders alone accounted for 38.1% of all net worth. Additional evidence shows that 78% of the wealth held by the richest 1% is owned by a scant one-half of 1% of all households (Mishel et al. 2003, 286).

In short, the degree of economic inequality in the United States is substantial by any measure. In fact, the United States now has a greater economic inequality than all industrial countries in Western Europe (Hacker 1997; Mishel et al. 2003; Wolff 2002). Moreover, the extent of this inequality is increasing. One standard measurement of the extent of inequality is the Gini Ratio, which measures the extent of discrepancy between the actual distribution of income and a hypothetical situation in which each quintile of the population receives the same percentage of income. Values of the Gini Ratio range between 0 and 1, where 0 indicates complete equality and 1 indicates complete inequality. Thus the higher the number, the greater the degree of inequality. The U.S. Census Bureau (2003) reports that the Gini Ratio has steadily and incrementally increased from .399 in 1975 to .460 in 2000—representing a 17% increase over a twenty-five-year span. An Internal Tax Revenue study (Balkovic 2003) further shows that the 400 American taxpayers with the highest adjusted gross income in 2000 accounted for 1.6% of all income in the United States, more than double their corresponding share in 1992. The average income of these 400 taxpayers was $174 million, nearly quadruple the average in 1992. These top 400 American taxpayers increased their incomes at fifteen times the rate of the bottom 90% of Americans. These figures actually underestimate the true amount of inequality, because adjusted gross income does not include nontaxable income earned on municipal bonds and other tax shelters. These data clearly show that the rich in America are getting richer. Another indication of income inequality is revealed by a comparison of pay for the chief executive officers of major corporations with that of rank-and-file employees. CEO pay as a ratio of average worker pay increased from 26 to 1 in 1965 to 310 to 1 in 2000, with much of the compensation package for CEOs coming in the form of stock options (Mishel et al. 2003).

Increases in wealth inequality are also dramatic. The percentage of net worth held by the richest 1% of Americans nearly doubled between 1976 and 1997, from approximately 20% to 35% (Wolff 2002, 83). In short, the gap between those who live off investments and the large majority of people who work for a living has considerably widened in recent decades.

Consideration of wealth as opposed to just income in assessing the total amount of economic inequality in society is critical for several reasons. First, the really big money in America comes not from wages and salaries but from owning property—particularly the kind of property that produces more wealth. If it "takes money to make money," those with capital to invest have a distinct advantage over those whose only source of income is wages. Apart from equity in owner-occupied housing, assets that most Americans hold are the kind that tend to *depreciate* in value over time—cars, furniture, appliances, clothes, and other personal belongings. Many of these items end up in used car lots, garage sales, and flea markets selling at prices much lower than their original cost. The rich, however, have a high proportion of their holdings in the kinds of wealth that *increase* in value over time. In 1998, for instance, the top 1% of households held 67.7% of all business equity, 50.8% of all financial securities, 54% of all trusts, 49.4% of all stocks and mutual funds, and 35.8% of all nonhome real estate (Wolff 2002, 26).

Second, wealth is especially critical with respect to inheritance. When people inherit an estate, for instance, they inherit accumulated assets—not incomes from wages and salaries. Inheritance of estates, in turn, is an important nonmerit mechanism for the transmission of privilege across generations. In strictly merit terms, inheritance is a form of getting something for nothing. Since only very large estates are taxed, there is no systematic accounting of the transfer of wealth across generations. As a result, data on the flow of these resources are skimpy. There are several factors that contribute to the lack of data on wealth transfers. Because wealth is highly concentrated, only a small percentage of Americans own great wealth. Since the number is small, wealthy people tend not to be sampled on random national surveys—the primary tool used in mobility research. In addition, the rich often conceal the full extent of their assets both for tax purposes and to avoid public scrutiny. Even when interviewed on surveys, most people are not fully aware of the current value of their assets and liabilities. In some studies, the lack of systematic and reliable data on wealth is addressed through the use of complex simulation models that take what wealth data are available and "fill in the gaps" with empirically derived estimates (see Keister 2000; Gokhale et al. 2001).

Despite these limitations, there is strong evidence that a substantial amount of accumulated wealth is passed on through inheritance. One source of information on wealth is the annual list of the 400 wealthiest Americans published by *Forbes* magazine. An early study of the *Forbes* list, for instance, showed that at least 40% of the 1982 *Forbes* list inherited at least a portion of their wealth, and the higher on the list, the greater the likelihood that wealth was derived from inheritance (Canterbery and Nosari 1985). A more recent study of the 1997 *Forbes* list showed that the majority of individuals on the list (56%) inherited a fortune of at least $50 million (Collins 1997). Among the ten wealthiest Americans on the 2003 *Forbes* list, five (ranked 4–8 respectively) are direct descendants of Sam Walton, founder of the Wal-Mart empire. The five Walton heirs have a combined estimated net worth of $102.5 billion (*Forbes* 2003).

Although there is some movement over time into and out of the *Forbes* list of the richest 400 Americans, this does not mean that those who fall off the list have lost or squandered their wealth. Most likely, when wealthy individuals fall off the *Forbes* list they have not lost wealth at all but rather have not gained it as fast as others. Although those who fall off the 400 list may have lost ground relative to others, they typically still have vast amounts of wealth and most likely remain within the upper 1% of the richest Americans. The richest 400 Americans represent less than two ten-thousandths of 1% of the total American population, but in 2001 they collectively accounted for 2.3% of all personal wealth (Kennickell 2003, 3). The top 1% of wealth holders represents approximately 2.9 million Americans—a considerable number but still far less than one-tenth of the approximately 33 million Americans living below the official poverty line.

Examining the population as a whole, the evidence suggests that as with income shares, shares of wealth ownership are very stable over long periods of time. Using a wealth simulation model spanning the period from 1975 to 1995, sociologist Lisa Keister reports that:

> The results indicate that there was little movement among wealth percentiles, even over a 20 year period. Sixty percent of those who started in the bottom 25% in 1975 were in the bottom 25% of wealth owners in 1995. Only 21% had moved to the second quartile, and 12% had moved to the third quartile. There was no movement over these 20 years from the bottom of the distribution to the very top, among these households. Downward movement was also rare. (2000, 79)

In short, most of those starting in the top of the wealth distribution remained there, with a few sliding down marginally. Continuity of income and wealth within households over time is not the same thing as continuity of wealth across generations. Unfortunately, the longitudinal data required to track transfers of wealth across specific parent-child households are not yet available. However, given that we know there is a high level of concentration of wealth in the United States and that wealth has great staying power over time, it is reasonable to assume that a substantial amount of concentrated wealth is available for transfer across generations. One study, for instance, estimates that sums in excess of $10 trillion will be available for transfer intergenerationally between 1990 and 2040 (Shapiro 1994). It is unlikely that these vast amounts of wealth simply evaporate between generations.

In the absence of longitudinal data on individual intergenerational wealth transfers, one team of university and private sector economists led by Jagadeesh Gokhale recently created an empirically derived simulation model of the transmission of wealth inequality through inheritance (Gokhale et al. 2001). This model, although sensitive to necessary assumptions built into it, shows high degrees of stability of wealth across generations. In this model, Gokhale and his associates estimate that nearly one-half of children whose parents are in the upper 20% of wealth holders at age sixty-six will themselves end up at age sixty-six in that bracket as well. At the other end of the wealth scale, children whose parents are in the bottom 6% of wealth holders have a 47% chance of remaining in the bottom 6% of wealth holders as adults. They have an additional 47% of ending up between the bottom 6% and 18% of wealth holders. In other words, 95% of children born to parents in the bottom 6% of wealth holders will end up poor as adults. Children who start out life born to parents in the bottom 6% of wealth holders have only one-half of 1% chance of ending up even in the upper half of wealth holders as adults. In short, this model, which simulates the intergenerational transmission of wealth—consistent with patterns of intragenerational income and wealth inequality—shows very little movement from rags to riches or riches to rags. Most people stay at or very close to where they started, with most of the movement occurring between immediately adjacent categories.

Despite the evidence of wealth stability over time, much is made of the investment "risks" that capitalists must endure as a justification for returns on such investments. And to some extent, this is true. Most investments involve some measure of risk. The superwealthy, however, protect themselves as much as possible from the vicissitudes of "market forces"—most have diversified investment portfolios that are professionally managed. As a result, established wealth has great staying power. In short, what is good for America is, in general, good for the ownership class. The "risk" endured, therefore, is minimal. Instead of losing vast fortunes overnight, the more common scenario for the superrich is for the *amount* of their wealth to fluctuate with the ups and downs of the stock market as a whole. Given the very high levels of aggregate and corporate wealth concentration in the economy, the only realistic scenario in which the ownership class goes under is one in which America as a whole goes under.

THE CUMULATIVE ADVANTAGES OF WEALTH INHERITANCE

Inheritance is more than bulk estates bequeathed to descendants; more broadly defined, it refers to the total impact of initial social class placement at birth on future life outcomes. Therefore, it is not just the superwealthy who are in a position to pass advantages on to children. Advantages are passed on—in varying degrees—to all of those from relatively privileged backgrounds. Even minor initial advantages may accumulate during the life course. In this way existing inequalities are reinforced and extended across generations. As Harvard economist John Kenneth Galbraith put it in the opening sentence of his well-known book, *The Affluent Society:* "Wealth is not without its advantages and the case to the contrary, although it has often been made, has never proved widely persuasive" (1958, 13). Specifically, the cumulative advantages of wealth inheritance include the following.

Childhood Quality of Life

Children of the privileged enjoy a high standard of living and quality of life regardless of their individual merit or lack of it. For the privileged, this not only includes high-quality food, clothing, and shelter but also extends to luxuries such as entertainment, toys, travel, family vacations, enrichment camps, private lessons, and a host of other indulgences that wealthy parents and even middle-class parents bestow on their children. Children do not "earn" a privileged life style—they inherit it and benefit by it long before their parents are deceased.

Knowing with Which Fork to Eat

Cultural capital refers to what one needs to know to function as a member of the various groups to which one belongs. All groups have norms, values, beliefs, ways of life, and codes of conduct that identify the group and define its boundaries. The culture of the group separates insiders from outsiders. Knowing and abiding by these cultural codes of conduct is required to maintain one's status as a member in good standing within the group. By growing up in privilege, children of the elite are socialized into elite ways of life. This kind of cultural capital has commonly been referred to as "breeding," "refinement," possessing the "social graces," having "savoir faire" or simply "class" (meaning upper class). Although less pronounced and rigid than in the past, these distinctions persist into the present. In addition to cultivated tastes in art and music ("high brow" culture), cultural capital includes—but is not limited to—manners, etiquette, vocabulary, and demeanor. Those from more humble backgrounds who aspire to become elites must acquire the cultural cachet to be accepted in elite circles, and this is no easy task. Those born to it, however, have the advantage of acquiring it "naturally" through inheritance, a kind of social osmosis that takes place through childhood socialization.

Having Friends in High Places

Everybody knows somebody else. Social capital refers to the "value" of who you know. For the most part, rich people know other rich people, and poor people know other poor people. Another nonmerit advantage inherited by children of the wealthy is a network of connections to people of power and influence. These are not connections that children of the rich shrewdly foster or cultivate on their own. The children of the wealthy travel in high-powered social circles. These connections provide access to power, information, and other resources. The difference between rich and poor is not in knowing people; it is in knowing people in positions of power and influence who can do things for you.

Early Withdrawals on the Family Estate

Children of the privileged do not have to wait until their parents die to inherit assets from their parents. Inter vivos transfers of funds and "gifts" from parents to children can be substantial, and there is evidence suggesting that such transfers account for a greater proportion of intergenerational transfers than lump-sum estates at death (Gale and Scholz 1994). Inter vivos gifts to children provide a means of legally avoiding or reducing estate taxes. Currently, an individual can transfer $10,000 per recipient tax-free each year. In this way, parents can "spend down" their estates during their lives to avoid estate and inheritance taxes upon their deaths. In 2001, however, the federal government enacted legislation that will ultimately phase out the federal estate tax. Many individual states have also reduced or eliminated inheritance taxes. The impact of these changes in tax law on intergenerational transfers is at this point unclear. If tax advantages were the only reasons for inter vivos transfers, we might expect that parents would slow down the pace of inter vivos transfers. But it is unlikely that the flow of such transfers will be abruptly curtailed because they serve other functions. Besides tax avoidance, parents also provide inter vivos transfers to children to advance their children's current and future economic interests—especially at critical or milestone stages of the life cycle. These milestone events include going to college, getting married, buying a house, and having children. At each event, there may be a substantial infusion of parental capital—in essence an early withdrawal on the parental estate. One of the most common current forms of inter vivos gifts is paying for children's education. A few generations ago, children may have inherited the family farm or the family business. With the rise of the modern corporation and the decline of family farms and businesses, inheritance increasingly takes on more fungible or liquid forms, including cash transfers. Indeed, for many middle-class Americans education has replaced tangible assets as the primary form by which advantage is passed on between generations.

What Goes Up Doesn't Usually Come Down

If America were truly a meritocracy, we would expect fairly equal amounts of both upward and downward mobility. Mobility studies, however, consistently show much higher rates of upward than downward mobility. There are two key reasons for this. First, most mobility that people have experienced in America in the past century—particularly occupational mobility—was due to industrial expansion and the rise of the general standard of living in society as a whole. Sociologists refer to this type of mobility as "structural mobility," which has more to do with changes in the organization of society than with the merit of individuals. A second reason why upward mobility is more prevalent than downward mobility is that parents and extended family networks insulate children from downward mobility. That is, parents frequently "bail out" or "rescue" their adult children in the event of life crises such as sickness, unemployment, divorce, or other setbacks that might otherwise propel adult children into a downward spiral. In addition to these external circumstances, parents also rescue children from their own failures and weaknesses, including self-destructive behaviors. Parental rescue as a form of inter vivos transfer is not a generally acknowledged or well-studied benefit of inheritance. Indirect evidence of parental rescue may be found in the recent increase in the number of "boomerang" children—adult children who leave home only to later return to live with parents. Demographers report that in 1980, one in thirteen adults between the ages of twenty-five and thirty-four lived with parents; by 1990, one in eight did (Edmondson and Waldrop 1993). The reasons for adult children returning to live at home are usually financial—adult children may be between jobs, between marriages, or without other viable means of self-support. Such living arrangements are likely to increase during periods of high unemployment, which in 2003 topped 6% of the civilian labor force.

If America operated as a "true" merit system, people would advance solely on the basis of merit and fail when they lack merit. In many cases, however, family resources prevent or at least reduce "skidding" among adult children. One of the authors of this book recalls that when he left home as an adult, his parents took him aside and told him that no matter how bad things became for him out there in the world, if he could get to a phone, they would wire him money to come home. This was his insurance against destitution. Fortunately, he has not yet had to take his parents up on their offer, but neither has he forgotten it. Without always being articulated, the point is that this informal familial insurance against downward mobility is available—in varying degrees—to all except the poorest of the poor, who simply have no resources to provide.

Live Long and Prosper

From womb to tomb, the more affluent one is, the less the risk for injury, illness, and death (Cockerham 2000; National Center for Health Statistics 1998, Smith 1999). Among the many nonmerit advantages inherited by those from privileged backgrounds is higher life expectancy at birth and greater chances of better health throughout life. Data from the Internal Revenue Service show that those who file estate tax returns have on average lived three years longer than the average life expectancy for the general population as a whole, which is itself likely to be much higher than the life expectancy of those living below the poverty line (Johnson and Mikow 1999). In addition to longer life expectancy, the wealthier are also healthier throughout their lives. Americans who report that they are in excellent health have 74% more wealth than those who report being in fair or poor health (Smith 1999). There are several possible reasons for the strong and persistent relationship between socioeconomic status and health. Beginning with fetal development and extending through childhood, increasing evidence points to effects of "the long reach of early childhood" on adult health (Smith 1999). Prenatal deprivations, more common among the poor, for instance, are associated with later life conditions such as retardation, coronary heart disease, stroke, diabetes, and hypertension. Poverty in early childhood is also associated with increased risk of adult diseases. This may be due in part to higher stress levels among the poor. There is also evidence that cumulative wear and tear on the body over time occurs under conditions of repeated high stress. Another reason for the health-wealth connection is that the rich have greater access to quality health care. In America, access to quality health care is still largely for sale to the highest bidder. Under these conditions, prevention and intervention are more widely available to the more affluent. Finally, not only does lack of income lead to poor health, poor health leads to reduced earnings. That is, if someone is sick or injured, he or she may not be able to work or may have limited earning power.

Overall, the less affluent are at a health disadvantage due to higher exposure to a variety of unhealthy living conditions. As medical sociologist William Cockerham points out:

> [P]ersons living in poverty and reduced socioeconomic circumstances have greater exposure to physical (crowding, poor sanitation, extreme temperatures), chemical

> and biochemical (diet, pollution, smoking, alcohol, and drug abuse), biological (bacteria, viruses) and psychological (stress) risk factors that produce ill health than more affluent individuals. (1998, 55)

Part of the exposure to health hazards is occupational. According to the Department of Labor, the occupations with the highest risk of being killed on the job are—listed in order of risk—fishers, timber cutters, airplane pilots, structural metal workers, taxicab drivers, construction laborers, roofers, electric power installers, truck drivers, and farmworkers. With the exception of airline pilots, all the jobs listed are working-class jobs. Since a person's occupation is strongly affected by family background, the prospects for generally higher occupational health risks are in this sense at least indirectly inherited. Finally, although homicides constitute only a small proportion of all causes of death, it is worth noting that the less affluent are at higher risk for being victims of violent crime, including homicide.

Some additional risk factors are related to individual behaviors, especially smoking, drinking, and drug abuse—all of which are more common among the less affluent. Evidence suggests that these behaviors, while contributing to poorer health among the less affluent, are responsible for only one-third of the "wealth-health gradient" (Smith 1999, 157). These behaviors are also associated with higher psychological as well as physical stress. Indeed, the less affluent are not just at greater risk for physical ailments; research has shown that the less affluent are at significantly higher risk for mental illness as well (Cockerham 2000). Intriguing new evidence suggests that—apart from material deprivations—part of the link between wealth and health may be related to the psychological stress of relative deprivation, that is, the stress of being at the bottom end of an unequal social pecking order, especially when the dominant ideology attributes being at the bottom to individual deficiencies.

Despite the adage that "money can't buy happiness," social science research has consistently shown a tendency for happiness and subjective well-being to be related to the amount of income and wealth people possess (Frey and Stutzer 2002). This research shows that people living in countries that are wealthier (and more democratic) tend to be happier and that rates of happiness are sensitive to overall rates of unemployment and inflation. In general, poor people are less happy than others, although greater amounts of income beyond poverty only slightly increase levels of happiness. Beyond a certain threshold, additional increments of income and wealth are not likely to result in additional increments of happiness. Thus, although money may not *guarantee* a long, happy, and healthy life, a fair assessment is that it aids and abets it....

YOU CAN'T TAKE IT WITH YOU

Whatever assets one has accumulated in life that remain at death represent a bulk estate. Bequests of such estates are what is usually thought of as "inheritance." Because wealth itself is highly skewed, so are bequests from estates. Beyond personal belongings and items of perhaps sentimental value, most Americans at death have little or nothing to bequeath. There is no central accounting of small estates, so reliable estimates on total number and size of estates bequeathed are difficult to come by. Only about 20% of all households report ever having received a bequest (Joulfaian and Wilhelm 1994; Ng-Baumhackl et al. 2003). As the wealth of households increases, however, so does the likelihood of reporting having received a bequest. When bequests are made, they often involve substantial amounts. Recent data from the IRS show that less than 2% of all estates are subject to the federal estate tax (Qohnson and Eller 1998). To be subject to federal estate tax, estates had to have a value of over $600,000 after allowable deductions and exemptions. In short, although few are likely to inherit great sums, bequests from estates are nevertheless a major mechanism for the transfer of wealth across generations (McNamee and Miller 1989, 1998; Miller et al. 2003).

Some may argue that those who receive inheritances often deplete them in short order through spending sprees or unwise investments and that the playing field levels naturally through merit or lack of it. Although this may occur in isolated cases, it is not the general pattern—at least among the super-wealthy. Although taking risks may be an appropriate strategy for acquiring wealth, it is not common strategy for maintaining wealth. Once secured, the common strategy is to protect wealth by playing it safe—to diversify holdings and to make safe investments. The super-wealthy usually have teams of accountants, brokers, financial planners, and lawyers to "manage" portfolios for precisely this purpose. One of the common ways to prevent the quick spending down of inheritances is for benefactors to set up "bleeding trusts" or "spendthrift trusts," which provide interest income to beneficiaries without digging into the principal fund. Despite such efforts to protect wealth, estates may, in some families, be gradually diminished over generations through subdivision among descendants. The rate at which this occurs, however, is likely to be slow, especially given the combination among the wealthy of low birthrates and high rates of marriage within the same class. Even in the event of reckless spending, poor financial management, or subdivision among multiple heirs, the fact remains that those who inherit wealth benefit from it and are provided opportunities that such wealth provides—for as long as it lasts—regardless of how personally meritorious they may or may not be.

REFERENCES

Balkovic, Brian, 2003. "High-Income Tax Returns for 2000." *SOL Bulletin* (Spring): 10–62.

Canterbery, E. Ray, and Joe Nosari. 1985. "The Forbes Four Hundred: The Determinants of Super-Wealth." *Southern Economic Journal* 51: 1073–83.

Cockerham, William. 1998. *Medical Sociology*. 7th ed. Upper Saddle River, N.J.: Prentice Hall.

——— 2000. *Medical Sociology*. 8th ed. Upper Saddle River, N.J.: Prentice Hall.

Collins, Chuck. 1997. *Born on Third Base: The Sources of Wealth of the 1997 Forbes 400*. Boston: United For A Fair Economy.

Edmondson, Brad, and Judith Waldrop. 1993. "Married with Grown Children." *American Demographics* 15: 32.

Forbes. 2003. "The Forbes 400." 172, no. 7 (October 6): special issue.

Frank, Robert H. 1999. *Luxury Fever: Why Money Fails to Satisfy in an Era of Excess*. New York: Free Press.

Frey, Bruno S., and Alois Stutzer. 2002. *Happiness and Economics: How the Economy and Institutions Affect Well-Being*. Princeton, N.J.: Princeton University Press.

Galbraith, John Kenneth. 1958. *The Affluent Society*. New York: Mentor Press.

Gale, William G., and John Karl Scholz. 1994. "Intergenerational Transfers and the Accumulation of Wealth." *Journal of Economic Perspectives* 8: 145–60.

Gokhale, Jagadeesh, Laurence J. Kotlikoff, James Sefton, and Martin Weale. 2001. "Simulating the Transmission of Wealth Inequality Via Bequests." *Journal of Public Economics* 79: 93–128.

Hacker, Andrew. 1997. *Money: Who Has How Much and Why*. New York: Touchstone.

Johnson, Barry E., and Martha Britton Eller. 1998. "Federal Taxation of Inheritance and Wealth Transfers." Pp. 61–89 in *Inheritance and Wealth in America*, edited by Robert K. Miller Jr. and Stephen J. McNamee. New York: Plenum Press.

Johnson, Barry W., and Jacob M. Mikow. 1999. "Federal Estate Tax Returns, 1995–1997." *Internal Revenue Service, Statistics of Income Bulletin* 19, no. 1: 69–129.

Joulfaian, D., and M. O. Wilhelm. 1994. "Inheritance and Labor Supply." *Journal of Human Resources* 29: 1205–34.

Keister, Lisa A. 2000. *Wealth in America: Trends in Wealth Inequality*. Cambridge, U.K.: Cambridge University Press.

Kennickell, Arthur. 2003. *A Rolling Tide: Changes in the Distribution of Wealth in the U.S., 1989–2001*. Washington, D.C.: Federal Reserve Board.

McNamee, Stephen J., and Robert K. Miller Jr. 1989. "Estate Inheritance: A Sociological Lacuna." *Sociological Inquiry* 38: 7–29.

Miller, Robert K., Jr., Jeffrey Rosenfeld, and Stephen J. McNamee. 2003. "The Disposition of Property:

Transfers between the Living and the Dead." pp. 917–25 in *Handbook of Death and Dying*, edited by Clifton D. Bryant. Thousand Oaks, Calif.: Sage.

Mishel, Larence, Jared Bernstein, and Heather Boushey. 2003. *The State of Working America, 2002/2003.* Ithaca, N.Y.: Cornell University Press.

National Center for Health Statistics. 1998. *Health, United States, 1998 with Socio-economic Status and Health Chartbook.* Hyattsville, Md.: National Center for Health Statistics, U.S. Department of Health and Human Services.

Ng-Baumhacki, Mitj, John Gist, and Carlos Figueiredo. 2003. *Pennies from Heaven: Will Inheritances Bail Out the Boomers?* Washington, D.C.: American Association of Retired Persons Public Policy Institute.

Shapiro, H. D. 1994. "The Coming Inheritance Bonanza." *Institutional Investor* 28: 143–48.

Smith, James P. 1999. "Healthy Bodies and Thick Wallets: The Dual Relation between Health and Economic Status." *Journal of Economic Perspectives* 13: 145–66.

Thurow, Lester C. 1999. *Building Wealth: The New Rules for Individuals, Companies and Nations in a Knowledge-Based Economy.* New York: HarperCollins.

U.S. Census Bureau. 2003. Table H-4, "Gini Ratios of Households, by Race and Hispanic Origin of Householder: 1967–2000." In *March Current Population Survey.* Washington, D.C.: U.S. Government Printing Office.

Wolff, Edward N. 2002. *Top Heavy: The Increasing Inequality of Wealth in America and What Can Be Done about It.* New York: The New Press.

DISCUSSION QUESTIONS

1. The table in this selection divides income and net worth into separate columns. As you examine this table, what conclusions do you come to? Does the table illustrate a social problem, in your opinion?
2. Should society limit the wealth that people can accumulate? If you agree, how can it be done? What might be some negative consequences? If you do not agree, do you believe that inequality in society will be more or less limited in the future?
3. In your opinion, does economic inequality as described in this selection undermine a democratic society?
4. What is just society? Is it one where hard work should be rewarded the most? Is it one where the best educated or most talented should be rewarded the most? Is it one where those who contribute to society are rewarded the most? Is it one where those who are most intelligent are rewarded the most? Or is a society only a place where everyone must simply compete without seeking "justice"?

PART IV

Social Problems: Work and Unemployment

For many people work is an important problem. Because work is so individualistic, some people may find it hard to believe that work is a *social* problem. Probably, however, work has been a social problem for a long time, perhaps in all societies, even though people may not have always defined it as such.

In the first selection, from the book *Changing Contours of Work: Jobs and Opportunities in the New Economy*, Stephen Sweet and Peter Meiksins provide a troubling assessment of the ways the new economy is exacerbating class, gender, racial/ethnic, and international chasms. In order to address the many inequalities of the new economy, the authors delineate important roles that organized labor, employers, and the government must play.

The middle class historically has been central to the economic and political health of the United States. Griff Witte from the *Washington Post* addresses issues that he sees as negatively affecting middle-class jobs in the United States.

Finally, Barry Schwartz tackles the problem of meaningful work. Work should be a calling, not a job, not simply a career. Unfortunately, an economic system that relies on the market causes people to regard work as a means to the "good life" rather than a part of the good life itself. This is why, Schwartz believes, so many people are disappointed by their work. He also believes that this is a social problem, not just an inevitability.

10

Reshaping the Contours of the New Economy

STEPHEN SWEET AND PETER MEIKSINS

The Four Questions

1. What is the problem Sweet and Meiksins describe?
2. Is this a *social* problem? Who is harmed by this problem?
3. What are the causes of this problem?
4. How do Sweet and Meiksins think we can solve it?

Topics Covered

New economy
Jobs and opportunity
Class inequalities
Gender inequalities
Unemployment
Social movements
Organized labor
Unionism
Corporate
Social responsibility
Fair labor standards

There is little evidence that the new economy is moving in a direction that will ensure that everyone has opportunities to engage in meaningful work, will earn a comfortable income, or will have the resources to construct satisfactory lives outside of work. On the contrary, employer practices and institutional arrangements continue to sustain—and sometimes even deepen—the chasms that separate workers from opportunity....

OPPORTUNITY CHASMS

Class Chasms

One of the major problems in the new economy is that so many workers find it difficult or impossible to find satisfying and secure jobs that pay. There are both old and new aspects to this problem. Although workers across the class spectrum face problems of strained schedules, insecure jobs, and uncertain futures, these concerns are felt most strongly by those working in the numerous "McJobs" that emerged as a result of efforts to deskill work in the old economy. Today, legions of workers labor in jobs that offer low pay, few benefits, and few opportunities for growth. These workers can expect scant rewards from their jobs and slim prospects that diligent efforts will result in upward mobility. And because women and minorities tend to be funneled into these low-end jobs, this class divide contributes to gender and racial inequalities.

Although bad jobs continue to exist at the bottom of the American economy, there also have been efforts to chip away at the securities and leisure time available to those in the middle class. Insecurity extends to nearly all segments of the workforce, and most workers contend with strong

SOURCE: From Stephen Sweet and Peter Meiksins, *Changing Contours of Work: Jobs and Opportunities in the New Economy*, Thousand Oaks, CA: Pine Forge Press, 2008, pp. 165–186. Reprinted by permission.

prospects that their (or their spouse's) employer will close shop, lay them off, or restructure their jobs away. Their careers commonly require working long hours and affect their ability to form rewarding family lives. These workers' lives remain very different from those laboring at the bottom, but a feature of the new economy is that many "good" jobs have acquired features once associated with less desirable employment.

At the same time, the transition to the new economy witnessed the richest members of society becoming even wealthier and further separating their lives from the rest of society. In contemporary America, the significant gaps are not just between the middle class and the poor, but also between the affluent and everyone else. It is hard not to conclude that the good fortunes of those at the top of the class structure have come from the bad fortunes of those at the bottom and the increasingly fragile middle.

What can be done? One of the most pressing concerns is to address the economic problems faced by low-wage workers and to reshape the terms under which their jobs operate. There needs to be a national dialogue, and action, to rebalance the equation of a fair day's effort and a fair day's pay, including issues of compensation, scheduling, and security. But beyond those issues, resources and institutional arrangements will also need to be restructured in ways that enable workers to move from dead-end, low-skilled positions into jobs that offer greater rewards. This will require, among other things, designing and supporting educational opportunities that fit the structure of the new economy. And because the returns on work are so lopsidedly allocated, the lion's share of the financial burden of paying for these changes must be borne by the affluent.

Issues of job insecurity, unmanageable schedules, tenuous access to health care, limited (and sometimes nonexistent) vacations, summary dismissal, and a variety of other concerns cut across class lines. Creating entitlements to reasonable treatment and reinforced safety nets will help all workers, not just those laboring in bad jobs. Launching initiatives to address these concerns will require abandoning cultural mindsets that divide groups of workers from one another. As long as low-wage workers see themselves as pitted against those with education, or the middle class sees itself as being pitted against those receiving welfare, the effort to mobilize support for collective resources will be hampered. As was the case in bringing about the Fair Labor Standards Act and Social Security legislation in the wake of the Great Depression, catalyzing change will almost necessarily require recruiting the fragile middle class into the struggle.

Gender Chasms

In the old economy, social policies, organizational cultures, job designs, and personal expectations were shaped by the assumption that men would be breadwinners and women would be homemakers. Within this gender regime, women's efforts in the home were defined as something other than real work, and care work was not given its economic due. In the new economy, women are nearly as likely as men to work outside of the home and to aspire to meaningful careers. And it is not simply women's desires that influenced their increased integration into the paid labor force—most families need two earners to make ends meet. Inequalities between men and women persist in no small degree because of the persistence of ideas and structures that place most of the burden of care work on women and impede their access to the most financially rewarding jobs.

Gendered work standards, modeled on what men used to be able to bring to the workplace, are now structured into job designs and employers' definitions of who are "ideal workers" (Williams 2000). Workers have been socialized to accept a career mystique, an unsustainable set of expectations promoting an intense commitment to work (Moen and Roehling 2005). These structures and beliefs chafe against the resources available to the new workforce, one composed increasingly of dual-earner and single-parent families. How are workers responding? Most commonly by laboring long hours, shifting the timing of life events (such as marriage and childbirth), doing without, and shouldering the burden of increased stress. Their ability to manage also is often predicated on their ability to employ low-wage workers to care for children or aging parents.

Recognizing these facts highlights the need to dismantle a gender regime that continues to assign men and women to different jobs inside and outside the home. But beyond countering the forces of socialization and interpersonal discrimination, structural changes will have to occur as well. Compensation will need to be recalibrated so that care work receives equitable returns as a form of labor. And career and job templates that assume workers have the capacity to work like men with full-time, stay-at-home spouses will need to be reconfigured.

Workers and employers in the new economy need to develop new definitions of how much work employees should be expected to perform and to redesign jobs to be reasonably compatible with what workers can provide across their careers. Efforts to create a "one-size-fits-all" approach to work and family strains are unlikely to succeed, as the potential contributions of most workers vary over the life course. A more fruitful direction for change involves fostering new templates for flexible careers (Moen and Sweet 2004). These can include opportunities to scale back work hours or to take time-outs from the labor force. And individuals and couples will need to have the resources to be able to effectively plan careers so that they can land on their feet when jobs are lost.

The issues surrounding gender and work are commonly cast as "women's problems," but the reality is that these are family problems that affect men, women, their children, and their aging parents. Accelerating the pace and extending the range of reform will require bringing men into the movement to humanize work by persuading them of the benefits to be reaped from reconfigured gender roles and resources. This is actually an easy argument to make. Which stressed worker would not want to see the establishment of reasonable work schedules, expanded vacations, and opportunities to secure time-outs from his job? If more men are convinced that they lose out with gendered divisions of labor—particularly in their access to time to spend with their children—the clamor for change will increase, and coalitions for change will strengthen.

Racial and Ethnic Chasms

Although Civil Rights–era legislation imposed legal barriers against discrimination, racial and ethnic ties continue to influence access to jobs. Today, race and ethnicity play a role in determining where one lives, the resources families have to pass from one generation to the next, and the attitudes gatekeepers hold about individuals from different backgrounds. Like the gender chasm, the race chasm is maintained through social structures that differentially allocate resources and opportunities, as well as cultural orientations that shape expectations.

If the new economy is to become truly race-blind, it will need to address—but also go beyond—sources of interpersonal discrimination, such as the widespread use of stereotypes that label potential employees as suspect. The disadvantages minority group members face stem from being raised in families that lack the economic resources to afford college tuition, the social connections to facilitate entry into good jobs, and the cultural capital that provides the soft skills needed for success in the new economy. But more than family disadvantages and interpersonal discrimination contributes to racial inequities. They also result from community-level resource deficits. The forces that concentrate members of underprivileged racial and ethnic minority groups into poor neighborhoods make it likely that they will be deprived of quality education, information about work and careers, and the types of social ties that link them to jobs.

Racial tensions continue to divide the U.S. workforce, with controversies over illegal immigration, affirmative action programs, and welfare reform all exhibiting strong racial overtones. The dialogue about whether illegal immigrants are "taking jobs from American workers" or "doing jobs that no one else will perform" distracts from the larger question of how to regulate low-wage work. A crucial step in reorienting this debate will be to shift the focus away from blaming the victims of globalization and economic restructuring, and toward analyzing the conditions of work—at home and abroad. It is interesting to note that virtually absent from the illegal immigration and welfare reform debates are discussions of the exploitative conditions under which most low-wage work occurs or of the role of employers who provide jobs to undocumented workers. Building racial coalitions is going to be a great challenge because prejudice and discrimination are etched so deeply into the culture and structure of American society. But these barriers can be dismantled by groups that share common interests (which we discuss later in this chapter), and as this happens, prospects for opening opportunities for underrepresented minority groups will be enhanced.

International Chasms

The new economy's global reach is profoundly affecting work opportunities around the world. It is impossible to ignore the remarkable transformations occurring in China, India, and many other developing nations. Longer life expectancies, as well as access to medical care, educational opportunities, and consumer goods, reflect a variety of positive outcomes of globalization. But at the same time, globalization has contributed to a number of new problems in the developing world, including pollution, overpopulation of cities, and the creation of hazardous work environments. Although some societies are advancing, other places in the global economy—especially in sub-Saharan Africa—have been largely left behind.

The flow of work opportunities into the developing world and the movement of employment away from its traditional locations in the United States and Europe have contributed to American workers' feelings of insecurity and fueled international antagonisms. Some American workers express hostility toward lower-priced workers overseas, who are perceived as stealing Americans' jobs. At the same time, the United States is increasingly viewed negatively overseas, not only because of recent political debacles, but also because its power is perceived as coming from the exploitation of workers abroad. The cheap consumer goods available in American stores often come at a high cost to workers in foreign lands—and these facts are well recognized within the developing world. As a result, the globalization of work may be catalyzing radical responses to what is perceived as American cultural, political, and economic domination. And even in countries that are making economic strides, vast numbers of people are unable to achieve a standard of living even remotely like that enjoyed by an average American.

The solutions to these problems are not likely to come from the free market. As Karl Marx observed, the forces that make capitalism so productive are intertwined with the drive to exploit labor, and to do so around the globe:

> The need of a constantly expanding market for its products chases the bourgeoisie over the whole surface of the globe. It must nestle everywhere, settle everywhere, establish connections everywhere ... All old-established industries have been destroyed or are daily being destroyed. They are dislodged by new industries, whose introduction becomes a life and death question for all civilized nations, by industries that no longer work up indigenous raw material, but raw material drawn from the remotest zones; industries whose products are consumed, not only at home, but in every quarter the globe. In place of the old wants, satisfied by the production of the country, we find new wants, requiring for their satisfaction the products of distant lands and climes. (Marx and Engels 1972[1848])

The current system enables unfettered mobility of capital and creates incentives for employers to move jobs to locations where labor and the environment can be most easily exploited. Developing societies will not necessarily benefit from this movement because the same mechanisms that enable capital to move in also enable shifting the profits to other locations. Noted analysts such as Thomas Friedman (2005) optimistically assert that globalization will "flatten" the divides between developed and developing countries. Although this may be true in cultural terms, in economic terms the divides between developed and many developing countries are expanding (United Nations 2005). This is not to say that life will not continue to improve in most developing nations, but it will not be nearly as good as it could be if reasonable checks and controls, designed to ensure more equitable returns from work, were placed on free-market capitalism. This needs to be done not only for ethical reasons, but also because it is in the interests of workers in the United States to do so (Kochan 2005).

THE AGENTS OF CHANGE

... Our argument is that refashioning the new economy will require the concerted efforts of multiple agents, ranging from individual consumers to activist groups, unions, employers, government, and international organizations. However, not every agent will carry the same level of force, and when their efforts are misapplied (or misrepresented), they can actually impede change rather than accelerate it.

The Role of Individuals

Efforts to reshape the workings of the new economy, and to remedy its problems, often encourage a "do-it-yourself" approach. But can the cumulative efforts of disconnected individuals lead to the types of changes needed to substantially refashion the contours of work and opportunity? Consider, for example, the advice manual *The Better World Handbook: From Good Intentions to Everyday Actions* (Jones, Haenfler, Johnson, and Klocke 2002). This book (like a number of others) suggests that the path to reform leads through self-reflection and conscious efforts to "do the right thing." Toward that end, consumers are advised to limit their trade with companies that are less than environmentally sensitive or labor friendly, as well as to treat coworkers and subordinates with the dignity that they deserve. Investors are encouraged to buy stocks and invest in companies that have a positive track record on issues of public concern. And, like innumerable other self-help books, it offers advice on the best ways to "balance" work and family.

Though intuitively appealing these "do-it-yourself" solutions are unlikely to create truly meaningful change by themselves. One problem with the focus on the isolated consumer is that individuals face the Herculean challenge of identifying those brands and companies that are "responsible." To do so requires not only understanding the actions of distributors, but also the various companies that are linked to them in vast global supply chains. And companies can have mixed histories—for example, they may be strong on labor concerns but much weaker on environmental concerns. Socially responsible consumers also experience considerable difficulty determining which products to buy because the information they receive is often confusing and misleading (Nestle 2003; Seidman 2007). And even if consumers did have access to the relevant information, options to purchase responsibly do not always exist. This is a classic example of the exit/voice dilemma—individual consumers may possibly influence policy through exit (by avoiding choices they reject) but lack voice; that is, they cannot *create* options that correspond to their preferences. Of equal concern is the observation that consumers are reluctant to purchase fair trade products that cost much more than standard goods. In practice, people tend to place higher premiums on issues of style and price than on ethical business practices (Iwanow, McEachern, and Jeffrey 2005; Pelsmacker, Driesen, and Rayp 2005).

The expectation that individuals will provide solutions for collective problems has a long history in American culture. For conservatives, individualism reflects an American virtue, and they reference Tocqueville (1969 [1836]) in their observations of the ways their country's successes hinged on Americans' embrace of autonomy and volunteerism. Conservatives argue the merits of an "ownership society" that allocates as many resources as possible to individuals. Thus, rather than invest in a national health care system or Social Security, the individualistic approach is to offer health "savings accounts," and personally managed retirement portfolios. Nor is it only conservatives who place individualistic perspectives at the forefront of social policy; so do many liberals. This can be witnessed, for example, in President Bill Clinton's efforts to end "welfare as a way of life" and to introduce strict limits on support through the Personal Responsibility and Work Opportunity Reconciliation Act. In all of these cases, reforms were presented as empowerment, but actually had the consequence of shifting the responsibility for managing risk to individuals. As Jared Bernstein (2006) argues, these programs rest on an underlying philosophy that says "you're on your own" and that individuals have no right to expect support or protection from hardship (see also Hacker 2006).

Numerous cultural critics have argued that individualism results in a diminished capacity to empathize and understand how one's fortunes (or misfortunes) are connected to larger social processes (Bellah, Madsen, Sullivan, Swidler, and Tipton 1985; Mills 1959; Putnam 2000; Reisman 2001 [1961]). Even when sympathy exists, notice how individual volunteers are assigned the task of ameliorating hardship, supplanting bolder initiatives to challenge oppressive structural arrangements. For example, consider two of the most prominent volunteer and semi-volunteer groups in America today—Habitat for Humanity and Teach for America. Habitat for Humanity International, with a nearly all-volunteer membership, has erected approximately 200,000 homes in America and abroad since its inception. In 2005, Teach for America had 3,500 dedicated young teachers working in impoverished school districts. Without disparaging these accomplishments, it is important to place them in context. In 2006, in the United States alone, nearly *8 million* families lived in poverty and *1 in 10* children did not graduate from high school. As well-intentioned as volunteers are, they commonly do little (if anything) to eliminate structural barriers or establish a more equitable distribution of social resources.

In sum, individualistic solutions are hampered by three fundamental concerns. First, people may not have the inclination or the capacity to behave in a manner that will transform the new economy. Second, individualistic efforts generally leave untouched the underlying forces that shape the contours of work and opportunity. Third, as a cultural framework, individualism reinforces the shift of risks to individuals, rather than building on the strength of collectivities. This does not mean that individuals cannot make a difference by trying to be socially conscious or that these efforts cannot create good outcomes. But these efforts alone will not be sufficient to address the root sources of the problems or make a significant impact on the operations of the new economy (Bernstein 2006).

The Role of Activist Groups

If individuals are not going to be the answer, how about activist groups? These groups comprise individuals who band together to exert pressure on governments, employers, and consumers. Their goals are to use collective action, primarily directed at the local level, to influence employment and trade practices, not only at home in America, but also abroad in developing countries. Here, we see greater cause for optimism, but the impact of their efforts will necessarily depend on recruitment, their alliances with other organizations, and their ability to identify and target institutions that can effect change.

Consider the successes of activist groups. such as United Students Against Sweatshops and United Students for Fair Trade, which have influenced their own educational institutions to sever contracts with producers with histories of labor abuse (Crawford 2003; Glover 2003). Among their accomplishments is increased public awareness of exploitative labor practices and their links to the production of well-known consumer items, such as those endorsed by P. Diddy, Kathie Lee Gifford, and Michael Jordan. By putting pressure on institutions from within, and by mobilizing public opinion, these groups engage in a "name and shame" strategy to persuade employers to change their business practices (Seidman 2007)....

Although these successes provide reason for hope, it is important to assess them relative to the magnitude of the problems. Even the most successful grassroots initiatives tend to have limited reach in sparking change in employment practices in the new economy. In part, they are hampered by limited resources and fluctuating memberships. When they do effect change, gains can be fleeting, as companies remain free to move facilities to "employer-friendly" locations (Armbruster-Sandoval 2005). Another problem is related to the strategy of focusing change at the local level, which can sometimes be trumped by more macrolevel initiatives. For example, companies can influence the creation of legislation at the state level to restrict the right of local communities to set their own labor standards. And focusing on the local level tends to produce changes that are confined to small groups of workers in particular locations. Careful study of living wage initiatives, for example, reveals that comparatively few low-wage workers have been affected (most studies estimate the number to be less than 100,000, a small fraction of today's workforce) (Freeman 2005).

Arguably, the biggest success of activist groups has not been in local reforms, but in increasing public awareness of labor and environmental abuses. In response, companies have become more conscious of their public images, which in turn has influenced some major employers to reform labor and trade practices. For example, once the widespread use of child labor gained public recognition, companies such as Nike, Reebok, and Adidas agreed to monitor their suppliers' employment practices. Coffee distributors such as Starbucks have signed agreements with "preferred suppliers," who have promised to pay better wages, not to employ child labor, and to have their operations monitored (Schrage 2004). Still, critics, such as the Organic Consumers Association, claim that Starbucks makes only limited use of "fair trade" coffee beans and that the company's commitment to economic justice is superficial. As we discuss later, employers have developed a variety of strategies to maintain the impression of being socially responsible, while often avoiding being truly responsible (Seidman 2007). The question that we will return to concerns the most strategic means for harnessing the collective power of these grassroots initiatives to force change.

The Role of Organized Labor

The American union movement has been in decline for much of the past 50 years and efforts to reverse this trend have proven largely unsuccessful.... Unions face a variety of problems in the new economy, including the decline of manufacturing employment, a legal framework that makes organizing difficult, capital mobility, the proliferation of smaller workplaces and subcontractors, and many more. These problems facing organized labor are compounded by public relations concerns. Many American workers have ambivalent feelings toward unions, which are commonly perceived as the reason why jobs move overseas, as protecting deadbeat employees, as corrupt, and as undermining the principle that individual workers (rather than classes of workers) should be compensated according to their efforts (Lichtenstein 2002). Only one in five Americans (20%) thinks that unions are "excellent" or "very good" for the country, one in three (36%) believes that unions block economic progress, and two in three (69%) think unions have enough or too much power.[1]

Although unions have diminished power in the new economy, it is important to recognize that they still hold considerable sway. Unions play a vital role in influencing job contracts for those who are members or who work in unionized workplaces. Today, even though smaller proportions of the labor force are organized, they remain a substantial political force, particularly within the Democratic Party (Dark 2001). Fully 1 in 10 workers is a union member, and many more Americans belong to a family in which at least one person is a union member....

Several recent examples of successful organizing campaigns give proponents hope that the resurgence of union activity has already begun. In some cases, conventional unions have succeeded in organizing previously unorganized groups of workers. The Justice for Janitors campaign has made remarkable progress in organizing and winning better conditions for building maintenance workers (Milkman and Voss 2004). Similarly, the Hotel and Restaurant Workers Union succeeded in negotiating labor contracts with most of the full-service hotels in San Francisco (Wells 2000). Unions are even attracting the interest of professional workers, such as graduate students, who successfully gained union representation at several major universities (Lafer 2003). Surprisingly, some of the most successful organizing campaigns have been conducted by some of the oldest (and allegedly most "conservative") unions, those once affiliated with the American Federation of Labor (AFL). Ruth Milkman (2006) argues that this is because they are concentrated in growing sectors of the economy, in industries unaffected by capital mobility (such as hotels), and because their approaches are better suited to a volatile economy.[2]...

A resurgent union movement will have to confront globalization and develop a means of collaborating and organizing workers on an international level. This would not be an altogether new idea. Even in the late 19th century, labor "internationals" existed, and many 20th century unions belonged to a variety of international union federations. Most recently, an attempt to consolidate these international organizations led to the formation in 2006 of the International Trade Union Confederation (ITUC), which claims more than 300 affiliates in a wide range of countries. However, most international labor organizations, including the ITUC, do not engage in organizing efforts, but rather serve as clearinghouses for information, protest various forms of injustice against workers around the world, and pressure national and international organizations (such as the World Bank) on behalf of workers. Today, there are other opportunities for cross-national cooperation, such as an American union working together with a sister union at a branch plant overseas. Alternately, workers in one country could be organized to act in sympathy with workers in the same industry overseas. This happened in 1998, when American workers refused to unload ships affected by an Australian dock strike.

Unions will be better positioned to promote change when they integrate themselves with other activist groups, such as the anti-sweatshop campaigns and living wage efforts mentioned earlier. In the past, unions and new social movements have not been receptive to one another. During the Vietnam War era, unions were often perceived (sometimes accurately) as opposed to the student movement. Nevertheless, when unions work in concert with other grassroots organizations, an approach that has been called "social movement unionism," the results are often superior to when they operate alone (Gordon 2005; Milkman 2006). Successful living wage campaigns illustrate this well; so do the much-discussed demonstrations in Seattle in 2001, in which labor groups cooperated with church groups, environmentalists, and others to protest the meetings of the World Trade Organization and its policies regarding global trade (Clawson 2003). And the remarkably successful organizing drives in Las Vegas hotels and casinos also owe much to a conscious strategy of working together with other community organizations (Fantasia and Voss 2004).

In short, labor organization could, again, become a powerful force for workplace reform. However, there is no guarantee that this will happen. Obstacles to union organization remain powerful and the gains discussed here are fragile. The New York University (NYU) graduate students' union, which organized in 2002 and made substantial gains in wages, benefits, and rights to overtime, was essentially shattered in 2005 (Arenson 2005; Epstein 2005). Worker centers, too, often find their gains to be temporary or unenforceable and struggle both to forge links with nonimmigrant groups and to grow beyond a few hundred members (Gordon 2005). So, although it may be premature to announce the death of American unions, unions are unlikely suddenly to become powerful enough to reshape American workplaces by themselves.

The Role of Employers

The tensions between work and family life, and the growing awareness of the problems of global trade, are creating pressures to implement more socially responsible designs for workplace operations. But are these pressures sufficient to motivate employers to advance meaningful change from within? What is the likelihood that "internal" actions by the leading agents in the new economy will foster new approaches to work that serve the interests of workers and the communities in which they live?

One strategy to make employers take ownership of change relies on demonstrating how responsive policies can actually benefit companies. Work–family scholars call this the "dual agenda," and posit that what is good for working families, or the wider society, does not necessarily come at a cost to employers. Some family-responsive policies, such as flexible work arrangements (when applied in specific types of settings to specific types of workers) can actually work in the employer's interests (Pavalko and Henderson 2006). Fulfilling the dual agenda requires demonstrating the potential for positive "returns on investment" that can be measured by increased productivity, greater profits, better employee retention, consumer loyalty, or other indicators of success. This business case for reform shows employers how their companies can benefit from altering production practices and taken-for-granted ways of working (Kossek and Fried 2006).

In many industries, a business case can be made to enhance labor standards through increased wages, benefits, or the creation of flexible work arrangements. For instance, recognition as a "best employer" in publications such as *Working Mother* magazine can help a company market itself to consumers and potential employees. Similarly, a business case can be made to respond to brand tarnishing that results from "naming and shaming" pressures brought to bear by activist groups. A marketing business case can be made for developing products that fill the emerging niche for "fair trade" products. And opening opportunities for bridge jobs or flexible employment can provide a means to retain the talents, experience, and knowledge (of an aging workforce (DeLong 2004). Researchers at the Massachusetts Institute of Technology (MIT) have worked with employers to design collaborative experiments to test how dismantling and replacing entrenched ways of organizing work affect the bottom line (Bailyn, Bookman, Harrington, and Kochan 2005). Other researchers, in collaboration with employers, are documenting the extent to which companies are implementing worker-responsive practices and then publicizing findings from which companies can "benchmark" their own performance relative to that of the wider corporate community (Harrington and James 2006).

Although these approaches show some promise, the employers in these studies tend to be those who want to retain and focus the energies of skilled workers. It is much harder, and sometimes impossible, to identify a business case to improve the conditions of work for those engaged in low-end jobs at home or abroad. If one traces the history of work from industrialization forward, it becomes apparent that the business case is often strongest for *dismantling*, rather than implementing, worker-responsive practices (such as providing high wages, secure jobs, health insurance, retirement benefits, and safe working environments—all of which have been eroded in the new economy). This is not to say that a dual agenda cannot sometimes be satisfied, especially for flexible work arrangements for highly skilled workers. But it is important to recognize the limitations of this approach because it is unlikely to close many of the major opportunity chasms, such as those that channel the lowest strata of the workforce into poorly paid, low-skilled work. For those workers, the business case (established by history and numerous current cases) is to degrade work and opportunity.

Beyond the business case strategy, others suggest that employers may move the new economy forward as they embrace an ethos of corporate citizenship or corporate social responsibility (CSR). In this case, the internal force is an open commitment, by corporate leadership, to understand their roles as stewards of the new economy (Blowfield 2005). CSR requires managers not to direct their actions simply to expand profit within the narrow confines of the law, but also to follow normative standards of professional conduct that go beyond what is obligatory or customary (Carter 2004). Perhaps most familiar to American consumers is Ben & Jerry's ice cream, a company that maintains that community-mindedness is a core component of the brand.[3] Beyond this company, the adoption of corporate social responsibility codes has been demonstrated to have a positive impact in the textile, leather tanning, and shoe industries in South Asia and in the wine industry in South Africa (Luken and Stares 2005; Nelson, Martin, and Ewert 2005; Winstanley, Clark, and Leoson 2002). And individual socially responsive leaders can make a difference in the lives of their workers. For example, because of the challenges of being a single parent following his wife's death, CEO Lewis Platt created sweeping changes at Hewlett-Packard, including the introduction of expansive flextime, flexplace, and family leave policies (Abelson 1999)....

Is corporate leadership likely to guide companies to be "good citizens" in the new economy? Or are they more likely to continue to engage in "a race to the bottom" and seek ways to maximize profit at the expense of workers and the communities in which they live?... Considerable evidence suggests that the latter is more likely to occur. It is important to remember that capitalism itself discourages social responsibility because there are clear financial motivations to employ populations that can be most easily exploited and to move to locations where environmental protections are weakest. The case of Wal-Mart, the world's largest employer and retailer, amply demonstrates this proposition. In the wake of widespread condemnation of its employment and trade practices, Wal-Mart has directed far greater attention to its strategies for managing public opinion than it has to reforming its labor practices, and its profits and power continue to increase (Goldberg 2007).

The Role of Government

Political conservatives commonly caution against the interference of "big government" and the impediments regulation creates for the functioning of the economy. This perspective on government's role in the regulation of the new economy has a number of problems. First, government is *always* involved in economic activity. As the economic historian Karl Polanyi (1944) argued, the market itself could not exist unless the government had been there to help create it and continued to provide the political, legal, and military framework within which it operates. Second, government has a long history of making trade possible, and of shielding workers from abusive work conditions. Most of the protections workers have today are the result of government intervention: these include child labor laws, unemployment insurance, environmental regulations, social security, minimum wage standards, rights to overtime pay, workplace safety oversight, the right to unionize, prohibitions against discrimination, and the right to family medical leave.... Those who espouse laissez-faire capitalism often overlook these necessary "intrusions" into the economy (Bernstein 2006; Galbraith 1976).

....What role should the American government play in the new economy? The types of action we advocate will require reenvisioning the regulation of work, in much the same way as happened during the New Deal. This reenvisioning should consider specific workplace regulations as well as the collective responsibility to provide for the basic needs of citizens, irrespective of their attachment to the labor force. We suggest three types of policy initiatives. First, there must be a revisiting of the issue of what constitutes reasonable conditions for work, and a revision of the standards enacted under the Fair Labor Standards Act (and other legislation) to consider both the nature of jobs and the composition of the labor force in the new economy. Second, there needs to be a national discussion of the means by which affluence and opportunities can be more equitably distributed. Third, although much attention has been focused on the U.S. government's role in promoting the interests of employers and protecting jobs at home, there needs to be a forceful discussion of its role in regulating the terms under which global supply chains operate and the role of the United States in fostering *positive* development abroad.

A Fair Labor Standards Act for the New Economy When the Fair Labor Standards Act was enacted in 1938, it offered Americans, for the first time, the right to overtime pay and a minimum wage. Since its passage, its provisions have received modest updates and some new worker protections have been instituted. However, the regulation of work today largely fails to address the needs of many workers in the new economy. Today, employers can expect workers to labor for below poverty-level wages; employers can impose long, short, or unpredictable schedules that interfere with employees' family responsibilities, and they are not obligated to provide information sufficient to enable employees to plan lives and careers. Redefining what constitutes the minimum standards of fair treatment is long overdue.

In early 2007, the federal minimum wage remained at a scant $5.15 per hour, unchanged since 1996, and held a real value that was lower than it was in 1951. Although teenagers are commonly considered to be the typical minimum wage workers, the reality is that nearly one in two (47%) minimum wage workers is older than age 25, and two in five (39%) work full time (Haugen 2003; Shulman 2005). As we write … a newly elected Democratic Congress passed a bill that will raise the minimum wage to $7.25 per hour by 2009. This standard approximates that advocated by 650 leading economists (including five Nobel Laureates) (Economic Policy Institute 2006). However, in contrast to the recommendations offered by these leading economists, the bill does not tether the increased minimum standard to inflation adjustments, so the minimum wage will continue to be subject to the vagaries of political will.

During the Great Depression, the problem of unemployment was exacerbated because those who did have jobs were expected to work long hours, which in turn deprived others of the opportunity to work. The Fair Labor Standards Act responded by implementing overtime provisions that penalized employers for working their employees beyond 40 hours per week. By mandating time-and-a-half compensation for overtime, the act stimulated a reduction in work hours, as well as dispersed work opportunity. But because full-time employees are typically the only ones who receive benefit, and because the costs of benefits like health care have increased substantially, the penalty for overworking employees has decreased in the new economy. And because of changing opportunity structures, many more employees in the new economy are exempt from overtime provisions, as they labor in salaried positions and within organizational cultures that treat long hours as desirable and normal.

Today, the 40-hour work week plus time-and-a-half pay for overtime is a taken-for-granted arrangement, but one that needs to change. Although the current threshold may have been a reasonable expectation for a husband/breadwinner–wife/homemaker arrangement, today many dual-earner couples work a combined 80 to 100 hours per week (and sometimes more), leaving them frazzled and exhausted. One approach to fixing this problem can be found in France, a country that reduced the threshold for overtime eligibility to 35 hours per week. The result of this legislation has been to increase work opportunity and to decrease the level of stress shouldered by working families. One study of the impact of the French law found that nearly two in three workers responded that it had made it easier for them to combine work life with their family life (Fagnani and Letablier 2004). In 2007, the newly elected Nicolas Sarkozy, who ran for president on a *conservative* platform, stated no plans to fundamentally rework the 35-hour labor standard. Alternately, rather than lowering the threshold at which overtime begins, the penalty for overworking employees could be increased—perhaps from time-and-a-half to double-time compensation.

A Fair Labor Standards Act for the New Economy needs to address the issue of overwork in other ways as well. For example, to bring the country more in line with the treatment of workers in Europe, Americans would need the opportunity to take 4 to 5 weeks of paid vacation a year. Current labor standards need to go far beyond the 12 weeks of unpaid leave provided by the Family and Medical Leave Act and at a minimum, the current standard needs to be modified to accommodate paid leaves of absence (which in turn will make family leave available to all workers). The variety of family leave and family supports enacted in Europe in the past 30 years provide valuable lessons to inform efforts to reshape American family leave policy (see Gornick and Meyers 2003; Hobson 2002; Kelly 2006; Pfau-Effinger 2004).

Finally, a Fair Labor Standards Act for the New Economy needs to address the issue of job insecurities. In part, adjustments of family leave and overtime provisions may help to reduce insecurity by discouraging under-employment and forced "voluntary" exits from the labor force. Again, lessons from Europe may help to craft this legislation, as well as anticipate its impact on worker performance. In France, for example, complex "licenciement" laws regulate the conditions under which employees can be fired. To terminate a worker, companies commonly have to demonstrate that the employee could not be retained, show that the company cannot afford to keep the job in existence, or prove that the employer is incompetent. As a result, when employees underperform, rather than firing them outright (as is commonly done in America), French employers sometimes pay them to disappear graciously, and even go so far as to keep problematic employees on the books, but not assign them important work (Smith 2006). Although this type of legislation could have the result of keeping greater numbers of "deadbeat" employees on the payroll, French workers are actually more productive (as measured by productivity per hour worked) than American employees are (Krugman 2005). Whether American culture could embrace such commitments remains to be seen, but the key to winning the debate will be concerted effort to shift the discussion from employer rights to worker rights (Shaiken 1984), and from "you're on your own" to "we're in this together" (Bernstein 2006)....

CONCLUSION

... The new economy has many new characteristics but it has also integrated old practices into the design of jobs and the allocation of resources. Some of the problems evident today are new, but many reflect unresolved problems that emerged in the old economy, failed approaches to those problems, and lagging responses to emerging concerns. Today, America faces challenges analogous to those it confronted at the turn of the 20th century. Old means of working are being discarded, as new technologies and organizational practices are introduced. Previously dominant sectors of the economy are in decline, as new jobs and industries are emerging. And the workforce is changing, as are communities of workers. Along with the new challenges these changes present, there remain persistent concerns—of inequalities, overwork, unemployment, insecurity, and strained schedules.

If left unchanged, the contours of work in the new economy will be all too familiar. The future of work will be characterized by divided opportunities, with workers segregated on the basis of class, gender, race, and nationality. Some workers, particularly women and members of minority groups, will have careers dislodged or never make it onto career tracks at all. Large portions of the workforce will labor in alienating, low-skilled, low-wage jobs. Some will have careers disrupted when they have children; others will keep their careers intact by foregoing having children altogether. Large numbers of workers will experience stress from having too much work, whereas many others will be work-poor and labor in jobs with unpredictable schedules, if they can find work at all. And most workers will labor under conditions of uncertainty and have inadequate resources to plan careers and to weather job loss.

Economics should work for people, not the other way around. The shape of the new economy is not immutable, nor is its future inevitable. One of the lessons learned by looking at work in historical perspective is that people have been able to effect positive change—even when faced with formidable resistance....

Essential to solving all of the issues presented in this book is recognizing that the problem does not rest in a scarcity of wealth or resources. It rests in the ways resources are allocated in a tremendously productive economy. There is no reason to accept the premise that good jobs and a good life will inevitably be beyond the reach of the hardworking men and women who make it all possible.

NOTES

1. Authors' analysis of the General Social Survey.
2. For example, Milkman notes that the AFL tries to get all (or most) of the employers in an area to agree to labor standards so that a single employer's threat to move or close down is less effective.
3. Ben & Jerry's was, however, recently purchased by an international conglomerate. There has been an effort to preserve the company's socially conscious image, but it remains to be seen whether this can be sustained over the long haul.

REFERENCES

Arenson, Karen. 2005. "N.Y.U. Moves to Disband Graduate Students Union," *New York Times* (June 17), p. 2.

Armbruster-Sandoval, Ralph. 2005. "Workers of the World Unite? The Contemporary Anti-Sweatshop Movement and the Struggle for Social Justice in the Americas." *Work and Occupations* 32: 464–485.

Bailyn, Lotte, Ann Bookman, Mona Harrington, and Thomas Kochan, 2005. "Work-family Interventions and Experiments: Workplaces, Communities, and Society." In *The Work and Family Handbook: Multi-Disciplinary Perspectives, Methods, and Approaches*, edited by Marcie Pitt-Catsouphes, Ellen Ernst Kossek, and Stephen Sweet. Mahwah, NJ: Erlbaum.

Bellah, Robert, Richard Madsen, William Sullivan, Ann Swidler, and Steven Tipton. 1985. *Habits of the Heart: Individualism and Commitment in American Life.* New York: Harper & Row.

Bernstein, Jared. 2006. *All Together Now: Common Sense for a Fair Economy.* San Francisco: Berrett-Koehler.

Blowfield, Michael. 2005. "Corporate Social Responsibility: Reinventing the Meaning of Development?" *International Affairs* 81: 515–524.

Carter, Craig R. 2004. "Purchasing and Social Responsibility: A Replication and Extension." *Journal of Supply Chain Management: A Global Review of Purchasing & Supply* 40: 4–16.

Clawson, Dan. 2003. *The Next Upsurge: Labor and the New Social Movements.* Ithaca, NY: ILR Press.

Crawford, Elizabeth. 2003. "Good to the Last Drop." *Chronicle of Higher Education* 49: A8.

Dark, Taylor. 2001. *The Unions and the Democrats: An Enduring Alliance*, 2nd ed. Ithaca, NY: Cornell University Press.

DeLong, David. 2004. *Lost Knowledge: Confronting the Threat of an Aging Workforce.* New York: Oxford University Press.

Economic Policy Institute. 2006. "Hundreds of Economists Say: Raise the Minimum Wage." Retrieved from http://www.epinet.org/content.cfm/minwagestmt2006

Fagnani, Jeanne, and Marie-Therese Letablier. 2004. "Work and Family Life Balance: The Impact of the 35 Hour Laws in France." *Work, Employment and Society* 18: 551–72.

Fantasia, Rick, and Kim Voss. 2004. *Hard Work: Remaking the American Labor Movement.* Berkeley: University of California Press.

Freeman, Richard. 2005. "Fighting for Other Folks' Wages: The Logic and Illogic of Living Wage Campaigns." *Industrial Relations* 44: 14–31.

Freidan, Betty. 1963. *The Feminine Mystique.* New York: Norton.

Friedman, Thomas. 2005. *The World Is Flat: A Brief History of the Twenty-First Century.* New York: Farrar, Straus and Giroux.

Galbraith, John Kenneth. 1976. *The Affluent Society*, 3rd ed. Boston: Houghton Mifflin.

Glover, Katherine. 2003. "No Swear in Minneapolis." *Dollars and Sense* 249: 8–9.

Goldberg, Jeffrey. 2007. "Selling Wal-Mart: Annals of Spin." *New Yorker*, pp. 32–40.

Gordon, Jennifer. 2005. *Suburban Sweatshops: The Fight for Immigrant Rights.* Cambridge, MA: Harvard University Press.

Gornick, J.C., and M. K. Mevers. 2003. *Families That Work: Policies for Reconciling Parenthood and Employment.* New York: Russell Sage Foundation.

Hacker, Jacob. 2006. *The Great Risk Shift: The Assault on American Jobs, Families, Health Care and Retirement and How You Can Fight Back*. New York: Oxford University Press.

Harrington, Brad, and Jaquelyn James. 2006. "The Standards of Excellence in Work-Life Integration: From Changing Policies to Changing Organizations." In *The Work and Family Handbook: Multi-Disciplinary Perspectives, Methods, and Approaches*, edited by Marcie Pitt-Catsouphes, Ellen Ernst Kossek, and Stephen Sweet. Mahwah, NJ: Erlbaum.

Haugen, Steven. 2003. "Characteristics of Minimum Wage Workers in 2002." *Monthly Labor Review* 126: 37–40.

Hobson, Barbara. 2002. *Making Men Into Fathers: Men, Masculinities and the Social Politics of Fatherhood*. Cambridge, UK: Cambridge University Press.

Iwanow, H., M. G. McEachern, and A. Jeffrey, 2005. "The Influence of Ethical Trading Policies on Consumer Apparel Purchase Decisions." *International Journal of Retail and Distribution Management:* 371–387.

Jones, Ellis, Ross Haenfler, Brett Johnson, and Brian Klocke. 2002. *The Better World Handbook: From Good Intentions to Everyday Actions*. Gabriola Island, BC: New Society.

Kelly, Erin. 2006. "Work-Family Policies: The United States in International Perspective." Pp. 99–124 in *The Work and Family Handbook: Multi-Disciplinary Perspectives, Methods, and Approaches*, edited by Marcie Pitt-Catsouphes, Ellen Ernst Kossek, and Stephen Sweet. Boston: Erlbaum.

Kochan, Thomas. 2005. *Restoring the American Dream: A Working Families' Agenda for America*. Cambridge, MA: MIT Press.

Kossek, Ellen Ernst, and Alyssa Fried. 2006. "The Business Case: Managerial Perspectives on Work and Family." In *The Work and Family Handbook: Multi-Disciplinary Perspectives, Methods, and Approaches*, edited by Marcie Pitt-Catsouphes, Ellen Ernst Kossek, and Stephen Sweet. Boston: Erlbaum.

Krugman, Paul. 2005. "French Family Values." *New York Times* (July 29). p. 1.

Lafer, Gordon. 2003. "Graduate Student Unions: Organizing in a Changed Academic Economy." *Labor Studies Journal* 28: 25–43.

Lichtenstein, Nelson. 2002. *State of the Union: A Century of American Labor*. Princeton, NJ: Princeton University Press.

Luken, Ralph, and Rodney Stares. 2005. "Small Business Responsibility in Developing Countries: A Threat or an Opportunity?" *Business Strategy and the Environment* 14: 38–53.

Marx, Karl, and Friedrich Engels. 1972 [1848]. "Manifesto of the Communist Party." In *The Marx-Engels Reader*, edited by Robert C. Tucker. New York: Norton.

Milkman, Ruth. 2006. *L.A. Story: Immigrant Workers and the Future of the U.S. Labor Movement*. Berkeley: University of California Press.

Milkman, Ruth, and Kim Voss. 2004. *Rebuilding Labor: Organizers and Organizing in the New Labor Movement*. Ithaca, NY: ILR Press.

Mills, C. Wright. 1959. *The Sociological Imagination*. New York: Oxford University Press.

Moen, Phyllis, and Patricia V. Roehling. 2005. *The Career Mystique*. Boulder, CO: Rowman & Littlefield.

Moen, Phyllis, and Stephen Sweet. 2004. "From 'Work-Family' to 'Flexible Careers': A Life Course Reframing." *Community, Work and Family* 7: 209–226.

Nelson, Valerie, Adrienne Martin, and Joachim Ewert. 2005. "What Difference Can They Make? Assessing the Social Impact of Corporate Codes of Practice." *Development in Practice* 15: 539–545.

Nestle, Marion. 2003. *Food Politics*. Los Angeles: University of California Press.

Pfau-Effinger, Birgit. 2004. *Development of Culture, Welfare States and Women's Employment in Europe*. Burlington, VT: Ashgate.

Pavalko, Eliza, and Kathryn Henderson. 2006. "Combining Care Work and Paid Work: Do Workplace Policies Make a Difference." *Research on Aging* 28: 359–374.

Pelsmacker, Patrick De, Liesbeth Driesen, and Glenn Rayp. 2005. "Do Consumers Care About Ethics? Willingness to Pay for Fair Trade Coffee." *Journal of Consumer Affairs* 39: 363–386.

Polanyi, Karl. 1944. *The Great Transformation*. New York: Farrar and Rinehart.

Putnam, Robert. 2000. *Bowling Alone: The Collapse and Revival of American Community*. New York: Simon & Schuster.

Riesman, David, Nathan Glazer, and Reuel Denney. 2001 [1961]. *The Lonely Crowd: A Study of the Changing American Character.* New Haven, CT: Yale University Press.

Schrage, Elliot. 2004. "Supply and the Brand." *Harvard Business Review* 82: 20–21.

Seidman, Gay. 2007. *Beyond the Boycott: Labor Rights, Human Rights, and Transnational Activism.* New York: Russell Sage Foundation.

Shaiken, Harley. 1984. *Work Transformed: Automation and Labor in the Computer Age.* New York: Lexington Books.

Shulman, Beth. 2005. *The Betrayal of Work: How Low-Wage Jobs Fail 30 Million Americans and Their Families.* New York: New Press.

Smith, Craig. 2006. "Letter From Paris: 4 Simple Rules for Firing An Employee in France." *International Herald Tribune* (March 29), p. 1.

Tocqueville, Alexis de. 1969 [1836]. *Democracy in America.* New York: Doubleday.

United Nations Statistics Division. 2005. "Demographic and Social Indicators: Statistics and Indicators on Women and Men." Retrieved from http://unstats.un.org/unsd/demographic/products/indwm/wwpub.htm

Wells, Miriam. 2000. "Immigration and Unionization in the San Francisco Hotel Industry." *In Organizing Immigrants: The Challenge for Unions in Contemporary California*, edited by Ruth Milkman. Ithaca, NY: Cornell University Press.

Williams, Joan, 2000. *Unbending Gender: Why Family and Work Conflict and What To Do About It.* New York: Oxford University Press.

Winstanley, D., J. Clark, and H. Lesson. 2002. "Approaches to Child Labour in the Supply Chain." *Business Ethics: A European Review* 11: 210–223.

DISCUSSION QUESTIONS

1. Life-long jobs are a thing of the past in the new economy, and American workers are competing with workers all across the globe for "good jobs." This is the globalization of labor. But is this a *social* problem? Why or why not?
2. Who benefits from the globalization of work? Who suffers? What is the government's responsibility to those who cannot make a living in the new economy? What is the individual's responsibility?
3. What can activist groups and social movements do about the problems stemming from the new economic realities of globalized work?
4. Only about 10% of American workers are union members. Can unions really have a serious impact on the many problems of the new economy? Explain your answer.
5. What responsibility does the corporation have to its employees? To society?
6. Do you agree with the authors' call for a "Fair Labor Standards Act for the New Economy," or do you think that corporations are able to act in the best interest of their employees without government legislation? Explain your response.

11

The Vanishing Middle Class

GRIFF WITTE

The Four Questions

1. Is there a problem here? What exactly is it?
2. Is it a *social* problem?
3. What are some of the factors causing this problem?
4. Does Witte give some way to alleviate the problem?

Topics Covered

Middle class

Work

Job loss

Unemployment

Economic inequality

Scott Clark knows how to plate a circuit board for a submarine. He knows which chemicals, when mixed, will keep a cell phone ringing and which will explode. He knows how to make his little piece of a factory churn hour after hour, day after day.

But right now, as his van hurtles toward the misty silhouette of the Blue Ridge Mountains, the woods rising darkly on either side and Richmond, Va., receding behind him, all he needs to know is how to stay awake and avoid the deer.

So he guides his van along the center of the highway, one set of wheels in the right lane and the other in the left. "Gives me a chance if a deer runs in from either direction," he explains. "And at night, this is my road."

It's his road because, at 3:43 a.m. on a Wednesday, no one else wants it. Clark is nearly two hours into a workday that won't end for another 13, delivering interoffice mail around the state for four companies—none of which offers him health care, vacation, a pension or even a promise that today's job will be there tomorrow. His meticulously laid plans to retire by his mid-fifties are dead. At 51, he's left with only a vague hope of getting off the road sometime in the next 20 years.

Until three years ago, Clark lived a fairly typical American life—high school, marriage, house in the suburbs, three kids and steady work at the local circuit-board factory for a quarter-century. Then in 2001 the plant closed, taking his $17-an-hour job with it, and Clark found himself among a segment of workers who have learned the middle of the road is more dangerous than it used to be. If they want to keep their piece of the American dream, they're going to have to improvise.

Figuring out what the future holds for workers in his predicament—and those who are about to be—is key to understanding a historic shift in the U.S. workforce, a shift that has been changing the rules for a crucial part of the middle class.

This transformation is no longer just about factory workers, whose ranks have declined by 5 million in the past 25 years as manufacturing moved to countries with cheaper labor. All kinds of jobs that pay in the middle range—Clark's $17 an hour, or about $35,000 a year, was smack in the center—are vanishing, including computer-code

SOURCE: From Griff Witte, "As Income Gap Widens, Uncertainty Spreads," *The Washington Post*, September 20, 2004.

crunchers, produce managers, call-center operators, travel agents and office clerks.

The jobs have had one thing in common: For people with a high school diploma and perhaps a bit of college, they can be a ticket to a modest home, health insurance, decent retirement and maybe some savings for the kids' tuition. Such jobs were a big reason America's middle class flourished in the second half of the 20th century.

Now what those jobs share is vulnerability. The people who fill them have become replaceable by machines, workers overseas or temporary employees at home who lack benefits. And when they are replaced, many don't know where to turn.

"We don't know what the next big thing will be. When the manufacturing jobs were going away, we could tell people to look for tech jobs. But now the tech jobs are moving away, too," says Lori G. Kletzer, an economics professor at the University of California at Santa Cruz. "What's the comparative advantage that America retains? We don't have the answer to that. It gives us a very insecure feeling."

The government doesn't specifically track how many jobs like Clark's have gone away. But other statistics more than hint at the scope of the change. For example, there are now about as many temporary, on-call or contract workers in the United States as there are members of labor unions. Another sign: Of the 2.7 million jobs lost during and after the recession in 2001, the vast majority have been restructured out of existence, according to a study by the Federal Reserve Bank of New York.

Each layoff or shutdown has its own immediate cause, but nearly all ultimately can be traced to two powerful forces that reinforce each other: global competition and rapid advances in technology.

Economists and politicians—including the presidential candidates—are locked in a vigorous debate about the job losses. Is this just another rocky stretch of the U.S. economy that, if left alone, will foster new industries generating millions of as-yet-unimagined jobs, as it has during other times of upheaval? Or is the workforce hollowing out permanently, with those in the middle forced to slide down to low-paying jobs without benefits if they can't get the education, credentials and experience to climb up to the high-paying professions? ...

Some of the consequences are already evident: The ranks of the uninsured, the bankrupt and the long-term unemployed have all crept up the income scale, proving those problems aren't limited to the poor. Meanwhile, income inequality has grown. In 2001, the top 20 percent of households for the first time raked in more than half of all income, while the share earned by those in the middle was the lowest in nearly 50 years.

Within the middle class, there has been a widening divide between those in its upper reaches whose jobs provide the trappings of the good life, and those in the lower rungs whose economic fortunes are less secure.

The growing income gap corresponds to a long-term restructuring of the workforce that has carved out jobs from the center. In 1969, two categories of jobs—blue-collar and administrative support—together accounted for 56 percent of U.S. workers, according to an analysis by economists Frank Levy of MIT and Richard J. Murnane of Harvard. Thirty years later the share was just 39 percent.

Jobs at the low and high ends have replaced those in the middle—the ranks of janitors and fast-food workers have expanded, but so have those of lawyers and doctors. The problem is, jobs at the low end don't support a middle-class life. And many at the high end require special skills and advanced degrees. "However you define the middle class, it's a lot harder now for high school graduates to be in it," Levy says. College graduates aren't immune, either. In places like Richmond, the overall health of the economy masks layoffs that have snared not only blue-collar workers like Clark, but also thousands of office workers at companies like credit card giant Capital One Financial Corp. and high-tech retailer Circuit City Stores Inc. Those cutbacks have educated even those with bachelor's degrees in the new ways of a volatile economy.

A University of California at Berkeley study last year found that as many as 14 million jobs are vulnerable to being sent overseas. Many economists,

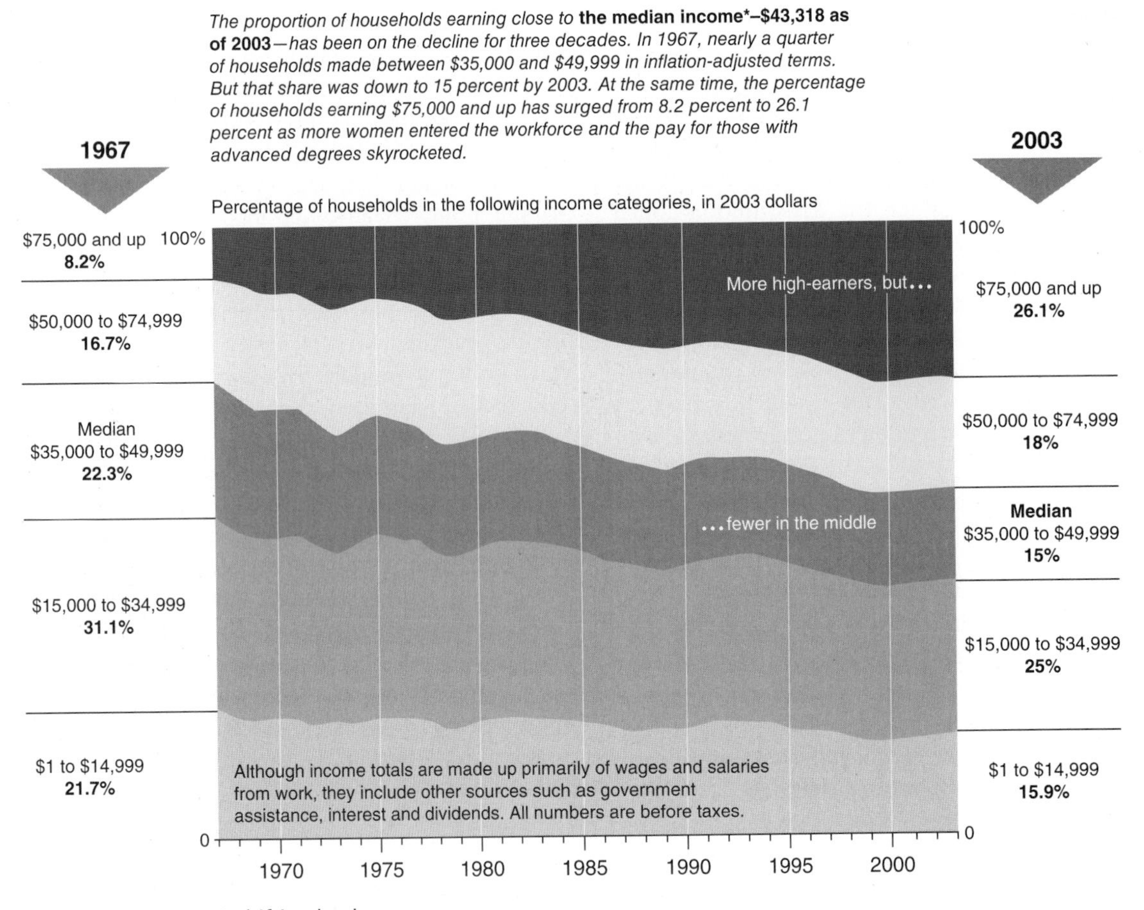
The proportion of households earning close to the median income*–$43,318 as of 2003—has been on the decline for three decades. In 1967, nearly a quarter of households made between $35,000 and $49,999 in inflation-adjusted terms. But that share was down to 15 percent by 2003. At the same time, the percentage of households earning $75,000 and up has surged from 8.2 percent to 26.1 percent as more women entered the workforce and the pay for those with advanced degrees skyrocketed.
1967
2003
Percentage of households in the following income categories, in 2003 dollars
$75,000 and up 8.2%
100%
$50,000 to $74,999 16.7%
Median $35,000 to $49,999 22.3%
$15,000 to $34,999 31.1%
$1 to $14,999 21.7%
More high-earners, but...
...fewer in the middle
Although income totals are made up primarily of wages and salaries from work, they include other sources such as government assistance, interest and dividends. All numbers are before taxes.
0
1970
1975
1980
1985
1990
1995
2000
$75,000 and up 26.1%
$50,000 to $74,999 18%
Median $35,000 to $49,999 15%
$15,000 to $34,999 25%
$1 to $14,999 15.9%

FIGURE 11.1 A shifting landscape.

though, say offshoring is more opportunity than threat because it allows companies to make and sell goods for less, and offer even better jobs than those that are lost. "Offshoring can't explain job loss. It can only explain job switch," says David R. Henderson, a Hoover Institution economist.

Henderson says the middle class is thriving, and by many measures, he's right. As a group they're earning more money than they have before, and their ranks have swollen with members who can afford the DVDs, SUVs and MP3s now seen by many families as part of the essential backdrop to modern life. Whereas Census numbers show the median household earned $33,338 in 1967 when adjusted for inflation, that number was up by $10,000 in 2003.

But when compared with those at the top, the middle has lost much ground. And many in the middle have dropped well behind their peers.

The gaps are likely to widen, according to Robert H. Frank, a Cornell economist. He says that as more people worldwide become available to do routine work for less money and as computers take on increasingly complex functions, the demand for those Americans whose skills are easily duplicated could drop. "The new equilibrium," Frank says, "may be a little meaner and more unpleasant than it was before." In the Washington area, the federal government and its contractors have cushioned the impact of the change in the workforce. But you don't have to travel far for evidence of the shift: Just two hours south on 1-95, to Richmond.

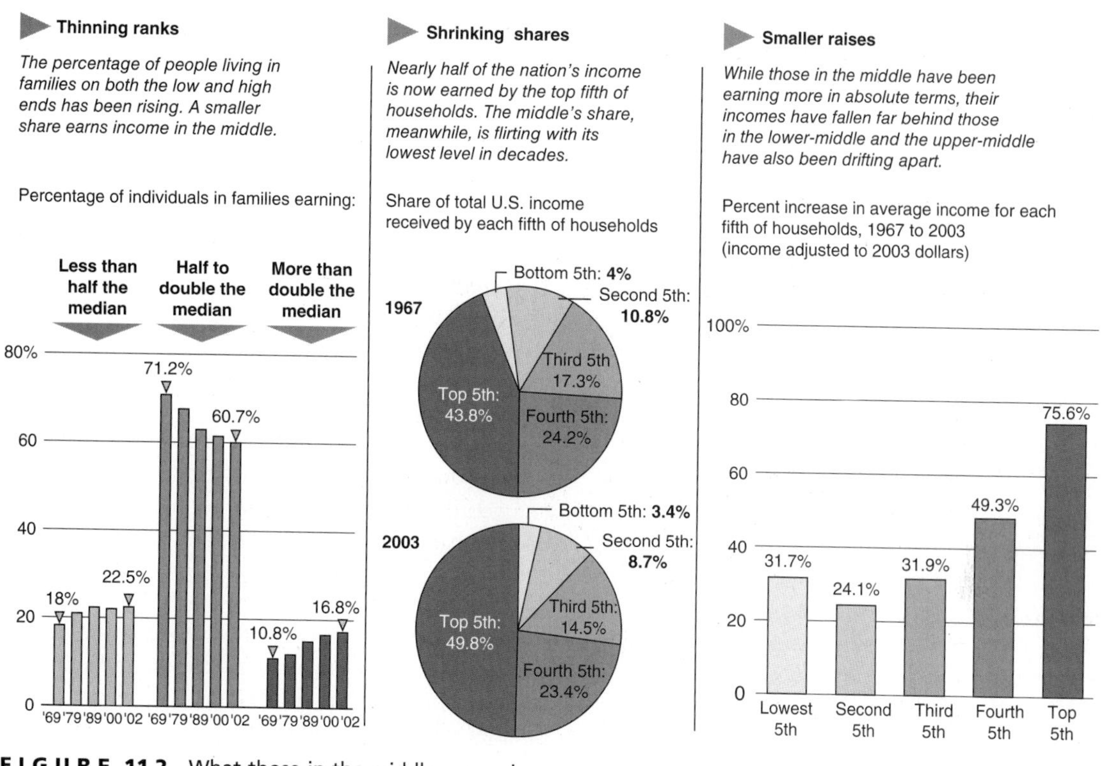

FIGURE 11.2 What those in the middle are seeing.

*Median is the level at which exactly half the households earn more and half earn less.

SOURCE: The source for all charts except for "Thinning Ranks" is the U.S. Census Bureau. The source for "Thinning Ranks" is the Economic Policy Institute based on government data.

From a distance, like many parts of the United States, Richmond looks like a place where the middle class should thrive. As its economy evolved over the past century from agriculture to manufacturing to services and, finally, to technology, it hung on to some aspects of each phase. That diversity keeps the jobless rate below the national average. Paychecks for professionals are growing. Major corporations such as Philip Morris USA are adding staff. A biotech park has taken root in downtown. Two new malls recently opened in the suburbs. And yet, for some who lack the right skills to match employers' demands, Richmond has less to offer than it used to.

"I think we're tending not to see any growth in the middle," says Michael Pratt, a Virginia Commonwealth University economics professor, "but I don't know anywhere in America where you are."

It wasn't always that way.

When Fred Agostino moved to suburban Richmond to head the Henrico County Economic Development Authority in the mid-1980s, employers wanted semi-skilled workers they could train for half a day and hire for life at a decent wage with benefits. Now companies looking to relocate to Richmond just want to know what percentage of the local population has a PhD. "They have to have educated, skilled, world-class people," Agostino says.

Meanwhile, the lifetime jobs were cut short.

The Viasystems Inc. circuit board factory was once known as "Richmond Works," and it provided good pay for people who didn't get past high school—like Scott Clark. He was also among the 2,350 people who lost their jobs in 2001, when the plant shut for good.

Today Clark is a driver-for-hire, willing to work virtually any schedule, and drive any route for less than anyone else. His old factory job was outsourced to workers in China, Canada or Mexico. But now he benefits from outsourcing, doing work that once might have been someone else's full-time job with benefits. A former proud union man, he has become part of the steady exodus from the labor movement, which now represents just under 13 percent of the workforce. Instead, he's part of another nearly 13 percent of the workforce that has grown, not shrunk—those who do jobs that are temporary, contract or on-call.

At least the work's not going anywhere. A real person in America, he reasons, has to drive American roads to get things from one place to another. There's security in that.

Clark used to feel the same security about work at the factory. When he started there in the mid-1970s, it was a new Western Electric plant, part of the Ma Bell family. When managers called him for an interview and he got the job, he could hardly believe it: "I said, 'It's funny you called me. My girlfriend's got college, and you ain't called her.' They said, 'What kind of college?' I said, 'She's taking biology and chemistry and all that stuff.' Before I got home, they called her and I had to turn around and bring her back up."

His girlfriend, Kathy, dropped out of school immediately. They started work the same day in 1976, making less than $10 an hour between them. Marriage followed.

Clark, a big, profane man, makes his way through Virginia yelling at other drivers, yelling at talk radio, and, occasionally, singing along to a sweet, sad blue-grass tune.

He doesn't have much patience for politicians. When Sen. John E. Kerry of Massachusetts, the Democratic presidential nominee, comes on the radio to talk about the economy, proclaiming, "I believe in building up our great middle class," Clark sneers, "Yeah, right." When President Bush's voice echoes through the cab a little later, Clark dubs him "a liar."

Clark has few nice things to say about corporations, either, but he concedes that the factory—for most of his years there—was run pretty well. He enjoyed the work, putting copper plating on circuit boards that would power phones, computers and even a few submarines for the Navy. Working in the chemical division was a dirty job. But because it was dirty, managers stayed away. Amidst the fumes, working long into the night on the second shift, the workers forged deep friendships. Clark and three buddies played the lottery religiously, with a vow that if one hit the jackpot, they would split the winnings and all retire on the spot.

"It was a real close-knit group of people," says Kathy Clark, who also worked the second shift for years. "We grew up there. We had our families there." But in 1996, the plant was sold by Lucent

Many of the occupations the government predicts will lose the most jobs by 2012 pay a middle-class wage and require no postsecondary education. The jobs with the largest projected growth are divided between high-paying jobs that require college and low-paying jobs that do not.

Key
Bar color indicates quartile rank by 2002 median annual earnings
Top 25% (Highest earners)
Third 25% (includes median wage)
Second 25%
First 25% (Lowest earners)
Postsecondary education or training:
Associate's degree
Bachelor's degree
Doctoral degree
Short on-the-job training
Medium on-the-job training
Long on-the-job training

Biggest Projected Gainers
Selected occupations with largest projected, 2002–2012 (from the top 30)

Occupation	Number of jobs to be added	Postsecondary education or training
Registered nurses	623,000	Associate's degree
Postsecondary teachers	603,000	Doctoral degree
Retail salespeople	596,000	Short on-the-job training
Customer service reps	460,000	Medium on-the-job training
Cashiers, except gaming	454,000	Short on-the-job training
Janitors and cleaners*	414,000	Short on-the-job training
Waiters and waitresses	367,000	Short on-the-job training
Security guards	317,000	Short on-the-job training
Home health aides	279,000	Short on-the-job training
Computer systems analysis	184,000	Bachelor's degree

FIGURE 11.3 The future job market.

Technologies Inc., which had inherited it from AT&T Corp. Although the union made a bid, the victor was a start-up called Viasystems.

Many of the workers, Scott Clark included, had a feeling Viasystems was not invested in the plant for the long term. The reality was hard to ignore: By 2001, few companies still made circuit boards in the United States. They could earn a bigger profit producing them where business costs were lower, and where the workers would not demand overtime or sick leave. Scott Clark was not surprised on the day Viasystems announced the factory would shut down.

"They point-blank told us.... 'You could work for nothing and we would still close this plant,'" Kathy Clark says.

On the plant's final day, the workers were told to throw their ID passes and beepers into a box in the auditorium. Scott Clark wouldn't do it. Instead he broke into a meeting of managers, and placed his pass on the table. "When I walked into this plant, they handed me that pass," he told them. "They were proud to give it to me, and I was proud to take it." Now he was giving it back. He turned, and left the plant for the last time.

A handful of employees stayed behind to remove the machines so they could either be shipped overseas or sold for scrap. In the end, Richmond Works was just a shell. The building still sits vacant off the side of Interstate 64 just outside Richmond, a 700,000-square-foot tan tombstone in a weedy field.

Kathy Clark was unemployed for a year after the plant closed. Scott Clark lost time to training as he began his second career on the road. With their savings all but evaporated, the Clarks have spent the past two years starting over.

Working 15-hour days, Scott Clark has been pulling in good money. He won't say exactly how much for fear that competitors will undercut him, but in the Richmond area, he says, a courier can make $800 a week for doing routes less time-consuming than his. That's more than his base pay at the factory, though his new job lacks any benefits and he has to pay for the van and the gas.

Kathy Clark, meanwhile, got a full-time job this summer after two years of temp work. But they still have a lot of ground to make up. Had the plant stayed open, they would have been ready for retirement in just a few more years.

Now, "I feel like I'm 18 years old again," said Kathy Clark, as she sat in a rocking chair in her living room, strands of light gray overtaking the dark brown of her short hair. The Clarks know they have it better than many of their friends from the plant. They have frequent, impromptu reunions at Wal-Mart, where the talk inevitably turns to who has found work and who hasn't.

Raffael Toskes Sr. has, but only for $11 an hour. He rides around each day in an armored car, a gun strapped to his side. "I consider myself a middle-class person," says Toskes, who made $17 an hour at the plant. "But right now, I'm probably a lower-middle-class person."

Selected occupations with the largest projected job decline, 2002–2012 (from the top 30)

	Number of jobs to be lost	Postsecondary education or training
Computer operators	–30,000	■ ■
Telephone operators	–28,000	■
Postal service mail sorters	–26,000	■
Loan interviewers and clerks	–24,000	■
Brokerage clerks	–11,000	■ ■
Eligibility interviewers, government programs	–11,000	■ ■
Prepress technicians	–10,000	■ ■ ■
Meter readers, utilities	–8,000	■
Chemical plant and system operators	–7,000	■ ■ ■
Mixing and blending machine setters, operators and tenders	–7,000	■ ■

FIGURE 11.4 Biggest projected losers.

Lawrence Provo has given up on trying to find a job. He was out of work for nearly two years after the plant closed. "That was probably the worst time in the world to become unemployed. Everybody was downsizing. Everybody was laying off," he says.

Provo and his wife cut back on expenses and sold their car, furniture and jewelry. They even sold their home, and moved in with Provo's mother-in-law. But it was not enough. They had come to rely on his factory wage, and now their debts spiraled into the tens of thousands. They declared bankruptcy, joining a record 1.6 million who filed last year.

Provo finally got a job through a temp agency for $8.50 an hour, less than $18,000 a year and a little more than a third of his pay at Viasystems. He was just getting his life back together when, in November last year, his heart failed him. "My doctor told me, 'You've got a choice: You can work or you can live,'" he says.

Robert Boyer retrained in computers after the plant closed. But tech companies told him they wanted five years' experience, not a certificate from a six-month course. So he works for $11.50 an hour at Home Depot, using the wisdom of four decades as plant electrician to help customers pick light bulbs for their remodeled kitchens.

Boyer turns angry at any suggestion that the jobs picture is not that bad. "When these guys get on the boob tube and say there's jobs out there, you just gotta go out there and get them, it makes me want to go out there and grab them by the throat and say, 'Where? Where are the jobs at?'"

Ask Richmond's leaders, and they'll say the jobs are in infotech, biotech, nanotech and other kinds of tech yet to be conceived. "People have the impression that Richmond is a good-old-boy town. And we do have some old money here. But that money is going to build the new economy," says Robert J. Stolle, executive director of the Greater Richmond Technology Council. "Tech is the backbone of the Richmond economy."

One home-grown company seems to capture in its name Richmond's most deeply held ambitions: Circuit City. Born in 1949 to sell television sets to the masses, its existence attests to the enduring strength of the middle class. And all those sales of computers and video games have created a lot of jobs. With a local staff of 3,072, the chain is one of the Richmond area's largest employers.

But the work has a tendency to disappear. In the eight years after he moved to Richmond to take an offer at Circuit City, Chuck Moore lost his job in that company three times, proving that a white collar and a college degree are no protection from the forces that have shifted the ground under blue-collar workers like Clark. At 35, Moore spent the first nine months of 2004 desperate for a job as he watched his grip on the middle class slipping away. His story complicates the idea that to be comfortable in America today, all you need is a little more education.

Moore's roots are solidly blue-collar: His father worked as an electrician for the same company for 40 years. His stepfather drove a truck. His brother went to work at the Georgia Pacific plant. His mother still manages the local Shoney's. No one in his family had ever graduated from college.

For nine years after his high school graduation, he and his wife, Terry, worked full time to pay for Chuck to complete his degree at the Savannah College of Art and Design. With a knack for electronics and an artistic eye, he wanted to animate movies or video games. "I thought that walking out that door with that degree in my hand, I wouldn't have to look. I would have people coming to me," Moore says.

But while Moore was in school—designing animation by day, manning a hotel desk by night—the technology had continued to improve and so had employers' capacity to hire artists anywhere on earth. A bachelor's degree might have been enough before; now you needed a master's or even a doctorate.

Moore started looking for computer jobs instead. He and Terry both had luck at Circuit City.

Moore's first job disappeared when the company closed a tech support center and began moving its call center operations to India. His second job—designing ads for the recruitment division—evaporated when the tech bubble burst. His last job there ended in January when the database he built to manage marketing projects worked so well that the company no longer needed the help of a human.

Until a week ago, his job search had gone like this: 320 resumes sent out, six calls back. Three interviews. No offers. At first, he had put his old salary on his résumé: $40,000. Later he switched to, "negotiable."

"I've already been willing to go down 10 [thousand dollars]. And if it goes much longer, I might have to go down 15. For a guy with a bachelor's degree to take $25,000, I might as well be working at McDonald's," Moore said in August. "There's something not right about that."

Yet on Sept. 18, when an animal hospital offered him work as a veterinary assistant—for half what he had been making in his old job and no benefits—he accepted immediately. He started last Monday, cleaning out kennels and, he hopes, learning how to use the X-ray machines or work in the lab so he can add to his repertoire of skills.

Moore has thought of going back for his master's degree. But that's hardly an option when he has a 3-year-old son, not to mention a mortgage and student loans.

Instead, to help make ends meet, he's been teaching computer basics at J. Sargeant Reynolds Community College, where his students can identify with their teacher's plight. One is a 20-year Army veteran who found that the best he could do without college was become a salesman at Lowe's, the home-improvement store. He was taking Moore's class so he could go to a four-year college in the fall.

"The job market for people like me is not that good," says the man, Albert DiCicco. "Maybe it is for people with bachelor's degrees."

Lately, DiCicco's predicament has been on the mind of Federal Reserve Chairman Alan Greenspan. In June, Greenspan warned that a shortage of highly skilled workers and a surplus of those with fewer skills has meant wages for the lower half of the income scale have remained stagnant, while the top quarter of earners sprints away. Greenspan said the skills mismatch "can and must be addressed, because I think that it's creating an increasing concentration of incomes in this country and, for a democratic society, that is not a very desirable thing to allow to happen."

But it already has happened. The gap between the wages of a 30-year-old male high school graduate and a 30-year-old male college graduate was 17 percent as of 1979, according to analysis by Harvard's Murnane and MIT's Levy in their book, "The New Division of Labor." Now it tops 50 percent, with an even larger differential for women. Real wages for both high school graduates and high school dropouts have actually fallen since the 1970s. Meanwhile, wages for college graduates—who make up only about a quarter of the adult population—have soared upward.

The trend seems poised to continue. The list of the 30 jobs the Labor Department predicts will grow the most through 2012 includes high-paying positions such as postsecondary teachers, software engineers and management analysts. But nearly all require a college degree. There are also plenty of jobs that demand no college—including retail sales and security guard—but they pay a low wage.

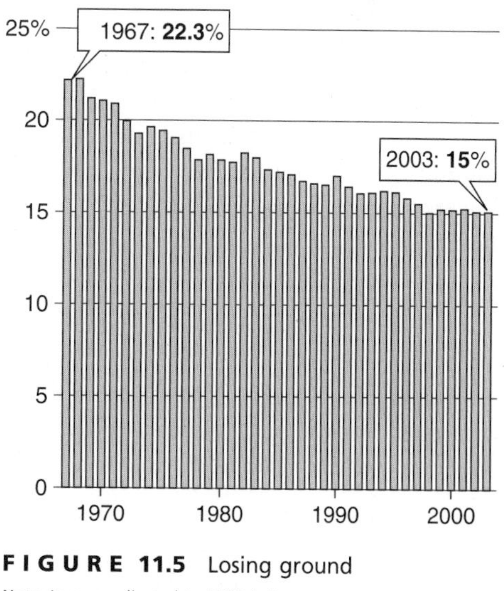

FIGURE 11.5 Losing ground

Note: Income adjusted to 2003 dollars.

SOURCE: U.S. Census Bureau.

And yet, as Moore's situation shows, a college diploma offers a porous shield when demand for a certain skill evaporates. College graduates have, in recent years, become an increasingly large percentage of the long-term unemployed. When they find new work, their salary cuts have been especially deep.

The optimists among economists—and there are many—point to trends that could help mitigate the pain of job losses and lead to future growth. One is the coming mass retirement of baby boomers, which could leave plenty of openings for those trying to break into the workforce. Economists tend to believe, too, that trade and technology will ultimately create new efficiencies that produce far more jobs than they destroy and leave everyone, on average, better off.

Scott Clark isn't sure if he will emerge better off. Spending day and night in the cab of a van was not exactly how he planned to live out his fifties and sixties, but he'll get by. He's even managed to save enough money to begin cutting his hours from 15 down to 11.

It's the end of the day now and as Clark battles the Richmond evening rush hour, his thoughts are turning to home. He's already fulfilled his part of the American dream, doing better than his parents did. "Everybody tells me I'm low class," Clark says, chuckling faintly. "But we're middle class. We're definitely middle class."

Yet his kids—his son is 26 and his twin daughters are 21—still live at home because they can't afford places of their own. None of them went to college, although his daughters had 3.8 grade-point averages in high school and his son aced the SATs. They're saving to go back to school—eventually. In the meantime, they work. His son lays carpet and his daughters stock shelves in a warehouse.

Will they be able to move up the economic ladder, just like he did? Clark ponders the question. After a long day, he is showing the strain, getting sleepy with his regular bedtime of 6:30 p.m. fast approaching.

"I really don't know. It's just too uncertain. It really is. There's nothing there," he says, turning completely serious for the first time all day. "There's nothing you can just count on. I wish there was."

DISCUSSION QUESTIONS

1. What basic trends does Witte describe that characterize the U.S. middle class?
2. If Witte is right, how might these trends affect your future? Can you improve your chances of success through knowing these trends?
3. If we treat the decline of the middle class as a major cause of social problems in the future, what are some of the major probable problems for our future society?
4. If you were a policy maker, what types of social policies might you suggest to alleviate the plight of the shrinking middle class?
5. What advice would you give other students in the university concerning plans for their future? Should they worry? Should they focus on developing certain skills? Should they ignore the trends Witte describes and hope for the best?
6. Who in the United States benefits if middle-class jobs become less important to our economy?
7. If Witte is wrong, if the middle class is not vanishing, what exactly do you think is happening in the American class structure?

12

The Demeaning of Work

BARRY SCHWARTZ

The Four Questions

1. What is the problem Schwartz describes?
2. Is this a *social* problem? Who is hurt? Is society hurt? Is the cause social? Are many people concerned about it?
3. What causes this problem?
4. How does Schwartz think we might deal with it?

Topics Covered

Work
Job
Career
Calling
Teachers
Market system
Division of labor
Alienation

What would we do if we didn't have to work, if we inherited substantial wealth—if, as my mother likes to say, we were fortunate enough to have our parents born before us? Would we spend all of our time consuming? Would we live lives of leisure—playing tennis today, golf tomorrow; visiting Aspen this week, Palm Springs next? I don't think so, at least not most of us. While it is true that work is a means to many of the things we value, it is also true that work is valuable in itself. While it is true that we work to live, it is also true that we live to work. Repeatedly over the years, when people in a wide range of occupations have been asked in surveys about what they want and expect out of work, material rewards have emerged as only one of several significant aspirations—and typically not the most important one. People want work that is interesting, challenging, and socially valued. They want to work with people they like. And they want work in organizations that treat them with respect. People continue working even when they no longer have to—indeed, even when they could make out better financially by retiring. Small-business owners, for example, characteristically work longer hours than they would if they worked for a similar income for a corporation, and they keep the business going even though they could do better by selling it and investing the proceeds.[1]

Despite the evidence that people want more from work than their paychecks, according to what might be called the "official ideology" of our culture, people wouldn't work if they didn't have to. Work is regarded as inherently unpleasant—as a cost, a burden. And an elaborate set of incentives, coupled with vigilant supervision, operates in most work settings to get people to do what they're being paid to do. Advertisements repeatedly tell us that we live for weekends and that, if we could get away with it, we would make every day a weekend. Just as the official ideology tells us we must "economize on love" because we can't count on people to treat one another well out of a sense of care, concern, and respect, it also tells us we must "economize on

SOURCE: From Barry Schwartz, *The Costs of Living: How Market Freedom Erodes the Best Things in Life* (Philadelphia, PA: Xlibris, 2000); originally published in 1994 by W.W. Norton and Co. Reprinted by permission of the author.

work" because we can't count on people to do their jobs well out of a sense of pride, responsibility, and commitment. Just as we can get people to treat one another well only if, in some sense, we make it pay, we can get them to do their jobs well only if we make that pay.

Most of us know that this "official ideology" is false. In our own work, we aspire to more than just a healthy paycheck. We actually want to work, and we want work that is meaningful, challenging, and fulfilling. Obvious though this may be, the official ideology denies it. And in denying it, it encourages the creation of working environments in which meaning, challenge, and fulfillment are increasingly difficult to find.

... It is useful to distinguish three different attitudes that people might have toward their work. Some may view their work as a *job,* some as a *career,* and some as a *calling.* People who have jobs don't expect them to have much meaning. They do the jobs for the material benefits they bring, and nothing more. Work is simply a means to consumption. I imagine that, for practically everyone who does it, slinging or serving hamburgers at a fast-food restaurant is just a job. Without close supervision and/or a system of incentives that rewards diligence, people with jobs can be expected to do as little as they can get away with. The jobs aren't challenging, and they are of little social value. So who cares?

For people with careers the situation is different. While people with jobs are just working for a wage, people with careers are going somewhere. There is a clear trajectory of advancement that defines success, and people derive satisfaction from advancing over and above the larger paychecks that it brings. So people with careers can be energized not just by the prospect of high wages but by the prospect of moving up the ladder. They will work hard and take initiatives, even without constant supervision, in the hope that such efforts will bring advancement. The hamburger slinger who hopes one day to manage a store of her own may see her work as a career.

The trouble with careers is that not everyone can keep advancing. The fast-food restaurant has many burger makers and servers, but only one manager. Owing to the pyramidal structure typical of workplace organizations, many people with careers are continually driven to do more and better work than their peers. The pressure to keep up is relentless, and time devoted to career advancement eats into time that might otherwise be available for other kinds of activity. One result of this pressure to work... is that it competes with the time required to be a good spouse, lover, friend, or parent. A second result is that because people are forced to work longer and harder than they want, they get less satisfaction out of their work than they otherwise might. And despite these great efforts, and neglect of friends and family, most people embarked on career paths will fail to reach the heights to which they aspire. By the time most people get to be about forty, if not before, the realization hits that they are stuck. They won't be getting to the top, as they might have imagined in their youthful, ambitious fantasies. They face twenty-five years or more of just holding on, with nowhere to go but down. As the realization sets in that they have gone about as far as they can go, diligence wanes, initiative-taking disappears, and close supervision is required to see to it that their work meets even minimal standards of acceptability. When a person's reason for doing her best is the prospect of advancement, as that prospect dims, enthusiasm dims with it.

For people with a calling, the situation is still different. For them, it is the concrete products of their work, and not just personal advancement, that provide meaning and satisfaction. People with a calling are doing something of value. It will not lose its value even if they are stuck doing it, with no prospect of moving up, for twenty-five years or more. The very unusual hamburger slinger who sees herself as providing nourishing food at moderate prices in clean, attractive surroundings to ease the pressure on modern, two-job families with modest incomes has neither a job nor a career but a calling. She can go to work each day knowing that just by meeting her responsibilities, she will be doing something of value for other people. People with a calling don't require careful supervision or an elaborate system of incentives. Their motivation to do their work well comes from inside. The work is just too important to be done carelessly or lazily....

Let's consider the work of teaching.... For some people, in some settings, teaching is only about salary and benefits: short school days, long vacations, summers off, excellent retirement and health plans. The trick, for people like this, is to meet the requirements of the job while doing as little as possible. It goes without saying that these teachers, for whom teaching is a job, will not be very good at it, unless their performance is carefully monitored and various incentives can be brought to bear on it.

For other people, teaching is a career. In addition to wages and benefits, there is tenure, promotion, and perhaps administrative authority to be sought. There may also be various forms of recognition for excellence. People for whom teaching is a career may be fine teachers, especially if the criteria that determine advancement are actually connected to classroom performance. They may often do far more than the minimum required for them to keep their jobs. They may stay long after school is out to give students individual assistance. They may accompany kids on trips and supervise extracurricular activities. But it is very important to them to know that someone in authority is watching, that they will get credit for the extra work. These teachers' eyes are primarily on the next step on the ladder, not on the children in their classes. If the requirements for advancement and the requirements for classroom effectiveness ever start to pull in different directions, we know which direction these teachers will take.

Finally, there are some people for whom teaching is a calling. Their eyes are firmly focused on the real aim of what they do. They are interested in doing whatever they can to help their students learn. They do extra work, but they don't regard it as "extra," since the point of their work is to help students learn. They concentrate on the task and let the recognition and promotion take care of themselves.

We may be tempted to say to ourselves that teaching *ought* to be a calling, and to hold the teachers for whom it is a merely a job or a career responsible for their misguided attitude toward their work. It is, after all, interesting work, with new challenges arising almost every day. And it is important, socially valued work, since teachers are entrusted with the future of our children and thus of our society. But neither interest nor importance can be taken for granted. Much of what teachers do can be extremely repetitive and uninteresting. They teach the same material, and correct the same mistakes, year after year. Furthermore, efforts by administrators to develop a uniform curriculum and set of teaching methods can undermine the desires of teachers to be creatively involved in both the setting of educational goals and the development of methods to achieve them. If the structure of the work is made routine and repetitive, the people who value creative and challenging engagement will be driven away. People who want to use their intelligence in the pursuit of excellence will seek other kinds of work. We will be left at best with careerists and at worst with people who are just putting in time. And as to the importance of teaching, what is our typical response when we learn that someone we've just met is a teacher? Do we respond with respect and admiration? Or do we ask ourselves why he's *only* a teacher?

So there is nothing that is obvious or automatic to influence teachers to regard their work as a calling. And the same is true of other occupations. Treating the same illnesses year after year, or writing the same wills and contracts, or farming the same crops, or selling the same stereos—whether these activities are jobs, careers, or callings depends upon whether their products are generally regarded as valuable. Doctors, lawyers, salespeople, and farmers can keep boredom from the door, and be content doing the same thing over and over again, only if they can be convinced that what they do is important....

Whether work is a job, a career, or a calling is determined by the attitudes that the people doing the work bring to it. Give two people the same work, side by side, and one of them will be enthusiastic, dedicated, and responsible while the other will just be working for a wage. For example, there was a man who worked for several years in my neighborhood as a church janitor and a school crossing guard. Neither of these tasks strikes me as being either especially challenging or highly valued socially. Nevertheless, this man put himself into his work with a level of passion, care, and devotion that would be rare and admirable in a brain surgeon. He watched his corner like a hawk, stopping rush hour traffic fearlessly to allow the young children to cross the street safely. He knew all the little kids whose journey to school he protected, and engaged them in conversation regularly. He even knew the cars and faces of many of the commuters who drove by his corner every day, and never failed to give them a warm wave hello. And he maintained the church grounds as though they were an arboretum. Why this enthusiasm and commitment? Perhaps he had the view that *all* work is meaningful, that everyone contributes to society by doing his job well, whatever it is. Perhaps it was just his nature to find something valuable in any situation. Perhaps he thought that all people are called by God to do something, and that God does not distinguish between "significant" tasks and "menial" ones. Whatever the reason, my temptation, in watching him, was to want to bottle what's inside him and spread it around....

Without denying what is undeniable—that different people approach the same work differently—focusing on changing the attitudes of individuals might be the wrong way to proceed. It is easier to find meaning and challenge in some kinds of work than in others. And the work people do ought to have meaning and challenge. So, rather than finding ways to adapt people to work that has little challenge or social value, we could be looking for ways to adapt work to people who want and need challenge and social value. The people will be better for it, and so will the work.

WORK AND "MARKET REALISM"

I have made three claims ... that create a puzzle when taken together. First, work can be viewed as a job, a career, or a calling. Second, any given type of work can, in theory, belong in any one of these categories. Third, when work is regarded as a calling, the work is done better and the people doing it are more satisfied. What these claims suggest is that society would be well served if work were structured so that as many people as possible saw what they did as a calling....

Why have so many of us allowed ourselves to be put in a position where we spend half our waking lives doing what we don't want to do in a place we don't want to be? Is there something about work as just a means to an end that serves the market system very well and thus has encouraged those in positions of power and authority to organize work in this way? To address this question, let's examine how it helps the market system to have a collection of workers who work only for the wages they receive.

Market activity is about buying and selling things. For the market to function effectively, it must be possible for the exchange of things to occur freely and easily. What makes this possible is a medium of exchange that all participants can use: money. But the market is also about buying and selling labor. The market requires that labor be mobile, that people be willing to move around from job to job as fluctuations in consumer demand increase the call for some kinds of work and decrease the call for other kinds. Unless people are willing to stop working in a hula hoop factory and go to work in a computer game factory if changes in consumer demand require it, manufacturers will not be able to switch their resources from hula hoops to computer games, and supply will not keep up with demand. And just as a medium of exchange is essential for the marketing of things, it is essential for the marketing of labor. What makes labor mobile is the same medium of exchange that works for goods—namely, money. If you work simply for the wage, you're a mercenary. You can easily be induced to change jobs for a better wage. Thus the market can induce the car mechanic to switch from fixing cars to fixing bicycles if consumer demand requires it. But if you work at least in part because of the meaning and significance of the particular work that you do, inducing you to change jobs is much, much harder. Meaning and significance simply can't be dispensed like money.

If, in general, people do work simply in exchange for money, then labor will be mobile and the market will run smoothly. The value of people's time and work will be equal to the price they fetch on the market and nothing more. People won't care about the character of the work they do, since work will simply be the means to consumption. People will sell their labor to the highest bidder, to maximize the rate of return on labor time and thus maximize opportunities to consume. And employers will dictate what the workers do so as to maximize their rate of return, or profit. So both workers and employers will strive to get the greatest return for the least effort.

If, in contrast, people do work that has meaning and significance, so that there is value in the activities themselves, how can we induce them to change their work? How can we induce them to do whatever we want them to do in exchange for a wage? People are not indifferent enough toward work that is challenging and meaningful to be willing to sell their time to the highest bidder. So, to establish a market system that functions effectively, work must be organized in such a way that people are willing to trade it on the market.

And there is another requirement that must be met for labor to be mobile. The training, knowledge, and skill demanded by the work must be minimal. To see why, think about a worker whose job requires a great deal of training and knowledge. Such a worker might not become productive until after a year or more of training. During that training period, her company is paying her a salary, but getting little or no production in return. The company finances the training as an investment in the worker's future productivity. But once the worker is trained, she can hold the company hostage, threatening to take her skills to a competitor unless her salary demands are met. What can the company do? It can't afford to let the worker leave and be faced with the expensive training of a replacement. Thus it must meet her demands. And even if she isn't making demands, when times are slack, the company will be reluctant to let her go, since all the time and money invested in her will then be wasted.

To respond to this problem, and keep labor mobile, companies can try to structure work so that it requires as little training, knowledge, and skill as possible. By doing this, the company will have only an insignificant investment in each worker. It can then afford to let workers come and go as they, and the market, please. But as a consequence of doing this, the company will deprive its workers of opportunities to use their intelligence in the workplace. All jobs will be as menial as possible, as interchangeable as possible. With a workplace organized like this, it is hardly surprising that most people will be driven to seek satisfaction in what their work buys them, rather than in the work itself....

It is interesting to realize, as we read modern cries of alarm about the "deskilling" of labor, that it didn't happen by accident. A century ago, the deskilling of labor was the explicit objective of management. And as more and more of the U.S. economy has shifted from production industries to service industries, the routinized, deadening methods of the factory have been imported into the service sector. People who work for airline companies, telephone companies, mail-order houses, and government agencies, and who spend their days talking to customers on the telephone, are being subjected to a degree of performance monitoring and regimentation that would have made Taylor [an important researcher who studied efficiency] proud. Thanks to the modern computer, it is now possible to monitor automatically how long such workers take on each call, as well as when, and for how long, they are away from their stations. In one company, the computers keep track of how much time workers take between calls (they are allowed seven seconds), how much time they take for lunch away from their stations (they are allowed thirty minutes), how much time they take on their breaks (two fifteen-minute breaks per day), and even how much time they spend in the bathroom. Each employee gets a grade (automatically computed) each week. Good grades bring workers small bonuses (an hour off with pay), while bad grades bring them penalties (an hour of extra work with no pay). Supervisors can listen in on calls, to make sure that the proper service is being provided and no time is being wasted. There may be as many as twenty million American workers whose work is being micromonitored in this way. They report a much higher incidence of work-related ailments (back pain, eyestrain, and the like), and a much higher incidence of depression, than workers doing similar jobs who are not being micromonitored and managed.[2]

A defender of modern workplace organization might concede that it didn't *have* to go the way it did, but argue that the astonishing productivity and efficiency it made possible made it far superior to previous forms of work organization. As a result, it represents a kind of cultural evolution. The cultures that develop practices and modes of organization that best meet the demands imposed on them by the environment are the ones that survive and flourish. Modern industrial society is no cultural accident. It is the most efficient and productive system ever known. The industrial revolution is still with us because nothing can match it.

Is it true that the modern workplace replaced its predecessors because it was so much more efficient and productive? Economist Stephen Marglin has attempted to show that the case *cannot* be made that workplace organization evolved on the basis of efficiency and productivity. He has argued that key developments in early industrialization had almost nothing to do with productivity and efficiency. Instead, what industrialization and division of labor did was give the boss much greater control over the production process than he had previously had. And it was this increase in control, along with the opportunity for profit it provided, that was the point of industrialization....

This point seems obvious. The more anyone knows about all aspects of an operation, the more likely she is to keep it afloat when things go wrong, or to design significant improvements in it. But few modern American factories are organized to take advantage of this obvious fact. Nor, for that matter, are most white-collar bureaucracies. They seem increasingly to operate on a "need to know" basis, jealously guarding information and meting it out as if it were gold, one speck at a time....

Even Adam Smith, [an economist who described the free market system,] at the same time that he was extolling the efficiency of the division of labor, recognized that it had a dark side, a deadening effect on the capacity of workers to exercise judgment and intelligence:

> The man whose whole life is spent in performing a few simple operations ... has no occasion to exert his understanding, or to exercise his invention.... He naturally loses, therefore, the habit of such exertion, and generally becomes as stupid and ignorant as it is possible for a human creature to become.[3]

There is one final respect in which the minute social division of labor is inefficient. When work is organized in this way, it is deprived of almost all challenge, meaning, and significance. As a result, it is nothing but a job. If the sole point of a job is to get as much as possible for doing as little as possible, then people will be tempted to cut corners—to lie, to cheat, to steal, to break the law—in the service of increased profitability or decreased effort. Why not? There is nothing—no pride in craft and no social responsibility—to hold them back. And the temptation to cheat requires a host of (unproductive and inefficient) inspectors and regulators to make sure that quality is being preserved and that customers are being treated honestly and fairly. Some economists who are critical of the current character of workplace organization have estimated that as many as 34 million people in the American workforce are engaged in monitoring, supervising, and policing production workers who would rather not be doing what they do for a living. This amounts to about one policer for every 2.3 productive workers.

Why would anyone willingly submit to a life of work that is menial and deadening? Marglin points out that most early factory workers really had no effective choice. They were either country people who had been driven off their land, or paupers, or disbanded soldiers, or women and children who had been offered to the factory by their husbands and fathers. Eventually, the problem of inducing workers to put up with the conditions of the factory disappeared. As Marglin notes, recruiting the first generation of factory workers was the key problem. For this generation's progeny the factory was part of the natural order, perhaps the only natural order. Once grown to maturity, fortified by the discipline of church and school, the next generation could be recruited to the factory with probably no greater difficulty than the sons of colliers are recruited to the mines or the sons of career soldiers to the army.[4]

So it is that what for one generation is intolerable becomes for the next "only natural." So it is that ways of being that are forced upon people by the social and material circumstances in which they happen to live come to be viewed as "human nature"—eternal, inevitable, and unchangeable. The historical deskilling of work is another manifestation of what I [have] referred to … as idea technology, or ideology. Once we realize that there is nothing inevitable about the character of work as we know it, we have to ask ourselves what work *should* be like, and how work *should* be organized.

What, then, is to be done? How can meaning and significance be restored to the work that people do? Turning back the clock is not an option. We can't go back to the days of independent householding family units. The large-scale technology of the modern workplace is here to stay. The genie is out of the bottle. If we are to reintroduce meaning and significance to work, we must do so within the structures of society that currently exist. But even within these constraints, there are significant possibilities for change.

MAKING WORK A CALLING

If the only point of your work is to make as much money as possible, whether for yourself as an employee or for the company as an employer, concerns about product or service quality, consumer satisfaction, consumer safety, and the like will fade into the background. These aspects of your work will be relevant only to the extent that they contribute to profit. They will simply be means to an end. If you can find other means that are more effective, you will leave concern for customers and quality behind. When work comes to be defined or framed strictly in terms of the making of money, rather than the making of things, then sources of satisfaction with a job well done, a product well made, a customer well served erode. Those people who continue to care about the quality of what they do find themselves subverted by management decisions that sacrifice quality for profit. Under conditions like these, the only satisfaction that remains available to the worker is the satisfaction that comes from earning and consuming a large salary.

So the temptation to view work as a job must be eased. The way to do this is to reintroduce challenge, meaning, and significance into work—to reemphasize the service to the community that work can provide. Saying this is one thing; doing it is quite another….

There are plenty of local businesses, of all types, that seek to serve rather than deceive customers. They do so partly because they know the people with whom they deal. Could you imagine picking up a piece of chicken that had fallen on the filthy floor, dusting it off, and serving it to your mother or your best friend? Could you imagine being careless about installing brakes in a new car if the car was going to be driven by your daughter? Well, it becomes similarly difficult to do such things to people you get to see every day and to know, people who come to depend on you for whatever good or service you provide…. [F]amily relations are being assimilated to the model of the market, with literal contracts replacing metaphorical ones…. [W]e might all be better off if the reverse occurred, if market relations were assimilated to the model of the family. The repeated interactions between workers of all types and their regular customers is an approximation to the family model of doing business.

Can we afford it? Can shopkeepers afford to take the time to see to the well-being of their customers?

Can they afford to pass up or discourage potential sales because they don't think the purchases will really serve their customers' interests? I think it depends on how they do the accounting. If customers realized how much they are cheated and exploited in the markets in which they are treated anonymously, they might be willing to pay the extra few cents that the local merchant requires for the personalized service he provides. If customers realized how much time they spend making sure they are not being taken advantage of, they might be willing to pay the extra few cents and place their trust in the local merchant. And if shopkeepers could count on customer loyalty, they might be able to stop wasting money on advertising and spend it instead to keep the place clean, keep the stock fresh, and serve their customers while keeping their prices down.

There is a model of the kind of retail setting in which the proprietors and the customers see themselves as on the same side. It is the consumer cooperative. In a consumer co-op, the customers *are* the proprietors. They own shares in the co-op, and any profits are either distributed to the shareholders or put back into the business. Employees in the coop truly are there to serve the customers. Since the customers are the owners, no one is served by selling stale food, or by price gouging. Accomplishment is measured by finding good-quality products at reasonable prices, and by keeping customers as informed as possible about what they are buying.

The co-op does not require people of exceptional honesty. It does not require extensive government protection of consumers. It does not demand of each customer that she be an expert on nutrition, or drugs, or automobile repairs before she enters. Instead, it creates a structure in which people of ordinary honesty and sophistication can behave honestly, selling what they claim to sell and getting what they pay for, because buyers and sellers are not engaged in a competitive, zero-sum game. Buyers and sellers can both come out winners.

A similar kind of structure can make for dramatic changes on the production side of business as well. In company after company, management control has shifted out of the hands of people who know about production and into the hands of people who know about finance. The result has been a shift in company mission—from making things to making money. When making money becomes a company's sole mission, concerns about the product fade. You make it well only if your advertising department can't find a way to sell it when it's made poorly.

When the company's sole mission is to make money, the work of the people who make the product is slowly but surely degraded. When workers see their efforts to make quality products subverted, diligence, initiative, creativity, and care on the shop floor are undermined. If the company is in it only for the money and the boss is in it only for the money, what can we expect of the people who actually make the product?

Putting control of companies back in the hands of people who know about production is a step toward reintroducing concerns about quality to the workplace. But it's too small a step. It still leaves the problem of a workforce whose intelligence and talent are underutilized, and whose ability is not respected. More is required. People on the shop floor need to become involved at all levels of production. They need to feel that the product they are making is *their* product. If they are given the opportunity to use their talent and intelligence on the job, they will start getting more out of their work than just a paycheck. When they can legitimately take responsibility for a product well made, a job well done, they will begin to care that the product is well made. And, as a result, well-made products will become more common than they now are—without extensive government regulation or plant supervision.[5]

Steps along these lines have been taken in various forms of producer cooperatives. What producer cooperatives have in common is that they give the workers a stake in the company. But there are many different ways to do this, and not all of them will be equally effective in changing the character and quality of the work that is done. Least effective is the approach that gives workers the opportunity to share in the profits, either through cash bonuses or through the award of company stock. Under arrangements like this, the more money a company makes, the more money the workers make. But nothing about what the workers do and how they do it is affected. Managers still make all the decisions, and the talents of-workers continue to be underutilized. Meaningless work may be more tolerable if, when the company prospers, you prosper, but if the company's prosperity is not tied to the quality of its products, an incentive system like this is not going to produce better-quality products. The interests of the workers, like those of any other shareholders, will be tied to profit, not to production.

Far more powerful is an arrangement in which workers take on a significant share of decision making, quality control, and supervision. They sit on the board of directors, participate in long-range planning, and contribute to decisions about how their work should be organized and executed, and what standards it should meet. In the limiting case, in which the company is wholly owned by the workers (a true producers' cooperative), they don't merely sit on the board; they make up the board. But even when the company continues to be owned by the shareholders, and continues to have a level of senior management that is removed from the shop floor, workers can be given a fair measure of control over what they do.

How do companies that are partly or entirely run by workers fare? How does breaking the barrier between labor and management affect product quality and company profitability? There is no universal agreement on the answers to these questions. Some worker-managed firms fail; others succeed. In some worker-managed firms, product quality is not enhanced; in others it is. In some worker-managed firms, worker satisfaction does not go up; in others it does. So the turning of control over to the workers is no guarantee of anything. It has to be done right, and what it means to do it right may vary from case to case. But when it is done right, worker engagement and satisfaction rise dramatically, and so does the quality of what is produced. And it is becoming increasingly clear that, worker satisfaction aside, some kind of worker participation in the running of companies is essential if American industry is to be successful in an increasingly competitive global economy. As one group of commentators on the current world economic scene put it, it is vital to the well-being of the American economy that employers "replace the iron fist with the sustained handshake, improving economic performance by giving workers a stake in their enterprises and economy rather than clubbing them into submission."[6]

It is unlikely that producer cooperatives can achieve the same results that consumer cooperatives do. In consumer cooperatives, the adversarial relation between buyer and seller vanishes; both are after the same thing. In producer cooperatives, this isn't true. Especially when workers share in company profits, their financial interests are served by selling as much of the product as possible for as big a profit as possible. However, what producer cooperatives can do is introduce interests to the workforce in addition to financial ones. Having responsibility for the product, and the opportunity to make it well, encourages workers to care not just about how much they make but about how they make it.

So restructuring the way in which commercial activity occurs—both the making and the selling of goods—can go a long way toward bringing challenge, meaning, and significance back to work. It can increase the satisfaction that people get out of making the things they make, which in turn will increase the satisfaction that people get out of buying the things they buy. It just isn't true that "business is business," nor is it true that work is inherently dissatisfying. Business can be the ruthless pursuit of self-interest, or it can be in the service of the public interest. Were it to become the latter, everyone would be better off. Work can be meaningless and alienating, or it can be meaningful and involving. And were it to become the latter, everyone would be better off.

But we have to be realistic. We should heed Marglin's warning about how quickly people can come to view meaningless, alienating work as the natural order of things. Undoing this perception will be a laborious and time-consuming process. It should probably begin not in the workplace but in the school, the institution in which we get our first instruction about the nature and meaning of work.

ENDNOTES

1. For a recent discussion of work motivation, see E. A. Locke and G. P. Latham, *A Theory of Goal Setting and Task Performance* (Englewood Cliffs, N.J.: Prentice-Hall, 1990). On the small-business owner's nonmaximizing decision to keep at his business, see J. Fallows, "What Can Save the Economy?" *New York Review of Books*, April 23, 1992, pp. 12–17.
2. F. W. Taylor, *Principles of Scientific Management* (1911) (New York: Norton, 1967). For more detailed discussion, see B. Schwartz, R. Schuldenfrie, and H. Lacy, "Operant Psychology as Factory Psychology," *Behaviorism* 6 (1978): 220–54.
3. A. Smith. 1776. *The Wealth of Nations*, pp. 734–35. On the proportion of the U.S. workforce devoted to policing the work of others, see Bowles, Gordon, and Weisskopf "Economic Strategy."
4. S. Marglin, "What Do Bosses Do?" in *Division of Labour*, edited by A. Gorz (London: Harvester, 1976), pp. 38–39.
5. For a review of some evidence on how worker managed firms perform, see D. R. Fusfeld, "Labor-Managed and Participatory Firms: A Review Article," *Journal of Economic Issues* 17 (1983): 769–89.
6. For an argument about the need for worker-manager cooperation to produce a viable American economic future, see I. Bluestone and B. Bluestone, *Negotiating the Future: A Labor Perspective on American Business* (New York: Basic, 1993). The quotation is from S. Bowles, D. M. Gordon, and T. E. Weisskopf, "Economic Strategy for Progressives," *The Nation*, February 10, 1992, pp. 145, 163–65.

DISCUSSION QUESTIONS

1. Do you think that people are really supposed to enjoy their work?
2. What is a "job"? What is a "career"? What is a "calling"? Is one better than the others?
3. Unions protect workers' rights and enhance workers' benefits. Does this in itself undermine work as a calling? Apply this to teaching.
4. Can working as a janitor be a calling? Can being a social worker be a job?
5. Describe the conditions necessary for enjoyable work.
6. If you wanted your employees to enjoy their work, what steps would you take? Would giving them good pay help or hinder?
7. What does it mean to "control one's own work"? Does anyone control his or her own work today?

PART V

Social Problems: Racial and Ethnic Inequality

The United States is a society of widely diverse racial and ethnic groups. Many people would argue this is not a problem at all, but a strength. However, the stratification of these groups creates many problems. It is one thing to be a society of a multitude of groups; it is another to be a society in which people are ranked, discriminated against, treated as outsiders, segregated, and/or oppressed. This tends to be the sociological view of diversity in America: It is diversity tied to inequality.

Racism is a problem if one is committed to democratic values. It is a social problem because it is a result of societal rather than individual patterns, because it affects large numbers of people, because it has negative consequences for society, and because many people agree that it is wrong and should be addressed.

The problems associated with racism that are described in the four selections are not the only ones that exist. Indeed, many sociologists and other academics would argue that no social problem is as important to our society as racism and that most of our other social problems have their origin in racism.

Lawrence Bobo and Ryan Smith describe modern racism, which includes a subtle racist philosophy which argues that even though we are finally operating as a democratic society, there is something wrong with racial minorities themselves that keeps them from being successful in American life. Attitude surveys show that most Americans are open to the idea of an equal society and claim they are against racism. Yet, the authors maintain, the surveys also reveal that Americans are not prepared to do anything about changing society so that minorities can succeed. When policy questions are asked, people are less supportive of creating an equal society. There is an inconsistency between what Americans say they want and what they are willing to do.

Roberto Suro also describes members of the Latino community as outsiders, strangers, and foreigners. Suro examines immigration, its expectations and its setbacks. The number of young Latinos is growing faster than any other segment of the population, and this will inevitably bring to our society great challenges and great change.

John Farley and Gregory Squires are most interested in the processes that keep the United States segregated. They examine segregated housing and the ways it impacts those who are discriminated against in education, health care, jobs, and the criminal justice system.

The final selection is by Amy Wax from her book *Race, Wrongs, and Remedies: Group Justice in the 21st Century*. Wax assesses the link between racial inequality and deficits in human capital, focusing specifically on the deficits in education, employment rates, and in family structure and home environments.

You may find some ideas in these selections controversial because racism is an emotional issue. However, analyzing them will help you to better understand this important problem.

13

Laissez-Faire Racism

LAWRENCE D. BOBO AND RYAN A. SMITH

The Four Questions

1. According to Bobo and Smith, what specifically is the problem?
2. In what sense do they consider this a *social* problem?
3. Do they emphasize individual cause or societal cause? What do they identify as the cause?
4. Bobo and Smith criticize society more than they offer clear ways to deal with the problem they identify. Do they imply any sensible ways to improve this problem?

Topics Covered

The American dilemma

Racism

African-Americans

Jim Crow racism

Laissez-faire racism

Racial attitudes

Group position in society's structure

The Swedish economist and social reformer Gunnar Myrdal arrived in the United States on September 10, 1938. He had come at the request of the Carnegie Corporation, which had commissioned him to head a comprehensive study of the status of African Americans. Among his first undertakings was a tour of the American South. This journey brought the energetic Swede face-to-face with Jim Crow segregation and discrimination against blacks. It also impressed on him the backwardness of the southern economy and the extreme poverty of most people in the region, especially but not only blacks. The journey convinced Myrdal of the importance of his mission for the nation as a whole.[1] With these stark images of a caste society and economic underdevelopment foremost in his mind, Myrdal and a distinguished staff and team of research collaborators began the research for *An American Dilemma: The Negro Problem and American Democracy*.[2]

The book was two impressive volumes. Throughout most of its pages, *An American Dilemma* provided a detailed account of discrimination against blacks in every domain of American life, debunked claims of innate black inferiority, and examined in detail black institutions (e.g., the church and political organizations). *An American Dilemma* provided the most comprehensive and shocking portrayal of the status of blacks ever assembled. Yet the legacy of Myrdal was not, in the main, the conditions he documented. Myrdal's legacy is to be found in the interpretive context in which he set "the Negro problem in American democracy."

Myrdal's analysis declared that above all else the race problem was a moral dilemma. He suggested that the United States, more than any other industrial society, possessed an explicit and popularly understood political culture that extolled the values

SOURCE: From Lawrence D. Bobo and Ryan A. Smith, "From Jim Crow Racism to Laissez-Faire Racism: The Transformation of Racial Attitudes," in *Beyond Pluralism: The Conception of Groups and Group Identities in America*, ed. Wendy F. Katkin, Ned Landsman, and Andrea Tyree.

of freedom, individual rights, democracy, equality, and justice. The status and treatment accorded African Americans by their fellow white citizens, however, stood in sharp contrast to what Myrdal viewed as the national religion or, more fittingly, the "American Creed."

Most white Americans, in his judgment, faced an "ever-raging conflict" between their general values, as expressed in the American creed, and their specific attitudes and behaviors toward blacks. The "American dilemma" was the inherent moral discomfort white Americans experienced in their relation to blacks.

An American Dilemma decisively reshaped how educated and liberal whites, especially those in the North, understood the race problem in American society. It is difficult to overestimate the impact of the book in this regard. According to the historian David Southern, "Myrdal's book played a significant role in changing the thought patterns and feelings of a people. For twenty years the Swede's authority was such that liberals simply cited him and confidently moved on."[3] Myrdal's biographer, Walter Jackson, wrote that *An American Dilemma* "established a liberal orthodoxy on black-white relations and remained the most important study on race issues until the middle 1960s."[4]

Indeed, Myrdal's work was a genuine cultural input to the coalescence of what has been called America's Second Reconstruction. The Second Reconstruction was a short but critical era from roughly the late 1950s to the mid-1960s, when the U.S. Supreme Court, the Congress, and the White House appeared to act in unison to protect the basic citizenship rights of black Americans.[5] The reach of Myrdal's influence is perhaps most clearly seen in explicit reference to *An American Dilemma* in the landmark 1954 *Brown v. Board of Education* ruling—the still controversial footnote 11—and the subsequent denunciation of Myrdal by southern defenders of segregation and other extreme right-wing groups.[6]

His influence had been seen earlier. The report of President Truman's Committee on Civil Rights, *To Secure These Rights,*[7] adopted Myrdal's theme of the contradiction between democratic values and the conditions of blacks. Truman's committee also borrowed one other notion from Myrdal, namely, his faith that American social values would win out over the customs, interests, and prejudices that had to that point combined to subjugate blacks in the postslavery American South.

FROM OPTIMISM TO PESSIMISM

Myrdal had been optimistic about the course future events would take. He anticipated positive change because the nation had much to gain from modernizing the southern economy; because levels of education were rising, particularly for blacks, who were increasingly migrating to urban and northern areas; and because changes had been induced by the wartime mobilization. The core, deeply rooted commitment to the American creed, along with these other inducements and opportunities, prompted him to adopt the optimistic assessment that the American dilemma would be resolved in favor of equality and integration.

Yet generating optimism about the course of black-white relations is perhaps harder now than at any other point in the post–World War II period. To be sure, a quarter of a century ago the Kerner Commission warned us: "Our nation is moving toward two societies, one black, one white, separate and unequal."[8] In the wake of the Simi Valley police brutality verdict and the rebellions in Los Angeles in 1992, even these words seem pallid and inadequate to capture the enormous gulf in perception, social standing, and identity that apparently still separates black and white Americans from one another. Myrdal's optimism now seems too naive. It is perhaps fitting then that Andrew Hacker's more recent book, *Two Nations: Black and White, Separate, Hostile, Unequal,* updates and provides an even bleaker assessment of race relations in the United States than the Kerner Commission did.[9]

At bottom, Hacker's point is that white-dominated society and institutions have never intended full inclusion for blacks and do not now show any real inclination toward bringing it about. An equally bleak depiction of race relations was offered in *Faces at the Bottom of the Well: The Permanence of Racism,* by Derrick Bell, a black legal scholar.[10] For Bell, each wave of racial change, reform, and apparent progress, in the end, merely reconstitutes black subordination on a new plane. The underlying racial hierarchy in the United States has not fundamentally changed. Although the Kerner Commission shared Hacker's and Bell's belief that white racism was the central cause of the oppressive conditions in which black Americans lived, it stressed that the rift between black and white could be reduced through "new attitudes, new understanding, and above all, new will" to address the racial divisions in the United States. Much of the recent scholarship and dialogue on race doubts the potential for genuine transformation of the type once envisioned by Myrdal and, to a degree, even the Kerner Commission.

The purpose of this [selection] is to assess whether these new attitudes have emerged or show any sign of emerging. Have racial attitudes genuinely improved, and are there grounds for optimism? Or is Hacker's prophecy that the United States faces "a huge racial chasm ... and there are few signs that the coming century will see it closed" the more accurate forecast?[11] Although many positive changes in racial attitudes have taken place, we believe that racism is the core problem affecting black-white relations and that it remains a disfiguring scar on the American body politic....

FROM JIM CROW RACISM TO LAISSEZ-FAIRE RACISM

Along with Howard Schuman and Charlotte Steeh, Lawrence Bobo, the senior author of this [selection], wrote a book assessing broad patterns of change in American racial attitudes.[12] Writing in 1985, we concluded that whites' attitudes toward blacks had undergone a dramatic positive transformation. A key aim of this [selection] is to delimit the scope and meaning of that transformation more precisely. Specifically, we suggest that in the post-World War II period the predominant pattern of racial attitudes among white Americans has shifted from Jim Crow racism to a modern-day laissez-faire racism. We have witnessed the virtual disappearance of overt bigotry, demands for strict segregation, advocacy of governmentally enforced discrimination, and adherence to the belief that blacks are the categorical intellectual inferiors of whites. Yet Jim Crow racism has not been replaced by an embracing and democratic vision of the common humanity, worth, dignity, and equal membership of blacks in the polity. Instead, the tenacious institutionalized disadvantages and inequalities created by the long slavery and Jim Crow eras are now popularly accepted and condoned under a modern free-market or laissez-faire racist ideology.

Laissez-faire racism blames blacks themselves for the black-white gap in socioeconomic standing and actively resists meaningful efforts to ameliorate America's racist social conditions and institutions. These racial attitudes continue to justify and explain the prevailing system of racial domination, even while a core element of racist ideology in the United States has changed. Jim Crow racism was premised on notions of black biological inferiority; laissez-faire racism is based on notions of black cultural inferiority. Both serve to encourage whites' comfort with and acceptance of persistent racial inequality, discrimination, and exploitation....

If the nature and causes of this transformation from the once dominant ideology of Jim Crow racism to the currently dominant ideology of laissez-faire racism fit the data we discuss below, then Hacker's and Bell's pessimism may be solidly grounded. Neither the decline of Jim Crow racism nor the emergence of laissez-faire racism can be attributed to the goodwill of the American people or to the gradual ascendancy of the American creed of freedom, equality, justice, and democracy.[13] On the contrary, both of these epochal ideologies appear to involve support for specific forms of racial domination. These forms of domination each fit the different economic and political conditions of their eras.[14]

WHY CALL IT RACISM?

For those who may doubt that the United States, which is legally committed to an antidiscrimination policy, still is a racially dominative society, we review a few facts.[15] First, the black-white gap in socioeconomic status remains enormous. Black adults remain two-and-one-half times as likely as whites to be unemployed. Strikingly, this gap exists at virtually every level of the educational distribution.[16] If one casts a broader net to ask about "underemployment"—those who have fallen out of the labor force entirely, are unable to find full-time work, or are working full-time at below poverty-level wages—then the black-white ratio in major urban areas has over the past two decades risen from the customary 2 to 1 disparity to very nearly 5 to 1.[17] Conservative estimates show that young, well-educated blacks who match whites in work experience and other characteristics still earn 11 percent less annually.[18] Studies continue to document direct labor market discrimination at both low-skill, entry-level positions[19] and more highly skilled positions.[20] A growing number of studies indicate that even highly skilled and accomplished black managers encounter "glass ceilings" in corporate America,[21] prompting one set of analysts to suggest that blacks will never be fully admitted to the power elite.[22]

Judged against differences in wealth, however, the huge black-white gaps in labor-force status and earnings seem absolutely paltry.[23] The average differences in wealth show black households lagging behind white ones by nearly twelve to one. For every one dollar of wealth in white households, black households have less than ten cents. In 1984 the median level of wealth held by black households was around $3,000; for white households the figure was $39,000. Indeed, white households with incomes of between $7,500 and $15,000 have "higher mean net worth and net financial assets than black households making $45,000 to $60,000."[24] Whites near the bottom of the white income distribution have more wealth than blacks near the top of the black income distribution.

Wealth is in many ways a better indicator of likely quality of life than earnings are. When we pose a few hypothetical questions, the reasons for this claim become clear. If we envision an "average" black family with about $3,000 in wealth and an average white family with about $39,000 in wealth we might then ask: which of these families is best equipped to send a child to college for four years? Which of these families could best survive a four-month period of unemployment? Which of these families could pay for costly medical treatment? Which of these families can attempt to start a business of its own? Indeed, which of these families might be able to do all of these things, and which one might be unable to do any? The gaping disparity in accumulated wealth is the real inequality in standard of living produced by three hundred plus years of systematic and pervasive racial discrimination.

Second, blacks are far and away the group from which whites maintain the greatest social distance.[25] The demographers Douglas Massey and Nancy Denton concluded that it makes sense to describe the black condition as "hypersegregation." Blacks are the only group, based on 1980 census data for large metropolitan areas, to rank as "hypersegregated" on four or more measures, and this was true for sixteen areas covering nearly a quarter of all blacks.[26] Housing audit studies continue to show high levels of direct racial discrimination in the housing market.[27] Middle-class blacks have enormous difficulty translating their economic gains into residential mobility, which has been a critical pathway to assimilation into the economic and social mainstream for other groups. Residential segregation has social consequences. As we all know, neighborhoods vary in services, school quality safety, and levels of exposure to a variety of unwanted social conditions.[28]

Third, the value this society places on black life appears to be in steady decline. This is seen in how blacks and black life are treated by the criminal justice system as well as in overall figures in life expectancy. A 1990 study showed that fully 42 percent of black males between the ages of eighteen and twenty-four in the nation's capital are in jail, on probation, or have warrants out for their arrest. Blacks are seven times more likely than whites to die as victims of homicide. Blacks who kill whites are more severely punished than whites who kill blacks.[29] When blacks kill whites, prosecutors are forty times more likely to request the death penalty than when blacks kill other blacks. Such profound differences prompted retiring Supreme Court justice Harry Blackmun to publicly repudiate the death penalty.

Looking beyond violent crime and the criminal justice system, black life expectancy at birth declined for four years in a row between 1985 and 1989, although this was a period of modest but continuing increase in life expectancy for whites. Most stunning, the decline in 1988 reached such a level that it brought down the overall national average. Yet our national leadership conveyed no sense of real emergency about this shocking set of social statistics.

We could go on, but the severity of the disparities and the extent to which they cut across class lines in the black community are sufficiently clear to establish a strong prima facie case for maintaining that the United States society still has a system of racial domination....

PATTERNS OF CHANGE IN RACIAL ATTITUDES

The longest trend data from national sample surveys may be found for racial attitude questions that deal with matters of racial principles, the implementation of those principles, and social distance preferences. Principle questions ask whether American society should be integrated or segregated and whether individuals should be treated equally without regard to race. Such questions do not raise issues of the practical steps that might be necessary to accomplish greater integration or to ensure equal treatment. Implementation questions ask what actions, usually by government, especially the federal government, ought to be taken to bring about integration, to prevent discrimination, and to achieve greater equality. Social distance questions ask about the individual's willingness to personally enter hypothetical contact settings in schools or neighborhoods that vary from virtually all white to heavily black.[30]

Transformation of Principles

Questions on racial principles provide the most consistent evidence on how the attitudes of white Americans toward blacks have changed. From crucial baseline surveys conducted in 1942, trends for most racial principle questions show whites increasingly support the principles of racial integration and equality. Whereas a solid majority 68 percent, of white Americans in 1942 favored segregated schools, only 7 percent took such a position in 1985 (see Figure 13.1). Similarly, 55 percent of whites surveyed in 1944 thought whites should receive preference over blacks in access to jobs, compared with only 3 percent who offered such an opinion as long ago as 1972. Indeed, so few people were willing to endorse the discriminatory response to this question on the principle of race-based labor market discrimination that it was dropped from national surveys after 1972. On both these issues, then, majority endorsement of the principles of segregation and discrimination have given way to overwhelming majority support for integration and equal treatment.

This pattern of movement away from support for Jim Crow toward apparent support for racial egalitarianism holds with equal force for those questions dealing with issues of residential integration, access to public transportation and public accommodations, choice among qualified candidates for political office, and even interracial marriage. It is important to note, however, that the high levels of support seen for the principles of school integration and equal access to jobs (both better than 90 percent) do not exist for all questions on racial principles. Despite improvement from extraordinarily low levels of support in the 1950s and 1960s, survey

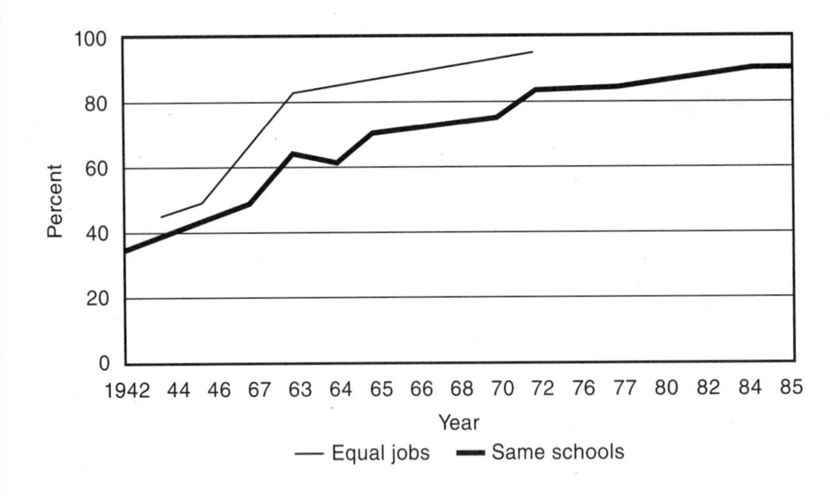

FIGURE 13.1 Trends in Racial Principle Questions among Whites, 1942–85. (Based on data in Howard Schuman, Charlotte Steeh, and Lawrence Bobo, *Racial Attitudes in America: Trends and Interpretations* [Cambridge, Mass.: Harvard University Press, 1985], 74–75)

data continue to show substantial levels of white discomfort with the prospect of interracial dating and marriage, for instance....

There has been a sweeping transformation of attitudes about the rules that should guide black-white interaction in the more public and impersonal spheres of social life. Those living outside the South, the well-educated, and younger people led the way on these changes. However, change has usually taken place in all categories of people. Schuman, Steeh, and Bobo characterized this change as a fundamental transformation of social norms regarding race. Robert Blauner's in-depth interviews with blacks and whites over nearly three decades led him to a very similar conclusion: "The belief in a right to dignity and fair treatment is now so widespread and deeply rooted, so self-evident that people of all colors would vigorously resist any effort to reinstate formalized discrimination. This consensus may be the most profound legacy of black militancy, one that has brought a truly radical transformation in relations between the races."[31] In short, a tremendous progressive trend has been evident in white racial attitudes where the broad issues of integration, equality, and discrimination are concerned.

Those who believe that America is making progress toward resolving the "American dilemma"point to this evidence as proof that Americans have taken a decisive turn against racism. As Richard G. Niemi, John Mueller, and Tom W. Smith argued, "Without ignoring real signs of enduring racism, it is still fair to conclude that America has been successfully struggling to resolve its Dilemma and that equality has been gaining ascendancy over racism."[32] If anyone doubts the validity of this transformation, it is noteworthy that even former Klansman David Duke felt compelled to assert that he was no longer a bigot and had shed parts of his past during his failed bid to become governor of Louisiana. Whether his claim is true is less important than the fact that Duke had to take such a public position. Some ideas—support for segregation, open discrimination, and claims that blacks are inherently inferior to whites—have fallen into deep public disrepute. Surveys have documented the speed, social location, and breadth of this transformation.

RESISTANCE TO POLICY CHANGE

If trends in support of progressive racial principles are the optimistic side of the story of the transformation of racial attitudes, the patterns for implementation questions are the pessimistic side of the story. It should be noted that efforts to assess how Americans feel about government efforts to bring about greater integration and equality or to prevent discrimination really do not arise as sustained matters of inquiry in surveys until the 1960s. To an important degree, issues of the role of government in bringing about racial change could not emerge until sufficient change involving the basic principles had actually occurred.

There are sharp differences between support for racial principles and support for policy implementation. This is not surprising insofar as principles, viewed in isolation, need not conflict with other principles, interests, or needs that often arise in more concrete situations. However, the gaps between principle and implementation are large and consistent in race relations. In 1964, for example, surveys showed that 64 percent of whites nationwide supported the principle of integrated schooling; however, only 38 percent thought that the federal government had a role to play in bringing about greater integration (see Figure 13.2). The gap had actually grown larger by 1986, when 93 percent supported the principle, but only 26 percent endorsed government efforts to bring about school integration. We return to this point later.

Similar patterns emerge in the areas of jobs and housing. Support for the principle of equal access to jobs stood at 97 percent in 1972. Support for federal efforts to prevent job discrimination, however, had reached only 39 percent. Likewise in 1976, 88 percent supported the principle that blacks have the right to live wherever they can afford, yet only 35 percent said they would vote in favor of a law requiring homeowners to sell without regard to race.

There are not only sharp differences in absolute levels of support when moving from principle to implementation but also differences in trends. Most striking, there is a clear divergence of trends in the area of school integration. During the 1972 to 1986 time period, when support for the principle of integrated schooling rose from 84 percent to 93 percent, support for government efforts to bring about integration fell from 35 percent to 26 percent. It should be noted that this decline is restricted almost entirely to those living outside the South .... By 1978 there was virtually no difference between college-educated whites outside the South and southern whites who had not completed high school when it came to supporting federal efforts to help bring about school integration. To put it colloquially, Bubba and William F. Buckley increasingly found themselves in agreement on this issue.

Two complexities are worthy of note. First, a couple of implementation issues do show positive trends. The most clear-cut case involves a question on whether the government has a role to play in assuring blacks fair access to hotels and public accommodations. This may be the only instance where parallel questions on principle and implementation show parallel positive change. A somewhat similar pattern is found for the principle of residential integration and support for an open or fair housing law. However, even as recently as 1988 barely 50 percent of white Americans endorsed

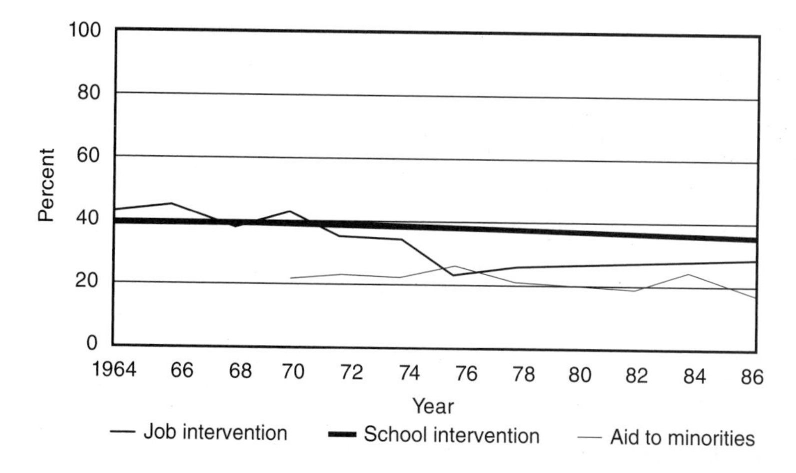

FIGURE 13.2 Trends in Implementation Questions among Whites, 1964–86. (Based on data in Howard Schuman, Charlotte Steeh, and Lawrence Bobo, *Racial Attitudes in America: Trends and Interpretations* [Cambridge, Mass.: Harvard University Press, 1985], 88–89.)

a law that would forbid racial discrimination in the sale or rental of housing.

It should be borne in mind that antiblack animus is not the only source of opposition to government involvement in bringing about progressive racial change. Howard Schuman and Lawrence Bobo have shown that whites are equally likely to oppose open housing laws whether the group in question is black, Japanese American, or another minority.[33] There appears to be an important element of objection to government coercion in this domain that influences attitudes. At the same time, however, Schuman and Bobo also found that whites express a desire for greater distance from blacks than they do from other groups.

Second, opposition to implementation is widespread and is not substantially affected by the usual socioeconomic characteristics of respondents, including education, region, and age. Weak to nonexistent effects of education and age in particular suggest that we are unlikely to see much change in the future.

Unfortunately, comparatively few survey trend questions speak directly to affirmative action policies. Many different questions have been asked beginning in the mid-1970s. Affirmative action is a much maligned and misunderstood concept. Affirmative actions can range from advertising and special recruitment efforts to preferential treatment requiring quotas. Support for affirmative action varies dramatically, depending on exactly which type of policy is proposed.[34] Policies that mainly aim to increase the human capital attributes of blacks are comparatively popular.[35] Policies that lean in the direction of achieving equal outcomes, as powerfully symbolized by the term *quotas,* elicit overwhelming opposition among whites....

BASIC ROOTS OF THE SHIFT

... [W]hat accounts for the momentous changes that occurred in whites' racial attitudes? We believe that structural changes in the American economy and polity that reduced the importance of the Jim Crow system of exploited black agricultural labor to the overall economy lie at the base of the positive change in racial attitudes. In short, the structural need for Jim Crow ideology disappeared. Correspondingly, though slowly and only in response to aggressive and innovative challenge from the black civil rights movement, political and ideological supports for Jim Crow institutions yielded. The defeat of Jim Crow ideology and the political forms of its institutionalization (e.g., segregated schooling and public facilities, voting hindrances) was the principal accomplishment of the civil rights movement.

We submit that there are inevitable connections between economic and political structures, on the one hand, and patterns of individual thought and action, on the other hand. As the structural basis of long-standing patterns of social relationships changes, there is a corresponding potential for change in the ways of thinking, feeling, and behaving that had previously been commonplace.

Our argument is similar to Myrdal's. His optimism about the future course of race relations in the United States rested explicitly on a set of ideas about economic interests and needs, demographic trends, and the wartime mobilization, which he thought would all work in the direction of more fully integrating blacks into American society. We part company with Myrdal, however, when he argued that the American creed was a fundamental impetus to changing conceptions of the place of African Americans. Instead, we are impressed with how long many white Americans have been comfortable with conditions in the black community and in the daily lives of African Americans that constitute profound violations of the high moral purposes articulated in the American creed....

Structural Change and Changing Attitudes

The declining importance of cotton to the U.S. economy and as a source of livelihood for blacks opened the door to tremendous economic and political opportunity for blacks. The product of these opportunities, stronger churches, colleges, and political organizations, culminated in a sustained movement of protest for racial justice. The movement and the organizations it created had indigenous leadership, financing, and a genuine mass base of support. Through creative, carefully designed, and sustained social protest, this movement was able to topple a distinct, epochal form of racial oppression that was no longer essential to the interests and needs of a broad range of American political and economic elites.

Widespread cultural attitudes endorsing elements of the Jim Crow social order, quite naturally then, began to atrophy and wither under a steady assault by blacks and their white allies. Segregationist positions were under steady assault and increasingly lacked strong allies. The end product of these forces, the decline of Jim Crow racism, is the broad pattern of improvement seen in whites' racial attitudes in the United States.

The effectiveness of the NAACP's legal strategy challenging segregation, the passage of the Civil Rights Act of 1964, and the passage of the Voting Rights Act of 1965 amounted to an authoritative legal and political rebuke of the Jim Crow social order. This rebuke, however, did not directly alter the socioeconomic status of blacks, especially those living in the northern urban areas. This rebuke also did not directly alter entrenched patterns of racial residential segregation that existed nationwide.[36] Nor did widespread attitudes of hostility toward blacks suddenly disappear.[37] The enormous and far-reaching successes of the civil rights movement did not eliminate stark patterns of racial domination and inequality that existed above and beyond the specific dictates of the distinctly southern Jim Crow system. Instead of witnessing genuine racial comity, we saw the rise of laissez-faire racism.

THE SENSE OF GROUP POSITION AND CHANGING RACIAL ATTITUDES

Students of prejudice and racial attitudes may have misunderstood the real "object" of racial attitudes. The attitude object, or perceptual focus, is not really the social category "blacks" or "whites," whether as groups or individuals. It is not neighborhoods or schools of varying degrees of racial mixture. Instead, as Herbert Blumer argued forty years ago,[38] the real object of "prejudice," what we are really tapping with our questions, is attitude toward the proper relation between groups: that is, the real attitude object is relative group positions. This sense of group position is historically and culturally rooted, socially learned, and modifiable in response to new information, events, or structural conditions so long as these factors contribute to or shape contexts for social interaction among members of different groups.

What does this "group position" view of racial attitudes mean in the context of all that we have reviewed to this point? First, attitudes toward "integration" or toward "blacks" are, fundamentally statements about preferred positional relations among groups. They are not simply or even mainly emotional reactions to groups, group symbols, or situations. Nor are they best understood as statements of simple feelings of like or dislike of minority groups and their members. Nor are they simply perceptions of group traits and dispositions. Instead, racial attitudes capture preferred group positions and those patterns of belief and feeling that under-gird, justify, and make understandable a preference for relatively little group differentiation and inequality under some social conditions or for a great deal of differentiation and inequality under others.

In the case of changing white racial attitudes in the United States, increasing openness to the principle of integrated education does not mean a desire for greater contact with blacks or even an attachment to integrated education. From the vantage point of group position theory, it means declining insistence on forced group inequality in educational institutions. Declining support for segregated public transportation does not signal a desire for more opportunities to interact with blacks on buses, trains, and the like. Instead, it means a declining insistence on compulsory inequity in group access to this domain of social life.

Second, the group position view sees change in political and economic structures as decisively shaping the socially constructed and shared sense of group position. The sources of change in attitudes—changes in preferred group positions—are not found principally in changing feelings of like and dislike. Changes in the patterns of mass attitudes reflect changes in the structurally based, interactively defined and understood needs and interests of social groups. To put it differently, to have meaning, longevity, and force in people's everyday lives, the attitudes individuals hold must be linked to the organized modes of living in which people are embedded.[39] A demand for segregated transportation, segregated hotels, and blanket labor market discrimination increasingly rings hollow under an economy and polity that have less need for—in fact may be incurring heavy costs because of—the presence of a superexploited, black labor pool. When the economic and political needs of significant segments of a dominant racial group no longer hinge on a sharp caste system for effective functioning, the ideology that explained and justified such a caste system should weaken. It becomes vulnerable to change; its costs should become increasingly apparent—and be rejected.

Third, a key link between changing structural conditions and the attitudes of the public are those prominent social actors who articulate, and frequently clamor over and debate, the need for new modes of social organization.[40] The claims and objectives of leaders presumably spring from their conceptions of the interests, opportunities, resources, and needs of the group at a particular point in time. Readily appreciated examples of the role of leaders include the justices' 1954 *Brown* decision, President Kennedy's speech following the effort to enroll two black students at the University of Mississippi, President Johnson's invocation of the civil rights slogan "We shall overcome," and perhaps most memorably Dr. Martin Luther King's "I have a dream" speech.

Of course, not all leadership statements and actions were supportive of positive change. There were White Citizens' Councils, Ku Klux Klan rallies, and a wide variety of other forms of resistance to change. Indeed, Kennedy's speech, which the historian Carl Brauer credits with launching the Second Reconstruction,[41] followed on the heels of Alabama governor George Wallace's "Segregation now! Segregation tomorrow! Segregation forever" declaration. In addition, there were powerful voices and forces on the Left activated by the civil rights movement that were demanding greater change than either Kennedy or Johnson was ready to accept. Our point is that the direction and tenor of change is shaped in the larger public sphere of clashes, debate, political mobilization, and struggle.

CONCLUSION

Racism Old and New

Can we now share the faith and optimism that Gunnar Myrdal expressed in 1944? Or are the bleak depictions offered by Andrew Hacker and Derrick Bell more accurate analyses? We cannot share Myrdal's optimism, although we resist pessimism and despair.

The long and unabated record of sweeping change in racial attitudes that national surveys document cannot be read as a fundamental breakdown in either racialized thinking or antiblack prejudice. Instead, we have witnessed the disappearance of a racial ideology appropriate to an old social order, that of the Jim Crow South. A new and resilient laissez-faire racism ideology has arisen in its place. As a result, America largely remains "two nations," with African Americans all too often viewing the world from the "bottom of the well." …

The end product of these conditions and processes is a new racialized social order with a new racial ideology—laissez-faire racism. Under this regime, blacks are blamed as the cultural architects of their own disadvantaged status. The deeply entrenched cultural pattern of denying societal responsibility for conditions in many black communities continues to foster steadfast opposition to affirmative action and other social policies that might alleviate race-based inequalities. In short, many Americans have become comfortable with as much racial segregation and inequality as a putatively nondiscriminatory polity and free-market economy can produce. Such individuals also tend to oppose social policies that would substantially improve the status of blacks, hasten the pace of integration, or aggressively attack racial discrimination. Enormous racial inequalities thus persist and are rendered culturally palatable by the new laissez-faire racism.

ENDNOTES

1. Walter A. Jackson, *Gunnar Myrdal and America's Conscience* (Chapel Hill: University of North Carolina Press, 1990).
2. Gunnar Myrdal, *An American Dilemma: The Negro Problem and Modern Democracy*, 2 vols. (New York: Random House, 1944).
3. David W. Southern, *Gunnar Myrdal and Black-White Relations: The Use and Abuse of 'An American Dilemma,"* 1944–1969 (Baton Rouge: Louisiana State University Press, 1994), xvi.
4. Jackson, *Gunnar Myrdal and America's Conscience*, xviii.
5. Carl M. Brauer, *John F. Kennedy and the Second Reconstruction* (New York: Columbia University Press, 1977); C. Vann Woodward, *The Strange Career of Jim Crow*, 3d rev. ed. (New York: Oxford University Press, 1974).
6. See Southern, *Gunnar Myrdal and Black-White Relations*, 155–86.
7. U.S. President's Committee on Civil Rights, *To Secure These Rights* (New York: Simon and Schuster, 1947).
8. National Advisory Commission on Civil Disorders, *Report of the National Advisory Commission on Civil Disorders* (New York: Bantam Books, 1968), 1.
9. Andrew Hacker, *Two Nations: Black and White, Separate, Hostile, Unequal* (New York: Macmillan, 1992).
10. Derrick Bell, *Faces at the Bottom of the Well: The Permanence of Racism* (New York: Basic Books, 1992).
11. Hacker, *Two Nations*, 219.
12. Howard Schuman, Charlotte Steeh, and Lawrence Bobo, *Racial Attitudes in America: Trends and Interpretations* (Cambridge, Mass.: Harvard University Press, 1985).
13. Compare Myrdal, *American Dilemma*.
14. Herbert Blumer, "Race Prejudice as a Sense of Group Position," *Pacific Sociological Review* 1, no. 1 (1958): 3–7.
15. Prominent legal scholars have pointed to the persistence of racism despite the enactment of antidiscrimination laws. See Charles R. Lawrence, "The Id, the Ego, and Equal Protection: Reckoning with Unconscious Racism," *Stanford Law Review* 39 (January 1987): 317–52; and Kimberle W. Crenshaw, "Race, Reform, and Retrenchment: Transformation and Legitimation in Antidiscrimination Law," *Harvard Law Review* 101 (May 1988): 1331–87.
16. Gerald D. Jaynes, "The Labor Market Status of Black Americans: 1939–1985," *Journal of Economic Perspectives* 4, no. 4 (1990): 9–24; Franklin D. Wilson, Marta Tienda, and Lawrence Wu, "Race and Unemployment: Labor Market Experiences of Black and White Men, 1968–1988," *Work and Occupations* 22 (Summer 1995): 245–70.
17. Daniel T. Lichter, "Racial Differences in Unemployment in American Cities," *American Journal of Sociology* 93 (January 1988): 771–92; Roderick J. Harrison and Claudette E. Bennett, "Racial and Ethnic Diversity," in *State of the Union: America in the 1990s*, vol. 2, ed. R. Farley (New York: Russell Sage, 1995), 141–210.
18. Reynolds Farley, *Blacks and Whites: Narrowing the Gap?* (Cambridge, Mass.: Harvard University Press, 1984), 80.
19. Joleen Kirschenman and Kathryn M. Neckerman, "We'd Love to Hire Them, But. ..: The Meaning of Race for Employers," in *The Urban Underclass*, ed. C. Jencks and P. E. Peterson (New York: Brookings Institution, 1991), 203–31; Margery A. Turner, Michael Fix, and Raymond J. Struyk, *Opportunities Denied, Opportunities Diminished: Racial Discrimination in Hiring*, Urban Institute Report 91—9 (Washington, D.C.: Urban Institute Press, 1991); Roger Waldinger and Thomas Bailey, "The Continuing Significance of Race," *Politics and Society* 19 (September 1991): 291–329.
20. Joe R. Feagin and Melvin P. Sikes, *Living with Racism: The Black Middle-Class Experience* (Boston: Beacon, 1994).
21. John P. Fernandez, *Black Managers in White Corporations* (New York: John Wiley, 1975); Edward W. Jones, "Black Managers: The Dream Deferred," *Harvard Business Review* 64 (May–June 1986): 84–93; Ryan A. Smith, "Race, Income and Authority at Work: A Cross-Temporal Analysis of Black and White Men (1972–1994)," *Social Problems* 44 (February 1997): 19–37.

22. Richard L. Zweigenhaft and G. William Domhoff, *Blacks in the White Establishment: A Study of Race and Class in America* (New Haven, Conn.: Yale University Press, 1990).
23. Gerald D. Jaynes and Robin M. Williams, *A Common Destiny: Blacks and American Society* (Washington, D.C.: National Academy Press, 1989); Melvin L. Oliver and Thomas M. Shapiro, *Black Wealth/White Wealth: A New Perspective on Racial Inequality* (New York: Routledge, 1995).
24. Paul Starr, "Civil Reconstruction: What to Do without Affirmative Action," *American Prospect* (Winter 1992): 12.
25. Lawrence Bobo and Camille L. Zubrinsky, "Attitudes on Residential Integration: Perceived Status Differences, Mere In-group Preferences or Racial Prejudice?" *Social Forces* 74 (March 1996): 883–909; Camille L. Zubrinsky and Lawrence Bobo, "Prismatic Metropolis: Race and Residential Segregation in the City of Angels," *Social Science Research* 25 (December 1996): 335–74; Martin Sanchez Jankowski, "The Rising Significance of Status in U. S. Race Relations," in *The Bubbling Cauldron: Race, Ethnicity, and the Urban Crisis*, ed. M. P. Smith and J. R. Feagin (Minneapolis: University of Minnesota Press, 1995), 77–98.
26. Douglas S. Massey and Nancy S. Denton, *American Apartheid* (Cambridge, Mass.: Harvard University Press, 1993).
27. Diana M. Pearce, "Gatekeepers and Homeseekers: Institutional Patterns in Racial Steering," *Social Problems* 26 (February 1979): 325–42; Margery A. Turner, "Discrimination in Urban Housing Markets: Lessons from Fair Housing Audits," *Housing Policy Debates* 3, no. 2 (1992): 185–215.
28. D. S. Massey, A. B. Gross, and M. L. Eggers, "Segregation, the Concentration of Poverty, and the Life Chances of Individuals," *Social Science Research* 20 (December 1991): 397–420.
29. General Accounting Office, *Death Penalty Sentencing: Research Indicates Patterns of Race Disparities*, Report to the Senate and House Committees on the Judiciary (Washington, D.C.: Government Printing Office, 1990), 5–6.
30. Schuman, Steeh, and Bobo, *Racial Attitudes in America.*
31. Robert Blauner, *Black Lives, White Lives: Three Decades of Race Relations in America* (Berkeley: University of California Press, 1989), 317.
32. Richard G. Niemi, John Mueller, and Tom W. Smith, *Trends in Public Opinion: A Compendium of Survey Data* (New York: Greenwood, 1989), 167.
33. Howard Schuman and Lawrence Bobo, "Survey Based Experiments on Whites' Racial Attitudes toward Residential Integration," *American Journal of Sociology* 94 (September 1988): 273–99.
34. James R. Kluegel and Eliot R. Smith, *Beliefs about Inequality: Americans' View of What Is and What Ought to Be* (New York: Aldine de Gruyter, 1986); Seymour Martin Lipset and William Schneider, "The Bakke Case: How Would It Be Decided at the Bar of Public Opinion?" *Public Opinion* (March/April 1978): 38–48.
35. Lawrence Bobo and James R. Kluegel, "Whites' Stereotypes, Social Distance, and Perceived Discrimination toward Blacks, Hispanics, and Asians: Toward a Multiethnic Framework" (paper presented at the annual meetings of the American Sociological Association, Cincinnati, August 25, 1991); Lawrence D. Bobo and James R. Kluegel, "Opposition to Race-Targeting: Self-Interest, Stratification Ideology and Racial Attitudes," *American Sociological Review* 58 (August 1993): 443–64.
36. Karl E. Taeuber and Alma F. Taeuber, *Negroes in Cities: Residential Segregation and Neighborhood Change* (Chicago: Aldine, 1965).
37. Paul B. Sheatsley, "Whites' Attitudes toward the Negro," *Daedalus* 95 (Winter 1966): 217–38.
38. Blumer, "Race Prejudice as a Sense of Group Position."
39. Earl Raab and Seymour Martin Lipset, "The Prejudiced Society," in *American Race Relations Today*, ed. Earl Raab (New York: Doubleday, 1962), 29–55.
40. Blumer, "Race Prejudice as a Sense of Group Position."
41. Brauer, *John F. Kennedy and the Second Reconstruction*, 259–64.

DISCUSSION QUESTIONS

1. What is the "American dilemma"? Do you believe that it exists? What do the authors believe?
2. How is it possible that attitudes have changed considerably in the United States, yet we still have segregation and inequality?
3. What is laissez-faire racism? Is it really racism, or is it something else?
4. What exactly is the problem identified in this selection? What is its cause?
5. Put yourself into the mind of someone who is attracted to laissez-faire racism as an explanation of racial inequality. What would you argue? How would the authors react to your arguments? What do you really think?
6. The authors emphasize the importance of group position rather than prejudice as a cause of racism. What does this mean?
7. Do the authors suggest ways to deal with the problem they identify? How would they approach the problem of segregation and inequality if they were asked?

14

Latino Lives in a Changing America

ROBERTO SURO

The Four Questions

1. What is the problem that is described?
2. Is this a *social* problem? Does it have a social cause? Are a lot of people affected? Is society affected? Is there a consensus that there is a problem?
3. What are the causes of this problem?
4. What do the authors recommend for dealing with the problem?

Topics Covered

Latinos Mexican-Americans

Education

Work

Immigration

Poverty

Racism

On Imelda's fifteenth birthday, her parents were celebrating everything they had accomplished by coming north to make a new life in the United States. Two short people in brand-new clothes, they stood in the driveway of their home in Houston and greeted relatives, friends, and neighbors, among them a few people who had come from the same village in central Mexico and who would surely carry gossip of the party back home. A disc jockey with a portable stereo presided over the backyard as if it were a cabaret instead of a patch of grass behind an overcrowded bungalow where five people shared two bedrooms. A folding table sagged with platters of tacos and fajitas. An aluminum keg of beer sat in a wheelbarrow atop a bed of half-melted ice cubes. For Imelda's parents, the festivities that night served as a triumphant display of everything they had earned by working two jobs each. Like most of the other adults at the party, they had come north to labor in restaurants, factories, warehouses, or construction sites by day and to clean offices at night. They had come to work and to raise children in the United States.

Imelda, who had been smuggled across the Rio Grande as a toddler, wore a frilly dress ordered by catalog from Guadalajara, as befits a proper Mexican celebrating her *quinceañera,* which is the traditional coming-out party for fifteen-year-old Latin girls. Her two younger sisters and a little brother, all U.S. citizens by birth, wore new white shirts from a discount store. Their hair had been combed down with sharp, straight parts and dabs of pomade.

When it came time for Imelda to dance her first dance, her father took her in his arms for one of the old-fashioned polkas that had been his favorite when a band played in the town square back home. By tradition, boys could begin courting her after that dance. Imelda's parents went to bed that night content they had raised their children according to proper Mexican custom.

The next morning at breakfast, Imelda announced that she was pregnant, that she was dropping out of school, and that she was moving in with her boyfriend, a Mexican-American who did not speak Spanish and who did not know his father. That night, she ate a meal purchased with food stamps and cooked on a hot plate by her boyfriend's mother. She remembers the dinner well. "That night, man, I felt like an American. I was free."

This is the promise and the peril of Latino immigration. Imelda's parents had traveled to Texas on a wave of expectations that carried them from the diminishing life of peasant farmers on a dusty *rancho* to quiet contentment as low-wage workers in an American city. These two industrious immigrants had produced a teenage welfare mother, who in turn was to have an American baby. In the United States, Imelda had learned the language and the ways. In the end, what she learned best was how to be poor in an American inner city.

Latino immigration delivers short-term gains and has long-term costs. For decades now, the United States has engaged in a form of deficit spending that can be measured in human lives. Through their hard work at low wages, Latinos have produced immediate benefits for their families, employers, and consumers, but American society has never defined a permanent place for these immigrants or their children and it has repeatedly put off considering their future. That future, however, is now arriving, and it will produce a reckoning. The United States will need new immigration policies to decide who gets into the country. More importantly, the nation will need new means of assuring political equality and freedom of economic opportunity. Soon Americans will learn once again that in an era of immigration, the newcomers not only demand change; they create change.

When I last met Imelda, she was just a few weeks short of her due date, but she didn't have anything very nice to say about her baby or her boyfriend. Growing up in Houston as the child of Mexican immigrants had filled her with resentment, especially toward her parents, and that was what she wanted to talk about.

"We'd get into a lot of yelling and stuff at home because my parents, they'd say, 'You're Mexican. Speak Spanish. Act like a Mexican girl,' and I'd say, 'I'm here now and I'm going to be like the other kids.' They didn't care."

Imelda is short and plump, with wide brown eyes and badly dyed yellow hair. She wore a denim shirt with the sleeves ripped off, and her expression was a studied pout. Getting pregnant was just one more way of expressing anger and disdain. She is a dimestore Madonna.

Imelda is also a child of the Latino migration. She is a product of that great movement of people from Latin America into the United States that is older than any borders but took on a startling new meaning when it gradually gained momentum after the 1960s and then turned into something huge in the 1980s. Latino immigrants were drawn north when America needed their services, and they built communities known as barrios in every major city. But then in the 1990s, as these newcomers began to define their permanent place here, the ground shifted on them. They and their children—many of them native-born Americans—found themselves struggling with an economy that offered few opportunities to people trying to get off the bottom. They also faced a populace sometimes disconcerted by the growing number of foreigners in its midst. Immigration is a transaction between the newcomers and the hosts. It will be decades before there is a final tally for this great wave of immigration, but the terms of the deal have now become apparent.

Imelda's story does not represent the best or the worst of the Latino migration, but it does suggest some of the challenges posed by the influx. Those challenges are defined first of all by demography. No other democracy has ever experienced an uninterrupted wave of migration that has lasted as long and that has involved as many people as the recent movement of Spanish-speaking people to the United States. Twelve million foreign-born Latinos live here. If immigration and birth rates remain at current levels, the total Hispanic population will grow at least three times faster than the population as a whole for several decades, and Latinos will become the nation's largest minority group, surpassing the size of the black population a few years after the turn of the [twenty-first] century. Despite some differences among them, Latinos constitute a distinctive linguistic and cultural group, and no single group has ever dominated a prolonged wave of immigration the way Latinos have for thirty years. By contrast, Asians, the other large category of immigrants, come from nations as diverse as India and Korea, and although the Latino migration is hardly monolithic, the Asian influx represents a much greater variety of cultures, languages, and economic experiences. Moreover, not since the Irish potato famine migration of the 1840s has any single nationality accounted for such a large share of an immigrant wave as the Mexicans have in recent decades. The 7 million Mexican immigrants living in the United States in 1997 made up 27 percent of the entire foreign-born population, and they outnumbered the entire Asian immigrant population by more than one million people. Latinos are hardly the only immigrants coming to the United States in the 1990s, but they will define this era of immigration, and this country's response to them will shape its response to all immigrants.

Latinos, like most other immigrants, tend to cluster together. Their enclaves are the barrios, a Spanish word for neighborhoods that has become part of English usage because barrios have become such a common part of every American city. Most barrios, however, remain a place apart, where Latinos live separated from others by custom, language, and preference. They are surrounded by a city but are not part of it. Imelda lived in a barrio named Magnolia Park, after the trees that once grew along the banks of the bayou there. Like other barrios, Magnolia is populated primarily by poor and working-class Latinos, and many newly arrived immigrants start out there. Magnolia was first settled nearly a hundred years ago by Mexicans who fled revolution in their homeland and found jobs dredging the ship channel and port that allowed Houston to become a great city. Latinos continued to arrive off and on, especially when Houston was growing. Since the 1980s, when the great wave of new arrivals began pouring into Magnolia, it hasn't mattered whether the oil city was in boom or bust—Latinos always find jobs, even when they lack skills and education. Most of Magnolia is poor, but it is also a neighborhood where people go to work before dawn and work into the night.

Like other barrios, Magnolia serves as an efficient port of entry for Latino immigrants because it is an easy place to find cheap housing, learn about jobs, and keep connected to home. Some newcomers and their children pass through Magnolia and find a way out to more prosperous neighborhoods where they can leave the barrio life behind. But for millions like Imelda who came of age in the 1990s, the barrios have become a dead end of unfulfilled expectations.

"We could never get stuff like pizza at home," Imelda went on, "just Mexican foods. My mother would give me these silly dresses to wear to school. No jeans. No jewelry. No makeup. And they'd always say, 'Stick with the Mexican kids. Don't talk to the Anglos; they'll boss you. Don't run around with the Chicanos [Mexican-Americans]; they take drugs. And just don't go near the *morenos* [blacks] for any reason.'"

Imelda's parents live in a world circumscribed by the barrio. Except for the places where they work, the rest of the city, the rest of America, seems to them as remote as the downtown skyline visible off in the distance on clear days. After more than a dozen years, they speak all the English they need, which isn't much. What they know best is how to find and keep work.

Imelda learned English from the television that was her constant childhood companion. Outside, as Magnolia became a venue for gangs and drug sales, she learned to be streetwise and sassy. Growing up fast in Magnolia, Imelda learned how to want things but not how to get them.

Many families like Imelda's and many barrios like Magnolia are about to become protagonists in America's struggles with race and poverty. Latino immigrants defy basic assumptions about culture and class because they undermine the perspective that divides the nation into white and nonwhite, a perspective that is the oldest and most enduring element of America's social structure. Are Latinos white or nonwhite? There is only one correct answer, though it is often ignored: They are neither one nor the other. This is more than a matter of putting labels on people. Americans either belong to the white majority or to a nonwhite minority group. That status can determine access to social programs and political power. It decides the way people are seen and the way they see the world. White and nonwhite represent two drastically dissimilar outcomes. They constitute different ways of relating to the United States and of developing an American identity. Latinos break the mold, sometimes entering the white middle-class mainstream, often remaining as much a group apart as poor blacks.

Most European immigrants underwent a period of exclusion and poverty but eventually won acceptance to the white majority. This process of incorporation occurred across generations as the immigrants' economic contributions gained recognition and their American-born children grew up without foreign accents. Too many Latinos are poor, illegal, and dark-skinned for that path to serve as a useful model.

African-Americans traveled an even greater distance to achieve levels of material and political success unthinkable fifty years ago, but as a racial group, they remain juxtaposed to the white majority. Blacks have formally become part of the body politic, but they remain aggrieved plaintiffs. Latino immigrants lack both the historical standing and the just cause to win their place by way of struggle and petition. And these newcomers are not likely to forge an alliance with blacks, but instead, these two groups are already becoming rivals.

Neither the European ethnics nor the African-Americans were free to choose the means by which they became part of American society. Their place in this country is a product of history, and in each case it is a history of conflict. After centuries of slavery and segregation, it took the strife and idealism of the civil rights era to create a new place for African-Americans within the national identity. The Irish, the Italians, and other European ethnics had been coming here for decades but did not win full acceptance until after the Great Depression and World War II reforged and broadened the American identity to include them. Now the Latinos stand at the gate, looking for a place in American society, and the conflict that will inevitably attend their arrival is just beginning to take shape.

Latinos are different from all other immigrants past and present because they come from close by and because many come illegally. No industrialized nation has ever faced such a vast migration across a land border with the virtual certainty that it will continue to challenge the government's ability to control that border for years to come. No immigrant group has carried the stigma of illegality that now attaches itself to many Latinos. Unlike most immigrants, Latinos arrive already deeply connected to the United States. Latinos come as relations, distant relations perhaps, but familiar and connected nonetheless. They seem to know us. We seem to know them, and almost as soon as they are in the house, they become part of our bedroom arguments. They are newcomers, and yet they find their culture imbedded in the landscape of cities that have always had Spanish names, such as Los Angeles and San Antonio, or that have become largely Spanish-speaking, such as Miami and New York. They do not consider themselves strangers here because they arrive to something familiar.

They come from many different nations, many different races, yet once here they are treated like a pack of blood brothers. In the United States, they live among folk who share their names but have forgotten their language, ethnic kinsmen who are Latinos by ancestry but U.S. citizens by generations of birthright. The newcomers and the natives may share little else, but for the most part they share neighborhoods, the Magnolias, where their fates become intertwined. Mexican-Americans and Puerto Ricans account for most of the native-born Latino population. They are the U.S.-made vessel into which the new immigration flows. They have been Americans long enough to have histories, and these are sad histories of exploitation and segregation abetted by public authorities. As a result, a unique designation was born. "Hispanics" became a minority group. This identity is an inescapable aspect of the Latino immigrant experience because newcomers are automatically counted as members of the group for purposes of public policy and because the discrimination that shaped that identity persists in some segments of the American public. However, it is an awkward fit for several reasons. The historical grievances that led to minority group designation for Latinos are significant, but compared to slavery or Jim Crow segregation they are neither as well known nor as horrible. As a result, many Americans simply do not accept the idea that Latinos have special standing, and not every native Latino embraces this history as an inescapable element of self-concept. Moreover, Latinos do not carry a single immutable marker, like skin color, that reinforces group identity. Minority group status can be an important element of a Latino's identity in the United States, but it is not such a clear and powerful element of American life that it automatically carries over to Latino immigrants.

"Hispanic" has always been a sweeping designation attached to people of diverse cultures and economic conditions, different races and nationalities, and the sweep has vastly increased by the arrival of immigrants who now make up nearly 40 percent of the group. The designation applies equally to a Mexican-American whose family has been in Texas since before the Alamo and a Mexican who just crossed the Rio Grande for the first time. Minority group status was meant to be as expansive as the discrimination it had to confront. But now for the first time, this concept is being stretched to embrace both a large native Latino population with a long undeniable history of discrimination and immigrants who are just starting out here. The same is occurring with some Asian groups, but the Latino phenomenon has a far greater impact because of the numbers involved. Latino immigrants are players in the old and unresolved dilemma of race in America, and because they do not fit any of the available roles, they are a force of change.

Like all other newcomers, Latino immigrants arrive as blank slates on which their future course has yet to be written. They are moving toward that future in many directions at once, not en masse as a single cohesive group. Some remain very Latino; others become very American. Their skin comes in many different colors and shades. Some are black, and some of them can pass very readily as white. Most Latinos arrive poor. Some stay poor, many do not. Latino immigrants challenge the whole structure of social science, politics, and jurisprudence that categorizes people in terms of lifetime membership in racial or ethnic groups. The barrios do not fit into an urban landscape segregated between rich and poor, between the dependent and the taxed.

Latino immigrants come in large numbers. They come from nearby. They join fellow Latinos who are a native minority group. Many arrive poor, illegally, and with little education. Those are the major ingredients of a challenge unlike any other....

More than a third of all Latinos are younger than eighteen years old. This vast generation is growing faster than any other segment of the population. It is also failing faster. While dropout rates among Anglos and African-Americans steadily decline, they continue to rise among Latino immigrants, and mounting evidence suggests that many who arrive in their teens simply never enter American schools at all. A 1996 Rand study of census data found that high school participation rates were similarly high—better than 90 percent—for whites, blacks, and Asians, native and immigrant alike, and for native Latinos, as well. Latino immigrants, especially from Mexico, were the only group lagging far behind, with less than 75 percent of the school-age teens getting any education. Only 62 percent of the Mexican immigrant seventeen-year-olds were in school, and these young people are the fuel of U.S. population growth into the twenty-first century.

Dropout rates are only one symptom. This massive generation of young people is adapting to an America characterized by the interaction of plagues. Their new identities are being shaped by the social epidemics of youth homicides, pregnancy and drug use, the medical epidemic of AIDS, and a political epidemic of disinvestment in social services. These young Latinos need knowledge to survive in the workforce, but the only education available to them comes from public school systems that are on the brink of collapse. They are learning to become Americans in urban neighborhoods that most Americans see only in their nightmares. Imelda and a vast generation of Latino young people like her are the victims of a vicious bait and switch. The United States offered their parents opportunities. So many of the children get the plagues.

For the parents, movement to the United States almost always brings tangible success. They may be poor by U.S. standards, but they measure their accomplishments in terms of what they have left behind. By coming north, they overcome barriers of race and class that have been insuperable for centuries in Latin America. Meanwhile, the children are left on the wrong side of the barriers of race and class that are becoming ever more insuperable in the United States. With no memory of the *rancho,* they have no reason to be thankful for escaping it. They look at their parents and all they see is toil and poverty. They watch American TV, and all they see is affluence. Immigrant children learning to live in this dark new world face painful challenges but get little help. Now, on top of everything else, they are cursed by people who want to close the nation's doors against them. The effects are visible on their faces.

"I can tell by looking in their eyes how long they've been here," said the Reverend Virgil Elizondo, former rector of San Fernando Cathedral in San Antonio, Texas. "They come sparkling with hope, and the first generation finds that hope rewarded. Their children's eyes no longer sparkle. They have learned only to want jobs and money they can't have and thus to be frustrated."

The United States may not have much use now for Imelda's son, but he will be eighteen and ready to join the labor force in the second decade of the next century, just as the bulk of the baby-boom generation hits retirement age. Then, when the proportion of elderly to young workers is going out of whack, this country will have a great need for him and the other children born in the barrios, who will contribute financial sustenance in the form of their payroll deductions and other taxes. This is already an inescapable fact because of the relatively low birth rates among [the] U.S.'s whites and African-Americans for the past several decades. Women of Mexican ancestry had fertility rates three times higher than non-Hispanic women in the 1990s (and they were the least educated mothers of any group). Mexican immigrant women account for more than a quarter of all the births in California and nearly a third of the births to teenage mothers. The United States may not care about the children of the barrios, but it must start to address their problems now. If it lets them fail, there will be a great price to pay....

Given their relatively low levels of education when they arrive here, it is remarkable that most

Latino immigrants do fairly well economically in the United States. Through sheer exertion and determination they earn enough to enjoy their own version of the American dream even when that requires working an extra job at night. Millions of Latinos who came north with minimal work skills have firmly established their positions in the American economy and have found homes in suburbs. But others have suffered a far different fate. About a third of all recent Latino immigrants live below the official poverty line. More than a million and a half Mexicans who entered the country legally and illegally since 1980—43 percent of the total—were officially designated as poor in 1994. With little education and few skills, most have nowhere to start but low on the economic ladder, and in America today, people who start low tend to stay low and their children stay low as well unless they get an education....

Latino poverty will not be remedied by the welfare-to-work programs that are now virtually the sole focus of U.S. social policy, and it will not be fixed by trying to close the nation to further immigration. The Latino poor are here and they are not going to go away. Unless new avenues of upward mobility open up for Latino immigrants and their children, the size of America's underclass will quickly double and in the course of a generation it will double again. That second generation will be different than the first. It will not only suffer the economic and political disenfranchisement that plagues poor blacks today but it will also be cut off from the American mainstream in even more profound and dangerous ways....

The latest wave of immigrants has come to the United States only to find the ladder broken. Their arrival has coincided with changes in the structure of the U.S. economy that make the old three-generation formula obsolete. The middle rungs of the ladder, which allowed for a gradual transition into American life, are more precarious because so many jobs disappeared along with the industrial economy of smokestacks and assembly lines. In addition, the wages paid at the bottom of the labor force have declined in value since the early 1980s....

Starting at the bottom has usually been an immigrant's fate, but this takes on a new meaning in an increasingly immobile and stratified society. Skills and education have come to mark a great divide in the U.S. workforce, and the gap is growing ever broader. The entire population is being divided into a two-tier workforce, with a college education as the price of admission to the upper tier. In the new knowledge-based economy, people with knowledge prosper. People without it remain poor. These divisions have the makings of a new class system because this kind of economic status is virtually hereditary. Very few Latino immigrants arrive with enough education to make it into the upper tier of the workforce. Their children, like the children of all poor people, face the greatest economic pressures to drop out and find work. When they do stay in school, the education they receive is, for the most part, poor....

Fears often reflect preexisting conditions in the mind of the victim, and fear of foreigners is no different. Immigrants served as emblems for perils that had already begun to gnaw away at this country's sense of confidence. The seemingly unregulated flow of people struck many Americans as another irrational product of feckless Washington. The immigrants themselves were seen as unworthy beneficiaries of American largesse, arriving unbidden to take advantage of jobs, welfare programs, and much else. Because they are nonwhite and because Hispanic civil rights groups had pushed relentlessly for more open admissions, Latino immigrants also became associated in the minds of some whites with the era of minority-group activism and fears of "reverse discrimination." The case with which illegal aliens flaunted border controls haunted those who believed that the United States exists in a world full of unworthy but vexing adversaries....

Devising effective immigration control is an important challenge because without a credible immigration policy the American people are unlikely to make the kind of effort necessary to ensure the successful integration of Latino immigrants and their children. Illegal immigration and high dropout rates in barrio schools may seem like unrelated problems, but in fact it will be difficult to muster the political will and the resources necessary to head off the looming crisis in the barrios without first gaining control of the borders. Over the next few decades, despite efforts to close the nation's doors, immigrants will continue to come and, along with the millions already here, many will form a new class of outsiders. No one knows where these new people are supposed to fit into American society, and yet their story has become an American one....

Latinos are a people in motion. Coming from many different places, they are headed in many different directions, and it is the recent immigrants who travel the farthest and the fastest. America was changing when they got here, and they became part of that change. The new Latino immigration is the story of people struggling to adapt to an economy undergoing a prolonged and profound transformation. It is the story of communities trying to find their place in a society suffering confusion and complacency. Now, because of their energy and their numbers, Latino immigrants are helping determine where an era of change will take the nation.

Latinos are rapidly becoming the nation's largest minority group at a time when that term is quickly losing its meaning. Latino immigration can prompt the creation of a new civil rights framework that distinguishes between two distinct tasks—redressing the effects of past discrimination and providing protection against new forms of bias—and undertakes both tasks aggressively.

Latinos are also rapidly adding to the ranks of the working poor at a time when the nation is redefining the role of its lower classes. The divisions between rich and poor, between the knowledgeable and the unskilled, grow greater even as a broad political consensus favors reducing services and benefits for the poor. Understanding recent Latino immigrants, however, involves appreciating a very distinct kind of poverty. The ambition and optimism of the Latino poor could sour in the future, especially if the second generation gets nowhere, but in the meantime Latino immigration offers this country a chance to revise its attitudes toward the poor. Understanding the poverty of hard work will carry Americans beyond the common misperception that the poor are no more than an unsightly appendage to an affluent society. Instead they will be viewed as an integral part of the larger whole, one that must have opportunities to escape poverty in order for the whole to prosper.

These changes can occur, however, only if Latinos alter some attitudes of their own. Long-term residents of the barrios—natives and immigrants alike—must realize that they, more than anyone else, suffer the ill effects of illegal immigration and that it is in their self-interest to turn illegals away from their communities. Latinos must also take a new approach to language. Instead of preserving Spanish as a way to redress past grievances with the education system, English-language training should be pursued as a means of securing a successful future in a new land.

Finally, Latino immigration will cause the United States to rethink the connection between the issues of race and poverty. For too long, the two have been linked in an easy but false equation that renders the problems of the poor as the problems of African-Americans and vice versa. This constitutes a form of prejudice and, like all prejudice, it is blinding. The arrival of Latino immigrants tangibly breaks the connection.

Addressing these challenges will require a cohesion and purposefulness that the United States has sorely missed for many years. By their numbers alone, the Latinos will require the country to find a place for them. Along the way, there is a chance that America might find itself again.

In the meantime, they will keep coming.

DISCUSSION QUESTIONS

1. What are some of the reasons that so many Mexicans come to the United States? Normally, we should be able to identify "push factors" from the society of origin and "pull factors" from the society of destination.
2. What are Americans afraid of as far as Mexican immigration is concerned?

3. Imagine yourself to be Imelda living in the United States. What kind of inner conflicts would you face?
4. In what ways do you believe Mexican-Americans are in a different position than other minorities in the United States?
5. Suppose you were living in poverty in Mexico. Suppose the only way you could get your family to the United States would be to come here illegally. Would you try? Explain your answer.
6. What is the future going to be, given the large numbers of Mexican-Americans immigrating to the United States? What will have to change, and how will it change?
7. Do you believe that racism against Mexican-Americans is a problem in the United States?

15

Fences and Neighbors: Segregation in 21st-Century America

JOHN E. FARLEY AND GREGORY D. SQUIRES

The Four Questions

1. What is the problem?
2. Is it a *social* problem?
3. What are the causes of the problem?
4. What ways do Farley and Squires identify for dealing with the problem?

Topics Covered

Racial inequality

Segregation

Discrimination

Index of dissimilarity

Fair housing

Costs of segregation

Community Reinvestment Act

"Do the kids in the neighborhood play hockey or basketball?"

—ANONYMOUS HOME INSURANCE AGENT, 2000

America became less racially segregated during the last three decades of the 20th century according to the 2000 census. Yet, despite this progress, despite the Fair Housing Act, signed 35 years ago, and despite popular impressions to the contrary racial minorities still routinely encounter discrimination in their efforts to rent, buy, finance, or insure a home. The U.S. Department of Housing and Urban Development (HUD) estimates that more than 2 million incidents of unlawful discrimination occur each year. Research indicates that blacks and Hispanics encounter discrimination in one out of every five contacts with a real estate or rental agent.

SOURCE: From John E. Farley and Gregory D. Squires, "Fences and Neighbors: Segregation in 21st-Century America," in *Contexts,* Vol. 4, No. 1, 33–39. © 2005 by the American Socialogical Association. Permession by the University of California Press.

African Americans, in particular, continue to live in segregated neighborhoods in exceptionally high numbers.

What is new is that fair-housing and community-development groups are successfully using antidiscrimination laws to mount a movement for fair and equal access to housing. Discrimination is less common than just ten years ago; minorities are moving into the suburbs, and overall levels of segregation have gone down. Yet resistance to fair housing and racial integration persists and occurs today in forms that are more subtle and harder to detect. Still, emerging coalitions using new tools are shattering many traditional barriers to equal opportunity in urban housing markets.

SEGREGATION: DECLINING BUT NOT DISAPPEARING

Although segregation has declined in recent years, it persists at high levels, and for some minority groups it has actually increased. Social scientists use a variety of measures to indicate how segregated two groups are from each other. The most widely used measure is the index of dissimilarity (labeled *D* in Figure 15.1) which varies from 0 for a perfectly integrated city to 100 for total segregation.[1] Values of *D* in the 60s or higher generally represent high levels of segregation.

Although African Americans have long been and continue to be the most segregated group, they are notably more likely to live in integrated neighborhoods than they were a generation ago. For the past three decades, the average level of segregation between African Americans and whites has been falling, declining by about ten points on the *D* scale between 1970 and 1980 and another ten between 1980 and 2000. But these figures overstate the extent to which blacks have been integrated into white or racially mixed neighborhoods. Part of the statistical trend simply has to do with how the census counts "metropolitan areas." Between 1970 and 2000, many small—and typically integrated—areas "graduated" into the metropolitan category, which helped to bring down the national statistics on segregation. More significantly, segregation has declined most rapidly in the southern and western parts of the United States, but cities in these areas, especially the West, also tend to have fewer African Americans. At the same time, in large northern areas with many African-American residents, integration has progressed slowly. For example, metropolitan areas like New York, Chicago, Detroit, Milwaukee, Newark, and Gary all had segregation scores in the 80s as late as 2000. Where African Americans are concentrated most heavily, segregation scores have declined the least. As Figure 15.1 shows, in places with the highest proportions of black population, segregation decreased least between 1980 and 2000. Desegregation has been slowest precisely in the places

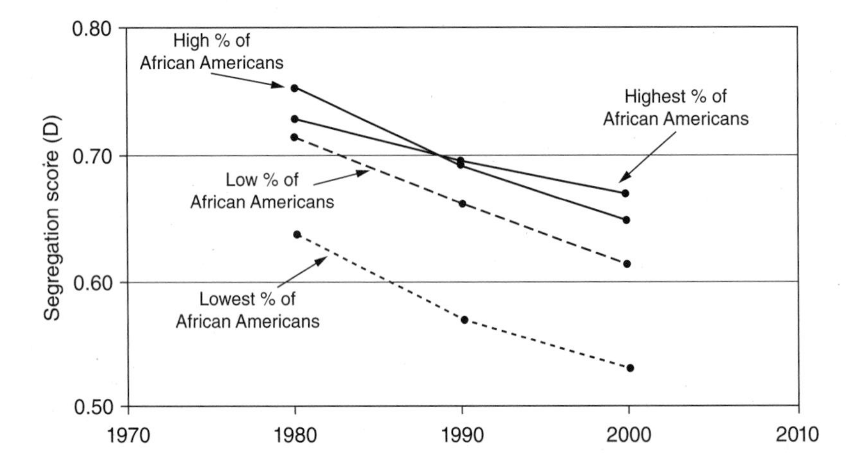

FIGURE 15.1 Declining segregation of African Americans in four groups of metropolitan areas.

SOURCE: U.S. Census Bureau, *Housing Patterns—Racial and Ethnic Residential Segregation in the United States: 1980–2000*. Online http://www.census.gov/hhes/www/housing/resseg/tab5-2.html

African Americans are most likely to live. There, racial isolation can be extreme. For example, in the Chicago, Detroit, and Cleveland metropolitan areas, most African Americans live in census tracts (roughly, neighborhoods) where more than 90 percent of the residents are black and fewer than 6 percent are white.

Other minority groups, notably Hispanics and Asian Americans, generally live in less segregated neighborhoods. Segregation scores for Hispanics have generally been in the low 50s over the past three decades, and for Asian Americans and Pacific Islanders, scores have been in the low 40s. Native Americans who live in urban areas also are not very segregated from whites (scores in the 30s), but two-thirds of the Native Americans who live in rural areas (about 40 percent of their total population) live on segregated reservations. Although no other minority group faces the extreme segregation in housing that African Americans do, other groups face segregation of varying levels and have not seen a significant downward trend.

CAUSES OF CONTINUING SEGREGATION

Popular explanations for segregation point to income differences and to people's preferences for living among their "own kind." These are, at best, limited explanations. Black-white segregation clearly cannot be explained by differences in income, education, or employment alone. Researchers have found that white and black households at all levels of income, education, and occupational status are nearly as segregated as are whites and blacks overall. However, this is not the case for other minority groups. Hispanics with higher incomes live in more integrated communities than Hispanics with lower incomes. Middle-class Asian Americans are more suburbanized and less segregated than middle-class African Americans. For example, as Chinese Americans became more upwardly mobile, they moved away from the Chinatowns where so many had once lived. But middle-class blacks, who have made similar gains in income and prestige, find it much more difficult to buy homes in integrated neighborhoods. For example, in 2000 in the New York metropolitan area, African Americans with incomes averaging above $60,000 lived in neighborhoods that were about 57 percent black and less than 15 percent non-Hispanic white—a difference of only about 6 percentage points from the average for low-income blacks.

Preferences, especially those of whites, provide some explanation for these patterns. Several surveys have asked whites, African Americans, and in some cases Hispanics and Asian Americans about their preferences concerning the racial mix of their neighborhoods. A common technique is to show survey respondents cards displaying sketches of houses that are colored in to represent neighborhoods of varying degrees of integration. Interviewers then ask the respondents how willing they would be to live in the different sorts of neighborhoods. These surveys show, quite consistently, that the first choice of most African Americans is a neighborhood with about an equal mix of black and white households. The first choice of whites, on the other hand, is a neighborhood with a large white majority. Among all racial and ethnic groups, African Americans are the most disfavored "other" with regard to preferences for neighborhood racial and ethnic composition. Survey research also shows that whites are more hesitant to move into hypothetical neighborhoods with large African-American populations, even if those communities are described as having good schools, low crime rates, and other amenities. However, they are much less hesitant about moving into areas with significant Latino, Asian, or other minority populations.

Why whites prefer homogeneous neighborhoods is the subject of some debate. According to some research, many whites automatically assume that neighborhoods with many blacks have poor schools, much crime, and few stores; these whites are not necessarily responding to the presence of blacks per se. Black neighborhoods are simply assumed to be "bad neighborhoods" and are avoided as a result. Other research indicates that "poor schools" and "crime" are sometimes code words for racial prejudice and excuses that whites use to avoid African Americans.

These preferences promote segregation. Recent research in several cities, including Atlanta, Detroit, and Los Angeles, shows that whites who prefer predominantly white neighborhoods tend to live in such neighborhoods, clearly implying that if white preferences would change, integration would increase. Such attitudes also imply tolerance, if not encouragement, of discriminatory practices on the part of real estate agents, mortgage lenders, property insurers, and other providers of housing services.

HOUSING DISCRIMINATION: HOW COMMON IS IT TODAY?

When the insurance agent quoted at the beginning of this article was asked by one of his supervisors whether the kids in the neighborhood played hockey or basketball, he was not denying a home insurance policy to a particular black family because of race. However, he was trying to learn about the racial composition of the neighborhood in order to help market his policies. The mental map he was drawing is just as effective in discriminating as the maps commonly used in the past that literally had red lines marking neighborhoods—typically minority or poor—considered ineligible for home insurance or mortgage loans.

Researchers with HUD, the Urban Institute, and dozens of nonprofit fair housing organizations have long used "paired testing" to measure the pervasiveness of housing discrimination—and more recently in mortgage lending and home insurance. In a paired test, two people visit or contact a real estate, rental, home-finance, or insurance office. Testers provide agents with identical housing preferences and relevant financial data (income, savings, credit history). The only difference between the testers is their race or ethnicity. The testers make identical applications and report back on the responses they get. (Similar studies have exposed discrimination in employment; see "Is Job Discrimination Dead?" *Contexts,* Summer 2002.) Discrimination can take several forms: having to wait longer than whites for a meeting; being told about fewer units or otherwise being given less information; being steered to neighborhoods where residents are disproportionately of the applicant's race or ethnicity; facing higher deposit or down-payment requirements and other costs; or simply being told that a unit, loan, or policy is not available, when it is available to the white tester.

In 1989 and 2000, HUD and the Urban Institute, a research organization, conducted nationwide paired testing of discrimination in housing. They found generally less discrimination against African Americans and Hispanics in 2000 than in 1989, except for Hispanic renters (see Figure 15.2). Nevertheless, discrimination still occurred during

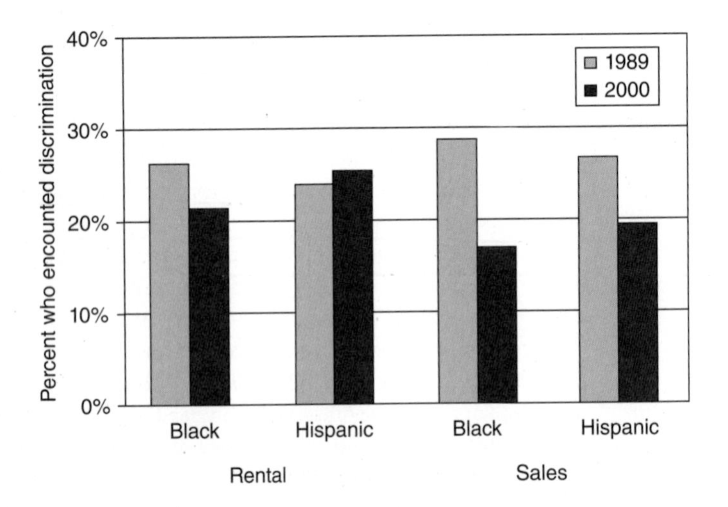

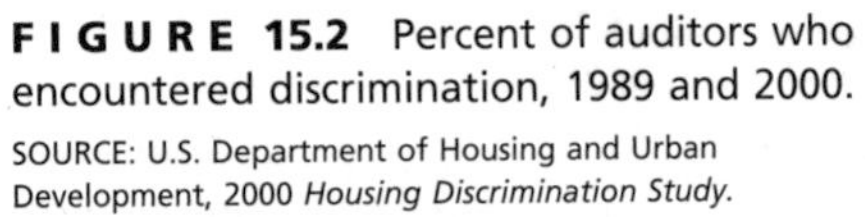
FIGURE 15.2 Percent of auditors who encountered discrimination, 1989 and 2000.

SOURCE: U.S. Department of Housing and Urban Development, 2000 *Housing Discrimination Study.*

17 to 26 percent of the occasions when African Americans and Hispanics visited a rental office or real-estate agent. (The researchers found similar levels of discrimination against Asians and Native Americans in 2000; these groups were not studied in 1989.)

In 2000, subtler forms of discrimination, such as invidious comments by real estate agents, remained widespread. Even when whites and non-whites were shown houses in the same areas, agents often steered white homeseekers to segregated neighborhoods with remarks such as "Black people do live around here, but it has not gotten bad yet"; "That area is full of Hispanics and blacks that don't know how to keep clean"; or "(This area) is very mixed. You probably wouldn't like it because of the income you and your husband make. I don't want to sound prejudiced."

Given the potential sanctions available under current law, including six- and seven-figure compensatory and punitive damage awards for victims, it seems surprising that an agent would choose to make such comments. However, research shows that most Americans are unfamiliar with fair housing rules, and even those who are familiar and believe they have experienced racial discrimination rarely take legal action because they do not believe anything would come of it. Most real estate professionals do comply with fair housing laws, but those who work in small neighborhoods and rely on word of mouth to get clients often fear losing business if they allow minorities into a neighborhood where local residents would not welcome them. In a 2004 study of a St. Louis suburb, a rental agent pointed out that there were no "dark" people in the neighborhood to a white tester. She said that she had had to lie to a black homeseeker and say that a unit was unavailable because she would have been "run out of" the suburb had she rented to a black family.

Discrimination does not end with the housing search. Case studies of mortgage lending and property insurance practices have also revealed discriminatory treatment against minorities. White borrowers are offered more choice in loan products, higher loan amounts, and more advice than minority borrowers. The Boston Federal Reserve Bank found that even among equally qualified borrowers in its region applications from African Americans were 60 percent more likely to be rejected than those submitted by whites. Other paired-testing studies from around the country conclude that whites are more likely to be offered home insurance policies, offered lower prices and more coverage, and given more assistance than African Americans or Hispanics.

THE CONTINUING COSTS OF SEGREGATION

Beyond constricting their freedom of choice, segregation deprives minority families of access to quality schools, jobs, health care, public services, and private amenities such as restaurants, theatres, and quality retail stores. Residential segregation also undercuts families' efforts to accumulate wealth through the appreciation of real estate values by restricting their ability both to purchase their own homes and to sell their homes to the largest and wealthiest group in the population, non-Hispanic whites. Just 46 percent of African Americans owned their own homes in 2000, compared to 72 percent of non-Hispanic whites. In addition, recent research found that the average value of single-family homes in predominantly white neighborhoods in the 100 largest metropolitan areas with significant minority populations was $196,000 compared to $184,000 in integrated communities and $104,000 in predominantly minority communities. As a result of the differing home values and appreciation, the typical white homeowner has $58,000 in home equity compared to $18,000 for the typical black homeowner. Segregation has broader effects on the quality of neighborhoods to which minorities can gain access. In 2000, the average white household with an income above $60,000 had neighbors in that same income bracket. But black and Hispanic households with incomes above $60,000 had neighbors with an average income of under $50,000. In effect, they lived in poorer neighborhoods, and this gap has widened since 1990.

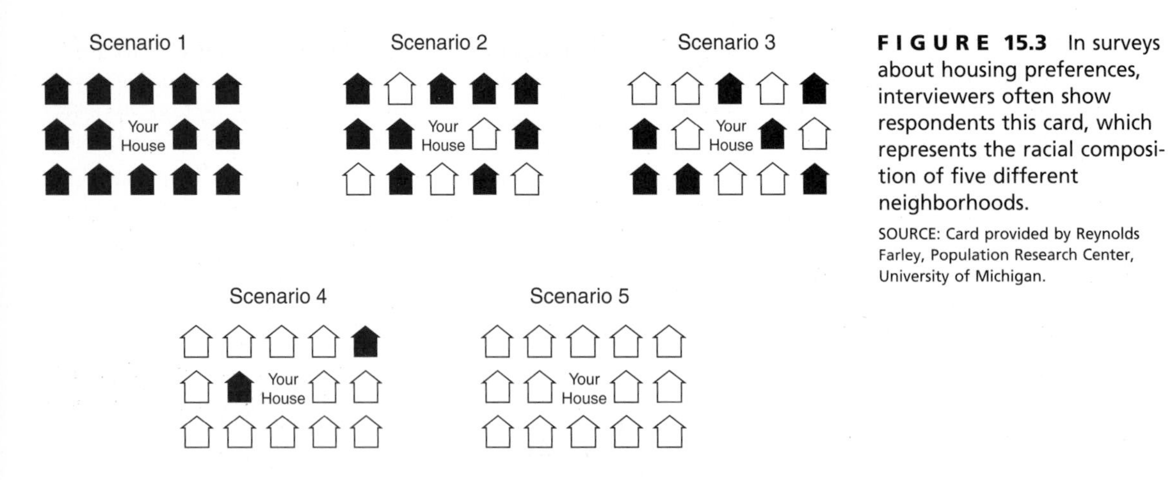

FIGURE 15.3 In surveys about housing preferences, interviewers often show respondents this card, which represents the racial composition of five different neighborhoods.

SOURCE: Card provided by Reynolds Farley, Population Research Center, University of Michigan.

Segregation restricts access to jobs and to quality schools by concentrating African Americans and Hispanics in central cities, when job growth and better schools are found in the suburbs. Amy Stuart Wells and Robert Crain found, for example, that black children living in St. Louis who attend schools in the suburbs are more likely to graduate and to go on to college than those attending city schools. Yet only busing makes it possible for these students to attend suburban schools, and America has largely turned away from this remedy to segregation. According to research by Gary Orfield at the Harvard Civil Rights Project, our nation's schools are as segregated today as they were 35 years ago. Most job growth also occurs in suburban areas, and difficulty in finding and commuting to those jobs contributes to high unemployment rates among African Americans and Latinos.

The risks of illness and injury from infectious diseases and environmental hazards are also greater in minority neighborhoods, while resources to deal with them are less available than in mostly white areas. For example, in Bethesda, Md., a wealthy and predominantly white suburb of Washington, D. C, there is one pediatrician for every 400 residents compared to one for every 3,700 residents in Washington's predominantly minority and poor southeast neighborhoods. As John Logan has argued, "The housing market and discrimination sort people into different neighborhoods, which in turn shape residents' lives—and deaths. Bluntly put, some neighborhoods are likely to kill you." (See "Life and Death in the City," *Contexts,* Spring 2003).

Finally, segregation helps perpetuate prejudice, stereotypes, and racial tension. Several recent studies show that neighborhood-level contact between whites and African Americans reduces prejudice and increases acceptance of diversity. Yet with today's levels of housing segregation, few whites and blacks get the opportunity for such contact. More diverse communities generally exhibit greater tolerance and a richer lifestyle—culturally and economically—for all residents.

A GROWING MOVEMENT

In 1968, the U.S. Supreme Court ruled that racial discrimination in housing was illegal, characterizing it as "a relic of slavery." In the same year, Congress passed the Fair Housing Act, providing specific penalties for housing discrimination along with mechanisms for addressing individual complaints of discrimination. These legal developments laid the groundwork for a growing social movement against segregation that has brought limited but real gains. Members of the National Fair Housing Alliance, a consortium of 80 nonprofit fair housing organizations in 30 cities and the District of Columbia, have secured more than $190 million for victims of housing discrimination since 1990 by using the Federal Fair Housing Act and equivalent state and local laws. In addition, they have negotiated legal settlements that have transformed the marketing and underwriting activities of the nation's largest property insurance companies, including State Farm, Allstate, Nationwide, American Family, and Liberty. The key investigative technique members of the alliance have used to secure these victories is paired testing.

Community reinvestment groups have secured more than $1.7 trillion in new mortgage and small business loans for traditionally underserved low- and moderate-income neighborhoods and minority markets since the passage of the Community Reinvestment Act (CRA). The CRA was passed in order to prevent lenders from refusing to make loans, or making loans more difficult to get, in older urban communities, neighborhoods where racial minorities are often concentrated. Under the CRA, third parties (usually community-based organizations) can formally challenge lender applications or requests by lenders to make changes in their business operations. Regulators who are authorized to approve lender applications have, in some cases, required the lender to respond to the concerns raised by the challenging party prior to approving the request. In some cases, just the threat of making such challenges has provided leverage for community organizations in their efforts to negotiate reinvestment agreements with lenders. Community groups have used this process to generate billions of new dollars for lending in low-income and minority markets. Sometimes, in anticipation of such a challenge, lenders negotiate a reinvestment program in advance. For example, shortly after Bank One and JP Morgan Chase announced their intent to merge in 2004, the lenders entered into an agreement with the Chicago Reinvestment Alliance, a coalition of Chicago-area neighborhood organizations. The banks agreed to invest an average of $80 million in community development loans for each of the next six years. Research by the Joint Center for Housing Studies at Harvard University indicates that mortgage loans became far more accessible in low-income and minority neighborhoods during the 1990s and that the CRA directly contributed to this outcome.

Many housing researchers and fair-housing advocates have criticized fair-housing enforcement authorities for relying too heavily on individual complaints and lawsuits to attack what are deeper structural problems. Currently, most testing and enforcement occurs when individuals lodge a complaint against a business rather than as a strategic effort to target large companies that regularly practice discrimination. Reinvestment agreements recently negotiated by community groups and lenders illustrate one more systemic approach. More testing aimed at detecting what are referred to as "patterns and practices" of discrimination by large developers and rental management companies would also be helpful. Such an undertaking, however, would require more resources, which are currently unavailable. Despite the limits of current enforcement efforts, most observers credit these efforts with helping to reduce segregation and discrimination.

Resistance to fair housing and integration efforts persists. For example, lenders and their trade associations continually attempt to weaken the CRA and related fair-housing rules. Yet fair-housing and community-reinvestment groups like the National Fair Housing Alliance and the National Community Reinvestment Coalition have successfully blocked most such efforts in Congress and among bank regulators. As more groups refine their ability to employ legal tools like the CRA and to litigate complex cases under the jurisdiction of the Fair Housing Act, we can expect further progress. The struggle for fair housing is a difficult one, but with the available tools, the progress we have made since 1970 toward becoming a more integrated society should continue.

ENDNOTE

1. *Segregation* refers to the residential separation of racial and ethnic groups in different neighborhoods within metropolitan areas. When a metropolitan area is highly segregated people tend to live in neighborhoods with others of their own group, away from different groups. The index of dissimilarity (*D*), a measure of segregation between any two groups, ranges from 0 for perfect integration to 100 for total segregation. For segregation between whites and blacks (imagining, for the sake of the example, that these were the only two groups), a *D* of 0 indicates that the racial composition of each neighborhood in that metropolitan area is the same as that of the entire area. If the metropolitan area was 70 percent white and 30 percent black, each neighborhood would reflect those percentages. A *D* of 100 would indicate that every neighborhood in the metropolitan area was either 100 percent white or 100 percent black. In real metropolitan areas, *D* always falls somewhere between those extremes. For example, the Chicago metropolitan area is 58 percent non-Hispanic white and 19 percent non-Hispanic black. Chicago's *D* was 80.8 in 2000. This means that 81 percent of the white or black population would have to move to another census tract in order to have a *D* of 0, or complete integration. On the other hand, in 2000 the Raleigh-Durham, N.C. metropolitan area, which is 67 percent non-Hispanic white and 23 percent non-Hispanic black, had a *D* of 46.2—a little more than half that of Chicago.

RECOMMENDED RESOURCES

John Iceland and Daniel H. Weinberg, with Erika Steinmetz. "Racial and Ethnic Residential Segregation in the United States: 1980–2000." *Census 2000 Special Reports*, 2002. Online, http://www.census.gov/hhes/www/housing/resseg/front_toc.html. This study summarizes housing segregation in the year 2000 and trends in segregation since 1980.

Keith R. Ihlanfeldt and Benjamin Scafidi. "Whites' Neighborhood Preferences and Neighborhood Racial Composition in the United States: Evidence from the Multi-City Study of Urban Inequality." *Housing Studies* 19 (2004): 325–359. The authors demonstrate whites' preference for predominantly white neighborhoods and how that promotes segregation in three metropolitan areas.

John R. Logan, Brian J. Stults, and Reynolds Farley. "Segregation of Minorities in the Metropolis: Two Decades of Change." *Demography* 41. (2004): 1–22. This study reviews patterns of desegregation since 1980. One critical finding is that black-white segregation did not decline more in areas where income gaps were reduced than in other areas.

Douglas S. Massey and Nancy Denton. *American Apartheid: Segregation and the Making of the Underclass* (Harvard University Press, 1993). The most thorough and comprehensive account of racial housing segregation in the United States.

Margery Austin Turner, Stephen L. Ross, George C. Galster, and John Yinger. *Discrimination in Metropolitan Housing Markets: National Results from Phase I HDS 2000. Final Report* (U.S. Department of Housing and Urban Development, 2002). Online. http://www.huduser.org/Publications/pdf/Phasel _Report.pdf. This report presents findings from the most recent national study of discrimination in the sale and rental of housing.

DISCUSSION QUESTIONS

1. What exactly is segregation?
2. Is segregation a social problem? If no, then why? If segregation is a social problem, who is harmed by it, and who is responsible for its continuation?
3. Do you believe that fair housing laws are really fair to the sellers?
4. How would you feel if you were discriminated against in housing for being nonwhite? Should

whites care? Should the government get involved?

5. Segregation also occurs by "subtle discrimination," according to the authors. What is this? Do you believe it occurs? Will it always occur?

6. The criticism many make against ending segregation is that people should have the right to live among their own kind. Are they right, and if so, why? If you disagree with their criticism, then what can a democratic society do to encourage residential desegregation.

16

Group Justice in the 21st Century

AMY L. WAX

The Four Questions

1. What is the problem identified by Wax?
2. Is this a *social* problem? Is this problem caused by social factors? Who does this problem affect?
3. What is the cause of this problem?
4. What can be done about this problem?

Topics Covered

Racial capital

Educational deficits

Employment deficits

Family Structure deficits

Racial Inequality

Much of the last century saw steady improvement in blacks' educational achievement, earnings, and job status as reflected in the growth of the black middle class.[1] In the past few decades, that progress has slowed or leveled off, with significant gaps remaining and even widening on some measures. As a result, black Americans continue to lag behind other racial and ethnic groups in indicators of social and economic well-being....

DEFICITS IN EDUCATION

From the 1950s until the mid 1980s, blacks posted gradual but steady gains in both educational attainment (as reflected in years of schooling completed) and educational achievement (as measured by test scores) relative to whites nationwide.[2] In the past two decades, that trend has faltered, with progress slowing significantly or ceasing entirely on many measures of educational success.[3] For example, blacks' math achievement test scores remained steady after the late 1980s; in reading, black students made modest gains from 1975 to the mid 1980s, but then lost ground. Disparities between black and white school-age children in these basic skills remain great, with black children in fourth to twelfth grades testing between 0.8 and 1 standard deviation behind whites on the National Assessment of Educational Progress (NAEP) in this

SOURCE: From Amy L. Wax, *Race, Wrongs, and Remedies: Group Justice in the 21st Century*. New York: Roman & Littlefield Publishers, 2009, pp. 47–93. Reprinted by permission.

decade. As summarized by economist Derek Neal, "overall, black-white math and reading gaps in 2004 among 9- to 13-year-olds are quite similar to the gaps observed in the late 1980s."[4] Disparities by race among school-age children persist into the teenage years, with recent SAT scores revealing a large and widening gap. In 2003, the average combined scores for incoming college freshmen (who represent the most able group among all test-takers) were 1063 for whites and 1083 for Asians, as compared with 857 for blacks and 905 for Mexican-Americans. The overall 206-point white-black gap in 2003 is up from 187 points in 1993.[5] A substantial difference in the scores of blacks and whites, representing one standard deviation, is also found on the Armed Forces Qualification Test (AFQT), a test of cognitive skills administered to military recruits.[6] Average score differences mask more pronounced disparities among top scorers, with little or no recent improvement in black under-representation among the most able contingent of test-takers.[7] Blacks also lag in educational attainment. In a study conducted in 2000, 34 percent of whites received at least a bachelor's degree, as compared with 17 percent of blacks.[8]

Why have gaps stopped closing or even widened, and why do significant disparities remain? The potential explanations divide roughly into two types: those that look to external factors independent of, and beyond the immediate control of, the students and families—what social scientists call exogenous factors—and those that, although influenced by external circumstances, are caught up with the behavioral choices of students and their parents. These are sometimes designated as endogenous factors because the end-points being measured (such as student learning, achievement, or completed years of schooling) are not solely the product of external forces. Rather, they are substantially within the control of the subjects themselves.

Economists and sociologists have long linked racial disparities in educational outcomes to socioeconomic factors such as parental education and resources. In general, students from more affluent and higher-status families outscore those who are poor, and black families are on average less affluent and educated than others. The precise mechanism through which family background differences produce disparate student achievement is contested, with some researchers emphasizing family resources and others pointing to child-rearing practices that correlate with income and education. But regardless of how the observed gaps actually arise, it is now well documented that differences in family income and education fail to explain most of the black-white achievement gap. Significant racial disparities exist even among students with similar economic backgrounds. Black students from families earning more than $70,000 per year have lower SAT scores than white students from families with less than $30,000 in income. White students with parents who are high school graduates outscore black students whose parents attended or graduated from college. These data indicate that a significant portion of blacks' poor educational performance is not due to poverty or parents' lack of education.[9]

Likewise, the theory that observed racial differences stem from institutional failures, such as poor schools, unequal educational spending, or lack of funds for higher education, does not hold up well despite voluminous attempts to link these variables with deficits in learning. School quality is often invoked to explain the race achievement gap. This is a slippery concept that can encompass a range of factors, including caliber of instruction, expenditures per pupil, and student socioeconomic background, that are sometimes hard to measure and that operate through a variety of mechanisms. In particular, it is important to recognize that a school's quality is not wholly a function of exogenous variables. Rather, it depends critically on endogenous elements, such as the characteristics, behavior, and education-related attitudes of the students themselves. These very factors are notoriously resistant to outside manipulation.

There is evidence that, in some limited respects, the schools that blacks attend are not quite as good as those in which whites are enrolled. In general, however, measured differences are small and of uncertain or undocumented importance to outcomes. Overall, there is little reason to believe that black students' schools are significantly inferior on the parameters that have been demonstrated to matter to quality of education. Along traditional dimensions of school quality (such as class size, student-teacher ratios, curriculum, and computer resources), the schools that blacks and whites attend became more similar during the 1980s and 1990s and now differ little on average.[10] To be sure, expenditures per pupil tend to track average family income, which varies by district. These resource differences mean that schools in districts with many poor (and black) students tend to be somewhat more crowded, older, less well maintained, and less varied in course offerings.[11] Nonetheless, some majority black city districts spend heavily on their public schools, so expenditures per pupil are similar by race overall, with the average per student amount for whites actually lower than for nonwhites.[12] In general, however, the relationship between spending and learning is not well established, and research has failed to demonstrate a clear link.[13] Between 1972 and 1992, when increased state-level spending significantly reduced interdistrict disparities, race gaps in test scores consistently failed to narrow.[14]

Likewise, key factors that affect the classroom experience do not seem to explain black-white differences in attainment. From the 1970s through the 1990s, teacher quality (as measured by teachers' educational background, quality of schools attended, and qualifications for subject of teaching) deteriorated across the board in all school districts, regardless of affluence and racial composition. Although poor and nonwhite districts improved relative to whiter and wealthier districts in teacher-to-student ratios, they continued to lag somewhat in teacher training and experience.[15] The differences were not large, however, and none of these variables has been reliably linked to student outcomes.

Researchers have also sought to examine other institutional variables that might bear on learning but are ordinarily not captured in standard studies of school quality. They have attempted for decades to investigate the claim that low teacher expectations depress black student achievement. The theory is that low expectations create "self-fulfilling prophecies": If teachers expect black students to do poorly, students respond by failing to achieve and learn. But the literature fails to demonstrate that self-fulfilling prophecies depress black achievement.[16] In the few studies suggesting that lower expectations might influence minority children, measured effects are "small, fragile, and fleeting."[17] In addition, most research in this area suffers from the serious methodological flaw of failing to distinguish between low teacher expectations as a cause of black students' poor performance or as a response to those students' actual underachievement.[18]

Yet another hypothesis attributes low black achievement to a race mismatch of students and teachers: Students do poorly because they lack role models or because their teachers don't understand or respond to their learning style. There is little support for this view. Although one recent Tennessee study detects somewhat better performance by students with same-race teachers, the authors themselves admit that the effects have "at best been limited and qualified."[19] Overall, empirical studies have found little or no evidence that being taught by black teachers raises black student achievement.

In a similar vein, achievement differences by race are frequently blamed on the effects of "stereotype threat." The theory is that black students fall short because they are afraid tests and other performance measures will be used to confirm negative views about their group's abilities. Despite much publicity, the impact of stereotype threat has been greatly exaggerated. There is currently no evidence that the effects attributed to stereotype threat account for more than a negligible portion of the overall black-white gap in achievement and test scores.[20]

In sum, the data reveal that the schools black children attend lag a bit on some measures of school quality, such as teacher experience, physical condition of the school, and number of advanced or enrichment courses, but are better on others, including student-teacher ratios and spending per student. The measured effects are not large in either direction. The key questions are whether these observed disparities are the driving force behind racial gaps in achievement and whether there is reason to believe that erasing these differences (if that could be achieved) would eliminate existing discrepancies in student performance.

The critical issue is whether the observed minor differences in schooling quality are causing the present significant gaps in educational achievement. For teacher training and certification, for example, one study found that predominantly minority schools have 7.6 percent fewer teachers with master's degrees than schools with mostly nonminority students.[21] But the significance of even that small disparity is questionable: Most studies show little or no connection between teacher certification and student learning. Similarly, teachers in predominantly black schools tend to have somewhat less teaching experience. But the effect of small differences in experience has not been well established.[22] Although there is evidence that teachers do vary in qualities that affect student learning, research also shows that measurable credentials (such as possession of a master's degree or years of teaching) are poor proxies for good teaching. Rather, there is general agreement that direct classroom monitoring is needed to assess teacher quality.[23] But reliable, large-scale, observational information about what teachers actually do in their classrooms is not currently available, so there is as yet no basis for attributing the black-white achievement gap to poor teaching.

DEFICITS IN EMPLOYMENT

Blacks and whites differ in employment rates, earnings, and occupational attainment. Although racial gaps in workplace success narrowed after World War II and the enactment of civil rights legislation, progress has slowed or stalled in recent decades. Discrimination is often cited as an important cause of lingering disparities. Yet, social scientists and economists have increasingly turned their attention to so-called supply-side factors—those bearing on job performance and the development of human capital. The list of proposed sources of such deficits includes resource constraints (such as insufficient funds for higher education and underfinancing of schools), private endowments and inputs (such as inadequate parental investments in children), and personal behaviors (such as criminal activity, lack of academic self-development, and choices reflecting poor socialization or work ethic).

A consensus has now emerged in labor economics that supply-side deficits account for almost all remaining racial disparities in employment outcomes. Much of the racial gap in jobs and earnings can be traced to differences in learning and academic achievement. A recent summary states unequivocally that "black-white skills gaps are the driving force behind black-white differences in labor market outcomes among adults."[24] Investigations of the contribution to job outcomes of various human capital inputs—including cognitive skill, educational achievement, and actual learning—yield a remarkably consistent picture: Personal and behavioral attributes related to productivity are by far the most important predictors of job market success, regardless of race.[25] Most studies reveal that blacks and whites with similar levels of cognitive skill (as measured by tests of intelligence and educational aptitude) and educational attainment (as measured by quantity and quality of education, basic math and reading proficiency, and overall learning) receive similar job market rewards, including salaries and promotions.[26] Where unexplained racial disparities persist, they are typically small and fail to apply across the board. Although the data do not rule out the possibility that these residual differences are due to old-fashioned discrimination, other factors—such as poor job networking, geographic mismatch, and disparities in degree and type of work experience—could also explain the observed results.[27] Moreover, whereas less-educated black men lag a bit behind their white counterparts in earnings, black male college graduates earn as much or more than white college graduates of similar aptitude. All in all, there is little evidence that observed job market patterns are due to discrimination. Skill, ability, experience, and productivity count far more.

Poorer job outcomes and the expectation of job market discrimination are sometimes cited as a cause of lower educational achievement in blacks. One contention is that blacks' failure to invest in education is a rational strategy because employers can be expected to undervalue their skills.[28] The data show that this logic is flawed. Because "the relationship between basic skills and eventual earnings is stronger among black men than white men," and because earnings grow faster for blacks than whites as educational level increases, blacks have every reason to stay in school and invest in self-development.[29] Although black men may harbor and act on the belief that they are discriminated against in the job market, the evidence reveals that this reaction is unjustified and dysfunctional. Failing to pursue an education is not a rational response to actual patterns of labor market rewards.

To be sure, the econometric findings do not mean that racial discrimination has altogether disappeared. There are some data suggesting that employers are leery of less-educated blacks, especially men. For example, employers tend to hire white over black ex-convicts[30] and to disfavor job candidates with black-sounding names.[31] Studies with job testers reveal a mild preference in some markets for white job candidates over blacks with similar objective credentials.[32] In general, however, studies like these involve small samples of subjects. In contrast, econometric analyses showing that discrimination is minimal examine large population trends. In addition, the effects attributed to discrimination, even in these smaller-scale studies, are modest. They often barely reach statistical significance.

The patterns of residual racial disparity are often suggestive of so-called rational or statistical discrimination.[33] Statistical discrimination represents an employer's response to real average gaps in performance between members of different groups. When groups differ systematically in productivity-relateds skills or behaviors and it is difficult or expensive to observe individual characteristics directly, employers may use group attributes such as race as a proxy for job-related traits. If the average black male performs less well than the average white with apparently similar credentials, employers may "read between the lines" to devalue all black male employees, regardless of individual ability to do the job.

Because easily measured qualifications, such as years of schooling, are not entirely reliable indicators of actual knowledge or skill, statistical discrimination can be a cost-effective, albeit imperfect, strategy for screening prospective employees. Paper credentials sometimes mask real racial differences relevant to job performance. For example, blacks graduate from high school with markedly lower average reading and math test scores than whites.[34] This evidence suggests that, for jobs that draw on academic ability, blacks are on average less qualified than whites with similar years of education. There may also be group-related differences in noncognitive or "soft" attributes—such as social skills, work-related attitudes, or elements of social background (including family structure and stability of personal relationships)—that may influence or correlate with productivity but are not always revealed by conventional screening methods. Although there is some evidence of racial differences along these lines, behavioral contrasts of this type can be hard to demonstrate systematically.[35] Nonetheless, many employers harbor the perception that soft skills are not uniform by race.[36] Such perceptions may lead employers to disfavor blacks generally or to be wary of distinct subgroups, such as black ex-convicts or those with black-sounding names. The situation is complicated by the fact that race in our society correlates with social or economic background, which, regardless of paper credentials, is linked with job-related behaviors. In fact, persons with distinctly black names tend to come from less educated and less affluent homes, and there is evidence that employers are sensitive to this.[37]

Statistical discrimination, when and if it occurs, does pose an obstacle to employment by making the struggle for economic success even harder for blacks. Nonetheless, the impact of this phenomenon, by any measure, is small. The data fail to reveal a significant role for this sort of rational discrimination or for discrimination of any kind. Even if some racial differences in job outcomes remain unexplained, it is far from clear that race-based discrimination, as opposed to other factors, account for them. Employers may not be using race at all. Rather, they may be responding *directly* to so-called "unobservable" or "hidden" variables—that is, real differences in credentials or productivity-related attributes that are apparent to managers or supervisors and that correlate with race but are difficult for social scientists to assess. In any event, investigations of employer bias focus almost exclusively on the screening and hiring stages. There is virtually no hard evidence that race-based generalizations affect decisions about pay raises or job advancement. And there is even less reason to believe that statistical discrimination would come into play for those decisions because evaluations of existing employees are based on a richer body of information, including direct observations of on-the-job performance, than is available at initial hiring.

DEFICITS IN FAMILY STRUCTURE AND HOME ENVIRONMENT

The black family has long been in disarray. Compared with whites and other races, blacks marry less often, divorce more, and have far higher rates of extramarital childbearing.[38] Yet the situation was not always so dire. Earlier in the twentieth century, black marriage rates were high and extramarital childbearing was the exception. That did not last. Although the postwar years have seen periods of black progress on other fronts, the black family has steadily deteriorated.

The growth in family structure differences by race has occurred against a background of divergence by education and economic class.[39] Although marriage rates have declined for all sociodemographic groups, the trend is most pronounced among the poor and least skilled. Class differences are particularly stark among whites, with the affluent and well educated still marrying and staying married at high rates. In contrast, marriage has become less common among blacks in every social class, with the steepest decline among poorer blacks.[40] Overall, marriage rates are significantly lower, and divorce and extramarital birth rates higher, for blacks than for whites at all levels of education and income.

One consequence of these patterns is that relatively fewer black adults and children reap the benefits of living in stable, conventional nuclear families. In particular, the decline in marriage among blacks has important implications for the setting in which children grow up. Today there are many more black than white children living with only one parent. Although fractured and fatherless families prevail among the poor, they are also common among middle-class blacks. Overall, more than two-thirds of black children are born out of wedlock. Multiple-partner fertility—or the practice of having children by more than one partner—is significantly more common among blacks than other American ethnic groups, and more often involves extramarital children.[41] In contrast, about one-quarter of white children are born to unmarried mothers, with almost all such births in this group confined to low-income and less-educated women. Very few white women with a college degree bear children outside of marriage, and relatively few become single mothers through divorce.[42]

Do blacks' low rate of marriage and higher rate of out-of-wedlock childbearing matter? These patterns have important consequences because family structure is linked to deficits that hold blacks back. A growing body of research shows that children who grow up with single or unmarried parents are less well-off on many measures. In addition to having lower educational achievement and completing fewer years of schooling, they experience more behavioral and psychological problems throughout life and have less stable adult relationships.[43] This is partly due to fewer resources: Single parent and nonmarital families have less money than families headed by married couples. But these effects are also observed when parents are matched for income and education. That is, growing up without two married parents in itself produces worse outcomes.[44] The relatively high rates of divorce and cohabitation among blacks also add to risks for children. Recent research suggests that children do best when raised by married, biological parents and that children in blended or step-parent families fare no better than children raised by comparable single or divorced parents.[45] Households with biologically unrelated males are particularly detrimental for children.[46]

Although social scientists disagree on how much black family fragility contributes to educational and labor market disparities by race, the consensus is growing that its influence is significant. Building on prior work by economist James Heckman, Derek Neal identifies black-white differences in early childhood experiences as a critical cause of persistent racial gaps in labor market outcomes.[47] Family structure is thought to affect investments in children. These investments in turn contribute to cognitive as well as noncognitive attributes that are crucial to educational attainment and job success.[48] Black family structure and reproductive patterns also harm children indirectly by weakening neighborhoods and communities. Responsible, married fathers are exceptional, and resident biological fathers are uncommon even among the black middle class. The absence of fathers undermines the supervision and proper socialization of children, and loosely attached adult males create a potentially disruptive presence.

In sum, the evidence suggests that the continuing deterioration of the black family has contributed significantly to widening disparities in economic position and human capital development between blacks and other racial groups. Nonetheless, fractured families are only one source of racial gaps in human capital development that bear on economic success. Cultural differences in parenting practices also appear to play role.[49] Although black children on average come from poorer and less-educated homes, significant racial differences in children's school readiness and subsequent performance are observed even for families with similar income, parental education, and family structure. In attempting to account for these differences, social scientists have investigated such factors as home environment and routines, parental behavior, children's activities, and family attitudes toward learning. Findings suggest that young black children are exposed to lower levels of cognitive and emotional stimulation than whites and Asians, even in families matched for income, education, and IQ. Black children watch more TV, read fewer books, are taken on educational outings less often, and are subjected to more erratic routines. Developmental research on these elements of upbringing, although not definitive, suggests they are important predictors of lifelong outcomes.[50] Although measured differences by race are reflected most immediately in school readiness and performance, they have potential consequences beyond the childhood years for adult social adjustment and occupational success.

The conclusion that home environment and private social life exert a pivotal influence finds additional support in a growing body of literature on the relationship between noncognitive attributes and life outcomes. Although achievement in school is important to economic position, the evidence suggests that neither educational attainment nor subsequent occupational success depends exclusively on cognitive skills. Also crucial are noncognitive aspects of personality and character, including industriousness, perseverance, trustworthiness, honesty, cooperativeness, agreeableness, conscientiousness, and future orientation.[51] These traits predict school success and also translate into the "soft skills" that employers seek. These same characteristics may be important to staying out of trouble with the law and to the ability to sustain an orderly, harmonious and stable family life.

There is some evidence for a race gap in these noncognitive attributes. Black students outscore whites, even controlling for socioeconomic background, on some measures of anti-social behaviors.[52] In one study, a preemployment screening personality test administered by a large American company revealed differences by race, with blacks scoring lower on two key noncognitive attributes, agreeableness and conscientiousness, that the employer at issue deemed desirable.[53]

Why might these types of traits be unevenly distributed? There has been relatively little attention paid to date to the origins of those noncognitive skills that are most crucial to life success. More research is needed to identify the attributes, habits, inclinations, and behaviors that contribute to effective performance and to determine how best to foster them. Existing evidence suggests, however, that families are vital to proper socialization. Private, informal institutions have proved most effective in inculcating these characteristics.[54]

Across the board, the social science reveals that observed racial differentials in educational attainment, economic position, and labor market outcomes no longer can be ascribed to the "usual suspects." First and foremost, there is no straightforward relationship between present, ongoing discrimination based on race and observed patterns, and thus little reason to believe that eliminating such discrimination will work a significant change. This does not entail denying a connection between current patterns and past oppression. But the fact that persistent shortfalls are the outgrowth of past mistreatment tells us little about how current inequalities can effectively be erased....

ENDNOTES

1. See Stephen Thernstrom and Abigail Thernstrom, *America in Black and White* (New York: Simon & Schuster, 1997).
2. For a detailed account of trends among elementary and high school students, see Derek Neal, "Why Has Black-White Skill Convergence Stopped?" in *The Handbook of Economics of Education*, Vol. I, Eric. A. Hanushek and Finis Welch, eds., (Amsterdam: North-Holland, 2006). See also Derek Neal, "Black-White Labour Market Inequality in the United States," in the *New Palgrave Dictionary of Economics*, 2007; Melanie Phillips and Christopher Jencks, *The Black-White Test Score Gap* (Washington, D.C.: Brookings Institution Press, 1998); Kathryn Neckerman, ed., *Social Inequality* (New York: Russell Sage Foundation Press, 2004), 468–473. For a review of black educational trends and attainment since the 1950s, see Orley Ashenfelter, ed., Special Issue on Brown vs. Board of Education, *American Law and Economics Review* 8 (Summer 2006).
3. Meredith Phillips and Tiffani Chin, "School Inequality: What Do We Know?" in *Social Inequality*, Kathryn Neckerman, ed., (New York: Russell Sage Foundation Press, 2004), 468–473. See also Alan Krueger, Jesse Rothstein and Sarah Turner, "Race, Income and College in 25 years: Evaluating Justice O'Connor's Conjecture," *American Law and Economics Review* 8 (Summer 2006): 285.
4. Neal, "Black-White Labour Market Inequality in the United States" Racial gaps are not confined to scores in reading and math but affect other important areas of learning. Recent testing of basic science knowledge in eighth graders nationwide showed black students lagging well behind whites even within the same school districts. See Diana Jean Schemo, "Most Students in Big Cities Lag Badly in Basic Science," *The New York Times,* 16 November 2006, A22.
5. See June Kronholz, "SAT Scores Are Highest Since 1974, "*The Wall Street Journal,* 27 August 2003, A2.
6. Melanie Phillips and Christopher Jencks, *The Black-White Test Score Gap*, 482.
7. See Krueger, Rothstein, and Turner, "Race, Income and College in 25 Years," documenting the paucity of blacks among top scorers on the SATs. A black student is about six times more likely than a white student to finish below the fifth percentile in a standardized math test but only about one-twentieth as likely as a white student to finish above the ninety-fifth percentile. See Phillips and Jencks, *The Black-White Test Score Gap*, 158–159, 174–175. Twenty percent of white twelfth-graders were rated as proficient or advanced on the math section of the 2000 NAEP, as compared with fewer than five percent of blacks. See Abigail Thernstrom and Stephan Thernstrom, *No Excuses: Closing the Racial Gap in Learning* (New York: Simon & Schuster, 2003), 16.
8. Phillips and Chin, "School Inequality: What Do We Know?," at 471.
9. For data on test scores by race, education, and income, see sources cited in notes 6 & 7 *supra.*
10. See Sean Corcoran et al., "The Changing Distribution of Education Finance, 1972–1997" in *Social Inequality*, ed. Kathryn Neckerman, ed., (New York:

Russell Sage Foundation Press, 1998), 433–465; Phillips and Chin, "School Inequality: What Do We Know?" 467–519, reviewing data on school spending and quality from the 1980s and 1990s.
On disparities in school and teacher quality, see also Richard Rothstein, *Class and Schools: Using Social, Economic, and Educational Reform to Close the Black-White Achievement Gap* (Washington, DC: Economic Policy Institute, New York: Columbia University Teachers College, 2004), 100–102.

11. See, Corcoran et al., "The Changing Distribution of Education Finance, 1972–1997," 433–465; Phillips and Chin, "School Inequality: What Do We Know?' " 467–519.
12. Phillips and Chin, "School Inequality: What Do We Know?," 510. See also Corcoran et al., "The Changing Distribution of Education Finance, 1972–1997," 440. The ratio of spending per white student to spending per nonwhite student was 1.02 in 1972, 0.99 in 1982, and 0.97 in 1992.
13. See, e.g., Michael A. Rebell, "Poverty, 'Meaningful' Educational Opportunity, and the Necessary Role of the Courts," *North Carolina Law Review* 85 (June 2007), 1467–1543, at 1479–1482 (reviewing the debate over whether "money matters"); Michael Heise, "Litigated Learning, Law's Limits, and Urban School Reform Challenges," *North Carolina Law Review* 85 (June 2007), 1419–1446, (presenting evidence against the connection between education funding levels and differential achievement); see also id. at 1446–1450 (observing that urban public schools, including those with mostly black students, have been comparatively well-funded in recent decades).
14. Corcoran et al., "The Changing Distribution of Education Finance, 1972–1997," 440–442.
15. Phillips and Chin, "School Inequality: What Do We Know?," 508–510.
16. Lee Jussim and Kent D. Harber, "Teacher Expectations and Self-Fulfilling Prophecies: Knowns and Unknowns, Resolved and Unresolved Controversies," *Personality and Social Psychology Review* 9, 2 (2005): 131–155; see also Lee Jussim, Jacquelynne Eccles, Stephanie Madon, "Social Perception, Social Stereotypes, and Teacher Expectations: Accuracy and the Quest for the Powerful Self-Fulfilling Prophecy" *Advances in Experimental Social Psychology* 28 (1996); 281–288.
17. Jussim and Harber, "Teacher Expectations," 151.
18. *Ibid.*
19. Thomas S. Dee, "Teachers, Race, and Student Achievement in a Randomized Experiment," *Review of Economics and Statistics* 86 (2004): 195. Race matching may actually be detrimental. Minority teachers tend to lag on average in other attributes—such as knowledge of subject matter and measures of cognitive ability—that have been shown to matter to student learning. The tension between race-matching and quality teaching was noted in the 1960s by the sociologist James Coleman. See Linda Seebach, Editorial: "Dissenters Often Offer Better Glimpse of the Truth," *Contra Costa Times* 16 Feb. 1997, A19, stating that "In the 1970s, when sociologist James Coleman was doing research of inequalities of educational opportunity, he discovered two things: that children's success was strongly affected by their teachers' verbal ability as revealed on standardized tests and that black teachers, many of them trained in segregated schools, did badly on such tests. But the potential implications of the idea that black children might do just as well or better with white teachers were simply unthinkable, and Coleman and his colleagues did not pursue it. Looking back on the episode, Coleman admitted it could be true that 'we aided in the sacrifice of educational opportunity for many children, most of whom were black, to protect the careers of black teachers.' "); Chuck Forrester, Editorial: "Merger Model Could Work with the Right Elements," *Greensboro News and Record* 26 May 1991, B3 stating that "[s]ome years ago the state of North Carolina did a survey to determine how good students are produced. To everyone's surprise class size and per pupil expenditures mattered little, if at all. The quality of teachers was found to be most important. James Coleman had found the same thing in 1964: a direct link between teachers' vocabularies and student achievement.").
20. See Paul R. Sackett et al., "On Interpreting Stereotype Threat as Accounting for African American-White Differences on Cognitive Tests," *American Psychologist* 59, 1 (January 2004): 7–13; Amy L. Wax, "The Threat in the Air," *The Wall Street Journal*, 13 April 2004, A20. For a comprehensive critique of stereotype threat (ST) as an explanation for performance differences by race and gender, see Amy L. Wax, "Stereotype Threat: A Case of Overclaim Syndrome?" in Christina Hoff

Sommers, ed., *The Science on Women in Science* (AEI Press, forthcoming April 2009).

21. See Phillips and Chin, "School Inequality: What Do We Know?," 476–479.
22. See Phillips and Chin, "School Inequality: What Do We Know?" 476–479. See also Rothstein, *Class and Schools*; Thomas J. Kane, Jonah Rockoff, and Douglas Staiger, "What Does Certification Tell Us About Teacher Effectiveness? Evidence from New York City," unpublished paper. http://papers.nber.org/papers/w12155.pdf.
23. See Kane, Rockoff, and Staiger. Kane and colleagues claimed to find a link between directly observed teacher quality and student test scores in a study conducted in New York City. They did not look specifically at whether minority children had poorer teachers, however. There is some modest evidence for a relationship between a teacher's cognitive ability (as measured, for example, by IQ tests) and knowledge (as reflected in quality of undergraduate education) and student outcomes. But there is no consistent, reliable data on whether, in general, minority students have teachers who are less learned or less smart. See, e.g., Rob Greenwald, Larry V. Hedges, and Richard D. Laine, "The Effect of School Resources on Student Achievement," *Review of Educational Research* 66 (1996): 361–396; Linda Darling-Hammond, *Teacher Quality and Student Achievement: A Review of State Policy Evidence* (Seattle: Center for the Study of Teaching and Policy, 1999).
24. Derek Neal, "Black-White Labour Market Inequality in the United States." There is evidence that blacks on average fall short of whites in performance on the job. See Patrick F. McKay and Michael A. McDaniel, "A Re-examination of Black-White Mean Differences in Work Performance: More Data, More Moderarors," *J. of Applied Psych. 91* (2006): 538–554 presenting data on "mean racial disparities in work performance." As the authors of this paper note, these observations raise the question of "which aspects of performance underlie these disparities." Id. At 538. Although the answer is complex, the research suggests that differences in work-related skills and abilities account for almost all performance gaps. See also Devah Pager, "The Use of Field Experiments for Studies of Employment Discrimination: Contributions, Critiques, and Directions for the Future," *Annals of the American Academy of Political and Social Sciences* 609 (January 2007): 108 noting numerous "influential studies" showing that "when relevant individual characteristics—in particular, cognitive ability—have been accounted for, racial disparities in wages among young men narrow substantially or disappear"; James Heckman, "Detecting Discrimination," *J. of Econ. Perspectives* 12 (1998): 101 noting that "[m]ost of the disparity in earnings between blacks and whites in the labor market of the 1990s is due to differences in skills they bring to the market, and not to discrimination within the labor market," and describing employment discrimination as "the problem of an earlier era."
25. Derek Neal, "Why Has Black-White Skill Convergence Stopped?" 512–576; see also June O'Neill, "The Role of Human Capital in Earnings Differences Between Black and White Men," *The Journal of Economic Perspectives* 4,4 (Autumn 1990): 25–45; George Farkas et al. "Cognitive Skill, Skill Demands of Jobs, and Earnings Among Young European American, African American, and Mexican American Workers," *Social Forces* 75 (1997): 913; George Farkas and Keven Vicknair, "Appropriate Tests of Racial Wage Discrimination Require Controls for Cognitive Skill: Comment on Cancio, Evans, and Maume," *American Sociological Review* 61 (August 1996): 557–660; Derek Neal and William Johnson, "The Role of Pre-market Factors in Black-White Wage Differences." *J. of Political Economy* 104 (1996): 869–895; Barry T. Hirsch and David A. Macpherson, "Wages, Sorting on Skill, and the Racial Composition of Jobs," *Journal of Labor Economics* 22, 1 (2004): 189–210; James Heckman and Dmitri Masterov, "Labor Market Discrimination and Racial Differences in Premarket Factors," in *Handbook of Research on Employment Discrimination, Rights and Realities*, Laura Beth Nelson and Robert Nelson, eds. (New York: Springer, 2005).
26. Derek Neal, "Why Has Black-White Skill Convergence Stopped?"; Richard J. Murnane, John B. Willet, and Frank Levy, "The Growing Importance of Cognitive Skills in Wage Determination," *Review of Economics and Statistics* 77 (1995): 251–266.
27. William R. Johnson and Derek Neal, "Basic Skills and the Black-White Earnings Gap," in *The Black-White Test Score Gap*, Christopher Jencks and

Meredith Phillips, eds. (Washington, DC.: Brookings Institution Press, 1998), 480–497.

28. See Stephen Coate and Glenn C. Loury, "Will Affirmative-Action Policies Eliminate Negative Stereotypes?" *American Economic Review* 83 (1993): 1220–1240.
29. Johnson and Neal, "Basic Skills and the Black-White Earnings Gap." 495.
30. Devah Pager and Lincoln Quillian, "Walking the Talk? What Employers Say Versus What They Do," *American Sociological Review* 70 (2005): 355–380. See also Devah Pager, *Marked: Race, Crime, and Finding Work in an Era of Mass Incarceration* (Chicago: University of Chicago Press, 2007).
31. Marianne Bertrand and Sendhil Mullainathan, "Are Emily and Greg More Employable Than Lakisha and Jamal? A Field Experiment on Labor Market Discrimination," *The American Economic Review* 94, 4 (September 2004): 991–1013.
32. P. A. Riach and J. Rich, "Field Experiments of Discrimination in the Market Place," *The Economics Journal* 112 (November 2002): F480–F518.
33. On rational discrimination, see Richard Epstein, *Forbidden Grounds: The Case Against Employment Discrimination Laws* (Cambridge: Harvard University Press, 1992). See also Gary Becker, *The Economics of Discrimination* (Chicago: University of Chicago Press, 1971).
34. See Sam Dillon, "Schools Slow in Closing Gaps between Races," *The New York Times,* 20 November 2006, A1, noting that "African American and Hispanic students in high school can read and do arithmetic at only the average level of whites in junior high school"; Rene Sanchez, "Colleges Compete for Minority Students by Helping Them Achieve," *The Washington Post,* 28 Dec. 1996, A1, explaining that "In California, for example, only 5 percent of black graduates from public schools in the state, and only 4 percent of Hispanic graduates, now meet the university system's admissions standards. That compares with 13 percent of whites."
35. See notes 51 and 53, this chapter, infra.
36. See Joleyn Kirschenman and Kathryn Neckerman, "We'd Love to Hire Them, But…" in *The Urban Underclass,* Christopher Jencks and Paul E. Peterson, eds. (Washington, DC.: The Brookings institution, 1992); Elijah Anderson, *Code of the Streets: Decency, Violence, and the Moral Life of the Inner City* (New York: Norton, 1999); Tanya Mohn, "Sometimes the Right Approach Is Putting the Best Face Forward," *The New York Times,* 7 May 2006, sec. 10, p. 1 (noting that some inner-city black men develop the habit of wearing a "game face," or menacing expression, that employers find intimidating).
37. Roland G. Fryer, Jr. and Steven D. Levitt, "The Causes and Consequences of Distinctively Black Names" *The Quarterly Journal of Economics* 119, 3 (August 2004): 767–805, explaining negative employer responses to black names as a reaction to those names' association with lower social class, which is in turn linked to lower job productivity.
38. David T. Ellwood and Jonathan Crane, "Family Change Among Black Americans: What Do We Know?" *Journal of Economic Perspectives* 4, 4 (Autumn 1990): 65. For a comprehensive overview of the most recent data on family structure by race and class, see U.S. Census Bureau, Current Population Survey (CPS) Reports, *America's Families and Living Arrangements:* 2008, at http://www.census.gov/population/www/socdemo/hh-fam.html.
39. David T. Ellwood and Christopher Jencks, "The Uneven Spread of Single-Parent Families: What Do We Know? Where Do We Look for Answers?" in *Social Inequality,* Kathryn M. Neckerman, ed. (New York: Russell Sage Foundation, 2004), 3–77; David T. Ellwood and Christopher Jencks, "The Spread of Single-Parent Families in the United States Since 1960," in Daniel P. Moynihan, Timothy M. Smeeding, and Lee Rainwater, eds., *The Future of the Family* (New York: Russell Sage Foundation, 2004), 25–65; Sara McLanahan, "Diverging Destinies: How Children Are Faring Under the Second Demographic Transition," *Demography* 41, 4 (November 2004); 607–627. For a review of recent trends in family structure and possible explanations, see Amy L. Wax, "Engines of Inequality: Class, Race, and Family Structure," *Family Law Quarterly* (Fall 2007), 567–599.
40. In 1960, 80 percent of black women and 66 percent of black men aged 20–34 were married at least once. By 1990, those figures had declined to 46 percent and 38 percent, respectively. See Robert G. Wood, "Marriage Rates and Marriageable Men: A Test of the Wilson Hypothesis," *Journal of Human Resources* 30 (1995): 163.

41. See Cassandra Logan et al., *Men Who Father Children with More Than One Woman: A Contemporary Portrait of Multiple Partner Fertility* (Washington, D.C.: Childtrends Publication (2006–10, 2006), showing that black men are twice as likely as white men to have children by more than one woman. See also Maria Cancian and Daniel R. Meyer, "The Economic Circumstances of Fathers with Children on W-2," *FOCUS* 22 (Summer 2002), 19, 21–23; Karen B. Guzzo and Frank F. Furstenberg, Jr., "Multi-partnered Fertility Among American Men," *Demography* 44 (August 2007), 583–601 (presenting data that black men are significantly more likely than others to have children by more than one woman). For an overview of developments within the black family, see Kay S. Hymowitz, *Marriage and Caste in America* (Chicago: Ivan R. Dee, 2006).
42. See Sara McLanahan, "Diverging Destinies: How Children Are Faring Under the Second Demographic Transition," *Demography* 41, 4 (November 2004): 607–626, noting that the out-of-wedlock childbearing rate to white college-educated women in the 1990s remained well under 5 percent. See also Steven P. Martin, "Growing Evidence for a 'Divorce Divide'? Education and Marital Dissolution Rates in the U.S. Since the 1970s," *Demography Research* 15 (2006): 537–560, noting declining divorce rates among well-educated whites.
43. See Kristin Anderson Moore et al., "Marriage from a Child's Perspective: How Does Family Structure Affect Children?" *Childtrends Research Brief* (2002); Wendy Sigle-Rushton and Sara McLanahan, "Father Absence and Child Well-Being: A Critical Review," in *The Future of the Family*, Daniel Patrick Moynihan et al., eds. (New York: Russell Sage Foundation. 2004), 116–155; Abigail Thernstrom and Stephan Thernstrom, *No Excuses: Closing the Racial Gap in Learning* (New York: Simon & Schuster, 2003), 132; McLanahan, "Diverging Destinies," 607–627.
44. McLanahan, "Diverging Destinies," 607–627.
45. See Kathryn Edin and Maria Kefalas, *Promises I Can Keep: Why Poor Women Put Motherhood Before Marriage* (Berkeley: University of California Press, 2005), 215, noting that *"living apart from either biological parent at any point during childhood is what seems to hurt children"* (emphasis in original). See also Donna K. Ginther and Robert A. Pollak, "Family Structure and Children's Educational Outcomes: Blended Families, Stylized Facts, and Descriptive Regressions," *Demography* 41 (November 2004): 671–696; Sandra Hofferth, "Residential Father Family Type and Child Well-Being: Investment Versus Selection," *Demography* 43 (Feb. 2006): 53–77. For a recent summary of the social science evidence on family structure and child well-being to date, see Kristin Anderson Moore et al., "Marriage from a Child's Perspective: How Does Family Structure Affect Children and What Can We Do About It?" *Child Trends Research Brief* (June 2002).
46. Patricia Schnitzer and Bernard Ewigman, "Child Deaths Resulting from Inflicted Injuries: Household Risk Factors and Perpetrator Characteristics," *Pediatrics* 116 (November 2005): e687–e689.
47. See Derek Neal, "Why Has Black-White Skill Convergence Stopped?" 512–576.
48. See Valerie E. Lee and David T. Burkam, *Inequality at the Starting Gate: Social Background Differences in Achievement as Children Begin School* (Washington, D.C.: Economic Policy Institute, 2002); Petra Todd, "The Production of Cognitive Achievement in Children: Home, School and Racial Test Score Gaps," unpublished ms., at http://www.econ.brown.edu/econ/events/revpaper.pdf.
49. See Valerie E. Lee and David T. Burkam, *Inequality at the Starting Gate.*
50. *Ibid.*
51. For noncognitive attributes as predictors of subsequent outcomes on a broad range of social indicators, see James J. Heckman, Jora Stixrud, and Sergio Urzua, "The Effect of Cognitive and Noncognitive Abilities on Labor Market Outcomes and Social Behavior," *Journal of Labor Economics* 24, 3 (2006): 411–482; James Heckman and Yona Rubinstein, "The Importance of Noncognitive Skills: Lessons from the GED Testing Program," *American Economic Review* 91 (2001): 145–149; See also Angela Lee Duckworth, *Intelligence Is Not Enough: Non-IQ Predictors of Achievement* (unpublished doctoral dissertation, chapter 3, University of Pennsylvania); Angela L. Duckworth, Christopher Peterson, Michael D. Matthews and Dennis R. Kelly, "Grit: Perseverance and Passion for Long-Term Goals," *Journal of Personality and Social Psychology* 92 (2007): 1087–1101.
52. Richard Rothstein, *Class and Schools*, 100–102.

53. See David Autor and David Scarborough, "Will Job Testing Harm Minority Workers?" *Quarterly Journal of Economics* 123, 1 (Feb. 2008): 219–77 presenting data that reveal group differences on personality tests of job-related attributes like agreeableness and conscientiousness. See also Lewis R. Goldberg et al., "Demographic Variables and Personality: The Effects of Gender, Age, Education, and Ethnic/Racial Status on Self-Descriptions of Personality Attributes," *Personality and Individual Differences* 24, 3 (1998): 393–403, finding some average personality differences between whites, blacks, and Hispanics in a large national sample.
54. See James J. Heckman and Alan B. Krueger, *Inequality in America: What Role for Human Capital Policies?* (Cambridge: MIT Press, 2003).

DISCUSSION QUESTIONS

1. What factors do you believe might account for the current deficits in educational attainment for African Americans when compared to whites and Asians?
2. What are the social consequences of deficits in family structure for children according to Amy Wax? How do deficits in family structure and one's home situation contribute to the broader problem of racial disparities?
3. What public policies might address the deficits to family structure?
4. Can the government, through political interventions, address deficits to family structure, (or, is this a *personal* problem that only individuals can solve)? Explain your response.

PART VI

Social Problems: Immigration

Our society is built on a long history of immigration, of welcoming people from all over the world. It was also once built on slavery and the long history of racist ideas and institutions that continue to create a number of social problems today. This section focuses primarily on the problems faced by immigrants to the United States, those who come legally, those who come illegally, and those who are a forced to come under practices of *new slavery*.

The problems described in these five readings are by no means the only quandaries faced by immigrants coming to the United States, but they represent some of the more intractable ones. The readings cover a range of topics from the history of settlement and theories of immigrant incorporation, to the racism that immigrant groups experience and, sometimes, perpetuate. This section also raises questions about the economic costs of immigration and the role of government entitlements programs in addressing the wellbeing of immigrants to the United States.

John Iceland discusses historical settlement patterns of various immigrant groups to the United States, and theories that explain the ways these groups were incorporated into the nation's culture.

Mark Krikorian's essay, from the book *The New Case Against Immigration: Both Legal and Illegal,* raises some difficult questions about the social costs of immigration on government spending, such as whether the U.S. can maintain its current level of immigration, both legal and illegal, given strained government spending on various public entitlements, such as welfare, healthcare, education, and criminal justice programs.

Kevin Bales takes on a global social problem that has a compelling link to immigration: the problem of *new slavery* in the global marketplace. He provides a gripping example of a young migrant who was brought to the United States to work as a slave. But more than providing gripping tales about the link between new slavery and the global economy, Bales shows that entrenched poverty, more than ever, is the driving force behind this social problem.

The next selection deals with the tensions faced by young Hmong Americans in a nation that remains polarized by race. *The Ideological Blackening of Hmong American Youth* is a sagacious essay that challenges the "model minority" stereotype for Asian Americans by exposing the failure of the assimilation model in explaining the immigrant experience of Hmong Americans.

At some point in their experience with their new country, immigrants will encounter racism and racist ideology. Some will arrive with perceptions of racism from their home countries, and others will come in holding to the belief that racism does not exist in the United States—that America and the American Dream is based wholly on the meritocracy ideal. The final article in this section is by Beth Frankel Merenstein, and it deals with the problem of racism: immigrants' perceptions of, and experiences with, this intractable American dilemma.

17

Immigration and Race in the United States

JOHN ICELAND

The Four Questions

1. What is the problem identified by Iceland?
2. Is this a *social* problem? Are people aware of this issue?
3. What is the cause of the problems faced by immigrants?
4. What can be done about these problems?

Topics Covered

Immigration
Theories of immigration
Settlement patterns
Assimilation paradigm
Spatial assimilation
Human capital
Ethnic disadvantage
Transnationalism
Place stratification
Segmented assimilation

Immigration has continually reshaped the racial and ethnic character of the United States. The earliest settlers from Europe to America migrated to a fairly sparsely populated land and eventually took from the Native Americans by force what they could not obtain otherwise. From the start of the colonial period in 1607, to the adoption of the Constitution in 1789, close to a million people came to the United States—about 600,000 from Europe and perhaps 300,000 Africans.[1] At the time of the first U.S. census, in 1790, about 60 percent of the population was from England, but many were also from Scotland, Germany, the Netherlands, and France. Voluntary migrants came to the colonies for a variety of religious, economic, and political reasons, such as the anticipation of economic opportunity or freedom from religious persecution.[2]

During the colonial period, migrants from England not only impressed the largest demographic stamp on the area that eventually became the United States but also exerted the greatest cultural, social, and political influence. As Philip Martin and Elizabeth Midgley note: "They [the English colonists] built communities at Jamestown and Plymouth, seized control from the Dutch in New York, and overran various French and Spanish settlements. These colonists established English as the public language and England's common law as the model for the U.S. legal system."[3]

While the English imprint was clearly the most significant in these early years, Americans developed a distinct identity over time. This identity was in large part a function of the mixing of cultures, as was noted by observers.[4] This notion of an *American* identity, then, reflects not simply the assimilation of new immigrants into an English culture but also the emergence of a culture influenced by the population's multiple origins.

At the same time, immigrants and their children retained elements of their previous cultures, if to differing degrees. Even as immigrants became acculturated to their new environment, they still often lived

SOURCE: From John Iceland, *Where We Live Now: Immigration and Race in the United States*. Berkeley, CA: University of California Press: 2009, pp. 14–30.

in distinct communities with particular institutions and economic niches. Irish immigrants in the nineteenth century, for example, were concentrated in large cities such as Boston and New York, were over-represented as common laborers and domestic servants and later on in government occupations (such as the police force), and established numerous Irish Catholic churches.[5] These twin and often competing forces in the residential sphere—assimilation and ethnic retention—are the focus of this chapter.

HISTORICAL SETTLEMENT PATTERNS

Regional Concentrations

It has long been noted that immigrants of different origins are often concentrated in different regions of the United States. For example, in their book, *Immigrant America,* Alejandro Portes and Ruben Rumbaur describe that it is hardly an accident that European immigrants first settled in states along the Atlantic seaboard, while Asian immigrants populated the Pacific Rim states, and Latin American immigrants concentrated in the Southwest.[6] Settling in areas with greater proximity to the country of origin lowered both the cost of moving to the United States and the cost of the return trip home.

In colonial times there were also distinct regional differences in settlement patterns among European immigrants. The English were particularly well represented in New England states, although they were present in large numbers in all of the colonies. Germans were concentrated in Pennsylvania, the Dutch in New York and New Jersey, and the relatively small number of Swedes in Delaware. Black slaves, who were of course mainly involuntary migrants, were overwhelmingly concentrated in the South. For example, they constituted over two-fifths of the populations of South Carolina and Virginia in 1790.[7]

In the nineteenth century the Irish tended to settle in the large cities, mainly in New England, New York, and New Jersey, though not exclusively. In contrast, German immigrants in the nineteenth century were considerably less urban and much more concentrated in the Midwest, in states such as Wisconsin and Ohio. Scandinavian immigrants flocked to midwestern and north-central states, particularly Minnesota. Polish immigrants of the late nineteenth and early twentieth centuries settled in large eastern and midwestern cities such as Chicago, New York, Pittsburgh, and Buffalo. A vast majority of both Chinese and Japanese immigrants in the nineteenth century lived along the Pacific coast, with a strong concentration in California. Mexican immigrants of the early 1900s not only settled in border states but also moved to cities in the Midwest such as Chicago, Milwaukee, and Gary, Indiana.[8]

Regional concentrations, then, stemmed not only from proximity to one's home country but also from established ethnic concentrations and the economic conditions of the communities they settled in at the time of entry.[9] Ethnic networks drew people to established communities, where compatriots helped the newcomers find housing and jobs. The growing industrial power of midwestern cities in the nineteenth and early twentieth centuries drew immigrants of many origins to those cities.

Residential Patterns within Metropolitan Areas

Examining *regional* settlement patterns is informative for a couple of reasons. Ethnic identity is more likely to be reinforced by living in a region with many members of one's own ethnic group (known as *coethnics)* than by living in one with few. In addition, when a particular group occupies an area for many generations, that area may become associated with distinct cultural traits.[10] However, there are limitations to drawing conclusions about the spatial assimilation of immigrants based on regional patterns alone. Groups may be overrepresented in different regions but still display relatively little social and cultural difference from each other, especially over time. For example, while Poles remain concentrated in the Northeast and Midwest today, they show low degrees of neighborhood-level segregation from other white ethnic groups in the metropolitan areas where they reside.[11] Conversely, other groups may live in the same metropolitan areas as one another, but may still live worlds apart. Even though blacks and whites are highly represented in many of the same metropolitan areas, neighborhood-level black-white segregation is on average quite prevalent, which is indicative of the level of racism, and quality of black-white relations more generally, in the United States.[12]

Sociologists grew increasingly interested in the social organization and functioning of communities within metropolitan areas in the early decades of the twentieth century. The neighborhood was seen as shaping the opportunities and constraints individuals faced in their everyday life and was the basis for both political participation and control.[13] In 1925, Robert Park wrote about how neighborhood residential segregation and racial and ethnic divisions more generally reinforced immigrant group solidarity:

> The isolation of the immigrant and racial colonies of the so-called ghettos and areas of population segregation tend to preserve and, where there is racial prejudice, to intensify the intimacies and solidarity of the local and neighborhood groups. Where individuals of the same race or of the same vocation live together in segregated groups, neighborhood sentiment tends to fuse together with racial antagonisms and class interests. Physical and sentimental distances reinforce each other, and the influences of local distribution of the population participate with the influences of class and race in the evolution of the social organization.[14]

Robert Park, Ernest Burgess, and other sociologists at the University of Chicago were particularly interested in how new migrants to Chicago became incorporated into the city's life. They were largely responsible for developing the *assimilation paradigm* for understanding this incorporation process. They saw assimilation as the "process of interpenetration and fusion in which persons and groups acquire the memories, sentiments, and attitudes of other persons and groups and, by sharing their experience and history, are incorporated into the mainstream of American life."[15]

These sociologists held that immigrants were likely to assimilate into the host society over the long run. In 1964 Milton Gordon provided a systematic discussion of the assimilation concept and identified seven dimensions of assimilation in his widely cited book, *Assimilation in American Life*. In his view there was a critical difference between "acculturation" and "structural assimilation." He referred to acculturation as the minority group's adoption of the cultural patterns of the host society, whereas structural assimilation was the entry of members of an ethnic minority into primary-group relationships, such as close friendships and intermarriage, with the majority group. Gordon discussed at length the importance of prejudice and discrimination in contributing to social distance between groups. Nevertheless, he argued that American-born children of immigrants were by and large "irreversibly" on the path to complete acculturation, though not necessarily structural assimilation.[16] He did, however, note that the barrier between blacks and whites was more formidable, a fact also long-recognized by other commentators....

THEORIES OF IMMIGRANT INCORPORATION AND RESIDENTIAL SEGREGATION

Three common theoretical perspectives are used to explain how immigrants and their children become incorporated into society: assimilation, ethnic disadvantage (or ethnic retention), and segmented assimilation.

Assimilation

Assimilation refers to the decline of ethnic distinctions between groups. Classic assimilation theory posits that ethnic immigrant groups experience integration with a society's majority groups through the adoption of mainstream attitudes and culture, and through educational and work experiences. Contemporary assimilation theorists emphasize that assimilation need not be a one-way street, where minority members become more like the majority group members, which in the United States consists of non-Hispanic whites. Rather, assimilation involves a general convergence of social, economic, and cultural patterns.[17]

Spatial assimilation refers to the convergence of residential patterns in particular. Upon their arrival, immigrants may be residentially segregated from other groups for a variety of reasons. Social networks, both kin and community, are key factors shaping where immigrants live.[18] Immigrants often feel more comfortable living with, and welcomed by, their fellow coethnics. In addition, the low socioeconomic status of many immigrants may mean that such individuals may simply not be able to afford to live in the same neighborhoods as more affluent whites. People with little education or few skills—attributes usually referred to as "human capital"—may be particularly dependent on their ethnic communities. In contrast, immigrants who are professionals, such as scientists and engineers, are likely to rely less on ethnic networks and more on the ties they develop with a particular employer.[19]

According to the spatial assimilation model, immigrants are more likely to move out of ethnic communities as they become acculturated with the host society and as they achieve socioeconomic upward mobility. For example, as immigrants become more familiar with local norms and as their English-language ability improves, they may become more comfortable interacting with others and living outside of their ethnic enclave. Immigrants may also become more familiar with the amenities of alternative neighborhoods, such as good schools and clean streets, and, if their own socioeconomic standing allows it, they may be more likely to move to those areas, which often contain members of other ethnic groups. The result is a dispersion of immigrant group members and desegregation over time.[20]

Richard Alba and Victor Nee, in their discussion of assimilation theory, explain how assimilation is not necessarily a universal outcome for all groups. Moreover, assimilation is a lengthy process that typically spans generations: "To the extent that assimilation occurs, it proceeds incrementally as an intergenerational process, stemming both from individuals' purposive action and from the unintended consequences of their workaday decisions. In the case of immigrants and their descendants who may not intentionally seek to assimilate, the cumulative effect of pragmatic decisions aimed at successful adaptation can give rise to changes in behavior that nevertheless leads to eventual assimilation."[21]

Spatial assimilation theory offers clear and testable hypotheses concerning the residential patterns of immigrants. Residential exposure to the majority group is expected to increase the longer immigrants are in the host country and across generations. The quality of neighborhoods of residence is likewise expected to improve with socioeconomic mobility and acculturation.[22]

There is considerable evidence that the descendants of European immigrants of the nineteenth and early twentieth centuries have largely assimilated into U.S. society. While certainly some groups remain concentrated in particular regions of the country, white ethnic groups tend to share many of the same neighborhoods within the metropolitan areas where they reside.[23]

However, just because white immigrants of the previous great wave of immigration have assimilated does not mean that post-1965 immigrants will have the same experience. Commentators have pointed to a number of differences in the conditions under which these assorted waves arrived in the United States. For example, some argue that the hiatus in immigration that began in the 1920s, following the passage of tough immigration restrictions by Congress, served to weaken ethnic communities. Ethnic enclaves no longer received new members who could replace individuals departing to outlying areas. With regard to the more recent wave of immigration, Alba and Nee note that despite the very important exception of Mexico, there has been a shift in immigrants' countries of origin ever since 1965, such as a decrease in the number of immigrants from Korea and Taiwan and increases in the number from other countries, such as El Salvador and Guatemala. Such continual shifts in flows result in a very diverse immigration stream that tends to undermine the formation of large ethnic enclaves.

Another feature of current immigration that some argue works against assimilation is the racial distinctiveness of new immigrant groups. The argument is that while the previous wave of immigrants hailed from many different countries, they were more racially homogeneous. The counterargument is that race is largely a social construct, and a reading of historical accounts indicates that many groups of the previous wave of immigration, including the Irish, Jews, and Italians, were usually perceived as racially distinct from the majority of native-born Americans. As historian Roger Daniels writes: "However curious it may seem today, by the late nineteenth century many of the 'best and brightest' minds in America had become convinced that of all the many 'races' (we would say 'ethnic groups') of Europe one alone—variously called Anglo-Saxon, Aryan, Teutonic, or Nordic—had superior innate characteristics. Often using a crude misapplication of Darwinian evolution, which substituted these various 'races' for Darwinian species, historians, political scientists, economists, and, later, eugenicists discovered that democratic political institutions had developed and could thrive only among Anglo-Saxon peoples."[24]

The notion that immigrants from different European countries constituted different races dissipated only over time as various groups achieved political, social, and economic mobility. In addition, eugenic theories that were in style at the time lost favor in the wake of Hitler's defeat in World War II, Hitler having been a champion of eugenics.[25]

There are other possible differences in the conditions immigrants face that may affect patterns of assimilation today. Some argue that high levels of income inequality will hinder the socioeconomic mobility of recent immigrants and their children. However, levels of income inequality were also high in the 1920s. Others mention differences in the ideological climate regarding ethnic diversity, such as greater support for multiculturalism and less pressure to assimilate in recent decades, which may in turn weaken the assimilation process. Some counter that the ideological climate may actually facilitate the incorporation process. The ideal of tolerating the cultural differences between groups may serve to reduce tensions and the social distance between groups over the longer run.

Still others argue that the emergence of transnationalism, where immigrant communities may span national borders, makes it less likely that immigrants will assimilate in the host country. Yet ties have long existed between immigrants in the host country and their kin and communities in the sending country. Daniels estimates that emigration from the United States to countries of origin varied considerably across groups. For example, between 30 to 50 percent of Italian immigrants returned to Italy, compared with fewer than 5 percent of Jewish immigrants. Greeks also had strong ties to Greece, and perhaps a little over half emigrated back to Greece.[26] Alba and Nee, in an extended discussion of the distinctions between eras of immigration, conclude that, while not unimportant, such differences are not as clear-cut as often thought, and likely do not undermine the predictive power of assimilation theory.[27] Others do not share this view.

Ethnic Disadvantage

Assimilation was the dominant theory of immigrant incorporation until the 1960s and 1970s, when it came under attack for empirical and political reasons. Empirically, some researchers pointed to the persistent subordination of some groups in American society—particularly African Americans. Politically, this was a time of celebration of group differences and rebellion against the politics of conformity. Assimilation theory was taken as a criticism of the unique contributions of minorities and immigrants.[28] Indeed, even the term *assimilation*—which to some implies the repudiation of one's culture and origins—has been out of fashion in the academic literature on immigrant *incorporation* or *integration*.

In contrast to assimilation theory, the ethnic disadvantage (or ethnic retention) model holds that increasing knowledge of the language of the new country and familiarity with its culture and customs often do not lead to increasing assimilation. For one thing, lingering discrimination and structural barriers to opportunities often hamper the assimilation process. Differential access to wealth, power, and privilege affect how well immigrants fare in the host country and may impede their full incorporation into American society.[29]

In research on ethnic residential patterns, the offshoot of the ethnic disadvantage theory is termed *place stratification*. This perspective likewise emphasizes that lingering prejudice and discrimination by the dominant group (non-Hispanic whites in the U.S. context) prevent neighborhood-level integration. Segregation is the tool used by whites to maintain social distance from minority groups. The effects of structural barriers are thought to be greatest for blacks in the United States because blacks have historically been perceived in the most unfavorable terms.[30]

The desire of minority group members to live with their coethnics plays a role in continuing segregation. In studies of individuals' preferences to live in neighborhoods with varying degrees of integration, sociologist Camille Charles describes how respondents of all races tend to express a desire to live near people of the same ethnicity. However, whites tend to exhibit a stronger preference than Hispanics, blacks, and Asians for same-race neighbors and would be less comfortable as a numerical minority. Since each minority group has a preference for a greater number of coethnic neighbors than most whites could tolerate in their own neighborhood, this likely often leads to "tipping" toward a majority-race makeup rather than a stable neighborhood mix. In these studies of racial preferences, whites also tend to be the most desirable out-group (group other than a certain group's own), while blacks are the least desirable, suggesting a racial rank ordering of residential preferences.[31]

Discriminatory practices in the housing market against African Americans in particular, as well as Hispanics and Asians, have been well documented.[32] As described earlier, over the years these discriminatory practices have included real-estate agents steering racial groups to certain neighborhoods, providing less information and assistance to minority home seekers, the provision of unequal access to mortgage credit, and neighbors' hostility.[33] Research has indicated both a willingness of whites to live in more integrated neighborhoods and a decline in discrimination in the housing market in recent years. Changing attitudes in society, the rising economic status of minority customers, and the continuing effect of the Fair Housing Act on the real-estate industry and the law's enforcement all likely play a role in these trends.[34] However, despite some declines in discrimination, many believe that both it and white avoidance of mixed or minority neighborhoods still play a central role in shaping the residential patterns of various ethnic groups in the United States.[35]

Segmented Assimilation

A third theory of immigrant incorporation is segmented assimilation. This perspective focuses on divergent patterns of incorporation among contemporary immigrants. It grew out of studies that found that many of the disadvantages faced by poor immigrant families were sometimes reproduced or even enhanced in the next generation. For example, the prevalence of female heads of household increased across generations among many Latin American nationality groups in the United States. This suggests that acculturation may do little to reduce disadvantage, and in some cases may exacerbate it if the ethnic group members adopt norms of the host society, such as nonmarital childbearing, that do not enhance the well-being of the next generation.[36]

According to the segmented assimilation perspective, the host society offers uneven possibilities to different immigrant groups. Recent immigrants are being absorbed by different segments of American society, ranging from affluent middle-class suburbs to impoverished inner-city ghettos. Individual- and structural-level factors play key roles in affecting the incorporation process, and there is an important interaction between the two levels. Individual-level factors include education, career aspiration, English-language ability, place of birth, age on arrival, and length of residence in the United States. Immigrants who arrive at a young age and whose parents have high levels of education and English-language ability, for example, are more likely to achieve socioeconomic success as adults and to assimilate into "mainstream" society. Children of immigrants with low levels of education who themselves arrived as teenagers are less likely to achieve the same kind of socioeconomic success. In response to a perceived lack of opportunities, immigrants may cultivate ties with their ethnic communities in order to achieve upward mobility rather than try to integrate into the mainstream.[37]

Structural-level factors that likewise affect patterns of incorporation include racial stratification and the range of economic opportunities available in a particular place at a particular time. As discussed earlier, racial discrimination may diminish the opportunities available to non-white immigrants. Increasing income inequality in the United States has also had a negative impact on the earnings of immigrants with low levels of education. Thus, according to the segmented assimilation model, we should expect to see considerable differences in residential patterns for various immigrant groups, with some groups experiencing no decline in their residential segregation from non-Hispanic whites over time, while others will more readily assimilate into the mainstream.

These three theoretical models—assimilation, ethnic disadvantage, and segmented assimilation—provide us with alternative ways of understanding immigrant incorporation in the United States. The following chapters review the evidence regarding which is the most useful for understanding the residential patterns of immigrants. As a result, we will get a better sense of not only what patterns we see today in U.S. metropolitan areas but also what we might expect in the future, in terms of both integration and the possible trajectory of the American color line. If segregation is low and declining among some groups, this would suggest that the social boundaries between groups are becoming less distinct—and less important—over time. However, if we see little evidence of residential integration, the implication is that social boundaries between groups will likewise remain strong in the coming years.

ENDNOTES

1. Roger Daniels (2002), *Coming to America*, 2nd ed. (New York: Perennial), 20.
2. Philip Martin and Elizabeth Midgley (2003), "Immigration: Shaping and Reshaping America," *Population Bulletin* 58, 2 (June): 12.
3. Ibid., 11.
4. A number of observers have commented on the uniqueness of American culture. See for example, J. Hector St. John Crevecoeur (1792), *Letters from an American Farmer* (repr., New York: Albert and Charles Boni, 1925), 54–55; Alexis de Tocqueville (1835), *Democracy in America* (New York: Penguin Putnam); and Israel Zangwill (2006), *From the Ghetto to the Melting Pot: Israel Zangwill's Jewish Plays; Three Playscripts* (Detroit: Wayne State University Press). Zangwill's play coined the famous phrase "melting pot."
5. Daniels (2002, 128–45).
6. Alejandro Portes and Ruben G. Rumbaut (2006), *Immigrant America: A Portrait*, 3rd ed. (Berkeley: University of California Press), 38–40.
7. Daniels (2002, 66–69).
8. Daniels (2002, 136–37, 149–50, 220, 240, 250); Portes and Rumbaut (2006, 39–41).
9. Portes and Rumbaut (2006, 38).
10. Ibid., 62.
11. Ibid., 62; Michael J. White (1987), *American Neighborhoods and Residential Differentiation* (New York: Russell Sage), 97.
12. Douglas S. Massey and Nancy A. Denton (1993), *American Apartheid: Segregation and the Making of the Underclass* (Cambridge: Harvard University Press), 9–10.
13. Robert E. Park (1925), "The City: Suggestions for the Investigation of Human Behavior in the Urban Environment" in *The City*, ed. Robert E. Park and Ernest W. Burgess (Chicago: Chicago University Press), 7.
14. Ibid., 9–10.
15. Robert E. Park and Ernest W. Burgess (1921), *Introduction to the Science of Sociology* (repr., Chicago: University of Chicago Press, 1969), 735.

16. Milton Gordon (1964), *Assimilation in American Life: The Role of Race, Religion, and National Origins* (New York: Oxford University Press), 244.
17. Alba and Nee (2003, 11).
18. Portes and Rumbaut (2006, 41); Alba and Nee (2003, 49).
19. Portes and Rumbaut (2006, 41).
20. Alba and Nee (2003, 29).
21. Ibid., 38.
22. Ibid., 255–56.
23. White (1987, 96–98).
24. Daniels (2002, 276).
25. Alba and Nee (2003, 131–32).
26. Daniels (2002, 189, 201–2).
27. Alba and Nee (2003, 126–57).
28. Charles Hirschman, Philip Kasinitz, and Josh DeWind, eds. (1999), *The Handbook of International Migration: The American Experience* (New York: Russell Sage), 129.
29. Bean and Stevens (2003, 98–99).
30. Camille Zubrinsky Charles (2003), "Dynamics of Racial Residential Segregation," *Annual Review of Sociology* 29, I: 167–207; Lawrence Bobo and Camille Zubrinsky (1996), "Attitudes on Residential Integration: Perceived Status Differences, Mere In-Group Preference, or Racial Prejudice?" *Social Forces* 74, 3:883–909; Reynolds Farley, Charlotte Steeh, Maria Krysan, Tara Jackson, and Keith Reeves (1994), "Stereotypes and Segregation: Neighborhoods in the Detroit Area," *American Journal of Sociology* 100, 3:750–80.
31. Camille Zubrinsky Charles (2000), "Neighborhood Racial-Composition Preferences: Evidence from a Multiethnic Metropolis," *Social Problems* 47, 3:370–407. See also Maria Krysan (2002), "Whites Who Say They'd Flee: Who Are They and Why Would They Leave?" *Demography* 39, 4:675–96.
32. Margery Austin Turner and Stephen L. Ross (2003), *Discrimination in Metropolitan Housing Markets: Phase 2—Asians and Pacific Islanders of the HDS 2000* (Washington, D.C.: U.S. Department of Housing and Urban Development): Margery Austin Turner, Stephen L. Ross, George Galster, and John Yinger (2002), *Discrimination in Metropolitan Housing Markets: National Results from Phase 1 of the Housing Discrimination Study (HDS)* (Washington, D.C.: U.S Department of Housing and Urban Development).
33. See, for example, John M. Goering and Ron Wienk, eds. (1996), *Mortgage Lending, Racial Discrimination and Federal Policy* (Washington, D.C.: Urban Institute Press); and John Yinger (1995), *Closed Doors, Opportunities Lost: The Continuing Costs of Housing Discrimination* (New York: Russell Sage).
34. Charles (2003, 182–91); Stephen L. Ross and Margery Austin Turner (2005), "Housing Discrimination in Metropolitan America: Explaining Changes between 1989 and 2000," *Social Problems* 52, 2:152–80.
35. See, for example, Camille Zubrinsky Charles (2005), "Can We Live Together?" in *The Geography of Opportunity*, ed. Xavier de Souza Briggs (Washington D.C.: Brookings Institution Press, 2005), 72–76; Margery Austin Turner and Stephen L. Ross (2005), "Racial Discrimination and the Housing Search," in *The Geography of Opportunity*, ed. Xavier de Souza Briggs (Washington, D.C.: Brookings Institution Press), 82; and Gregory D. Squires and Charis E. Kurbin (2006), *Privileged Places: Race, Residence, and the Structure of Opportunity* (Boulder, CO: Lynne Rienner).
36. Portes and Zhou (1993, 74–96); Min Zhou (1999), "Segmented Assimilation: Issues, Controversies, and Recent Research on the New Second Generation," in *The Handbook of International Migration: The American Experience,* ed. Charles Hirschman, Philip Kasinitz, and Josh DeWind (New York: Russell Sage), 196–211.
37. Zhou (1999, 211).

REFERENCES

Alba, Richard D., and Victor Nee. 2003. *Remaking the American Mainstream: Assimilation and Contemporary Immigration.* Cambridge: Harvard University Press.

Bean, Frank D., and Gillian Stevens. 2003. *America's Newcomers and the Dynamics of Diversity*. New York: Russell Sage.

Bobo, Lawrence, and Camille Zubrinsky. 1996. "Attitudes on Residential Integration: Perceived Status Differences, Mere In-Group Preference, or Racial Prejudice?" *Social Forces* 74, 3:883–909.

Charles, Camille Zubrinsky. 2000. "Neighborhood Racial-Composition Preferences: Evidence from a Multiethnic Metropolis." *Social Problems* 47, 3:370–407.

———. 2003. "Dynamics of Racial Residential Segregation." *Annual Review of Sociology* 29, 1:167–207.

———. 2005. "Can We Live Together?" In *The Geography of Opportunity,* edited by Xavier de Souza Briggs. Washington, D.C.: Brookings Institution Press.

Crevecoeur, J. Hector St. John. 1782. *Letters from an American Farmer.* Reprint, New York: Albert and Charles Boni, 1925.

Daniels, Roger. 2002. *Coming to America.*, 2nd ed. New York: Perennial.

Farley, Reynolds, Charlotte Steeh, Maria Krysan, Tara Jackson, and Keith Reeves, 1994. "Stereotypes and Segregation: Neighborhoods in the Detroit Area," *American Journal of Sociology* 100, 3:750–80.

Goering, John M., and Ron Wienk, eds. 1996. *Mortgage Lending, Racial Discrimination and Federal Policy.* Washington, D.C.: Urban Institute Press.

Gordon, Milton. 1964. *Assimilation in American Life: The Role of Race, Religion, and National Origins.* New York: Oxford University Press.

Hirschman, Charles, Philip Kasinitz, and Josh De Wind, eds. 1999. *The Handbook of International Migration: The American Experience.* New York: Russell Sage.

Krysan, Maria. 2002. "Whites Who Say They'd Flee: Who Are They, and Why Would They Leave?" *Demography* 39, 4:675–96.

Martin, Philip, and Elizabeth Midgley. 2003. "Immigration: Shaping and Reshaping America." *Population Bulletin* 58, 2 (June); 1–44.

Massey, Douglas S., and Nancy A. Denton. 1993. *American Apartheid: Segregation and the Making of the Underclass.* Cambridge: Harvard University Press.

Park, Robert E., and Ernest W. Burgess. 1921. *Introduction to the Science of Sociology.* Reprint, Chicago: University of Chicago Press, 1969.

Park, Robert E., Ernest W. Burgess, and Roderick D. McKenzie. 1925. *The City.* Chicago: Chicago University Press.

Portes, Alejandro, and Ruben G. Rumbaut. 2001. *Legacies: The Story of the Immigrant Second Generation.* Berkeley: University of California Press.

———. 2006. *Immigrant America: A Portrait.* 3rd ed. Berkeley: University of California Press.

Portes, Alejandro, and Min Zhou. 1993. "The New Second Generation: Segmented Assimilation and Its Variants among Post-1965 Immigrant Youth." *Annals of the American Academy of Political and Social Science* 530 (November): 74–96.

Ross, Stephen L., and Margery Austin Turner. 2005. "Housing Discrimination in Metropolitan America: Explaining Changes between 1989 and 2000." *Social Problems* 52, 2:152–80.

Squires, Gregory D., and Charis E. Kurbin. 2006. *Privileged Places: Race, Residence, and the Structure of Opportunity.* Boulder, CO: Lynne Rienner.

Tocqueville, Alexis de. 1835. *Democracy in America.* Reprint, New York: Penguin Putnam, 2004.

Turner, Margery Austin, and Stephen L. Ross. 2003. *Discrimination in Metropolitan Housing Markets: Phase 2—Asian and Pacific Islanders of the HDS 2000,* Washington, D.C.: U.S. Department of Housing and Urban Development.

———. 2005. "Racial Discrimination and the Housing Search." In *The Geography of Opportunity,* edited by Xavier de Souza Briggs. Washington, D.C.: Brookings Institution Press.

Turner, Margery Austin, Stephen L. Ross, George Galster, and John Yinger. 2002. *Discrimination in Metropolitan Housing Markets: National Results from Phase 1 of the Housing Discrimination Study (HDS).* Washington, D.C.: U.S. Department of Housing and Urban Development.

———. 1987. *American Neighborboods and Residential Differentiation.* New York: Russell Sage.

Zinger, John. 1995. *Closed Doors, Opportunities Lost: The Continuing Costs of Housing Discrimination.* New York: Russell Sage.

Zangwill, Israel. 2006. *From the Ghetto to the Melting Pot: Israel Zangwill's Jewish Plays; Three Playscripts.* Detroit: Wayne State University Press.

Zhou, Min. 1999, "Segmented Assimilation: Issues, Controversies, and Recent Research on the New

Second Generation." In *The Handbook of International Migration: The American Experience*, edited by Charles Hirschman, Philip Kasinitz, and Josh DeWind, 172–95. New York: Russell Sage.

DISCUSSION QUESTIONS

1. According to the author, why is racial distinctiveness working against assimilation for many new arrivals to the United States? Do you agree with this proposition? Why or why not?
2. Why should assimilation be the goal of new immigrants to the United States? Why is it necessary to give up their unique cultural practices and traditions, including their native language, in order to be fully accepted—or assimilated—as "American"?
3. In what ways have discriminatory practices worked to undermine the goal of assimilation for many immigrants to the United States?
4. Which of the three theories on the experience of immigration (assimilation, ethnic disadvantage, and segmented assimilation) provides the best explanation in your opinion?

18

The New Case Against Immigration

MARK KRIKORIAN

The Four Questions

1. What is the problem that is described in this article?
2. Is this a *social* problem? Who does this problem affect? Is society affected by this problem?
3. What are the root causes of this problem?
4. What does the author recommend doing about this problem?

Topics Covered

Legal immigration

Illegal immigration

Government spending and debt

Immigrant welfare costs

Immigrant health care care costs

Immigrant education costs

Net cost of immigration

Immigration policy

Government Spending

The conflict between mass immigration and modern America may be most evident when studying

SOURCE: From "Government Spending," from *The New Case Against Immigration: Both Legal and Illegal* by Mark Krikorian,

government services. The combined spending by federal, state, and local governments accounts for almost one third of our total economy, a figure many times larger than during prior waves of immigration. This includes welfare spending, of course, but also spending on schools, roads, criminal justice, and other tax-supported activities.

The problem this poses was summed up by Nobel Prize–winning economist Milton Friedman: "It's just obvious that you can't have free immigration and a welfare state."[1]

Importing millions of poor people with large families means that, by definition, they will pay relatively little in taxes but make heavy use of government services. This is true not because of any moral defect in the immigrants or any meaningful differences between today's immigrants and those of the past. Instead, it is *we* who have changed—our modern society embraces a larger role for government, expecting it to underwrite a broad system of social provision for the poor, education for the young, support for the elderly, a large portion of the nation's medical care, and many other functions.

There is no way to avoid this conflict between mass immigration and the functions of modern government—the welfare state and other aspects of big government are inherent to modern society. Efforts to trim government and make welfare policy less morally problematic may well succeed to some degree; the 1996 welfare reform, for instance, is generally acknowledged to have been a success, and perhaps other such measures are possible.

But whatever measures we might take in the future to limit the size and scope of government, big government in some form or another is never going away. As one welfare scholar has put it, "Transfer or redistribution policies are a pervasive, if not predominant, government activity in all modern societies."[2] And that means that mass immigration can never stop being a drain on public coffers.

The fiscal threat that immigration now poses did not exist in earlier eras, when government was much smaller. In 1901, for instance, the entire federal budget totaled $525 million, equivalent to about $13.5 billion in 2006 dollars.[3] By 2006, the nation's population had quadrupled, but the federal budget reached $2.7 *trillion,* 200 times larger, in real terms, than in 1901.[4]

That means that a century ago, during the previous wave of mass immigration, the federal government spent roughly $178 a year per American, in today's dollars, while today it spends about $9,000—a fifty-fold increase in the amount of government per person.

This huge growth in government is apparent in other statistics as well. For instance, federal spending accounted for less than 3 percent of the economy in 1900 but grew to 20.1 percent in 2005. Government has also grown at the state and local levels; the combined spending of all levels of government—federal, state, and local—has grown from about 8 percent of the total economy in 1900 to 31 percent in 2005.[5]

WELFARE

A large part of this growth in government has been the development of the welfare state—the extensive system of government support for the poor. There was no welfare state during prior waves of mass immigration—in fact, it wasn't until well after the end of the last great wave of immigration in 1924 that the institutions of the welfare state began to develop.

As the late journalist Richard Estrada put it, "From the late 1920s through the mid-1960s, the period when the welfare state went from swaddling clothes to its Sunday best, the levels of immigration were virtually negligible. Today, they are at their highest levels in U.S. history."[6]

The Social Security Act of 1935 established old-age pensions (what we usually mean when we talk about Social Security), as well as unemployment assistance and Aid to Families with Dependent Children (which was renamed in 1996 as Temporary Assistance to Needy Families, or TANF). The current Food Stamp program was established in 1964. Medicare and Medicaid were established in 1965. The Child Nutrition Act of 1966 created the WIC program, short for the Special Supplemental Nutrition Program for Women, Infants and Children.

Then in 1974 came Supplemental Security Income (SSI) for the indigent elderly, blind, and disabled.

Whatever concerns many conservatives have with these New Deal and Great Society programs, the costs of some of them would have remained more manageable had the poor population not increased. Instead, the United States saw a surge in immigration, mainly of poor and unskilled people with big families. That makes them just like the immigrants of the past, of course, but also just the kind of people this new welfare system was designed to subsidize.

The low level of education among immigrants is clear. In 2005, about 30 percent of all immigrants in the workforce lacked a high-school education, nearly quadruple the rate for native-born Americans.[7] Well over half (55 percent) had no more than a high school diploma. The largest immigrant group by far—Mexicans—had the lowest levels of education, with 62 percent of working-age Mexican immigrants having less than a high-school education, and only 5 percent having a college degree.[8]

In a modern economy, people with very little education generally earn very low wages, with limited opportunities for advancement, and this holds true for immigrants. In 2005, immigrants were nearly half again as likely to be in poverty than natives—17 percent of immigrants had incomes below the official poverty line (about $20,000 for a family of four), compared with 12 percent of native-born Americans.[9] When immigrants' U.S.-born young children are added (which makes sense, since they live in the same household), the immigrant poverty rate is more than 18 percent.

Nearly half of immigrant households—45 percent—were in or near poverty, meaning their income was below 200 percent of the poverty level, compared with 29 percent of native-headed households. The near-poverty benchmark is important because people whose incomes are below that level generally don't pay federal income taxes and are usually eligible for means-tested welfare programs.

Mexicans—the largest and least-educated immigrant group—are also the poorest, with nearly two thirds living in or near poverty and more than one in four actually below the poverty line.

The result of this widespread immigrant poverty is widespread immigrant welfare use. Among households headed by native-born Americans, 18 percent use at least one major welfare program. That's already quite high, of course, but among immigrant-headed households, the rate is half again as high, at almost 29 percent.[10] Immigrant use of cash assistance programs[11] is only a little higher than among Americans, while the gap is larger for use of food stamps and government-owned or -subsidized housing. The biggest differences are in WIC and subsidized school lunches, which immigrants are more than twice as likely to use, and Medicaid (health insurance for the poor), which immigrants are about two thirds more likely to use.

These welfare programs combined cost federal taxpayers some $500 billion a year, most of which is for Medicaid. This is significant because it is in Medicaid—the biggest and most expensive welfare program—where the gap between immigrants and the native born is the biggest. Use of most of the other welfare programs, whether by Americans or immigrants, is mostly in the single digits; but Medicaid is used by fully 15 percent of native-born households and an astounding 24 percent of immigrant households (and 37 percent of Mexican immigrant households).

Underlining the mismatch between mass immigration and the modern welfare system is the extraordinarily high immigrant eligibility for the Earned Income Tax Credit (EITC). The EITC is available only to people who work, and it functions as a kind of negative income tax—those who file a tax return have their EITC check calculated automatically by the IRS, based on their income and family size. In its surveys, the Census Bureau estimates whether people are eligible for the EITC, so the number of people actually receiving it is probably somewhat lower, but the figures are quite high nonetheless. More than 15 percent of native-headed households are eligible for this $30 billion program, but nearly twice as large a share (30 percent) of immigrants qualify, including fully half of Mexican-immigrant households.

The EITC was designed as a way to encourage enterprise rather than dependency among the poor.

For the same reasons, the 1996 welfare-reform legislation introduced work requirements for certain programs. At the same time, other welfare programs are targeted at helping children, such as WIC, subsidized school lunches, and the State Children's Health Insurance Program (a part of Medicaid). In other words, the modern American welfare system is designed mainly to provide support for the working poor with children.

Who are immigrants but the working poor with children? Immigrants work (in 2001, almost 80 percent of immigrant households using welfare had at least one person working[12]), they're poorer than Americans (see above), and they have more children than Americans (the average immigrant woman has 2.7 children in her lifetime, versus 2.0 for native-born women). In other words, mass immigration is almost perfectly designed to overwhelm modern America's welfare system.

In the words of two welfare scholars, "This very expensive assistance to the least advantaged American families has become accepted as our mutual responsibility for one another, but it is fiscally unsustainable to apply this system of lavish income redistribution to an inflow of millions of poorly educated immigrants."[13]

HEALTH CARE

A specific aspect of assistance to the poor warrants separate treatment. The United States spent more than $2 trillion on health care in 2006, nearly half of the expenditures coming from government at all levels, including Medicare (for the elderly), Medicaid (for the poor), and other costs.[14] Health-care expenditures have been growing by more than 7 percent a year, much higher than inflation, and one of the issues this has brought to the fore is the number of people who lack medical insurance.

Universally available, high-quality medical care is an important value for a modern society, which is why the problem of the uninsured has given rise to many proposed solutions, from a single-payer, government-run system suggested by the Left to expanded "health savings accounts" offered by the Right. It's all the more curious, then, that there's little discussion of how mass immigration subverts our efforts to address this issue. In 2005, immigrants were two and half times more likely to be uninsured than the native born, almost 34 percent versus about 13 percent.[15] In other words, about one out of three immigrants in the United States has no health insurance.

If the number of immigrants were small, this might not make much difference. But because of their huge numbers, immigrants account for a large part of the uninsured problem. One in four people in the United States without health insurance is an immigrant, and among children who are uninsured, one out of three is either an immigrant or the young child of an immigrant.

And immigrants got to be such a large share of the uninsured population because they are responsible for most of its growth. Immigrants and their U.S.-born children account for nearly three quarters of the growth in the uninsured population from 1989 to 2005.[16] Over a different time period, another study found that immigrants accounted for 86 percent of the growth in the uninsured from 1998 to 2003.[17] Looking at only one portion of the immigration population, yet another study estimated that illegal aliens alone accounted for one third of the growth in the number of uninsured adults from 1980 to 2000.[18] It would not be too much to say that the crisis of the uninsured is a creation of our immigration policy.

This large population of uninsured immigrants would be even larger, of course, except that so many immigrants are on Medicaid and thus directly supported by taxpayers. In fact, nearly half (47 percent) of people in immigrant families are either uninsured or on Medicaid, nearly double the rate for native families.

The cost to taxpayers of the Medicaid program is obvious, but immigrants without any insurance at all also impose significant costs on government and on consumers. After all, even the uninsured get sick, and no modern society is going to allow them to go entirely without treatment. This treatment usually happens at emergency rooms; in the words of one recent study, emergency rooms "have become one of the nation's principal sources of care for patients with limited access to other providers, including the 45 million uninsured Americans."[19] (Actually, as we've seen, a large portion of the uninsured are noncitizens, rather than Americans.)

Taxpayers bear some of the costs incurred by the emergency departments of these "safety net" hospitals. Medicaid sends "disproportionate share hospital" payments to hospitals serving the poor, including immigrants; and in the Medicare prescription drug bill, Congress allotted payments of $250 million per year (for four years, through 2008) to the states for treatment provided to illegal aliens.[20] In addition, "A number of states also provide additional support to emergency and trauma systems through general revenues or special taxes."[21]

In California, where households headed by immigrants (legal and illegal) make up the *majority* of the uninsured, [22] emergency departments statewide lost nearly half a billion dollars in 2001–2002, up 58 percent from just two years before.[23] In Texas, where immigrant households account for nearly four in ten of the uninsured, the state's twenty-one trauma centers lost $181 million in 2001 due to unreimbursed costs.

There is no estimate of total health-care costs borne by taxpayers due to immigration overall, but there are a number of estimates regarding just illegal aliens, who represent about one third of the foreign-born population. One estimate of the cost to emergency providers of treating illegal aliens is $1.45 billion, and the cost to counties along the Mexican border nearly a quarter of a billion dollars.[24] Other state-specific estimates: nearly $600 million in uncompensated care for illegals provided by New York taxpayers in 2006;[25] in Florida in 2005, $165 million;[26] in Texas in 2004, $520 million;[27] in California in 2004, more than $1.4 billion;[28] and in Arizona in 2004, $400 million.[29]

These estimates do not include the costs of care provided to uninsured *legal* immigrants—an important point, since the main reason for immigrant poverty, and thus the lack of health insurance, is lack of education, not legal status.[30]

But not all uncompensated care is covered by government, and hospitals must thus write off a portion of the costs of treating the uninsured. Hospitals then shift these costs onto paying patients and their insurance carriers, resulting in higher premiums for those who *do* have health insurance. One report estimated that each American family with insurance through a private-sector employer paid more than nine hundred dollars extra in premiums in 2005 due to the cost of treating the uninsured.[31]

The Emergency Medical Treatment and Active Labor Act (EMTALA) requires all emergency rooms nationwide to screen and stabilize any and all comers, whether they can afford it or not—so when hospitals can no longer shift enough of the costs for uncompensated care to others, they simply close their emergency rooms. This doesn't create a direct monetary cost for consumers, but it does hold the potential to levy the ultimate tax—death—on Americans in need of emergency care. From 1993 to 2003, the number of hospitals with emergency rooms declined by 9 percent, even though the number of emergency room visits increased 26 percent, double the rate of increase in the population.[32] In Los Angeles, more than sixty hospitals have closed their emergency rooms over the past decade.[33]

A 2005 report bleakly summarized the situation: "To the degree that immigration continues to increase, it is likely that the uninsured will also continue to increase as a proportion of the population."[34]

EDUCATION

Another very costly service provided by the government is education. Total expenditures nationwide on all levels of education reached $866 billion in the 2003–2004 school year, accounting for 7.9 percent of the gross domestic product.[35] An indication of how this huge investment in education is characteristic of our modern society is seen in the fact that even as late as 1949, education spending accounted for only 3.3 percent of our GDP.

The role of immigration in increasing the cost of education is especially large because immigrant women are more likely to be in their child-bearing years, and immigrants generally have larger families. So while immigrants constitute about 12 percent of our total population, the children of immigrants (some born here, some immigrants themselves) comprise 19 percent of the school-age population (five to seventeen years old) and 21 percent of the preschool population (four and below).[36] In California, nearly half the children in both age groups are from immigrant families. While more than two thirds of these children are native born, their use of American public schools is a direct result of our immigration policy allowing their parents to enter.

Total school enrollment nationwide reached a post-Baby Boom low of 44.9 million in 1984, and is now at about 55 million.[37] This means that the 10.3 million school-age children in immigrant families account for all—100 percent—of the growth in elementary and secondary school enrollment nationwide over the past generation. Obviously this isn't the case in every school or every town or even every state. But in the nation as a whole, the surge in school enrollments is due entirely to the federal government's immigration policies.

This immigration-driven growth in enrollment has caused overcrowding at many schools. The U.S. Department of Education has found that 22 percent of public schools are overcrowded, with 8 percent of the total being overcrowded by more than 25 percent of capacity. The overcrowding is worst in precisely the kind of schools immigrant children are likely to attend: large-capacity schools in central cities, especially in the West, especially those with more than 50 percent minority enrollment, where the majority of students receive free or reduced-price school lunches (i.e., are from poor families).[38]

The education expenses incurred by this immigration-driven surge in enrollment are huge. One study, which looked at only part of the impact of immigration, found that education of illegal-alien students cost the states $12 billion a year, and when U.S.-born children of illegal aliens are added, the cost more than doubles to $28.6 billion.[39] California was estimated to have spent $7.7 billion on education for the children of illegal aliens, nearly 13 percent of the state's total education budget for 2004–2005.

And this cost estimate did not take into account the extra expense of educating immigrant students as opposed to native-born students; a study conducted some twenty-five years ago found that even then, bilingual education programs cost from $100 to $500 more per pupil, while a more recent look at California found that supplementary programs for "limited English proficient" students cost an extra $361 per student.[40] Another more recent study, of Florida, found that "ESOL [English for speakers of other languages] students cost $153 million in fiscal year 2003–04 beyond what they would have if enrolled in basic programs."[41] Across the nation, this adds up, since about 10 percent of all public-school students are considered limited English proficient, otherwise known as "English language learners."[42]

CRIMINAL JUSTICE

Total spending on justice—including police, courts, and prisons—by all levels of government was $185 billion in 2003, up more than 400 percent from 1982, a growth rate sixteen times greater than the rate of increase in the population.[43]

How much of this cost is attributable to immigration is hard to gauge. The multiplicity of jurisdictions involved means that different information is collected and tallied in different ways. In the words of a recent government report, there is "no reliable population and incarceration cost data on criminal aliens incarcerated in all state prisons and local jails."[44]

Despite the stereotype of immigrants contributing disproportionately to crime, the skimpy evidence that exists suggests that immigrants—both today and a century ago—are no more likely to be involved in crime than the native born, and perhaps less so.[45] But given that there are more than 36 million immigrants, and they are disproportionately young, single men, they will inevitably have a large presence in the criminal justice system and create large costs. This is all the more a concern, given research that finds the likelihood of an immigrant being incarcerated grows with longer residence in the United States[46] and that the U.S.-born children of immigrants are dramatically more likely to be involved in crime than their immigrant parents. For instance, native-born Hispanic male high-school dropouts are eleven times more likely to be incarcerated than their foreign-born counterparts.[47]

The costs at the federal level are easier to calculate, consisting of expenses incurred housing aliens in federal prisons plus the grants made under the State Criminal Alien Assistance Program (SCAAP), which reimburses certain states and localities for a small part of the costs they incur when incarcerating certain illegal aliens. About one quarter of federal prisoners are aliens, and when combined with the reimbursement program, the Government Accountability Office estimates they cost the federal government nearly $6 billion for 2001–2004.[48] The amount would have been larger had Congress not consistently cut the SCAAP program, narrowing its scope (now only those illegals held four days or more can qualify) and cutting the number of cents on the dollar it reimbursed.

There are some estimates of state costs for illegal aliens; New York, for instance, was estimated to have spent $165 million,[49] Florida $60 million,[50] Texas $150 million,[51] California $1.4 billion,[52] and Arizona $80 million.[53] Large as the costs are, they include only prison costs, not jail costs (prisons hold offenders for more than one year; jails are the local facilities where people are kept for shorter periods of time), nor the costs incurred by police or the courts, nor the monetary losses suffered by the victims. At least as important is the fact that these estimates count the incarceration only of illegal aliens, not the rest of the immigrant population.

NET COSTS

Of course, all these costs might not matter if all immigrants were highly productive, earning high wages, paying large amounts in taxes, and making only modest demands on government services. Immigrants, legal and illegal, pay taxes, of course—income and payroll taxes, sales and excise taxes, property taxes (included in their rent), and many others. But given their low levels of education and consequent low incomes, and their large families and heavy use of services, it is a mathematical inevitability that today's immigrants create a burden on taxpayers—not because they're especially different from yesterday's immigrants, but because *we* are from our predecessors.

Heritage Foundation scholar Robert Rector has calculated that in 2004, low-skill immigrant households (those headed by an immigrant with less than a high-school education), paid more than $10,000 in all taxes (federal, state, and local) but received in services more than $30,000, representing a net burden on American taxpayers of nearly $20,000 per household per year.[54] In fact, he found that since the average earnings of low-skill immigrant households in that year were just under $29,000, "the average cost of government benefits and services received by these households not only exceeded the taxes paid by these households, but actually exceeded the average earned income of these households." Rector calculated the total tax burden created by such households at more than $89 billion each year. Furthermore, the average lifetime cost to the taxpayer of *each* low-skilled immigrant household is $1.2 million.

The high stakes for taxpayers are made clear by this one fact: "It takes the entire net tax payments (taxes paid minus benefits received) of one college-educated family to pay for the net benefits received by one low-skill immigrant family. Each extra low-skill immigrant family which enters the U.S. requires the taxes of one college-educated family to support it."[55]

An earlier study of this issue was *The New Americans,* a landmark 1997 report by the National Research Council.[56] Although the numbers are certainly bigger now, the report found that immigrants of all kinds (legal and illegal, high-skill and low-skill) create a net burden on government at all levels of between $11 billion and $22 billion a year, swamping the presumed net economic benefit that immigrants create (by lowering the wages of the poor) of between $1 billion and $10 billion per year. The report also calculated the costs incurred by immigrants in two very different states, California and New Jersey. The average immigrant-headed household in California used almost $3,500 more in state and local services than it paid in taxes, amounting to an extra tax burden for each native-headed household of nearly $1,200 per year. New Jersey, with a more ethnically diverse and more highly educated immigrant population, still saw the average immigrant family consume nearly $1,500 more in services than it paid in taxes, increasing the tax burden for each native-born family by more than $200 per year.

The report also estimated the lifetime cost of an immigrant based on different educational levels. It found that the average immigrant high-school dropout would cost American taxpayers a total of $89,000 over his lifetime, and that an immigrant with only a high-school degree would still cost taxpayers $31,000. An immigrant with education beyond high school, though, was estimated to create a fiscal benefit of $105,000 over his lifetime. Given the mix of immigrants the United States takes in, the report estimated a net lifetime cost to taxpayers of $3,000 per immigrant.

In addition to these real-world cost estimates, the National Research Council report also included a kind of thought experiment, projecting the net fiscal effect not only of the immigrant but also of his descendants for three hundred years into the future. Even after three centuries, the balance sheet of taxes paid versus services used for an immigrant high-school dropout and his posterity would *still* be negative, to the tune of $13,000. This mathematical game did find that other immigrants and their descendants would be a plus, adding up to a total benefit to taxpayers over a period of three hundred years of $80,000 for the average immigrant.

Although supporters of mass immigration latched onto the three-hundred-year projection's net plus for taxpayers, the report's authors warned, "It would be absurd to claim that the projections into the 23rd century are very reliable." As another scholar has written, "We cannot reasonably estimate what taxes and benefits will be even 30 years from now, let alone 300."[57]

Notwithstanding this parlor trick, the lesson of *The New Americans* is the same as Milton Friedman's—"It's just obvious that you can't have free immigration and a welfare state." Other, more narrowly focused, research has made the same point. For instance, a 2005 study found that "in Florida the net burden on state and local governments from immigrants is on the order of $2,000 per immigrant household."[58] A nationwide study looking only at illegal aliens found that they cost federal taxpayers about $10 billion more in services than they paid in taxes, and the gap is even larger at the state and local levels, since that's the source of most services used by immigrants.[59] Estimates of the net cost of illegal immigration at the state level (just for public education, health care, and incarceration) point to the same problem: a net cost to New York taxpayers of $4.5 billion in 2005; Florida, nearly $1 billion; Texas, $3.7 billion; California, nearly $9 billion; and Arizona, more than $1 billion.[60]

A final aspect of the issue of government services is different from what's been discussed above. There is a huge existing investment in public infrastructure that immigrants immediately benefit from without ever paying in to—like joining a club without a buy-in fee. These public assets—built over the years with the taxes of Americans—include roads, navigable waterways, ports, subway systems, and airports, plus things like water treatment plants and sewage systems, national parks, public beaches, and other public lands, schools, universities, and libraries, as well as government buildings and military bases.

In a smaller America with a smaller public sector, immigrants weren't inheriting quite as much when they showed up. In fact, the previous period of mass immigration in our nation's history—about one lifetime's worth of mass inflows from the late 1840s to the early 1920s—was precisely the period when our country underwent the urbanization and industrialization (which immigrants contributed massively to) that led to the creation of such a large public infrastructure in the first place. But now that America has completed that process and reached maturity, continued mass immigration simply represents a gift of billions of dollars in infrastructure to new immigrants, a kind of inheritance tax on

Americans, lessening the value of their share of the public assets bequeathed them by their ancestors.

WHAT TO DO?

Given that current immigration policy ensures that immigrants will be a fiscal burden, is there a way out? Is there a way a modern nation can undertake mass immigration without soaking the taxpayer?

The most simplistic suggestion comes from utopian libertarians, whose slogan is "Immigration *sí*, welfare *no*!" In other words, if there's a problem with immigrants using welfare, the solution is to abolish welfare. In the words of libertarian journalist Tom Bethell, "a few million more 'undocumented' newcomers may just about finish off the welfare state: one more argument for an open border."[61]

Looking at the actual numbers, the Heritage Foundation estimated that "in order for the average low-skill household to be fiscally solvent (taxes paid equaling immediate benefits received), it would be necessary to eliminate Social Security and Medicare, all means-tested welfare, and to cut expenditures on public education roughly in half."[62]

Needless to say, this isn't going to happen, because despite widespread desire for reform, the American people don't want to abolish the welfare system (let alone cut education spending in half). A detailed examination of public-opinion surveys shows that Americans have embraced the idea of some kind of government system of social provision for the poor.[63] In a 1939 article accompanying a poll on public expectations of government, *Fortune* magazine wrote, "In 1929, beneficence would probably not have been accepted as a proper function of government. But today it is emphatically held desirable."

There has been no change in this view over the generations. The compilers of the survey of polls wrote, "The stability of opinion on the central issues in this debate is nothing short of astonishing." When presented with the statement "It is the responsibility of the government to take care of people who can't take care of themselves" in almost two dozen surveys from 1987 to 2001, a majority of respondents agreed in every survey, from a low of 56 percent agreement, up to 74 percent.

The "astonishing" persistence of support for some sort of welfare system demonstrates that it is an inherent part of modern society. (And pollsters don't even bother to ask the public whether they support abolition of the public schools or public highways, since the answers are obvious.) Despite wide support for work requirements, concerns about illegitimacy, and ambivalence about the effectiveness of welfare programs, the idea of abolishing the welfare system altogether is nothing more than a fantasy.

Well, if we don't abolish big government, would shrinking it break the connection between mass immigration and huge tax burdens? Again, the answer is no. Grover Norquist, one of the most active advocates for smaller government (and, improbably enough, a vocal supporter of open borders), laid out a plan in 2000 to cut the size of government's role in the economy by half in twenty-five years.[64] The success of such an extraordinarily ambitious plan would reduce government from one third of the economy to one sixth. Put aside the inconvenient fact that, nearly one third of the way along Norquist's twenty-five-year timetable, government has only gotten bigger; but if by some miracle this plan were to come to fruition, we would still have a multitrillion-dollar government sector—and Milton Friedman's basic insight would still be valid.

Perhaps instead of relying on utopian schemes to abolish government as a way of limiting the fiscal fallout of immigration, we might leave government services essentially intact but simply wall off the immigrants, denying them access. This is the heart of the idea of turning away immigrants who are likely to become a "public charge," a principle in American immigration law dating back to colonial times.[65] This principle holds that foreigners who cannot support themselves should be kept out of the country, or if already here, deported. In reality, the public-charge provision is most useful in denying people access to the United States in the first place, rather than removing those already here; some 10 percent of visa applicants are rejected on public charges grounds, whereas during the entire decade of the 1980s, only twelve aliens were deported on those grounds.[66]

Two laws passed in 1996 sought to reinvigorate the principle that immigrants shouldn't be using welfare. The Illegal Immigration Reform and Immigrant Responsibility Act increased the qualifications to sponsor an immigrant (i.e., promising to support the immigrant if he falls on hard times, rather than letting him use welfare), and made the sponsorship agreement, called the "affidavit of support," a legally enforceable contract.

Meanwhile, the immigration-related parts of the big 1996 welfare-reform law (the Personal Responsibility and Work Opportunity Reconciliation Act) tried to limit immigrant welfare eligibility. The law kicked immigrants already here off SSI and food stamps (though that was never fully enforced), and future immigrants were barred for five years from most means-tested benefits.

Welfare reform in general appears to have worked in shrinking some programs and encouraging work. But the immigrant-specific provisions did nothing to reduce immigrant welfare use. First of all, there were numerous exceptions in the laws (they did not apply to refugees and asylum recipients, for instance) and Congress almost immediately rolled back certain provisions (eliminating the ban on SSI for immigrants here before the law's passage and allowing food stamps for the elderly, children, and the disabled). Also, many states (particularly those where immigrants are concentrated) took up the slack, since it was a state option to provide Medicaid and TANF to noncitizens—resulting in nothing but a shifting of the processing of the welfare benefits from federal to state bureaucrats. Finally, the national-origin groups most likely to receive welfare before the law's passage saw the biggest increases in naturalization rates after its passage, because once an immigrant attains citizenship, the alien-specific welfare limits no longer apply.[67]

The result of all this was that the experiment in excluding immigrants from the modern welfare system was a failure. After briefly falling in the late 1990s, welfare use was back to 1996 levels within five years; in 1996, 22 percent of immigrant-headed households used at least one major welfare program, and in 2001 the figure was 23 percent.[68] Use of TANF and food stamps did decline, to a level only a little higher than that of the native born, but this resulted in almost no cost savings because it was offset by higher costs for Medicaid, the biggest and most expensive welfare program of all. Five years after welfare reform, the number of immigrant families using welfare had actually *increased* by 750,000, accounting for 18 percent of households on welfare, up from 14 percent in 1996.

In a broader sense, it's simply out of the question for our modern society to do what it takes to prevent fiscal costs from immigration. Immigrants don't just cross a physical border when entering the United States; they also cross a moral border, entering a nation that will not tolerate the kind of premodern squalor and inhumanity that is the norm in much of the rest of the world. Are Americans prepared to allow people to die on the hospital steps because they're foreigners? No. Are we going to deny immigrants access to the WIC program? Say the name of the program out loud and you'll get your answer: Americans are not going to deny nutrition to women, infants, and children.

And even barring illegal-alien children from the schools makes the public uncomfortable. A 2006 *Time* magazine poll, for instance, found that only 21 percent of respondents thought illegals should be allowed to obtain government services like health care or food stamps, only 27 percent thought illegals should get driver's licenses, but fully 46 percent thought illegal alien children should be allowed to attend public schools.[69]

Walling immigrants off from government benefits once we've let them in is a fantasy. The 1996 welfare reform was a vast social experiment which taught us two things: First, welfare *can* be made less harmful to the recipient and to society; and second, fine-tuning welfare policy to limit the costs of immigration is doomed to fail.

Perhaps then the answer is amnesty? After all, if illegal aliens (the least educated and poorest part of the immigrant population) were legalized, their wages and tax payments would increase, perhaps solving our fiscal problem.

Again, no. It's true that amnesty would increase the income of illegal aliens, by perhaps 15 percent, which would translate to an increase in their average tax payments to the federal government of 77 percent as not only their incomes rose but also as more of them were paid on the books.[70] But they would remain poor—with an average household income one third below others'—and now, as legal immigrants, they would be eligible for many more government programs and less reluctant to make use of them. The result would be that an amnesty for illegal immigrants would nearly triple the fiscal burden they place on the federal budget, from $10.4 billion a year to $28.8 billion. Robert Rector of the Heritage Foundation has written that, far from saving money, an amnesty "would be the largest expansion of the welfare state in 35 years."[71]

The last option for limiting the fiscal fallout of mass immigration might be to bar the immigration of the uneducated and instead permit mass immigration only of highly educated foreigners. Were this possible, it might well be a fiscal plus; the National Research Council report found that immigrants with more than a high-school education represented a net lifetime fiscal plus of $105,000.

While there is no doubt that educated immigrants are not a fiscal burden like the uneducated, even highly educated immigrants make much heavier use of public services than comparable natives and thus are not the fiscal boon that they could be. Among those with at least a college degree, average incomes are very close—$45,000 for the native born and $42,000 for immigrants. But even at that high level of education, immigrants are still more than twice as likely as the native born to use welfare (13 percent versus 6 percent) and to be uninsured (17 percent versus 7 percent).[72]

Also, the idea of a massive flow composed exclusively of highly educated immigrants is as fantastical as the "immigration *sí*, welfare *no*" mantra. It's true that there are a significant number of highly educated immigrants—about 10 percent have graduate or professional degrees, the same proportion as among the native born. But that cannot become a mass flow, because immigrants are people who are dissatisfied with conditions in their homelands—and highly educated people are the most likely to have opportunities that would keep them in their homelands. What's more, immigration is not an atomized process, with unconnected individuals randomly seeking to move here; instead, immigration takes place through networks of family and friends, networks that are not based on education levels or work skills.

More important, the fiscal effect is only one part of the conflict between mass immigration and modern society. Different immigrants will conflict with different goals and characteristics of a modern society—the highly educated are not likely to represent a fiscal burden nor drive down the wages of the poor, but they would, for instance, still be part of the social-engineering project that uses mass immigration to artificially increase our population (see the discussion of skilled immigration in the previous chapter). Also, they might well represent a greater assimilation challenge, not because they won't learn English, but because they are far more likely than the uneducated to have acquired a fully developed, modern national consciousness in their home country—and thus are more resistant than peasants to the adoption of a new American sense of nationhood and more likely so pursue dual citizenship.

ENDNOTES

1. Peter Brimelow, "Milton Friedman at 85," *Forbes*, Dec. 29, 1997.
2. Robert E. Rector, "Setting the Record—and the Research—Straight: Heritage Responds (Again) to the *Wall Street Journal*," June, 12, 2007, http://www.heritage.org/Press/Commentary/ed061207b.cfm.
3. Using the *Columbia Journalism Review* inflation calculator (http://www.cjr.org/tools/inflation/) results

in $525 million of 1901 dollars becoming $11.9 billion in 2002 dollars. Then the Bureau of Labor Statistics inflation calculator (http://data.bls.gov/cgi-bin/cpicalc.pl) converted that amount into $13.524 billion in 2006 dollars.

4. The historical data is from *Historical Tables, Budget of the United States Government, Fiscal Year 2007* (Washington, DC: U.S. Government Printing Office, 2006), http://www.gpoaccess.gov/usbudget/fy07/pdf/hist.pdf.
5. Chris Edwards, "Downsizing the Federal Government." Cato Institute, Nov. 2005.
6. Richard Estrada, "Immigration Magnets' Power," *Dallas Morning News*, July 22, 1994.
7. Steven A. Camarota, "Immigrants at Mid-Decade: A Snapshot of America's Foreign-Born Population in 2005," Center for Immigration Studies, Dec. 2005. http://www.cis.org/articles/2005/back1405.pdf, table 6.
8. Ibid., table 14.
9. Ibid., table 10.
10. Ibid., table 13.
11. Cash assistance includes state-run general-assistance programs, TANF, and SSI.
12. Steven A. Camarota, "Back Where We Started: An Examination of Trends in Immigrant Welfare Use Since Welfare Reform," Center for Immigration Studies. Mar. 2003. http://www.cis.org/articles/2003/back503release.html.
13. Robert E. Rector and Christine Kim, "The Fiscal Cost of Low-Skill Immigrants to the U.S. Taxpayer." Heritage Foundation, special report no.14, May 22, 2007, http://www.heritage.org/Research/lmmigration/sr14.cfm.
14. Office of the Actuary. "National Health Care Expenditures Projections: 2005–2015," Centers for Medicare and Medicaid Services, 2006, http://www.cms.hhs.gov/NationalHealthExpendData/downloads/proj2005.pdf.
15. Camarota, "Immigrants at Mid-Decade," table 11.
16. Ibid., p. 15.
17. Paul Fronstin, "The Impact of Immigration on Health Insurance Coverage in the United States," *Employee Benefit Research Institute Notes*, June 2005, http://www.ebri.org/pdf/notespdf/EBRI_Notes_06–2005.pdf.
18. Dana P. Goldman, James P. Smith, and Neeraj Sood, "Legal Status and Health Insurance Among Immigrants." *Health Affairs*, vol. 24, no. 6, Nov.–Dec. 2005, pp. 1640–53.
19. Committee on the Future of Emergency Care in the United States Health System, *Hospital-Based Emergency Care: At the Breaking Point* (Washington, DC: National Academies Press, 2006), prepublication copy, http://newton.nap.edu/catalog/11621.html, p. 2.
20. U.S. General Accounting Office, *Undocumented Aliens: Questions Persist About Their Impact on Hospitals' Uncompensated Care Costs*, GAO–04–472. May 2004. http://www.gao.gov/new.items/d04472.pdf, pp. 3–4.
21. Committee on the Future of Emergency Care, *Hospital-Based Emergency Care*, p. 41.
22. Camarota, "Immigrants at Mid-Decade," table l8.
23. Committee on the Future of Emergency Care, *Hospital-Based Emergency Care*, p. 44.
24. Ibid., p. 43.
25. Jack Martin, *The Costs of Illegal Immigration to New Yorkers*, Federation for American Immigration Reform, Sept. 2006, http://www.fairus.org/site/DocServer/NYCosts.pdf?docID=1161.
26. Jack Martin and Ira Mehlman, *The Costs of Illegal Immigration to Floridians*, Federation for American Immigration Reform, Oct. 2005, http://www.fairus.org/site/DocServer/fla_study.pdf?docID=601.
27. Jack Martin and Ira Mehlman, *The Costs of Illegal Immigration to Texans*, Federation for American Immigration Reform, Apr. 2005, http://www.fairus.org/site/DocServer/texas_costs.pdf?docID=301.
28. Jack Martin and Ira Mehlman, *The Costs of Illegal Immigration to Californians*, Federation for American Immigration Reform, Nov. 2004, http://www.fairus.org/site/DocServer/ca_costs.pdf?docID=141.
29. Jack Martin and Ira Mehlman, *The Costs of Illegal Immigration to Arizonans*, Federation for American Immigration Reform, 2004, http://www.fairus.org/site/DocServer/azcosts2.pdf?docID=101.
30. Steven A. Camarota, *The High Cost of Cheap Labor: Illegal Immigration and the Federal Budget*, Center for Immigration Studies, paper no. 23, Aug. 2004, http://www.cis.org/articles/2004/fiscal.pdf, p. 5.

31. Families USA, *Paying a Premium: The Added Cost of Care for the Uninsured*, publication no. 05–101, June 2005, http://www.familiesusa.org/assets/pdfs/Paying_a_Premium_rev_July_13731e.pdf.
32. Committee on the Future of Emergency Care, *Hospital-Based Emergency Care*, p. 29.
33. Ibid., p. 44.
34. Fronstin, "The Impact of Immigration on Health Insurance Coverage."
35. National Center for Education Statistics, *Digest of Education Statistics 2004*, U.S. Department of Education, http://nces.ed.gov/programs/digest/d04/index.asp, ch. 1, table 29.
36. Camarota, "Immigrants at Mid-Decade," table 16.
37. National Center for Education Statistics, *Digest of Education Statistics 2004*, ch. 2, table 3.
38. National Center for Education Statistics, *Condition of America's Public School Facilities: 1999*, U.S. Department of Education, June 2000, http://nces.ed.gov/pubs2000/2000032.pdf.
39. Jack Martin, *Breaking the Piggy Bank: How Illegal Immigration Is Sending Schools into the Red*, Federation for American Immigration Reform, June 2005, http://www.fairus.org/site/PageServer?pagename=research_researchf6ad.
40. Patricia Gandara, *Review of Research on the Instruction of Limited English Proficient Students: A Report to the California Legislature*, University of California Linguistic Minority Research Institute, Feb. 1999, http://lmri.ucsb.edu/publications/97_gandara.pdf, p. 14.
41. David Denslow and Carol Weissert, "Tough Choices: Shaping Florida's Future," LeRoy Collins Institute, University of Florida. Oct. 2005, http://www.bebr.ufl.edu/system/files/Tough_Choices.pdf. p. 216.
42. Rafael Lara-Alecio, et al., *Texas Dual Language Program Cost Analysis*, Jan. 2005. http://ldn.tamu.edu/Archives/CBAReport.pdf, p. 4.
43. Kristen A. Hughes, *Jusitce Expenditure and Employment in the United States, 2003*, Bureau of Justice Statistics, NCJ 212260, rev., May 10, 2006, http://www.ojp.usdoj.gov/bjs/abstract/jeeus03.htm.
44. U.S. Government Accountability Office, *Information on Criminal Aliens Incarcerated in Federal and State Prisons and Local Jails*, GAO-05-337R, Apr. 7, 2005, http://www.gao.gov/new.items/d05337r.pdf, p. 2.
45. Carl F. Horowitz, *An Examination of U.S. Immigration Policy and Serious Crime*. Center for Immigration Studies. Apr. 2001, http://www.cis.org/articles/2001/crime/toc.html.
46. Ruben G. Rumbaut, Roberto G. Gonzales, Golnaz Komaie, and Charlie V. Morgan. "Debunking the Myth of Immigrant Criminality: Imprisonment Among First- and Second-Generation Young Men." *Migration Information Source*, June 2006, http://www.migrationinformation.org/Feature/display.cfm?id=403.
47. Ruben G. Rumbaut and Walter A. Ewing, "The Myth of Immigrant Criminality and the Paradox of Assimilation: Incarceration Rates Among Native and Foreign-Born Men," Immigration Policy Center, Spring 2007, http://www.ailf.org/ipc/special_report/sr_022107.pdf.
48. U.S. Government Accountability Office, *Information on Criminal Aliens*, p. 3.
49. Martin, *The Costs of Illegal Immigration to New Yorkers*.
50. Martin and Mehlman, *The Costs of Illegal Immigration to Floridians*.
51. Martin and Mehlman, *The Costs of Illegal Immigration to Texans*.
52. Martin and Mehlman, *The Costs of Illegal Immigration to Californians*.
53. Martin and Mehlman, *The Costs of Illegal Immigration to Arizonans*.
54. Rector and Kim, "The Fiscal Cost of Low-Skill Immigrants."
55. Rector, "Setting the Record—and the Research—Straight."
56. James P. Smith and Barry Edmonston, eds., *The New Americans: Economic, Demographic, and Fiscal Effects of Immigration*, (Washington, DC: National Academies Press, 1997), http://newton.nap.edu/catalog/5779.html, ch. 6–7.
57. Robert Rector, "Amnesty and Continued Low Skill Immigration Will Substantially Raise Welfare Costs and Poverty," Heritage Foundation, backgrounder no. 1936. May 12, 2006, http://www.heritage.org/Research/Immigration/bg1936.cfm.
58. Denslow and Weissert, "Tough Choices: Shaping Florida's Future," p. 385.
59. Camarota, *The High Cost of Cheap Labor*.

60. Martin and Mehlman, *The Costs of Illegal Immigration to New Yorkers*, and comparable reports for other states.
61. Tom Bethell, "Immigration, Si; Welfare, No," *American Spectator*, Nov. 1993.
62. Rector and Kim, "The Fiscal Cost of Low-Skill Immigrants."
63. Karlyn H. Bowman, "Attitudes About Welfare Reform," *AEI Studies in Public Opinion*, American Enterprise Institute, Mar. 6, 2003, http://www.aei.org/publications/pubID.14885/pub_detail.asp.
64. Grover G. Norquist, "Reducing the Government by Half: How and Why We Can Cut the Size and Cost of Government in Half in One Generation—the Next Twenty-five Years." *Heritage Insider*, May 2000, http://www.atr.org/content/html/2000/may/000501op-govt_in_half.htm.
65. James R. Edwards Jr., "Public Charge Doctrine: A Fundamental Principle of American Immigration Policy," Center for Immigration Studies, May 2001. http://www.cis.org/articles/2001/back701.html, p. 2.
66. Ibid., p. 4.
67. George J. Borjas, *The Impact of Welfare Reform on Immigrant Welfare Use*. Center for Immigration Studies, Mar. 2002, http://www.cis.org/articles/2002/borjas.htm.
68. Camarota, "Back Where We Started."
69. *Time*/SRBI poll, Mar. 31, 2006, http://www.srbi.com/TimePoll_Final_Report-2006-03-31.pdf.
70. Camarota, *The High Cost of Cheap Labor*.
71. Robert Rector, "Amnesty and Continued Low Skill Immigration Will Substantially Raise Welfare Costs and Poverty," Heritage Foundation, backgrounder no. 1936, May 12, 2006, http://www.heritage.org/Research/Immigration/bg1936.cfm.
72. Camarota, "Immigrants at Mid-Decade," table 15.

DISCUSSION QUESTIONS

1. In light of the evidence presented in this reading, respond to Milton Friedman's quote: "It's just obvious that you can't have free immigration and a welfare state." Do you agree or disagree with the renowned economist? Why or why not?
2. What immigration policy should the United States support in view of our unprecedented debt levels and the unsustainable growth in entitlement programs?
3. Do you agree with the presented evidence that support the argument that the current immigration level is a "fiscal burden" on the government? If yes, what policies would you advocate and why? If no, what supporting evidence can you find to make the counter argument that immigration is a net benefit for the country?
4. What should comprehensive immigration reform entail in your opinion?
5. "Immigration *sí*, Welfare *no!*" What do think about the libertarian approach to this social problem? Can we really have immigration without copious amounts of needed social services, government entitlement programs, and subsidized public education for poor and vulnerable immigrants? Explain your response.

19

Understanding the World of New Slavery

KEVIN BALES

The Four Questions

1. What is the problem Bales identifies in this selection?
2. Is it a *social* problem?
3. What is the cause of this problem? Who does this problem affect? Who benefits from this problem?
4. What can be done about this problem?

Topics Covered

New slavery
Immigration
Cost of slavery
Human trafficking
Global poverty
Absolute poverty
Globalization
Political corruption

Louis works for the phone company near Washington, D.C. He also frees slaves. When he got together with family and friends for the Thanksgiving holiday a few years ago, Louis did what everyone does—he got out his video camera. As he recorded the holiday gathering, he noticed something strange: in the large group of family and friends, one teenage girl always tried to hide when he turned on the camera.

> And I asked myself, you know, what's wrong with the young lady? At first I asked her, where do you come from? She told me she was visiting from Indiana. That is, she was staying with my cousin there. But something stuck in my mind.

Louis had visited his cousin in Indiana several times and didn't remember ever meeting this young woman. Driving home from the party, he asked his wife what she knew about the girl. His wife had heard that the girl had eloped and was now hiding with their cousin. Yet this seemed a little odd as well, because normally a girl seeking refuge would look first to her own family.

Louis had come to the United States from Cameroon, West Africa, in 1985 and eventually became a U.S. citizen. With a degree in management from an American university, he had a good job and served as an elder in his Presbyterian church. Living in the suburbs in Virginia, Louis and his family were pursuing the middle-class dream of stability and security. By his late forties, Louis had achieved much of his dream. But he found himself deeply unsettled by this strange and troubled girl in the middle of his family's Thanksgiving holiday. Things just weren't adding up, so a few days later he drove over to visit a relative, where the young woman was staying.

> And then she began to tell me the story. And felt so bad about it.... I mean everyone has a human feeling, if you hear

> a story which is so terrible, you are moved, being human. I began to put down in writing the stories she told me, and probably if you read the whole report that I wrote, you would come to the conclusion that something was really, really wrong.

The more Louis heard, the more sickened he became. There was no elopement in this girl's past. He listened as, confused, isolated, and still in shock, she painfully recounted years of slavery in the suburbs of Washington, D.C.

Her name was Rose. Back in Cameroon, at the age of fourteen, she had just begun her school's summer vacation when a friend of her aunt stopped by her house. This woman explained that a Cameroonian family in the United States needed someone to help around the house. In exchange, the family would help Rose go to school in America. It sounded like a great opportunity to Rose and to her parents. They talked it over carefully and agreed that she could go right away. Since the summer vacation had just begun, she would have time to settle in before starting school in the United States. Away from her parents, Rose was introduced to the family she would be working for. They bought her air ticket and escorted her through customs and immigration, passing her off as one of their family when they reached the United States. Everything seemed normal until they reached Rose's new home in America. Then the trap closed.

The husband and wife showed Rose the jobs they wanted her to do. Soon the jobs filled her day completely, rapidly taking control of her life. Up at six in the morning, Rose had to work until long past midnight. When she began to question her treatment, the beatings began. "They used to hit me," Rose said. "I couldn't go for three days without them beating me up." The smallest accident would lead to violence. "Sometimes I might spill a drink on the floor by mistake. They would hit me for that," she said.[1] In a strange country, locked up in a strange house far from home, Rose was cut off from help. If she tried to use the phone, she was beaten; if she tried to write a letter, it was taken away from her. "It was just like she was lost in the middle of a forest," said Louis: "she was completely isolated."

Under the complete control of others, subject to physical abuse, paid nothing, working all hours, this fourteen-year-old schoolgirl had become a slave. The promise that she could go to school in America was just the bait used to hook her. In Cameroon her parents received no word from her, only occasional reassuring messages from the family who had enslaved their daughter. The beatings and constant verbal attacks broke Rose's will, and her life dissolved into a blur of pain, exhaustion, work, abuse, and fear. Rose lived in slavery for two and a half years.

Someday we may know the details of what happened in these years, but probably not. Until recently Rose was often nervous and withdrawn, still suffering from the trauma of her enslavement. Demonstrating remarkable resilience, she has moved on to a new life, and it would be understandable if she never wants to revisit that period of unspeakable pain. Her mind has deeply buried her memories. What we do know is that not long after she had turned eighteen, Rose was found trying to talk to a neighbor by the woman who controlled her. Dragging her away, "the woman started yelling at me, started cursing me," Rose explains, "and I couldn't take it anymore. I just had to run away." Later that day she ran to the home of a friend of her "employer." She pleaded for help, but this woman called the family who had enslaved her. When Rose realized the betrayal, she ran again, this time to the parking lot of a nearby K-Mart. Her only hope was a Cameroonian man she had met in her employers' home. He had seemed nice, and she had learned his phone number. Begging change from a stranger to make a phone call, she managed to leave a message for the man, asking him to pick her up. Without a coat, with no other place to go in the cold November night, she waited in the parking lot outside a store. Four hours later, at nearly 11:00 P.M., she was picked up and taken home to the man's family. This man was Louis's cousin.

Although in safety, Rose was still in limbo. The family she was staying with simply did not know what to do with her, and she feared that in time her "employers" would try to take her back. Then came Thanksgiving and the meeting with

Louis. As Louis gently drew out Rose's story, he was shocked and saddened:

> I felt terrible, I mean, I felt really terrible, because I couldn't imagine, not even in my slightest imagination, that in this day and age someone would treat somebody's child the way she was treated. It made me sick in my stomach.

Soon Louis took Rose to stay with his own family, and as she opened up to him, more shocking facts came tumbling out. It occurred to Louis to ask Rose if she knew any more girls in the same situation. "Oh, yes," she replied. Following up on what Rose could remember, he found two more young women in slavery, and by himself, with real daring, he liberated them. One of them, Christy, had been brought to the United States at seventeen and had spent five years as a domestic slave. Sally had been brought at age fifteen and had spent three years in bondage.[2] Now Louis had three young women staying with his family, with their care and support coming from his own pocket. His first job, he decided, was to reassure the women's families, so he took videos of all three relaying messages to their families and then traveled to Cameroon. He showed the video footage to the girls' families, who were shocked but overjoyed, as Louis explains:

> They were very happy to see me, and especially the fact that I took the video of their children, they were extremely happy, because even if they now saw what their daughters had gone through, at least they had firsthand information from their children. I felt good about it because it was like a conclusion to me that I had done the right thing. I could see their faces, and I could see that they realized at least that someone was concerned about the lives of their children.

More than a year before, Sally's family had been told that their daughter had died in America, and their emotions at seeing her alive are hard to imagine. Meanwhile Louis was investigating the connections that had smuggled the girls into the United States. He found a network that recruited girls from poor families by promising education and jobs. One woman provided a house where the girls were taken after leaving their families and were prepared for the trip to America. Some respected members of the Cameroonian community in the United States were involved, and Louis began to understand that he was up against something big.

With the help of lawyers from an organization called CASA Maryland, the girl's "employers" have now been prosecuted in both criminal and civil courts. Rose bravely went through the ordeal of being cross-examined and having her slavery and abuse exposed in open court. The trial resulted in a conviction, and the couple who enslaved her were sentenced to nine years in prison and ordered to pay her $100,000 in restitution. Of course, being awarded restitution and actually getting it are two different things, and Rose will probably never see the money. She'll also never see her parents again; they both died in 2002.

Christy's "employers" got five years in prison and were ordered to pay $180,000 in back wages. So far, Christy has received about $2,000. She and Rose live together now, sharing an apartment in the Washington, D.C., area. Both are working as nursing assistants and dream of being nurses. For the moment they can't afford to go to college, because much of what they earn is sent back to Cameroon to support their families. Christy's remittances are building a house for her parents and are paying the school fees of her younger brothers and sisters.[3] Louis supported the three girls as long as they needed help. Although he slowly used up his savings, he is convinced that he did the right thing. "People treat dogs better than these girls were treated," he told me. "Anyone who cares about other people would do what I did."

If the story of these young women were unique, it would be shocking enough. That there are slaves in the suburbs of Washington, D.C., the capital of the land of the free, might cause us to question our assumption about this country. Our questions become more troubling when we face the fact that Rose's story is one of many such tales in the Washington, D.C., area, and one of thousands in the United States. The U.S. Department of State estimates that as many as 17,500 people are brought into the country *each year* and forced into agricultural work, prostitution, domestic service, or sweatshop labor. These are not poorly paid migrant workers; these people are slaves. According to conservative estimates, there are tens of thousands of slaves in America today.[4] And the thousands in the United States and other developed countries are just a fraction of the total across the globe. The estimated twenty-seven million slaves in the world today equals more than twice the number of people taken from Africa during the 350 years of the Atlantic slave trade.[5] But given that legal slavery no longer exists, how can we call these people *slaves*?

WHAT MAKES A SLAVE?

Slavery has been with us since the beginning of human history. When people began to congregate in Mesopotamia and make the first towns around 6800 B.C., they built strong external walls around their towns, suggesting the occurrence of raiding and war. We find the first depiction of slavery in clay drawings that survive from 4000 B.C.; the drawings show captives of battle being tied, whipped, and forced to work by the Sumerians. Papyrus scrolls from 2100 B.C. record the ownership of slaves by private citizens in Egypt and list the first documented price of a slave: eleven silver shekels. When money began to be used, slave trading became a way of making a living, and we see records showing slave-raiding expeditions from Egypt capturing 1,554 slaves in Syria in one season. Around 1790 B.C., the first written laws introduced the legal status and worth of slaves. These Babylonian law codes clearly stated that slaves are worth less than "real people," a principle that is repeated for the next four thousand years. The ancient code is gruesomely clear: a physician making a fatal mistake on a patient, for example, is ordered to have his hands cut off, unless the patient is a slave, in which case he only has to replace the slave.

In past centuries, people had no problem understanding who was a slave and who wasn't, even given the existence of temporary enslavement. Slaves then and now share one central condition: violence is freely used to control them or punish them. The Babylonian code again: "If a slave strike a free man, his ear shall be cut off"; and the Louisiana Slave Code of 1724: "The slave who will have struck his master, his mistress, the husband of his mistress, or their children, either in the face or resulting in a bruise or the outpouring of blood, will be punished by death." For nearly four thousand years the right to inflict violence on a slave was enshrined in law. When the legal ownership of humans ended, as it did in the United States in 1865, many people thought that slavery had ended as well.

But even when people in the United States no longer owned slaves legally, they often continued to control them—by restricting their housing and food supply, refusing them education, limiting their movements, and threatening them with violence. This fact does not diminish the great achievements of the abolitionists and the slaves who fought for their freedom; if there are tens of thousands of slaves in the United States today, it is worth remembering that there were once four million. Still, no matter how many laws were passed against it, de facto slavery never stopped. Throughout history, slavery has meant taking total control of a person and exploiting that person's labor. The essence of slavery is neither legal ownership nor the business of selling people; the essence of slavery is controlling people through violence and using them to make money. Before laws gave one person the right to own another person, even before the invention of money as a means of exchange, slavery was part of human life. Today the laws allowing slavery have been repealed, but people around the world are still brutalized and broken and reduced to slavery through violence. Their free will is taken away. Their labor, their minds, and their lives are consumed by someone else's greed. Slavery, at its most fundamental, has just three elements: control through violence, economic exploitation, and the loss of free will. Slavery is not about race, color, or ownership. Any one or all of these may be used to justify slavery, but they are not essential for its existence.

If people are not legally recorded as being slaves, how can we really call them slaves? The answer is relatively simple, though as with most human conditions, there are always cases that defy clear definition. We can start by asking, Can this person walk away from the situation without fear of violence? If the answer is no, if the person is beaten when trying to leave, then you have one indication of slavery. Another question we can ask: is this person paid nothing, or at a level that barely keeps the person alive from one day to the next? Look again at Rose. She couldn't leave because of the threat of violence, she was paid nothing, she was given only enough food to keep her alive, and she was economically exploited. Her ability to exercise free will was taken from her. Rose was a slave. The newspapers might have called her condition "virtual slavery," or said she lived in "slavelike" conditions, but make no mistake: like women in bondage in ancient Babylonia or the antebellum American South, Rose was a slave.

Slavery is also not a matter of duration. The fact that Rose spent "only" two and a half years in bondage does not make her any less a slave. Slavery isn't necessarily a permanent condition. That was never the case, even when slavery was legal. The ancient Babylonian law and the Louisiana Slave Code both allowed for temporary enslavement. For thousands of years people have been captured, snared, coerced, tricked, sold, kidnapped, drugged, arrested, swindled, seduced, assaulted, or brutalized into slavery. A fortunate few have then managed to make their way out again through any number of exits. Some were released when their health and strength broke down and they were no longer useful. Some managed to escape after decades and others after just weeks. For some families of slaves, it took generations. On rare occasions a master would free a slave as a gift, but that did not change the fact that the person had been a slave. The same is true today.

We have to put behind us the picture of slavery most of us hold in our minds, that of slavery in the antebellum South. Contemporary slavery shares with the slavery of the past the essentials of violence and exploitation, but today it is not a legal institution, a key part of any country's economy, or a relationship crucially dependent on race or ethnicity. Today, as in the past, slavery exists in many different forms around the world. But modern slavery has two key characteristics that make it very different from slavery of the past: slaves today are cheap, and they are disposable.

A CATASTROPHIC FALL IN PRICE

This new variant of slavery arrived with the twenty-first century. Today slaves are cheaper than they have ever been. The enslaved fieldworker who cost the equivalent of $40,000 in 1850 costs less than $100 today. This dramatic fall in price has forever altered the basic economic equation of slavery. When the price of any commodity drops radically, the balance of supply and demand is fundamentally changed. Today there is a glut of potential slaves on the market. That means they cost very little but can generate high returns, since their ability to work has not fallen with their price. The return to be made on slaves in 1850 Alabama averaged around 5 percent. Today returns from slavery start in the double figures and range as high as 800 percent. Even when they are used in the most basic kinds of work, slaves can make back their purchase price (however that acquisition occurred) very quickly.

Slaves were far more costly in the past. Although finding a clear equivalent between modern dollars and the currency used in ancient Babylonia or Rome is impossible, we can look at how much slaves cost in terms of things that don't change very much over time. An understanding of slave prices from the past is based on three measures: the value of land, the annual wages paid to a free agricultural worker, and the price of oxen. For all of human history (and still today), people have maintained records of the price of land and the cost of keeping a worker in the fields. And for thousands of years, until the Industrial Revolution, oxen were used as a power source to get food out of the ground and onto the table. (When you read *oxen*, think *tractor*. Today an American farmer may pay upward of $100,000 for a tractor; in the past oxen were expensive.) For the price of three or four oxen, a farmer could buy a productive field big enough to support a family and then some, or pay the annual salaries of two or three agricultural workers. Three or four oxen were a big capital investment. Yet, on average, they were worth the price of only one healthy slave. Or consider that for the price of one slave in the Deep South in the 1850s, a person could buy 120 acres of good farmland. It should not be any surprise that slaves were expensive; after all, slaveholders were buying the complete productive capacity of a human being, all the work they could squeeze out for as long as they could keep the slave alive.

In India today you can still buy slaves as well as farmland and oxen, which remain the essential "tractors" that keep farms going. But when we compare the price of a slave to the modern prices of land, labor, or oxen in India, the slave costs, on average, 95 percent less than in the past. This precipitous collapse, unprecedented in all of human history, has dire consequences for slaves.[6] If you could buy a fully equipped, brand new car for $40, do you think your relationship to your car would change? If your car were that cheap, you would begin to treat it as something to be used and then discarded. Why even fix a flat tire if the whole car costs less than the repair? Most slaveholders feel that way about slaves today.

The inexpensiveness of slaves is good for the slaveholder and great for the bottom line but disastrous for slaves. A low purchase price means that a slave does not represent a large investment requiring special care; the slave is easily replaced. Slaves today are treated like cheap plastic ballpoint pens, the kind we all have in our desk drawers or pockets. No one worries about the care and maintenance of these pens or about keeping a careful record of their whereabouts. No one files a deed of ownership for these pens or sends out a search party if one goes missing. No one takes out insurance on these pens. These pens are disposable, and, because they are so cheap, so are slaves.

If slaves get ill, are injured, outlive their usefulness, or become troublesome to the slaveholder, they are dumped—or worse. The young woman enslaved as a prostitute in Thailand is thrown out on the street when she tests positive for HIV. The Brazilian man tricked and trapped into slavery making charcoal is tossed out when the forest is razed and no trees are left to cut. The boy in India who spends all day rolling bidi cigarettes is dumped or sent back to his family if he is injured or ill, and the slaveholder will try to take another child in his place. The young women in "ritual slavery" in Ghana, who has been exploited, sexually abused, and impregnated again and again by a *trokosi* priest, will be sent back to her parents when the priest tires of her or her health breaks down. Enslaved domestic workers around the world will be discarded when their "family" moves to another city or country. Like plastic pens or paper cups, slaves and potential slaves are so numerous that they can simply be used up and thrown away. Rose, for example, was fairly expensive as slaves go today, yet her slaveholders paid nothing for her. Her acquisition cost involved only the time needed to spin a web of lies to her and her parents and the expense of bringing her to America.

WHERE DO ALL THESE SLAVES COME FROM?

How do we explain a world in which twenty-seven million people are in slavery? Where did all these slaves come from? And how did these people become enslaved? Basically, three factors that converged after World War II gave birth to the resurgence of slavery. The first factor is the world population explosion. After 1945 the world population grew like never before in human history. This growth was the product of many positive developments: the control of infectious diseases, better health care for children, and a prosperity that provided sustenance for the coming billions. World population exploded from two billion to over six billion in about fifty years, with most of this growth occurring in the developing world. Figure 19.1, plots world population growth along with the fall in the prices of slaves. What becomes clear is that the population explosion helped create a glut of potential slaves flooding the market and leading to a crash in prices.

Population growth helps to explain the drop in prices, but it doesn't necessarily explain the growth in numbers of slaves. Simply having a lot of people doesn't make them into slaves; other things had to happen to lead them or force them into slavery.

The second factor pushing these growing millions toward slavery is a collection of dramatic social and economic changes, many of which were supposed to make those people's lives better. Like the world population, the global economy boomed

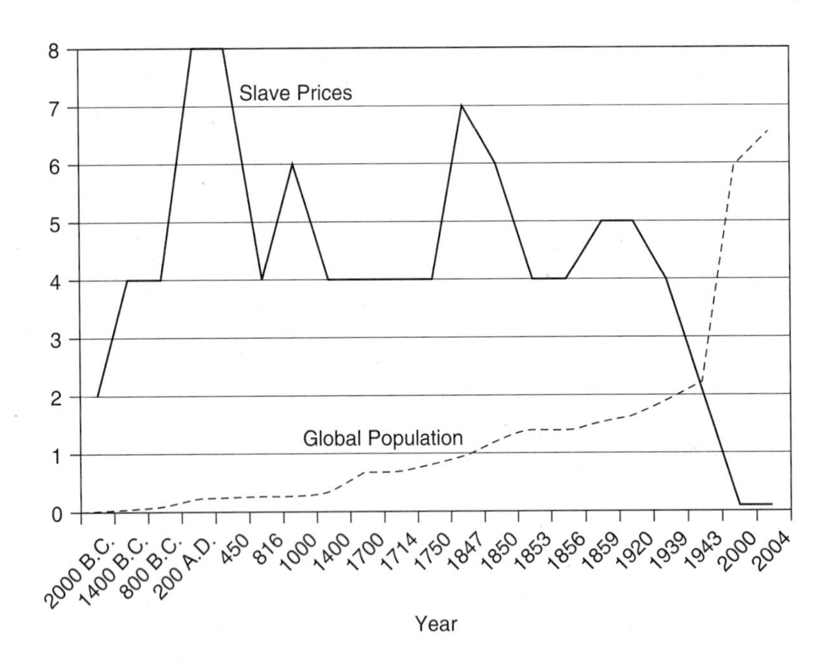

FIGURE 19.1 World population and the price of slaves, 2000 B.C. to A.D. 2004. The Y-axis represents both the global population in billions and the Slave Price Index, which is composed of equivalency measures for oxen, land, or agricultural wages and which varies from 0 to 8. The irregularity of the years along the X-axis is due to the lack of any regular data on slave prices; the years given reflect those years in which slave price information can be obtained.

after 1945. As colonies in the developing world gained their independence, many of the new countries opened up to Western businesses. In the 1950s people remarked on the spread of Ford cars and Coca-Cola all over the world; by the 1960s the rapid economic changes in the developing world were seen as commonplace; by the 1990s no one thought it surprising that teenagers in India, Malaysia, or the Ukraine were eating the same McDonald's hamburgers and humming the same tunes from MTV as teenagers in Chicago, Tokyo, or London. As the economy became global and grew exponentially, its benefits were shared in many parts of the world. But other parts of the world did not take part in that growth. Whole populations were left behind, stuck in the subsistence poverty of the past, or worse.

Poverty is often thought of as having two levels, though this is just a rough guideline. At the bottom level are more than one billion people who live on $1 a day or less. All these people are living in the developing world, and for the most part, they are living outside towns and cities living out the same hand-to-mouth existence that was the rule for most of human history. What does it mean to be this poor? It means you are always hungry and that access to medical care and education for your children is pretty much out of the question. It means you are unlikely to have the basic needs of life: clean water, a roof to keep out the rain, adequate clothing, or even a pair of shoes.

This is life without options. Every action must be aimed at day-to-day survival, and even that survival is nor assured. Desperation is the norm, and families are ready to do anything to survive. These families are found especially in rural South Asia and rural Africa, areas where slavery thrives. Later in this book you will meet families in India who are trying to live on forty cents a day. These are families whose children are regularly harvested into slavery.

One step up from this extreme poverty, again as a guideline, are families living on approximately $1 or $2 a day, a level that is sometimes called "moderate poverty."[7] Many of these families are living in the vast shantytowns that surround the major cities of the developing world. For example, Mexico City has a population of twenty million, and about half of the residents live in shacks and lean-tos of cardboard and scrap wood that lack basic amenities. Many of these families are economic refugees from rural areas where family farms have been converted into plantations growing cash crops for export. Their lives as subsistence farmers, based around the village and church, were shattered when they were dispossessed of the land they had worked, often for generations. Searching for jobs, they migrated to the city, only to find themselves competing with millions of other campesinos. In the shantytown they have lost the neighbors, the church, and the customs of the rural village. Criminal gangs control much of the shantytown areas. The government has little time or attention to give these poor and disenfranchised people, relegating them to second-class status. This pattern is repeated across the developing world, and the result, whether in Rio, New Delhi, Manila, or Bangkok, is extreme vulnerability. The police do not protect you, the law is not your shield, you can't buy your way out of problems, and any weapon you have is no match for those of the gangs and the police.

In fact, if we compare the level of poverty and the amount of slavery for 193 of the world's countries, the pattern is obvious.[8] The poorest countries have the highest levels of slavery. The relationship would be exact except for the effects of global human trafficking in which the vulnerable are enslaved and transported from poor countries to rich ones, with the result that the richest countries have significant pockets of slavery.

This link between poverty and slavery holds almost any way you measure it. For example, the United Nations has classified thirty-eight countries of the world as being "high-debt countries," which means that these countries are carrying a crippling load of debt owed to international lenders. A high-debt country has to use what little income or taxes it can gather to service debt rather than to invest in its own people. This is often called a *debt overhang*. Debt from the past bears down on a country, paralyzing it and preventing any growth in the future.[9] The types of investments a high-debt country is *not* able to make—schools, law enforcement, economic growth, and so on—are exactly the ones that are most likely to reduce the amount of slavery. If you look across all the countries in the world, those with the largest amount of debt overhang also have some of the highest levels of slavery.[10]

One of these poor countries is Cameroon, where Rose came from. Cameroon has all the usual markers for serious poverty: half the population is poor, the economy carries a high debt load, infant mortality is high, HIV/AIDS is rampant, and life expectancy is around fifty years and falling. This level of poverty creates many hardships, but just having a very large number of poor people still doesn't make those people slaves. Rapid population growth and the impoverishing impact of globalization, epidemics, natural disasters, war and civil conflict, kleptocratic governments, and the international arms trade all support the emergence of slavery but do not cause it. To turn the poor into slaves requires violence, and violence needs the right conditions to grow unchecked.

ON THE TAKE

Corruption, especially police corruption, is the third force that drives the growth of slavery. For slavery to exist, the slaveholder must be able to keep the slave where the law can't protect them. Rose, enslaved in the United States, had to be isolated and locked up. But in many parts of the world, a simple payment to the local police allows a slaveholder to use violence without fear of arrest. Sometimes the police themselves will provide violence for an extra fee. When governments fail to protect their citizens and to maintain the rule of law, those citizens can become slaves. And because slavery is now illegal everywhere, the complicity of crooked police is a fundamental requirement for slavery to take root. In Western Europe, Canada, and the United States, slavery happens *in spite* of the efforts of the police, but in many countries slavery flourishes *because of* the work of the police. Almost everywhere you find slavery in the developing world, you find police or government officials on the take, turning a blind eye.

It is not difficult to understand the pressure on police to join forces with the slaveholders. If a police officer's salary is $10 or $20 a month, the opportunity to bring in an extra $100 a month is the difference between poverty and being able to feed a family, send the children to school, and have electricity. Taking the bribe is even easier when the police officer is urged to do so by the boss. Landlords, moneylenders, and businesspeople, the solid citizens of the town or village, are likely to use slavery in their businesses. Does a policeman really want to jeopardize his job, put his family at risk, and alienate the most powerful members of his local community just to protect people no one seams to care about? Since the enslaved are often migrants from somewhere else, or members of a lower caste, lower class, or different ethnic or religious group, serving their interests will offer few rewards and carry many penalties. Again, if you examine most of the world's countries in a systematic way, you find a strong relationship between slavery and corruption.[11]

The pattern is strong and clear: more corruption means more slavery. This is a special challenge when corruption becomes institutionalized. The bribes pass up the chain of command and into the hands of politicians and government officials. Soon law enforcement is dedicated to protecting systematic law violation. In Thailand, for example, lucrative police commands are sold to the highest bidder, and the regular payments from slaveholders join the flow of money from other criminals into the pockets of police and government officials. Russia is now a major exporter of trafficked women; in Moscow a single monthly payment providing protection from government taxes, police investigation, fire, theft, vandalism, safety inspections, and parking tickets is deposited directly into a U.S. bank account.[12] The size of the payment depends on the size of the business and whether or not it is legal. The bribes required for smuggling drugs or people are high, but then these crimes make huge profits.

If this sort of corruption is widespread, the national government faces an enormous task. Pakistan, for example, enacted a strong law against debt bondage slavery in 1988 and revised it in 1992, but in spite of a large number of cases coming to light, not a single offender has been convicted. The enslaved may be freed, usually through the actions of human rights organizations, but because of corruption the slaveholders are never punished. Instead, the ex-slaves and their liberators are at great risk of violent retaliation and persecution.[13] More than twenty-five years ago India enacted an excellent law against debt bondage slavery, setting a three-year prison sentence and a fine for anyone convicted of forcing someone into, or keeping them in, bondage. To my knowledge, of the hundreds of cases prosecuted, no convicted slaveholder has ever served prison time. Today fines of just 100 rupees (less than $2) are common for those convicted, making a mockery of the law. The linchpin of slavery in many countries is government indifference or, worse, complicity. When corruption is widespread, governments must do more than just pass tough laws; they must also root out corruption and give protection to those who have come out of slavery. At international meetings, many countries make statements of heartfelt concern for the enslaved, only to forget about them as soon as their representatives return home. In any event, slaves are voiceless, and unless they have powerful friends, their cause is forgotten....[14]

NOTES

UNDERSTANDING THE WORLD OF NEW SLAVERY

1. Quotations from Rose were gathered in interviews for the film *Slavery: A Global Investigation*, TrueVision Productions, London, 2001.
2. "Sally" didn't want her real name to be used. I have changed the names of all the people, enslaved or freed, whom I interviewed for this book, with the exception of those people whose stories have already been publicized, activists and abolitionists with public roles, and public officials. The safety of some antislavery workers and freed slaves depends on their anonymity. Those still enslaved are under enough threat without my adding to it.
3. The information on the outcomes of the trials was gathered in interviews for the film *Dreams Die Hard*, Free the Slaves, Washington, D.C., 2005.
4. The U.S. Department of State estimates that around 17,500 people are trafficked into the United States each year. We know from work done by Free the Slaves and the Human Rights Center, University of

TABLE 19.1 Levels of Poverty and Levels of Slavery for 193 Countries

	Level of Slavery in Country (%)					
	No slavery	Rare or very little slavery	Persistent low level of slavery	Regular slavery in a few sectors	Slavery in many sectors	Total % (nations)
Extreme poverty	0	3.2	48.4	32.3	16.1	100 (31)
Moderate poverty	13.8	17.2	24.1	17.2	27.6	100 (29)
Low income	21.2	33.3	39.4	1.5	4.5	100 (66)
Middle income	30.0	55.0	10.0	5.0	0	100 (47)
Rich nations	20.2	30.1	30.1	10.9	8.8	100 (20)

TABLE 19.2 Levels of Debt and Levels of Slavery for 204 Countries

	Level of Slavery in Country (%)					
Countries	No slavery	Rare or very little slavery	Persistent low level of slavery	Regular slavery in a few sectors	Slavery in many sectors	Total % (nations)
High-debt	2.6	5.3	42.1	26.3	23.7	100 (38)
All other	25.3	36.1	25.9	7.8	4.8	100 (166)

California, Berkeley, that the average period that trafficking victims are enslaved in the United States is between three and five years. For more on slavery in the United States, see Free the Slaves and Human Rights Center, University of California, Berkeley, *Hidden Slaves: Forced Labor in the United States*, 2004, available at http://freetheslaves.net/files/Hidden_Slaves.pdf; or at "Hidden Slaves: Forced Labor in the United States," *Berkeley Journal of International Law* 23, no. 1 (2005): 47–111.

5. For an explanation of the estimate that there are about twenty-seven million slaves in the world, see Kevin Bales, *Disposable People: New Slavery in the Global Economy* (Berkeley: University of California Press, 1999); or "International Labor Standards: Quality of Information and Measures of Progress in Combating Forced Labor, *Comparative Labor Law and Policy* 24, no. 2 (Winter 2004).
6. A slave in India costs less than 1 percent of the price of a productive field, or about 10–15 percent of the annual wages of a farm laborer (one of the lowest-paid workers in India), and around 17 percent of the price of an ox. In the Ivory Coast today, a slave costs just under 4 percent of the annual wages of a poor farm worker. Antislavery field workers in India collected the prices of land, oxen, agricultural ages, and slaves in 2005 and 2006.
7. These categories, of course, cannot portray the realities of poverty around the world and are used as guidelines. There is justified criticism of this structure ($1 a day equals abject poverty, and $1 to $2 equals moderate poverty) because it may be much harder to live on $2 a day in a shantytown than it is to live on less than $1 in a place where people grow their own food. The suffering that accompanies poverty may not be diminished just because someone has crossed from $1 to $2 a day. As we will see, vulnerability to slavery often has more to do with lack of access to productive assets than it has to do with income.
8. Table 19.1 shows the relationship between poverty and slavery in the world today. The table groups 193 countries according to poverty as measured by their Gross Domestic Product (GDP) and shows the amount of slavery that occurs for each group of countries. The relationship between these two measures is statistically significant at greater than the .001 level (Chi-square = 76.44, df = 16), and the Spearman's correlation is −.496, also significant at greater than the .001 level.

TABLE 19.3 Levels of Corruption and Levels of Slavery for 177 Countries

	Level of Slavery in Country (%)					
Countries	No slavery	Rare or very little slavery	Persistent low level of slavery	Regular slavery in a few sectors	Slavery in many sectors	Total % (nations)
Low corruption	48.3	53.7	0	0	0	100 (29)
Medium corruption	32.8	34.4	21.9	6.3	4.7	100 (64)
High corruption	3.6	19.0	44.0	19.0	14.3	100 (84)

9. For a further discussion of debt overhang see Jeffrey Sachs, *The End of Poverty* (New York: Penguin, 2005).
10. Table 19.2 shows the link between international debt and slavery for 204 countries. The relationship between these two measures is statistically significant at greater than the .001 level (Chi-square = 42.62, df = 4), and the Spearman's correlation is .433, also significant at greater than the .001 level.
11. Table 19.3 shows the link between slavery and government corruption for 177 counties. An annual report by the organization Transparency International scores most of the countries in the world on their level of corruption. I have grouped these countries into those with high, medium, and low levels of corruption. The relationship between these two measures is statistically significant at greater than the .001 level (Chi-square = 66.68, df = 8), and the Spearman's correlation is .589, also significant at greater than the .001 level.
12. I once owned part of a business in Moscow, and another businessman there explained to me in detail the nature of the corruption he faced.
13. Human Rights Commission of Pakistan, *State of Human Rights in 2006* (Lahore, Pakistan, 2007), pp. 239 and 251.
14. The eight Millennium Development Goals (MDGs)—which range from halving extreme poverty to halting the spread of HIV/AIDS to providing universal primary education, all by the target date of 2015—form a blueprint agreed to by all the world's countries and all the world's leading development institutions. The MDGs have galvanized unprecedented efforts to meet the needs of the world's poorest. However, it has to be said that at the beginning of 2007, the world was already falling behind on the timetable to achieve these goals. The plan to end extreme poverty by 2025 is explained in Jeffrey Sachs, *The End of Poverty*.

DISCUSSION QUESTIONS

1. "There are no slaves in the United States!" What would you say to a friend who held this belief, and what evidence would you present to counter this assertion?
2. What factors make modern-day slavery unique and less detectable than the slavery of the American antebellum south?
3. What is the connection between poverty and slavery? And what should be done to address this relationship?
4. Why is slavery a social problem? And why should Americans care about the plight of slaves when the institution of slavery has been a part of human society since recorded history?
5. Should we extend the "slave" label to individuals who labor in multinational sweatshops, at low wages, under dirty and dangerous conditions, with forced overtime, and with the constant threat of termination if they refuse to comply with orders and/or attempt to form unions? Are these individuals "slaves," or does the fact that they have "choice" disqualify them from the label? Explain your position.

20

Hmong American Youth

STACEY J. LEE

The Four Questions

1. What is the problem that Lee identifies in this article?
2. Is it a *social* problem? Who is this problem harming? Are people aware of this problem?
3. What is the cause of this problem?
4. What can be done about this problem? Does the author offer suggestions?

Topics Covered

Hmong assimilation
"Model Minority" stereotype
Racial "blackening"
Hmong American families
Racial disparities culturation
Racial hierarchy "perpetual foreigners" label
Assimilation

I was in junior high school when I first heard my mother explain that *money whitens*. As a Chinese-American growing up in Mississippi during the 1940s and 1950s, my mother's social experiences were shaped by the black and white discourse on race. My mother explained that as her parents, grocery store owners, became more economically successful the whites in the town were "friendlier" to them. Most significantly, my mother recalls that there were social advantages, including attending white schools, which came with being whitened. For Chinese-Americans in the Mississippi Delta, the process of being whitened simultaneously involved being de-blackened. Although my mother recognizes that she experienced significant privileges as someone who was whitened, she has always pointed out that she and other Chinese-Americans were never viewed as being actually white. In Tuan's (1998) language, Chinese-Americans in the Mississippi Delta achieved the status of "honorary whites." As "honorary whites," Chinese-Americans were always subject to the scrutiny of whites. My mother's tenuous status as a whitened Chinese-American was driven home when her high school principal denied her the right to make the graduation speech, a right she had earned as the salutatorian of her graduating class.

My mother's story is not entirely unique to her or to Chinese-Americans in the segregated south. Various scholars have made similar observations about the significance of social class and the position of Asian-Americans in the racial hierarchy (Okihiro, 1994; Ong, 1996). Today the experiences of newer Asian-American immigrants are similarly influenced by the largely black and white discourse that continues to shape understandings of race in the United States (Lee, 2005; Lei, 2003; Ong, 1996; Reyes, 2007). Ong (1996) asserts, "non-white immigrants in the First World are simultaneously, though unevenly, subjected to two processes of normalization: an ideological whitening or blackening that reflects dominant racial oppositions and an assessment of cultural competence based on imputed human capital and consumer power in the minority subject" (p. 737). The educational and economic

SOURCE: From Stacey J. Lee, "The Ideological Blackening of Hmong American Youth," in *The Way Class Works: Readings on School, Family, and the Economy*. Edited by Lois Weis, pp. 305–314.

success of many East Asian and South Asian immigrants has perpetuated the stereotype that Asian-Americans are model minorities (i.e., near whites). On the other hand, the high rates of poverty and low rates of educational attainment among many Southeast Asian immigrants have led to their ideological blackening. In her ethnographic study of Southeast Asian youth, for example, Reyes (2007) observed that the youth "have fallen prey to these familiar stereotypes traditionally assigned to African-Americans, as they settle in impoverished urban areas across the United States and participate in gang culture...." (p. 12).

As it was during my mother's childhood, the racist discourse on race associates whiteness with all that is valued in dominant society (e.g., economic self-sufficiency and self-reliance) and blackness with failure and laziness. Groups who are identified as being like whites earn social privilege and economic opportunities. On the other hand, being associated with blackness has potentially negative social and economic consequences. Although economic self-sufficiency is certainly tied to the process of whitening, money alone does not lead to the process of whitening. In order to be viewed as "near white" (Okihiro, 1994) or "honorary whites" (Tuan, 1998), Asian-Americans must adopt cultural characteristics associated with the dominant white middle class, and distance themselves from Asian (i.e., foreign) cultural norms. In my maternal grandparents' case, for example, this meant becoming Southern Baptists.

Similarly, the process of blackening also involves more than an evaluation of a group's economic circumstances. If the process of blackening were simply linked to poverty, working-class and poor Chinese-American residents of Chinatowns would be subject to ideological blackening. Instead, these individuals are identified as perpetual or "forever foreigners" (Tuan, 1998). Chinatowns are seen as exotic locations where the residents speak "foreign" languages and eat "foreign" foods. Segregated from other ethnic and racial groups, Chinese residents of US Chinatowns are viewed as foreign others who exist outside the category of citizens (Lowe, 1996). In contrast to Chinatown residents who remain outside of the black and white discourse of race, Vaught (2006) found that Samoan youth in the Pacific Northwest are blackened in the eyes of their teachers and their peers. In the case of the Samoan youth, issues of poverty, low academic achievement, and the fact that they lived in predominantly black neighborhoods appeared to contribute to their blackening. While poverty is certainly associated with blackness, these examples suggest that the process of racialization is complex and multifaceted.

For immigrant students, there are serious educational consequences that come with being whitened or blackened. Asian-American groups who have achieved a model minority or "near white" status are held to high academic expectations and are often favored by teachers (Lee, 1996). Although this may appear to be an advantage, members of this group who fail to achieve model minority success may be overlooked by teachers, and may suffer from shame. Like their African-American peers who face racial barriers in education, Asian-American groups who are blackened face low expectations and marginalization in schools (Lee, 2005; Vaught, 2006).

The Hmong American students' experiences illustrate the role that schools play in the racialization of immigrant youth. The specific focus in this chapter will be on the ways high rates of poverty within the Hmong community, Hmong cultural expressions of family, Hmong American youth styles, student achievement, and issues of space intersect in the process of blackening that Hmong American youth undergo. The data in this chapter come from my ethnographic study on Hmong American high school students living in a mid-sized Midwestern city that I call Lakeview (Lee, 2005). This study explored the ways race, class, and gender informed the academic and social experiences of Hmong American high school students at a school I call University Heights High School (UHS). I conducted fieldwork (e.g., participant observation of Hmong American students inside and outside of school, interviews of students and staff) from January 1999 until June 2000. Although located in a predominately white, middle to upper-middle-class neighborhood, UHS draws students from the south side of the city where the majority of the Hmong American and African-American students live in lower-income housing.

THE HMONG: HISTORICAL BACKGROUND

The first Hmong arrived in the United States as refugees from Laos thirty years ago. During the Vietnam War the Hmong were US allies in the "secret war" against communism in Laos. The Hmong suffered tremendous casualties during the war, and many more died in refugee camps before being resettled in the United States and around the world (Quincy, 1995). Early Hmong refugees faced significant linguistic, cultural, and economic barriers in their adjustment to life in the United States. Hmong culture in Laos was primarily oral, and children learned cultural traditions and day-to-day skills by observing their elders' daily routines. Few boys, and even fewer girls, had access to formal schooling. Scholars estimate that up to 70 percent of the Hmong refugees were non-literate when they arrived in the United States (Takaki, 1989). Early research on the educational experiences of Hmong refugees highlighted the problems that Hmong students faced, including high dropout rates from middle and high school (Goldstein, 1985; Trueba, Jacobs & Kirton, 1990).

According to the 2000 census, there are 186,310 people who identify themselves as Hmong living in the United States (US Census Bureau, 2000). The largest Hmong American communities are in California, Minnesota, and Wisconsin. As is the case in other immigrant communities, the Hmong American community is young, with a median age of 16.1. The Hmong community continues to suffer from high rates of poverty. According to the 2000 Census, 37.8 percent of Hmong lived under the poverty line, compared to 12.4 percent of the US population overall. Half of the Hmong who lived under the poverty line were individuals under the age of eighteen (Pfeifer & Lee, 2004).

HMONG AMERICAN STUDENTS: "AT RISK"

UHS enjoys an excellent reputation in the city and throughout the state of Wisconsin. Despite this reputation, however, there is a significant achievement gap at UHS that reflects race and class lines. Few Southeast Asian, African-American, or Latino students are on the honor roll or in the higher academic tracks. In conversations with teachers, guidance counselors, and administrators at UHS I found that they used similar language to talk about Hmong American and African-American students. Hmong American students, for example, were described as experiencing high rates of poverty that put them "at risk" for academic failure and for becoming members of the "new underclass," both terms historically associated with black people (Lipman, 1998). The language used to describe Hmong and other Southeast Asian students stood in stark contrast to the characterization of East Asian American students. While Hmong students were typically described in deficit terms, East Asian students were described as model minorities. As high-achieving model minorities who were in the advanced classes and participated in high status extracurricular activities, East Asian students at UHS achieved a whitened or near white status. One of the vice-principals, for example, asserted that many East Asian students at UHS were "outstripping white kids in terms of attendance and probably achievement" (Lee, 2005, p. 48).

UHS educators suggested that the neighborhoods where Hmong American students lived put them "at risk" for underachievement and deviant behavior. Although the city of Lakeview has become increasingly diverse in the last two decades, the neighborhoods are highly segregated by race and class, with most African-Americans and Southeast Asians living in low-income neighborhoods where there were few whites. Most of the East Asian students were middle class, and lived in predominately white neighborhoods, a fact that may have contributed to their "near white" status. The great majority of Hmong American students, on the other hand, qualified for free or reduced lunch and lived in low-income housing on the south side of the city, an area that teachers described as being economically disadvantaged and dangerous. Not insignificantly, the south side of Lakeview had been identified as a low-income black neighborhood long before the first Southeast Asians settled there in the mid-1970s. Educational researchers have highlighted the relationship between social space and identity (Perry, 2002; Weis, 2004). In her longitudinal ethnography of white working-class adults, for example, Weis (2004) clearly demonstrates the fact that white working-class adults believe that their distinct identities as whites depends on living in white neighborhoods. At UHS, it appears that the fact that Hmong American students lived in poor black neighborhoods played a role in how educators perceived the Hmong community, specifically the way educators racialized them.

The fact that most of the second-generation Hmong American students have adopted hip-hop styles of dress and speech that the UHS teachers associated with African-American youth was used as evidence that Hmong American youth were Americanizing in "bad ways" and "at risk" for academic failure and delinquency. UHS educators assumed that hip-hop culture was inherently dangerous and anti-school. The process of blackening intersects in particularly significant ways for Hmong American boys. Like African-American boys, Hmong American boys were viewed as being potentially dangerous and therefore in need of control (Ferguson, 2000). Many UHS educators expressed fear that Hmong American boys were "gang involved." Not insignificantly, teachers interpreted Hmong American boys' hip-hop style clothing as "evidence" of gang involvement. While there were white, middle-class boys who dressed in hip-hop clothing, they were not identified as being in gangs. Despite the teachers' assumptions regarding the prevalence of gang activity among Hmong Americans, I did not find any evidence of a gang problem among Hmong youth. When I asked Hmong American youth about gangs most laughed and said that the Hmong students at UHS were simply gang "wannabees." Significantly, the teachers who were most convinced that Hmong American youth were involved with gangs were the ones who admitted that they didn't have much contract with Hmong students.

Interestingly, some non-Hmong students at UHS characterized Hmong American students as "acting black." In the following quote, a student explains the ways Hmong American students are stereotyped by their non-Hmong peers.

> Like the stereotypes that people have are like a lot of the Hmong—the Hmong stereotype is that they're all gangsters and they follow, like, the "black path" of wearing baggy clothes and being cool and forming gangs and not coming to school, and being truant, you know, all the time.

Not insignificantly, contained within the stereotype of Hmong American youth are problematic and racist stereotypes of African-Americans.

BLAMING FAMILIES

Ethnic culture, race, and social class have all been found to shape the nature of family life (Fine & Weis, 1998; Heath, 1983; Lareau, 2003; Valdes, 1996; Weis, 2004). In her ethnography of working-class and middle-class families, Lareau (2003) captures the many ways social class informs "critical aspects of family life: time use, language use, and kin ties" (p. 236). Furthermore, Lareau (2003) uncovers the way class-based differences in childrearing practices play out in schools and contribute to the reproduction of inequality. Hmong American families share many of the characteristics identified with working-class and poor families, including hierarchical relationships between parents and children, and emphasis on extended family relationships. The UHS educators' middle-class ideas regarding family and about parental involvement in schools led them to view Hmong families as being dysfunctional.

While middle-class family life revolves around the nuclear family, working-class and poor families have been found to emphasize the centrality of extended family (Lareau, 2003). For the Hmong, extended family and clan are central to identity. The importance of extended family was reflected in the fact that Hmong families will often travel great distances, and youth will miss school, in order to attend family funerals and/or weddings. UHS educators appeared to view these family obligations as being barriers to Hmong students' academic achievement. In discussing Hmong American students' academic problems, one teacher sighed as she explained that "Hmong students will miss school for several days to attend funerals." Although this teacher understood the significance of funerals and weddings in Hmong culture, she suggested that these "traditional" practices placed Hmong youth "at risk" academically.

Echoing the cultural deprivation language that originally targeted African-American families, another UHS educator observed:

> African-American kids, Hmong kids, Hispanic kids ... Some families, you know I shouldn't generalize, but some families are ... some push those kids to go and get education and are very pro-education. Some families have a kind of culture of not being involved in schools and not valuing education. So it's kind of sometimes put on a back burner ... the immediate needs of the family come first. Some kids might have to get a job, they might have to baby-sit sometimes when they are the only option, so they might miss school.
>
> (Lee, 2005, p. 46)

Both of these UHS educators concluded that Hmong parents simply don't place a high enough value on education. The educators' individualistic values prevented them from seeing the potential value of family-based decisions.

Within the Hmong community, the extended family serves as an important safety net and a form of control. One practice that Hmong families have adopted in the United States in order to control recalcitrant teens is sending them to live with family members in other communities. One of my research participants, for example, lived with her aunt because her parents wanted her to get away from "bad kids." Although this student was doing well in school, UHS educators who knew her viewed her living situation as being less than ideal. Her social studies teacher, for example, whispered when she said, "I'm not sure who she lives with, but I think she lives with her aunt." Although this teacher did not directly criticize the family, the fact that she lowered her voice and shared this information when discussing her concerns for the girl suggests that this family arrangement was understood to be inherently dysfunctional.

Most UHS educators identified early marriage and early childbearing as the two biggest challenges facing the Hmong community, particularly the girls. Although an increasing number of Hmong American girls are postponing marriage until they complete high school, the general perception among UHS educators is that Hmong girls are pushed into early marriage. The subject of early marriage within the Hmong culture has received a lot of attention from academics and journalists, and the practice of early marriage has typically been understood as evidence of the traditional and vastly different culture (i.e., foreign) that Hmong immigrants/refugees have brought to the United States. In the conversations about early marriage, teachers slipped back and forth between the language of blackening and language that positioned Hmong students as perpetual foreigners. UHS teachers, guidance counselors, and other staff viewed early marriage as being quintessentially foreign. They described Hmong families that advocated early marriage as being "traditional." Most UHS educators were convinced that if the Hmong community did not abandon early marriage they would be "at risk" economically. The few girls who were married and had children were simultaneously described as victims of a traditional culture and as "teen moms" who were destined to become part of the "new underclass."

Finally, the fact that Hmong parents did not exhibit the kind of parental involvement valued by schools was seen as further evidence that Hmong families did not really care about education. UHS educators regularly complained that they had "trouble getting African-American and Hmong parents involved in the PTO [Parent Teacher Organization]." These UHS educators reasoned that the problems that Hmong students faced in school were linked to their parents' "lack of involvement" in educational matters. Although Hmong parents did not participate in their children's educations in the ways identified by UHS educators as "involved," Hmong American students all reported that parents consistently stressed the importance of education. Like other low-income parents of color, most Hmong parents lack the time, resources, and/or knowledge to participate in their children's educations in ways valued by the school.

BLACKENING AND EDUCATIONAL OPPORTUNITIES

For Hmong American youth at UHS there were negative educational consequences to being ideologically blackened. The racist assumptions regarding black students were extended to include Hmong American students. As such, Hmong students were held to low expectations and given little encouragement to take upper level courses or to pursue higher education. One guidance counselor explained his perspective on Southeast Asian students, including Hmong American students, in an interview.

> A lot of them are not intellectually motivated. These are Southeast Asians I'm talking about now. They are polite, they're nice, they never tell me what, where to go and that sort of thing, I'm a counselor. But they don't have a background of working hard academically, and they don't feel like it now.
>
> (Lee, 2005, p. 45)

While there were individual teachers at UHS who worked hard to advocate for Hmong American students and other students of color, most UHS educators appeared to have given up on them. Labeled "at risk" (i.e., blackened), Hmong American students were understood to be beyond the responsibility of the school. Their academic struggles were identified as being rooted in their social class, families, neighborhoods, cultures, and identities. By blaming the students and their families, these UHS educators were able to relieve themselves of the responsibility to serve Hmong American students.

PERSPECTIVES OF HMONG AMERICAN STUDENTS

Hmong American students' understandings of race and the racial hierarchy were shaped by their experiences at UHS. Hmong American youth were painfully conscious of the fact that whites were positioned at the top of the racial hierarchy at UHS and in the larger society. Furthermore, they were aware that non-Hmong people looked down on the Hmong culture and community. Significantly, Hmong American youth equated whiteness, middle class-ness, and American-ness. Hmong American students appeared to view whites as the only "real" Americans, as evidenced by the fact that they reserved the term "American" to describe whites, while using ethnically specific terms to describe themselves and other people of color.

Hmong American youth have internalized racial stereotypes about whites, African-Americans, and East Asians. When I asked Hmong American youth to describe "Americans," they typically described the stereotypic blue-eyed and blond-haired white person. Hmong American youth reported that "Americans" (i.e., whites) and East Asian Americans were "good" students. Based on their observations of the middle and upper-middle-class white students at UHS, and what they have learned from popular culture, Hmong American youth concluded that Americans (read: whites) are all wealthy. One Hmong American girl described her image of the typical white family in the following quote.

> When I think of the mainstream I think of a White family, I guess. As both parents working ... have really good jobs and maybe one kid or two kids, three at most. And the kids are doing house chores and everything, they like have good grades and even when the girl grows up, the woman, the mom has a good job like a doctor or something.
>
> (Lee, 2005, p. 71)

Although Hmong American girls and boys dreamed of "being rich," and many of the girls admired the gender equity they assumed exists in white families, most Hmong American youth could not identify with whites and did not want to be like white people. Hmong American youth asserted that whites were "selfish" and could not be trusted. Specifically, Hmong American students didn't trust white people to treat Hmong people with respect. One student explained,

> For me, I feel, I just feel like some White people neglect me. I mean as much as I try to be nice to them, give them respect, they don't give it back to me. Why should I even bother with them? Because I feel like I really don't need people like that...
>
> (Lee, 2005, p. 68)

Hmong students' relationship with whites influenced their attitudes towards blacks. Specifically, Hmong students identified with African-American students at UHS because both groups were subordinate to whites, both groups lived in the "ghetto," and both groups did poorly in school. One Hmong American student asserted that it was "the Hmong way" to be at least one year behind in school.

Hmong American students, however, were not simply passive victims of racialization. Their participation in hip-hop culture expressed resistance to the racial and class inequality they faced at school and in the larger society. Hmong American youth were well aware of the fact that hip-hop language and clothes were associated with African-Americans, and often teased each other for talking and dressing "ghetto" (code for black). Here, "ghetto" appears to be a space of oppositional racial power. Similarly, Reyes (2007) found that low-income Southeast Asian youth in Philadelphia were deeply involved with hip-hop culture, and often teased each other for "acting black." Although Hmong American youth identified with blacks, they maintained a distinct Hmong American identity. Hmong American youth who expressed a hip-hop aesthetic were quick to assert the importance of "Hmong pride" and lyrics from Hmong hip-hop songs often drew on Hmong cultural themes and encouraged ethnic pride.

UHS educators failed to see Hmong American students' adoption of hip-hop style as a legitimate form of social critique. Like many members of the dominant culture, UHS educators viewed hip-hop culture as inherently dangerous. The fact that Hmong American youth were adopting hip-hop styles confirmed UHS educators' belief that the Hmong are like blacks, and therefore "at risk." Thus, the students' resistance helped to confirm the process of blackening.

CONCLUSIONS

My research joins the work of other scholars who have identified schools as powerful sites of racialization (e.g., Fine, 1991; Fordham, 1996; Lei, 2003; Lipman, 1998; Olsen, 1997). Significantly, the process of racialization is intimately linked to ideas about social class. Teachers and other educators at UHS interpreted Hmong American students' family cultures, neighborhoods, clothing styles, academic achievement, and social class through the lens of race. The ideological blackening experienced by Hmong American youth demonstrates that ideas about race and class are conflated in the minds of the dominant group. In the dominant imagination, whiteness and blackness are both classed positions. Ideal whiteness is implicitly associated with middle-class status, and blackness is associated with poverty. As in my mother's youth, Asian-Americans are judged by the standards of whiteness (i.e., white, middle class) and are subject to ideological blackening if they are deemed to fall short of the criteria for being whitened. In other words, Asian-Americans are blackened when they lack the human capital and cultural capital associated with middle-class whites. Far from being neutral, the process of racialization involves privileging whiteness and denigrating blackness. Hmong Americans and other groups who are ideologically blackened are subject to racial bias and unequal educational opportunities that African-Americans have historically faced.

When I told my mother that I was writing a piece about blackening, and was writing about her childhood experiences with whitening, she responded, "Money does whiten, but the other day some stranger commented on how well I speak English. What nerve! The idiot was surprised I didn't have an Asian accent." My mother, clearly irritated, had once again encountered the stereotype that Asians are essentially foreign. My mother's recent experience suggests that social class plays a central role in determining whether Asian-Americans are whitened or blackened, but middle-class status does not protect Asian-Americans from being cast as perpetual foreigners. My data suggest that while Hmong American youth are blackened, they are also subject to being identified as perpetual foreigners. At times, UHS educators interpreted Hmong cultural norms to be simultaneously foreign and deficient. It appears that Asian-Americans are simultaneously subject to being judged by the black and white discourse of race, and subject to being seen as foreign.

REFERENCES

Ferguson, A. (2000). *Bad boys: Public schools in the making of black masculinity*. Ann Arbor, MI: University of Michigan Press.

Fine, M. (1991). *Framing dropouts: Notes on the politics of an urban public high school*. Albany: State University of New York Press.

Fine, M., & Weis, L. (1998). *The unknown city*. Boston, MA: Beacon Press.

Fordham, S. (1996). *Blacked out: Dilemmas of race, identity, and success at Capital High*. Chicago: University of Chicago Press.

Goldstein, B. L. (1985). Schooling for cultural traditions: Hmong girls and boys in American high schools. PhD dissertation, Department of Educational Policy Studies, University of Wisconsin-Madison.

Heath, S. B. (1983) *Ways with words: Language, life, and work in communities and classrooms*. Cambridge, England: Cambridge University Press.

Lareau, A. (2003). *Unequal childhoods: Class, race, and family life*. Berkeley, CA: University of California Press.

Lee, S. (1996). *Unraveling the model minority stereotype: Listening to Asian-American youth*. New York: Teachers College Press.

——— (2005). *Up against whiteness: Race, school, and immigrant youth*. New York: Teachers College Press.

Lei, J. (2003). (Un)necessary toughness? Those "loud black girls" and those "quiet Asian boys," *Anthropology & Education Quarterly, 34*(2), 158–81.

Lipman, P. (1998). *Race, class, and power in school restructuring*. Albany: State University of New York Press.

Lowe, L. (1996). *Immigrant acts: On Asian American cultural politics*. Durham, NC: Duke University Press.

Okihiro, G. (1994). *Margins and mainstreams: Asians in American history and culture*. Seattle: University of Washington Press.

Olsen, L. (1997). *Made in America: Immigrant students in our public schools*. New York: New Press.

Ong, A. (1996). Cultural citizenship as subject-making: Immigrants negotiate racial and cultural boundaries in the United States. *Current Anthropology, 37*(5), 737–62.

Perry, P. (2002). *Shades of white: White kids and racial identities in high school*. Durham, NC: Duke University Press.

Pfeifer, M., & Lee, S. (2004). Hmong population, demographic, socioeconomic, and educational trends in the 2000 census. In Hmong National Development Inc. & Hmong Cultural and Resource Center (Eds.), *Hmong 2000 census publication: Data and analysis* (pp. 3–11). Washington, DC: Hmong National Development, Inc.

Quincy, K. (1995). *Hmong: History of a people.* Cheney: Eastern Washington University Press.

Reyes, A. (2007). *Language, identity, and stereotype among Southeast Asian American youth: The other Asian.* Mahwah, NJ: Lawrence Erlbaum.

Takaki, R. (1989). *Strangers from a different shore.* Boston, MA: Little, Brown and Co.

Trueba, H. T., Jacobs, L., & Kirton, E. (1990). *Cultural conflict and adaptation: The case of Hmong children in American society.* Bristol, PA: Falmer Press.

Tuan, M. (1998). *Forever foreigners or honorary whites? The Asian ethnic experience today.* New Brunswick, NJ: Rutgers University Press.

US Census Bureau (2000). Race alone or in combination for American Indian, Alaska native, and for selected categories of Asian and of Native Hawaiian and other Pacific islander. At http://factfinder.census.gov/.

Valdes, G. (1996). *Con respeto: Bridging the distances between culturally diverse families and schools—an ethnographic portrait.* New York: Teachers College Press.

Vaught, S. (2006). *The peculiar institution: Racism, public schooling, and the entrenchment of whiteness.* Unpublished PhD dissertation, University of Wisconsin-Madison.

Weis, L. (2004). *Class reunion: The remaking of the American white working class.* New York: Routledge.

DISCUSSION QUESTIONS

1. What does the process of ideological-blackening entail? Can you think of other ethnic groups that have experienced the process of ideological blackening?
2. Do you agree with the author's binary description of the process of American "racialization," being either ideological "blackening" or "whitening"?
3. Stacey Lee argues that the process of ideological whitening involves key decisions on the part of Asian-Americans. What are these choices, and how do these strategic actions on the part of some ethnic groups in America relate to the idea of assimilation?
4. What social factors have contributed to the ideological blackening of Hmong youths, and in what ways are Hmong families blamed for their ideological blackening according to the author?

21

Immigrants' Preconceptions of Race

BETH FRANKEL MERENSTEIN

The Four Questions

1. What is the problem that Merenstein describes in this article?
2. Is this a *social* problem? Is it a personal problem? Who is harmed by this issue?
3. What is the cause of this problem?
4. What can be done about this problem?

Topics Covered

Racism
Immigration
Perception of racism
U.S. racial hierarchy
Assimilation
Racial ideology
American dream
Social class
Discrimination
Transnationalism

IMMIGRANTS' PERCEPTIONS OF RACISM

... Immigrants entering the United States today are largely from Latin American and Asian countries. Obviously, however, the large number of immigrants entering and residing in the United States means they differ on a number of important levels. Significantly, immigrants vary in terms of their income, educational, and occupational levels. For example, although we may lump together Asian immigrants, there is a vast difference between an upper-income, professional Japanese immigrant coming for a top-level business position and a lower-income, far less educated Vietnamese immigrant. Clearly, these class differences will affect their assimilation patterns ... such as where they settle and the schools their children attend.

Additionally, immigrants vary a great deal—often in direct relation to their socioeconomic status—in terms of their knowledge and understanding of English. This, too, will affect their perceptions of the racial social structure in the United States. Several other factors will also likely affect their positions once they enter the country: their gender can have a pronounced influence on their assimilation and acculturation (Hondagneu-Sotelo 1994): their age and family structure (whether they immigrate alone or with family members); and, importantly, their status—whether they immigrate legally, as a undocumented resident, or as a refugee....

Perceptions of Racism in the Home Countries

Although my argument is that immigrants acquire racial knowledge after immigrating to the United States, I also readily acknowledge that they arrive

SOURCE: From Beth Frankel Merenstein, *Immigrants and Modern Racism: Reproducing Inequality*. Boulder, Colorado: Lynne Rienner Publishers, 2008, pp. 39–51. Reprinted by permission.

here with their own understandings about race and racism, often based on how racial identities and categories were defined in their home countries. For example, in Latin America, racial constructions are more fluid and based on things like class and phenotype (color, hair texture, facial features); Latinos bring these conceptions with them, which influence how they view their own and others' racial identities (Rodriguez 2000). And in Latin America, just like in the United States, these racial constructions have a relation to racism and power.

What became readily apparent was that most of those immigrants denying the existence of racism in this country made a point to contrast it with the racism they said existed in their countries of origin. For example, Lisa, from Colombia, mentioned several times during her interview that one of the things she liked so much about the United States was that "everyone is treated the same," in contrast to her view of racism toward blacks in Colombia.

> LISA: There is a lot of racism towards blacks. There is a lot of racism toward the black race. {*In Colombia?*} In Colombia. {*Not in the US?*}. Here it seems everyone is treated the same. This is from what I see.... In my experience and in how long I have been here. I see that blacks, whites, or whatever they are, they are all treated the same.... [In Colombia,] there is a lot of racism toward the black race (and) the highest positions, the best jobs belong to white people. The black race does not appear to rise.

.... Some immigrants from Latin America said that there was little division in their countries based on race and that any existing distinctions often had more to do with class or politics. When it was about phenotype or color, they often said (as Laura did, above) that divisions had more to do with an individual's geographic location within the country. This was particularly the case for those immigrants from Peru, who said that most of the differences among groups depended on regional distinctions. All the Peruvians I spoke with repeated that there were three distinct regions in Peru: the coast, the mountains, and the jungle. They explained that people from each area looked down on the people from other areas. Occasionally they acknowledged that on some level racism did exist in their country, citing such examples as blacks not being allowed into certain private clubs. As Manny said, "I should also comment that in my country, black people distinguish themselves in everything—in the music, in sports—but are always discriminated against because of their color." Some of the Peruvian respondents were very clear and specific about the racism and in a similar way contrasted it with the lack of racism existing in the United States. For example, Karen said that she was specifically told by her Peruvian-American friend that one of the great things about the United States was that there was no racism, as there was in Peru.

Those who did discuss racism in their country often said race and class went "hand in hand." They claimed that whites held most of the positions of power in their countries and that often, even if a black person has money, he or she will still not be fully accepted. They would then consistently contrast the situation in their own countries with the situation in the United States, saying that whites did not hold all the power here and blacks of higher social classes were fully accepted in this country. Again, we see the strong relationship between the American Dream ideology and the belief that racism does not exist in the United States. For these immigrants, believing that the same situation exists here as in their countries, in terms of open social mobility or the lack thereof, would greatly negate their reasons for coming here. The strong belief in the American Dream is at odds with any belief in obstacles to that dream, such as racism.

Most of the immigrants I spoke with either claimed that racism did not exist in their home countries (and there were few distinctions in terms of a racial hierarchy), or, if it did, distinctions were more about politics or geography. For example, all the West Indians I talked with said that racism did not exist in their countries and that divisions were based more on class or politics. As Lee, from Jamaica, explained:

> LEE: Yeah, [Jamaica's] a place where we never, we don't know about racial things, 'til I come here. No, we don't know them things there, until I come to America, that's where I know about people prejudice against people. 'Cause white company back home, blacks basically run them and you know, it's white company, everybody get along, we hang out in the club. Back home we with white girls like nothing. You see. Up here it's like, you go into some white area with a white girl, and like, Ehh, what's he doing? [*laughs*]

Nevertheless, later in his interview, Lee did acknowledge that he had occasionally encountered racism, or racists, in Jamaica, but he insisted that black Jamaicans just ignore it when they experience it.

At times, these immigrants' portrayals of their countries contradicted what some scholars have documented. For example, the two Cuban men I spoke with insisted that there was hardly any racism in Cuba, which is not quite the same argument scholars studying race and racism in Cuba make (de la Fuente 2001, Logan 1999). Additionally, Audrey, from Brazil, claimed there was very little racism in Brazil and that the racism that existed was directed at poor blacks and was alleviated by the organizations that existed to defend black people. Furthermore, she maintained that the racism in the United States was worse than that in Brazil. Similar to the discrepancy between the perceptions of the Cuban interviewees and those of scholars writing on race in Cuba, this report of minimal racism in Brazil is not necessarily consistent with what some scholars have described regarding race and racism in Brazil, where they argue that often racism is only more covert than it is in the United States (Marx 1998, Premdas 1995).

We also need to look carefully not just at what people say about race, but also at the context and words they use to discuss the various concepts (van Dijk 1993, 1987, 1985). Examining the discourse in this way provides insight into how individuals think about a certain situation, when their previous words may contradict these beliefs. For example, Stacy (Peru) described how, while she was not racist, she knew racism existed to some extent in Peru. She provided an example of the type of racism that might exist in terms of employment: "Even in the area of employment. Many times you could see that in the jobs.... Discrimination to a certain extent, it wasn't that obvious. Hiring a pretty, white receptionist versus a dark-skinned, fat receptionist." In the example she gave here, we must question why she contrasted the image of "pretty and white" with "dark-skinned and fat"....

Preconceptions of the United States and Its Racial Hierarchy

There are a variety of possible ways for immigrants to begin learning about race and racial ideology in the United States. Obviously, exposure to US media is one such way immigrants may be exposed to US culture and ideologies before entering this country. Additionally, engaging in transnationalism, those who have already arrived here maintain ties to their home countries, thereby sharing information about what life is like here. If this is the case, we would expect information to be shared in both directions, and some of that information returning to the home country possibly would include mention of racial categories and racism in the new society. However, according to some scholars, such as Elliot Barkan (2006), this transnationalism is overstated and is, in fact, more like a Bell curve. "Mo[illegible]grants fall into the middle range, main[illegible] limited, intermittent, episodic, finan[illegible]en ties" (2006:15), which Barkan labels "[illegible]ocalism." Further, as other scholars have noted (Mahler 1995), many immigrants do not want to admit anything negative about this new society they have worked hard to come to, preferring to paint their new home as desirable and worthy of the effort to immigrate....

All of our major institutions have helped contribute to the ongoing racial ideologies. In particular, our media has consistently contributed to the dominant ideologies of racism in this country. Additionally, the media has had a greater impact on individuals around the world than have all other institutions.

Consistent with what we might think about the pervasiveness of US media outlets, about half of the immigrants I interviewed said they had watched a few US movies, watched news shows, and occasionally watched US television shows. For some, it was through the news in their own countries that they learned about issues in the United States. When asked whether she had ever heard that there was racism in the United States, Audrey, from Brazil, explained.

> AUDREY: Yes. On TV. I'd seen it on TV. In Brazil. I knew that was a very big issue between black and white people in the United States, bigger than in Brazil. So it's black people in their area and white people in their area, so it's very separate.

However, the impact of the US media was not quite as pervasive as I expected, as some of the immigrants claimed they had never watched US movies or television and rarely read US newspapers before coming to the United States. This distinction definitely seemed to exist on more of a class level, with the more-middle-class immigrants seeming to have had more access to US media sources before migrating. According to all the immigrants, however, it appeared that the predominant informational resource for learning about the United States before their arrival was family and friends.

Some immigrants said they had heard and learned about the United States durin[illegible] educa-tion. In the way some of the immigra[illegible]ed what they had learned, the informatio[illegible]erica [illegible] to have been very general.

> CRAIG (Colombian): When we were studying in the university we saw something about the history of the US. When you look at the history of America you see how this country developed. You see what groups came—they were basically the Europeans, the English, the Irish, the Scottish. Then there were the, well, the Indians were already here and then the blacks arrived.

In the way this immigrant explained it, the Europeans alone developed this country, and blacks just happened to arrive here, too....

Most of these immigrants' education about the United States occurred at colleges where they took courses on world history or United States history. Some of the immigrants from Europe (specifically, France and Spain), Peru, Colombia, and Cuba mentioned professors talking about US history in their college and high school classes. As Lori, from France, explained, what she learned and what she saw were not always the same thing:

> LORI: I remember studying the civil rights movement, and the Black Panthers, and Martin Luther King, so I had assumed it was a very divided country where black people were poor and struggling. And here we come and the [black] gardener [in California] was driving this huge, black car and seems to be quite well off. And that completely threw me. But then after that, that same summer, my cousin's wife took me to Watts and so then I had a very different impression. And that was more in keeping with what I'd imagined. But not that different than what I'd experienced in France with North Africans ... Although in France I don't think it's as divided as here.

Finally, some respondents talked about receiving information on the United States via government resources. Concerning the immigrants from Cuba, this source of knowledge was most often the case.

> ANDY (Cuba): *{How did you know there was discrimination against blacks in the US?}* In class, we had a class in universal history. There the professor explains about the racial systems in Africa, then we jump to the US when the KKK existed. When Martin Luther King and all those things we learned in class regarding blacks.

One of the basic sources from which immigrants receive information about the United States are previous immigrants, such as family members and friends. Unfortunately, both sources tend to provide false or misleading, and often fantasy, information (Mahler 1995). Freddy gave a nice example of seeing this firsthand and it's [illegible] repeating at length.

> FREDDY (Peru): A lot of times when you have a poor person here when they write their homes, "Oh, I'm a dishwasher, I have a car, I have a TV, I have a house," and they believe that and I know by experience, because a lot of times when I used to go back to Lima, they used to tell me that, "Oh, as a dishwasher or cleaning tables in a restaurant you can make a lot of money." Say no, as a dishwasher or a cleaning lady you're gonna starve, you're gonna starve to death, you not gonna have a nice apartment. Oh no, but so and so and so and so and so and so, they doing that and they telling me they live like kings. And then they look at me like I'm lying. So after a couple years I was telling them the truth, I say forget it. They don't want to listen to the truth, they want to listen to what they want to hear.... Well, a lot of people come here with that idea that they gonna live like kings without working too hard. When they come here and they find out that it's not like that, well then, rather than say the truth, back home, well, it's not like that, you have to work your tail off to get something back. If you work hard you're gonna enjoy. But so when they come here thinking they gonna work very little and live like kings and when they find out the difference, that's part of the protest, not the country, not the way of living.

The majority of the people I interviewed claimed that any information they received about the United States primarily came from letters and phone calls with family and friends who had previously immigrated to the United States....

For the majority of immigrants, it was often one particular family member or friend who had provided the basis for information. For example, Laura, from Colombia, spoke at length about an aunt who had immigrated previously and who had subsequently helped convince her to come to the United States. As she explained, although she was scared to come here, "I knew that I was going to be with my aunt in one area that would not be a problem." Similarly, Theresa (Cambodia) spoke about a sister who had immigrated and had urged Theresa and her mother to do so also. Theresa described how, at first, she was confused about what she saw and heard about the United States, but then her sister came here and wrote to her about it in a positive way.

> THERESA (Cambodia): I tell you that the people they know about like some TV, when we saw the picture, about the house, about the car, about the everything grow, you know, like the computer, light, phone, I am so confused. Like in my country we don't have that. My mother, she don't wanna come here, she said. "You can take your sister, your brother go there, oh I don't wanna go." Then my sister, she [come here] with her husband, explain about it, she took all the pictures, she had fun.

For some groups of immigrants, such as the Russian and Ukranian immigrants, their children who had immigrated before them were their primary source of information. They talked at length about their children's writing them letters and sending pictures of their homes and cars and how beautiful the United States always looked.

A major aspect of incorporating the current racial ideology is believing in the American Dream. The very reason most immigrants come to the United States is because of the power of this dream. They firmly believe that if they come here and work hard, boundless opportunities will exist and they will find a better economic future. This idea of meritocracy is firmly entrenched in modern racism, and was a concept discussed by many of the immigrants in my study.

Previous Knowledge of the American Dream

Interestingly, many of the immigrants I spoke with claimed to have had similar preconceptions of the United States as did the immigrants at the turn of the last century who wanted to immigrate to the "land paved with gold."…

With regard to learning about opportunities in the United States, family and friends, again, were clearly the primary sources of information. Regardless of whether the immigrants had watched US television shows popular in foreign countries, like *Baywatch* and *Dallas,* which portray inaccurate, sensationalized, and fantasy images of the United States, the most powerful source of immigrants' belief in becoming successful in the United States was from those who had immigrated previously. As Lisa, from Colombia, said, "[My friends] would say it was really good here, it was easy to get money. In a short period of time you could have the power to attain things—things that you've always wanted—easily." Another immigrant described what her relatives told her:

> ARIEL (Dominican): They always painted it as the best place in the world. There were many opportunities and they say there was a lot of work. Here you would work and return back home in a year and enjoy your money and buy a house. These were the plans of every normal person. They would say [the United States] was fabulous and when was I planning to apply for a visa to come here. That you can succeed here and that was the only way for me to get ahead, to have a family and live differently.

And in words we might have heard from Irish or Italian immigrants in the early 1900s, Claire, from Jamaica, said. "Men used to come here and they used to come back and they make it [illegible] you can just come here and pick money off the tree." Often, it was the success of those who had previously immigrated that predicted the belief in the dream. For many of the Russian and Ukrainian immigrants, as well as several older Asian immigrants I talked with, their children and grandchildren had succeeded in obtaining good jobs, relatively (to their previous lifestyle) large homes, and often college educations.

Other times it was the lack of success of those left behind that inspired someone to leave and come to the United States. For example, Freddy, from Peru, previously a jockey, explained:

> FREDDY: Your future over there is very limited as a rider. You only have one, maybe two, major tracks, and if you don't do well in those two tracks then you are going to struggle … while here in the States, the purses were very good all over the country.… Opportunities much, much better.… I am, I make that move, 'cause when I go back to Lima and go to the racetrack I see exriders that did extremely well and I can tell they have a rough life, so here your opportunities—after you stop riding you have so many opportunities, not perhaps to make the same kind of living when you were riding horses but you are not going to be in the poorhouse. While over there, most of them, they're struggling. Financially, not that well.

Aside from the more general questions I asked about previous knowledge and awareness of the United States. I also asked about diversity, discrimination, or race problems in the United States. Some immigrants said they had had no prior idea about the different racial and ethnic groups here, as explained by Audrey.

> AUDREY (Brazil): *{Would your aunt tell you anything about how diverse it was here?}* No, I didn't know, I was surprised. She did tell me that here they have a lot of Hispanic people. But I didn't know it was anything like that.

Other immigrants talked about being told of the extent of diversity and said they appreciated the variety of people from all over the world. For almost all of those I spoke with, this diversity was very different from what they had experienced in their home countries.

> KAREN I (Peruvian): Before I arrived I was very aware that this country is very diverse ethnically in regards to culture as well as the fact that this country has a lot to offer. It gives people the opportunity to come here and get a job, therefore you find people from all nationalities, all races, all religious beliefs. This country is very diverse.... [My boyfriend told me] that it was very diverse. The country is host to people from diverse racial backgrounds from all over the world. In the US, specifically in New York, you can find a diverse number of cultures—Chinese, Japanese, Peruvian, Colombians, Uruguayans, Europeans, Spanish, everything. You find everything.

What is important about this awareness of racial and ethnic diversity is its connection to the American Dream ideology. Hearing about and seeing all the different people in the United States reinforced for these immigrants the idea that anyone can make it here, regardless of his or her background. As Karen went on to say, "[My boyfriend would] say that he saw that in the US any person of color, Indians, Cholo or black, if professional, could accomplish any goal. This was something he did not see in Peru." Jesse also discussed hearing about the diversity in the United States when he was still in Mexico.

> JESSE: From all the countries—at every corner you would find someone from one country and at another corner you'd find someone from another country. You could find people from all the races in this country. They referred to it as the capital of the world. That's what they called it, the capital of the world. There was everything in the US.

For Jesse, his friends and family were making a connection between diversity and the potential for success.

Some of the interviewed immigrants said they heard about problems with discrimination in general, and with racism more specifically. However, some of those who heard about problems with racism claimed that it was not true. Some of Lisa's friends and family—who had been to the United States previously, or who had friends and family here already—told her not to come because of the discrimination here. However, Lisa (Colombian) said they told her the discrimination was toward Colombians specifically, not toward blacks; then she went on to say that she herself had never experienced any discrimination.

Occasionally problems with racism in the United States were heard about, not from friends or family, but from political figures. Both of the Cuban men I interviewed talked about how the Cuban government promoted the idea that racism existed in the United States. Aaron, however, believed the government's discourse about racism toward blacks in the United States was used to scare people from coming here and to bias them against the US political system.

> AARON: In my country, they talk a lot and say there is a lot of racism in the US. Cuba, and this is well known throughout the world, attacks the US regarding the racial problems. They say blacks are discriminated against here in the US, and that is not true. Politics in Cuba are based on saying that there is racism in the US. But of course, one who has never left Cuba cannot see that, they can't grasp it. A lot of people know it is not so because they receive visits from the US in Cuba or they get mail and they learn that it is not so.

Aaron claimed that the Cuban government's views are propaganda and Andy, also from Cuba, seemed to concur:

> ANDY: I knew it existed because I learned about it in school and I always paid attention

> to those things. I knew they exaggerated a little bit when they'd say there was discrimination in the US; they'd say that you couldn't go out to the streets late at night, they'd throw the dogs at you, they'd do things to you. I knew there was discrimination but only to a certain extent and I was able to understand. I'd say, I know there is discrimination but not as much, not in this day and age. Not at that level.

Andy said he learned this from government–sponsored school programs, and, like Aaron, he believed this racism does not exist to the extent he was told.

Occasionally an immigrant would talk specifically about being warned of the relationship between her or his own group and black Americans. Jesse, from Mexico, said he was warned by other Mexicans that "the blacks always had problems with us. *{With Mexicans?}* Yes, they would say that [blacks] would rob us and there were problems. They would warn me and say that it was good, but there also a lot of bad. *{Would they only mention the problems with the blacks? Did they ever mention anything about whites?}* No, they never mentioned anything about whites." Manny, from Peru, also said he was told black people believed they were discriminated against by anyone who looked at them and so he should be wary of interactions with them.

CONCLUSION

The immigrants I interviewed received knowledge about life in the United States from family and friends who had come before them; and although they often discovered falsehoods in these letters and phone calls, these sources of information were what they had to rely on before their arrival. Once they did arrive ..., they discovered a world full of segregation—in neighborhoods, schools, employment, and friendships. We can begin to see how previous racial knowledge, in combination with discovery of a new racial hierarchy (learned, in part, from agents of organizations they encounter), affect new immigrants' various racialization processes....

REFERENCES

Barkan, Elliot. 2006. "Introduction: Immigration, Incorporation, Assimilation, and the Limits of Transnationalism." *Journal of American Ethnic History* 25(2/3): 33–47.

de la Fuente, Alejandro. 2001. "The Resurgence of Racism in Cuba." *NACLA: Report on the Americas* 34(6): 29–34.

Hondagneu-Sotelo, Pierrette. 1994. *Gendered Transitions: Mexican Experience of Immigration.* Berkeley: University of California Press.

Logan, Enid. 1999. "El Apostol y el Comandante en Jefe: Dialectics of Racial Discourse and Racial Practice in Cuba, 1890–1999." *Research in Politics and Society* 6: 195–213.

Mahler, Sarah J. 1995. *American Dreaming: Immigrant Life on the Margins.* Princeton, NJ: Princeton University Press.

Marx, Anthony. 1998. *Making Race and Nation: A Comparison of the United States, South Africa, and Brazil.* Cambridge: Cambridge University Press.

Premdas, Ralph R. 1995. "Racism and Anti-Racism in the Caribbean." Pp. 241–260 in *Racism and Anti-Racism in World Perspective.* Edited by Benjamin Bowser. Thousand Oaks, CA: Sage Publications.

Rodriguez, Clara. 2000. *Changing Race: Latinos, the Census, and the History of Ethnicity in the United States.* New York: New York University Press.

van Dijk, Teun A. 1987. *Communicating Racism: Ethnic Prejudice and Thought and Talk.* Newbury Park, CA: Sage Publications.

——. 1993. *Elite Discourse and Racism.* Newbury Park, CA: Sage Publications.

——. ed. 1985. *Handbook of Discourse Analysis.* Vol. 4. London: Academic Press.

DISCUSSION QUESTIONS

1. How does belief in the American Dream ideology, particularly the idea of meritocracy, influence immigrants' convictions that America is a nation-state without racism?
2. What do immigrants believe about racism in America and its potential to impede on their goal of achieving the American Dream?
3. The American Dream, that anyone can "make it" if they are willing to work hard to achieve their goals, remains a major lure for attracting immigrants to the United States. On a whole, is the American Dream ideology a good thing for immigrants to adhere to, and if so, why? (Or, does the American Dream set immigrants up for disappointment by its unrealistic expectations of monetary success and material comfort?)
4. New immigrants don't understand the complexities of racism, sexism, and classism in the United States, so they are blissfully optimistic about their prospects of "making it" and shockingly ignorant in their assessment of the nation's "lack of racism." Respond to this statement. Do you agree? Why or why not?

PART VII

Social Problems: Gender Inequality and Issues in Sexual Orientation

As the selections in Part VII emphasize, sociologists usually understand the existence of gender inequality in the context of either structure/position/inequality/power or interaction/socialization. Among sociologists, it is unusual to find anything other than a social approach to this problem. Whatever biological differences exist between men and women do not explain the inequality that exists.

Sharon Hays considers many of the accompanying problems that welfare reform precipitates. Policies such as family cap and paternity requirement make the experience of receiving aid all the more onerous for recipients and their families.

Louise Marie Roth's detailed study *Selling Women Short: Gender Inequality on Wall Street* is based primarily on interviews of women on Wall Street. Her purpose was to understand the extent of inequality as well as the dynamics of discrimination. Discrimination, Roth concludes, is much more than obvious intentional actions by individuals; it is the result of the hierarchy, the culture, and the institutions in the interplay of career and family.

Susan Faludi began researching her book on the American man believing that he is the source of the inequality between men and women through his need to control. Instead, she came to a more sympathetic view: Men, like women, are faced with a society in which neither "are the masters of their fate." In fact, Faludi describes the reality of men rather than women in society as more of a serious social problem.

Jody Miller looks closely at the gendered dimension of inner-city violence, a social problem that is hidden from view of many middle class, suburban Americans. She describes in vivid and powerful first-person accounts the "pervasive sexualization" of day-to-day harassments and assaults that threaten the wellbeing of women and girls in the inner-cities of America.

Michael Bronski asks us to examine the source of America's condemnation of homosexuality. Instead of calling the fear of homosexuality irrational, Bronski maintains that it is "a completely rational fear," having to do with our cultural definition of sex. The problem for Bronski is the attack on homosexuals and the related question of citizenship in a democracy.

22

Flat Broke with Children

SHARON HAYS

The Four Questions

1. What is the problem identified by Hays?
2. Is it a *social* problem? Does it affect many people? Is it caused by social factors? Does it have importance to society?
3. What is the cause of this problem?
4. Does Hays suggest some way to alleviate the problem?

Topics Covered

Gender inequality

Poverty

Family

Welfare

Childcare

Child support

Ever since the inception of government-funded programs for the poor, policymakers have believed that the giving of benefits comes with the right to interfere in the family lives of the poor.[1] This is a notable exception to our strong cultural and constitutional prohibitions against state interference in private lives, particularly familial behavior. As political theorist Gwendolyn Mink argues in *Welfare's End,* the legal guarantee of privacy has been central to protecting basic family rights: to marry or not to marry, to make choices about reproduction and childrearing, and to determine living and custodial arrangements. Exceptions made in the case of the welfare poor have historically allowed, for instance, proper home requirements to determine the "suitability" of living arrangements and childrearing practices, "man-in-the-house" rules prohibiting men from sharing homes with welfare mothers, and attempts to control the reproductive behavior of poor women, including cases of forced sterilization. As Mink and others have noted, these policies often operated to discriminate against women (disproportionately nonwhite women) who were considered lacking in moral "virtue." All the most blatantly discriminatory policies have been struck down by the courts.[2] A few questionable family regulations have remained, however, and welfare reform has strengthened those and added more, reasserting the right of government to interfere in the familial life of the poor.

Welfare caseworkers therefore dutifully enforce a series of policies aimed at compelling responsible maternal behavior. The least intrusive is the demand that all children be properly vaccinated—verification is mandatory if the mother is to receive government aid. More onerous are the truancy requirements: all mothers with children aged 5 to 18 are held responsible for their children's school attendance. If a child misses three days in a row, five days in a month, or seven days over a longer period, the child is considered truant. The mother is called into the welfare office, where she must provide an account of the problem and formally outline a "plan" for getting the child to go to school. If the mother fails to control her child's school attendance, the child's portion of the family's welfare

SOURCE: From Chapter 3, "Promoting Family Values," from *Flat Broke with Children: Women in the Age of Welfare Reform* by Sharon Hays; copyright © 2003 by Sharon Hays. Used by permission of Oxford University Press, Inc.

benefits will be cut. Yet another policy, this one newly instituted with TANF, requires that all welfare recipients who are teenage parents must maintain full-time school attendance and remain under the watchful eye of more responsible adults, living with their own parents or another adult relative.

Although these policies may not seem particularly unreasonable on the surface, they are absolute and strictly enforced, with such stiff penalties that there is no room for difficult circumstances or individual choice. If a 16-year-old child is truant and beyond his mother's control, for instance, there is no exception available. If a mother would like to avoid the vaccination of her children (as some contemporary mothers do), she cannot. Or if a 17-year-old teenage mother wanted to drop out of school to seek work, hoped to move in with her boyfriend, or needed to escape a difficult family situation, she would be deemed ineligible for welfare benefits. Like the rules of the Work Plan, the inflexibility of these familial regulations belies the claim that welfare reform is aimed at the promotion of "personal responsibility."

Especially significant and debatable are those policies created by reform to stem the tide of out-of-wedlock childbearing that the law holds responsible for the "crisis in our nation." Three federal regulations directly address this issue; a fourth is offered as a state "option." First, the Personal Responsibility Act sets aside $50 million per year to subsidize state programs of "abstinence education" to teach "the social, psychological, and health gains to be realized by abstaining from sexual activity." These programs, Congress declares, will serve to remind the "at-risk" population that sex is only appropriate in the context of heterosexual, monogamous, marital relationships.[3] Next, our legislators direct attention to "predatory men," calling on the Justice Department to conduct a study of "the linkage between statutory rape and teenage pregnancy, particularly by predatory older men committing repeat offenses." In line with this, law enforcement officials are to be educated on the "prevention and prosecution" of statutory rape.[4]

The culmination of these efforts is rewarded by the third directive attacking unwed parenting—the so-called illegitimacy or antiabortion bonus. Every year, $100 million will be shared by the five states that do the best job of decreasing the number of out-of-wedlock children born *without* raising the abortion rate.[5] Significantly, the law does not include the slightest hint of further funding for birth control or family planning education, even though scholars have consistently demonstrated that affordable, easily accessible birth control is the one public policy that has proven most effective in lowering rates of nonmarital childbirth.[6] From a literal reading of the law, one would be forced to surmise that Congress imagines abstinence training and the prosecution of statutory rapists will do the trick. While this policy decision cleverly avoids controversy over birth control, most members of the American public are thoughtful enough to recognize that hunting down men who have sex with younger women and suggesting to the young and poor that they should avoid sexual activity altogether are insufficient solutions to the historical rise in the number of single parents.[7]

There is, however, one provision of the Personal Responsibility Act that addresses nonmarital births in an immediate and forceful way. This provision, which directly impacts life in the welfare office, is the "family cap"—barring from welfare receipt all children born to mothers who are already on welfare. This policy had been mandatory in the Republican version of welfare reform, but was made a state "option" in the version that finally passed the legislature. Almost half of the states, including Arbordale's, have instituted this plan.[8] The underlying logic is that women who consider becoming pregnant while on welfare will know that their progeny would be ineligible for benefits and hence will think twice before having more children.

The reality of this policy is faced by women like Joanne, a 29-year-old mother I met at a follow-up interview in the Arbordale welfare office. Joanne received welfare benefits for herself and her five-year-old daughter, Amanda. Her six-month-old son Tony, however, was conceived while she was on welfare—he was therefore a capped child.

Joanne was physically abused by Amanda's father, so she left him when her daughter was a toddler (and when she began to realize that the abuse was affecting her baby girl as well as herself). While on welfare Joanne met a new man who seemed to offer stability: he had a steady job and he did not beat her. When she became pregnant with their son, she was hopeful that the child would solidify the relationship, bring the family together, and get her off welfare. Initially, this seemed to be working—Tony's father moved in, and Joanne got off the rolls for a brief time. But less than a month after Tony was born his father left them, apparently unable to face the responsibility of supporting the whole family. Joanne and the two kids had since moved back in with her parents, sleeping in their living room.

Tony was a curious and alert little boy, with smiling blue eyes and dark curly hair. On the day I met him, he was playing with a stuffed panda as the caseworker talked to his mother. Both the caseworker and Joanne were referring to him as "the capped child," as if he wore a little beanie on his head embossed with the word "ineligible." That cap is not insignificant. Not only will he receive no benefits, but by welfare policy he doesn't actually exist, and hence the 18-month exemption from work requirements that is generally provided to new mothers in Arbordale does not apply to Joanne.

She was in the office that day because she had refused to work when Tony was still an infant; she was therefore sanctioned. She hadn't quite understood the reason for the sanction at the time, but she had obediently appeared in the welfare office as a result of a letter she received telling her that her one-month ineligibility was complete and she should report in. The caseworker explained the reasons for the sanction and told her she must find a job. Joanne asked instead to have her welfare case closed and her benefits discontinued. She said that Tony was just too young to be left in childcare; she did not feel that he would be safe. Given that there are no exceptions in welfare policies for mothers who are worried about the safety of their children, Joanne's case was closed and her benefits terminated. It was unclear to me how she and her children would survive.

There is no question that the family cap's impact on poor mothers and children is extremely harsh. For those who are less concerned about its cruelty, the family cap has two additional problems. First, as Gwendolyn Mink points out, this policy is arguably unconstitutional in that it systematically operates to penalize women for exercising their right to reproductive choice.[9] Second, the family cap provision is relatively ineffectual in meeting its stated goal of teaching poor women to control their fertility in that it suffers from both procedural inadequacies and a failure to understand the life circumstances of poor women.

The procedural failure of the family cap policy that is obvious to anyone who has spent any time in a welfare office is the implicit assumption that all welfare mothers know welfare rules, keep track of them, and plan their lives accordingly. As I have discussed, the rules and regulations of welfare are so complex that most welfare caseworkers cannot remember them all, and even the most conscientious and compliant welfare clients tend to keep track of only those that have an immediate impact or assert an immediate obligation. To correct this procedural flaw, one might recommend massive and ongoing national advertising campaigns to establish the necessary awareness, but even this would be inadequate. To the extent that welfare mothers are aware of welfare regulations, they generally make reasoned choices regarding the rules with which they will comply. Those choices, as well as the lack of prior knowledge, begin to explain why this provision has had no demonstrable impact on the rate of out-of-wedlock child-bearing in the one state where the family cap program was in effect long enough to study outcomes.[10]

Yet, more important than the procedural barriers to the success of the family cap is a much broader cultural barrier. *Welfare mothers have children for reasons other than the desire to get a few extra dollars in their welfare check.* As the tension between the Work Plan and the Family Plan illuminates, the idea of a calculated weighing of children against dollars is an apples and oranges equation. It is as culturally offensive to most Americans as the practice of seeking personal profit through the selling of babies on the open market.[11] On this score, most welfare mothers are no different from most Americans. Welfare mothers' decisions to have children arise from many sources, but they are generally not based on self-interested analyses measuring the size of their welfare checks against the value of raising a child. Research has repeatedly shown that rising rates of out-of-wedlock births have very little to do with rates of welfare benefits. In fact, national and international studies consistently demonstrate that attempts to lower rates of single parenting by lowering welfare benefits have been, and will be, generally ineffectual.[12]

Overall, the immunization rules, truancy requirements, teen supervision, abstinence education, and family cap policies are a mixed bag of attempts to impose standards of familial behavior on the poor. Meant to enforce an idealized middle-class model of the appropriate approach to family life, these policies effectively preclude a fundamental principle of middle-class family practices—the right to make *choices* regarding reproduction, parenting, and living arrangements. The problem here is not just one of denying welfare mothers the right of individual choice, it is also a problem of denying them the social *inclusion* that is implied by choice.

When we have finished enforcing this particular set of policies, we would be hard pressed to proclaim them the harbingers of a new era of familial commitment. Imagine a visitor from some far-off land trying to "read" these provisions for a vision of how American family life is appropriately ordered. The implicit message, so far, would simply state this: individual mothers are solely responsible for the health, education, and welfare of their children; all women without the financial resources, marketable skills, and stamina necessary to raise children alone should be assigned a life of celibacy. If these edicts were all we had to show as a representation of the family values we wish to champion, our nation and our families might well seem to this visitor cruel, unjust, or at least grossly underdeveloped....

SUBSIDIZING MARKET-BASED CHILDCARE

... The federal support of paid childcare for the poor makes obvious one of the central cost-benefit contradictions in the logic of welfare reform. *The costs of subsidizing childcare for the poor far outstrips the state and federal costs of paying a welfare mother to raise her own children.* It is cheaper, by far, to give a mother her monthly welfare check than it is to subsidize her childcare at market rates. In Arbordale, where welfare benefits are $354 a month for a family of three and $410 a month for a family of four, it costs $904 per month on average to purchase care for two children, $1,356 for the care of three.[13] Even if we include the supplementary benefit program of food stamps, placing welfare children in subsidized care for 40 hours per week is far more expensive than (inadequately) supporting their mothers to care for them 24 hours a day, seven days a week. The good news for taxpayers, and the bad news for poor single mothers and their children, is that the majority of welfare clients never actually receive childcare subsidies....

In Sunbelt City, only about one-quarter of eligible clients receive childcare subsidies; in Arbordale the figure is approximately 40 percent. Nationwide, less than one-third of eligible families actually receive the subsidy. In many locales there are simply not enough childcare facilities to go around. In situations where there are sufficient facilities, states keep running out of money to serve the seemingly endless stream of customers lining up for this valuable service.[14] Thus, in most places, welfare clients face long waiting lists either for the available childcare slots or for the subsidies.

It is important to remember that for these welfare mothers the question is not whether they *want* to put their children in childcare, they *must* do so in order to meet federal work requirements. Although states have the option of "exempting" from work requirements those mothers who are unable to find childcare (if their children are under age 6), neither Arbordale nor Sunbelt have used exemptions for this purpose. Given that federal regulations require

50 percent of welfare mothers to be in work or training, even the most generous state could not come close to exempting the two-thirds (or more) of mothers who are without a childcare subsidy.* Yet, as is true of nearly all things welfare related, the problems at ground level are even more troubling than any broad statistical accounting can convey.

The three-quarters of Sunbelt City mothers who are without subsidized childcare face a number of roadblocks, even though the welfare office, technically speaking, has no one on the waiting list for subsidies. The state has contracted out the administration of childcare to a private, nonprofit organization. One effect of this contractual system (a system used by many states for many of the services that welfare reform is obligated to provide) is that some welfare mothers are never properly referred for services, and some others don't fully understand that they are eligible. For the larger group who do seek out childcare help, they must travel long distances with poor public transportation just to get to the childcare office. Once they arrive, they are confronted with a bureaucratic process that rivals their experience of welfare.

The childcare administering agency requires an extensive application process, with piles of forms to be completed, many of which demand that clients provide the very same documentation they have already provided at the welfare office, including birth certificates and vaccination records. Clients must also produce, among other things, certification of eligibility and compliance from the welfare office, a separate letter from their employer stating their hours and pay-rate, and proof of a full physical examination and medical records for each of their children. Generally, this application process takes a number of trips and a number of weeks for clients to complete, and some are never able to provide all the documentation required. Those who are certified and do make use of the subsidy are then subject to the same kind of reporting obligations they face at the welfare office (change of hours, change of jobs, change of living situation, and so on). If they fail to comply, their subsidy will end.

Furthermore, the Sunbelt City childcare system allows children to be placed only in childcare facilities certified by the state. There are relatively high standards for state certification: extensive record checks on the provider, health and safety training, childcare training, and an on-grounds inspection of the home or facility. These requirements sound like a good thing, designed to protect children and ensure that they are well cared for. Unfortunately, these requirements present a number of problems for welfare mothers. First, as studies show, poor mothers (more than middle-class mothers) tend to want to have their children cared for by family members and friends, yet their family and friends are very unlikely to have the resources required for certification.[15] Second, although there are no waiting lists to receive subsidies, the demanding certification requirements limit the number and availability of childcare options and therefore mean that there are waiting lists for childcare providers, especially the most desirable and convenient ones. Hence, the childcare placements that do remain available to Sunbelt mothers tend to be at distant or odd locations. If mothers take one of these providers, they are often forced to spend an extra hour or more in transportation each day to manage the trip from home to childcare to work and back again—and their children spend that much more time in a childcare center that was not the mother's preferred provider in the first place.

Sunbelt mothers, therefore, tend to drop out of this program at one of four points—either they don't know they are eligible, they never make it through the initial application process, they can't keep up with the reporting requirements, or they have too much trouble accepting or managing the available childcare slots. And, as one might guess, it is just those mothers who are the most needy, the least educated, the least emotionally stable, the least

*As much as it would be impossible to exempt from work requirements all those welfare mothers who are without childcare, to exempt even 50 percent would mean that all those recipients with newborn infants, disabilities, and serious barriers to employment could not be exempted. (The Bush administration's 2002 proposals for higher work rates, if passed, will make this situation even more desperate [see Parrot et al. 2002].)

physically able, with the most difficult children who are the most likely to drop out of the program. Thus, Sunbelt is putting lots of mothers in a position where they must refuse to comply with the work requirements (and therefore be sanctioned), find some alternative childcare arrangements and manage to pay for it themselves, or leave their children at home alone during the workday. Over half of Sunbelt welfare mothers are without any form of childcare at all, subsidized or not. Since caseworkers in the Sunbelt welfare office don't handle childcare placements themselves, they tend to be unaware of these problems, and in any case, are relatively powerless to change them....

There is a large group of mothers for whom the system is completely unable to provide service, since no services exist. There are, for instance, no childcare centers to serve mothers who work swing shifts or graveyard shifts. There are no childcare services available for children with serious mental, emotional, or physical problems. I encountered welfare children with major physical disabilities, severe cases of attention deficit disorder, Down's syndrome, and crack cocaine addiction. I saw others who were truant and wild, and could not be controlled by their mothers, let alone a standard childcare provider. The mothers of all of these children must find work according to the rules of the Personal Responsibility Act. Those who lose their jobs for failure to appear at work on time, failure to be properly energetic while they are at work, or failure to concentrate single-mindedly on their work, will be sanctioned. Worries about children, in other words, are considered an unacceptable (unbusinesslike) distraction. Many caseworkers are sympathetic to their problems, but none can offer solutions that do not exist.

These multiple difficulties with the welfare childcare system point to the larger problems that *all* women in our society face in combining work and motherhood. The implicit federal answer has everywhere been the same: "This is a family matter. It is therefore your problem, and your personal responsibility." In the context of the welfare office, however, where government interference in family life is so readily apparent, the contradictory (and unjust) nature of this response is also apparent.

MAKING FATHERS PAY

...Child support enforcement, as the House Ways and Means Committee points out, serves a symbolic as well as practical purpose.[16] It sends a message to fathers that they cannot shirk their responsibilities. And many welfare clients strongly agree with this symbolic logic, and many are also very happy when their child support checks finally begin to arrive. The trouble is that for a relatively large proportion of welfare mothers this system actually generates greater financial hardship, or serious emotional distress, or even physical abuse....

One reason that some mothers would like to avoid enforcing child support is the reality of the paltry financial gains that mothers can expect from this process. As recipients quickly learn, for as long as they are on welfare the most they can expect to receive is a $50 per month "pass-through" of child support collected; the rest will go to the state to cover the costs of welfare.[17] Mothers must be very patient, and lucky, to get even this much. If things are running smoothly (which they rarely do), it takes about six months just to get a court order for child support. And even if the fathers have jobs, even if they are willing to pay, and even if they actually pay the full amount, the child support payments welfare mothers are likely to receive are generally inadequate to care for children. On one caseload, fathers who did pay child support were paying the following monthly amounts: $93, $38, $127, $5, $54, and $172. In 1996, the *average* amount received by single welfare parents nationally was $68 per month. And only about one-fifth of welfare mothers receive any child support at all.[18]

These figures give you an inkling of the status of the men who are the fathers of children on welfare: primarily poor men, some are in jail or prison (10 to 20 percent, caseworkers tell me), many are unemployed, others are employed only intermittently. In a number of locales, judges are ordering these men to pay back (previously unpaid) child support. This demand for back payment accounts for the tens of thousands of dollars owed by those fathers on the Sunbelt wanted poster. In many cases, this is support owed not to children but to the state in return for welfare benefits previously paid to those children. This situation means that many fathers owe far more than they can ever hope to pay.** Many of these absent fathers already offer more or less regular (under the table) support to their children, providing money, gifts, and services whenever they are able. Others have additional women and children that they support, and they are trying their best to fulfill those multiple obligations.[19] A number of them are very unhappy when they learn that the money they owe will actually go to the state rather than their own children. All are at risk for being jailed for their failure to pay....

In the Sunbelt City welfare office I met Cassandra, whose position highlights some of the problems involved. She was a 27-year-old mother of three children, all of them fathered by Eric, who had since moved in with another woman and fathered two more children. Eric was providing for this second family. He continued to visit Cassandra and the kids regularly and offered them money and gifts when he could. Yet Cassandra knew that if child support enforcement personnel were able to find him and order him to pay, Eric would flee, and the visits to her children as well as the gifts and occasional support would end. Cassandra and Eric had remained good friends, and Cassandra also knew his current girlfriend and children. She was painfully aware that he couldn't afford to take care of all five kids. And she knew that there were two children who would be left without his help at all if he should feel forced to run from the burden of regularized support payments. So Cassandra wished that she could avoid the enforcement process altogether. But she had no choice.

This trade-off between good relations with the father and gifts, visits, and intermittent aid for the family on the one side, as against regular child support payments on the other, is often a central catch-22 for welfare mothers.[20] One of the many ways that mothers make ends meet on a welfare check is by getting that additional (under the table) help from the children's fathers, or from the father's family. At least half the welfare mothers I encountered received such support. In a larger study of recipients' finances, nearly 60 percent received either regular (unreported) payments or in-kind support from absent fathers.[21] Many women are therefore resentful of child support enforcement rules. They are not only resentful that this process will ultimately result in a net financial loss, they are also resentful of the law's interference with their ability to maintain family ties....

There is a group of women who are *afraid* to comply with child support requirements. These mothers worry that violent or drug abusing or criminally dangerous men will be brought back into their lives. At the heart of this problem is the prevalence of domestic violence. It is estimated that approximately 60 percent of welfare mothers were, at one time or another, sexually or physically abused. The U.S. Department of Health and Human Services estimates that 15 to 34 percent of welfare mothers are *current* domestic violence victims.[22] In this context, court orders regarding child support may incite further violence. Although the law technically allows a "good cause" exemption from child support enforcement for mothers who are in this situation, such exemptions not only require mothers to disclose their circumstances, but in some cases they also require legal proof that a dangerous situation exists. Of the welfare mothers who told me they had suffered domestic violence, only half had disclosed the abuse to their welfare caseworkers; of those who needed legal proof, caseworkers estimated that less than a quarter were able to provide it....

**One study, conducted by a program serving low-income fathers, found that the fathers faced child support arrears averaging $2,000—even though the average total earnings of these men was only $2,800 over the preceding nine months (Sorenson et al. 2000). See also Edin (1995) and Garfinkel et al. (2001).

THE TANGLED WEB OF MATERNALISM, BUREAUCRACY, FAMILY VALUES, AND THE WORK ETHIC

One might be tempted to argue that the Family Plan and the congressional claim that "marriage is the foundation of a successful society" are just so much hot air. The family-related provisions of the Personal Responsibility Act appear relatively ineffectual in stemming the tide of single parenting and promoting happy and stable nuclear family relations. Indeed, in some cases these provisions seem more effective in driving a wedge between parents rather than drawing them together. Following the same pattern, even though the legislature makes two-parent families eligible for welfare, it has also retained its old "marriage penalty" in income eligibility requirements and created an additional penalty by establishing stricter and even more onerous work requirements for the few that qualify.[23]

Another apparent indication that family values are merely a decorative facade for the Personal Responsibility Act is the fact that the work requirements and the time limits for attaining self-sufficiency are far and away the primary values promoted in the welfare office. As I've noted, the vast majority of caseworkers I met did not see the regeneration of traditional family arrangements as part of their job description. They did not see themselves as couples counselors and did not urge their clients to go out and find marriageable partners. They did not read the rules of welfare reform as generating strong familial relations. In fact, almost none were aware that the law's preamble names marriage as foundational. Although many recognized the rise of single parenting as a serious social problem, few imagined that the practices of the welfare office provide a family-oriented solution.

In this light, it appears that the social tension between the values of work and family have been resolved in favor of work. To the extent that welfare reform's Work Plan is the only model that is left, the more worthy goals of work—its promise of independence, citizenship, and valued contributions to the collective good—have been debased or discarded. What remains is the individualistic ethic of self-sufficiency and an image of the "good society" as one full of unfettered individuals busily pursuing their daily bread in the marketplace, fending for themselves without a care or concern for others. After all, we could interpret the child-care subsidies as simply market-driven mechanisms provided solely for the purpose of pressing poor mothers into employment. The attack on deadbeat dads could be read as just another form of subsidy supplementing poor mothers' paychecks in order to keep them out of the welfare office and in their jobs. The provisions for vaccinations, truancy, and teen education could similarly be interpreted as edicts directed toward an image of mothers who must efficiently minimize their concern for children so that they can concentrate more fully on developing their earning potential. The abortion bonus and abstinence education are, in this vision, attempts to minimize the number of (always expensive and troublesome) children born....

As much as the complete triumph of the Work Plan would offer a disturbing vision of selfish individualism, the alternative offered by the Family Plan is equally narrow and disturbing. Marriage is pictured as little more than an economic transaction, and one where women are necessarily economic dependents. As is true of much of the logic of reform, the burden of creating new nuclear families is placed squarely on the shoulders of individual women: it is their job to use whatever skills they may have to seek out suitable men and enlist their aid in the support and rearing of children.

Nonetheless, there is good reason to believe that women who are sanctioned off the welfare rolls, who have been suffering through low-wage jobs and bouts of unemployment, or who have already reached their initial time limits on welfare receipt, will, in fact, seek out the help of men. After all, many of them have long relied on the help of boyfriends and the fathers of their children—and with welfare reform, their need for such help becomes increasingly pressing.[24] But larger questions remain as to whether women's efforts to find financially stable, marriage-ready mates will be any more successful than they were in the past, and whether reform can ultimately both decrease the rate of unwed childbearing and create warm, happy, and sustainable families.[25]

ENDNOTES

1. Abramovitz (1996), Gordon (1994), Horowitz (1995), Katz (1986), Mink (1998).
2. See Mink (1998), especially pages 93–98. See also Solinger (1999) on reproductive rights and welfare; see Fineman and Karpin (1995) on the general importance of the constitutional right to privacy to all women and families.
3. The law specifically suggests that this abstinence education "focus on those groups which are most likely to bear children out-of-wedlock." One cannot be sure how states will interpret that directive. See U.S. Congress (1996, PL 104–193, Title IX, Section 912).
4. U.S. Congress (1996, PL 104–193, Title IX, Section 906; see also Section 905).
5. U.S. Congress (1996, PL 104–193, Title I, Section 403 [a] [2]). See also Cornell (2000).
6. As sociologist Kristin Luker (1996) points out, "Federal provision of family planning to poor women has been one of the most significant, and least heralded, public policy successes of the past half-century" (p. 60), and the enormous success of this program "has been rewarded by having its funding cut almost in half" (p. 184).
7. None of this is to mention the problems involved in the condemnation of homosexuality that is implicit in the Personal Responsibility Act's suggestion that sex is "only appropriate in the context of heterosexual ... relationships." (See above.)
8. See Mink (1998) for the history of this provision. See Gallagher et al. (1998) and National Governors' Association (1999) for the states making use of it.
9. Mink (1998).
10. Regarding the much-debated outcomes of the New Jersey family cap, see Blau (1999: 148), Laracy (1994), Preston (1998).
11. See Hays (1996) and Zelizer (1985) for the cultural tensions between childrearing and the calculation of personal profit.
12. See especially Jencks (1997), Moffit (1992), Harris (1997), and Chapter 5.
13. Nationally, the average welfare family receives $357 a month in benefits (U.S. Department of Health and Human Services 1999A). Given that welfare mothers have an average of two children, this amounts to a $1.15 per hour, per child, rate for a 40-hour work week.
14. In 1997, only 1.25 million of 10 million low-income families nationwide received the childcare subsidies for which they were federally eligible. Studies suggest that welfare mothers are perhaps somewhat more likely to receive such subsidies than other low-income families, citing rates of 10 percent to 30 percent. See U.S. Department of Health and Human Services (1999B), Boushev and Gunderson (2001), Hernandez (1999), Weinstein (2001).
15. U.S. House of Representatives (1998), Edin (1995), Legler (1996).
16. U.S. House of Representatives (1998).
17. According to the Personal Responsibility Act, states are no longer required to "pass through" even $50 of child support collections, and many do not. Arbordale and Sunbelt, however, both allow clients to receive the $50 per month—that is, if a full $50 is collected in the first place. (At this writing, politicians are debating whether to loosen these rules and allow welfare mothers to receive a larger proportion of their child support checks.)
18. Sorenson and Zibman (2000), Edin (1995).
19. Edin (1995), Newman (1999).
20. Edin (1995).
21. Edin and Lein (1997). For further stories of how this works, see also Newman (1999), Garfinkel et al. (2001).
22. See U.S. Department of Health and Human Services (1999B), Bassuk et al. (1996), Raphael (1999), Kurz (1999), and Chapter 7.

23. Two-parent families' eligibility for welfare was federally mandated in 1988 (and had been a state option since 1961). Originally titled Aid to Families with Dependent Children—Unemployed Parent, this provision is now called Temporary Assistance to Needy Families—Unemployed Parent (or TANF—UP). The "unemployed parent" here referred to is, of course, the father—the language of the provision thus assumes that fathers are the primary breadwinner. The marriage penalty has been a part of this program since it was first installed in 1961 as an optional program: from "the beginning the eligibility criteria made it more sensible (economically speaking) to be a single mother than a married one. An additional penalty was instituted by the Personal Responsibility Act in that work participation rates require TANF-UP parents to find jobs faster and work more hours than is true of single parent cases. (See Sorenson et al. 2000, U.S. House of Representatives 1998.)"
24. Recent studies have, in fact, indicated that there is an increase in cohabitation among the poor (e.g., Cherlin and Fomby 2002).
25. See Garfinkel et al. (2001) and Edin (2000, 2001).

REFERENCES

Abramovitz, Mimi. 1996. *Under Attack, Fighting Back: Women and Welfare in the United States*. New York: Monthly Review Press.

Bassuk, Ellen L., Angela Browne, and John C. Buckner. 1996. "Single Mothers and Welfare." *Scientific American* 275(4): 60–67.

Blau, Joel. 1999. *Illusions of Prosperity: America's Working Families in an Age of Economic Insecurity*. New York: Oxford University Press.

Boushev, Heather and Bethney, Gunderson. 2001. *When Work Just Isn't Enough: Measuring Hardships Faced by Families after Moving from Welfare to Work*. Washington, DC: Economic Policy Institute.

Cherlin, Andrew J. and Paula Fomby. 2002. *A Closer Look at Changes in Children's Living Arrangements*. Welfare, Children, and Families; A Three-City Study, Working Paper 02-01. Baltimore, MD: Johns Hopkins University.

Edin, Kathryn. 1995. "Single Mothers and Child Support: The Possibilities and Limits of Child Support Policy." *Children and Youth Services Review* 17: 203–230.

Edin, Kathryn. 2000. "What Do Low-Income Single Mothers Say About Marriage?" *Social Problems* 47(1):112–133.

Edin, Kathryn. 2001. "Statement of Kathryn Edin: Testimony Before the Subcommittee on Human Resources of the House Committee on Ways and Means." http://waysandmeans.house.gov/humres/107cong/5-22-01/5-22edin.htm.

Edin, Kathryn and Laura Lein. 1997. *Making Ends Meet: How Single Mothers Survive Welfare and Low-Wage Work*. New York: Russell Sage Foundation.

Fineman, Martha Albertson and Isabel, Karpin. 1995. *Mothers in Law: Feminist Theory and the Legal Regulation of Motherhood*. New York: Columbia University Press.

Gallagher, L. Jerome, Megan Gallagher, Kevin Perese, Susan Schreiber, and Keith Watson. 1998. *One Year After Federal Welfare Reform: A Description of State Temporary Assistance for Needy Families (TANF) Decisions as of October 1997*. Washington, DC: Urban Institute.

Garfinkel, Irwin, Sara S. McLanahan, Marta Tienda, and Jeanne Brooks-Gunn. 2001. "Fragile Families and Welfare Reform: An Introduction." *Children and Youth Services Review* 23 (4/5): 277–301.

Gordon, Linda. 1994. *Pitied but not Entitled: Single Mothers and the History of Welfare*. Cambridge, MA: Harvard University Press.

Harden, Blaine. 2002. "Finding Common Ground on Poor Deadbeat Dads." *New York Times*, February 3: A3.

Harris, Kathleen Mullan. 1997. *Teen Mothers and the Revolving Welfare Door*. Philadelphia: Temple University Press.

Hays, Sharon. 1996. *The Cultural Contradictions of Motherhood*. New Haven, CT: Yale University Press.

Hernandez, Raymond. 1999. "Millions in State Child Care Funds Going Unspent in New York." *New York Times*, October 25: A29.

Horowitz, Ruth. 1995. *Teen Mothers: Citizens or Dependents?* Chicago: University of Chicago Press.

Jencks, Christopher. 1997. "The Hidden Paradox of Welfare Reform." *The American Prospect* 32 (May-June): 33–40.

Katz, Michael B. 1986 [1996]. *In the Shadow of the Poorhouse: A Social History of Welfare in America.* Revised edition. New York: Basic Books.

Kurz, Demie. 1999. "Women, Welfare, and Domestic Violence," pp. 132–151 in *Whose Welfare?* edited by Gwendolyn Mink. Ithaca, NY: Cornell University Press.

Laracy, Michael. 1994. *The Jury Is Still Out: An Analysis of the Purported Impact of New Jersey's AFDC Child Exclusion Law.* Washington, DC: Center for Law and Social Policy.

Legler, Paul K. 1996. "The Coming Revolution in Child Support Policy: Implications of the 1996 Welfare Act." *Family Law Quarterly* 30 (Fall): 519–563.

Luker, Kristin. 1996. *Dubious Conceptions: The Politics of Teenage Pregnancy.* Cambridge, MA: Harvard University Press.

Mink, Gwendolyn. 1998. *Welfare's End.* Ithaca, NY: Cornell University Press.

Moffitt, Robert A. 1992. "Incentive Effects of the U.S. Welfare System: A Review." *Journal of Economic Literature* 30(1) March: 1–61.

National Governors' Association for Best Practices. 1999. *Round Two Summary of Selected Elements of State Programs for Temporary Assistance for Needy Families.* Washington, DC: National Governors' Association.

Newman, Katherine S. 1999. *No Shame in My Game: The Working Poor in the Inner City.* New York: Vintage Books.

Preston, Jennifer. 1998. "Births Fall and Abortions Rise Under New Jersey Family Cap." *New York Times,* November 3: B8.

Raphael, Jody. 1999. "Keeping Women Poor: How Domestic Violence Prevents Women from Leaving Welfare and Entering the World of Work," pp. 31–44 in *Battered Women, Children, and Welfare Reform,* edited by Ruth A. Brandwein. Thousand Oaks, CA: Sage.

Solinger, Rickie. 1999. "Dependency and Choice: The Two Faces of Eve," pp. 7–35 in *Whose Welfare?* edited by Gwendolyn Mink. Ithaca, NY: Cornell University Press.

Sorensen, Elaine and Chava, Zibman. 2000. *Child Support Offers Some Protection Against Poverty.* Assessing the New Federalism, Series B, No. B-10. Washington, DC: Urban Institute.

U.S. Congress. 1996. *Personal Responsibility and Work Opportunity Reconciliation Act of 1996.* Public Law 104-193, H.R. 3734.

U.S. Department of Health and Human Services. 1999A. *Characteristics and Financial Circumstances of TANF Recipients.* Washington, DC: U.S. Government Printing Office.

U.S. Department of Health and Human Services. 1999B. *Temporary Assistance for Needy Families Program: Second Annual Report to Congress.* Washington, DC: U.S. Government Printing Office.

U.S. House of Representatives, Committee on Ways and Means. 1998. *Green Book: Overview of Entitlement Programs.* Washington, DC: U.S. Government Printing Office.

Uttal, Lynet. 1996. "Custodial Care, Surrogate Care, and Coordinated Care: Employed Mothers and the Meaning of Child Care." *Gender and Society* 10(3): 291–311.

Weinstein, Deborah. 2001. *The Act to Leave No Child Behind (S. 940/H.R. 1990): TANF and Food Stamp Reauthorization and Provisions to Ensure that Children and Families Receive Support to Promote Work and Reduce Poverty.* Washington, DC: Children's Defense Fund.

Zelizer, Viviana A. 1985. *Pricing the Priceless Child.* New York: Basic.

DISCUSSION QUESTIONS

1. How do procedural policies (like the family cap) impact the lives of welfare recipients? Is this itself a social problem?
2. What is paradoxical about the government's subsidized market-based childcare programs? Are there other solutions to childcare?
3. What values is the American welfare system promoting? Do you agree with these values? What problems does Hays see in these values?

4. What are the problems associated with welfare's paternity requirements. Do these problems themselves become a social problem? What recommendations would you make to improve the delivery of welfare?
5. In general, what responsibility does a democratic society have to poor families? What, if anything, is the cost for society if it ignores any responsibility at all?

23

Selling Women Short

LOUISE MARIE ROTH

The Four Questions

1. What is the problem?
2. Is it a *social* problem? Is it widespread? Is it socially caused? Does it harm society?
3. What are the causes of the problem?
4. Does the author suggest ways to alter the situation?

Topics Covered

Career and family

Success on Wall Street

Workaholic culture

Biased institutions

Gender division of labor

Child care

Gender discrimination

THE GENDER-PARENTHOOD WAGE GAP

It is well known from other settings that mothers earn less and fathers more than their childless counterparts.[1] In this respect, Wall Street was no exception. Figure 23.1 illustrates differences in pay for childless women, childless men, mothers, and fathers, while accounting for differences in undergraduate major, Wall Street experience prior to business school, weekly hours, rank, and firm prestige.[2] This figure clearly shows that mothers earned substantially less than their childless female peers while fathers received a large earnings bonus. But why do men's earnings increase and women's decrease when they have children? The common wisdom is that women become less committed to their careers when they have children, and make the choice to stay home or to reduce their

SOURCE: From Louise Marie Roth, *Selling Women Short: Gender and Money on Wall Street*, pp. 118–147, © 2006 by Princeton University Press. Reprinted by permission of Princeton University Press.

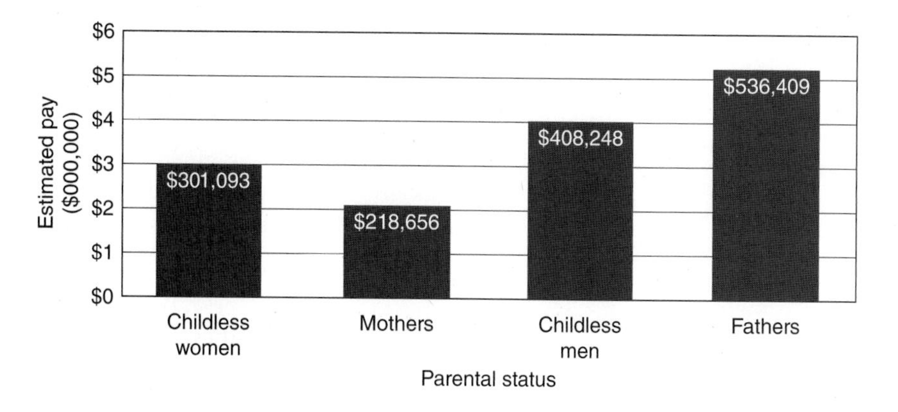

FIGURE 23.1 Estimated pay by parental status.

work commitment to care for their children. At the same time, as primary breadwinners, fathers become more career committed as they acquire more financial dependents. While reality is more complicated than this "commonsense" assumption implies, it suggests that fathers will and mothers will not maintain the devotion required to succeed on Wall Street.

AROUND-THE-CLOCK DEVOTION

Most Wall Street jobs required long hours and offered little flexibility. As a result, the ideal worker on Wall Street was a workaholic who was willing to work very long hours and had no external obligations. Workers without outside lives had better access to accounts and deals and were evaluated as better performers. Of course, a workaholic culture is not unique to Wall Street. Some scholars have argued that all workers increasingly work longer hours, sometimes by choice.[3] Others assert that managers and professionals work an increasing number of hours although many would prefer to work less, while many other workers are underemployed.[4] But all agree that inflexibility, in the form of long hours, face-time, and rigid schedules, is especially entrenched in high-paying jobs where workers demonstrate career commitment by centering their life on their work.[5]

Within this segment of the labor force, financial service professionals have experienced particularly large increases in work hours in the last few decades because consolidation, globalization, and new technologies have inflated competitive pressures. As a result, many Wall Street jobs require a willingness to work around-the-clock, especially in the career-building years. Caroline, a highly paid and highly satisfied investment banker, said,

> This is a job where you have to be prepared to subordinate everything else in your life to the job. It's very rare that you can say that you're going to take the weekend off and not have anything be there waiting for you, [and very common to] feel that you are somehow compromising something that you are working on because you didn't devote some time to it. There's always more than you could possibly handle in terms of things you need to do. Everything is always time sensitive, so a lot of things on a personal level have to take a back seat.

Junior workers in particular had to put up with having social plans and vacations canceled at the last minute. Many Wall Street workers, like Caroline, accepted this as part of the job, but others became more resistant to the workaholic nature of the business as time wore on. For example, Allison was a single woman who had recently changed jobs to improve her control over her personal life. But she still found the time demands of her job unreasonable.

> The fact [is] that up until recently, I could have a week-end planned and something could come up [such] that ... I would have to be in the office all weekend and I would have to cancel everything.... Even now, I can make plans to meet somebody at nine o'clock at night—nine o'clock at night—and I can't tell you how many times I have to cancel. That's ridiculous! That's nine o'clock at night!

As Allison suggested, most securities jobs, especially those in investment banking, required long and often unpredictable hours.

The bull market of the 1990s escalated the pressures for continuous work. Dave described the revolving door of business deals, remarking that client firms might do one important deal every decade or two, but that investment bankers worked on one after another. "The downside of course is that you work a lot of hours and there's not a lot of rest because when the company does an IPO they take the summer off, but you're on the next idea or the next transaction." So while the client firm might have a break after completing a large transaction, the investment bankers who orchestrated it would already be working on their next IPO, merger, or acquisition....

Most investment bankers thought that the job simply required someone who could maintain work as his or her *primary* priority and that anyone with different priorities could be replaced. Forbes blamed this on clients rather than managers, but the punch line was the same. When asked about family-friendly policies, he said that they simply were not feasible in investment banking because of the kind of service that clients expect.

> You can't sue the client if they hire a guy from [another firm] who is working ninety hours a week. So that's who you're competing against doing business. And if that's who your competition is, you either do what he does or you get out of the business.... And some people try to wrestle with that and try to legislate against the firm or try to argue for some position but clients dictate. It's only the clients who force us to work ninety hours a week. It's not that we choose to.... It's just—clients pay us ridiculous sums of money. They expect continuous work for that.

Clients expected Wall Street firms to deliver on their deadlines, and these firms promised more than their existing staff could produce without working very long hours....

Embedded within Wall Street's workaholic culture is an assumption that the ideal worker has no responsibilities outside of work. Family scholars have often observed that the definition of an ideal worker in advanced capitalist societies assumes a support person to take care of all nonwork aspects of life and is based on a breadwinner-homemaker division of labor in the family.[6] This is what most sociologists call an "institutionalized" assumption, or one that involves habitual and legitimized patterns of action that span organizations and are resistant to change.[7] Modern organizations have institutionalized employment practices that treat workers as though they have no family responsibilities, implicitly assuming that most families have a breadwinner-homemaker division of labor even though this is no longer the statistical norm.[8]...

In some ways, these time demands may seem gender neutral, because both men and women who were willing to work very long hours could succeed, but they had a disparate impact on women because gender profoundly shapes the structures of work and family life. Assumptions about the gender division of labor in the family inform the organization of work and the treatment of workers as men and as women, putting women at a disadvantage in the workplace because employers, workers, men with families, and even women themselves define care work as women's responsibility. In a broader culture that assigns all responsibility for childcare to women and demands that women's devotion to family trump all other commitments, family formation is guaranteed to have differential effects on male and female workers.[9]...

In an industry that demands around-the-clock devotion, most men and women viewed career reductions for family reasons as "just a personal thing." But the desire to work fewer than sixty hours per week is relevant only because of the way that Wall Street organizes work, which is hardly about "personal choice." In defining it as a choice, most workers did not challenge the underlying assumption that workers had no family responsibilities and, ideally, had stay-at-home spouses to manage all nonwork concerns. This posed problems for workers with childcare responsibilities.

THE ALLOCATION OF CHILDCARE

Given that most securities jobs conflicted with involvement in family life, how did parents who worked on Wall Street manage childcare? And how did their childcare arrangements fit with the workaholic culture of the industry? To address these questions, I asked workers with children how they expected becoming a parent to affect their careers before they had children, how being a parent had affected their careers, and how they managed childcare. I asked childless men and women if they hoped or planned to have children in the future and, if so, how they expected becoming a parent to affect their careers and how they expected to manage childcare. These workers envisioned only two options for managing childcare: a breadwinner-homemaker family in which the wife/mother is the primary caregiver, or hiring full-time childcare. Both men and women on Wall Street made enough to support a homemaker or to pay for full-time, high-quality childcare. But differential childcare arrangements by gender had a disparate impact on men's and women's careers because women took on responsibility for childcare in all cases, either doing all of it or arranging for another woman to do it.

Table 23.1 illustrates the approaches that parents took to childcare. Over four-fifths of the fathers had breadwinner-homemaker families, and 80 percent of the mothers hired caregivers. In fact, while breadwinner-homemaker families have become less common in the general population, they remained the norm for men on Wall Street and this norm was embedded in Wall Street culture. Maureen remarked on the uniformity of Wall Street men's family trajectories.

TABLE 23.1 Childcare Arrangements

Childcare Arrangement*	Mothers	Fathers	Total
Paid childcare	12	3	15
	(80.0%)	(17.6%)	(46.9%)
Breadwinner-homemaker	3	14	17
	(20.0%)	(82.4%)	(53.1%)
Total	15	17	32

*This reflects the childcare arrangement that the respondent used at the time of the interview. Some changed from one approach to another: in three cases, they switched from paid care to a breadwinner-homemaker division of labor.

> I think that in this environment, there is very much a standard of men at a certain age being married and, with few exceptions, whose wives don't work. Certainly, they don't work after they have children and everybody has had their first child at a certain step—when the men turn thirty—when they are about a third-year associate.

As Maureen indicated, the breadwinner-homemaker family pattern was an informally institutionalized norm for men on Wall Street. The support of stay-at-home wives meant that these men were able to devote most of their energy to their career and were perceived as more committed and stable when they married and had children.

Fathers with traditional wives were able to dedicate themselves to their career and were assumed to perform better because they had almost no family responsibilities aside from work. When asked how he and his wife managed childcare, Jorge said,

> For the first three years it was pretty much her by herself with help once a week or

> something like that. [Now] have somebody who comes four times a week and on the weekend for babysitting on Saturday night. But I really don't do anything at home. I would say we have more of a standard, old-fashioned [arrangement]…. I go to work and she takes care of everything in the house.

Like many male Wall Street professionals, Jorge had a full-time homemaking wife in addition to some paid domestic assistance, taking no responsibilities for caregiving himself.

Because of this, men with traditional wives had career advantages because they could put extremely long hours into their career that many other workers could not. Nick described a particularly strong devotion to his career, wearing a pager and checking his voice mail every hour over the weekend. But with three small children, he could only sustain this with the support of a stay-at-home wife. When asked if anything detracted from his satisfaction with his job, he said,

> No. Maybe time for my family, but that is also self-inflicted. I want to spend time working…. I don't have dinner with my kids during the week. I'm not the dad who reads them a story or gives them a bath. I also see less of my wife during the week. But I love it. She's supportive. She's independent and acknowledges how important it is for me.

In many ways, Nick was the ideal Wall Street worker because he put work ahead of other priorities.[10] Fifty-three percent of the fathers were able to focus on their careers with this intense level of devotion because they had traditional wives.[11] The large number of these men in this industry meant that the definition of the ideal worker did not have to shift to accommodate different family realities.[12]

This ideal worker notion disadvantaged workers who did not have stay-at-home wives. Caitlin, a single woman, said,

> Most of the vice presidents I work with are married, and they go home where they have a stay-at-home wife who has picked up their dry cleaning and maybe hired a maid, made their weekend plans. And sometimes it's tough because it's not that I want to have a wife at home, but there are advantages to that.

The norm of a stay-at-home spouse for men on Wall Street put women who were single or whose spouses worked in equally demanding careers at a relative disadvantage in living up to the demands of the workaholic culture….

Since they lacked the support of a homemaker, mothers who continued to work on Wall Street hired full-time childcare. Eighty percent of mothers and 18 percent of fathers hired childcare providers for their children so that both parents could have fulltime careers. But women in these families took on more responsibility than men for arranging childcare and for caregiving when they were not at work, so that their career opportunities were still restricted in ways that men's were not.

Men whose families hired childcare had wives who wanted to remain employed. They all earned substantially more than their wives and considered themselves to be the primary breadwinner.... The advantages of the breadwinner-homemaker arrangement are clear when comparing the average income of men with stay-at-home wives to the few with employed wives. Fathers with employed wives had average incomes similar to those of childless men ($375,000), rather than the higher average income of sole breadwinners ($628,846), suggesting that traditional wives dramatically improved men's opportunities, evaluations, and pay.

For women, hiring childcare was the only child-rearing solution that allowed them to continue their careers. Seventy-five percent of the mothers who hired childcare providers said that they were equal or primary breadwinners. But they were married to men with similarly demanding careers or to men with less demanding careers but who nonetheless worked full time. And because of broader cultural assumptions, mothers were responsible for childcare when paid caregivers were not present and did more childcare even when their husbands had less demanding and lower-paying jobs.... Childcare concerns that extended beyond the purview of the paid care provider typically fell upon the mother's shoulders, regardless of how demanding her career was. As other research has found, mothers felt responsible for spending time with and caring for their children in ways that fathers did not. Ultimately both men and women allocated responsibility for childcare to women.[13]

This responsibility could negatively affect women's careers. For example, Barbara discovered that unforeseen changes in her family life forced her to seek new childcare arrangements and made it impossible to balance work and family.

> I would say, in the beginning, it worked fine. Also, my husband, at the time, was in a fixed-income research job, and he had very similar hours to me, if not better. I worked basically from 7:30 to 5:30, and he worked 8:00 to 5:00. Or, that's when he had to work. He could do work on the computer at home. We had a nanny who came during the days to our apartment, so we were able to basically share relieving the nanny. This changed the same month I became pregnant with my second child. My husband was recruited by [another Wall Street firm].... Then he no longer had the same amount of free time as he had before.... More of the childcare responsibilities fall on me now that my husband has a much more demanding job. Really I think that that's what makes working hard. If you share all of the responsibilities equally, then you can do it. But if you both have the same type of demanding job, it's hard when you have to carry more of the responsibilities.

Barbara was considering taking a leave of absence so that she could devote the necessary energy to finding a new nanny. Because mothers take on more responsibility for arranging childcare and for filling the gaps when paid care providers are not on duty, even when their jobs are as demanding as those of their partners, surrogate caregivers did not support women's careers the way that stay-at-home wives supported men's careers.

PARENTHOOD AND WORK HOURS

Given childcare arrangements on Wall Street, it is clear that men and women were differently situated with respect to work-family conflict and that women had greater difficulty exhibiting the requisite career devotion. But whether or not men and women really differed in their commitment and productivity after having children is still an empirical question. While commitment and work effort are difficult to measure accurately and hours are not a perfect measure of productivity, one would expect women's family responsibilities to lead to reductions in hours and that compensation differences would roughly correspond to differences in hours. Figure 23.2 compares fathers, mothers, and childless workers by average hours per week, revealing that mothers do not work substantially less than childless women or fathers and that fathers work fewer hours than childless workers.

As this figure illustrates, women with children worked the fewest average hours, although not by much. But mothers' earnings penalty could not be attributed to differences in their hours, since fathers worked less than childless men or women and they made the most money. Mothers received an average of 53 percent as much money as fathers while working 92 percent as many hours.[14] Of course, some workers are more productive than others within the same amount of time, but mothers were much more likely to say that they had become more efficient at work so that they could work fewer hours while still performing as well. Meanwhile, fathers worked 90 percent as many hours as childless men, but received 122 percent as much money.[15] Table 23.2 illustrates differences in average pay, average hours, and average pay per hour, revealing that mothers received the least pay per hour and were less well compensated for their work effort.[16]

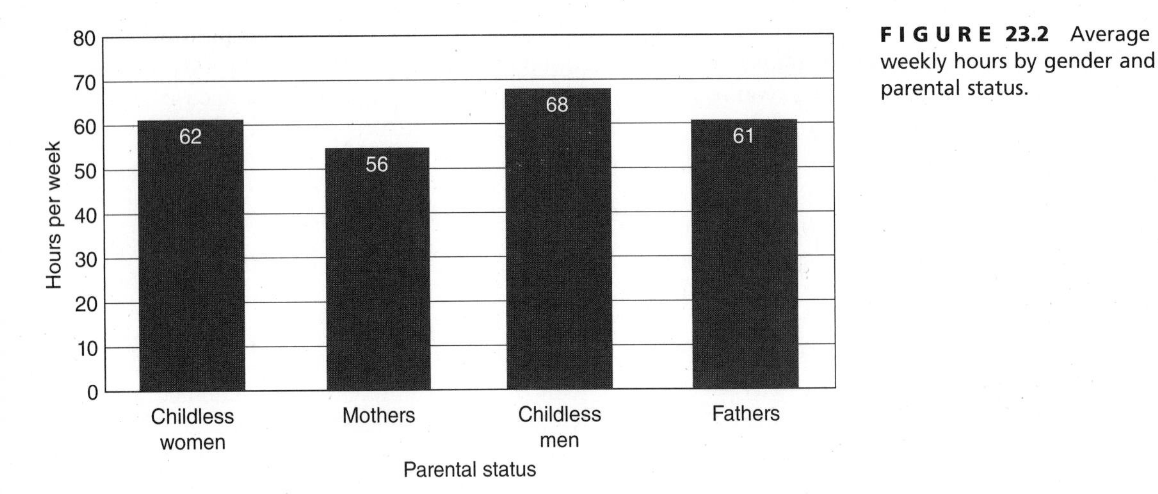

FIGURE 23.2 Average weekly hours by gender and parental status.

TABLE 23.2 Hours and Pay by Gender and Parental Status

	Average Pay (Raw Mean)	Average Hours/Week	Average Annual Pay/Weekly Hour
Childless women	$356,944	62	$5757.16
Childless men	$482,857	68	$7100.84
Mothers	$314,357	56	$5613.52
Fathers	$590,625	61	$9682.38
Total sample	$431,434	62	$6966.48

So how was it that women suffered an earnings penalty over and above their change in hours?

A MATTER OF PERCEPTION

While mothers did not substantially reduce their hours, coworkers' and managers' perceptions of women were influenced by gendered assumptions about work and family roles that spilled over into the workplace. Gutek first framed the term "sex-role" spillover to refer to the way that assumptions about gender permeate definitions of occupations.[17] She argued that predominantly female jobs like flight attendant or secretary take on stereotyped feminine characteristics, such as nurturing or being attractive to others, which then become part of the job description. Similarly, male-typed jobs like construction worker or litigator take on stereotypically masculine traits like aggressiveness, competitiveness, and predatory behavior.[18] Family-role spillover, the assumed gender division of labor in the family, is a specific form of sex-role spillover that leads co-workers and managers to presume that mothers are more committed to children's hands-on care while fathers are economic providers. This form of sex-role spillover is one way that organizations are invisibly gendered, with differential impact on men's and women's careers that is cloaked in the rhetoric of "personal choice."[19]...

Marriage improved men's image as stable and mature workers but had the opposite effect for women. In fact, women's image could be tarnished by the knowledge that they were married because they then seemed closer to becoming mothers and therefore less committed.

Jacqueline was warned by career counselors to avoid disclosing her marital status in job interviews. After she landed an equity research position, she found out that her manager would not have hired her if she had known that she was married.

> At [business school]... you're told that being married is a disadvantage when you're interviewing for a job. Because if

> you go in with a wedding band, they sort of view you as a temporary employee and it's a negative because they think you're just going to get a job, work for a few years, have a family, and leave.... I didn't wear my wedding band and I made no mention that I was married and you're obviously not allowed to ask the question. And she felt like I deceived her, but she also had made it a point that she might not have hired me if she knew I was married because the woman [who] was my predecessor had left her....

Assumptions that workers should not have responsibilities outside work and should make their career their central life focus led managers in Wall Street firms to offer more opportunities to workers who did not have and *were not expected to acquire* obligations to family or anything else. Because of the cultural assumption, shared by both men and women, that women's primary life focus is childbearing and caregiving, women on Wall Street found that they were equated with work-family conflict even when they did not have children. This led managers and coworkers to view all women as less effective workers, leading to harsher scrutiny and worse performance evaluations even when their performance was actually similar. They also assumed that all women would have children, even if they were childless and showed no signs of starting a family. Maureen, who was single, childless, and thirty-nine years old at the time of the interview, described comments that coworkers had made to her.

> It's nothing I ever think about and certainly don't talk about. But people make comments. I've had someone say to me that of all the people they know at the firm, in their mind, it would not surprise them that I would have a child on my own and be a single mother. Where did that come from? I've never even talked about children, so it's very random.

While she had no desire to have children and had never expressed such a desire to others, at least some workers still viewed her as a potential mother. As this comment suggests, women were always under suspicion of desiring a family even if they were single and approaching the end of most women's childbearing years....

Because of the view of all women as mothers or potential mothers, women's commitment to the workaholic standard was always suspect. After they had children, women often found that evaluations of their performance worsened and their bonuses declined because of the perception of them as less career committed. Todd had married another investment banker and described how she was treated after their first child was born: "I will say that my wife was penalized when she worked in investment banking. Her bonus was lower, and the perception of her changed. They perceived her as a mom rather than as a banker." In this and many other cases, mothers were automatically defined as uncommitted to their careers. On Wall Street, mothers who were career committed and fathers with desires for work-life balance did not fit the implicit model for family....

Discrimination on the basis of pregnancy was especially common. Thirty-six percent of women talked about pregnancy discrimination against themselves or other women. Renee described an instance of pregnancy discrimination that created a hostile work environment for her even though she was childless.

> I was at a meeting, for example, where the research analyst didn't show up because she was pregnant. Go figure. That is fine, but when the client caught on that everyone else had their research analyst except [my firm], the managing director said, "Damned women. Always getting pregnant." Of course, I was the only woman in the room when this happened. And I had worked the hardest on the presentation for this guy. All the eyes turned to me. I could see they felt bad for me, that my own firm was demeaning women like that. In fact, this client said to the senior MD, "Your behavior is

> offensive and we don't want to do business with you." So he got the message the hard way.

Renee was thankful that the client told her manager that his behavior was inappropriate and refused to work with him again. But her experience illustrates the blatant hostility that managers could have toward pregnant workers and how it created an inhospitable work environment for all women. Sometimes the hostility was less obvious, but women were concerned nonetheless about how they would be treated during pregnancy.

They had good reason to be concerned about how managers and coworkers would treat them when they were pregnant. Barbara worked as a trader during her first pregnancy.

> When I was pregnant, people I sat immediately with, they just did not think that I should come back to work. So they were mad at me if I did. One of my bosses and another MD would say "Who's going to raise your children? Don't you think that you're the best person to raise your children?" They would give me lectures every day. "This is a serious thing. You're bringing someone into the world." On and on and on. "You never know. You don't know." Every time something came on the radio about a nanny, child abuse, or cigarette burns, they'd say, "Why would you subject your child to that?" I heard that kind of thing all the time. Also, toward the end of my pregnancy, one of the MDs did not want me to sit in his chair because he was afraid my water was going to break. So he absolutely forbade me from sitting in his chair. Sometimes I forgot, and I sat, and he said, "Get out of my chair! Get out of my chair! Don't you sit in my chair!" It was like a joke.

Women's efforts to combine career and family often failed to fit a mold for "family" that their male colleagues could fully understand. After her maternity leave, Barbara moved into a sales position at another firm so that she could escape this hostile environment.

Even if they did not experience obvious discrimination during their pregnancies, many women were treated differently after giving birth. Managers often expected that women would not return to work or would quit soon after their maternity leave. This made them skeptical about mothers' commitment, leading to more negative evaluations....

These women attempted to use work-family policies but were penalized because they conflicted with the workaholic ethic of Wall Street culture. The women who became homemakers were further motivated by the fact that they were married to high-earning men, which gave them the financial option to stop working. So while they undoubtedly also wanted to spend more time raising their children, the decision to leave their careers was complex and at least partly fueled by a mismatch between performance and evaluations after they became mothers.

CONCLUSION

In Wall Street's male-dominated and workaholic culture, assumptions about the gender division of labor in the family led to beliefs that women would not live up to the demands of Wall Street careers and were less-than-ideal workers. Women's greater responsibilities for childcare may have affected their actual productivity and performance, but motherhood could also bias evaluations of performance over and above any actual changes in performance.[20]...

The broader cultural division of labor in the family had a disparate impact on women's careers and pay on Wall Street because all women were expected to quit when they had children—and all women were expected to have children. But women who quit were actually few in number. Four women quit to become homemakers, three of whom did not leave entirely voluntarily.[21] Contrary to stereotypes about women preferring the "mommy track" or "opting out" of careers, these women were pushed to leave by obstacles in the workplace. Of course, by leaving, they reinforced stereotypes that women want to cut back or stay home after they have babies. This perpetuated discrimination against women, while the workplace culture further reinforced the gender division of labor in the family.

Sometimes the influence of family responsibilities is portrayed as a supply-side influence on gender differences, meaning that it involves workers' characteristics or choices rather than "demand-side" aspects of the work environment like discrimination.[22] But the workaholic culture and demands for long hours cannot fairly be portrayed as worker preferences or choices. They are structural aspects of the work environment that, coupled with broader cultural injunctions, constrain women's success and produce systematic gender inequality. The combination of workplace culture and the broader culture surrounding family roles forced many women to view career and family as an either/or choice where men had a both/and option.

Of course, women's opportunities might have improved if the relentless time demands of Wall Street had lightened up and could better accommodate caregiving responsibilities. But Wall Street's workaholic culture defined careers and motherhood as incompatible, and the worker with a stay-at-home spouse was the norm as well as the ideal. Because of the spillover of the gender division of labor in the family into the workplace, this workaholic culture was highly gendered. As Martin has argued, Wall Street's work environment was *built out of* cultural beliefs about gender, including the breadwinner-homemaker family model.[23] These cultural beliefs were so widely held that they were invisible to many Wall Street workers, leading them to conclude that women's disadvantages were caused by their personal choices.

At the same time, securities jobs could have been managed differently while allowing workers with children to be effective employees.[24]

ENDNOTES

1. E.g., Budig and England 2001; Waldfogel 1997.
2. To construct this figure, I ran multiple regression models that calculated the effects of background characteristics (marital status, undergraduate major, and previous experience on Wall Street), average hours per week, rank (below or above vice president), and firm prestige on total compensation in 1997 for each parental status. I then computed the estimated pay by entering the group mean for each characteristic into the regression equation. The numbers presented in Figure 23.1 reflect group differences when the effects of these variables are accounted for. I did not include area of the firm largely for empirical reasons. The number within each parental status group was too low to account for all of these influences simultaneously. Also, mothers often changed areas because of work-family conflict or discrimination on the basis of pregnancy or motherhood, so that area appeared to often be a consequence of the processes that I explore in this chapter rather than a cause. Raw averages for each group exhibit a similar pattern, as shown in Table 23.2. In an analysis of variance, the difference in raw means across groups was statistically significant.
3. Hochschild 1997; Schor 1991. Hochschild's analysis emphasizes workers' incentives to spend more time at work and disincentives to spend time at home. Schor is more ambivalent about workers' preferences to work, viewing managers as the more important force in the trend toward long hours. She argues that employers pay a premium to overwork employees, which most workers accept even if many might prefer to work less.
4. Clarkberg and Moen 2001; Harrison and Bluestone 1988; Jacobs and Gerson 2004. These scholars argue that jobs have increasingly been divided between high-paying jobs that require long hours and low-paying jobs that do not provide fulltime work. As a result, workers at the top of the labor market, like Wall Street professionals, are overworked while many at the bottom are underemployed.
5. Bailyn 1993; Blair-Loy 2003; Blair-Loy and Wharton 2004; Jacobs and Gerson 2004.

6. Bernard 1981; Gerson 1985; Skolnick 1991; J. Williams 2000. This gendered division of labor fails to fit the reality of most contemporary families in the United States, in which women are in the labor force. Single-parent homes make up approximately one-third of families with children under eighteen.
7. DiMaggio and Powell 1991; Friedland and Alford 1991; Jepperson 1991.
8. Blair-Loy 2003; Clarkberg and Moen 2001; Moen 1992.
9. Blair-Loy 2003; Hays 1999. Blair-Loy called this the "family-devotion schema."
10. Nick's experience fits well with Hochschild's (1997) argument that workers prefer to spend time at work than at home. Hochschild claims that this preference is common because contemporary workplaces "value the internal customer" and provide rewards, encouragement, and a warm social environment. At the same time, she claims that as family life has become increasingly complex and homemakers are less common, the home requires work and involves emotional stress that many people prefer to avoid.
11. Nine of the seventeen fathers were extremely devoted to their careers and made large sacrifices in their personal lives. They all had homemaker wives, as did five other men who worked fewer hours and made time with their families a higher priority.
12. Wall Street workers were unusual in two noteworthy ways. First, only one of the workers I interviewed was a single parent—one mother was divorced. Second, the breadwinner-homemaker family has declined to approximately 10 percent of families within American society, largely due to economic necessity. But because Wall Street workers earned enough to support their families without a second income, the breadwinner-homemaker family was common in this industry. Men on Wall Street all married women who earned less than they did, and most married women who worked in low-paying occupations (e.g., teacher, flight attendant, social worker). Three men married another professional, but their wives ended their careers to become homemakers.
13. Hays 1999; Hertz 1997.
14. This figure and Table 23.2 use raw averages for pay and hours for each gender and parental status, rather than the estimated pay figures presented in Figure 23.1. The estimated pay in Figure 23.1 already accounts for differences in hours.
15. Childless men were much more likely to arrange a group dinner in the office or to horse around while they worked. This time was likely included in their estimations of their hours because it constituted face-time even though it was not productive time.
16. Hours were not significantly related to pay after accounting for all background characteristics, area, rank, and firm.
17. Gutek 1985.
18. Regarding litigators, see Pierce 1995.
19. Acker 1990; Martin 2001.
20. Budig and England 2001. After assessing a number of other theories for the wage gap, including explanations based on education, training and work experience (or "human capital"), and sex segregation, Budig and England argued that the residual wage difference between mothers and nonmothers could be a result of productivity differences, employer discrimination, or some combination of both.
21. In the remaining case, Kim left the industry after she got married but before she had children. Her primary reason for leaving was to relocate for her husband's job. She also expected to have children soon and to be a full-time mother for at least a few years.
22. Becker 1971; England 1992; Marini 1989. "Supply side" generally refers to characteristics that workers bring to the labor force and the choices that workers make. "Demand side" refers to employers' preferences, tastes, or requirements, which affect the jobs available to workers and the terms of those jobs.
23. Martin 2001.
24. In fact, it could be argued that workers with a better work-life balance are likely to be healthier and happier, and therefore more effective when they are at work.

DISCUSSION QUESTIONS

1. Do you believe there is a problem here? What values are violated? What goals are thwarted?
2. What is the point of Roth's discussion of "workaholic culture"? Is there a bias here simply because both men and women are expected to spend their primary efforts in the world of Wall Street? Is this expectation unfair to women?
3. If you were a president of a Wall Street business organization who came to believe the problem in this selection is important, what might you do to alleviate it?
4. What, if anything, should men come to realize in order to understand gender inequality in the workplace and the family?

24

The Betrayal of the American Man

SUSAN FALUDI

The Four Questions

1. What is the problem that Faludi identifies? What values are being violated or what goals thwarted?
2. Is this a *social* problem? In what sense?
3. What are the causes of this problem? Are men themselves the problem, according to Faludi?
4. How does Faludi approach meeting this problem?

Topics Covered

Gender inequality

Gender roles

Males

Masculinity

American culture

AT GROUND ZERO OF THE MASCULINITY CRISIS

...Wednesday evenings in a beige stucco building a few blocks from the freeway in Long Beach, Calif, I attended a gathering of men under court order to repent the commission of an act that stands as the emblematic masculine sin of our age. What did I expect to divine about the broader male condition by monitoring a weekly counseling session for batterers? That men are by nature brutes? Or, more optimistically, that the efforts of such a group might point to methods of "curing" such beastliness?

Either way, I can see now that I was operating from an assumption both underexamined and dubious: that the male crisis in America was caused by something men were doing unrelated to something being done to them. I had my own favorite

SOURCE: Excerpt from *Stiffed: The Betrayal of the American Man* by Susan Faludi. Copyright ©1999 by Susan Faludi. Reprinted by permission of HarperCollins Publishers.

whipping boy, suspecting that the crisis of masculinity was caused by masculinity on the rampage. If male violence was the quintessential expression of masculinity run amok, then a domestic-violence therapy group must be at the very heart of this particular darkness.

I wasn't alone in such circular reasoning. I was besieged with suggestions along similar lines from journalists, feminists, antifeminists and other willing advisers. Women's rights advocates mailed me news clips about male office stalkers and computer harassers. That I was not ensconced in the courtroom for O. J. Simpson's murder trial struck many of my volunteer helpers as an appalling lapse of judgment. "The perfect case study of an American man who thinks he's entitled to just control everything and everybody," one of them suggested.

But then, I had already been attending the domestic-violence group for several months—the very group O. J. Simpson was, by coincidence, supposed to have attended but avoided with the promise that he would speak by phone to a psychiatrist—and it was already apparent to me that these men's crises did not stem from a preening sense of entitlement and control. Each new member in the group, called Alternatives to Violence, would be asked to describe what he had done to a woman, a request that was met invariably with the disclaimer "I was out of control." The counselors would then expend much energy showing him how he had, in fact, been in control the entire time. He had chosen his fists, not a knife; he had hit her in the stomach, not the face. No doubt the moment of physical contact for these men had grown out of a desire for supreme control fueled by a need to dominate. I cannot conceive of a circumstance that would exonerate such violence. By making the abusive spouse take responsibility for his actions, the counselors were pursuing a worthy goal. But the logic behind the violence still remained elusive.

A serviceman who had turned to nightclub-bouncer jobs and pastry catering after his military base shut down seemed to confirm the counselors' position one evening shortly before his "graduation" from the group. "I denied it before," he said of the night he pummeled his girlfriend. "I thought I'd blacked out. But looking back at that night, I didn't black out. I was feeling good. I was in power, I was strong, I was in control. I felt like a man." But what struck me most strongly was what he said next: that moment of control had been the only one in his recent life. "That feeling of power," he said, "didn't last long. Only until they put the cuffs on. Then I was feeling again like I was no man at all."

He was typical in this regard. The men I got to know in the group had, without exception, lost their compass in the world. They had lost or were losing jobs, homes, cars, families. They had been labeled outlaws but felt like castoffs. There was something almost absurd about these men struggling, week after week, to recognize themselves as dominators when they were so clearly dominated, done in by the world.

Underlying all the disagreement over what is confusing and unnerving to men runs a constant line of thinking that blinds us—whatever our political beliefs—to the nature of the male predicament. Ask feminists to diagnose men's problems and you will often get a very clear explanation: men are in crisis because women are properly challenging male dominance. Ask antifeminists and you will get a diagnosis that is, in one respect, similar. Men are troubled, many conservative pundits say, because women have gone far beyond their demands for equal treatment and now are trying to take power away from men.

Both the feminist and antifeminist views are rooted in a peculiarly modern American perception that to be a man means you are at the controls at all times. The popular feminist joke that men are to blame for everything is the flip side of the "family values" reactionary expectation that men should be in charge of everything.

The man controlling his environment is today the prevailing American image of masculinity. He is to be in the driver's seat, the king of the road, forever charging down the open highway, along that masculine Mobius strip that cycles endlessly through a numbing stream of movies, TV shows, novels, advertisements and pop tunes. He's a man because he won't be stopped. He'll fight attempts to tamp him down; if he has to, he'll use his gun. But we forget that true Daniel Boone frontiersmanship was only incidentally violent, and was based on creating, out of wilderness, a communal context to which a man could moor himself through work and family.

Modern debates about how men are exercising or abusing their control and power neglect to raise whether a lack of mooring, a lack of context, is causing men's anguish. If men are the masters of their fate, what do they do about the unspoken sense that they are being mastered, in the marketplace and at home, by forces that seem to be sweeping away the soil beneath their feet? If men are mythologized as the ones who make things happen, then how can they begin to analyze what is happening to them?

More than a quarter century ago, women began to free themselves from the box in which they were trapped by feeling their way along its contours, figuring out how it was shaped and how it shaped them. Women were able to take action, paradoxically, by understanding how they were acted upon. Men feel the contours of a box, too, but they are told that box is of their own manufacture, designed to their specifications. Who are they to complain? For men to say they feel boxed in is regarded not as laudable political protest but as childish whining. How dare the kings complain about their castles?

What happened to so disturb the sons of the World War II GIs? The prevailing narrative that the sons inherited—fashioned from the battlefronts of Europe and the Pacific, laid out in countless newspapers, newsreels and movies—was a tale of successful fatherhood and masculine transformation: boys whose Depression-era fathers could neither provide for them nor guide them into manhood were placed under the benevolent wing of a vast male-run orphanage called the army and sent into battle. There, firm but kindly senior officers acting as surrogate fathers watched over them as they were tempered into men in the heat of a heroic struggle against malevolent enemies. The boys, molded into men, would return to find wives, form their families and take their places as adults in the community of a nation taking its place as a grown-up power in the world.

This was the story America told itself in dozens of war movies in which tough but tenderhearted commanding officers prepared their appreciative "boys" to assume their responsibilities in male society. It was the theme behind the 1949 film "Sands of Iwo Jima," with John Wayne as Sergeant Stryker, a stern papa molding his wet-behind-the-ears charges into a capable fraternity. "Before I'm through with you, you're gonna move like one man and think like one man," he tells them. "If I can't teach you one way, I'll teach you another, but I'm gonna get the job done." And he gets the job done, fathering a whole squad of youngsters into communal adulthood.

The veterans of World War II were eager to embrace a masculine ideal that revolved around providing rather than dominating. Their most important experiences had centered on the support they had given one another in the war, and it was this that they wished to replicate. As artilleryman Win Stracke told oral historian Studs Terkel in "The Good War," he came home bearing this most cherished memory: "You had 15 guys who for the first time in their lives could help each other without cutting each other's throat or trying to put down somebody else through a boss or whatever. I had realized it was the absence of competition and all those phony standards that created the thing I loved about the army."

The fathers who would sire the baby-boom generation would try to pass that experience of manhood on intact to their sons. The grunts who went overseas and liberated the world came home to the expectation that they would liberate the country by quiet industry and caretaking. The vets threw themselves into their federally funded educations, and later their defense-funded corporate and production-line jobs, and their domestic lives in Veterans Administration–financed tract homes. They hoped their dedication would be in the service of a higher national aim.

For their children, the period of soaring expectations that followed the war was truly the era of the boy. It was the culture of "Father Knows Best" and "Leave It to Beaver," of Pop Warner rituals and Westinghouse science scholarships, of BB guns and rocket clubs, of football practice and lettered jackets, of magazine ads where "Dad" seemed always to be beaming down at his scampy, cowboy-suited younger son or proudly handing his older son the keys to a brand-new convertible. It was a world where, regardless of the truth that lay behind each garden gate, popular culture led us to believe that fathers were spending every leisure moment in roughhouse play and model-airplane construction with their beloved boys.

In the aspiring middle-class suburb where I came of age, there was no mistaking the belief in the boy's pre-eminence; it was evident in the solicitous attentions of parents and schoolteachers, in the centrality of Cub Scouts and Little League, in the community life that revolved around boys' championships and boys' scores—as if these outposts of tract-home America had been built mainly as exhibition rings for junior-male achievement, which perhaps they had.

The speech that inaugurated the shiny new era of the 1960s was the youthful John F. Kennedy's address to the Democratic National Convention, a month before the launch of Echo. The words would become, along with his Inaugural oration, a haunting refrain in adolescent male consciousness. What Kennedy implicitly presented was a new rite of passage for an untested male generation. "The New Frontier of which I speak is not a set of promises," he told them. "It is a set of challenges." Kennedy understood that it was not enough for the fathers to win the world for their sons; the sons had to feel they had won it for themselves. If the fathers had their Nazis and "Nips," then Kennedy would see to it that the sons had an enemy, too. He promised as much on Inauguration Day in 1961, when he spoke vaguely but unremittingly of communism's threat, of a country that would be defined by its readiness to "pay any price" and "oppose any foe." The fight was the thing, the only thing, if America was to retain its masculinity.

The drumrolls promised a dawning era of superpower manhood to the boy born on the New Frontier, a masculine honor and pride in exchange for his loyalty. Ultimately, the boy was double-crossed. The fix was in from the start: corporate and cold-war America's promise to continue the World War II GI's wartime experience of belonging, of meaningful engagement in a mission, was never authentic. "The New Frontier" of space turned out to be a void that no man could conquer, let alone colonize. The astronaut was no Daniel Boone; he was just a flattened image for TV viewers to watch—and eventually, to be bored by. Instead of sending its sons to Normandy the government dispatched them to Vietnam, where the enemy was unclear and the mission remained a tragic mystery. The massive managerial bureaucracies of postwar "white collar" employment, especially the defense contractors fat on government largesse, produced "organization men" who often didn't even know what they were managing—and who suspected they weren't really needed at all. What these corporations offered was a secure job, not a vital role—and not even that secure. The postwar fathers' submission to the national-security state would, after a prosperous period of historically brief duration, be rewarded with pink slips, with massive downsizing, union-breaking and outsourcing. The boy who had been told he was going to be the master of the universe and all that was in it found himself master of nothing.

As early as 1957, the boy's diminished future was foreshadowed in a classic sci-fi film. In "The Incredible Shrinking Man," Scott Carey has a good job, a suburban home, a pleasure boat, a pretty wife. And yet, after he passes through a mist of atomic radiation while on a boating vacation in the Pacific, something happens. As he tells his wife in horror, "I'm getting smaller, Lou, every day."

As Carey quite literally shrinks, the promises made to him are broken one by one. The employer who was to give him lifetime economic security fires him. He is left with only feminine defenses, to hide in a doll house, to fight a giant spider with a sewing pin. And it turns out that the very source of his diminishment is implicitly an atomic test by his own government. His only hope is to turn himself into a celebrated freak and sell his story to the media. "I'm a big man!" Carey says with bitter sarcasm. "I'm famous! One more joke for the world to laugh at."

The more Carey shrinks, the more he strikes out at those around him. "Every day I became more tyrannical," he comments, "more monstrous in my domination of my wife." It's a line that would ring a bell for any visitor to the Alternatives to Violence group and for any observer of the current male scene. As the male role has diminished amid a sea of betrayed promises, many men have been driven to more domineering and some even "monstrous" displays in their frantic quest for a meaningful showdown.

THE ORNAMENTAL CULTURE

If few men would do what Shawn Nelson did one evening in the spring of 1995, many could relate. A former serviceman whose career in an army tank unit had gone nowhere, a plumber who had lost his job, a former husband whose wife had left him, the 35-year-old Nelson broke into the National Guard armory, commandeered an M-60 army tank and drove it through the streets of San Diego, flattening fire hydrants, crushing 40 cars, downing enough utility poles to cut off electricity to 5,000 people. He was at war with the domestic world that he once thought he was meant to build and defend. He was going to drive that tank he had been meant to command if it killed him. And it did. The police shot Shawn Nelson to death through the turret hatch.

If a man could not get the infrastructure to work for him, he could at least tear it down. If the nation would not provide an enemy to fight, he could go to war at home. If there was to be no brotherhood, he would take his stand alone. A handful of men would attempt to gun down enemies they imagined they saw in family court, employee parking lots, McDonald's restaurants, a Colorado schoolhouse and, most notoriously, a federal office building in Oklahoma. A far greater number would move their destruction of the elusive enemy to the fantasy realm to a clear-cut and controllable world of action movies and video combat, televised athletic tournaments and pay-per-view ultimate-fighting bouts.

But none of it would satisfy, because the world and the fight had changed....

Ornamental culture has proved the ultimate expression of the century, sweeping away institutions in which men felt some sense of belonging and replacing them with visual spectacles that they can only watch and that benefit global commercial forces they cannot fathom. Celebrity culture's effects on men go far beyond the obvious showcasing of action heroes and rock musicians. The ordinary man is no fool: he knows he can't be Arnold Schwarzenegger. Nonetheless, the culture reshapes his most basic sense of manhood by telling him that masculinity is something to drape over the body, not draw from inner resources; that it is personal, not societal; that manhood is displayed, not demonstrated. The internal qualities once said to embody manhood—surefootedness, inner strength, confidence of purpose—are merchandised to men to enhance their manliness. What passes for the essence of masculinity is being extracted and bottled and sold back to men. Literally, in the case of Viagra....

In a culture of ornament, manhood is defined by appearance, by youth and attractiveness, by money and aggression, by posture and swagger and "props," by the curled lip and flexed biceps, by the glamour of the cover boy and by the market-bartered "individuality" that sets one astronaut or athlete or gangster above another. These are the same traits that have long been designated as the essence of feminine vanity—the objectification and mirror-gazing that women have denounced as trivializing and humiliating qualities imposed on them by a misogynist culture. No wonder men are in such agony. At the close of the century men find themselves in an unfamiliar world where male worth is measured only by participation in a celebrity-driven consumer culture and awarded by lady luck.

The more I consider what men have lost—a useful role in public life, a way of earning a decent living, respectful treatment in the culture—the more it seems that men are falling into a status oddly similar to that of women at midcentury The '50s housewife, stripped of her connections to a wider world and invited to fill the void with shopping and the ornamental display of her ultrafemininity, could be said to have morphed into the '90s man, stripped of his connections to a wider world and invited to fill the void with consumption and a gym-bred display of his ultramasculinity. The empty compensations of a "feminine mystique" are transforming into the empty compensations of a masculine mystique, with a gentlemen's cigar club no more satisfying than a ladies' bake-off.

But women have rebelled against this mystique. Of all the bedeviling questions my travels and research raised, none struck me more than this: why don't contemporary men rise up in protest against their betrayal? If they have experienced so many of the same injuries as women, the same humiliations, why don't they challenge the culture as women did? Why can't men seem to act?

The stock answers don't suffice. Men aren't simply refusing to "give up the reins of power," as some feminists have argued. The reins have already slipped from most of their hands. Nor are men merely chary of expressing pain and neediness, particularly in an era where emoting is the coin of the commercial realm. While the pressures on men to imagine themselves in control of their emotions are impediments to male revolt, a more fundamental obstacle overshadows them. If men have feared to tread where women have rushed in, then maybe women have had it easier in one very simple regard: women could frame their struggle as a battle against men.

For the many women who embraced feminism in one way or another in the 1970s, that consumer culture was not some intangible force; they saw it as a cudgel wielded by men against women. The mass culture's portfolio of sexist images was propaganda to prop up the myth of male superiority, the argument went. Men, not the marketplace, many women believed, were the root problem and so, as one feminist activist put it in 1969, "the task of the women's liberation movement is to collectively combat male domination in the home, in bed, on the job." And indeed, there were virulent, sexist attitudes to confront. But the 1970s model of confrontation could get feminism only halfway to its goal.

The women who engaged in the feminist campaigns of the '70s were able to take advantage of a ready-made model for revolt. Ironically, it was a male strategy. Feminists had a clearly defined oppressive enemy: the "patriarchy." They had a real frontier to conquer: all those patriarchal institutions, both the old ones that still rebuffed women, like the U.S. Congress or U.S. Steel, and the new ones that tried to remold women, like Madison Avenue or the glamour and media-pimp kingdoms of Bert Parks and Hugh Hefner. Feminists also had their own army of "brothers": sisterhood. Each GI Jane who participated in this struggle felt useful. Whether she was working in a women's-health clinic or tossing her bottles of Clairol in a "freedom trash can," she was part of a greater glory, the advancement of her entire sex. Many women whose lives were touched by feminism felt in some way that they had reclaimed an essential usefulness; together, they had charged the barricades that kept each of them from a fruitful, thriving life.

The male paradigm of confrontation, in which an enemy could be identified, contested and defeated, proved useful to activists in the civil-rights movement, the antiwar movement, the gay-rights movement. It was, in fact, the fundamental organizing principle of virtually every concerted countercultural campaign of the last half century. Yet it could launch no "men's movement." Herein lies the critical paradox, and the source of male inaction: the model women have used to revolt is the exact one men not only can't use but are trapped in.

Men have no clearly defined enemy who is oppressing them. How can men be oppressed when the culture has already identified them as the oppressors, and when even they see themselves that way? As one man wrote plaintively to Promise Keepers, "I'm like a kite with a broken string, but I'm also holding the tail." Men have invented antagonists to make their problems visible, but with the passage of time, these culprits—scheming feminists, affirmative-action proponents, job-grabbing illegal aliens—have come to seem increasingly unconvincing as explanations for their situation. Nor do men have a clear frontier on which to challenge their intangible enemies. What new realms should they be gaining—the media, entertainment and image-making institutions of corporate America? But these are institutions already run by men; how can men invade their own territory? Is technological progress the frontier? Why then does it seem to be pushing men into obsolescence, socially and occupationally? And if the American man crushes the machine, whose machine has he vanquished?

The male paradigm of confrontation has proved worthless to men. Yet maybe that's not so unfortunate. The usefulness of that model has reached a point of exhaustion anyway. The women's movement and the other social movements have discovered its limits. Their most obvious enemies have been sent into retreat, yet the problems persist. While women are still outnumbered in the executive suites, many have risen in the ranks and some have achieved authoritative positions often only to perpetuate the same transgressions as their male predecessors. Women in power in the media, advertising and Hollywood have for the most part continued to generate the same sorts of demeaning images as their male counterparts. Blaming a cabal of men has taken feminism about as far as it can go. That's why women have a great deal at stake in the liberation of the one population uniquely poised to discover and employ a new paradigm—men.

DISCUSSION QUESTIONS

1. Do men truly face a problem, or is Faludi making it up?
2. Who is at fault? Men? Women? Society? What is Faludi's answer? What is yours?
3. As a man or woman, how do you react to Faludi's concerns? As a sociologist, how would you react? As someone who believes in democratic society, how would you react?
4. Are men worse off than women, according to Faludi? What is your opinion?
5. The answer for men is not to try to find out what it means to be a man but to find out what it means to be human. Is this a silly attempt to deal with the problem Faludi asks, or is it both possible and thoughtful on her part?

25

African American Girls, Inequality, Violence

JODY MILLER

The Four Questions

1. What is the problem Miller identifies?
2. Is this a *social* problem? Who is harmed by this problem? Who benefits? Are people aware of this problem?
3. What is the cause of this problem?
4. What can be done about this problem?

Topics Covered

Gender inequality

Gender-based violence

Sexism

Sexual harassment

Sexual assault

Sexual coecion

Gendered-risk-avoidance

Youth violence

Victim-blaming

...Urban disadvantage has important gendered dimensions. This is certainly evident in research on the urban street world, where researchers have provided consistent and extensive evidence of the salience and institutionalization of gender inequality, including violence against women.[1] It is also evident from nationally representative research, such as the National Crime Victimization Survey. The work of criminologist Janet Lauritsen indicates that, compared with other racial groups—where gender disparities in rates of victimization are pronounced—African American young women have rates of nonlethal victimization that are similar to those of their male counterparts, and these young women are more likely to be victimized by the people they know, including those in their neighborhoods. Lauritsen's area-level analyses reveal that the relationship between race and victimization risk is largely accounted for by those facets of extreme disadvantage that I documented in chapter 1.[2] This is why, though there is evidence of greater cultural support for violence against women in distressed urban neighborhoods, it is important to always keep its structural basis at the forefront. Robert Sampson refers to this as the *cultural structure* of such communities, to highlight the fact that cultural adaptations emerge in response to the structural conditions in which people—in this case urban Black adolescents—find themselves.[3]

...I turn the attention explicitly to the nature of violence against women in youths' neighborhoods, examining their experiences with sexual harassment and violence, as well as other public acts of violence against women.[4] I consider how gender structures neighborhood risks, the visibility of gender-based violence, and the strategies young women employ to maintain their safety and security while going about their daily lives.... I pay special attention to how youths interpret and make sense of violence against women in their communities.

GENDERED PERCEPTIONS OF NEIGHBORHOOD RISK

...Youths' neighborhood descriptions consistently included reports of various kinds of criminal activities, including drug sales, gang conflicts, and gun violence. Their accounts suggest that many have witnessed a great deal of community violence. As illustrated in table 25.1, this is confirmed by their responses to survey questions about exposure to violence.[5] For instance, nearly all of the youths had seen people being hit or physically assaulted. Though some of these incidents occurred in school or at home, fully two-thirds of the youths indicated that they had witnessed such events in neighborhoods. In fact, the more serious the violence witnessed, the larger the proportion of youths who reported that it had occurred in neighborhoods. A total of 36 youths reported having witnessed a stabbing; 33 of these incidents occurred in neighborhood contexts. Likewise, all of youths' reports of witnessing gunfire occurred in neighborhoods. It is especially striking that many had witnessed serious gun violence. Fully 60 percent of the girls and 70 percent of the boys had witnessed someone being shot. Of these 49 incidents, only two occurred in sites other than neighborhoods.[6] Likewise, more than one-third of the girls and nearly one-half of the boys had witnessed someone being killed. All 31 of these homicides occurred in the community.[7]

TABLE 25.1 Exposure to Violence

	Girls (N = 35)	Boys (N = 40)
Seen people hit	33 (94%)	40 (100%)
Seen physical assault	27 (77%)	39 (98%)
Seen guns shot	27 (77%)	38 (95%)
Seen someone shot	21 (60%)	28 (70%)
Seen robbery	16 (46%)	30 (75%)
Seen stabbing	15 (43%)	21 (53%)
Seen drive-by shooting	14 (40%)	17 (43%)
Seen someone killed	12 (34%)	19 (48%)

A growing body of research has examined the impact for youths of witnessing such extraordinary violence. This work indicates that witnessing violence is related to increased aggressive behavior, increased emotional and psychological distress, a heightened sense of vigilance, and increased risk of personal victimization.[8] Notably, several studies suggest that "repeated exposure to high levels of violence may cause children and adolescents to become uncaring toward others, and desensitized toward future violent events."[9]

What is missing from these studies, however, is an examination of the impact of exposure to violence against women, particularly in the community. Research has examined the consequences for children of witnessing violence among adults in their families,[10] but studies of community violence nearly always measure incidents such as those listed in table 25.1 and have not made distinctions between violence witnessed against women and other violent events.[11] This is because until recently, when scholars thought about urban violence, it was conceptualized primarily in gender-neutral terms or was assumed to be male on male. The consequence is that we have little information about how young women are affected by repeated exposure to male violence against women, including its psychological impact and its effects on girls' gender identities and their relationships with both males and other females. This is a critical gap in our knowledge, made all the more significant by my findings here, which suggest that not only do young women face gender-specific risks in public spaces but also that violence against women often takes on features of public spectacle....

Youths who believed that young men faced greater dangers focused on how young men's neighborhood activities structured risks against them. Specifically, they noted young men's participation in gangs and drug selling, along with the much greater likelihood for male violence to involve firearms. Young women, they believed, were insulated from such dangers because they were less involved in street action and rarely used guns; they believed that street conflicts over gangs and commodities from the drug trade tended to be the purview of young men.[12] For instance, Tisha explained, "it's probably safer for females, 'cause the males, they involved in them gangs and stuff. And I mean, you don't really hear too much about females shootin' each other the way you do males." Doug concurred: "A girl is cool, but the guys, it's just plain and simple. Like if we see you, and we don't know you in our neighborhood, we gonna jump you."

Alternatively, some youths believed that their neighborhoods were so dangerous that gender didn't insulate anyone. Curtis explained, "the neighborhood I'm living in now, it ain't safer for nobody." Likewise, Tommie said that "when all the drive-bys and stuff was goin' on" in his neighborhood, it was dangerous for "everybody. The kids, *everybody*." And Alicia explained, "I don't think it's safer for neither one, 'cause ... a girl can get shot just like a boy can." Nonetheless, youths often alluded to gender-specific risks, even within a broader sense of generalized danger. LaSondra said her neighborhood was safe "for nobody. Nobody." Clarifying, she explained, "females get raped, males get killed." Likewise, Dwayne said, "I don't think [my neighborhood's] safer for neither one. Anybody can get hurt." He continued, "if [girls] look good, somebody might try to touch 'em or something. And they might not want them to touch them and they might say something to 'em. And the dudes in my neighborhood, they might try to beat them up 'cause the girl wouldn't let them touch 'em."

As these last comments suggest, youths who emphasized the dangers posed to girls in their neighborhoods often focused on gender-specific issues. While risks for males were tied to gangs and offending, youths emphasized the dangers to girls caused by predatory male behavior. The threat of sexual violence was a common theme, and young women often spoke specifically about the dangers posed by nighttime. Britney explained:

> We face ... a lot of stuff. Males, if you don't want to give it up they'll probably try to take it. If you walking by yourself and in the dark they'll probably approach you and want to, you know what I'm saying, do something with you. That's why you don't never walk by yourself after dark. [I] always keep [protection] with me when I walk by myself.

Likewise, Gail noted: "I don't know if it's just me being paranoid. I'll be walkin' at night and people just be walkin' behind you.... [I] mean, I don't know, everybody get scared to walk at nighttime, so I guess it's just normal.... [I] just keep lookin' over my shoulder to make sure ain't nobody walkin' behind me." And LaSondra said:

> Over there, you have to watch. Especially when you walking or something, 'cause you never know who might be behind you. When I walk, I look around and sometimes when I'm walking I forget to look around. Next minute somebody's behind me. Like, where'd this person come from? I just start walking faster. That's scary. Especially if it's nighttime too....

Young men also emphasized sexual dangers in their discussions of girls' neighborhood risks. Antwoin said it was safer for males "'cause like the females, all the dudes be wanting to try to freak, you know, have sex with 'em, all that kinda stuff." Darnell explained, "you don't catch no girl walking, you know what I'm saying, on the northside. Not in no nighttime. Like my girlfriend, she stay like up the street from me.... I walk her home 'cause I don't trust her walking by herself, unless she got somebody with her." Asked what he thought might happen, Darnell explained:

> I don't know, I just can't trust things like that, you know. 'Cause things happen, things happen. Like you don't know if somebody'll jump out the bushes. Or wherever you going you can have somebody following you.... Like she could've been on her way to my house, somebody could've followed her, and I could've let her walk back home by herself and something would've happened to her.

Thus, nighttime posed different kinds of dangers for young women and young men. For young women, it signaled the potential for sexual danger; for young men, it was when the streets came alive with gang and drug violence. In both cases, though, perceived dangers came primarily at the hands of males.

In addition to their focus on sexual dangers, youths' perceptions of girls' neighborhood risks were tied to their beliefs about men's greater ability to protect themselves and to the vulnerabilities for women that emerged from perceptions of them as "weak." Jamellah explained, "it's always safer for males. They could protect theyself better than a female can. That's in any neighborhood." She continued:

> A girl could walk down the street and she hafta fear about somebody stoppin' and rapin' her and makin' her get in the car. And a boy could walk down the street and wouldn't hafta worry about that. 'Cause what woman gonna wanna rape a man for? And then a man is more likely to try to do something to a female anyway, 'cause we not as strong as a male.

Similarly, Britney believed males were less likely to be "messed with" because they were more likely to "carry a gun." In contrast, she said "dudes" believe females are "weak—won't do nuttin', won't say nuttin'."

As a consequence, Britney, like Jamellah, felt women were at greater risk for predatory behavior, including sexual assault, because they were deemed easy targets. April believed her neighborhood was safer for males for similar reasons:

> Ain't no nigga gon' mess with no 'n other nigga. They'll mess with a female 'fore they mess with another dude. 'Cause they more powerful than us. They know they can do, we like sensitive and all that stuff. We ain't gon' act like what no nigga do. We might try to fight back but we all for that hollerin' and ain't no dude gon' do that.... A dude'll pull a gun out on you. [With females], they can do something, knock it outta our hands or something, even if we had a gun.

A number of young men concurred. Curtis explained that "males more, I guess, rugged. Males more harder." Likewise, Kevin said, "women seem more vulnerable. So most people, like dope fiends and stuff, they'll rob women and stuff. They ain't gon' rob no man that's walking down the street 'cause they know he—especially a young man—probably got a gun or something on 'im." James explained, "dudes, I mean they can pretty much handle theyself and they ain't gotta worry about nuttin', they're safe. Females, somebody [can] overpower them.... I mean strength wise ... [it's] easy to take something from them 'cause there ain't too much they can do for theyself." Likewise, Maurice noted that "females can't protect theyself," explaining: "Who'll be the first person you rob, a male or female? It's like a big buff dude versus a feminine woman, you know what I'm saying, who the person gonna stick out to 'em? That woman."[14]

Thus, many youths culled from exaggerated notions of gender difference (males are buff, rugged, powerful; females are feminine, sensitive, vulnerable) in their explanations of young women's risks in dangerous neighborhoods. These widespread belief systems did not just affect youths' perceptions of gender and risk but contributed more broadly to a hierarchy on the streets in which females were situationally disadvantaged vis-à-vis males and therefore often viewed by males simply in terms of their sexual availability.[15] As Kenisha complained, "dudes get more respect than females.... It's just the way it is over there. The males, they have more authority than girls." Such gendered status hierarchies affected how young women were treated in public places and how others responded to incidents in which females were mistreated. Moreover, these inequalities limited the recourse available to young women for challenging gender-based violence.

In fact, gendered status hierarchies and the sexualization of young women meant that a number of youths looked to young women's behavior or dress in explaining their neighborhood risks. Kristy said that "dudes be safer than the gals ... 'cause I mean, they don't be walking around in hoochie clothes and stuff. They don't give people the wrong idea about what they want." Likewise, when Tisha was asked whether young women faced any particular dangers, she replied, "oh yeah, like when they out there wearin' all them tight clothes and all that, you know. They get a bad rep, they can be raped, anything like that." Asked whether girls faced particular dangers, Eugene said, "yeah, like people comin' and rape them or something, 'cause girls wear short stuff and … fellas think, well, dawg, you know, I'm fina get up on [that], I'm fina touch on her or something. And they get mad if somebody touch 'em. They gotta look [at] what they wearing and when they sending signals like that."…

In sum, youths who believed their neighborhoods were less safe for females than males emphasized both the sexual dangers facing women and the perception that women were weaker and thus easy targets.[16] In fact, regardless of whether they believed males or females faced greater danger, youths' explanations rarely deviated from gender-based interpretations of neighborhood risk. Their accounts provide further evidence of the extent to which public spaces operated as male-dominated terrains, providing avenues for male engagement (and, thus, victimization risk) and sites in which the presence of young women could be read as opportunities for sexual conquest.…

Witnessing Physical Violence against Women

Though scholars typically assume that intimate partner violence occurs primarily behind closed doors, many of the youths here described witnessing such incidents in public view. Youths' accounts of such incidents may seem somewhat peripheral to their own lives, particularly since much of the public violence they described involved adults and rarely included individuals they were intimately familiar with. In actuality, such incidents are quite meaningful, because they impart essential messages about how violence against women is to be interpreted and responded to. Thus, it is important to pay attention here not just to the descriptions that youths provided but also to the tone and implications of many of their comments. These provide evidence of the interpretive lenses brought to bear on such events, illustrating how youths were taught to think about violence against women.

Asked whether she had ever seen a man hurt a woman in her neighborhood,[17] Tisha responded:

> Yeah, it's this girl next door. I don't know if she a crackhead or what, her boyfriend just always beatin' her up. She always comin' [over], like, "Call the police for me!" But then when the police get there she don't wanna press charges or nothin' like that. I don't know, I guess she stuck on stupid. She like getting beat up I guess.

Tisha said that the woman "done let him [back] in [her home] over and over, and she know what he gon' do to her. So that's on her. We don't call the police for her no mo'. Just let her get beat up." Because, in Tisha's eyes, the woman had not taken sufficient action to extricate herself from the abusive relationship, she was seen as culpable and thus unworthy of further assistance.

In addition, Tisha speculated about whether the woman was a "crackhead." Other youths also referenced substance abusers in their descriptions of violence. Christal said she had seen "crackheads fightin' they boyfriends," and Tawanna described seeing a "prostitute standing around" who got beat up. Such labels functioned to distance the victims from the young women who witnessed the events and to suggest that the violence was deserved and the victim blameworthy because of her status. Gail said it was "lil' drunk, alcoholics [that] be fightin' up the street … just lil' drunkies, drunk people just gettin' drunk. Crackheads, drunks, you know.… We'll be laughin' at 'em or whatever." Thus Gail's response was unsympathetic. She described a recent incident:

> One time I seen a man hit a woman with a bottle. But again, they was drunk, you know.… Well we ain't really see it from the beginning. We just heard them arguin' [and went to look]. She was running down the street talkin' 'bout some, "He tryin' to kill me, he tryin' to kill me!" [*laughs*].… He was like, "Come here, come here."

> She was drunk, going back and stuff. And she talkin' 'bout, "You don't hit me." And he—bang—just stole her [hit her with the bottle].... [Then she] kept yellin, "Oh, he gon' kill me, he gon' kill me!" And knowin' us, we just laughin'.

Tami described a recent incident she witnessed involving the man across the street, who was routinely abusive toward his girlfriend:

> I don't know what they was arguing about. When he came outside he was telling her to get her stuff, and she don't run it, he run it. He be hittin' her upside her head and she be, "Why you hittin' me? Stop hittin' me!" and stuff. But she don't be fightin' him back or nothin'. He be hittin' her upside her ... head, she was just walkin' down the steps.

Asked how she reacted upon seeing the man hitting his girlfriend, Tami explained:

> We was laughin'. Then we was like, that's a shame and they shouldn't put up with that stuff. They should go'on, leave, but I guess they ain't got nowhere else to go so they just put up with the stuff.... But we figured it wasn't none of our business. It came outside [so] we was just lookin' to see what was goin' on.

Tami's comments suggest that her reaction to the event was complex. On the one hand, she and her friends found it an amusing public spectacle. On the other hand, Tami expressed some empathy for the woman she felt was likely trapped in the situation. Finally, she noted that it "wasn't none of our business," indicating the important norm of staying out of others' affairs....

Likewise, Dawanna saw "this man ... grabbin' this lady by her face and pulling her down the street." She explained, "she was yelling, but I ain't hear, you know, if it was like, 'Help' or anything like that. But when I walk down the street I was finding all her jewelry, you know, in a line going up to her house. [He was] pullin' stuff off her face, pullin' her face, hittin' her in the side." Asked whether she called the police or did anything else to intervene, Dawanna said, "I mean I didn't do nuttin' 'cause I mean, in my eyes that's their problems, that's not mine."

Shaun described participating in an altercation with a woman in his neighborhood that combined physical violence with sexual degradation:

> It was me and my homies, we was walking down the street, gonna get on the bus.... And my homie ... was talking to some lady, and she was real drunk and she was saying something. She was calling him a nigger or something like that. And he was like, he asked me if I wanted to see her strip, and I was like "Yeah." [So] we ripped her dress off. She was like, "Damn, you wanna see something?" She took her bra and her panties and her shoes off and she threw her shoes at him. And she was swinging on him and stuff. And he kicked her in her stomach and she fell and she got up and she was still swinging. And then his brother stole [punched] her in the face and she was knocked out, she was laying in the street.

Shaun and his friends then got on a passing bus and left. Kevin was one of the "homies" present, and recounted the story in similar detail. He said, "the bus came, so we had to get in the bus. [But] a bunch of [people at] the bus stop was staring and stuff." Fearing someone may have called the police, he and his friends rode until "another bus came, and we jumped off and got on the other bus just in case."

In fact, three young women did recount incidents in which they or someone in their family called the police. In two cases, an adult member of the young woman's family was present and took the initiative. In the third, the incident involved a woman who had been beaten and robbed by a stranger. Jamellah witnessed the incident as it happened near her house, explaining, "the dude got her by her shirt, just grabbin' her, punchin' her in her face. And she was like, 'Help, help, I'm getting robbed! Call the police!'" After the man took off,

Jamellah brought the woman to her house, called the police, and waited with her "for like thirty minutes." Since the police "ain't never show," Jamellah "waited on the bus stop with her ... and she caught the bus home."

Rennesha said that recently a young girl who lived down the street ran to their house and told them that her stepfather was beating her mother. She explained, "the lil' girl was cryin', telling me what happened ... [she] just looked so sorry." Though her family called the police, Rennesha said they "never did come." Finally, Kenisha saw a man brutally beating a woman across the street and "called my grandmother to the door. I was like, 'Look how he doin' her.' And my grandmother was like, 'Call the police,' so I called the police and [they] came." Following her grandmother's lead, Kenisha explained, "I felt like any man shouldn't hit a woman first of all. I don't care what the situation is. And the way he was doing her, I don't care how much money that she owed him or whatever the case was, he didn't have no business putting his hands on her." Kenisha said there were a number of people present when the incident happened, some of whom were "standing around." One young woman attempted to intervene, telling the man, "'It don't make no sense how you doing her. [If] it was a man, you wouldn't beat on no man like that!'" Nonetheless, Kenisha noted, "it was a good six people out there, and it was only that one girl who was trying to break it up."

There were times, then, when neighborhood residents attempted to intervene on incidents of violence against women. It appears that this was most likely to occur either when a responsible adult was present and took initiative or when the incident involved someone perceived as an innocent victim, as with the woman robbed by a stranger. Intimate partner violence, in particular, was defined as a private affair. Cherise, for instance, described an incident at the bus stop in which a young man got angry with her and "smacked me in my face." She explained, "there were people standing around [but] they didn't help. They jus' thought that was maybe my boyfriend or something, they didn't want to get involved." In addition, when youths were in the company of peers, incidents of violence against women took on a carnivalesque flavor, with group dynamics that encouraged watching and taking pleasure from the spectacle rather than intervening. Even when youths felt a pang of responsibility to do something, strong norms toward nonintervention won out.

In fact, Britney was explicit about this normative expectation. Residents at her housing project witnessed a lengthy assault in which a neighbor choked and beat his girlfriend when she tried to leave, then held her at gunpoint and forced her into his car. However, they refused to cooperate when the police arrived. She explained:

> People in the neighborhood, they be like, "We not gonna tell you nuttin', 'cause you the police," stuff like that. They was like, "You the police, we ain't fina tell you nuttin' about this person and that person." So the police couldn't get no information. So they walked away. And I couldn't tell 'em, the kids couldn't tell 'em. 'Cause they would've got on us for being a snitch.

Thus, the consequence of neighborhood norms that required people to mind their own business, coupled with distrust of the police and slow or poor police response, meant that many incidents took place in the public eye, but few interventions followed. As Britney noted, even if she wanted to provide information to the police, she believed her hands were tied as a "kid" among adults who refused to cooperate.

I noted earlier that public incidents of violence against women were significant because of the messages they taught youths about how such violence is to be interpreted and responded to. Central among these—which even young women learned and often internalized—was that women were to blame for their victimization. Moreover, they saw for themselves that intervention on behalf of female victims was rarely forthcoming. Such responses taught young women the profound lesson that they are likely to be on their own when it comes to dealing with gender-based violence....

Sexual Harassment in Neighborhoods

In addition to witnessing physical violence against women in their neighborhoods, a common problem described by young women and corroborated by young men was the routine sexual harassment of girls. While the issue of sexual harassment has received a great deal of scholarly attention in the institutions of school and the workplace, limited attention has been paid to harassment in public spaces such as neighborhoods.[18] In all, 89 percent of the girls we interviewed reported experiencing some form of sexual or gender harassment. While they were more likely to report having experienced harassment in school (77 percent), more than one-half of the girls (54 percent) described incidents that occurred in their neighborhoods.[19] Two kinds of neighborhood sexual harassment emerged in youths' accounts: sexual come-ons or comments by young men in the neighborhood, and harassment by adult men, which girls often found much more ominous.

Sexual Harassment by Adolescent Peers Young men's behavior was described as problematic in two situations: when they engaged in inappropriate sexual banter or touching with girls they were friends with, and when they approached young women they didn't know. Cherise complained about the behavior of some of the young men she hung out with in the neighborhood:

> When we be playing or something, they try to touch me on the fly. But we usually don't fight over it. We might, you know, I might just push him or something 'cause they might be cool.... I don't know what they might be thinking, but maybe they might try to take advantage of you or something.... I just know that they try to touch me and I don't like it.

Nicole exhibited a similar ambivalence when describing the behavior of her male friends in the neighborhood: "You gon' have to watch out for them boys though.... They just got problems, that's what I think. But they cool though, I hang out with 'em. But they'll say lil' [sexual comments], but they won't say nothin' nasty.... They be playin' with you." While Nicole appeared uncomfortable with the young men's sexual banter, she went along with some of it because she enjoyed their company. She continued, "they'll make you laugh with 'em, you know. They like to have fun or whatever, make you laugh." Thus, she vacillated between interpreting their behavior as "playful" and problematic.

In fact, Nicole's ambivalence was even more apparent when we consider the question that elicited her comments: "Do you think girls face any particular dangers in the neighborhood?" While she ultimately interpreted her friends' behavior as "playful," her discomfort was evidenced by the fact that her account was triggered by a question about the risks young women face in the neighborhood. Later in the interview, she was more explicit about her concerns. Asked how the neighborhood could be made safer for girls, she said:

> I think it could be made safer if the boys would think 'fore they talk. You know, you should always think.... Like you come up to a lady, you should approach her in a better way instead of being like, "Ah, yo' when we gon' do this? When we gon' fuck?" Naw, un-un. You ain't suppose to do that. You should just ask her, "Would you go with me? Would you be my lady?" You know, ask her in a appropriate way, you know, don't just come up to her like she a hooker or something.

When neighborhood friends were the ones engaging in this behavior, young women would typically negotiate the situation so as not to escalate to a conflict or damage the friendship, despite their discomfort with the young man's behavior. Young men used "play" claims to explain their behavior, which girls sometimes went along with, though seemingly uneasily. Jamal explained, "we don't try to take advantage of them.... We probably just have fun with them. Try to play with them." Jamal emphasized what he described as "joking" sexual banter, noting: "Taking advantage [is] like grabbing or touching them, we don't do that. You might catch one or two try to slap a girl on the butt, that's the most." Likewise, Raymond said, "the only time I see girls get hit on [is] when they playing around."

In addition to negotiating "play" banter with male friends, a number of girls complained about how boys approached them in the neighborhood, including those they didn't know or know well. Nicole's earlier comments are illustrative. Similarly, Alicia described what happened after a young man came up to her friend and "said, 'girl, you fine. Why don't you back that ass up?'" Alicia's friend "told him don't come to her like that, that's real disrespectful," but the young man replied, "'How is that disrespectful?'" In fact, some young men seemed oblivious to (or at least feigned ignorance of) girls' interpretations of such sexual comments. For instance, asked whether he knew of males in the neighborhood who tried to take advantage of girls, Tony replied, "naw, I don't know nobody that do that.... They [just] be tryin' to mack. Probably be like, 'Hey baby, can I get your number?' or something, or 'I want to get to know you.'" Reminded that during the survey he indicated that girls could be uncomfortable with such behaviors, Tony said, "oh yeah, sometimes when they just be playin'. But I don't think they be meanin' it though.... Probably be sayin, 'You ugly', callin' em B's [bitches] and stuff, then they'll laugh." Though he admitted that such incidents may feel serious "to the girl," Tony insisted that young men "don't be serious" when they behave in such ways.

While Tony downplayed the significance of such behavior, Dwayne was highly critical:

> It was a girl walking down the street, she had some little shorts on and a little top. And at first [the guy] try to talk to her, and she didn't want to talk to him. So he touched her and she told him not to put his hands on her no more or she gonna go get her big brother. He done slapped this girl. So she went to go get her big brother and stuff. That's how stuff be getting started. That's how people be getting killed ... putting they hands on people they don't even know.

In fact, Dwayne described intervening on behalf of his girlfriend to get the young men in his neighborhood to leave her alone. He explained, "like my girlfriend, she be like walking home and stuff and I be too busy at the house and I can't walk her home. They be trying to put they hands on [her] and stuff. So I be going around talking to them little cats about putting they hands on my girlfriend, so they don't be doing that stuff no more."

As Dwayne's earlier description illustrates, a number of youths reported that when a young woman rebuffed the advances of a young man she didn't know or know well, he often became angry or disrespectful toward her. Walking through her neighborhood, Anishika said, sometimes "one of them lil' thugs or something tryin' to talk to me.... They'll get mad or something if I don't say nothin' to them, they be like, 'Well fuck you then.' I don't say nothin' back 'cause ... I'm not even trying to get involved with nobody like that." And Katie said, "some boys I know, like if they see some new girls walkin' down the street or something, they'll try to talk to 'em or whatever. And then if the girls don't give 'em no play, they'll call 'em out of they name.... He'll curse her out or something, talk about her." Jermaine said that "like if gang members try to talk to the girls, and the girls don't want to talk to 'em, they like, they start arguing, like, 'I don't want to talk to you anyway, you ugly,' or something like that,"...

What we see, however, is that girls faced a double bind: not standing up for themselves meant that sexual harassment was likely to continue and would be seen as deserved or even desired, while standing up for themselves often prompted an angry response. Given male domination of public spaces, young women were in a lose-lose situation. None of the strategies available guaranteed positive outcomes. Held accountable for young men's sexualization of them if they appeared too friendly or compliant, they risked escalating into a conflict—or at least insults and name-calling—if they challenged or ignored young men's sexual advances....

Sexual Harassment by Adult Men Young women expressed greater apprehension and less ambivalence about sexual harassment by adult men in the community. These incidents, more so than incidents involving peers, heightened girls' sense of vulnerability and made them fear for their safety. Nykeshia explained:

> The one person I feel uncomfortable around in my neighborhood [is a man that] always askin' me to do something [sexual] with him. And I told my momma about it and she told me, "Just stay away from him, as far as you can, and I hope he stay away from you." She called the police about it 'cause he had asked me in front of my momma.... He just denied everything and they was like, "We don't have proof that he asked you that ... we [can't] do nothin' about that."

Nykeshia was fifteen when interviewed and said this man (who was four years her senior) had been harassing her "since I was about thirteen." She said he behaved this way toward other girls in the neighborhood as well. The police were less than helpful, and, to make matters worse, the officer who responded to the call was one that Nykeshia's friend said had sexually assaulted her. Nykeshia noticed that her friend, who "was over there" at the time, quickly "went back in the house" when the officer arrived. "I went in the house too, 'cause I was wondering what she was backing back for, she was just backing back." Thus, not only was the officer's response to the man's behavior ineffectual, but Nykeshia had evidence that the man called to protect her had sexually assaulted her friend. Since the police failed to intervene, she had little recourse but to try and avoid the man and hope he never hurt her.

Shauntell, who was just twelve, described a similar problem with a man in her neighborhood:

> He stay directly across the street from us. I'm real cool with his lil' brother. [But] he just got outta jail. And me [and] my best friend, we walked down the street after school and he comes across the street, stand there and he watched me go all the way up on my front until I get in the house. I ... go in the living room and looked out the window, he starin' straight at the window. ... I goes outside and he'll come from across the street and come sit on our steps and just sit up there and stare at me. I look at him, I be like, "Problem?" And he shook his head no. I go [sarcastically], "See something you like?" ... I looked like a lil' tomboy and I sit like [a tomboy] and he'll look between my legs. I closed my legs. Stuff like that, I just don't feel comfortable around him and he just got out of jail.

Despite her discomfort, Shauntell said her auntie told her, "don't worry about it." She continued, "I got too many people over there for him to think he gon' do something to me."

While Shauntell took some comfort in her belief that having relatives in the neighborhood would keep the man from assaulting her, Britney specifically called on her uncle to sanction a man who was menacing her:

> Older [women], they'll walk past, [and neighborhood men will] be like, "How you doing?" But like me, if I walk past, grown man say like, "Hey baby what's your name?" and I'm 14 years old, they can be 25. I be like, "I'm only 14 years old sir" And they'll be like, "Ain't no problem," you know what I'm saying. And that's not right. So one man approached me and he said, "Hey baby what's your name?" I say, "I'm 14 years old," and he said, "I ain't ask you [your age], I said what's your name." I keep on saying I'm 14. He said "I'm 25 and I don't care." I said, "Well I'm 14." Then he kept on saying it. So then my uncle had rolled up, [and] he didn't know that was my uncle. So my uncle got out the car and was like "What you saying to my niece?" He like, "I was asking her do she know where somebody stay at." I said, "No you didn't, you asked me what's my name and I kept telling you I'm 14, you a grown man." My uncle had

> said, "If you ever talk to my niece like that again [you will deal with me. So] he walked off.

As Britney's account suggests, girls felt that they were a particular target for adult men in the neighborhood because they were teenagers and thus seen as easy to take advantage of. Young women expressed great discomfort about being sexualized by adults and, consequently, tended to be especially leery of groups of men.

Young men also described the widespread nature of adult male harassment of adolescent girls. Andrew said in his neighborhood, "females get harassed so much." He believed it posed a real danger for young women:

> Like an older person come through and they'll see a young girl walking through. She can have a pretty shape, pretty face or whatever, and they'll see her and they'll try to dog [her].[20] They try to talk to her or whatever, and the female will tell them no, or that they're too young or they too old for them, or they not interested. They get mad and get to calling them bitches and ho's, disrespecting them. [Then the girls] get mad, they get to cussing them back out. Sometimes the dudes get mad and threaten 'em, saying, "I'll do this to you," or "I'll do that."... And a woman know they can't whoop a dude, so.

Andrew surmised that "older guys" targeted teenage girls because "I guess they feel like with a younger girl they can take advantage of them." He described being upset when his younger sister was targeted: "See like I got a little sister, and I feel bad when she come home and say a man came and did this to her and tried to do that or whatever.... It makes me mad to the point where if I see the dude that did it, it feel good to fight him. But it ain't gonna prove nothing, so ain't no point.... I mean, a man gonna be a man regardless."

Likewise, asked whether there were men in the neighborhood girls felt uncomfortable around, Ricky said yes, explaining:

> I don't mean to say it like this, but they act like typical niggas, you know what I'm sayin'. I mean you got them—the girls fear the guys that be out there constantly using the B-word [bitch] and constantly talkin' about sex and always wanting to sell drugs and smoke weed and drink. It's mainly those guys they fear. I mean, [the girls] know we not a threat 'cause we pretty much universal. We like to kick it, chill with females and stuff.... [But] it's some of the guys we hang around. We don't necessarily associate with 'em or nothin' but it's the particular older guys that's say from like 19 to 25.

He said a few of these men "go look for females to talk about and cuss out, and they look for females to give 'em a reason to hit 'em or say something very disrespectful, or try to find a reason to touch 'em or something." Ricky said such incidents occurred "on a everyday basis with the particular guys I'm talking about. I mean, they just love disrespecting women." While nonintervention was the norm, Ricky explained that "I would say something. Like, 'Man, you wouldn't want nobody to talk to your mother like that.'" In response, though, he would be told, "'Ole' punk ass nigga get out of here,' or something. I mean it's just they attitudes man." ...

The neighborhood-based harassment of young women, then, was particularly widespread and amplified by the fact that groups of men—young and old—often congregated in these spaces. Whether older men, perhaps unemployed and with time on their hands, sometimes drug-addicted, or young men actively involved in drug sales and gangs, these groups often looked on young women primarily as potential sexual conquests. Young women were routinely frustrated by the disrespectful sexual come-ons they received by young men, but they were often down-right threatened by the behavior of adults. While not all men or boys in their neighborhoods behaved in this way, it was a common enough problem to be a regular cause for concern and was tied directly to the structural inequalities that shaped their communities.

Sexual Assault and Coercion in Neighborhoods

Sexual assault was also an ongoing neighborhood danger for young women. In fact, the girls in this study reported very high rates of sexual victimization. More than one-half (54 percent) reported some form of sexual victimization, including rape (29 percent) or attempted rape (14 percent), or being pressured or coerced into unwanted sex (43 percent). Nearly one-third (31 percent) reported multiple sexual victimizations. These are staggering figures, especially considering that the average age of the young women we interviewed was just sixteen.[21] . . .

Sexual assaults and sexually coercive behaviors in youths' neighborhoods were often an extension of the broader sexualized treatment young women experienced. This is largely why young women were so leery of groups of men. As with sexual harassment, young women's risks for sexual assault and coercion were varied. Among peers, alcohol and drugs were often used to lower a young woman's level of awareness or incapacitate her. Rapes involving physical force more frequently involved adult men, though they happened with peers as well.

Recall Antwoin's earlier comment that the neighborhood was less safe for girls than boys because "like the females, all the dudes be wanting to try to freak, you know, have sex with 'em, all that kinda stuff." He continued, "females, they be having trouble [in the neighborhood], like boys be wanting to run a train on 'em[22] something, like two boys on one girl in a sandwich or something. One boy be hittin' her from the front, one boy he hittin' her from behind." Antwoin described such activities taking place fairly often around the neighborhood, and it appeared that young men preyed on particularly vulnerable young women. In fact, it is worth noting that Antwoin's earlier complaint about "stuck-up girls" followed directly after this commentary about running trains on girls. Stuck-up girls, apparently, were those who refused to engage when young men made sexual advances, yet it was these same advances that were used to lure girls to precisely the kinds of coercive situations he said made the neighborhood unsafe for them.

Shauntell described similar events in her neighborhood involving her male relatives and their friends: "I done known girls that done just walked down the block, and I ain't gon' even lie, my cousins, my brothers, snatch 'em up, take 'em around the corner and watch 'em and do some of I don't know what to 'em." She recounted a recent incident:

> The other day, two girls over there that just ran away from home, they came over here 'cause one of 'em use to go with my cousin. And they was all just over here doggin' 'em. All them boys took 'em around . . . the corner [to one of the boy's houses], he got his own house. . . . My cousin not givin' a care about the one girl [he used to go with]. He not carin'. They take her, take her sister, and do some of, I don't even wanna name it all. Then [the girls] come out cryin', sayin' "I'm fina go home," [They were doing] some of everything. Get 'em drunk, high, beat 'em, raped 'em, tortured 'em, everything.

Shauntell said that because one of the young men lived on his own, thus providing the boys with an unsupervised locale for their activities, "they do it all the time." Though she appeared to implicate her cousin for not caring about the girl he had previously been involved with, Shauntell also blamed the young women and was unsympathetic when they tried to talk to her about what had happened: "I come home from school and go over, [those girls] still there. Then they come to me, 'Shauntell—.' [I said], 'I don't care, 'cause you shoulda went home."

Sexually coercive behavior often appeared targeted toward young women who seemed easy to take advantage of. A common strategy, as Curtis described, was to "like get them drunk." Likewise, Kristy said some of the gang members she knew would "give 'em some GH, you know, B [GHB, gamma hydroxy butyrate, a common "date" rape drug], you know, like at a party." Ricky said young women in the neighborhood were particularly at risk in the context of parties: "They have to be extra careful about leaving. And they have to watch what they do. [Watch their drinking] and getting high. I mean, you got some smooth talkers in our neighborhood, so." Asked why he thought the guys in his neighborhood did that to girls, he explained:

> I think it's just to get a image, a name. To make theyselves look big.... I can't really explain it. A lot of guys do it just so other guys can be like, "Aw, man, he'll do this" or "He'll do that." Like for example, "We did this and we did that, and it was [so-and-so's] gal." Most of 'em just do it for a name, man, just for a image. Try to look like something they not.

Thus an important feature of girls' sexual abuse was the status rewards such behavior provided within male peer groups. In addition to Ricky's emphasis on young men's attempts to build a name for themselves through their coercive sexual conquests, Lamont suggested that disadvantaged neighborhoods gave rise to young women he defined as "freaks": "In the ghetto, every ghetto I can tell you ... you see all these ... freaks.... It's babies, little bitty babies walking around with coochie cutters [tight-fitting shorts] on. You know that they gonna grow up to be freaks. Just 'cause you know what they been influenced by." In some cases, the early sexualization of girls likely made them vulnerable to sexual abuses in adolescence.[23]...

LaSondra described being raped by a man as she walked home from her friend's house one night. She said "it was dark, nobody was outside." As a precaution, her friend "walk[ed] me half the way, and I crossed over to go home and she walked back the other way." LaSondra was passing through a gas station on her way home when the man approached her, "He was just looking at me, and he said, 'I seen you and your friend earlier.'" The man then got LaSondra into his car and raped her: "He was heavy-set, he be holding me down, like I can't run, I can't get away." Afterward:

> He just dropped me off in this one spot and then he just pulled off. It was so scary, I was shaking, I was crying, I was running, you know, trying to find somewhere to go. I seen some people that went to my school and they let me use they phone and I called my momma. I tell my momma everything that happened and so she took me to the hospital and she called the police. The detective peoples came over and stuff and she talk to them and they didn't find him.... I was crying, shaking like all night in the hospital bed having flashbacks.

While LaSondra's mother was very supportive of her, she was disappointed with the police. She explained, "police don't do nothing. Some of 'em just give up, they could care less.... That's how they are, they could just care less."

Janelle also described being raped by a young man in her neighborhood:

> I was over at a friend's house, this girl, and he came over there. And you know, he was sitting down and he was talking to me and conversating with me at first, you know. And then he got to trying to touch me and stuff, and I was like, "No, go away," pushing him and trying to avoid the whole situation. But me being a female, males have more strength. So he holds me down and there's no much I can do about it because, you know, I'm not as strong as him.

Janelle said the rape happened when "my friend had went to the store." Afterward, she said, "I was scared to tell somebody so I didn't tell anybody.... I guess I thought if he found out that I told somebody he would somehow try to do something else to me." She was also fearful of reporting the incident to the police: "I mean, after he got out [of jail] or whatever and they did charge him with it, if he would've got out he would have been mad about the fact that I got him locked up, so he might have did something more."...

As with sexual harassment, youths tended to hold victim-blaming attitudes toward young women who placed themselves in situations that made them vulnerable to sexual mistreatment, while taking as a given the "rewards of rape" for males in their communities.[24] Young women who described sexual victimization reported few resources for addressing these problems. The perceptual schemas youths brought to bear on incidents of sexual assault, coupled with norms of nonintervention, meant that young women's personal networks were not a promising avenue for support; moreover, police response was experienced as ineffectual or as an option that could not offer sustained protection.

GENDERED RISK-AVOIDANCE IN NEIGHBORHOODS

Time and again, we were struck by young women's proclamations that they simply avoided being outside in their neighborhoods. Alicia said she felt safe in her neighborhood, because "I don't be outside over there, I sit on the porch. I don't go nowhere." Likewise, Jackie explained, "I stay in the house.... I don't go outside at all when I'm at home. When I'm around my grandma's house I go outside sometimes. I know mostly everybody over there. But when I'm at home, I don't go outside unless I'm going to the store, but I don't talk to nobody." Likewise, Anishika said that while "I been staying [in the same neighborhood] all my life... I don't be on the block. I don't be outside like that." She clarified, "I might sit outside for a minute or something ... [but] I just don't be outside." And Nykeshia said she "rarely" went outside when she was at home, noting that she only went "on the front and back of the house. You know, I just don't trust my neighborhood." Finally, Tisha gave a lengthy response to the question of whether she felt safe in her neighborhood:

> Well, I don't be outside for real. So yeah, [I feel safe] 'cause I'm inside my house. 'Cause I don't wanna be involved in nothin' that goes on around there. I mean, I got my friends already, don't none of 'em live around there. I don't mess with nobody around there, I don't got no reason to go outside and conversate with them or nothin'. I know what they about and that's not what I'm about.

Though the most prevalent theme in girls' descriptions of risk-avoidance was to limit the time they spent on the streets, girls did spend some time in public spaces in their neighborhoods, including, in some cases, in social situations. At these times, they described either explicitly relying on others for protection or believing that their personal connections would protect them. LaSondra said, "I just don't walk at night by myself. If I'm gonna walk somewhere at night, I'm gonna call one of my friends on the phone or get my next door neighbor that's there so he'll walk with me." Likewise, April said, "if you walk somewhere at nighttime, you just gon' be with some dudes or walk and have something [a knife or mace] on you... I do both—walk with dudes and have something. 'Cause [the dudes] might run, you never know." Gail touched on multiple themes:

> I'm not really outside all the time. I might stay in the house and watch TV or whatever.... I don't really go nowhere by myself. I mean, I live down the street from a donut shop where it be a lot of grown men just sitting around and stuff. But I don't think they'll harm me, 'cause they like know my uncles and my daddy, and like, they know everybody we live around.

Though she expressed apprehension about groups of adult men in public spaces and described her tendency to avoid them, she also believed that her (male) family and community ties would help insulate her from becoming a target of men looking to take advantage of girls. Similarly, Shauntell said, "I got a lot of family [in the neighborhood] and I know they ain't gon' let nothin' happen to me."

In general, having extensive ties in the neighborhood provided young women with a sense of security, and these ties developed particularly through residential stability. Katie said she had been in the same home her whole life, and she believed that her relationships with others in her neighborhood would protect her: "I feel safe when there's a lot of people outside 'cause they know me and they wouldn't let nuttin' happen to me." Cherise said, "I get along with pretty much everybody, I know everybody. I've been staying in my neighborhood for about ten years." Jamellah articulated a similar set of beliefs, though with a bit less confidence than Cherise:

> I usually do feel safe in my [immediate] neighborhood. Because I been there, my mother grew up there, that's my grandfather's house, she grew up there. And we been staying there for like six years, so I know basically everybody. And it's all old people in this neighborhood, where I'm walkin' down the street and I'm saying, "Hi, how you doin'" to every house on the street.

However, Jamellah was more concerned in the broader neighborhood, where she had to travel daily, because "that's where all heroin users be at, and you know how they be over that drug—going crazy." There, she was concerned primarily about being robbed, but she hoped her long ties to the neighborhood would insulate her:

> I be lookin' nice, you know, I have on a nice jacket.... I don't never take off my necklace or my earrings, and I be like, I hope they don't do nothin' to me. I know 'em though, 'cause I grew up over there. But still, when a person on heroin, they mind go crazy. They don't care. You could be they momma, if ya got something they want, they gon' take it.... I usually keep a job and they know that. I'm just, I got respect over there, so. That what put me in the mind like, ain't nobody gon' mess with me over there 'cause they all been knowin' me since forever.

Finally, Cherise added that she insulated herself from neighborhood dangers through her lack of participation in criminal neighborhood activities: "I don't get affected by all these problems. Crack dealers get affected a lot ... the boys that sell drugs get affected. People get killed over gangs do, they get affected."

Thus, to the extent that girls felt themselves to have strong and longstanding ties in their neighborhoods, they believed they would be offered protection against more serious forms of community dangers. Having male relatives in the neighborhood could help place them "out of bounds" among adult males who targeted young women for sexual manipulation and violence. Likewise, the presence of an individual they could call on to walk with them in traversing the neighborhood—particularly at nighttime—gave young women confidence in their relative safety....

CONCLUSION

Three overarching themes emerge in youths' discussions of the gendered dimensions of neighborhood dangers. First, youths' accounts indicate that young men were much more likely to be active participants in neighborhood based street networks. Their neighborhood-descriptions (where boys spoke of "I" and "we," while girls spoke of "they"), their depictions of young men's (but not young women's) neighborhood risks being explicitly associated with gangs and drugs, and their accounts of girls' treatment on the streets all reveal that public community space was, in many ways, *male* space. Though this was particularly the case at night—when young women suggested that dangers lurked largest—male domination of public spaces was nonetheless ubiquitous.

Second, many facets of neighborhood risk were structured by gender, and youths brought gendered perceptual frameworks to their understandings of neighborhood violence. Like other forms of violence, violence against women was both widespread and highly visible. Youths recounted witnessing public violence against women around their neighborhoods, and young women expressed concern about the pervasive sexualization they faced, including sexual harassment by adolescent and adult men and their fears and experiences with sexual assault in their neighborhoods.

Third, neighborhood residents' (including adolescents') responses to violence against women were decidedly unhelpful to the victims, as were their experiences with the police. Victim-blaming attitudes were widespread, among both young women and young men, and norms favoring nonintervention often won out, frequently leaving female victims to fend for themselves. In some cases, such violence even became a source of entertainment, particularly when youths were in the company of their peers.

For adolescent young women, the public nature of violence against women likely had a meaningful impact. Its visibility created a heightened vigilance and awareness among girls of their own vulnerability, but it also resulted in coping strategies that included victim-blaming as a means of psychologically distancing themselves from such events. To the extent that violence could be seen as something that happened to *other* women and girls—those who could be understood as "deserving" of it—young women could construct a sense of self that they believed shielded them from such risks.[25] Moreover, in a setting where norms were not favorable for intervention into violent events, the result was that youths became desensitized to such violence. Because the ability to stand up for oneself is a key avenue for respect in disadvantaged communities,[26] girls who failed to do so—particularly when their actions could render them culpable in some way in the eyes of others—faced sanction, at the very least in the form of lack of empathy.

ENDNOTES

1. For example, Anderson, 1999; Bourgois, 1995; Maher, 1997; Miller, 2001; Steffensmeier, 1983; Steffensmeier and Terry, 1986; Wilson, 1996.
2. Lauritsen, 2003.
3. Sampson, 1997, p. 42.
4. As noted in chapter 1, youths' action spaces included both their own neighborhoods, neighborhoods they had recently moved out of, and the neighborhoods of relatives and friends. Except where specifically relevant to the story at hand, I do not make distinctions between youths' activities across these arenas and refer to them generally as youths' neighborhood contexts.
5. The totals in table 25-1 represent all reports of exposure to physical violence, nor differentiated by locale (neighborhood, school, home). However, as I detail in this section, most of the reports, particularly of more serious violence, occurred in neighborhood contexts, including youths' own neighborhoods and those they spent time in.
6. Both incidents occurred in respondents' homes.
7. In fact, the extent of violence witnessed by these young people is likely underestimated here, because youths were not asked how many violent incidents they had witnessed, only whether they had and where the most recent incident had occurred. In all likelihood, most youths had witnessed multiple violent events. On the difficulties in measuring exposure to chronic community violence, see Wolfer, 1999.
8. Studies that have linked violence exposure to increased aggression include Durant et al., 1994; Farrell and Bruce, 1997; and Fitzpatrick, 1997. For research on emotional and psychological stressors associated with exposure to violence, see Richters and Martinez, 1993; Shakoor and Chalmers, 1991 (but see Farrell and Bruce, 1997). Brown and Gourdine, 2001, document the heightened sense of vigilance among girls exposed to community violence. Lauritsen et al., 1992, and Sampson and Lauritsen, 1990, emphasize the significance of ecological proximity to violence as a risk for victimization.
9. Farrell and Bruce, 1997, p. 3; see also Garbarino et al., 1991.
10. Graham-Bermann and Edleson, 2001. Though it is beyond the scope of my focus here, it is worth noting that about 50 percent of the youths in this study had witnessed domestic violence. A sizeable

portion also reported having been abused by family members: 31 percent of girls and 28 percent of boys reported physical abuse, and 14 percent of girls reported having been sexually abused by a family member. In all, 66 percent of girls and 63 percent of boys reported some exposure to family violence—as witnesses, victims, or both.

11. This was the case even for the survey items I report here. During the survey, we did not ask youths specifically whether the incidents of physical violence they witnessed involved men using violence against women. However, in the in-depth interview, these issues were examined, as discussed in the next section.

12. Some youths also noted that males faced more dangers at the hands of the police (Brunson and Miller, 2006b). In contrast, several youths highlighted risks posed to girls who were involved in gangs and drug sales. If young women were visible players in offender networks within their neighborhood, they were believed to also be at risk for crime-related violence. For instance, Gail said gender "really don't matter, 'cause in my neighborhood the girls [are] tough as ... the boys." And Katie noted, "some girls that live on the block, some of 'em are in gangs, they be doin' the same things the boys do. Try to fit in with the boys." Marvin said girls face "the same things as the boys face. 'Cause I mean, girls can be like boys too. They can do the same stuff as we can." Likewise, Rennesha said the risks were "about equal," explaining, "'Cause we have some females in the neighborhood that is gang related, they call them thug queens of they block and stuff like that." She said conflicts typically involved "like girls on girls, and the guys on the guys." This was not a predominant theme in the interviews but is noteworthy nonetheless.

13. Though a number of youths specifically mentioned that girls were "easy" targets for robbery, their survey responses did not correspond with these gender-based beliefs. In all, 65 percent of the young men, but only 14 percent of the young women, said they had been the victim of a robbery. Their rates of personal victimization were more comparable for other forms of physical violence, including being hit by someone (75 percent and 71 percent, respectively, for boys and girls), having been jumped or beaten up (68 percent and 57percent, respectively), and having been stabbed (10 percent and 14 percent, respectively). Nevertheless, young men were more likely to have been threatened with a weapon (58 percent, versus 26 percent for girls) and to have been shot (15 percent, versus 3 percent).

14. For a discussion of women as situationally disadvantaged in public spaces, see Gardner, 1995. For an analysis of how gender is constructed through conversations about vulnerability to violence, see Hollander, 2001.

15. In addition, though not a common theme in girls' interviews when discussing their neighborhoods, some young men focused on dangers they believed young women created for themselves because of "their mouths." For instance, several young men said girls "talk too much" and thus "deserved" to be hit. These themes do come up in girls' discussions as well, primarily in their accounts of dating violence (see Chapter 5) and conflicts with other girls (Miller and Muffins, 2006). Because young women did not describe these problems as specific dangers they felt within their neighborhoods, I examine the issues in greater detail in these subsequent chapters.

16. An oversight in the in-depth interview guides meant that only young women were asked this question. In all, 43 percent described witnessing such incidents. Because we were interested in examining male peer support for violence against women, young men were asked only whether any of their friends engaged in such behaviors. Thus, broad accounts of violence against women in neighborhoods were not solicited from the young men. In hindsight, this was a disappointing omission.

17. The research literature on sexual harassment in schools is discussed in Chapter 3. On harassment in public places, see Gardner, 1995; MacMillan et al., 2000. In fact, MacMillan and his colleagues suggest that street harassment has profound consequences for women's fear of sexual victimization.

18. While "sexual harassment" refers to unwanted sexual attention the term "gender harassment" refers to hostile treatment toward women that includes sexist or degrading remarks or behavior. For a discussion, see Larkin, 1994.

    In all, 51 percent of the girls described incidents of gender harassment, 71 percent reported sexual comments that made them uncomfortable, and 49 percent described unwanted sexual touching or grabbing. Of the 18 girls who reported gender harassment, 67 percent described incidents that occurred in school, and 55 percent described incidents in neighborhoods. Likewise with sexual comments: 60 percent reported school incidents, and 48 percent reported neighborhood incidents. The largest disparity between sites was for unwanted touching or grabbing: of the 17 girls who reported such incidents, 65 percent described them occurring at school, and 35 percent in the neighborhood. These numbers do not include incidents of sexual assault.

Young men were also asked whether they had engaged in sexual or gender harassment toward girls, but they were not asked to distinguish where the incidents took place. In all, 70 percent of the boys said they had called girls names or said things to put them down; 48 percent said they had made sexual comments to girls that made them uncomfortable; and 38 percent reported having grabbed or touched girls in ways that made them uncomfortable.

19. Youths used the term "dog" or "dogging" to refer generally to disrespectful treatment of those considered subordinate. In this instance, the term is used to suggest the man is attempting to manipulate the young woman for sexual gain.
20. Consider, for example, that Mary Koss and her colleagues' (1987) groundbreaking self-report studies of sexual assault among college students found that 12 percent of college women reported having been raped. The adolescent girls in our study report a rate more than double this figure.
21. Youths used the term "running train" to refer to "gang bangs" or gang rapes. As I discuss in Chapter 4, nearly one-half of the young men interviewed reported having engaged in such behavior. Notably, they did not consider such incidents rape, though the three young women in our sample who had been victimized in this way did define what happened to them as a sexual assault.
22. An important part of this early sexualization for some girls is childhood sexual abuse. For instance, 20 percent of the girls in our study reported being victims of child sexual abuse. One of the subsequent outcomes of this is increased vulnerability to additional sexual coercion. I return to these issues in more detail in Chapter 4.
23. Scully, 1990.
24. Miller, 2001.
25. Anderson, 1999.

REFERENCES

Anderson, Elijah. 1999. *Code of the Street*, New York: Norton.

Benson, Michael L., Greer L. Fox, Alfred DeMaris, and Judy Van Wyk. 2003. "Neighborhood Disadvantage, Individual Economic Distress and Violence against Women in Intimate Relationships." *Journal of Quantitative Criminology* 19: 207–235.

Bourgois, Philippe. 1995. *In Search of Respect: Selling Crack in El Barrio*. Cambridge: Cambridge University Press.

Brown, Annie Woodley, and Ruby Gourdine. 2001. "Black Adolescent Females: An Examination of the Impact of Violence on Their Lives and Perceptions of Environmental Supports." *Journal of Human Behavior in the Social Environment* 4: 275–298.

Brunson, Rod K., and Jody Miller. 2006a. "Gender, Race, and Urban Policing: The Experience of African American Youths." *Gender and Society* 20: 531–552.

Carvalho, Irene, and Dan A. Lewis. 2003. "Beyond Community: Reactions to Crime and Disorder among Inner-City Residents." *Criminology* 41: 779–812.

Cobbina, Jennifer E., Jody Miller, and Rod K. Brunson. 2007. "Gender, Neighborhood Danger, and Risk-Avoidance Strategies among Urban African American Youth." Unpublished ms. University of Missouri–St. Louis.

Durant, Robert, Chris Cadenhead, Robert A. Pendergrast, Greg Slavens, and Charles W. Linder. 1994. "Factors Associated with the Use of Violence among Urban Black Adolescents." *American Journal of Public Health* 4: 612–617.

Farrell, Albert D., and Steven F. Bruce. 1997. "Impact of Exposure to Community Violence on Violent Behavior and Emotional Distress among Urban Adolescents." *Journal of Clinical Child Psychology* 26: 2–14.

Fitzpatrick, Kevin M. 1997. "Aggression and Environmental Risk among Low-Income African-American Youth." *Journal of Adolescent Health* 21: 172–178.

Garbarino, James, Kathleen Kostelney, and Nancy Dubrow. 1991. "What Children Can Tell Us about Living in Danger." *American Psychologist* 46: 376–383.

Gardner, Carol B. 1995. *Passing By: Gender and Public Harassment.* Berkeley: University of California Press.

Graham-Bermann, S. A., and J. L. Edleson eds. 2001. *Domestic Violence in the Lives of Children: The Future of Research, Intervention, and Social Policy.* Washington, D.C.: American Psychological Association.

Hollander, Jocelyn A. 2001. "Vulnerability and Dangerousness: The Construction of Gender through Conversation about Violence." *Gender and Society* 15: 83–109.

Ioe-Laidler, Karen A., and Geoffrey Hunt. 1997. "Violence and Social Organization in Female Gangs." *Social Justice* 24: 148–169.

Koss, Mary P., C. A. Gidycz, and W. Wisniewski. 1987. "The Scope of Rape: Incidence and Prevalence of Sexual Aggression and Victimization in a National Sample of Higher Education Students." *Journal of Consulting and Clinical Psychology* 55: 162–170.

Larkin, June. 1994. "Walking through Walls: The Sexual Harassment of High School Girls." *Gender and Education* 6: 263–280.

Lauritsen, Janet L. 2003. *How Families and Communities Influence Youth Victimization.* Juvenile Justice Bulletin. Washington, D.C.: U.S. Department of Justice.

Lauritsen, Janet L., and Robin J. Schaum. 2004. "The Social Ecology of Violence against Women." *Criminology* 42: 323–357.

MacMillan, Ross, Annette Nierobisz, and Sandy Welsh. 2000. "Experiencing the Streets: Harassment and Perceptions of Safety among Women." *Journal of Research in Crime and Delinquency* 37: 306–322.

Maher, Lisa. 1997. *Sexed Work: Gender, Race and Resistance in a Brooklyn Drug Market.* Oxford: Clarendon.

Marciniak, Liz Marie. 1998. "Adolescent Attitudes toward Victim Precipitation of Rape." *Violence and Victims* 13: 287–300.

Miles-Doan, Rebecca. 1998. "Violence between Spouses and Intimates: Does Neighborhood Context Matter?" *Social Forces* 77: 623–645.

Milier, Jody. 2001. *One of the Guys: Girls, Gangs and Gender.* New York: Oxford University Press.

Miller, Jody, and Christopher W. Mullins. 2006. "Stuck Up, Telling Lies, and Talking Too Much: The Gendered Context of Young Women's Violence." pp. 41–66 in *Gender and Crime: Patterns of Victimization and Offending,* edited by Karen Heimer and Candace Kruttschnitt. New York: New York University Press.

Pain, Rachel. 1991. "Space, Sexual Violence, and Social Control: Integrating Geographical and Feminist Analyses of Women's Fear of Crime." *Progress in Human Geography* 15: 415–431.

Richters, John E., and Pedro Martinez. 1993. "The NIMH Community Violence Project: Children as Victims of and Witness to Violence." *Psychiatry* 56: 7–35.

Sampson, Robert J., Stephen W. Raudenbush, and Felton Earls. 1997. "Neighborhoods and Violent Crime: A Multilevel Study of Collective Efficacy." *Science* 277: 918–924.

Scully, Diana. 1990. *Understanding Sexual Violence.* Boston: Unwin Hyman.

Shakoor, Bambade H., and Deborah Chalmers. 1991. "Covictimization of African-American Children Who Witness Violence and the Theoretical Implications of Its Effects on Their Cognitive, Emotional and Behavioral Development." *Journal of the National Medical Association* 83: 233–238.

Stanko, Elizabeth A. 1990. *Everyday Violence: How Women and Men Experience Sexual and Physical Danger.* London: Pandora.

Steffensmeier, Darrell J., and Robert Terry. 1986. "Institutional Sexism in the Underworld: A View from the Inside," *Sociological Inquiry* 56: 304–323.

Warshaw, Robin. 1988. *I Never Called It Rape.* New York: Harper and Row.

Wilson, William Julius. 1996. *When Work Disappears: The World of the New Urban Poor.* New York: Knopf.

Wolfer, Terry A. 1999. "'It Happens All the Time': Overcoming the Limits of Memory and Method for Chronic Community Violence Experience." *Journal of Interpersonal Violence,* 14: 1070–1094.

DISCUSSION QUESTIONS

1. The *'hood is an unsafe place for women and girls*; this is what the author of this selection suggests. What social factors are contributing to the high rate of gender-based violence?
2. What makes the violence against women both pervasive *and* highly public in the inner-city? Does the public nature of violence and harassment signal a lack of police involvement in the inner-city?
3. Why is victim-blaming such a common response when incidences of gendered-violence become public knowledge? What function does victim-blaming serve for those who witness extraordinary levels of violence against women and girls in the inner city?
4. Do misogynistic lyrics in popular music, especially some modes of "gangsta" rap and hip-hop, have anything to do with the culture of violence against women and girls in the inner-city? Explain your answer.
5. What possible treatment strategies would you suggest to assuage this social problem? What public policies should policy makers pursue to lower the rates of violence against women and girls in the inner-cities of the United States?
6. "*This is a social problem that begins and ends with violent men; they are the cause of this problem and the key to ending this dilemma.*" Respond to this statement.
7. What can inner-city women and girls do about this social problem, short of moving away from these violent neighborhoods?

26

Homosexuality and American Citizenship

MICHAEL BRONSKI

The Four Questions

1. What is the problem?
2. Is this a *social* problem, according to Bronski? Who is hurt? Is society hurt? Is the cause social? Is there widespread recognition of this as a problem?
3. What is the basic cause of this problem? This is what Bronski tries to get us to think about carefully.
4. What, if anything, does Bronski recommend we do about this problem?

Topics Covered

Culture wars

Homosexuality

Identity

Sexuality

Deviance

Citizenship rights

... The relationship between being an "American" and having the full rights and privileges of

SOURCE: From Michael Bronski, *The Pleasure Principle: Sex, Backlash, and the Struggle for Gay Freedom*, Bedford/St. Martin's, 1998.

legal citizenship is complicated. White women in the nineteenth century, for example, were considered American citizens even though they did not, in many states, have the right to own property, or to hold office, or even to vote. On the other hand, even though African Americans were released from slavery by the Emancipation Proclamation in 1863 and the Thirteenth Amendment in 1865, granted citizenship by the Fourteenth Amendment three years later, and (in the case of males) enfranchised by the Fifteenth Amendment in 1870, they still faced enormous legal and legislative battles to exercise those rights fully. And even full legal citizenship has not changed a prevailing attitude among many white citizens that whiteness is a prerequisite for being an authentic "American." Native Americans and residents of Asian descent, as well as African Americans, have suffered from this perception.[1]

In fact, the common definition of citizenship—as a relationship between an individual and a nation whereby individuals are granted clearly defined rights in exchange for certain duties, such as obeying laws, paying taxes, and serving in the military—is deceptively simple and inexact. Children, for example, are legally citizens but are denied the vote because their age is seen as an impediment to making informed decisions. White women had citizenship but were, for more than a century, denied basic rights. African Americans had citizenship rights they were barred from using. Yet both groups were expected to pay taxes, and black males were liable for military service.

This limited vision of citizenship has been an effective mechanism for curtailing minority groups' cultural and political freedom. While it secures rights and protections for some groups, it does not necessarily provide legal recourse when those rights are abridged. More important, it does not include the ability to live without the fear of violence or discrimination, or be assured of personal safety in public as well as private. It does not provide for minorities the same rights to public assembly, access to information, and self-expression that majority groups have. Ideally, full citizenship would mean the creation of a national social and political environment in which the cultural self-expression and determination of all minority groups would be respected.

The struggle over defining who is an "American," or a full citizen, is, at its heart, about power. It is about the power of the majority to set and maintain a political and cultural agenda and to create and regulate codes of personal, social, and sexual behavior that conform to the majority's standards. It is also about the struggles of those with less power to secure and protect their individual autonomy and dignity, as well as their right to a distinct, nonmajority identity.

The culture wars, in all of their manifestations, are the public forum where issues of what it means to be an "American" and a citizen are being debated and decided. Although many of the contemporary culture wars concern issues of race and ethnicity—immigration rights, bilingual education, Afrocentricism—a majority of them involve issues of sexuality and gender. Specifically, they revolve around reproductive rights, alternative definitions of the family, children's rights, feminism, pornography, and homosexuality. All of these issues are targeted by conservatives and the religious right as antithetical to "family values."

To the extent that this is true, these culture wars are actually about the restabilization and reaffirmation of heterosexuality and traditional gender norms in U.S. culture. Over the past century myriad factors have decentered or radically altered the institution of heterosexuality: increased freedom for women and children, the ability to regulate sexual reproduction, the effects of industrialization on the family, and the breakdown of established gender roles. But none of these are as deeply disruptive to the institution of heterosexuality as the emergence of homosexuality as a public identity and a political agenda—a "coming out" that forced the dominant culture to deal with homosexuality.

To understand why homosexuality is at the heart of the contemporary culture wars, it is important to examine the actual and metaphoric role that homosexuality plays in Western societies, and then to explore how mainstream culture's fear of homosexuality is constructed and manifested.

One of the most persistent myths of the gay rights movement, and of liberal thinking, is that the dominant culture's fear of homosexuality and hatred of homosexuals—what is commonly called "homophobia"—is irrational. This is untrue: It is a completely rational fear. Homosexuality strikes at the heart of the organization of Western culture and societies. Because homosexuality, by its nature, is nonreproductive, it posits a sexuality that is justified by pleasure alone. This stands in stark and, for many people, frightening contrast to the entrenched belief that reproduction alone legitimates sexual activity. This belief, enshrined in religious and civil law, is the foundation for society's limiting gender roles and the reason why marriage, traditionally, has been the only context recognized by society and by law for sexual relationships between men and women. It is the underpinning for the restrictive structure of the biological family unit and its status as the only sanctioned setting for raising children. It is the hidden logic determining many of our economic and work structures. In profound, if often unarticulated ways, this imperative view of reproductive heterosexuality has shaped our world.[2]

Although heterosexuals individually, throughout the centuries, have engaged in sex for pleasure, attempts to unseat the ideology of sex-as-reproduction have continually been resisted. Until recently, explicitly sexual material (that which portrays sex as simply pleasure) has routinely been censored, medical information about birth control has been banned, and access to contraceptives has been limited by law. Only in 1965 did the Supreme Court rule that married adults had a constitutional right to privacy in marriage that allowed them to use birth control. Even for heterosexuals, securing personal and sexual freedom has been a long and difficult battle.

The imperative of reproductive heterosexuality has had enormous effects on homosexuals as well. It is the basis for the labeling of homosexual activity as "unnatural," sinful, and sick. Historically, it has been the rationale for executing, castrating, and otherwise physically punishing male and female homosexuals—assaults that continue today under the rubric of "queer-bashing." It has prompted contemporary measures to curtail, regulate, and punish homosexual behavior and deny homosexual people such rights of citizenship as free association, freedom of expression, and the right to join the military or form legal marriages. It has fueled and inflamed attacks on gay people who fight for these freedoms.

These attacks occur because homosexuality and homosexuals present attractive alternatives to the restrictions that reproductive heterosexuality and its social structures have placed upon heterosexuals. The real issue is not that heterosexuals will be tempted to engage in homosexual *sexual* activity (although the visibility of such activity presents that option) but that they will be drawn to more flexible norms that gay people, excluded from social structures created by heterosexuality, have created for their own lives. These include less restrictive gender roles; nonmonogamous intimate relationships and more freedom for sexual experimentation; family units that are chosen, not biological; and new models for parenting. But most important, homosexuality offers a vision of sexual pleasure completely divorced from the burden of reproduction: sex for its own sake, a distillation of the pleasure principle.

The threat of homosexuality is thus so tremendous, and affects so significantly the basic structures of Western societies, that the sustained and often vicious antihomosexual response is not surprising. The construction (and manifestation) of antihomosexual fear and hatred, however, is a more complex phenomenon.

In his provocative analysis *Anti-Semite and Jew*, Jean-Paul Sartre argued that the fear and hatred of Jews was not a political or social "opinion" but a "passion," meaning not that it is emotional or irrational, but simply that it is the basis for a cohesive, sustained view that is held independently of material reality. It is "the *idea* of the Jew" that incites the anti-Semite, not necessarily any specific action or person.

So it is with homosexuality. It is the *idea*, the concept of homosexuality—that is, sexual pleasure without justification or consequences—that terrifies the gay-hater. As with Sartre's explanation of anti-Semitism, gay-hating derives less from a feeling about particular people than from a profound attachment to maintaining the existing social order. This helps explain why vocal antigay politicians are sometimes capable of maintaining cordial relationships with gay friends or family members.

The institutionalized hatred of homosexuality serves clear social functions. It brings together people of various races, religious beliefs, classes, and backgrounds in a unified vision and identity as heterosexual. In this way, it actually performs the function that traditional ethnic and religious "assimilation" has never achieved in the United States, uniting divergent groups, no matter what their genuine conflicts, under a common identity: heterosexuality. This identity then comes to delineate "American." Consequently, when homosexuals are denied specific legal rights and social opportunities, heterosexuals are then defined as authentic, full citizens.[3]

Like anti-Semitism, the hatred of homosexuality serves the function of valorizing the "normal"; it bonds heterosexuals, rallying them against a collective, and by implication more powerful, enemy. The creation of the "homosexual menace," not unlike the fantasy of a worldwide Jewish conspiracy, transforms the average man into a valiant defender of the "normal"—a process that celebrates and reinforces the sexual status quo while at the same time granting enormous importance to sexual differences. Just as the mythology of anti-Semitism magnifies the power of the Jew, gay-hating recasts the homosexual as sexually and socially dangerous. Yet because this fantasy homosexual also embodies a sexual freedom that stands in sharp relief to the restrictions of institutionalized heterosexuality, he is an object of envy as well as scorn.

But the creation of the homosexual menace—at once a threat and a temptation—is complicated by the fact that gay culture actually does, if its alternatives to institutionalized heterosexuality are embraced, present a threat to basic social structures. This poses a difficult dilemma for the dominant culture. If homosexuality is accepted, it challenges the status quo overtly; if it remains demonized, it presents a seductive vision of alternative possibilities.[4]

Western societies have dealt with this dilemma by curtailing and punishing homosexual behavior, by limiting the citizenship rights of homosexuals, and by creating a social milieu that rewards the closet and punishes visibility. In the United States, homosexuals' full citizenship is denied in many ways: from the denial of marriage rights to discrimination in the workplace, to physical violence. But these limitations on citizenship, which homosexuals share with other minority groups and subcultures, are indicative of a lack of standing as "Americans" as well.

The national debate, as manifested in the culture wars, raises the profound question of what it means to live in a pluralistic democracy. How are people of varying races, ethnicities, classes, sexual orientations, religions, and values all to live together? The ideal of "America" as the ultimate melting pot able to "assimilate" all difference and produce a unified national identity—the "American"—is a myth. Although U.S. culture is often split along ethnic, racial, and other identity lines, this is not a new phenomenon. The fantasy of a cohesive "American" national culture is a strategic, right-wing argument against minority and subcultural groups demanding full citizenship. The divisiveness of the culture wars is a replay of what has occurred throughout all of U.S. history....

The current culture wars provide a map of the dominant culture's discomforts and fears. Race, gender, and sexual orientation are hotly disputed factors in the exploration of the mostly unspoken question "Who is an American?" But for the dominant culture there is another, more frightening and complicated question lurking beneath this unarticulated inquiry: "Who are we?"

Still, although struggles for gender, ethnic, and racial equality have produced mixed results, an ongoing public debate has established these concerns as central to the broader discussion of social justice. This has been far less true of homosexuality, for several reasons. Whereas race and gender are seen as immutable, physically determined, and publicly discernible characteristics, sexual identity and behavior have historically been perceived as personal traits or decisions. Also, it is easier for homosexuals to "pass" than it was for minorities or women. And it was far more difficult to "pass" out of the strictures of race or gender, but relatively simple for homosexuals to "pass."

As a result, the threat that homosexuality poses to the dominant culture is substantially different from the threat posed by demands for racial and gender equality. The latter challenges come from outsiders whose otherness is clearly and physically defined. Homosexuality, on the other hand, is a far more complex, protean identity. It is rare that people are confused about their race or gender, but anyone can be a homosexual or engage in homosexual behavior. This ambiguity, combined with the homosexual's ability to "pass," challenges the presumed cohesion of mainstream heterosexual culture.[5]

The struggle over citizenship rights for homosexuals comes at a critical point in U.S. history. Campaigns for racial and gender equality have progressed far enough that the dominant culture believes that it has made enough concessions (or, from a conservative viewpoint, sustained enough losses). The dominant culture has become increasingly hostile to minority demands. While this animosity has affected the gay rights movement, particularly in the 1980s and 1990s, it accounts for only part of the antagonism that homosexuals face in the culture wars.

Struggles over racial and gender equality have deeply shaken the power structures in the United States, but the rise of a public and politicized homosexuality—in an organized gay rights movement and a visible gay culture—has proved even more threatening. Homosexuality offers a clear critique of the ideology of heterosexuality and attractive alternatives to heterosexuality as an institution. It challenges accepted ideas about sexual activity, gender roles, relationships, marriage, family, work, and child rearing. Most important, it offers an unstinting vision that liberates sex from the burden of reproduction and places pleasure at the center of sexual activity. This is what frightens Pat Buchanan and what is at the heart of the culture wars over sexuality, sexual orientation, and related issues.[6]

For the dominant culture, the threat of homosexuality is that it embodies and prioritizes the pleasure principle—the denial of which, we have always been told, is necessary for civilization, as we know it, to go on.

The question is this: Do we *want* civilization as we know it to go on? Or do we want to change it to meet our most basic needs?

ENDNOTES

1. The politics of woman suffrage campaigns have striking parallels to the current fight for gay rights. Ellen Carol DuBois's *Feminism and Suffrage* and Christine A. Lunardini's *From Equal Suffrage to Equal Rights* both detail how social "assimilationists" were pitted against political "activists." Eleanor Flexner's *Century of Struggle* is astute in its analysis of the tensions between a predominantly white suffrage movement and the fight for the black male vote, a situation similar to the gay rights movement and the African-American community today. Illuminating discussions of the complicated histories of Asian and Native American struggles for citizenship are found in Howard Zinn's *A People's History of the United States* and David H. Bennett's *The Party of Fear.*
2. "Homophobia" as a rational fear is a difficult idea for gay rights advocates to accept, since it is counter to the historical model of assimilation. George Weinberg coined the term "homophobia" in 1972 in his *Society and the Healthy Homosexual;* that work's now-simplistic analysis has been supplemented by works like Suzanne Pharr's feminist *Homophobia: A Weapon of Sexism*, and writings by people of color, such as James Baldwin's "Here Be Dragons," in *The Price of the Ticket*, and Barbara Smith's anthology *Home Girls.*
3. While the hatred of gay people reinforces male privilege, it also serves to unite, across a wide gender gap, heterosexual women and men. The idea that "heterosexuality" might function as a "nationalist" identity remains largely unexplored, but is touched upon in Jonathan Ned Katz's *The Invention of Heterosexuality* and George Mosse's *Nationalism and Sexuality.*

4. Weinberg's concept of "homophobia" does not address the complex interactions of gay culture with mainstream culture. Sartre's "ethnic" cultural/psychological model is reflected in Sander Oilman's *The Jew's Body*. Oilman's claim that anti-Semitism is fueled by sexual paranoia resonates with the psychological construction of the hatred of homosexuality.
5. Although "race" is a construct, most people identify as being a member of a specific "race." Recent discussions of the history and construction of "whiteness" include Richard Dyer's *White*, Theodore W. Allen's *The Invention of the White Race*, Alexander Saxton's *The Rise and Fall of the White Republic*, and David Roediger's *The Wages of Whiteness*.
6. For many politicians the culture wars are clearly just a political football, yet these battles often reflect real confusions over shifting moral and social values. William Bennett's *The Book of Virtues* and Gertrude Himmelfarb's *The De-Moralization of Society* are examples of marketing traditional "virtues" and "values" as conservative solutions to complicated social realities.

REFERENCES

Allen, Theodore W. *The Invention of the White Race*. London: Verso, 1994.

Bennett, David H. *The Party of Fear: From Nativist Movements to the New-Right in American History*. Chapel Hill: University of North Carolina Press, 1988.

Bennett, William. *The Book of Virtues: A Treasury of Great Moral Stories*. New York: Simon & Schuster, 1993.

DuBois, Ellen Carol. *Feminism and Suffrage: The Emergence of an Independent Women's Movement in America, 1848–1869*. Ithaca, N.Y.: Cornell University Press, 1978.

Dyer, Richard. *White*. New York: Routledge, 1997.

Flexner, Eleanor. *Century of Struggle: The Woman's Rights Movement in the United States*. New York: Atheneum, 1973.

Gilman, Sander. *The Jew's Body*. New York: Routledge, 1991.

Himmelfarb, Gertrude. *The De-Moralization of Society: From Victorian Virtues to Modern Values*. New York: Knopf, 1995.

Katz, Jonathan Ned. *The Invention of Heterosexuality*. New York: Dutton, 1995.

Lunardini, Christine A. *From Equal Suffrage to Equal Rights: Alice Paul and the National Woman's Party, 1910–1928*. New York: New York University Press, 1986.

Mosse, George L. *Nationalism and Sexuality: Respectability and Abnormal Sexuality in Modern Europe*. New York: Howard Fertig, 1985.

Pharr, Suzanne. *Homophobia: A Weapon of Sexism*. New York: Chardon Press, 1988.

Roediger, David R. *The Wages of Whiteness: Race and the Making of the American Working Class*. London: Verso, 1991.

Saxton, Alexander. *The Rise and Fall of the White Republic*. New York: Verso, 1990.

Smith, Barbara. *Home Girls: A Black Feminist Anthology*. Albany, NY: Kitchen Table—Women of Color Press, 1983.

Weinberg, George. *Society and the Healthy Homosexual*. New York: St. Martin's Press, 1972; Doubleday Anchor, 1973.

Zinn, Howard. *A People's History of the United States*. New York: Harper & Row, 1980.

DISCUSSION QUESTIONS

1. Does Bronski argue that homosexuality itself is a social problem? Do you? Does it hurt many people? Does it hurt society? Is its cause social? Is there general agreement that homosexuality is a social problem?

2. Does Bronski argue that societal reaction is the social problem? Do you? Does it hurt many people? Does it hurt society? Is its cause social?
3. Bronski argues that homosexuals constitute a minority in society different from ethnic and racial groups. To what extent do you believe this is true? In what ways are they all similar?
4. Bronski argues that the dominant group's fear of homosexuality is a rational fear. What does he mean by "rational"? What is his argument? Do you agree?
5. Does the existence of homosexuality in society threaten you personally? Does it threaten society?

PART VIII

Social Problems: Crime and Drugs

Most sociologists believe that crime and the criminal justice system are serious social problems in our society. It is frustrating to see these complex problems continue and seem to become worse, causing us to direct many resources in one direction rather than toward other problems we face. Why is there crime? How have we dealt with it as a society? Have we been successful? Can we do more? What are the links between poverty and crime, capitalism and crime, social power and crime, race and crime? The selections in this part explore many of these questions.

Elijah Anderson takes us into the inner city and describes why so many individuals are caught up in a world that encourages breaking the rules set by the dominant society. Anderson emphasizes poverty, hopelessness, anger, and alienation rather than blaming those who are caught in this web.

Stephen Rosoff, Henry Pontell, and Robert Tillman introduce white-collar crime as a social problem. Since white-collar crime is often the result of acts by people who are powerful and affluent, it is more difficult for society to define these acts as serious. The authors explore the causes and consequences of such crimes.

Arthur Benavie begins his essay on *Drugs: America's Holy War* by posing a powerful question: "*Who is the enemy in this war on drugs?*" Benavie offers a sophisticated treatment of the unintended consequences of America's long war on narcotics. Indeed, the article raises a profound question for the reader: Should society continue its current strategy on the "war on drugs"?

Todd R. Clear, writing with such passionate prose, asks us to consider the consequences of mass incarceration on individuals, families, and communities.

This article has the sociological audacity to suggest that the impact of incarceration on society is so formidable that at our current rates of imprisonment constitutes a unique form of violence, and is a *social* problem apart from the criminal acts themselves: "death by a thousand little cuts" indeed.

The final piece is by a sociologist with a unique combination of sociological acuity and empathy for a population of Americans that do not engender much compassion or acceptance: prisoners and ex-convicts. Devah Pager's *Blacklisted: Hiring Discrimination in an Era of Mass Incarceration* asks us to consider the severity of employment discrimination against ex-convicts. Pager poses troubling questions about notions of fairness and the punishment of stigmatizing labels that follow "ex-cons" long after they have supposedly "paid their debts to society." This is a difficult reading, but important for understanding the life-chances of millions of Americans.

27

Violence and the Inner-City Code

ELIJAH ANDERSON

The Four Questions

1. What is the problem Anderson identifies?
2. Is this a *social* problem? Who is hurt? Is society hurt? Is it socially caused? Are people aware of the problem?
3. What is the cause of this problem?
4. What can be done? Does Anderson give us any idea?

Topics Covered

Violence

Street crime

Code of the streets

Family

Neighborhood

Socialization

Poverty

Of all the problems besetting the poor inner-city black community, none is more pressing than that of interpersonal violence and aggression. This phenomenon wreaks havoc daily on the lives of community residents and increasingly spills over into downtown and residential middle-class areas. Muggings, burglaries, carjackings, and drug-related shootings, all of which may leave their victims or innocent bystanders dead, are now common enough to concern all urban and many suburban residents. The inclination to violence springs from the circumstances of life among the ghetto poor—the lack of jobs that pay a living wage, the stigma of race, the fallout from rampant drug use and drug trafficking, and the resulting alienation and lack of hope for the future.

Simply living in such an environment places young people at special risk of falling victim to aggressive behavior. Although there are often forces in the community which can counteract the negative influences—by far the most powerful is a strong, loving, "decent" (as inner-city residents put it) family committed to middle-class values—the despair is pervasive enough to have spawned an oppositional culture, that of "the streets," whose norms are often consciously opposed to those of mainstream society. These two orientations—decent and street—socially organize the community, and their coexistence has important consequences for residents, particularly for children growing up in the inner city. Above all, this environment means that even youngsters whose home lives reflect mainstream values—and the majority of homes in the community do—must be able to handle themselves in a street-oriented environment.

This is because the street culture has evolved what may be called a "code of the streets," which amounts to a set of informal rules governing interpersonal public behavior, including violence.[1] The rules prescribe both a proper comportment and the proper way to respond if challenged. They regulate the use of violence and so supply a rationale which allows those who are inclined to aggression to precipitate violent encounters in an approved way. The rules have been established and are enforced

SOURCE: From Elijah Anderson, "Violence and the Inner-City Code," in Joan McCord (ed.), *Violence and Childhood in the Inner City*, Cambridge University Press, 1997, pp. 1–30. Reprinted by permission.

mainly by the street-oriented, but on the streets the distinction between street and decent is often irrelevant; everybody knows that if the rules are violated, there are penalties. Knowledge of the code is thus largely defensive, and it is literally necessary for operating in public. Therefore, even though families with a decency orientation are usually opposed to the values of the code, they often reluctantly encourage their children's familiarity with it to enable them to negotiate the inner-city environment.

At the heart of the code is the issue of respect—loosely defined as being treated "right" or granted the deference one deserves. However, in the troublesome public environment of the inner city, as people increasingly feel buffeted by forces beyond their control, what one deserves in the way of respect becomes more and more problematic and uncertain. This situation in turn further opens the issue of respect to sometimes intense interpersonal negotiation. In the street culture, especially among young people, respect is viewed as almost an external entity that is hard-won but easily lost, and so it must constantly be guarded. The rules of the code in fact provide a framework for negotiating respect. Individuals whose very appearance—including their clothing, demeanor, and way of moving—deter transgressions feel that they possess, and may be considered by others to possess, a measure of respect. With the right amount, for instance, such individuals can avoid being bothered in public. If they are bothered, on the other hand, not only may they be in physical danger, but they will have been disgraced or "dissed" (disrespected). Many of the forms that dissing can take might seem petty to middle-class people (maintaining eye contact for too long, for example), but to those invested in the street code, these actions become serious indications of the other person's intentions. Consequently, such people become very sensitive to advances and slights, which could well serve as a warning of imminent physical confrontation.

This hard reality can be traced to the profound sense of alienation from mainstream society and its institutions felt by many poor inner-city black people, particularly the young. The code of the streets is actually a cultural adaptation to a profound lack of faith in the police and the judicial system. The police are most often seen as representing the dominant white society and as not caring enough to protect inner-city residents. When called, they may not respond, which is one reason many residents feel they must be prepared to take extraordinary measures to defend themselves and their loved ones against those who are inclined to aggression. Lack of police accountability has in fact been incorporated into the status system: The person who is believed capable of "taking care of himself" is accorded a certain deference, which translates into a sense of physical and psychological control. Thus, the street code emerges where the influence of the police ends and where personal responsibility for one's safety is felt to begin. Exacerbated by the proliferation of drugs and easy access to guns, this volatile situation results in the ability of the street-oriented minority (or those who effectively "go for bad") to dominate the public spaces.

This study is an ethnographic representation of the workings of this street code in the context of the socioeconomic situation in which the community finds itself.[2] The material it presents was gathered through numerous visits to various inner-city families and neighborhood hangouts and through many in-depth interviews with a wide array of individuals and groups; these interviews included sessions with adolescent boys and young men (some incarcerated, some not), older men, teenage mothers, and grandmothers. The structure of the inner-city family, the socialization of its children, the social structure of the community, and that community's extreme poverty which is in large part the result of structural economic change, will be seen to interact in a way that facilitates the involvement of so many maturing youths in the culture of the streets, in which violence and the way it is regulated are key elements.

THE ETHNOGRAPHIC METHOD

A clarifying note on the methodology is perhaps in order for those unfamiliar with the ethnographic method. Ethnography seeks to paint a conceptual picture of the setting under consideration, through the use of observation and in-depth interviews. The researcher's goal is to illuminate the social and cultural dynamics which characterize the setting by answering such questions as "How do the people in the setting perceive their situation?" "What assumptions do they bring to their decision making?" "What behavior patterns result from their choices?" and "What are the consequences of those behaviors?" An important aspect of the ethnographer's work is that it be as objective as possible. This is not easy since it requires researchers to set aside their own values and assumptions as to what is and is not morally acceptable—in other words, to jettison that prism through which they typically view a given situation. By definition, one's own assumptions are so basic to one's perceptions that it may be difficult to see their influence. Ethnographic researchers, however, have been trained to recognize underlying assumptions, their own and those of their subjects, and to override the former and uncover the latter (see Becker, 1970).

"DECENT" FAMILIES: VALUES AND REALITY

Although almost everyone in the poor inner-city neighborhood is struggling financially and therefore feels a certain distance from the rest of America, the decent and the street family in a real sense represent two poles of value orientation, two contrasting conceptual categories.[3] The labels decent and street, which the residents themselves use, amount to evaluative judgments that confer status on local residents. The labeling is often the result of a social contest among individuals and families of the neighborhood, and individuals of the two orientations can and often do coexist in the same extended family. Decent residents judge themselves to be so while judging others to be of the street, whereas street individuals may present themselves as decent, drawing distinctions between themselves and other even more street-oriented people. In any case, street is considered a highly pejorative epithet. In addition, there is quite a bit of circumstantial behavior among individuals—that is, one person may at different times exhibit both decent and street orientations, depending on the circumstances. Although these designations result from so much social jockeying, there do exist concrete features that define each conceptual category.

Generally, so-called decent families accept mainstream values more fully and attempt to instill them in their children. Whether a married couple with children or a single-parent (usually female) household, such families are generally working poor and so tend to be relatively better off financially than their street-oriented neighbors. The adults value hard work and self-reliance and are willing to sacrifice for their children. Because they have a certain amount of faith in mainstream society, they harbor hopes for a better future for their children, if not for themselves. Many of them go to church and take a strong interest in their children's schooling. Rather than dwell on the real hardships and inequities facing them, many such decent people, particularly the increasing number of grandmothers raising grandchildren sometimes see their difficult situation as a test from God and derive great support from their faith and from the church community....

Extremely aware of the problematic and often dangerous environment in which they reside, decent parents tend to be strict in their childrearing practices, encouraging children to respect authority and walk a straight moral line. They have an almost obsessive concern with trouble of any kind and remind their children to be on the lookout for people and situations which might lead to such difficulties. When disciplining their children, they tend to be liberal in the use of corporal punishment, but unlike street parents, who are often observed lashing out at their children, they tend to explain the reason for the spanking....

Consistent with this view, decent families also inspect their children's playmates for behavior problems, and they enforce curfews. For example, they may require their young children to come in by a certain time or simply to go no farther than the front stoop of their house, a vantage point from which the children attentively watch other children play unsupervised in the street; the obedient children may then endure the street kids' taunts of "stoop children." As the children become teenagers, these parents make a strong effort to know where they are at night, giving them an unmistakable message of their caring, concern, and love. At the same time, they are vigilant in guarding against the appearance of condoning any kind of delinquent or loose behavior, including violence, drug use, and staying out very late at night; they monitor not only their own children's behavior but also that of their children's peers, at times playfully embarrassing their own children by voicing value judgments in front of their children's friends....

Probably the most meaningful expression for describing the mission of decent families, as seen by themselves and outsiders, is instilling "backbone" in its younger members. In support of its efforts toward this goal, a decent family tends to be much more willing and able than the street-oriented family to ally itself with outside institutions such as school and church. The parents in such families have usually had more years of schooling than their street-oriented counterparts, which tends to foster in them a positive attitude toward their children's schooling. In addition, they place a certain emphasis on spiritual values and principles, and church attendance tends to be a regular family ritual, as noted earlier, although females are generally more so inclined than the males.

An important aspect of religious belief for the decent people is their conception of death. The intertwined ideas of fate, a judgment day, and an afterlife present a marked contrast to the general disorganization and sense of immediacy that is known to characterize street-oriented families and individuals. The situation of the street-oriented may then result in the tendency to be indifferent or oblivious to the probable consequences or future meaning of their behavior, including death. People imbued with a street orientation tend not to think far beyond the immediate present; their orientation toward the future is either very limited or nonexistent. One must live for the moment, for they embrace the general belief that "tomorrow ain't promised to you." For religious people, the sometimes literal belief in an afterlife and a day of reckoning inspires hope and makes life extremely valuable, and this belief acts to check individuals in potentially violent encounters. While accepting the use of violence in self-defense, they are less likely to be initiators of hostilities. Religious beliefs thus can have very practical implications. Furthermore, the moral feeling church-going instills in many people strengthens their self-esteem and underscores the sense that a positive future is possible, thus contributing to a certain emotional stability or "long fuse" in volatile circumstances.

In these decent families, then, there tends to be a real concern with and a certain amount of hope for the future. Such attitudes are often expressed in a drive to work to have something or to build a good life by working hard and saving for material things; raising one's children—telling them to "make something out of yourself" and "make do with what you have"—is an important aspect of this attitude. But the concern with material things, although accepted, often produces a certain strain on the young; often encouraged to covet material things as emblems testifying to social and cultural well-being, they lack the legitimate or legal means for obtaining such things. The presence of this dilemma often causes otherwise decent youths to invest their considerable mental resources in the street.

Involvement with institutions such as school and church has given decent parents a kind of *savoir faire* that helps them get along in the world. Thus, they often know how to get things done and use that knowledge to help advance themselves and their children. In the face of such overwhelming problems as persistent poverty, AIDS, or drug use, which can beset even the strongest and most promising families, these parents are trying hard to give their children as good a life—if not a better one—as they themselves had. At the same time, they are themselves polite and considerate of others, and they teach their children to be the same way. At home, at work, and in church, they work hard to maintain a positive mental attitude and a spirit of cooperation.

THE STREET

So-called street parents often show a lack of consideration for other people and have a rather superficial sense of family and community. Though they may love their children, many of them are unable to cope with the physical and emotional demands of parenthood, and they often find it difficult to reconcile their needs with those of their children. These families, who are more fully invested in the code of the streets than the decent people are, may aggressively socialize their children into it in a normative way. They believe in the code and judge themselves and others according to its values. In fact, the overwhelming majority of families in the inner-city community try to approximate the decent-family model, but there are many others who clearly represent the worst fears of the decent family. Not only are the financial resources of these other families extremely limited, but what little they have may easily be misused. Often suffering the worst effects of social isolation, the lives of the street-oriented are often marked by disorganization. In the most desperate circumstances, people often have a limited understanding of priorities and consequences, and so frustrations mount over bills, food, and, at times, drink, cigarettes, and drugs. Some tend toward self-destructive behavior; many street-oriented women are crack-addicted ("on the pipe"), alcoholic, or repeatedly involved in complicated relationships with men who abuse them. In addition, the seeming intractability of their situation, caused in large part by the lack of well-paying jobs and the persistence of racial discrimination, has engendered deep-seated bitterness and anger in many of the most desperate and poorest blacks, especially young people. The need both to exercise a measure of control and to lash out at somebody is often played out in the adults' relations with their children. At the least, the frustrations of persistent poverty shorten the fuse in such people—contributing to a lack of patience with anyone, child or adult, who irritates them....

The decent people firmly believe that it is this general set of cultural deficits—that is, a certain fundamental lack of social polish and commitment to norms of civility, which they refer to as "ignorance"—which makes the street-oriented quick to resort to violence to resolve almost any dispute. To those who hold this view, such street people seem to carry about them an aura of danger. Thus, during public interactions, the decent people may readily defer to the street-oriented, especially when they are strangers, out of fear of their ignorance. For instance, when street people are encountered at theaters or other public places talking loudly or making excessive noise, many decent people are reluctant to correct them for fear of verbal abuse that could lead to violence. Similarly, the decent people will often avoid a confrontation over a parking space or traffic error for fear of a verbal or physical confrontation. Under their breaths they may utter disapprovingly to a companion, "street niggers," thereby drawing a sharp cultural distinction between themselves and such persons. But there are also times when the decent people will try to approach the level of the "ignorant" ones by what they refer to as "getting ignorant" (see Anderson, 1990). In these circumstances, they may appear in battle dress, more than ready to face down the ignorant ones, indicating they have reached their limit, or threshold, for violent confrontation. And from this, an actual fight can erupt.

Thus, the fact that generally civilly disposed, socially conscious, and largely self-reliant men and women share the streets and other public institutions with the inconsiderate, the ignorant, and the extremely desperate puts them at special risk. In order to live and to function in the community, they must adapt to a street reality that is often dominated by the presence of those who at best are suffering severely in some way and who are likely to resort quickly to violence to settle disputes.

COMING UP IN STREET FAMILIES

In the street-oriented family, the development of an aggressive mentality can be seen from the beginning, even in the circumstances surrounding the birth of the child. In circumstances of persistent poverty, the mother is often little more than a child herself and without many resources or a consistent source of support; she is often getting by with public assistance or the help of kin. She may be ambivalent with respect to the child: On the one hand, she may look at the child as a heavenly gift; but on the other hand, as she begins to care for it, she is apt to realize that it is a burden, and sometimes a profound burden. These are the years she wants to be free to date and otherwise consort with men, to have a social life, and she discovers that the child slows her down (see Anderson, 1989). Thus, she sometimes leans on others, including family members, to enable her to satisfy her social needs for getting out nights and being with male and female friends.

In these circumstances, a woman—or a man, although men are less consistently present in children's lives—can be quite aggressive with children, yelling at and striking them for the least little infraction of the rules she has set down. Often little if any serious explanation follows the verbal and physical punishment. This response teaches children a particular lesson. They learn that to solve any kind of interpersonal problem, one must quickly resort to hitting or other violent behavior. Actual peace and quiet—and also the appearance of calm, respectful children that can be conveyed to her neighbors and friends—are often what the young mother most desires, but at times she will be very aggressive in trying to achieve these goals. Thus, she may be quick to beat her children, especially if they defy her law, not because she hates them but because this is the way she knows how to control them. In fact, many street-oriented women love their children dearly. Many mothers in the community subscribe to the notion that there is a "devil in the boy" that must be beaten out of him or that socially "fast girls need to be whupped." Thus, much of what borders on child abuse in the view of social authorities is acceptable parental punishment in the view of these mothers.

Many street-oriented women are weak and ineffective mothers whose children learn to fend for themselves when necessary, foraging for food and money and getting them any way they can. These children are sometimes employed by drug dealers or become addicted themselves. These children of the street, growing up with little supervision, are said to "come up hard." In the interest of survival, they often learn to fight at an early age, sometimes using short-tempered adults around them as role models. The street-oriented home may be fraught with anger, verbal disputes, physical aggression, and even mayhem. The children observe these goings-on, learning the lesson that might makes right. They quickly learn to hit those who cross them, and the dog-eat-dog mentality prevails. In order to survive, to protect oneself, it is necessary to marshal inner resources and be ready to deal with adversity in a hands-on way. In these circumstances, physical prowess takes on great significance.

In some of the most desperate cases, a street-oriented mother may simply leave her young children alone and unattended while she goes out. The most irresponsible women can be found at local bars and crack houses, getting high and socializing with other adults. Sometimes a troubled woman will leave very young children alone for days at a time. Reports of crack addicts abandoning their children have become common in drug-infested inner-city communities. Neighbors or relatives discover the abandoned children, often hungry and distraught over the absence of their mother. After repeated absences, a friend or relative, particularly a grandmother, will often step in to care for the young children, sometimes petitioning the authorities to send her, as guardian of the children, the mother's welfare check, if the mother gets one. By this time, however, the children may well have learned the first lesson of the streets: Survival itself, let alone respect, cannot be taken for granted; you have to fight for your place in the world.

THE SHUFFLING PROCESS

As indicated earlier, in order to carry on their everyday lives in poor inner-city neighborhoods, children from even the most decent homes must come to terms with the streets. This means that they must learn to deal with the various and sundry influences of the streets, including their more street-oriented peers. Indeed, as children grow up and their parents' control wanes, they go through a social shuffling process that can undermine, or at least test, much of the socialization they have received at home. In other words, the street serves as a mediating influence under which children may come to reconsider and rearrange their personal orientations. This is a time of status passage (see Glaser & Strauss, 1972), when social identity can become very uncertain as children sort out their ways of being. It is a tricky time because a child can go either way. For children from decent homes, for example, the immediate and present reality of the street situation can overcome the compunctions against tough behavior that their parents taught them, so that the lessons of the home are slowly forgotten and the child "goes for bad." Or a talented child from a street-oriented family may discover ways of gaining respect without unduly resorting to aggressive and violent responses—by becoming a rapper or athlete, for example, or, rarely, a good student. Thus, the kind of home a child comes from becomes influential but not determinative of the way he or she will ultimately turn out.

By the age of ten, all children from both decent and street-oriented families are mingling on the neighborhood streets and figuring out their identities. Here they try out certain roles and scripts—which are sometimes actively opposed to the wishes of parents—in a process that challenges both their talents and their socialization and may involve more than a little luck, good or bad. In this volatile environment, they learn to watch their backs and to anticipate and negotiate those situations that might lead to troubles with others. The successful outcomes of these cumulative interactions with the streets ultimately determine every child's life chances.

Herein lies the real meaning of so many fights and altercations, despite the ostensible, usually petty, precipitating causes, including the competitions over girlfriends and boyfriends and the "he say, she say" conflicts of personal attribution. Adolescents are insecure and are trying to establish their identities. Children from the middle- and upper-classes, however, usually have more ways to express themselves as worthwhile and so have more avenues to explore. The negotiations they engage in among themselves may also include aggression, but they tend to be more verbal in a way that includes a greater use of other options—options that require resources not available to those of more limited resources, such as showing off with things, connections, and so on. In poor inner-city neighborhoods, physicality is a fairly common way of asserting oneself. It is also unambiguous. If you punch someone out, if you succeed in keeping someone from walking down your block, "you did it." It is a *fait accompli*. And the evidence that you prevailed is there for all to see.

During this campaign for respect, through these various conflicts, those connections between actually being respected and the need for being in physical control of at least a portion of the environment become internalized, and the germ of the code of the streets emerges. As children mature, they obtain an increasingly more sophisticated understanding of the code, and it becomes part of their working conception of the world, so that by the time they reach adulthood, it comes to define the social order. In time, the rules of physical engagement and their personal implications become crystallized. Children learn the conditions under which violence is appropriate, and they also learn how the code defines the individual's relationship to his or her peers. They thus come to appreciate the give-and-take of public life, the process of negotiation.

The ethic of violence is in part a class phenomenon (see Wolfgang and Ferracuti, 1967). Children are more inclined to be physical than adults (because they have fewer alternatives for settling disputes), and lower-class adults tend to be more physical than middle- or upper-middle-class adults. Poor and lower-class adults more often find themselves in disputes that lead to violence. Because they are more often alienated from the agents and agencies of social control, such as the police and the courts, they are left alone more often to settle disputes on their own. And such parents, in turn, tend to socialize their kids into this reality.

But this reality of inner-city life is largely absorbed on the streets. At an early age, often even before they start school and without much in the way of adult supervision, children from such street-oriented families gravitate to the streets, where they must be ready "to hang," to socialize with peers. Children from these generally permissive homes have a great deal of latitude and are allowed to "rip and run" up and down the street. They often come home from school, put their books down, and go right back out the door. For the most severely compromised, on school nights, eight- and nine-year-olds remain out until nine or ten o'clock (and the teenagers come in whenever they want to). On the streets, they play in groups that often become the source of their primary social bonds. Children from decent homes tend to be more carefully supervised and are thus likely to have curfews and to be taught how to stay out of trouble.

In the street, through their play, children pour their individual life experiences into a common knowledge pool, affirming, confirming, and elaborating on what they have observed in the home and matching their skills against those of others. And they learn to fight. Even small children test one another, pushing and shoving others, and are ready to hit other children over circumstances not to their liking. In turn, they are readily hit by other children, and the child who is toughest prevails. Thus, the violent resolution of disputes—the hitting and cursing—gains social reinforcement. The child in effect is initiated into a system that is really a way of campaigning for self-respect.

There is a critical sense in which violent behavior is determined by situations, thus giving importance to the various ways in which individuals define and interpret such situations. In meeting the various exigencies of immediate situations, which become so many public trials, the individual builds patterns as outcomes are repeated over time. Behaviors, violent or civil, which work for a young person and are reinforced by peers whose reactions to such behavior come to shape the person's outlook, will likely be repeated.

In addition, younger children witness the disputes of older children, which are often resolved through cursing and abusive talk, and sometimes through aggression or outright violence. They see that one child succumbs to the greater physical and mental abilities of the other. They are also alert and attentive witnesses to the verbal and physical fights of adults, after which they compare notes and share their own interpretations of the event. In almost every case, the victor is the person who physically won the altercation, and this person often enjoys the esteem and respect of onlookers. These experiences reinforce the lessons the children have learned at home: Might makes right and toughness is a virtue; humility is not. In effect, they learn the social meaning of fighting. When it is left virtually unchallenged, this understanding becomes an ever more important part of a child's working conception of the world. Over time, the code of the streets becomes refined. Those street-oriented adults with whom children come in contact—including mothers, fathers, brothers, sisters, boyfriends, cousins, neighbors, and friends—help them along in forming this understanding by verbalizing the messages they are getting through experience: "Watch your back." "Protect yourself." "Don't punk out." "If somebody messes with you, you got to pay them back." "If someone disses you, you got to straighten them out." Many parents actually impose sanctions if a child is not sufficiently aggressive. For example, if a child loses a fight and comes home upset, the parent might respond, "Don't you come in here crying that somebody beat you up; you better get back out there and whup his ass. I didn't raise no punks! Get back out there and whup his ass. If you don't whup his ass, I'll whup yo' ass when you come home." Thus, the child obtains reinforcement for being tough and showing nerve....

SELF-IMAGE BASED ON JUICE

By the time they are teenagers, most youths have either internalized the code of the streets or at least learned the need to comport themselves in accordance with its rules, which chiefly have to do with interpersonal communication. The code revolves around the presentation of self. Its basic requirement is the display of a certain predisposition to violence. Accordingly, one's bearing must send the unmistakable if sometimes subtle message to "the next person" in public that one is capable of violence and mayhem when the situation requires it—that one can take care of oneself. The nature of this communication is largely determined by the demands of the circumstances but can include facial expressions, gait, and verbal expressions—all of which are geared mainly to detering aggression. Physical appearance, including clothes, jewelry, and grooming, also plays an important part in how a person is viewed; to be respected, you have to have the right look.

Even so, there are no guarantees against challenges, because there are always people around looking for a fight to increase their share of respect—or "juice," as it is sometimes called on the street. Moreover, if a male is assaulted, it is important, not only in the eyes of his opponent but also in the eyes of his "running buddies," for him to avenge himself. Otherwise, he risks being "tried" or "rolled on" (physically assaulted) by any number of others. To maintain his honor, he must show he is not someone to be "messed with" or "dissed." In general, the person must "keep himself straight" by maintaining his position of respect among others; this involves in part his self-image, which is shaped by what he thinks others are thinking of him in relation to his peers.

Objects play an important and complicated role in establishing self-image. Jackets, sneakers, and gold jewelry reflect not just a person's taste, which tends to be tightly regulated among adolescents of all social classes, but also a willingness to possess things that may require defending. A boy wearing a fashionable, expensive jacket, for example, is vulnerable to attack by another who covets the jacket and either cannot afford to buy one or wants the added satisfaction of depriving someone else of his. However, if a boy forgoes the desirable jacket and wears one that isn't hip, he runs the risk of being teased and possibly even assaulted as an unworthy person....

One way of campaigning for status is by taking the possessions of others. In this context, seemingly ordinary objects can become trophies imbued with symbolic value that far exceeds their monetary worth. Possession of the trophy can symbolize the ability to violate somebody—to "get in his face," to take something of value from him, to diss him, and thus to enhance one's own worth by stealing that which belongs to someone else. Though it often is, the trophy does not have to be something material. It can be another person's sense of honor, snatched away with a derogatory remark or action. It can be the outcome of a fight. It can be the imposition of a certain standard, such as a girl getting herself recognized as the most beautiful. Material things, however, fit easily into the pattern. Sneakers or a pistol—even somebody else's boyfriend or girlfriend—can become a trophy. When individuals can take something from another and then flaunt it, they gain a certain regard by being the owner, or the controller, of that thing. But this display of ownership can then provoke other people to challenge him or her. This game of who controls what is thus constantly being played out on inner-city streets, and the trophy—extrinsic or intrinsic, tangible or intangible—identifies the current winner.

An important aspect of this often violent give-and-take is its zero-sum quality. That is, the extent to which one person can rise depends on his or her ability to put another person down. This situation underscores the alienation that permeates the inner-city ghetto community. There is a generalized sense that very little respect is to be had, and therefore everyone competes to get what little affirmation is actually available. The craving for respect that results often gives people thin skins. It is generally believed that true respect provides an aura of protection. Thus, shows of deference by others can be highly soothing, contributing to a sense of security, comfort, self-confidence, and self-respect. Transgressions by others that go unanswered diminish these feelings and are believed to encourage further transgressions. Hence, one must be ever vigilant against the transgressions of others; one cannot even allow the *appearance* of transgressions to be tolerated. Among young people, whose sense of self-esteem is particularly vulnerable, there is an especially heightened concern with being "disrespected." Many inner-city young men in particular crave respect to such a degree that they will risk their lives to attain and maintain it.

The issue of respect is thus closely tied to whether a person has an inclination to be violent, even as a victim. In the wider society, people may not feel the need to retaliate physically after an attack, even though they are aware that they have been degraded or taken advantage of. They may feel a great need to defend themselves *during* an attack or to behave in such a way as to deter aggression (middle-class people certainly can and do become victims of street-oriented youths), but they are much more likely than street-oriented people to feel they can walk away from a possible altercation with their self-esteem intact. Some people may even have the strength of character to flee, without any thought that their self-respect or esteem will be diminished.

In impoverished inner-city black communities, however, particularly among young males and perhaps increasingly among females, such flight would be extremely difficult. To run away would likely leave one's self-esteem in tatters. Hence, people often feel constrained not only to stand up during and at least attempt to resist an assault but also to "pay back"—to seek revenge—after a successful assault on their person. This may include going to get a weapon. One young man described a typical scenario:

> So he'll [the victim] ask somebody do they got a gun for sale or somethin' like that. And they'll say yeah and they say they want a buck [a hundred dollars] for it or somethin' like that. So he'll go and get a hundred dollars and buy that gun off of him and then wait until he see the boy that he was fightin' or got into a argument [with] or somethin' like that. He'll sneak and shoot 'im or somethin' like that and then move away from his old neighborhood....
>
> Or if they already have a gun, they gonna just go get their gun and buy a bullet. And then they gonna shoot the person, whoever they was fightin', or whoever did somethin' to 'im. Then they'll probably keep the gun and get into a couple more rumbles and shoot people. And the gun'll probably have like nine bodies on it. Then they decide to sell the gun, and the other person'll get caught with it.

Or one's relatives might get involved, willingly or not. The youth continues:

> For instance, me. I was livin' in the projects [public housing] on Grant Street, right, and I think my brother is fightin' one of the other person's brother or cousin. Me and my brother look just alike—they thought I was my brother—and they try to throw me down the elevator shaft, but they threw me down three flights of steps. And I hit my face on a concrete rung, chipped my tooth, and got five stitches in my lip. And see, [later] my uncle killed one of them, and that's why he doin' time in jail now. Because they tried to kill me.

The very identity and self-respect—the honor—of many inner-city youths is often intricately tied up with the way they perform on the streets *during* and *after* such encounters. Moreover, this outlook reflects the circumscribed opportunities of the inner-city poor. Generally, people outside the ghetto have other ways of gaining status and regard, and thus, they do not feel so dependent on such physical displays.

BY TRIAL OF MANHOOD

Among males on the street, these concerns about things and identity have come to be expressed in the concept of "manhood." ... The operating assumption is that a man, especially a real man, knows what other men know—the code of the streets. And if one is not a real man, one is somehow diminished as a person, and there are certain valued things one simply does not deserve....

Central to the issue of manhood is the widespread belief that one of the most effective ways of gaining respect is to manifest "nerve." Nerve is shown when someone takes a person's possessions (the more valuable, the better), messes with someone's woman, throws the first punch, gets in someone's face, or pulls a trigger. Its proper display helps on the spot to check others who would violate one's person and also helps to build a reputation that works to prevent future challenges. But since such a show of nerve is a forceful expression of disrespect toward the person on the receiving end, the victim may be greatly offended and seek to retaliate with equal if not greater force. A display of nerve, therefore, can easily provoke a life-threatening response, and the background knowledge of that possibility has often been incorporated into the concept of nerve.

True nerve exposes a lack of fear of dying. Many feel that it is acceptable to risk dying over the principle of respect. In fact, among the hardcore street-oriented, the clear risk of violent death may be preferable to being dissed by another. The youths who have internalized this attitude and convincingly display it in their public bearing are among the most threatening people of all, for it is commonly assumed that they fear no man. As the people of the community say, "They are the baddest dudes on the street." They often lead an existential life that may acquire meaning only when faced with the possibility of imminent death. Not to be afraid to die is by implication to have few compunctions about taking somebody else's life. Not to be afraid to die is the quid pro quo of being able to take somebody else's life—for the right reasons, if the situation demands it. When others believe this is one's position, it gives one a real sense of power on the streets. Such credibility is what many inner-city youths strive to achieve, whether they are decent or street-oriented, both because of its practical defensive value and because of the positive way it makes them feel about themselves. The difference between the decent and the street-oriented youth is that the decent youth makes a conscious decision to appear tough and manly; in another setting—with teachers, say, or at his part-time job—he can be polite and deferential. The street-oriented youth, on the other hand, has made the concept of manhood a part of his very identity; he has difficulty manipulating it—it often controls him instead....

"GOING FOR BAD"

In the most fearsome youths, such a cavalier attitude toward death grows out of a very limited view of life. Many are uncertain about how long they are going to live and believe they could die violently at any time. They accept this fate; they live on the edge. Their manner conveys the message that nothing intimidates them; whatever turn the encounter takes, they maintain their attack—rather like a pit bull, whose spirit many such boys admire. The demonstration of such tenacity shows "heart" and earns their respect.

This fearlessness has implications for law enforcement. Many street-oriented boys are much more concerned about the threat of "justice" at the hands of a peer than at the hands of the police. According to one young man trying to lead a decent life, "When they shoot somebody, they have so much confidence that they gonna get away from the cop, you see. If they don't, then they be all mad and sad and cryin' and all that 'cause they got time in jail." At the same time, however, many feel not only that they have little to lose by going to prison but that they have something to gain. The toughening-up one experiences in prison can actually enhance one's reputation on the streets. Hence, the system loses influence over the hard core who are without jobs and who have little perceptible stake in the system. If mainstream society has done nothing *for* them, they counter by making sure it can likewise do nothing *to* them....

CONCLUSION

The attitudes of the wider society are deeply implicated in the code of the streets. Most people in inner-city communities are not totally invested in the code; but the significant minority of hardcore street youths who do embrace it have to maintain the code in order to establish reputations, because they have—or feel they have—few other ways to assert themselves. For these young people, the standards of the street code are the only game in town. The extent to which some children—particularly those who through upbringing have become most alienated and lack strong and conventional social support—experience, feel, and internalize racist rejection and contempt from mainstream society may strongly encourage them to express contempt for the more conventional society in turn. In dealing with this contempt and rejection, some youngsters will consciously invest themselves and their considerable mental resources in what amounts to an oppositional culture to preserve themselves and their self-respect. Once they do this, any respect they might be able to garner in the wider system pales in comparison with the respect available in the local system; thus, they often lose interest in even attempting to negotiate the mainstream system.

At the same time, many less-alienated young blacks have assumed a street-oriented demeanor as a way of expressing their blackness while really embracing a much more moderate way of life; they, too, want a nonviolent setting in which to live and raise a family. These decent people are trying hard to be a part of the mainstream culture, but the racism—both real and perceived—that they encounter helps to legitimate the oppositional culture; and so, on occasion, they adopt street behavior. In fact, depending on the demands of the situation, many people in the community slip back and forth between decent and street behavior.

A vicious cycle has thus been formed. The hopelessness and alienation that many young inner-city black men and women feel, largely because of endemic joblessness and persistent racism, fuel the violence they engage in. This violence serves to confirm the negative feelings many whites and some middle-class blacks harbor toward the ghetto poor, further legitimating the oppositional culture and the code of the streets in the eyes of many poor young blacks. Unless this cycle is broken, attitudes on both sides will become increasingly entrenched, and the violence—which claims victims black and white, poor and affluent—will only escalate.

ENDNOTES

1. This phenomenon is to be distinguished from that described by Wolfgang and Ferracuti (1967), who identified and delineated more explicitly a "subculture of violence." Wolfgang and Ferracuti postulated norms which undergirded or even defined the culture of the entire community, whereas the code of the streets applies predominantly to situational public *behavior* and is normative for only a segment of the community.
2. The ethnographic approach is to be distinguished from other, equally valid, approaches, most notably the social-psychological. A sensitive and compelling social-psychological analysis of the phenomenon of murder is to be found in Jack Katz's *Seductions of Crime: Moral and Sensual Attractions in Doing Evil* (New York: Basic Books, 1988). Katz's purpose is to make sense of the senseless, that is, to explain the psychic changes a person goes through to become, at a given moment, a murderer. In contrast, the analysis offered here focuses on conscious behavior, which, in the circumstances of the inner-city environment, is sensible (makes sense). Katz explores the moral dimension of violence, whereas I explore its practical aspect.
3. For comparisons in the ethnographic literature, see Drake and Cayton's discussion of the "shadies" and "respectables" in *Black Metropolis* (New York: Harper and Row, 1962, originally published

1945). Also, see the discussion of "regulars," "wineheads," and "hoodlums" in Anderson, *A Place on the Corner* (Chicago: University of Chicago Press, 1978). See also Anderson, "Neighborhood Effects on Teenage Pregnancy," in Jencks and Peterson, eds., *The Urban Underclass* (Washington, D.C.: Brookings, 1991).

REFERENCES

Anderson, E. (1978). *A place on the corner.* Chicago: University of Chicago Press.

———. (1989). Sex codes and family life among poor inner-city youths. In *The ghetto underclass: Social science perspectives*, ed. William Julius Wilson. Special edition of *The Annals of the American Academy of Political and Social Science* 501, 59–78.

———. (1990). *Streetwise: Race, class, and change in an urban community.* Chicago: University of Chicago Press.

———. (1991). Neighborhood effects on teenage pregnancy. In C. Jencks & P. E. Peterson (Eds.), *The urban underclass.* Washington, D.C.: Brookings Institution, pp. 375–98.

Becker, H. (1970). *Sociological work.* Chicago: Aldine.

Drake, S. C., & Cayton, H. (1962). *Black metropolis.* New York: Harper & Row.

Glaser, B. G., & Strauss, A. L. (1972). *Status passage.* Chicago: Aldine.

Katz, J. (1988). *Seductions of crime: Moral and sensual attractions in doing evil.* New York: Basic Books.

Wolfgang, M. E., & Ferracuti, F. (1967). *The subculture of violence.* London: Tavistock.

DISCUSSION QUESTIONS

1. Why does violent crime exist in the inner city, according to Anderson?
2. Place yourself in the situation described in this article. Would you be tempted to follow the street code? How would you try to escape conforming to the street code? If you were a parent in this situation, how would you try to assure that your child will be successful?
3. Is this a racial problem or a class problem, or both?
4. Would gun control make any difference in the situation described here? Is there any way for the school to work effectively as long as this street code is powerful? What can police do in this situation?
5. How do you think that violent crime in the inner city might be lessened?

28

White-Collar Crime

STEPHEN M. ROSOFF, HENRY PONTELL, AND ROBERT TILLMAN

The Four Questions

1. The problem is identified by the title of this selection. What is white-collar crime and why is in a problem?
2. Is this a *social* problem or an individual problem?
3. What causes this problem?
4. What can we do about this problem? Do the authors give us any suggestion? Is their suggestion an individual or a social attempt to deal with this?

Topics Covered

Crime
White-collar crime
Class
Capitalism
Society
Institutions
Organizations
Material success

Most people, most of the time, do *not* engage in fraud, embezzlement, bribery, and cover-ups. Rather, these acts involve persons occupying certain societal roles and transpire only under certain conditions. In this final chapter, we will consider some of the factors that give rise to white-collar criminality, along with some changes that need to be made in order to control it.

We will also reexamine both official and public responses to white-collar crime. Many commentators have suggested that white-collar crime flourishes because of a generally tepid reaction by law enforcement and other government agencies. Why does it seem that so many upperworld criminals are never brought to justice? Why do those few who are sanctioned seem to receive punishments so remarkably lenient? Moreover, why do convicted white-collar offenders suffer so little of the stigma and opprobrium that the citizenry attaches to "common" criminals? These are important questions that need to be addressed.

We will investigate, too, the consequences of white-collar crime....

CAUSES OF WHITE-COLLAR CRIME

... In attempting to identify the cause of white-collar crime we will focus on three basic levels of sociological analysis. We will see how *societies, institutions,* and *organizations* generate the opportunity, motives, and means for white-collar criminality.

Societal Causes

This is the level of analysis in which cultural explanations of white-collar crime are rooted. Of course, all criminal acts are committed by individuals. But as sociologists emphasize, individual decisions are

SOURCE: From *Honor: White Collar Crime and the Looting of America*, 5th Edition, © 2010, pp. 555–581. Reprinted by permission of Pearson Education, Inc. Upper Saddle River, NJ.

always made in larger societal contexts that present and provide support for particular options, making behavior that might otherwise seem unethical or unlawful appear as reasonable and legitimate. Ironically, when individuals engage in white collar crime, though they are breaking the law, they are often *conforming* to cultural values—such as the accumulation of wealth.

In 1938, the esteemed American sociologist Robert Merton published an article that was to become famous entitled "Social Structure and Anomie" in which he wrote: "Contemporary American culture continues to be characterized by a heavy emphasis on wealth as a basic symbol of success, without a corresponding emphasis on the legitimate avenues on which to march towards this goal."[1] At the same time, in Merton's view, American culture places negative sanctions on those who fail to achieve material success, attributing their failure to personal deficiencies and character flaws such as laziness. Merton observed that one response to this pressure is for certain individuals to turn to more "innovative" means to succeed, including, in some cases, criminal activities.[2] It is very significant that, in Merton's formulation, the standard of material success is not an objective one but exists on a sliding scale—so that even the relatively well-off may feel their aspirations are blocked and may thus resort to crime to achieve culturally defined goals.

In a similar vein, criminologist James W. Coleman has written more recently that the ultimate sources of white-collar criminality lie in the "culture of competition" that pervades American society. Coleman describes our industrial economy as one that includes a market structure which produces not only profits and losses but winners and losers.[3] The result is a "win at any cost" morality that encourages even the scrupulous entrepreneur or executive to bend the rules—or to engage in outright fraud and deception—in order to stay ahead of the competition. And while these pressures may be found in any society that produces "surplus wealth,"—that is, wealth beyond what is required for subsistence—they are felt most acutely in advanced capitalist societies like ours, where upward mobility is regarded as a right.

… White-collar offenders frequently employ what theorists Gresham Sykes and David Matza call "techniques of neutralization,"[4] mechanisms that allow them to annul any potential guilt or internal conflict stemming from conduct on their part that breaks laws to which they claim ostensible allegiance. A 1987 marketing study concluded that neutralization plays a major role in unethical business practices.[5]

The five commonly recognized neutralization techniques are:[6]

1. ***Denial of responsibility***: Offenders argue that they are not personally accountable for their actions because of overriding factors beyond their control. For example, ghetto price gougers typically justify usurious interest rates on the grounds that low-income consumers are "deadbeats."
2. ***Denial of injury***: Offenders contend that their misconduct causes no direct suffering. For example, producers of the rigged television quiz shows insisted that they merely had enhanced the dramatic appeal of innocuous programs designed only to entertain.
3. ***Denial of victim***: Offenders claim that violated parties deserved whatever happened. For example, many computer hackers maintain that anyone leaving confidential data files unprotected is really at fault for any subsequent break-ins.
4. ***Appeal to higher loyalties***: Offenders depict their conduct as the by-product of an attempt to actualize a higher value. For example, Richard Nixon and Oliver North portrayed their respective crimes as acts of patriotism—as did the CIA torturers at Abu Ghraib and Guantanamo.
5. ***Condemning the condemners***: Offenders seek to turn the tables on their accusers by claiming to be the true victims. For example, S&L crook Charles Keating repeatedly blamed regulators, prosecutors, politicians, journalists, and just about everyone, else—everyone that is, except himself.

In their avowals of innocence, corporate criminals also draw on what has been termed the

*Reminds me of article regarding domestic violence

"folklore of capitalism"[7] to recast themselves as misunderstood visionaries or persecuted dissidents. At the zenith of his career, when he was moving billions of dollars through the economy, junk bond impresario Michael Milken was fond of portraying himself as an agent of social change, a "social engineer" whose activities were restructuring society.[8]

In a similar manner, some white-collar criminals try to evoke images of the frontier. They liken themselves to pathfinders, blazing new trails through unchartered business terrain. This theme was sounded by a pair of notorious environmental criminals in an interview they gave from federal prison. Prior to their incarceration, the Colbert brothers ran a successful company which purchased toxic chemical wastes from American manufacturers and then illegally dumped them in Third World countries. In their interview, the Colberts steadfastly denied any guilt:

> We were, in a sense, innovators ahead of the times ... We're basically pioneers in the surplus chemical business, which is something that's a necessary business for society.[9] ...

Psychopathic Wealth

When the authors of this book were growing up in the 1950s, there was a generation entering the corporate arena far different than today's crop. Their personalities were shaped by the Great Depression and their experiences in World War II. They entered that arena with relatively modest expectations, especially when compared with their current counterparts. They wanted to participate in a burgeoning postwar economy. They wanted good jobs and economic security. They wanted to help build successful companies. Few of them likely expected to live like feudal lords, for there was far less wealth in American society then than there is now. They believed that if they paid their dues, worked hard, developed good products and services, were creative and took prudent risks they would be rewarded—as many of them surely were.

The corporate culture of the 1950s was very conformist. The "overnight" success stories, such as Xerox or Polaroid, were far more the exception than the rule. Indeed, the mid-20th century was not a time when great fortunes were being made overnight. Entrepreneurialism was not really a big part of the American business milieu at that time, for there was little venture capital available. With the Depression still gliding on the tightropes of their collective memories, Americans didn't feel comfortable taking big gambles; and it wasn't encouraged.

The hallmark of that generation of young executives was what has been termed ***patient wealth***. It was a time when almost nobody got rich quickly. Going public and making a billion dollars weeks later was not in their mentality or imagination.

Today's business leaders grew up in a very different America—an America that showcased enormous amounts of wealth. It now appears clear that many of them internalized a profound sense of entitlement. They expected things sooner, rather than later. They were the product of a much more materialistic society. "In your face" opulence that once would have been embarrassing had become the norm in their world. Not only do today's corporate elite seem to believe they should make fortunes quickly, but there are also far fewer restrictions on conspicuous consumption. One need only think of former Tyco CEO Dennis Koslowski's (Chapter* 7) $15,000 umbrella stand and $6,000 shower curtains, paid for by hapless shareholders. Greed thy name is Koslowski.

This dramatic transition has been fueled by a greed so out of control that even *Fortune* magazine—which once voted Enron the best company in the United States six years in a row—featured a cover in 2003 displaying pigs in Armani suits.[10] Former Treasury Secretary Robert Reich notes that the very rich now live in a "parallel universe" with its own separate rules.[11] This small elite class increasingly has become disconnected from the social mainstream and the responsibilities of good citizenship. A generation of executives who *make* things has been replaced by a generation who *take* things. The wealth they possess or aspire to possess is not a patient

*"Chapter" notes in Article 28 are related to the book *White Collar Crime*.

wealth. But to call it merely "impatient" wealth is to understate what has happened in corporate America. A more resonant term can be borrowed from the psychiatric lexicon, a term used to describe persons intensely selfish, conspicuously lacking in human empathy, and dispositionally unable to delay gratification. We seem to have entered an Age of **Psychopathic Wealth**.

The earnings gap between the CEOs of major corporations and their rank and file employees has widened steadily and dramatically. It is currently more than 500 to 1. While the average American is now working harder for less money, major CEOs not only enjoy huge salaries and incredibly lucrative stock option deals, but they also receive golden parachutes, forgivable loans, and a plethora of other financial benefits. The result is that the richest 1 percent of Americans now own more wealth than the bottom 95 percent—an unprecedented disparity. Historians Will and Ariel Durant concluded four decades ago that the gap between the wealthiest and poorest Americans has become greater than at any time since Imperial Rome.[12]

But evidently, all the aforementioned "legal" benefits aren't enough. As the recent corporate scandals reveal all too clearly, some greedy CEOs—along with other top executives—have illegally inflated the value of their firms' assets. This not only raises the value of the company's stock, but in particular increases the value of their own stock options, which have become the currency of senior management. Overstating earnings also makes exorbitant salaries, bonuses, and perquisites appear more justifiable.

It would be comforting to conclude that the problem is confined to a few highly publicized "bad apples"—the Enrons, WorldComs, Adelphia, Qwests, Global Crossings, and others. But no one can say this with any real confidence. Indeed, the evidence suggests otherwise. It suggests the emergence of a *kleptocratic* corporate culture—that is, a culture ruled by thieves. In the words of a former Securities and Exchange Commission (SEC) enforcement chief: "There is always greed and misconduct in the business world. But in today's society, more people tend to believe they can get away with it.[13]

Nearly 1,000 public corporations were forced to correct their financial statements between 1998 and 2003. The Enrons may be only the visible part of the financial reporting iceberg. As the economic collapse of 2008 revealed, what lay below the surface is so large that it is almost too scary to look.

Institutional Causes

While culture may provide the rationalizations for white-collar crime, it is specific institutions that provide the means and opportunities. Not all industries are contaminated by fraud. For instance, higher education, an institution with which the authors are intimately familiar, experiences very little fraud (with the exception of rare cases of falsified research noted in Chapter 11) simply because there is very little to be gained through deception. On the other hand, industries like the popular music business (Chapter 5), where sales for a single album may reach many millions of dollars, afford considerable incentive for bribery and corruption. At the same time, the unregulated character of the record industry and the individualistic basis of decision making by music broadcasters also provide ample opportunity for illegal conduct. Thus, certain industries are structured in ways that provide both the incentives and the opportunities for white-collar crime, and, therefore, one must examine that structure to discover root causes.

The concept of "criminogenic industries" is used to describe the factors that facilitate fraud in industries like the music business. Needleman and Needleman state that the defining feature of criminogenic industries is that "their internal structure—economic, legal, organizational and normative—play a role in generating criminal activity within the system, independent at least to some degree from the criminal's personal motives."[14] The automobile industry, for example has been described as "criminogenic" because it has a market structure characterized by a high concentration of manufacturers who exert control over the distribution process[15] and because excessive pressures from manufacturers on dealers to increase sales[16] create the conditions that give rise to extensive fraud and abuse at the retail level. Likewise, the largely unregulated and decentralized nature of automotive repair makes it another prime site for deceptive practices. As we saw in Chapter 2, numerous investigations have found that overcharging, billing for unnecessary work, and the sale of phantom parts are common practices at auto repair shops around the country—findings that may come as no surprise to any car owner who has ever left a mechanic's shop feeling more beat-up than the car.

And then there is the used car business. Fairly or otherwise, perhaps no figure has come to embody the fast-talking film-flam artist more than the used car dealer. One of the oldest and best known used car scams is odometer rollback, known as "clocking." This occurs when illegal changes are made to the mileage shown on a used vehicle's title and odometer in order to mask high mileage on a late-model car.[17] When the new car market struggles (as in the Recession of 2009), used car sales typically accelerate and more vehicles get clocked.[18] With the proliferation of leased and rental cars is recent years, odometer tampering has become a very lucrative criminal activity and is now recognized as a serious form of white-collar crime.[19] Clocking can add four or five thousand dollars to the price of a late-model used car.[20] In 2008, the National Highway Traffic Safety Administration estimated that more than 450,000 car buyers fell victim to odometer fraud.[21] The total cost to consumers has been estimated to be as high as $10 billion per year.[22] To cite one example, Robert "Bobby Cars" Fiorello, owner of several used car dealerships in New Jersey, pleaded guilty in 2008 to altering hundreds of odometers on late-model, high-mileage automobiles. Some odometers were rolled back as much as 100,000 miles.[23]

A more recent type of used car fraud is the illicit—and dangerous—sale of so-called "flood cars." In the aftermath of the three devastating 2005 Gulf Hurricanes—Katrina, Rita, and Wilma—more than 500,000 vehicles were submerged underwater.[24] When the waters receded, what remained were extremely hazardous "killer cars"—"automotive ticking time bombs that are basically rotting from the inside out."[25] Obviously, modern vehicles with their computer chips and sophisticated electrical systems suffer permanent damage when submerged for any length of time. "It's only a matter of time before the engines start to die in traffic without warning. The air bags may not inflate in a crash. The braking systems are likely to fail."[26] For those reasons, insurers usually "totaled" these cars—that is, took them from their owners and paid off insurance claims. One would assume this was the end of the road for the "Katrina cars," but that was not the case at all. According to a report from the Consumers for Auto Reliability and Safety organization in 2006: "Despite warnings they should all be crushed, most of them are headed for sale to unsuspecting used car buyers. Auto insurers have quietly shipped almost all of their 'total loss' Katrina flood cars to other states."[27] Flood cars soon appeared in used car lots far from the Gulf Coast.[28]

California, with its huge, lucrative market, was notorious as a favorite dumping ground for hurricane cars, thanks to a criminogenic chain of fraud. Insurers ship them to so-called "salvage pools," where they are auctioned off to the highest bidder. "[U]nlicensed rebuilders buy them, give them a power wash and replace the upholstery, and sell them to unscrupulous auto dealers, who in turn foist them off on unsuspecting used car buyers—often teenagers buying their first cars."[29] The insurers get a cut of the proceeds from the auction, reducing their losses by millions of dollars.[30]

One of the clearest illustrations of institutional criminogenesis was the corruption of Wall Street in the 1980s (Chapter 6). The "merger mania" of the era was orchestrated by investment banks and law firms, whose members were privy to valuable information about impending deals. The restricted nature of that information and the tremendous profits it could yield created both opportunities and incentives for insider trading.

Organizational Causes

In many white-collar crimes the motivation is straightforward—individual material gain. This would be the case, for example, when a Sheraton Corporation food buyer was sentenced to 15 months in prison for taking "hundreds of thousands of dollars" in kickbacks from produce wholesalers.[31]

It would also certainly be the case with many of the notorious criminals portrayed in these pages, such as Charles Ponzi, Ivan Boesky, or Jim Bakker.

However, in many of the other cases discussed in this book, individuals carried out the criminal schemes, but the ultimate benefit went to a larger organization, typically a corporation. A basic distinction in the analysis of white-collar crime is between what is known as "occupational crime" and what is labeled "corporate crime."

Clinard and Quinney defined occupational crime as "offenses committed by individuals in the course of their occupations and the offenses of employees against their employers" and corporate crime as "offenses committed by corporate officials for their corporation and the offenses of the corporation itself."[32] Occupational crime would include bank employees who embezzle, physicians who defraud health benefit programs, mechanics who overcharge or bill for needless services, and politicians who accept bribes and kickbacks.

Corporate crimes, on the other hand, usually are more complex and involve individuals who violate the law to advance the interests of the organization. One example would be James Donnell, former vice president at Affiliated Computer Services, who was convicted in 2004 of cheating the company's clients out of $3.8 million by inflating customer invoices.[33] Corporate crime already has been designated the "crime of choice" for the 21st century.[34] One author has characterized the modern corporation as a "pathological institution."[35]

In preceding chapters, we have looked at numerous instances of corporate crime. The illegal dumping of hazardous waste by companies well aware of the serious consequences, the knowing exposure of workers to unsafe conditions, and the deliberate sale of dangerous products are all examples of corporate crime, where the offenses frequently were carried out by supervisors and managers who claimed to be "just following orders."

More recently, the idea of corporate crime has been extended into more traditional categories of criminal law, such as homicide. In Chapter 3, we considered the infamous Pinto case in which the Ford Motor Company was charged with "reckless homicide" after three teenagers died when their Pinto burst into flames after being struck from the rear by an other vehicle. Although Ford eventually was acquitted—largely because of a procedural technicality—a significant legal precedent was set in the area of corporate criminal liability.

Some criminologists have sought to extend the concept of corporate liability to encompass organizations of all types, including government agencies and even nonprofit organizations. Schrager and Short distinguished "organizational crime" from individual offenses by defining it as illegal acts committed by "an individual or a group of individuals in a legitimate formal organization in accordance with the operative goals of the organization."[36] Other theorists have used similar terms, such as "organizational deviance"[37] and "elite deviance."[38]

Whatever the terminology, organizational crime also includes acts of governmental deviance described in Chapter 9, such as the FBI's unlawful surveillance and persecution of American citizens, the burglaries, dirty tricks, and cover-ups overseen by the White House in the Watergate scandal, and the abuses of power precipitated by the Iran-Contra Affair. To this list one could add the well-documented involvement of the CIA in drug trafficking around the world—particularly in Southeast Asia and Latin America.[39] These were all illegal activities undertaken by individuals on behalf of the covert agenda (but not the mandated goals) of official agencies and political administrations.

Why do some organizations promote "deviance" while others do not? Sometimes it is simply a matter of the greed of the individuals leading the organization. For example, Mickey Monus, co-founder and president of the 300-unit Phar-Mor drugstore chain, was convicted in 1995 of a mammoth fraud and embezzlement scheme that drove his own company into bankruptcy.[40]

Not all organizational deviance conforms to the "greedy leader" model exemplified by Mickey Monus, however. In many cases, it has more to do with the environments in which the organization operates. Some sociologists have carried Merton's notion of blocked aspirations to a higher level of analysis, suggesting that a particularly significant aspect of a criminogenic corporate environment is the "strain" placed on organizations. A case study of organizational misconduct by another drug store chain concluded that organizations, like individuals, may seek illegal means to achieve economic success when legitimate avenues are blocked and when opportunities are available to attain material goals unlawfully.

Executives of the chain reportedly ordered employees to falsify Medicaid claims after state officials had rejected earlier claims for administrative reasons. Those executives believed that reimbursement for the false claims was "owed" to the company.[41] A survey of retired middle managers from Fortune 500 companies reported that the most important cause of unlawful or unethical practices in their corporations, according to respondents, was the pressures applied by top management to show profits and reduce costs.[42]

... It may be very difficult to distinguish between acts that promote the goals of an organization from those that primarily benefit individuals, since, in many cases, "individuals and their organizations often reap mutual advantage from criminal conduct."[43] In fact, organizations can be used by members as "weapons" in the commission of crimes. As one study observed: "[O]ccupation and organization are to the world of white-collar crime what the knife and the gun are to street crimes."[44] Thus, elements like size, ownership form, and administrative structure can be used by organizational members to commit and later hide their crimes. This notion was applied to a study of fraud among insolvent S&Ls which found that "those institutions that were stock-owned ... were the sites and vehicles for the most frequent, the most costly and the most complex (as measured by the number of individuals involved) amounts of white collar crime."[45]

RESPONSES TO WHITE-COLLAR CRIME

Given the abundant opportunities that many persons have to engage in white-collar crime, it is interesting that more people do not. The most obvious reason for this is the fear of punishment. Despite the fact that relatively few individuals who cheat on their taxes are ever actually prosecuted, the knowledge that some have been (including the treasurer of the United States, as described in Chapter 1) is enough to keep most of us reasonably honest when we turn in our 1040s in April. In other words, many citizens do not cheat on their taxes simply because of strong moral compunctions but because the law *deters* them—especially when their personal reputations are at stake. For example, some experts have suggested "shaming" tax delinquents by publishing their names and pictures in local newspapers.[46] When the IRS reported $300 million in unpaid 2007 taxes, that amount included sizable bills owed by such celebrity tax delinquents as singer Dionne Warwick ($3.7 million), comedian Sinbad ($2.1 million), and former Clinton aide and current Fox News analyst Dick Morris ($1.5 million).[47] The argument goes that fear of embarrassment would deter tax evaders in general, but especially high-profile figures. Indeed, several persons nominated by President Obama in early 2009 to fill key positions in his administration were "shamed" into withdrawing their names because of revelations of unpaid back taxes.

A central concern about white-collar crime is that the risk-reward ratio is out of balance—that is, the potential rewards greatly outweigh the risks. Given the low probability of apprehension and the likelihood of light punishment, white-collar crime appears to be a "rational" course of action in many cases. The sometimes indifferent response to white-collar criminals stands in sharp contrast to the harsh punitive treatment customarily accorded to street criminals—a contrast summed up in Jeffrey Reiman's pithy observation that "the rich get richer and the poor get prison."[48] The comparative leniency shown white-collar offenders had been attributed to several factors related to their status and resources, as well as to the peculiar characteristics of their offenses. First, the relatively high educational level and occupational prestige of many white-collar offenders are seen as creating a "status shield" that protects them from the harsh penalties applied with greater frequency to "common" criminals. Many judges identify with defendants whose background and standing in the community are similar to their own. When three former executives of C.R. Bard Corporation, one of the world's largest medical equipment manufacturers, were convicted in 1996 of conspiring to test unapproved heart surgery catheters in human patients, the sympathetic judge reluctantly sentenced them to 18 months in federal prison. He almost seemed to be scolding the jury when he said: "I don't regard the defendants as being evil people or typical criminal types."[49] The U.S. attorney who prosecuted the case could not have disagreed more. He characterized the callous executives as "evil people doing evil things... for money."[50]

A second leniency factor punctuates one of the world's worst-kept secrets—that life is seldom fair. White-collar defendants' high incomes enable them to secure expensive legal counsel, whose level of skill and access to defensive resources are generally unavailable to lower-class defendants. In addition, the emerging field of "sentence consulting" of "post-conviction mitigation" has emerged, providing convicted white-collar offenders with assistance in sentence reduction such as demonstrations of charitable generosity or evidence of ailing health. Despite the "country club" mythology, there are no pleasant prisons; so these advisors also "pad the landing" when white-collar crooks fall from grace and land in a cell. One consultant, David Navel (mentioned earlier), says of his privileged clients: "They have absolutely no idea what it is they're facing. Unfortunately, they've heard a lot about 'Club Fed,' and at the same time they have seen movies like 'The Shawshank Redemption,' so they are not quite sure whether to bring their golf clubs or they are just in fear for their lives."[51]

Finally, white-collar crimes frequently involve complicated financial transactions in which the victims are either aggregated classes of unrelated persons, such as stockholders of large government agencies, such as the IRS—neither of which engender the kind of commiseration that individual victims of street crimes can elicit from judges and juries.

EFFECTS OF WHITE-COLLAR CRIME

The opening chapter noted that people often fail to express the kind of outrage over corporate crime that they do over street crimes because it is usually more difficult to visualize the damage wrought by white-collar offenses. Television often ignores or downplays corporate crime as well because those crimes lack the dramatic elements that fit the needs of the electronic media: clearly defined victims and villains; illegal actions that are easily understood and can be described in quick sound bytes; motivations like jealousy and rage that can be vividly portrayed; and heroes in the form of police or prosecutors, who apprehend and then punish those responsible. In contrast, white-collar crimes are frequently confusing; the perpetrators, because of their social status, are not easy to cast as lawbreakers; the injury caused by these crimes (as in the case of toxic dumping, for example) may take years to develop; and the resolution of the cases often occurs outside of criminal courts and away from television cameras in private negotiations between offenders and anonymous government officials. For these reasons, television often shows little interest in cases of corporate criminality, while at the same time deluging us with sensational stories of murder, mayhem, and madness.

As this book has stressed repeatedly, the effects of white-collar crimes are in many ways more serious than those of "common" crime. These harmful consequences fall into three categories: environmental and human costs; economic costs; and social costs. Let us consider each of these effects in more detail.

Environmental and Human Costs

The kinds of dire prophecies of environmental destruction issued by ecologists are usually shrugged off by most Americans as alarmist hyperbole—until disasters like the Exxon Valdez or Love Canal occur. Only then does the public recognize the terrible threat posed by industrial irresponsibility. Corporate apologists often depict these events as tragic but unpredictable "accidents" that no amount of legislation could prevent. After the Union Carbide catastrophe at Bhopal, India, in 1984 (Chapter 4), when a storage tank containing deadly gas burst and killed at least 2,500 people, some American magazines actually portrayed Union Carbide as the "victim"—expressing concern over the impact of the disaster on the company's reputation and future profitability. In contrast, Indian publications all focused on the true victims and consistently referred to the event as a "crime."[52] Indeed, on July 31, 2009, an Indian court issued an arrest warrant for the former CEO of Union Carbide (now owned by Dow Chemical) for his role in the Bhopal disaster 25 years earlier. At this writing, there is no indication that the U.S. will extradite the now 89-year-old man.[53]

The American media's unwillingness to see such disasters as manifestations of corporate crime was also evident in their response to the tragic 1991 fire at the Imperial chicken processing plant in North Carolina that killed 26 workers, who were trapped inside because management had chained the emergency exits shut to prevent employee pilferage. Despite the eventual manslaughter conviction of the plant's owner, the media tended not to focus on criminal culpability but on an alleged breakdown in government safety regulations, particularly the failure of OSHA to inspect the plant adequately.[54]

Economic Costs

One of the more immediate effects of corporate criminality is the drag it places on the economy. Chapter 1 pointed out that the annual costs to Americans from all white-collar crimes runs into hundred of billions of dollars; the total from personal fraud alone is more than $40 billion. In certain industries, the costs are especially high. For instance, the General Accounting Office has estimated that at least $100 billion (10 percent of the nearly $1 trillion spent by Americans on health care every year) is lost to fraud and abuse.[55]

Who pays these costs? Are they equally shared by all members of society? Should some groups—particularly those who may have profited from the crimes—absorb a larger share of the cost? The evidence indicates that the burden is *not* shared equally across society but is shifted to the middle- and lower-income segments of the population. Because these costs are typically passed on to taxpayers, they represent a "regressive tax" imposed on those individuals least able to pay. At the same time, companies victimized by white-collar crime or slapped with punitive fines for their own offenses pass the costs on to consumers.

Chapter 8 noted that the multibillion cost of bailing out failed S&Ls was paid for mainly through the issuance of government bonds, which eventually must be redeemed by taxpayers. In theory, the progressive nature of the American tax structure means that wealthy citizens would pay a higher proportion of their incomes to cover the bill. In reality, however, the loopholes and tax shelters enjoyed by high-income individuals mean that the bulk of the burden will fall on the middle class. The S&L bailout also forced reductions of government spending in other areas, notably the social services frequently needed by low-income persons—none of whom had benefited from the extraordinary high interest rates that corrupt thrifts paid to depositors during the I-got-mine years of the "greed decade." In contrast, the government chose to cover all lost deposits of the bankrupt S&Ls (not just those under $100,000, as the law required). This meant, for example, that many wealthy depositors were able to reap the benefits of the high interest rates offered by the S&L—without assuming any of the risks when they failed.[56] The economic elite, who never miss an opportunity to rail contemptuously against welfare payments to the poor, thus became the shameless beneficiaries of the biggest "entitlement" program ever implemented by the American government.

An illustration of how the economic costs of white-collar crime can generate a "ripple effect" is provided by the bankruptcy of Orange County, California. Because of fraud and gross mismanagement in one self-indulgent venue, every city, town, and school district in the United States will have to pay more to borrow money from now on.[57]

Other municipalities, most of them far less prosperous and guilty of no wrongdoing, have to share the bill for the Orange County debacle because municipal investors must now consider bankruptcy and default a feasible alternative to debt. In other words, Orange County has raised the level of risk for lenders, who will accordingly demand higher interest from *all* municipal bond issuers. This may be an affront to distributive justice, but it is no surprise. The diffusion of accountability is a common by-product of fiscal recklessness. Consider, for example, all the commercial banks that failed in the 1980s due to fraud and abuse. Federal regulators were able to cover the costs of those failures, while avoiding a direct S&L-type bailout, by raising the premium paid to federal deposit insurance funds by member institutions. The increased expenditure was then passed on by banks to customers in the form of higher interest rates on credit card balances—without any corresponding increase in the interest paid to depositors on their savings accounts. In effect, this transfer of debt represented a silent bailout of the banking industry funded by consumers.[58]

The same sort of buck-passing is also seen in a wide variety of other businesses, where the costs of white-collar crime are absorbed by consumers in the form of higher prices for goods and services. While malfeasant corporations always try to portray themselves somehow as "victims"—of unfair regulations, political persecution, or whatever—it is ultimately ordinary citizens who pay the tab. In this way, white-collar crime contributes to the widening economic gap between the "haves" and the "have nots" of American society.

Social Costs

In addition to the environmental damage, the human destruction, and the economic losses, there are less tangible but other very serious social consequences of white-collar crime. The impact of upperworld criminality tends to radiate, influencing people's attitudes toward society and each other. Chapter 1 observed how persistent, unpunished corporate and governmental corruption can produce feelings of cynicism among the public, remove an essential element of trust from everyday social interaction, delegitimate political institutions, and weaken respect for the law. One could identify many specific effects—as previous chapters have done—but a negative consequence still remaining for consideration is the relationship between white-collar offenses and other forms of crime. There is a connection, both direct and indirect, between "crime in the suites" and "crime in the streets."

White-collar criminality surely encourages and facilitates other types of crime. Indirectly, the existence of elite lawbreaking promotes disrespect for the law among ordinary citizens and provides ready rationalizations for potential street criminals seeking to justify their misconduct. Underclass youths who see or hear about local politicians taking bribes, police officers stealing drugs, or merchants cheating customers are able to minimize the harms caused by their own crimes by arguing that "everybody's doing it," or "we're no worse than anybody else," or "that's the way the system works." For many poor, young urbanites, crime among the rich and powerful affirms the futility of legitimate work and demonstrates the monetary and status rewards that accompany illegal activities. This cynical worldview was vividly articulated by a member of a Puerto Rican street gang in Chicago:

> We grew up at a time when people were making money and making it quick. You know, we saw on television people getting rich overnight. Then we saw, you know, you had the white-collar criminals—you know the guys who are just like us but never get caught, except that everybody knows that they are crooks. You had all these guys becoming filthy rich. And what do you think that's going to tell us? Shit, it didn't tell me to go and get a job at McDonalds and save my nickels and dimes.[59]

For such young offenders from the underclass, it seems, a life of hustling on the streets represents a "rational" alternative to the limited opportunities offered to them by legitimate society—just as Merton had proposed seven decades ago.

White-collar criminality also promotes other forms of crime in more direct ways. One of the allegations from the Iran-Contra hearings (Chapter 9), for example, was that a number of small banks with ties to organized crime were used by American officials to channel funds illegally to the Nicaraguan Contras.[60]

Organized crime, in fact, has long depended on corruption for its existence. In 1895, the Lexow Commission, investigating the New York Police Department, reported that "money and promise of service to be rendered are paid to public officials by the keepers or proprietors of gaming houses, disorderly houses or liquor saloons ... in exchange for promises of immunity from punishment or police interference."[61] Nearly 75 years later, the President's Commission on Law Enforcement and Administration of Justice reached the same conclusion: "All available data indicate that organized crime flourishes only where it has corrupted local officials."[62] From organized crime's point of view, this relationship is functional—bribery of cops and politicians enables it to conduct its illicit business free of official interference.[63] From the perspective of society and the victims of organized crime, however, this depraved symbiosis is wholly dysfunctional.

White-collar criminal activities support the illegal drug industry as well. Corrupt banks are central to the operations of the international cartels that import billions of dollars in cocaine and other drugs into the United States each year. Given the cash-heavy nature of drug sales, distributors and retailers are faced with the inevitable problem of converting huge amounts of cash into forms that can be more easily invested or transported overseas. Federal law requires that financial institutions must report all transactions of $10,000 or more. This creates a dilemma for drug dealers who may accumulate millions of dollars in cash. But for a price, unscrupulous bankers can "launder" that money, converting it into more legitimate forms without reporting it to the authorities. At a time when banks are coming under increased scrutiny, drug dealers are turning to other financial institutions, like insurance companies, to launder their cash.[64] Money laundering, of course, is not limited to drug dealers but can be found anywhere that a group has a need to disguise its sources of income—most notably with the operators of gambling casinos in the United States, who launder cash skimmed from profits in order to evade taxes.[65]

CONTROLLING WHITE-COLLAR CRIME

When the Republican majority won control of Congress in 1994, it quite naturally sought to impose its self-reliant philosophy—enunciated in its "Contract with America"—on the public. Many citizens, unhappy with the status quo, embraced this pact, but others managed to control their enthusiasm. Critics contend that beneath the Contract's seductive rhetoric, one finds the same vision espoused by Adam Smith and his disciples: a society dominated by the marketplace, a society in which the government has limited power to protect consumers, a society in which citizens have few legal recourses to challenge the ascendancy of corporations, and a society in which the people essentially must place their trust in the "good will" of bankers, manufacturers, brokers, and the other conductors of the free market.

The Contract was a blueprint for a *laissez-faire* economy. One of its clauses proposed a moratorium on new governmental regulations and would subject all existing regulations—including such areas as environmental pollution, employee safety, and consumer protection—to a cost-benefit analysis to determine if they impede commerce in anyway. Those rules determined to be too "costly" would be eradicated.[66]

Securities regulation was a major target of the Contract. Legislation has already been proposed (unsuccessfully) to downgrade the SEC's original watchdog function. Such an action would effectively have enfeebled much of the investor protection law that has underpinned U.S. capital markets for decades.[67] Another clause, approved by Congress in 1905 under the guise of "tort reform," made the recovery of money lost to securities fraud extremely difficult in federal court. The bill included a much higher burden of proof of intent and penalties for persons who sue and do not win.

The Contract also called for the capping of product liability awards by juries. Vocal critics of America's tort system, such as former Vice President Dan Quayle, contend that the willingness of juries to make excessive multimillion dollar awards to plaintiffs strangles the economy by discouraging companies from introducing new products. A common example used to support this argument is that of Monsanto, which has developed a patented phosphate fiber substitute for asbestos but has kept it off the market because it fears potential lawsuits. In fact, the Environmental Protection Agency (EPA) believes this asbestos substitute to be as carcinogenic as asbestos. So Monsanto's decision not to sell it may be based on sound corporate policy and not intimidation.[68] Nevertheless, Quayle and his cohorts seem intent on bestowing blanket immunity on manufacturers by exploiting the universal unpopularity of ambulance-chasing lawyers.

There was another notable aspect to the Contract with America. It called for the implementation of much harsher punishment for "street" criminals but said virtually nothing about the punishment of white-collar offenders.

Deregulation represents a movement in exactly the wrong direction in the fight against white-collar crime. When the corporate establishment—through its congressional surrogates—tells the American people, "We want to get government off *your* backs," they almost always mean, "We want to get government off *our* backs"—a very different proposition. It is impossible to imagine how the absence of an SEC could have prevented the insider trading epidemic or how the elimination of banking restrictions would thwart another savings and loan crisis, or how doing away with the EPA would tender future Love Canals less likely. Advocates of *laissez-faire* solutions do point out, with some justification, that government regulations have already failed to prevent the aforementioned calamities. Yet, calling for their abolishment because they have been less than entirely successful seems dangerous and absurd. If budget slashing forces regulatory agencies to cut back on enforcement, prosecutors "may well find that the cuffs are on them."[69]

But if deregulation moves us in the wrong direction, then what might the right direction be? A number of suggestions to curb white-collar criminality can be offered, requiring changes of three types: legal, institutional, and social.

Legal Changes

It has been proposed that new laws that impose tougher penalties on white-collar criminals might well deter some potential offenders. More punitive federal sentencing guidelines have resulted in more white-collar crime convictions.[70] Adhering to the new guidelines would also serve to redress the sentencing imbalance between white-collar and traditional "common" criminals.[71] A federal prosecutor has declared: "You deal with white-collar the same way as street crime. You try to raise the likelihood they will be caught and punished."[72]

But whether this new toughness serves as a general deterrent is not at all clear. Current laws likely fail to deter because white-collar offenders are aware of the frequent lack of vigorous enforcement and the relatively low probability that their crimes will be detected or punished severely.

For example, in 2008 movie star Wesley Snipes was convicted on three misdemeanor counts of failing to file tax returns.[73] He was probably the highest-profile tax evasion defendant since hotel empress Leona Helmsley, the so-called Queen of Mean, was convicted in 1989 of evading $1.2 million in federal income taxes.[74] Although Snipes did receive a three-year sentence, he had faced up to 16 years in prison but was acquitted on the more serious felony charges—despite never even filing a tax return between 1999 and 2004.[75] The government was clearly disappointed with the relative leniency shown to Snipes. The prosecuting attorney had told the court in a sentencing memorandum: "In the defendant Wesley Snipes, the court is presented with a wealthy, famous and inveterate tax scofflaw. If ever a tax offender was deserving of being held accountable to the maximum extent for his criminal wrongdoing—Snipes is that defendant."[76]

The system of regulatory codes and administrative agencies that monitor corporate conduct and respond to criminal violations is another important part of the legal apparatus. Some scholars believe that we do not need more regulation; rather, we need "smarter" regulation. Simply applying harsher laws to corporations and individuals, they argue, will only produce a subculture of resistance within the corporate community "wherein methods of legal resistance and counterattack are incorporated into industry socialization."[77] Regulation works best when it is a "benign big gun"—that is, when regulators can speak softly but carry big sticks in the form of substantial legal penalties.[78]

Joseph T. Wells, the founder and chairman of the Association of Certified Fraud Examiners, has offered a newer strategy: *executive transparency*. Wells argues for a law requiring corporate executives to open up their own personal bank accounts for scrutiny by auditors and regulators.[79] The rationale is that in many of the high-profile corporate fraud cases, the fraud is not discovered until after the money has been frittered away. Wells cites the huge "loans" Bernie Ebbers gave to himself so that he could buy hundreds of thousands of acres of timberland and the biggest cattle ranch in Canada; the profligate spending by the Rigas family, including the construction of their own private golf course; the millions of dollars embezzled by Mickey Monus to finance his personal basketball league; and the grotesque self-indulgence of Dennis Kozlowski. As Wells notes, major corporate fraud cases almost always begin at the top.[80]

> To head off the financial rape of public corporations, I would suggest a law that requires selected company insiders to furnish their individual financial statements and tax returns to independent auditors. They should also sign an agreement allowing access to their private banking information. The data should be available in cases where suspicions arise.[81]

Executives of public corporations have a fiduciary duty to act in the best interests of shareholders,[82] and Wells' call for "transparency" seems consistent with that duty.

A hierarchical structure of corporate sanctions also has been proposed, in which the first response to misconduct consists of advice, warnings, and persuasion; then escalates to harsher responses culminating in what is termed "corporate capital punishment" or the dissolution of the offending company.[83] The goal of this model is compliance: "Compliance is thus understood within a dynamic enforcement game where enforcers try to get commitment from corporations to comply with the law and can back up their negotiations with credible threats about the dangers faced by defendants if they choose to go down the path of non-compliance."[84] The strength of such a system is that it works at multiple levels and holds all the actors involved—"executive directors, accountants, brokers, legal advisers, and sloppy regulators"[85]—accountable for criminal misconduct.

Institutional Changes

Preceding chapters have highlighted numerous examples of "criminogenic industries"—industries whose structure and traditional practices seem to encourage, or even embrace, criminal behavior. Any successful preventive strategy must therefore seek to "deinstitutionalize" white-collar crime, that is, remove its institutional sources. This is by no means an easy task. About 20 percent of the 1,000 largest American corporations already have "ethics officers," who help formulate codes of proper conduct.[86] A *Wall Street Journal* reporter, however, has written that these codes "are little more than high-sounding words on paper."[87]

Sociologist Amitai Etzioni has suggested that a better way to encourage ethical business conduct is to "foster associations and enforce moral codes somewhat like those of lawyers and physicians" in the business community.[88] Such associations would lack legal authority, but they could discipline violators through public censure and other informal control mechanisms.

Another way to change the environment in which corporate organizations do business is to create internationally agreed upon standards of conduct. This would be particularly important in the case of American companies that operate in—and often export white-collar crimes to—foreign countries. The most significant step in this direction has been the United Nation's attempt to draw up a Code of Conduct for multinational corporations. That document sets standards of acceptable behavior for global firms. Among other things, the Code calls upon multinationals to abstain from corrupt practices, such as political bribery, and to carry out their operations in an environmentally sound manner.[89] With these same goals in mind, the Clinton administration has proposed a code of ethics for American firms operating overseas. Known as the Model Business Principles, this code encourages companies to adopt more scrupulous behavior abroad by "providing a safe and healthy workplace ... pursuing safe environmental practices ... and complying with U.S. laws prohibiting bribery."[90]

Neither of these proposed codes establishes any authority to sanction violators; they are merely guidelines. Enforcement is left up to individual nations. This could be a significant weakness. Some Third World countries (as we have seen) have shown a willingness to tolerate egregious abuses by American corporations—either because of official corruption or out of desperation for economic growth. Nevertheless, these efforts represent a first step toward battling white-collar crime in a global economy.

SOCIAL CHANGES

Earlier in this chapter, it was suggested that Americans tend to have very strong feelings about crime and criminals but are often ambivalent in their responses to those convicted of white-collar offenses. As noted, many corporate lawbreakers have received the support of their colleagues and other prominent members of their community. Even after their convictions, they were quickly accepted back into legitimate society, suffering few of the stigmas and resentments that other types of convicted felons routinely experience. Indeed, it almost seems that we extend a begrudging respect to those who are clever and bold enough to fleece us out of millions. But if we are ever to gain control over white-collar crime, these attitudes must change.

Criminologist John Braithwaite argues that the broader corporate milieu needs to be transformed. Braithwaite calls for the creation of a "communitarian corporate culture" in which organizations draw "everyone's attention to the failings of those who fall short of corporate social responsibility standards [shaming], while continuing to offer them advice and encouragement to improve [reintegration]."[91] A simple, yet very effective form of "shaming" that can be employed when corporations violate the law is adverse publicity—spreading information about misconduct to consumers, who could then express their disapproval by refusing to patronize the offending company.[92] Braithwaite's ideal corporation, "well integrated into the community and therefore amenable to the pressures of social control,"[93] stands in dramatic contrast to Friedman's model of an amoral corporation, divorced from obligations to the community and inevitably producing a criminogenic environment.

Many of the old assumptions about the utility of business ethics have been challenged. One such assumption is that "most people act unethically for two reasons: to make more money and to beat the competition."[94] But according to a number of newer studies, an ethical culture can actually make a business *more* successful. Walker Research, which tests employees' perceptions about their employers, reports "an attitudinal link between perceptions of ethics and a company's ability to hold on to employees."[95] According to a 1999 Walker report:

> [C]ompanies viewed as highly ethical by their employees were six times more likely to keep their staff members. On the other hand, 79 percent of employees who questioned their bosses' integrity said they felt trapped at work or uncommitted, or were likely to leave their jobs soon—sobering news in an economy where holding on to every good employee counts.[96]

Moreover, another recent study found that businesses with a "strong culture of shared values"[97] tend to be more profitable than other companies. "Companies with a strong sense of values see revenue grow four times faster, jobs get created seven times faster, and stock prices increase 12 times faster."[98]

Another way to change the culture of tolerance for white-collar crime is to alter the socialization of future captains of industry. One logical place to look is to the elite MBA programs where many American business leaders are trained.

A number of reformers have called for the integration of rigorous ethical analyses into the curricula of business schools.[99] The sad reality, however, is that by that point it may already be too late. Some MBA programs have begun to offer business ethics courses, either as electives or in some cases as requirements,[100] but whether such reform will have any measurable future impact on corporate morality seems a long shot at best. At one leading business school, an elective called Managing in the Socially Responsible Corporation is reportedly ridiculed by students as a useless "touchy-feely" course.[101] A former chairman of IBM has declared, "If an MBA candidate doesn't know the difference between honesty and crime, between lying and telling the truth, then business school, in all probability, will not produce a convert."[102]

Like death, taxes, and *I Love Lucy,* white-collar crime may always be with us. Measures such as those just outlined will not entirely extirpate the problem, but they can help to make it less commonplace and to minimize its fallout. If nothing else, raising the public's consciousness of the depth, dynamics, and disaster of white-collar crime—as this book has tried to do—hopefully can reshape people's view of events like the insider trading scandal and S&L crisis, the recent corporate collapses, the tragedy of carcinogenic pollution, and other licensed assaults on the quality of life in the United States.

ENDNOTES

1. Merton, Robert K. "Social Structure and Anomie." In Robert Merton, *Social Theory and Social Structure*. Glencoe, IL: The Free Press, 1957; 131–160.
2. *Ibid.*
3. Coleman, James W. *The Criminal Elite* (3rd Edition). New York: St. Martin's, 1994.
4. Sykes, Gresham and Matza, David. "Techniques of Neutralization." *American Sociological Review* 22, 1957: 664–670. For empirical applications of this concept to white-collar crime, see Rothman, Mitchell and Gandossy, Robert "Sad Tales: The Accounts of White-Collar Defendants and the Decision to Sanction." *Pacific Sociological Review* 25, 1982: 449–473; Benson, Michael, "Denying the Guilty Mind: Accounting for involvement in White-Collar Crime." *Criminology* 27, 1989: 769–794.
5. Vitell, Scott J. and Grove, Stephen J., "Marketing Ethics and the Techniques of Neutralization." *Journal of Business Ethics* 6, 1987: 433–438.
6. From Sykes and Matza, *op.cit.*
7. Thurmon, Arnold, *The Folklore of Capitalism*, New Haven, CT: Yale University Press, 1937.
8. Bailey, Fenton. *The Junk Bond Revolution*. London: Fourth Estate, 1991.
9. Center for Investigative Reporting, *Global Dumping Ground*, Washington, D.C.: Seven Locks Press, 1990; 37.
10. Useem, Jerry. "Oink! CEO Pay Is Still Out of Control," *Fortune* 147, April 28, 2003: 56–64.
11. Reich, Robert B., "Secession of the Successful," *New York Times Magazine*. January 20, 1991: 16.
12. Durant, Will and Durant, Ariel. *The Lessons of History*. New York: MJF Books, 1968.
13. Strauss, Gary. "How Did Business Get So Darn Dirty?" USATODAY.com, June 12, 2002.
14. Needleman, Martin and Needleman, Carolyn. "Organizational Crime: Two Models of Criminogenesis." *Sociological Quarterly* 20, 1979: 517
15. Leonard, William and Weber, Marvin. "Automakers and Dealers: A Study of Criminogenic Market Forces." *Law and Society Review* 4, 1970: 407–424.
16. Farberman, Harvey, "A Criminogenic Market Structure: The Automobile Industry," *Sociological Quarterly* 16, 1975: 438–457.

17. *washoelegalservices.org*. "Consumer Law: Odometer Fraud," December 29, 2002.
18. *businessweek.com*. "TCC Tip: Odometer Fraud," December 30, 2005.
19. Scripture, James E. "Odometer Rollback Schemes." *totse.com*, January 12, 2009.
20. *businessweek.com, op.cit.*
21. *autopi.com*, "Odometer Fraud," December 29, 2008.
22. *businessweek.com, op.cit.*
23. *nj.com*, "Used Car Dealer Admits to Odometer Scam," August 30, 2008.
24. Redding, Robert L., Jr. "Congress Assesses Flood Vehicle Problem." *asashop.org*, February 29, 2006.
25. Shahan, Rosemary, "Rotting Flood Cars Should Be Banned from California." *californiaprogressreport.com*, August 31, 2006.
26. *Ibid.*
27. *Ibid.*
28. Benton, Joe. "Congress Treads Water on 'Flood Car' Bill." *consumeraffairs.com*, December 4, 2006.
29. *Ibid.*
30. *Ibid.*
31. Foderaro, Lisa W. "Former Sheraton Food Buyer Is Sentenced for Kickbacks." *New York Times*, July 12, 1997: B25.
32. Clinard, Marshall and Quinney, Richard. *Criminal Behavior Systems* (2nd Edition), Cincinnati, OH: Anderson 1986: 189.
33. *bizjournals.com*. "Ex-VP of ACS Convicted of Fraud." January 20, 2004.
34. Johnston, Jeffrey L. "Following the Trail of Financial Statement Fraud," *Business Credit* 97, October 1995: 48.
35. Bakan, Joel. *The Corporation: The Pathological Pursuit of Profit and Power*. New York: Free Press, 2004.
36. Schrager, Laura and Short, James F. Ir. "Toward a Sociology of Organizational Crime." *Social Problems* 25, 1978: 411–412.
37. Ermann, David and Lundman, Richard. "Deviant Acts by Complex Organizations: Deviance and Social Control at the Organizational Level of Analysis." *Sociological Quarterly* 19, 1978: 53–67.
38. Simon, David and Eitzen, Stanley, *Elite Deviance*. Boston: Allyn and Bacon, 1982.
39. McCoy, Alfred. *The Politics of Heroin: CIA Complicity in the Global Drug Trade*. Brooklyn, NY: Lawrence Hill Books, 1991.
40. Collins, Glenn. "Ousted Phar-Mor President Found Guilty in $1 Billion Fraud." *New York Times*, May 26, 1995: D3.
41. Vaughan, Dianne, *Controlling Unlawful Organizational Behavior*, Chicago IL: University of Chicago Press, 1983.
42. Clinard, Marshall, *Corporate Ethics and Crime*. Beverly Hills, CA: Sage, 1983.
43. Wheeler, Stanton and Rothman, Mitchell. "The Organization as Weapon in White-Collar Crime." *Michigan Law Review* 80, 1982: 1405.
44. *Ibid.*, 1426.
45. Tillman, Robert and Pontell, Henry. "Organizations and Fraud in the Savings and Loan Industry." *Social Forces* 4, 1995: 1458.
46. Soled, Jay A. and Ventry, Dennis J. Jr. "A Little Shame Just Might Deter Tax Cheaters," *USA Today*, April 10, 2008: 11A.
47. Cauchon, Dennis. "Tax Delinquents: Who Owes—and How Much—May Be Surprise." *USA Today*, April 14, 2008: 1A, 4A.
48. Reiman, Jeffrey. *The Rich Get Richer and the Poor Get Prison: Ideology, Crime and Criminal Justice* (4th Edition). Boston, MA: Allyn and Bacon, 1995.
49. Quoted in Ranalli, Ralph. "Execs Get 18 Months in Med-Testing Flap." *Boston Herald*, August 9, 1996, 20.
50. Quoted in *Ibid.*, 20.
51. Robertson, Tatsha and Krasner, Jeffery. "Seeking Not-So-Hard Time." *boston.com*, April 9, 2004.
52. Lynch, Michael, Nalla, Makesh, and Miller, Keith. "Cross-Cultural Perceptions of Deviance: The Case of Bhopal." *Journal of Research in Crime and Delinquency* 26, 1989: 7–35.
53. Eltman, Frank. "Man Wanted in Deadly Gas Leak 'Haunted,' Wife Says." *Houston Chronicle*, August 2, 2009: A25.
54. Wright, John, Cullen, Francis, and Blankenship, Michael. "The Social Construction of Corporate Violence: Media Coverage of the Imperial Food Products Fire." *Crime & Delinquency* 41, 1995: 20–36.

55. Witkin, Gordon, Friedman, Doran, and Guttman, Monika. "Health Care Fraud," *U.S. News & World Report*, February 24, 1993: 34–43.
56. *New York Times*. "Who Should Pay for the S&L Bailout?" September 21, 1990: A31.
57. Petruno, Tom and Flanigan, James. "Orange County in Bankruptcy." *Los Angels Times*, December 7, 1994: D1.
58. U.S. Congress, House Committee on Banking, Finance, and Urban Affairs, Subcommittee on Consumer Affairs and Coinage. "H.R. 2440, Credit and Charge Card Disclosure Amendments of 1991." 102nd Congress, Second Session, October 9, 1991.
59. Quoted in Padilla, Felix. *The Gang as an American Enterprise*. New Brunswick, NJ: Rutgers University Press, 1992: 103.
60. Brewton, Pete. *The Mafia, CIA & George Bush*, New York: Time Books, 1994.
61. New York Senate. *Report and Proceedings of the Senate Committee Appointed to Investigate the Police Department of the City of New York*. Albany, NY: 1895: 6.
62. U.S. President's Commission on Law Enforcement and Administration of Justice. Task Force Report: Organized Crime. Washington, D.C.: U.S. Government Printing Office, 1967: 6.
63. Merton, Robert K. *Social Theory and Social Structure*, Glencoe, IL: The Free Press, 1957.
64. U.S. Congress, House Committee on Banking, Finance, and Urban Affairs. "Federal Government's Response to Money Laundering," May 25, 1993.
65. Block, Alan and Scarpitti, Frank "Casinos and Banking: Organized Crime in the Bahamas." *Deviant Behavior* 7, 1986: 301–312.
66. Gillespie, Ed and Schellhas, Bob (Eds.). *Contract With America*, New York: Time Books, 1994.
67. Cooke, Stephanie, "SEC Under Attack." *Furomoncy* 317, September 1993: 84–88.
68. Stein, Ben, "Insecurities Trading." *Los Angeles Magazine* 41, January 1996: 31–33.
69. *Economist*, "Business Not Guilty," Vol. 326. February 13, 1996: 63–64.
70. Lee, Charles S. "Bit Crime, Big Time," *Newsweek* 126, December 11, 1995: 59.
71. *Ibid.*
72. U.S. Code Congressional and Administrative News. 98th Congress, Second Session, 1984, Vol. 4. St. Paul. MN: West.
73. Quoted in Langberg, Mike. "White Collar Crime Erodes Faith in Business." *San Jose Mercury News*, February 12, 1989: IE.
74. *nytimes.com*, "Wesley Snipes Gets 3 Years for Not Filing Tax Returns." April 25, 2008.
75. Fried, Joseph. "U.S. Jury Finds Helmsley Guilty of Tax Evasion but Not Extortion." *nytimes.com*, August 31, 1989.
76. *MSNBC.com*, "Wesley Snipes Acquitted of Federal Tax Fraud," February 1, 2008.
77. Quoted in Goddard, Jacqui. "Wesley Snipes Given Three Years for Tax Evasion." *timesonline.co.uk*, April 25, 2008.
78. Ayres, Ian and Braithwaite, John. *Responsive Regulation*, Oxford: Oxford University Press, 1992: 20.
79. *Ibid.*
80. Wells, Joseph T. "Never a Better Time for 'Executive Transparency.'" *USA Today*, August 1, 2005: 11A.
81. *Ibid.*
82. *Ibid.*
83. *Ibid.*
84. Fisse, Brent and Braithwaite, John. *Corporations, Crime and Accountability*, Cambridge: Cambridge University Press, 1993.
85. *Ibid.*, 143.
86. *Ibid.*, 230.
87. Yenkin, Jonathan, "Ethics Officers Manage Companies' Morals." *Orange County Register*, August 29, 1993: Business Section. 1.
88. Rakstis, Ted I. "The Business Challenge Confronting the Ethics Issue." *Kiwanis Magazine*, September 1990: 30.
89. Etzioni, Amitai, *Public Policy in a New Key*. New Brunswick, NJ: Transaction, 1993: 103.
90. U.S. Congress, Senate Committee on Foreign Relations, Subcommittee on International Economic Policy, Trade Oceans and Environment. "UN Code of Conduct on Transnational Corporation." 101st Congress, Second Session, October 11, 1990.
91. *Los Angeles Times*, "White House Unveils Its Overseas Code of Corporate Conduct." March 28, 1995: D1.

92. Braithwaite, John, *Crime, Shame and Reintegration* New York: Cambridge University Press, 1989.
93. Coffee, John. "No Sound to Damn, No Body to Kick: An Unscandalized Inquiry into the Problem of Corporate Punishment." *Michigan Law Review* 9, 1981: 424–429.
94. Braithwaite, 1989, *op. cit.*, 144.
95. Galvin, *op. cit.*, 99.
96. *Ibid.*, 99.
97. *Ibid.*, 99.
98. *Ibid.*, 99.
99. Quoted in *Ibid.*, 99
100. See, for example: Etzioni, *op. cit.*
101. Fraedrich, John P. "Do the Right Thing: Ethics and Marketing in a World Gone Wrong." *Journal of Marketing* 60, January 1996: 122–123.
102. Marks, *op. cit.*

DISCUSSION QUESTIONS

1. Is white-collar crime harmful? Is it more harmful than other kinds of crime?
2. Why is it so difficult for people to regard white-collar crime as a serious problem?
3. Should people who engage in white-collar crime go to prison for what they do? If they do go to prison, is this good for society? Is it good for the perpetrators? Is it good for the victims? Is there any alternative?
4. Who is most likely to commit white-collar crime, in your opinion?
5. What are the worst types of white-collar crime, in your opinion?

29

Drugs: America's Holy War

ARTHUR BENAVIE

1. What is the problem identified by Benavie? Is it the individual's drug abuse? Is it society's response to drug-abusing individuals? Or is it something else?
2. Is this a *social* problem? Who is harmed by this problem? Does society recognize it as a social problem?
3. What is the cause of this problem?
4. How can society best address the problems discussed in this article?

Topics Covered

Illegal drugs

Illicit drugs

War on drugs

Drug laws

Problems of the drug war

Drug policy

History of drug use

WHY CERTAIN DRUGS ARE PROHIBITED

Who's the enemy in this war on drugs? It's a set of prohibited or "illicit" drugs along with those who possess or sell them. The federal Controlled Substances Act (1970) distinguished five categories or "schedules" of drugs based on their alleged susceptibility to abuse and their medical usefulness.[1]

A schedule I drug is defined as both dangerous and lacking any medical benefit. Drugs in this category cannot be possessed or sold legally and cannot even be prescribed by physicians. Examples are marijuana, heroin, LSD, and ecstasy.

A schedule II drug allegedly has a high potential for abuse but has some recognized medical value and can therefore be prescribed. Examples include cocaine, morphine, and methamphetamine.

The remaining categories include medically useful substances that are defined as decreasingly dangerous. For example, schedule III includes amphetamines and codeine, schedule IV Valium and Darvon, and schedule V other medicines containing narcotics.

Illicit drugs include all of those in schedule I along with the substances in the other four categories when used or sold outside of medical channels. The severest criminal penalties are applied to the possession or distribution of the drugs in schedules I and II.

Who decides which drug belongs in which schedule? You might think it would be the Federal Drug Administration (FDA) or the Surgeon General. Not so. It's the responsibility of the U.S. Drug Enforcement Administration (DEA) of the U.S. Department of Justice to make recommendations.[2] The Attorney General makes the final decision, one of many indications that drug abuse is considered a legal rather than a medical problem in the United States.

All the illicit drugs have two things in common—they are psychoactive (mood altering) and they give pleasure to the consumer. According to the National Survey on Drug Use and Health, in 2005, approximately 8.1 percent of Americans over the age of twelve (19.7 million) were "current users" of illicit drugs, that is, they had used an illicit drug in the month prior to the interview. Six percent of Americans over twelve were current users of marijuana in 2005 (14.6 million); 10 percent (2.4 million) used cocaine; 0.3 percent (0.72 million) crack; 0.1 percent (0.24 million) heroin; and 0.2 percent (0.48 million) methamphetamine.[3]

We are so often reminded of the harmful effects of these drugs, we can forget that people take them for a reason. Here I describe some of the pleasurable sensations people report along with some of the dangers.

Marijuana smokers report that they experience relaxation, euphoria, laughter, amusing distortions of space and time, heightened sensitivity to colors and sounds, increased sociability, and an enhanced enjoyment of music and sex.[4] *The Merck Manual of Medical Information* refers to "the sense of exaltation, excitement, and inner joyousness" that the smoker often experiences. Not everyone enjoys marijuana, however. Many try it a few times and stop. One common negative reaction is a feeling of paranoia. A major complaint, often made by parents, is that users lose the drive to achieve in school or work.[5] (We explore this in chapter 10.)

A laboratory study involving hypothetical choices among drugs by heroin addicts found that marijuana and heroin are "substitutes," that is, a rise in the price of one causes an increase in the consumption of the other.[6] This result probably reflects the fact that the high from heroin is similar to the intoxication marijuana smokers experience.[7] The heroin high is a warm, drowsy, and euphoric state that lasts four or five hours.[8] For those who inject the drug, the high is preceded by a "rush," which lasts only a few minutes. The rush has been likened to a sexual orgasm, and is described as a feeling of "great relief." Given the myths about heroin,

I was shocked to learn that, unlike alcohol or cocaine, it does not cause organic damage even to those who have become dependent, as long as they don't overdose. As *The Merck Manual* put it, "People who have developed tolerance [to heroin] may show few signs of drug use and function normally in their usual activities as long as they have access to drugs."[9] Many of the complications of heroin dependence are due to unsanitary conditions in the black market. But, the drug is physically addictive and carries the danger of overdosing, which can be life threatening.[10]

While marijuana depresses brain activity[11] and heroin is a depressant to the central nervous system,[12] cocaine is a stimulant, with effects similar to those produced by caffeine and amphetamine.[13] At moderate doses, users typically report that the drug combats fatigue, boosts energy, increases sociability, enhances sexual arousal, and creates feelings of euphoria and competence.[14] If the drug is injected or smoked, rather than inhaled, it produces a rush similar to heroin, followed by a high, which lasts less than a half an hour. High doses can lead to nervousness and agitation. Chronic users often suffer severe depression if they try to stop. Dependence on cocaine can develop rapidly if the drug is used frequently. Those who become dependent on it often suffer from "degeneration of the nasal mucous membranes ... digestive disorders, nausea, loss of appetite, weight loss, tooth erosion, brain abscess, stroke, cardiac irregularities, occasional convulsions, and sometimes paranoid psychoses and delusions of persecution."[15] Since the drug raises blood pressure, it can be lethal for someone with heart problems.[16]

Unlike marijuana, heroin and cocaine, amphetamines are not plant based. They were first synthesized in Germany in the 1880s. Among the drugs classified as amphetamines are ecstasy (MDMA) and methamphetamine (speed).[17]

Ecstasy is a mood elevator, which produces feelings of empathy, euphoria, and sociability, lasting three to six hours.[18] Users take the drug at "rave" dances, which often last all night. Ecstasy is not highly addictive and few deaths are associated with it (nine in 1998).[19] Some of those deaths are related to overheating, which is preventable if water is available at rave dances along with rooms where people can rest. Ecstasy raises the blood pressure and the heart rate. At high doses the drug can be dangerous, causing a large increase in body temperature that may result in muscle breakdown, kidney failure, and lethal cardiovascular effects in people who have heart disease.[20] Recent research has found a positive association between heavy use of MDMA and decreased performance in memory tasks and other cognitive functions.[21] It was made a schedule I drug in 1985.[22]

Methamphetamine, developed in Japan in 1919, has become the most frequently abused amphetamine in the United States.[23] Meth is a powerful stimulant, creating a rush of energy. Users report that it makes them "feel like Superman."[24] The drug can be eaten, snorted, injected, or smoked. When it's injected or smoked, it can, like cocaine, give the user a highly pleasurable rush that lasts a few minutes, followed by a euphoric high. As *The Merck Manual* puts it, "Amphetamines increase alertness (reduce fatigue), heighten concentration, decrease the appetite, and enhance physical performance. They may induce a feeling of well-being or euphoria."[25] Amphetamines temporarily relieve depression, improve athletic performance in the short run, help long-distance truck drivers stay awake, and bolster students putting in an all-nighter studying for exams.[26] From Vietnam up through Desert Storm, the Air Force has provided amphetamines to pilots.[27] Meth is prescribed on a limited basis for narcolepsy and attention-deficit hyperactivity disorder.[28] While the drug is rarely fatal, even small amounts can have potentially lethal effects, such as strokes, respiratory problems, cardiovascular collapse, anorexia, hyperthermia, convulsions, Parkinson-type diseases, insomnia, paranoia, and depression.[29]

It's widely believed that stimulants such as cocaine and the amphetamines are instantly addictive. Not true; but any substance that causes euphoria also creates a potential for abuse. As a highly respected psychopharmacology text put it: "Despite the popular view that drugs like cocaine and heroin are instantly and automatically addictive, that is not the case."[30] Drug researchers at the Duke University Medical School sum up the issue this way: "There are thousands of people, ranging from children with ADHD to truck drivers, who regularly use psychomotor stimulants but never develop a compulsive pattern of use … yet we know that the drive to use cocaine or amphetamine is considerably stronger than that for any of the other addictive drugs."[31]

Meth labs have been spreading rapidly in the rural areas of the Southeast.[32] Labs are found in homes, cheap hotel rooms, and backyards, and on deserted roadways. The drug can be cooked up in a few hours with a potentially explosive cocktail of household ingredients, such as farm fertilizer, anhydrous ammonia, lithium from car batteries, and pseudoephedrine from cold tablets. The waste from these meth labs is hazardous for humans and the environment.[33]

LSD is a hallucinogen, which induces the sensation of floating, along with visual and auditory distortions.[34] The user often experiences "sensory crossovers," that is, hearing colors, tasting sounds, and seeing music in colors. The LSD "trip" usually lasts six to twelve hours. It can be good or bad depending on the person's mood and the setting in which the drug is taken. A bad trip can cause anxiety and panic. A danger of using the drug is impaired judgment. For example, the user may believe they can fly and may try to do so. Chronic users may experience recurring hallucinations, called flashbacks, which usually disappear within a year after the last use, but may recur occasionally for up to five years. The existing data suggest that bad trips and flashbacks are rare.[35] There are no withdrawal symptoms, and I found no reports of deaths from the drug. Recreational use of LSD was banned in the United States in 1967.

Getting back to the question of why these drugs are prohibited, we might ask whether heroin, cocaine, and amphetamines are more life threatening than tobacco and alcohol. The answer is no. Tobacco and alcohol are more dangerous by far.[36] Together, they kill over a half million people a year: in 2000, tobacco was responsible for 435,000 deaths, and alcohol for 85,000 deaths.[37] The deaths per 100,000 users per year has been estimated at 650 for tobacco and 150 for alcohol.[38] In 2000, the consumption of *all* illicit drugs contributed to the deaths of around 17,000 people.[39] The deaths per 100,000 users per year have been estimated at eighty for heroin and four for cocaine.[40] Most of these deaths are due to contamination by black market suppliers and the government's restrictions on clean needles for drug injectors.[41] No death has ever been reported for marijuana.[42] According to a recent report by the European Monitoring Centre for Drugs and Drug Addiction, "Acute deaths related solely to cocaine, amphetamines or ecstasy are unusual despite the publicity they receive."[43]

A recent annual survey of hospitals and medical examiners conducted by the National Institute on Drug Abuse (NIDA), called DAWN (Drug Abuse Warning Network), found that the emergency room visits involving the so-called club drugs—which include methamphetamine, LSD and ecstasy—are "relatively rare," constituting 4 percent of drug-related emergencies. Some of these visits are caused by multiple drug use. Combinations involving alcohol are especially risky.[44]

Are the banned drugs the most addictive? No again. Legal drugs such as nicotine and alcohol have a higher addictive potential than many illicit drugs.[45] Ironically, except for heroin, the schedule I drugs—marijuana, LSD and ecstasy—are, according to studies, less likely to hook you than nicotine, alcohol, or even caffeine. In a National Survey on Drug Abuse, nicotine was found to be most addictive, with less than 20 percent of tobacco users managing to avoid getting hooked.[46] You may be surprised to learn that most heroin users consume only intermittently and do not become heavy users.[47] Of those who use cocaine, around 20 percent are dependent.[48] And about 10 percent of those who use marijuana smoke it daily.[49] In one survey, where 746 drug abuse researchers were asked to rank commonly used drugs on how addictive they were, nicotine and heroin ranked as the most addictive, followed in descending order by crack, meth, cocaine, alcohol, amphetamine, caffeine, marijuana, LSD, and ecstasy.[50] Contrary to conventional wisdom, most consumers of the prohibited drugs use them casually, suffer little harm, and stop within five years without coercion or treatment.[51]

Many people believe that the rationale for prohibiting certain drugs is that they induce violence, an accusation often leveled at crack in particular. The truth is that crack is not in the same league as alcohol in the violence department.[52] .... It's the drug war, along with alcohol, not the consumption of illicit drugs, that is responsible for most drug-related violence.[53]

As David Musto, an historian of American drug policy put it, "The history of drug laws in the United States shows that the degree to which a drug has been outlawed or curbed has no direct relation to its inherent danger."[54] So, how can we explain why certain drugs are prohibited while more dangerous ones are not? Researchers suggest that illicit drugs are associated in the minds of many with groups that are feared or hated, such as immigrants, African Americans, criminals, and rebellious young people.[55] The history of United States' drug policy illustrates how the criminalization of drug users has been fueled by racial, ethnic, and economic antagonisms.... Nineteenth-century anti-opium laws were aimed at Chinese laborers who were in competition with American workers and whose opium dens were an attraction to American youth. Marijuana first came to be seen as a problem in the early twentieth century, when it was the drug of choice of Mexican farm laborers who were thought to be taking scarce jobs away from Americans. Marijuana was further stigmatized by its connection with antiwar and anti-establishment protests during the Vietnam War. Cocaine and heroin came to be perceived as a danger to society as they were associated with blacks, juvenile gangs, and prostitutes.[56] The drug scene is frightening to mainstream society because it's seen as a cause of violence, disease, and social disorder.[57]

Those who study the drug war often claim that it's "rooted in a puritanical strain in American culture."[58] The strongest opponents of decriminalizing drugs appear to be evangelical Christiains, many of whom see drug use as sinful because they believe it's a threat to the authority of parents, the nation, and the social order. For example, S. K. Oberdeck claimed in a 1971 *National Review* article that marijuana represents a lifestyle that threatens to "bring down 'ordered life as we know it.'"[59]

Other researchers of the drug war argue that psychoactive drugs are perceived as a danger to our capitalistic economic system. The historian H. Wayne Morgan observed that illicit drugs produced an "indolence and reverie" that "seemed especially unsuitable to a modernizing industrial society whose success depended on hard work, rationality, and the mastery of complex facts."[60]

In sum: America's war on drugs is not driven by objective information concerning the effects of illicit drugs or the repercussions of the drug laws, but rather by emotional factors: racism, xenophobia, a fear of social disorder and disease, a perceived threat to the success of the economic system, and a religious crusade against "vice." As a result, the mainstream public and the government tune out overwhelming evidence that the drug war is causing irreparable damage to our society. As economist Thomas Sowell has written, "Policies are judged by their consequences, but crusades are judged by how good they make the crusaders feel."[61] The drug war is clearly a crusade.

ATTITUDES TOWARD DRUG LAW VIOLATORS

Some of the statements made by public figures about drug law violators are more suggestive of a religious crusade than a law enforcement issue. Consider a few examples.

Nancy Reagan has gone beyond "just say no"; she has stated that "The casual drug user cannot morally escape responsibility for the actions of drug traffickers and dealers. I am saying that if you are a casual drug user, you are an accomplice to murder."[62]

On September 5, 1990, the Los Angeles police chief Daryl Gates testified before the Senate Judiciary Committee that casual drug users "ought to be taken out and shot."[63] (That would polish off a third of our tenth and twelfth graders.)[64]

In August 1996, former House Speaker Newt Gingrich, while campaigning for presidential candidate and former Senator Robert Dole, told a crowd that drug dealers should be executed. (Even a high-school student selling pot to his friends?)[65]

In November, 1999, at a literary luncheon in Brisbane, Australia, "Judge Judy" suggested that, instead of attempting to control AIDS and hepatitis by providing clean needles to drug addicts, we should "give them all dirty needles and let them die."[66] (This pretty much reflects government policy.)

Drug czar William Bennett told a national radio audience that it was "morally" OK to behead drug traffickers.[67]

According to former Senator Paula Hawkins (R., FL), "Drug traffickers are mass murderers."[68]

The United States agreed to support Colombia and Peru in a policy to shoot down planes that were even suspected of carrying illegal drugs. In April 2001, a Peruvian jet fighter mistakenly shot down a private plane carrying American missionaries.[69]

Hardly a critical eyebrow was raised when the distinguished liberal Justice Thurgood Marshall told *Life* magazine, "If it's a dope case, I won't even read the petition. I ain't giving no break to no drug dealer."[70]

In a 1989 White House document, called the *National Drug Control Strategy*, drug czar William Bennett argued that casual users were more dangerous than hard-core users:

> The non-addicted casual or regular user ... is likely to have a still-intact family, social, and work life. He is likely still to "enjoy" his drug for the pleasure it offers. And he is thus much more willing and able to proselytize his drug use—by action or example—among his remaining non-user peers, friends, and acquaintances. A non-addict's drug use, in other words, is *highly* contagious.[71]

Under Bennett's reign as drug czar there was a dramatic increase in drug-related arrests and an accelerated expansion in the prison system.[72]

The loathing expressed for drug law violators seems out of all proportion to their crimes. The drug war appears to give people a license to say things about illicit drug users that they would never be allowed to say about the groups whom they associate with these drugs: African Americans, Latinos, and rebellious young people.

PUNISHMENTS FOR DRUG LAW VIOLATORS

Given these powerful emotions about drug law violators, it's not surprising that the punishments imposed on them often rival those for murder and rape. In our criminal justice system, the federal courts are designed to deal with the larger and more serious cases, such as interstate drug trafficking, while the state courts typically handle the less complicated street crimes.

Many drug offenses involve mandatory minimum sentences. For example, under federal law, a dealer, convicted for the first time of selling 5 grams of crack (a weight about equal to that of a quarter), will serve a minimum of five years in prison. A first-time peddler of powder cocaine would have to sell 500 grams to receive the same sentence.[73] This 100 to 1 punishment disparity between two forms of the same substance is widely seen as racist, since most crack dealers are African American, while peddlers of powder cocaine are usually white. The average federal sentence for a first-time, low-level crack dealer is 128 months, longer than the average term for rape (79 months) or for weapons offenses (91 months). The average sentence for murder is 153 months.[74]

Under federal law, as well as the laws of various states, drug offenders who have served their prison sentences can be denied public housing, college loans, food stamps, and welfare benefits, often for life.[75] Armed robbers, rapists, and even murderers, once out of prison, are still eligible for these benefits.[76]

State drug laws vary enormously. In many cases they are even harsher than the federal law. In Oklahoma, for a first-time offender, a conviction for the possession of any amount of cocaine less than 1 ounce leads to a minimum of two years in prison, while the possession of 1 ounce or more requires incarceration for ten years to life.[77] In Louisiana, being convicted of possessing 1 ounce of cocaine mandates imprisonment for a minimum of ten years and a maximum of sixty.[78]

While eleven states have decriminalized marijuana, most states, as well as the federal government, still have severe laws against the drug.[79] In Louisiana, the sale of 1 ounce of marijuana can lead to incarceration for twenty years, in Washington state, selling almost any amount results in a recommended prison sentence of five years.

Under New York's Rockefeller drug laws, enacted in 1973, the possession of 4 ounces or the sale of 2 ounces of a narcotic substance carries a mandatory penalty of fifteen years to life. This is as harsh as the sentence for murderers and kidnappers, and harsher than the penalty for rapists and arsonists. Here are a couple of examples.

- Amy Fisher was to serve four years and ten months for shooting a woman in the head, and Robert Chambers was serving five years for a Central Park strangling, while Lawrence V. Cipolione, Jr., was sentenced to fifteen years to life for selling 2⅓ ounces of cocaine to an undercover offcer.[80]
- Andre Neverson shot his girlfriend's uncle five times. The uncle lived and Neverson served five years in prison. Kenia Tatis was convicted of possessing 20 ounces of cocaine and is serving a mandatory sentence of fifteen years to life in a state prison. She had never been in trouble with the law before and no drugs were found on her. She was convicted by the testimony of a woman who in return received a lighter sentence for herself.[81]

In spite of widespread opposition to the Rockefeller drug laws, the politicians have so far not been able to agree on an overhaul. However, the law has been increasingly circumvented over the years to reduce some of its harshness.[82] Prosecutors have started to divert addicts into treatment programs instead of prison. Also, the governor and the legislature have agreed to allow convicted felons with no history of violence to earn reductions in their sentences with good behavior. Still, there are currently about 16,500 drug offenders incarcerated in the state of New York.

The drug war has led to an overcrowding of our prisons. As a result of the mandatory minimums, it has become routine for wardens to grant early releases to violent criminals to make room for nonviolent drug violators who must serve out their full sentences.[83]

THE GOAL OF A DRUG-FREE AMERICA

The stated goal of the war against drugs is to eliminate the consumption of these illicit substances—it's called zero tolerance. The phrase "drug-free America" pops up frequently. It was the title given to a White House conference in 1986, and showed up in legislation of that year which declared that it was the "policy of the United States Government to create a Drug-Free America by 1995."[84]

When 1995 rolled around, President Clinton's drug czar Lee Brown concluded that not only had America not been freed of drugs but that "drugs are readily available to anyone who wants to buy them. Cocaine and heroin street prices are low and purity is high—making use more feasible and affordable than ever."[85]

Facts notwithstanding, many of our political leaders still assert that zero tolerance is achievable if we just try harder. In 1998, House Speaker Dennis Hastert coauthored a plan to "help create a drug-free America by the year 2002."[86] The year 2002 was a reprise of 1995.

To state the obvious—that zero tolerance is an impossible dream—would be politically hazardous for any politician. Even examining the pros and cons of the drug war is dangerous territory. In December 1993, Bill Clinton's Surgeon General Jocelyn Elders said, "I do feel that we would markedly reduce our crime rate if drugs were legalized. But I don't know all the ramifications of this. I do feel that we need to do some studies."[87] She observed that in countries that had legalized drugs "there had been a reduction in the crime rate and there has been no increase in their drug use rate."[88] Her remarks caused a storm of protest. Eighty-seven Republican members of the House sent a letter to the president demanding that she be fired. Former drug czar William Bennett labeled her "nutty, just plain nutty." Elders continued to argue that the drug war should at least be studied. The White House sided with her critics! Press secretary Dee Dee Myers said, "The president is against legalizing drugs, and he's not interested in studying the issue."[89] President Clinton fired Elders in December 1994.

THE DRUG WAR IS A MISSION IMPOSSIBLE

It's no secret that we're losing the drug war. It's been raging for over eighty years with no end in sight. No political leader has explained how we would know that an armistice had arrived. Over the past twenty-five years, our government has dramatically accelerated spending on the war. Enormous quantities of contraband have been confiscated and our prison population has exploded, but the street prices of cocaine and heroin have declined,[90] and the consumption of illicit drugs by our teenagers is higher in 2005 than it was in the early 1990s.[91]

Listen to Joseph D. McNamara, the former police chief of both San Jose, California and Kansas City, Missouri: "It was my own experience as a policeman trying to enforce the laws against drugs that led me to change my attitude about drug-control policy ... I was a willing foot soldier at the start of the modern drug war, pounding a beat in Harlem ... We made many arrests but it did not take long before cops realized that arrests did not lessen drug selling or drug use."[92]

Scholars overwhelmingly agree with McNamara that Washington's war on drugs is a mission impossible. The reason? The laws of supply and demand, which even the most powerful government on the planet cannot repeal. There are two mechanisms at the heart of these laws. One is the *lure of profits*. Suppliers respond to profits like bees to honey, and the underground market for illegal drugs generates enormous profits. According to the United Nations, in 2001 a kilogram of heroin in Pakistan cost about $300 and sold on the streets of the United States for about $290,000, while a kilogram of coca base in Colombia cost about $400–600 and sold in the United States for about $110,000.[93] How could law enforcement stand up to a market with such a tax-free profit margin?[94]

The other mechanism is called the "balloon" (or "hydra," or "push-down-pop-up") effect; namely, whenever law enforcement cracks down on the production or distribution of illicit drugs in one area, these substances pop up in another.[95] For example, in 1989 the voters of Jackson County, Missouri (containing most of Kansas City), imposed a ¼ percent sales tax on themselves to be given to law enforcement to fight the war on drugs.[96] Fifteen months after the tax went into effect, the head of the Street Narcotics Unit of the Kansas City Police Department reported that the police were "holding even" and that, "At best, we may move [dealers] a few blocks."[97] Albert Riederer, the Jackson County Prosecuting Attorney, who had proposed and campaigned for the sales tax, said a few years later, "I got caught up in it [the idea that the police can stamp out drugs] and I probably believed it. It isn't true."[98]

... The billions we spend attempting to eradicate marijuana plants, coca plants, and opium poppies is money poured down a rat hole. The acreage required to satisfy demand is an infinitesimal fraction of that suitable for growing these crops.[99] When eradication reduces production in one area, new suppliers inevitably pop up elsewhere. In addition, our eradication campaign is inflicting economic, political, and health damage on the citizens of the countries that grow these plants.

We are also spending billions trying to prevent illicit substances from entering the country. Making even a dent in the inflow of these drugs is impossible.[100] Around 8 million shipping containers enter U.S. ports every year and our customs check only 2 percent. The United States has 5,500 miles of coastline and 7,000 miles of shared borders with Canada and Mexico.[101] There's no way to search the bowels of the millions of cars, planes, trains, buses, boats, and passengers that enter the country everyday.[102] Even President Ronald Reagan admitted the futility of stopping drugs from coming into the country, and likened interdiction to "carrying water in a sieve." But he concluded that it was important to keep trying.[103]

Using law enforcement to eliminate the domestic distribution of illicit drugs is also a hopeless undertaking. Since the transactions are between willing buyers and sellers, there's no injured party to bring charges. To convict someone, law enforcement agents have to go undercover to set up an illegal buy. Consequently, the likelihood of being convicted for drug peddling is very small. There is a much higher probability that the undercover police will be corrupted or injured or killed in performing this dangerous, futile, and expensive task. When one drug pusher is taken off the streets, another is waiting in line to take his place. Selling illicit drugs promises quick profits, and it costs little for a new peddler to set up shop.[104]

DAMAGE CAUSED BY THE DRUG WAR

Study the drug war from any perspective—economic, public health, civil liberties, violence, corruption—and you get a different dimension of the enormous damage caused by this crusade.

- **Violence** Since drug lords operate outside the law, they cannot appeal to the legal system to resolve disputes.[105] They're forced to take the law into their own hands. According to the FBI Uniform Crime Reports, the most violent episodes in the twentieth century coincided with the prohibition of alcohol and the escalation of the modern-day war on drugs.[106]
- **Drug contamination** When the government outlaws a drug, it surrenders the authority to regulate its quality. Consequently, black market drugs are frequently contaminated, sometimes with toxic substances. Most overdoses are a result of contaminants and ignorance about potency. Both would occur far less in a legal and regulated market.[107]
- **Property crimes** Drug prices are far higher in a black market than they would be in a legal market. In order to pay these artificially elevated prices, drug addicts commit property crimes to finance their habit. A study of prison inmates by the Bureau of Justice Statistics revealed that one in three robbers and burglars had committed their crimes to obtain money for drugs.[108]
- **Corruption** There is voluminous evidence that profits from the black market have corrupted many law enforcement officials. Corruption is inevitable where large amounts of cash are available to bribe low-paid police to look the other way in an undercover operation.[109] Judge James Gray observed that "Almost everyone in the legal profession knows someone who has succumbed to the temptation of large amounts of 'easy' drug money."[110] According to the FBI, over half of the police convicted of corruption between 1993 and 1997 were involved in drug-related offenses.[111]
- **The Fourth Amendment** The United States Constitution guarantees against "unreasonable searches and seizures." This protection has been gutted by the drug war. For example, police can get search warrants on the word of an anonymous informant and often don't even bother with a warrant. International travelers can be strip searched and held incommunicado without reasonable suspicion or probable cause.[112]
- **Asset-forfeiture** The drug laws have allowed, and in fact encouraged, state and federal law enforcement to seize and keep real and financial property, often without even bringing charges. In 2000, the federal law was finally changed so that asset forfeiture could occur only if evidence existed that the person was involved in a drug crime. But property can still be permanently confiscated under federal law without a conviction.[113]
- **Drug McCarthyism** Project DARE is an anti-drug program taught by police officers in the school systems of 5,000 communities. Every study evaluating this program has

concluded that it's ineffective, yet it continues.[114] In a number of cases, the program has resulted in students snitching on their parents, with serious legal consequences.[115] Drug czar William Bennett told middle-school students that exposing drug-using parents "isn't snitching … It's an act of true loyalty—of friendship."[116] Sergeant Robert Gates, DARE's national coordinator, defended snitching on parents, saying that "an arrest is the best thing that could ever happen to that parent … What may turn out to be negative for the parent is positive for society."[117]

- **The AIDS epidemic** Drug prohibition has made heroin more expensive, inducing addicts to get a bigger bang for the buck by injecting. Our government has refused to support the provision of clean needles, even though injection with dirty needles has been estimated by the Centers for Disease Control to be responsible for more than 250,000 HIV infections and more than half of the pediatric AIDS cases in the United States.[118] Numerous scientific studies, many funded by the government, have shown that providing clean needles would both save lives and, by bringing drug injectors into contact with the health care system, reduce drug addiction.[119] Needle exchange is strongly recommended by many organizations, including the National Academy of Science, the National Institutes of Health, and the American Medical Association.
- **Incarceration** We punish illicit drug users with prison, even though numerous studies have found that "the perceived certainty and severity of punishment are insignificant factors in deterring use."[120] What is worse, the prison environment may actually encourage drug use. Drugs are readily available in most prisons. A federal survey found that, of those inmates who had regularly used hard drugs, three-fifths had not done so until after they were first incarcerated.[121]

ENDNOTES

1. See, for example, James A. Inciardi, *The War on Drugs III*, pp. 314–15; or Erich Goode, *Drugs in American Society*, p. 104.
2. Inciardi, *The War on Drugs III*, p. 314.
3. www.oas.samsha.gov/NSDUH/2k5NSDUH/2k5results.htm#Ch2, p. 229.
4. Jerrold S. Meyer and Linda F. Quenzer, *Psychopharmacology*, p. 333; *The Merck Manual of Medical Information*, p. 491; and Inciardi, *The War on Drugs III*, p. 45.
5. Mitch Farleywine, *Understanding Marijuana*, pp. 114, 204–205.
6. Nancy M. Petry and Warren K. Bickel, "Polydrug abuse in heron addicts."
7. Robert J. MacCoun and Peter Reuter, *Drug War Heresies*, p. 361.
8. John Kaplan, *The Hardest Drug*, p. 23; *The Merck Manual of Medical Information*, p. 487; Meyer and Quenzer, *Psychopharmacology*, pp. 199, 248; and Cynthia Kuhn, Scott Swartzwelder, and Wilkie Wilson, *Buzzed*, pp. 182–3.
9. *The Merck Manual of Medical Information*, p. 487; and Kuhn et al., *Buzzed*, p. 190.
10. *The Merck Manual of Medical Information*, p. 488.
11. Ibid., p. 491.
12. National Institute on Drug Abesu, *InfoFacts: Heroin*.
13. John Morgan and Lynn Zimmer, "The social pharmacology of smokeable cocaine," pp. 137–8; and *The Merck Manual of Medical Information*, p. 492.
14. *The Merck Manual*, ibid.; and Meyer and Quenzer, *Psychopharmacology*, pp. 280–281.
15. Inciardi, *The War on Drugs III*, p. 142. See also Meyer and Quenzer, *Psychopharmacology*, pp. 286–9; and Kuhn et al., *Buzzed*, pp. 225–30.
16. Kuhn et al., ibid., p. 144; and *The Merck Manual of Medical Information*.
17. *The Merck Manual*, ibid., p. 491; and Inciardi, *The War on Drugs III*, pp. 51–2, 62–63.

18. Jerome Beck and Marsha Rosenbaum, *Pursuit of Ecstasy;* Meyer and Quenzer, *Psychopharmacology*, p. 296; and Kuhn et al., *Buzzed*, pp. 76–7.
19. Drug Abuse Warning Network, "Club Drugs," p. 4.
20. Kuhn et al., *Buzzed*, p. 79.
21. Meyer and Quenzer, *Psychopharmacology*, p. 299.
22. See Goode, *Drugs in American Society*, p. 265.
23. *The Merck Manual of Medical Information*, pp. 491–2.
24. Holly Hickman, "Meth use 'choking' western hills of North Carolina."
25. *The Merck Manual of Medical Information*, p. 491.
26. Inciardi, *The War on Drugs III*, p. 51.
27. Naval Strike and Air Warfare Center, "Performance maintenance during continuous flight operations: A guide for flight surgeons." NAVMED P-6410. January 1, 2000, p. 8, available online through the Virtual Naval Hospital of the University of lowa, at www.vnh.org/PertormMaint.
28. "Amphetamine dependence," *The Merck Manual of Diagnosis and Therapy*, section 15: Psychiatric Disorders, chapter 195: "Drug use and dependence."
29. Ibid., and National Institute on Drug Abuse, *InfoFacts: Methamphetamine.*
30. Meyer and Quenzer, *Psychopharmacology*, p. 191.
31. Kuhn et al., *Buzzed*, p. 226.
32. Hickman, "Meth use 'choking' western hills of North Carolina"; Robert E. Pierre. "Keeping pills away from drug abusers"; "Johnston sheriff fears new drug plague"; Martha Quillin, "Rural country is meth central"; "You take the high road."
33. See also www.cdc.gov/noish/npg/pgdstart.html.
34. Inciardi, *The War on Drugs III*, p. 49; *The Merck Manual of Medical Information*, pp. 493–4; and Meyer and Quenzer, *Psychopharmacology*, pp. 352–7.
35. Meyer and Quenzer, ibid., pp. 356–7.
36. MacCoun and Reuter, *Drug War Heresies*, p. 23.
37. Ali H. Mokdad, James S. Marks, and Donna F. Stroup, "Actual causes of death in the United States, 2000," cited at www.drugwarfacts.org/causes.htm.
38. United States Surgeon General, *Reducing the Health Consequences of Smoking*, p. 160, cited in Steven B. Duke and Albert C. Gross, *America's Longest War*, p. 77.
39. *Journal of the American Medical Association*, 293, no. 3, January 19, 2005, p. 298, cited at www.drugwarfacts.org/causes.htm.
40. United States Surgeon General, *Reducing the Health Consequences of Smoking*, p. 160, cited in Duke and Gross, *America's Longest War*, p. 77.
41. Ibid.
42. Earleywine, *Understanding Marijuana*, pp. 143–4.
43. Report by the European Monitoring Center for Drugs and Drug Addiction, quoted in "A survey of illegal drugs," p. 10.
44. www.DAWNinfo.samhsa.gov.
45. Earleywine, *Understanding Marijuana*, p. 32; SAMHSA National Survey on Drug Abuse, 1999, reported in "A survey of illegal drugs," p. 8; Philip J. Hilts, "Is nicotine addictive?"
46. SAMHSA National Survey on Drug Abuse, 1999, reported in "A survey of illegal drugs," p. 8.
47. Ibid.; Se also Duke and Gross, *America's Longest War*, pp. 61–2; Goode, *Drugs in American Society*, pp. 316–17; and Tom Carnwath and lan Smith, *Heroin Century*, p. 81.
48. SAMHSA National Survey on Drug Abuse, 1999, reported in "A survey of illegal drugs," p. 8; and J. Anthony, L. Warner, and R. Kessler, "Comparative epidemiology of dependence on tobacco, alcohol, controlled substances and inhalants," cited in MacCoun and Reuter, *Drug War Heresies*, p. 18.
49. Anthony et al., ibid., p. 19.
50. Communication from Professor Mitch Earlywine, March 31, 2004. For similar rankings, see Deborah Franklin, "Hooked—not hooked."
51. See P. A. Ebener et al., *Improving Data and Analysis to Support National Substance Abuse Policy;* and national survey results on drug use from the Monitoring the Future Study, available at www.isr.umich.edu/src/mtf, cited in MacCoun and Reuter, *Drug War Heresies*, p. 16.
52. Duke and Gross, *America's Longest War*, pp. 38–42; Goode, *Drugs in American society*, pp. 340–43 and Paul J. Goldstein, Henry H. Brownstein, Patrick J. Ryan, and Patricia A. Bellucci, "Crack and homicide in New York city."

53. See Paul J. Goldstein, Henry H. Brownstein, and Patrick J. Ryan, "Drug-related homicide in New York," cited in Jeffrey A. Miron and Jeffrey Zwiebel, "The economic case against drug prohibition," p. 176.
54. David Musto, *The American Disease*, p. 254.
55. See, for example, the excellent study by Diana R. Gordon, *The Return of the Dangerous Classes*.
56. David T. Courtwright, *Dark Paradise*, pp. 95–96.
57. Inciardi, *The War on Drugs III*, p. 48.
58. Eva Bertram et al., *Drug War Politics*, p. 63.
59. Ibid., p. 98.
60. H. Wayne Morgan, *Drugs in America*, p. 37.
61. Mark Thornton, "Prohibition vs. legalization."
62. Reported in the *New York Times*, February 29, 1988.
63. Bertram et al., *Drug War Politics*, pp. 115–116.
64. www.lib.ncsu.edu/congbibs/senate/101dgst2.htm; and http://monitoringthefuture.org, table 2.
65. Associated Press, "Gingrich wants drug dealers executed."
66. *Fresno Bee* [CA], November 30, 1999.
67. See, for example, the *Idaho Observer*, August, 2000, available at www.proliberty.com/observer/20000807.htm.
68. Paul Anderson, "Drug bill tough on smugglers," quoted in Steven Wisotsky, *Beyond the War on Drugs*, p. 200.
69. Larry Rohter, "Brazil carries the war on drugs to the air."
70. *Life* magazine, September, 1987, p. 105.
71. See Michael Massing, *The Fix*, p. 201.
72. Bertram et al., *Drug War Politics*, pp. 115–116.
73. United States Sentencing Commission, *Cocaine and Federal Sentencing Policy*, pp. 2–3. See also Eric E. Sterling, "Take another crack at that cocaine law"; Sterling was counsel to the House Judiciary Committee, principally responsible for anti-drug legislation.
74. United States Sentencing Commission, *Cocaine and Federal Sentencing Policy*, p. 150; *Sourcebook of Criminal Justice Statistics* 1996 (Washington, D.C.: Bureau of Justice Statistics, 1997), p. 476, table 5.58.
75. Ibid.; also Drug Policy Alliance, *State of the States*, p. 7.
76. See Greg Winter, "A student aid *ban for past drug use is creating a furor.*"
77. Andrews University, the MayaTech Corporation, and RAND Corporation, *Illicit Drug Policies*.
78. Ibid.
79. Eric Schlosser, "Make peace with pot."
80. James P. Gray, *Why Our Drug Laws Have Failed*. p. 32.
81. Bob Herbert, "The ruinous drug laws," p A23.
82. Schlosser, "Make peace with pot."
83. Gray, *Why Our Drug Laws Have Failed*, p. 36–7; and Bertram et al., *Drug War Politics*, p. 52.
84. Joshua W. Shenk, "America's altered states," p. 239; and Bertram et al., *Drug War Politics*, p. 139.
85. Testimony of Lee Brown before Judiciary Committee, U.S. Senate, February 10, 1994, quoted in Bertram et al., *Drug War Politics*, p. 12.
86. Shenk, "America's altered states," p. 255.
87. Christopher Connell, "Legalizing drugs would reduce crime rate," quoted in Dan Baum, *Smoke and Mirrors* p. 334.
88. Ibid,; and Bertram et al., *Drug War Politics*, pp. 160–161.
89. Bertram et al., ibid.
90. Drug Enforcement Agency, *Illegal Price/Purity Report*, March, 1991, pp. 3, 5, 7; April, 1994, pp. 1, 4, 6; and June, 1995, pp. 1, 4, 6; cited in Bertram et al., *Drug War Politics*, Appendix 2.
91. See www.monitoringthefuture.org. table 2.
92. Statement by Joseph D. McNamara, in William Buckley, Jr., et al., "The war on drugs is lost," repr. in Gray, *Busted*, p. 207.
93. "A survey of illegal drugs," p. 6.
94. Statement by Joseph D. McNamara, in Buckley et al., "The war on drugs is lost," repr. in Gray, *Busted*, p. 204.
95. Duke and Gross, *America's Longest War*, p. 222; and Bertram et al., *Drug War Politics*, pp. 18–19.
96. Gordon, *Return of the Dangerous Classes*, ch. 6.
97. Ibid., pp. 82–83.
98. Ibid., p. 84.
99. Bertram et al., *Drug War Politics*, pp. 14–20.
100. See the study on interdiction by the Government Accounting (now Accountability) Office (GAO), available at www.fas.org/irp/gao/nsi95032.htm.

101. Duke and Gross, *America's Longest War*, pp. 203–7.
102. Ethan Nadelmann, "Drug prohibition in the U.S."
103. Bertram et al., *Drug War Politics*, p. 22.
104. See, for example, Massing, *The Fix*, p. 67.
105. See Miron and Zwiebel, "The economic case against drug prohibition," pp. 177–8; and Jeffrey A. Miron, "Violence and U.S. prohibitions of drugs and alcohol."
106. See United States Census Data and FBI Uniform Crime Reports, *Murder in America*.
107. MaCoun and Reuter, *Drug War Heresies*, p. 125.
108. Bureau of Justice Statistics, *Special Report*, cited in Duke and Gross, *America's Longest War*, p. 110 and Inciardi, *The War on Drugs III*, p. 193.
109. Duke and Gross, *America's Longest War*, pp. 113–15; and Gray, *Why our Drug Laws Have Failed*, pp. 95–100.
110. Gray, ibid., p. 73.
111. United States General Accounting Office, *Report to the Honorable Charles B. Rangel, House of Representatives, Law Enforcement*, p. 35, cited at www.drugwarfacts.org/corrupt.corrupt.htm.
112. Duke and Gross, *America's Longest War*, p. 124.
113. Ibid., pp. 135–43; and Gray, *Why Our Drug Laws Have Failed*, pp. 117–22.
114. See, for example, the long-run study by researchers at the University of Kentucky, summarized at www.druglibrary.org/think/-inr/noeffct.htm.
115. Joseph Pereira, "The informants in a drug program," quoted in Duke and Gross, p. 174.
116. Times Wire Service, "Drug czar urges pupils to turn in parents, says it's not 'snitching,'" *Los Angeles Times*, May 18, 1989, quoted in Duke and Gross, *America's Longest War*, p. 174
117. Pereira, "The informants in a drug program," quoted ibid., p. 174.
118. Drug Policy Alliance, *State of the States*, p. 8; also Ethan Nandelmann, "Commonsense drug policy."
119. United States Surgeon General, *Evidence-Based Findings on the Efficacy of Syringe Exchange Programs: An Analysis from the Assistant Secretary for Health and the Surgeon General of the Scientific Research Completed since April 1998*, accessed May 11, 2005, at www. harm reducation.org/research/surgeongenrev/surgreview.html, cited at drugwarfacts.org/syringe.htm.
120. Patricia G. Erickson, "A public health approach to demand reduction," p. 565, cited in Bertram et al., *Drug War Politics*, p. 29; and Mathew Purdy, "Warehouse of addiction."
121. Purdy, ibid.; Elliot Currie, *Reckoning*, p. 170; and Daniel K. Benjamin and Roger Leroy Miller, *Undoing Drugs*, p. 105.

REFERENCES, BOOKS AND REPORTS

Andrews University, The MayaTech Corporation, and RAND Corporation. *Illicit Drug Policies: Selected Laws from the 50 States*, 2002, rev. 2003; available at www.andrews.edu/ipa/2004/charrhook/pdf.

Baum, Dan, *Smoke and Mirrors*, Boston: Little, Brown, 1996.

Bayer, Ronald, and Gerald M. Oppenheimer, eds., *Confronting Drug Policy*. New York: Cambridge University Press, 1993.

Beck, Jerome, and Marsha Rosenbaum, *Pursuit of Ecstasy: The MDMA Experience*. Albany: State University of New York Press, 1994.

Benjamin, Daniel K., and Roger Leroy Miller, *Undoing Drugs*. New York: Basic Books, 1991.

Bennett, William J., John J. DiIulio, Jr., and John P. Walters, *Body Count: Moral Poverty—and How to Win America's War against Crime and Drugs*. New York: Simon & Schuster, 1996.

Bertram, Eva, Morris Blachman, Kenneth Sharpe, and Peter Andreas, *Drug War Politics*. Berkeley: University of California Press, 1996.

Bonnie, Richard J., and Charles H. Whitebread II, *The Marijuana Conviction*. New York: Lindesmith Center, 1999.

Brands, H. W., *The First American*. New York: Doubleday, 2000.

Bureau of Justice Statistics, *Special Report: Drugs and Jail Inmates*, 1989. Washington, D.C.: U.S. Department of Justice, August 1991.

Commission on the Advancement of Federal Law Enforcement, *Law Enforcement in a New Century and a*

Changing World. Washington, D.C.: Commission on the Advancement of Federal Law Enforcement, 2000.

Courtwright, David T., *Dark Paradise: A History of Opiate Addiction in America*. Cambridge, MA: Harvard University Press, 2001.

CSDP (Common Sense for Drug Policy), *Revising the Federal Drug Control Budget Report: Changing Methodology to Hide the Cost of the Drug War*, 2003: available at www.csdp.org/research/ondcpenron.pdf.

Currie, Elliott, *Reckoning*. New York: Hill & Wang, 1993.

Davenport-Hines, Richard, *The Pursuit of Oblivion*. New York: W. W. Norton, 2001.

Drug Enforcement Agency, *Illegal Price/Purity Report*, Washington, D.C.: U.S. Department of Justice, March 1991.

——, *Illegal Price/Purity Report*. Washington, D.C.: U.S. Department of Justice, April, 1994.

——, *Illegal Price/Purity Report*. Washington, D.C.: U.S. Department of Justice, June, 1995.

——, *In the Matter of Marijuana Rescheduling Petition*, Washington, D.C.: U.S. Department of justice (docket #86-22), September 6, 1988.

Drug Policy Alliance, *State of the States: Drug Policy Reforms: 1996–2002*. September, 2003; available at www.drugpolicy.org/docUpload/sos_report2003.pdf.

Drummer, O. H., *Drugs in Drivers Killed in Australian Road Traffic Accidents*. (Report no. 0594), Melbourne, Australia: Monash University, Victorian Institute of Forensic Pathology, 1994.

Duke, Steven B., and Albert C. Gross, *America's Longest War*. New York: G.P. Putnam's Sons, 1993.

Durose, Matthew R., and Patrick A. Langan, *State Court Sentencing of Convicted Felons, 1998 Statistical Tables*, Washington, D.C.: U.S. Department of Justice, Bureau of Justice Statistics, December, 2001; available at www.ojp.usdoj.gov/bjs/abstract/scsc98st.htm.

Earleywine, Mitch, *Understanding Marijuana*. New York: Oxford University Press, 2002.

Ebener, P. A., J. P. Caulkins, S. A. Geshwind, D. McCaffrey and H.I. Saner, *Improving Data and Analysis to Support National Substance Abuse Policy: Main Report*. Santa Monica, CA: RAND Corporation, 1994.

Ehlers, Scott, *Policy Briefing: Asset Forfeiture*. Washington, D.C.: Drug Policy Foundation, 1999.

Eldredge, Dirk Chase, *Ending the War on Drugs*. Bridgehampton, NY: Bridge Works Publishing Company, 1998.

Everingham, Susan S., and C. Peter Rydell, *Modeling the Demand for Cocaine*. Santa Monica, CA: RAND Corporation, 1994.

Federal Bureau of Investigation, *Crime in the United States: FBI Uniform Crime Reports 2005*. Washington, D.C.: U.S. Government Printing Office, 2005.

Fish, Jefferson M., ed., *How to Legalize Drugs*. Northvale, NJ: Jason Aronson, 1998.

Friedman, Lawrence M., *American Law in the 20th Century*. New Haven, CT and London: Yale University Press, 2002.

Friedman, Milton, and Thomas S. Szasz, *On Liberty and Drugs: Essays on Prohibition and the Free Market*. Washington, D.C.: Drug Policy Foundation, 1992.

Goode, Erich, *Drugs in American Society*, 6th ed. New York: McGraw Hill, 2005.

Goodman, Louis, and Alfred Gilman, eds., *The Pharmacological Basis of Therapeutics*, 4th ed. New York: Macmillan, 1970.

Gordon, Diana R., *The Return of the Dangerous Classes*. New York: W. W. Norton, 1994.

Gray, James P., *Why Our Drug Laws Have Failed and What We Can Do about It*. Philadelphia: Temple University Press, 2001.

Gray, Mike, ed., *Busted: Stone Cowboys, Narco-Lords, and Washington's War on Drug*. New York: Thunder's Mouth Press/Nation Books, 2002.

——, *Drug Crazy: How We Got into This Mess and How We Can Get Out*. New York: Random House, 1998.

Grinspoon, Lester, and James B. Bakalar, *Cocaine: A Drug and its Social Evolution*, rev. ed. New York: Basic Books, 1985.

——, *Marihuana, The Forbidden Medicine*. New Haven and London: Yale University Press, 1997.

Hall, Kermit L., ed., *The Oxford Companion to the Supreme Court of the United States*. New York and Oxford: Oxford University Press, 1992.

Hall, W., R. Room, and S. Bondy, *WHO Project on Health Implications of Cannabis Use: A Comparative Appraisal of the Health and Psychological Consequences of Alcohol, Cannabis, Nicotine and Opiate Use.* Geneva, Switzerland: World Health Organization, March, 1998.

Hanson, David J., *Alcohol Beverage Consumption in the U.S.: Patterns and Trends,* State University of New York, Sociology Department, 2003: available at www2.potsdam.edu/hansondj/Controversies/1116895242.html.

Harrison, Paige M., and Allen J. Beck, *Prisoners in 2004.* Washington, D.C.: U.S. Department of Justice, Bureau of Justice Statistics, October 2005.

Husak, Douglas, *Legalize This! The Case for Decriminalizing Drugs.* New York: Verso, 2002.

Inciardi, James A., *The War on Drugs III: The Continuing Saga of the Mysteries of Intoxication, Addiction, Crime, and Public Policy.* Boston: Allyn & Bacon, 2002.

Institute of Medicine, *Dispelling the Myths about Addiction.* Washington, D.C.: National Academy Press, 1997.

Irwin, John, Vincent Schiraldi, and Jason Ziedenberg, *America's One Million Nonviolent Prisoners.* Washington, D.C.: Justice Policy Institute, March, 1999.

Joy, Janet E., Stanley J. Watson, Jr., and John A. Benson, Jr., *Marijuana and Medicine: Assessing the Science Base.* Washington, D.C.: National Academy Press, 1999.

Kaplan, John, *The Hardest Drug: Heroin and Public Policy.* Chicago: University of Chicago Press, 1983.

Karburg Jennifer C., and Doris J. James, *Substance Dependence, Abuse, and Treatment of Jail Inmates,* 2002, Washington, D.C.: U.S. Department of Justice, Bureau of Justice Statistics, July 2005.

Krauss, Melvin, and Edward P. Lazear, eds., *Searching for Alternatives: Drug-Control Policy in the United States.* Stanford, CA: Hoover Institution Press. 1991.

Kuhn, Cynthia, Scott Swartzwelder, and Wilkie Wilson, *Buzzed,* rev. ed. New York: W. W Norton. 2003.

Lyman, Michael D, and Gary W. Porter, eds., *Drugs in Society: Causes, Concepts, and Control.* Cincinnati, OH: Anderson, 1991.

Lynch, Timothy, ed., *After Prohibition: An Adult Approach to Drug Policies in the 21st Century.* Washington, D.C.: CATO Institute, 2000.

MacCoun, Robert J., and Peter Reuter, *Drug War Heresies.* Cambridge: Cambridge University Press, 2001.

Maruschak, Laura, *HIV in Prisons,* 2001. Washington, D.C.: U.S. Department of Justice, Bureau of Justice Statistics, January 2004.

Massing, Michael, *The Fix.* Berkeley: University of California Press, 1998.

Meierhoefer, B. S., *The General Effect of Mandatory Minimum Prison Terms: A Longitudinal Study of Federal Sentences Imposed.* Washington, D.C.: Federal Judicial Center, 1992.

The Merck Manual of Medical Information, 16th ed., 1992; available at www.merck.com.

The Merck Manual of Diagnosis and Therapy; available at www.merck.com.

Merrill, J. C., and K. S. Fox, *Cigarettes, Alcohol, Marijuana: Gateway to Illicit Drug Use,* New York: National Center on Addiction and Substance Abuse, Columbia University, 1994.

Meyer, Jerrold S., and Linda F. Quenzer, *Psychopharmacology: Drugs, the Brain, and Behavior.* Sunderland, MA: Sinauer Associates, 2005.

Miron, Jeffrey A., *Drug War Crimes.* Oakland, CA: Independent Institute, 2004.

Morgan. H. Wayne, *Drugs in America.* Syracuse, NY: Syracuse University Press, 1981.

Mumola, Christopher J., *Substance Abuse and Treatment, State and Federal Prisons, January* 1997, Washington. D.C.: U.S. Department of Justice.

Musto, David F., *The American Disease.* New York: Oxford University Press. 1999.

National Center on Addiction and Substance Abuse, *Behind Bars: Substance Abuse and America's Prison Population,* New York: National Center on Addiction and Substance Abuse, Columbia University, 1998.

National Commission on Acquired Immune Deficiency Syndrome, *The Twin Epidemics of Substance Use and HIV.* Washington, D.C.: National Commission on Acquired Immune Deficiency Syndrome, 1991.

National Institute on Drug Abuse, *InfoFacts: Heroin,* Rockville, MD: U.S. Department of Health and Human Services: available at www.nida.nih.gov/infofacts/heroin.html.

——, *InfoFacts: Methamphetamine.* Rockville, MD: U.S. Department of Health and Human Services; available at www.nida.nih.gov/infofacts/methamphet amine.html.

National Research Council. *Informing America's Policy on Illegal Drugs.* Washington, D.C.: National Academy Press, 2001.

Norris, Mikki, Chris Conrad, and Virginia Resner, *Shattered Lives: Portraits from America's Drug War.* El Cerrito, CA: Creative Xpressions. 2000.

O'Brien, Charles P., and Jerome H. Jaffe, *Addictive States.* Green Bay, WI: Raven Press. 1991.

Office of National Drug Control Policy, *National Drug Control Strategy*, Washington, D.C.: Office of National Drug Control Policy, 1995; 1997-1998; 2000.

——, *What Americans Need to Know about Marijuana.* Washington. D.C.: Office of National Drug Control Policy, 2003.

Reinarman, Craig, and Harry G. Levine. eds., *Crack in America.* Berkeley: University of California Press. 1997.

Reischauer, Robert, *The Andean Initiative: Objective and Support.* Washington D.C.: Congressional Budget Office, March 1994; available at www.cho.gov/showdoc.cfm?index=4885&sequence=0.

Royal College of Psychiatrists and Royal College of Physicians, *Drugs: Dilemmas and Choices.* London: Gaskell Press, 2000.

Rydell, Peter C., and Susan S. Everingham, *Controlling Cocaine: Supply versus Demand Programs.* Santa Monica, CA: RAND Corporation, 1994.

Satcher, David, *Evidence-Based Findings on the Efficacy of Syringe Exchange Programs: An Analysis from the Assistant Secretary for Health and the Surgeon General of the Scientific Research Completed since April 1998.* Washington, D.C.: U.S. Department of Health and Human Services, 2000: accessed May 11, 2005, at www.harmreduction.org/research/surgeongenrev/surgreview.html.

Schiraldi, Vincent, Barry Holman, and Phillip Beatty, *Poor Prescription: The Costs of Imprisoning Drug Offenders in the United States.* Washington, D.C.: Justice Policy Institute, July 2000; available at www.cjcj.org.

Schlosser, Eric, *Reefer Madness.* New York: Houghton Mifflin, 2003.

Shafer, Raymond P., et al., *Marihuana: A Signal of Misunderstanding.* Washington, D.C.: U.S. National Commission on Marihuana and Drug Abuse, 1972.

Siegel, Ronald K., *Intoxication.* New York: F. P. Dutton, 1989.

Skolnik, Jerome H., *Policing Drugs: The Cultural Transformation of a Victimless Crime.* Berkeley: Jurisprudence and Social Policy Program, University of California, 1986.

Spillane, Joseph F., *Cocaine.* Baltimore and London: Johns Hopkins University Press, 2000.

Tabbush, V., *The Effectiveness and Efficiency of Publicly Funded Drug Abuse Treatment and Prevention Programs in California: A Benefit Cost Analysis.* UCLA, March, 1986.

Terhune, K. W., C. A. Ippolito, and D. J. Crouch. *The Incidence and Role of Drugs in Fatally Injured Drivers.* (DOT HS Report No.808 065), Washington, D.C.: U.S. Department of Transportation, National Highway Traffic Safety Administration, 1992.

Trebach, Arnold S., *The Great Drug War.* New York: Unlimited Publishing, 2005.

——, *The Heroin Solution.* New York: Unlimited Publishing. 2006.

Trebach, Arnold S., and Kevin B. Zeese, eds., *Drug Prohibition and the Conscience of Nations.* Washington, D.C.: Drug Policy Foundation, 1990.

Uchtenhagen, Ambros, Felix Gurzwiller, and Anja Dobler-Mikola, *Programme for a Medical Prescription of Narcotics: Final Report of the Research Representatives.* Zurich: Addiction Research Institute, July 10, 1997; available at www.druglibrary.org/schaffer/heroin/programme.htm.

United Nations, *Economic and Social Consequences of Drug Abuse and Illicit Trafficking.* (United Nations International Drug Control Program, Technical Series Report #6), New York: UNDCP, 1998.

United States Census Data and FBI Uniform Crime Reports. *Murder in America, 1900–1998*; available at www.drugwarfacts.org/crime.htm.

United States Department of Health and Human Services. *Drug Abuse and Drug Abuse Research,* Rockville, MD: 1991.

——, *Result from the … National Survey on Drug Use and Health: National Findings.* Rockville, MD: 2005.

United States General Accountability Officer, *Federal Drug Offenses: Departures from Sentencing Guidelines and Mandatory Minimum Sentences. Fiscal Years 1999–2001.* Washington, D.C.: USGAO, October, 2003.

United States General Accounting Office, *Drug Control: Narcotics Threat from Colombia Continues to Grow.* Washington, D.C.: USGAO, 1999.

——, *Federal Drug Interdiction Efforts Need Strong Central Oversight*, Report by the Comptroller General, June 13, 1983. Washington, D.C.: USGAO, 1983.

——, *Law Enforcement: Information on Drug-Related Police Corruption: Report to the Honorable Charles B. Rangel, House of Representatives*. Washington, D.C.: USGAO, May 1998.

——, *Marijuana: Early Experiences with Four States, Laws that Allow Use for Medical Purposes*. Washington, D.C.: USGAO, November 2002.

United States Sentencing Commission, *Report to Congress: Cocaine and Federal Sentencing Policy*, May, 2007.

——, *Report to Congress: MDMA Drug Offenses*, May, 2001.

United States Surgeon General, *Reducing the Health Consequences of Smoking: 25 Years of Progress*. Washington, D.C.: U.S. Department of Health and Human Services, 1989.

Weil, Andrew, *The Natural Mind*. Boston: Houghton Mifflin, 1972.

Wisotsky, Steven, *Beyond the War on Drugs*. Buffalo, NY: Prometheus Books, 1990.

World Health Organization, *Cancer Pain Relief and Palliative Care: Report of a WHO Expert Committee*. Geneva: WHO, 1990.

——, *Effectiveness of Sterile Needle and Syringe Programming in Reducing HIV/AIDS among Injecting Drug Users*. Geneva: WHO, 2004; available at www.euro.who.int/document/e7777.pdf.

Ziedenberg, Jason, and Vincent Schiraldi, *The Punishing Decade: Prison and Jail Estimates at the Millennium*. Washington. D.C.: Justice Policy Institute, revised estimates, May, 2000.

Zimmer, Lynn, and John P. Morgan, *Marijuana Myths, Marijuana Facts*. New York and San Francisco: Lindesmith Center, 1997.

Articles

Abdel-Mahgoud. M., and M. K. Al-Haddad, "Patterns of heroin use among a non-treatment sample in Glasgow, Scotland," *Addiction*, 91, 1996, pp. 1859–64.

Abrams, Jim, "Interdiction hasn't stemmed drug flow, Congress is told," *Boston Globe*, February 26, 1993.

American Medical Association, "About the AMA position on pain management using opioid analgesics," 2004; available at www.ama-assn.org/ama/pub/category/11541.html.

Anderson, Paul, "Drug bill tough on smugglers," *Miami Herald*, May 26, 1984.

Andreas, Peter, "Profits, poverty, and illegality: The logic of drug corruption," *NACLA Report on the Americas*, 27, no. 3, 1993, pp. 22–8.

Anthony, J., L. Warner, and R. Kessler, "Comparative epidemiology of dependence on tobacco, alcohol, controlled substance and inhalants: Basic findings from the national comorbidity study," *Experimental and Clinical Psychopharmacology*, 2, 1994, pp. 244–68.

Buckley, William F., Jr., Schmoke, Kurt, McNamara, Joseph D., and Sweet, Robert W., "The war on drugs is lost," *National Review*, July 1, 1996; repr. in *Busted*, ed. Mike Gray, New York: Thunder's Mouth Press/Nation Books 2002, pp. 198–202.

Connell, Christopher, "Legalizing drugs would reduce crime rate: Elders." *Los Angeles Times*, December 10, 1994.

Coolidge, Sharon, "No jail for patient who grew his own marijuana," *Cincinnati Enquirer*, May 28, 2004.

"Dealing with drug use: Treatment (not jail time) saves lives," editorial, *San Francisco Chronicle*, November 15, 2004.

De Benedictis, Don J., "How long is too long." *ABA Journal*, 79, 1993, pp. 74–9.

Doblin, Richard, and Mark A. R. Kleiman, "Marijuana as antiemetic medicine: A survey of oncologists' experiences and attitudes." *Journal of Clinical Oncology*, July 9, 1991.

Drug Abuse Warning Network, "Club drugs," *The DAWN Report*, Washington, D.C.: Office of Applied Studies. SAMHSA (Substance Abuse and Mental health Services Administration), December 2000.

Duke, Steven B., "The drug war and the constitution," in *After prohibition*, ed. Timothy Lynch, Washington, D.C.: CATO Institute, 2000, pp. 47–51.

Duke, Steven B., and Albert C. Gross, "Forms of legalization," in *How to Legalize Drugs*, ed. Jefferson M. Fish, Northvale, NJ: Jason Aronson.

Dumond, Chris, "Attorneys say region has produced most meth busts," *Bristol Herald Courier*, August 22, 2003.

Duster, Troy, "Race in the drug war," in *Crack in America*, ed. Craig Reinarman and Harry G. Levine, Berkeley: University, of California Press, 1997.

Eckholm, Eric, "States are growing more lenient in allowing felons to vote," *New York Times*, October 12, 2006.

Ehrlich, Isaac, "The deterrent effect of capital punishment—A question of life and death," *American Economic Review*, 65, 1975, pp. 397–417.

Erickson, Patricia G., "A public health approach to demand reduction," *Journal of Drug Issues*, 20, no, 4. 1990.

Evans, Richard N., "What is 'legalization'? What are drugs?," in *How to Legalize Drugs*, ed. Jefferson M. Fish, Northvale, NJ: Jason Aronson.

"Exec seeks $20 million in bogus bust," *San Diego Union-Tribune*, December 2, 1992, available at paranoia.lycaeum.org/war.on.drugs/casualties/botched.raids.

Farrell, A. D., S. J. Danish, and C.W. Howard, "Relationship between drug use and other problem behavior in urban adolescents," *Journal of Consulting and Clinical Psychology*, 60, 1992, pp. 705–12.

Fish, Jefferson, "First steps toward legalization," in *How to Legalize Drugs*, ed. Jefferson M. Fish, Northvale, NJ: Jason Aronson.

Franklin, D., "Hooked—not booked: Why isn't everyone an addict?," In *Health*, 4, no, 6, 1990, pp. 39–52.

Goldstein, Paul J., Henry H. Brownstein, and Patrick J. Ryan, "Drug-related homicide in New York: 1984 and 1988," *Crime and Delinquency*, 38 October 1992, pp. 459–76.

Herbert, Bob, "The ruinous drug laws," *New York Times*, July 18, 2002.

Hickman, Holly, "Meth use 'choking' western hills of North Carolina," *News & Observer*, February 29, 2004.

Miron, Jeffrey A., and Jeffrey Zwiebel, "The economic case against drug prohibition," *Journal of Economic Perspectives*, 9, no. 4, 1995, pp. 175–92.

Mokdad, Ali H., James S. Marks, and Donna F. Stroup, "Actual causes of death in the United States, 2000," *Journal of the American Medical Association*, 291, 2004, pp. 1238, 1241–2.

Morgan, John P., and Lynn Zimmer, "The social pharmacology of smokeable cocaine: Not all it's cracked up to be," in *Crack in America*, ed. Craig Reinarman and Harry G. Levine, Berkeley: University of California Press, 1997, pp. 131–70.

Morganthau, Tom, "Why good cops go bad," *Newsweek*, December 19, 1994.

Murphy, Dean F., "Backers of medical marijuana hail ruling," *New York Times*, October 15, 2003.

Murphy, Sheigla B., and Marsha Rosenbaum, "Two women who used cocaine too much," in *Crack in America*, ed. Craig Reinarman and Harry G. Levine, Berkeley: University of California Press, 1997, pp. 98–112.

Nadelmann, Ethan, "Commonsense drug policy," *Foreign Affairs*, January–February 1998, pp. 111–26; repr. in *Busted*, ed. Mike Gray, New York: Thunder's Mouth Press/Nation Books, 2002, pp. 173–86.

——, "Drug prohibition in the U.S.," in *Crack in America*, ed. Craig Reinarman and Harry Levine, Berkeley: University of California Press, 1997, pp. 290–92.

——, "Europe's drug prescription," *Rolling Stone*, January 26, 1995, p. 38.

——, "The hospice raid and the war on drugs," *San Diego Union Tribune*, September 19, 2002.

——, "Legalization is the answer," *Issues in Science and Technology*, Summer 1990.

——, "Thinking seriously about alternatives to drug prohibition: Parts 1 and 2," *Daedalus*, 121, 1992. pp. 87–132; repr. in *How to Legalize Drugs*, ed. Jefferson M. Fish. Northvale, NJ: Jason Aronson, 1998, pp. 578–616; also available at www.drugpolicy.org/library/.

O'Connor, Ann-Marie, "No drug link to family in fatal raid, police say," *Los Angeles Times*, August 28, 1999.

Ostrowski, James, "Drug prohibition muddles along: How a failure of persuasion has left us with a failed policy," in *How to Legalize Drugs*, ed. Jefferson M. Fish, Northvale, NJ: Jason Aronson, 1998.

Petreira, Joseph, "The informants in a drug program: Some kids turn in their own parents," *Wall Street Journal*, April 20, 1992.

Petry, Nancy M., and Warren K. Bickel, "Polydrug abuse in heroin addicts: A behavioral economic analysis," *Addiction*, 93, 1998, pp. 321–35.

Pierre, Robert E., "Keeping pills away from drug abusers." *Washington Post*, national weekly edition, June 23–29, 2003.

Polen, M. R., "Health care use by frequent marijuana smokers who do not smoke tobacco." *Western Journal of Medicine*, 1993, pp. 596–60.

Purdy, Mathew, "Warehouse of addiction: Bars don't stop flow of drugs into the prisons," *New York Times*, July 2, 1995.

Rohter, Larry, "Bolivian leader's ouster seen as warning on U.S. drug policy." *New York Times*, October 23, 2003.

——, "Brazil carries the war on drugs to the air." *New York Times,* July 25, 2004.

Schlosser, Eric, "Make peace with pot," *New York Times*, April 26, 2004.

Schmidt, William E, "To battle AIDS, Scots offer oral drugs to addicts," *New York Times*, February 8, 1993.

Scott, David Clark, "New cooperation seen in anti-drugs strategy." *Christian Science Monitor*, March 16, 1993.

Shane, S., "Test of 'heroin maintenance' may be launched in Baltimore." *Baltimore Sun*, June 10, 1998.

Shane S., and G. Shields, "Heroin maintenance quickly stirs outrage." *Baltimore Sun*, June 12, 1998.

Shedler J., and J. Block, "Adolescent drug use and psychological health: A longitudinal inquiry." *American Psychologist*, 45, 1990, pp. 612–30.

Shenk, Joshua W., "America's altered states," in *Busted*, ed. Mike Gray, New York: Thunder's Mouth Press/Nation Books, 2002.

Thomas, Chuck, "Marijuana arrests and incarceration in the United States," *Drug Policy Analysis Bulletin*, no. 7, June, 1999.

Thornton, Mark, "Prohibition vs. legalization: Do economists reach a conclusion on drug policy?" *Econ Journal Watch*, April, 2004, p. 97: available at www.econjournalwatch.org.

Winter, Greg "A student aid ban for past drug use is creating a furor," *New York Times*, March 13, 2004.

DISCUSSION QUESTIONS

1. Do you agree with the author's assertion that the American "War on Drugs" is a crusade that is driven largely by emotions? If yes, what policies would you recommend to shift the current paradigm? If no, then make the case for staying the course in our long war against drugs.
2. Why should the U.S. legalize marijuana? And what are some possible unintended consequences of decriminalizing "pot"?
3. Consider the impressive list under the heading "Damage Caused by the Drug War." Based on the author's argument, which side is winning the "War on Drugs"? Who is bearing the bulk of this war's casualties? And what should be done to lessen or eradicate these damages, aside from legalization which is unlikely?
4. Is it possible to have a rational debate on the deleterious effects of the "War on Drugs" given the vested support for the status quo from politicians and the general public alike? Is a national conversation on the current approach even possible? Why or why not?

30

Death by a Thousand Little Cuts: Studies of the Impact of Incarceration

TODD R. CLEAR

The Four Questions

1. What is the problem?
2. Is this a *social* problem? What makes this problem a social one?
3. What is the cause of this problem? Who is responsible for it? Is it society? Is it the individual?
4. What, if anything, can be done to address this problem?

Topics Covered

Incarceration policy
Negative impact of incarceration
Marriage rates
Parenting
Male imprisonment
Community disruption
Intimate relations
Sexually transmitted diseases
Delinquency

CHILDREN AND FAMILIES

We were walking around the South City neighborhood of Tallahassee, the summer of 2001. As summers always are in the Florida Panhandle, it was steamy hot. People stay inside; there may be a threesome of pre-teens playing basketball in the schoolyard, or a couple of pre-schoolers yelling exuberantly, playing in the shade of a tree. But for the most part, the out-of-doors is empty. School's out, so it would be natural to wonder where the kids are. We enter a two-story housing project, go upstairs to knock on a door in the door of a household in our sample, and explain why we are there. We are invited in. It is dark inside, the television is blaring. Four children at various ages, maybe ranging from 4 to 9, are sitting—no, bouncing—on the couch. That is where the children are, we realize, inside. Hidden or hiding. Watching TV. It becomes a theme in almost every house we visit. Mothers, children, television, noise ... and no fathers.

Families are the building blocks of a healthy society, and family functioning is the key ingredient in child development. Adults in the family socialize their children about the normative rules and behavioral expectations of society. Family members connect one another—especially children—to networks of social supports that become the foundation for later social capital as adults. Families are the central mechanism of informal social controls, bolstering the limited capacity of formal social controls to shape behavior. And the interpersonal dynamics of families are the source of later psychological and emotional health (or maladjustment). There is no single institution that carries more importance in the well-being of children than the family, and the prospects for healthy social relations in adulthood rely heavily on the existence of a vibrant family life.

There are indications that family life in America is changing, especially among people who are poor.

SOURCE: From Todd R. Clear, *Imprisoning Communities: How Mass Incarceration Makes Disadvantaged Neighborhoods Worse*, pp. 93–117. © 2007 Oxford University Press. Reprinted by permission.

In this group, changes over the last 40 years have been devastating: divorce rates are one-third higher and births to unmarried mothers have doubled, as has the rate of households headed by single mothers (see Western, Lopoo, and McLanahan 2004). Small wonder that so much attention has recently been given to strengthening and sustaining the family life of poor families in America.

Incarceration policy has been a fellow traveler in the deterioration of poor American families. For example, almost three out of five African-American high-school dropouts will spend some time in prison, a rate five times higher than for equivalent whites (Pettit and Western 2004). Two-fifths of those African-American high-school dropouts are fathers who were living with their children before they entered prison (Western, Patillo, and Weiman 2004). One-fourth of *juveniles* convicted of crime have children (Nurse 2004); locking up these fathers increases the chances of divorce and damages their bonds with their children. Counting both adult and juvenile parents, there are probably close to 2 million children in the United States with a parent currently behind prison bars (extrapolating from Western, Pattillo, and Weiman 2004). Between 1991 and 1999, according to Murray and Farrington (forthcoming), the number of children with a mother in prison was up 98 percent, and the number with a father in prison rose by 58 percent. Half of these children are black, almost one-tenth of all black children (Western 2005: fig. 6.2).

That incarceration affects families and children in deleterious ways should be obvious. There is a long and rich literature showing that removal of a parent from the home has, on the average, negative consequences for the partner and the children who remain (see, for example, Bloom 1995 and Hairston 1998). This is the average picture, of course, which masks considerable variation in outcomes. Some families do well in the face of loss; others fare disastrously. What is not clear is the nature and extent of the disruption that follows an adult's incarceration, though numerous negative and positive effects can be posited from the literature (see Hagan and Donovitzer 1999). Phillips et al. (2006) point out that:

> There is evidence … that the arrest of parents disrupts marital relationships, separates children and parents, and may contribute to the permanent legal dissolution of these relationships. It may also contribute to the establishment of grandparent-headed households and, upon parents' return home from prison, to three-generation households. (103)

They go on to say that, "One must be careful… in attributing these family risks to parents' involvement with the [criminal justice system] because these same situations (e.g., divorce, parent-child separation, economic strain, instability, large households, and so forth) also occur when parents have problems such as substance abuse, mental illness, or inadequate education" (104). Thus, while it is known that the incarceration of a parent (especially the mother) increases the chances of foster care or other substitute-care placement, and that substitute (i.e., foster) care is associated with poorer long-term life outcomes, "we know remarkably little about whether children placed in substitute care fare better or worse than similar children remaining with their own parent or parents" (Johnson and Waldfogel 2004:100). And while there is a host of behavioral and emotional problems associated with being a child of an incarcerated parent (see Hagan and Dinovitzer 1999), there are but a few studies showing that the parent's incarceration *causes* this kind of distress.

Having an adult family member go to prison has been shown as a source of problems, not the least of which is increased risk of juvenile delinquency (Widom 1994). Myriad studies show that children and partners of incarcerated adults tend to experience other difficulties, as well, compared to children of nonincarcerated parents. These include school-related performance problems, depression and anxiety, low self-esteem, and aggressiveness (see Hagan and Dinovitzer 1999). The studies show that the negative psychological, behavioral, and circumstantial impact on children from the removal of a parent for incarceration is similar in form, though not always in degree, to that produced by removal owing to divorce or death….

Marriage

There is consistent evidence that poor neighborhoods in which there is a large ratio of adult women to men are places where female-headed, single-parent families are common, and that incarceration is one of several dynamics that have removed black males from their neighborhoods, producing this ratio (Darity and Myers 1994). This may be a race-specific effect; in a county-level analysis for 1980 and 1990, Sabol and Lynch (2003) found that *both* removals to and returns from prison increased the rate of female-headed households in the county.

Being incarcerated reduces marriage prospects for young men. Analyzing the National Longitudinal Survey of Youth (NLSY), Harvard economist Adam Thomas (2005) found that going to prison substantially reduces the likelihood of being married. The effects hold across all racial and ethnic groups, but are strongest for black males over 23 years old, whose likelihood of getting married drops by 50 percent following incarceration. Thomas concluded that "past imprisonment is associated with a lower probability of marrying not only in the near term but also over the long run... [and] it is certainly the case that, among blacks, the relationship cannot easily be explained away by casually controlling for economic, family background characteristics, and neighborhood effects" (2005:26–27). He does find, however, that men with past incidents of incarceration who do become involved with women are more likely instead to cohabit without marriage. These unconventional living arrangements contribute to intergenerational family dysfunction. Cohabitation is associated with previous parental divorce, suggesting an intergenerational pattern, and it carries the risk for future abuse or neglect (called "troubled home"). Western's (2006) analysis of the NYSL confirms these patterns, and his analysis of the Fragile Families Survey of Child Well-Being estimates that going to prison cuts the rate of marriage within a year of the birth of a child by at least one-half and about doubles the chance of separating in that same year (figs. 6.8, 6.9). It is thus not surprising that Lynch and Sabol (2004b) have estimated that 46 percent of the prison population are currently divorced, compared to 17 percent of nonimprisoned adults.

The reduction in the rate of marriage is important in several respects. Families formed by marriage have more longevity than those defined by cohabitation, on average. Mothers in marriages expect and receive more support from their male partners than those in cohabitation relationships (Gibson, Edin, and McLanahan 2003, cited in Western, Lopoo and McLanahan 2004). With increases in family disruption and reduced male involvement in the home there are increased risks of poor school performance by children, of domestic violence, and of contact with the juvenile justice system (Western 2006, esp. figs. 6.10, 6.11).

For the men, there are also consequences of incarceration on marriage. Laub, Nagin, and Sampson (1998) have shown that stable marriages promote life-style changes in adults who were previously criminally active; marriage can thus serve as a turning point in their criminal careers (see also Sampson and Laub 1993). Men who do not get married tend to find it harder to form pro-social relationships and identify the positive social bonds that promote an end to criminal activity. It follows that reduced marriage prospects resulting from a term in prison are a risk factor in recidivism.

Parenting

While they are locked up, many men maintain contact with their children; about half receive mail and/or phone calls, and one-fifth receive visits while in prison (U.S. Department of Justice 1997, cited in Western, Pattillo and Weiman 2004: table 1.5). But the rate at which mothers dissolve their relationships with their children's father during the latter's imprisonment is very high, even for fathers who were active in their children's lives prior to being arrested (Edin, Nelson, and Paranal 2004). In general, incarceration has "a deleterious effect on relationships between former inmate fathers and their children" (Nurse 2004:90). In their longitudinal study of the Fragile Families Survey and Child Well-Being Study data, Western, Lopoo, and McLanahan show that "men who have been incarcerated are much less likely to be married to or cohabiting with the mother of their children twelve months after the birth of their children than men who have not been incarcerated" (2004:39–40). There are several reasons for this. For women who live in poor communities, "the decision to marry or remarry depends in part on the economic prospects, social respectability and trustworthiness of their potential partners" (Western, Lopoo, and McLanahan 2004:23). Against these criteria, many (or most) ex-prisoners do not fare well.

Even when mothers retain their relationships with the incarcerated father of their children, incarceration diminishes the capacity for effective parent-child relationships. Edin, Nelson, and Paranal point out that "incarceration often means that fathers miss out on ... key events that serve to build parental bonds and to signal ... that they intend to support their children both financially and emotionally.... The father's absence at these crucial moments ... can weaken his commitment to the child years later, and the child's own commitment to his or her father" (2004:57).

Researchers point out that this general pattern has to be understood in the context of two caveats. First, as a group, young fathers who are poor and have marginal human capital typically struggle to maintain good relationships with their children and their young mothers, and incarceration may not be a cause of further difficulties so much as a correlate of the personality and situational factors that produce these difficulties. Under the best of circumstances, the bonds between criminally active fathers and their children are often quite fragile. As a result, sometimes young mothers who made progress when their male partners went to prison suffer setbacks when they return (Cohen 1992). Second, for some fathers who have had little contact with their children before imprisonment, the descent into prison is a life-changing moment that opens a door to renewal of those bonds in ways that are not only beneficial for the child but for the father as well (Edin, Nelson, and Paranal 2004).

There are, after all, over 600,000 men who enter prison in a year, and the range of parental interests and patterns for such a group must run the gamut. On the average, however, having one's parent go to prison is not a positive life experience. It disrupts the family and damages parenting capacity.

Family Functioning

Lynch and Sabol (2004b) have estimated that between one-fourth and one-half of all prisoners disrupt a family when they are removed for incarceration. Joseph Murray's (2005) excellent review lists a dozen studies of the way incarceration of a male parent/spouse (or partner) affects the functioning of the family unit he left behind. The most prominent impact is economic—spouses and partners report various forms of financial hardship, sometimes extreme, that result from the loss of income after the male partner's incarceration. This "loss of income is compounded by additional expenses of prison visits, mail, telephone calls ... and sending money to [the person] imprisoned" (2005:445). Because most families of prisoners start out with limited financial prospects, even a small financial impact can be devastating. Phillips et al. (2006) longitudinal study of poor, rural children in North Carolina found that having a parent get arrested led to family break-up and family economic strain, both of which, they point out, are risk factors of later delinquency.

After the male's imprisonment, the family responds to his incarceration in a variety of ways. In order to deal with changed financial circumstances, prisoners' families often move, leading to family disruptions that may include the arrival of replacement males in the family, reduced time for maternal parenting owing to secondary employment, and so on (Edin, Nelson, and Paranal 2004). Moves may also result in more crowded living conditions (especially when the prisoner's family moves in with relatives) and changes in educational districts that may produce disruptions in schooling.

There are also relationship problems. Female partners who find a male replacement for the man who has gone to prison often face the psychological strains that accompany the arrival of a new male in the household. Prisoners' spouses and partners report strains in relationships with other family members and neighbors. Carlson and Carvera (1992) showed that women often have to rely on family and friends to fill the hole left by the incarcerated husband, providing money, companionship, and babysitting and generally straining those ties. Strains in the relationships with children are also reported, often resulting from emotional and functional difficulties that spouses and partners encounter when a male partner goes to prison....

Child Functioning

Incarceration has an effect on the child that is both direct and indirect (Murray 2005). The mother's incarceration has been shown to produce "a significant worsening of both reading scores and behavioral problems" (Moore and Shierholz 2005:2). Likewise, male parent incarceration has been shown to lead to later antisocial behavior by children in an English sample (Murray and Farrington 2005). This latter study compared children whose parents were incarcerated during their early childhoods to children without parental separation, children for whom parental separation was due to other factors (i.e., death), and children whose parents went to prison and returned before they were born. Outcomes included delinquency, antisocial personality measures, and life successes. The children of incarcerated parents were between 2 and 13 times more likely to have had various negative outcomes than any of the comparison groups. These effects appear to be direct, in part because they survive in the face of statistical controls for social class and other family demographics.

Indirect effects stem from the way incarceration undermines family stability. Changes in parental working conditions and family circumstances often a result from incarceration and are known to affect children's social adjustment and norm transmission across generations (Parcel and Menaghan, 1993). Among the problems suffered by children during a parent's incarceration are: "depression, hyperactivity, aggressive behavior, withdrawal, regression, clinging behavior, sleep problems, eating problems, running away, truancy, and poor school grades" (Murray 2005:466). Studies have also shown that parental incarceration is a risk factor in delinquency (Gabel and Shindledecker 1993), emotional maladjustment (Kampfner 1995), and academic performance problems (Phillips and Bloom 1998). In her summary of the literature on childhood loss of a parent, including parental incarceration, Marcy Viboch (2005) points to a range of common reactions to the trauma, including depression, aggression, drug abuse, and running away. Murray and Farrington's (forthcoming) systematic review of studies of parental incarceration on children find evidence of both direct and mediated effects on anti-social behavior, school performance, mental health, drug abuse, and adult unemployment. They conclude that "parental imprisonment may cause adverse child outcomes because of traumatic separation, stigma, or social and economic strain." (1)

The potential negative impact of incarceration on school performance is particularly important. School success is linked to family structure, which has an effect independent of social class and parenting style in impoverished families (Vacha and McLaughlin 1992). Behavioral problems in school are also correlated with problems in parental relationships, including child abuse—an early family dynamic that contributes to later delinquency. Psychologist Cathy Spatz Widom (1989, 1994), following a cohort of children from early school years to adulthood, has observed that victims of early childhood abuse had earlier criminal activity, increased risk of an arrest during adolescence (by more than 50%), and, when they became adults, twice as many arrests as controls....

Community Dynamics

At a most basic level, the absence of males restricts the number of adults available to supervise young people in the neighborhood. The presence of large numbers of unsupervised youth is predictive of various aspects of community-level disorder, including serious crime (Sampson and Groves 1989). It is also known that the existence of "adverse neighborhood conditions" as rated by mothers, tended to be associated with a decrease in the self-control of youth who live in disadvantage (Pratt, Turner, and Piquero 2004). Under these conditions, the informal social control over youths that might have been exercised by family members or neighbors often fails to materialize. These attenuated informal social controls are the springboard for crime, and the effects can be felt in adjacent neighborhoods as well (Mears and Bhati 2006).

In the face of community disruptions, some families isolate themselves from neighbors. In a series of interviews in the South Bronx, Andres Rengifo (2006) has observed that many residents seek to withdraw from their impoverished surroundings. One housing project resident, a single mother with four children (one of whom was attending Yale University and two of whom were in the prestigious Bronx High School of Science public school) said that although she had lived in the projects for seven years, "this place is a dump. I don't talk to anyone, I don't know anyone. That's how we made it here."

While it is commonly assumed that criminally active adults are less capable or less willing guardians, there is no evidence to support this. In fact, Venkatesh (1997) reports that although many problems within the housing project he studied were gang related, gang members involved in criminal activity tended to be accepted because they contributed to the well-being of the community in a variety of ways. For instance, they acted as escorts or protectors, renovated basketball courts, and discouraged truancy. These factors eroded perceptions of them as social deviants, partly because their roles as sons and brothers helped residents view them as "only temporarily" bad, and partly because the gang helped the community in tangible ways.

Intimate (Sexual) Relations

The incarceration of large numbers of parent-age males restricts the number of male partners available in the neighborhood. This means that mothers find more competition for intimate partners and to serve as parents for their children. In the context of more competition for male support, mothers may feel reluctant to end relationships that are unsuitable for children, partly because prospects for suitable replacements are perceived as poor. Thus, even when men who have been sent to prison were abusers, if they are replaced by men who are also abusive, the tradeoff is negative. Likewise, men living with advantageous gender ratios may feel less incentive to remain committed in their parenting partnerships. When the remaining family unit is forced to choose from a thinning stock of males, the options may not be attractive. For those women who end abusive relationships and live alone, the neighborhood implications may also be problematic.

Citing these dynamics, epidemiologists James Thomas and Elizabeth Torrone (2006) investigated the role of high rates of incarceration on sexual behavior in poor neighborhoods. They argued that the pressures on men for safe sex and monogamy are reduced as the ratio of women to marriageable men gets very high. Analyzing North Carolina counties and communities, they found that incarceration rates in one year predicted later increases in rates of gonorrhea, syphilis, and chlamydia among women. They also found that a doubling of incarceration rates increased the incidence of childbirth by teenage women by 71.61 births per 100,000 teenage women. They conclude that "high rates of incarceration can have the unintended consequence of destabilizing communities and contributing to adverse health outcomes: (2006:1)."

This latter finding is notable because teenage births are associated with numerous problematic outcomes for both the mothers and their children. For mothers, teenage births are more likely to lead to a life plagued with lower wages, underemployment, reliance upon welfare, and single parenthood. For *all* children of mothers who have their first child at a very early age, there is an increased likelihood that those children will be arrested for delinquency and violent crime (Pogarsky, Lizotte, and Thornberry 2003). Not surprisingly, rates of out-of-wedlock births also predict higher levels of incarceration across time in the United States (Jacobs and Helms 1996).

Incarceration also seems to explain at least part of the higher rate of HIV among African-American men and women. Johnson and Raphael (2005) analyzed data on AIDS infection rates, provided by the U.S. Centers for Disease Control and Prevention, from 1982 to 2001, combined with national incarceration data for the same period. As did Thomas, they posit that "male incarceration lowers the sex ratio (male to female), abruptly disrupts the continuity of heterosexual relationships, and increases the exposure to homosexual activity for incarcerated males—all of which may have far-reaching implications for an individual or group's AIDS infection risk" (2005:2). What is important about their perspective is that they assume *both* removal and reentry as potential destabilizing factors in sexual relations. They find "very strong effects of male incarceration rates on both male and female AIDS infection rates [and] that the higher incarceration rates among black males over this period explain a large share of the racial disparity in AIDS between black women and women of other racial and ethnic groups" (3).

Incarceration and Families: A Summary

Available studies, listed above, show that incarceration imposes a long list of costs on families and children. Children experience developmental and emotional strains, have less parental supervision, are at greater risk of parental abuse, and face an increased risk of having their own problems with the justice system. Mothers find it harder to sustain stable intimate relationships with men who have gone to prison, and they have an increased risk of contracting sexually transmitted diseases. Families are more likely to break up, and they encounter economic strains. Girls raised in these high-imprisonment places are more likely to become pregnant in their teen years; boys are more likely to become involved in delinquency. Residents feel less positively about their neighbors, and they may tend to isolate themselves from them. The descriptions of these challenges to family life in impoverished places are not surprising, as we have long known that poor neighborhoods are problem settings for family life. What is new is to see, demonstrated in the data, the role incarceration plays in creating these challenges by disrupting social networks and distorting social relationships.

To be sure, incarceration is not the sole cause of these situations. Studies find that parental problems such as drug abuse and mental illness also contribute directly and separately to these situations (Phillips et al. 2006). Addressing the problem of incarceration alone will not be sufficient to ameliorate these problems. By the same token, trying to overcome the many family deficits experienced by these at-risk children without considering the effect of incarceration is not likely to work.

REFERENCES

Bloom, Barbara. 1995. Imprisoned mothers. In Katherine Gabel and Denise Johnston, eds., *Children of Incarcerated Parents*. New York: Lexington.

Carlson, Bonnie, and Neil Cervera, 1992. *Inmates and Their Wives*. Westport, CT: Greenwood Press.

Cohen, Barbara E. 1992. *Evaluation of the Teen Parents Employment Demonstration*. Washington, DC: Urban Institute.

Darity, William A., and Samuel L. Myers, Jr. 1994. *The Black Underclass: Critical Essays on Race and Unwantedness*. New York: Garland.

Edin, Kathryn, Timothy Nelson, and Rechelle Paranal. 2004. Fatherhood and incarceration as potential turning points in the criminal careers of unskilled men. In Mary Pattillo, David Weiman, and Bruce Western, eds., *Imprisoning America: The Social Effects of Mass Incarceration*, 46–75. New York: Russell Sage.

Gabel, Stewart, and Richard Shindledecker. 1993. Characteristics of children whose parents have been incarcerated. *Hospital and Community Psychiatry* 44(7):543–559.

Gibson, Christina, Katheryn Edin, and Sara McLanahan, 2003. High hopes but even higher expectations: The retreat from marriage among low-income couples. Working Paper 03-066-FF. Princeton, NJ: Center for Research on Child Wellbeing, Princeton University.

Hagan, John, and Ronit Dinovitzer. 1999. Collateral consequences of imprisonment for children, communities and prisoners. In Michael Tonry and Joan Petersilia, eds., *Prisons*, 121–162. Chicago: University of Chicago Press.

Hairston, Creasie. 1998. The forgotten parent: Understanding the forces that influence incarcerated fathers' relationships with their children. *Child Welfare* 5:617–39.

Jacobs, David, and Ronald E. Helms. 1996. Toward a political model of incarceration: A time series examination of multiple explanations for prison admission rates. *American Journal of Sociology* 102:323–57.

Johnson, Rucker C, and Steven Raphael. 2005. The Effects of Male Incarceration on Dynamics of AIDS Infection Rates among African-American Women and Men. Unpublished paper presented to the incarceration study group of the Russell Sage Foundation, July.

Kampfner, Christina Jose. 1995. Post-traumatic stress reactions of children of incarcerate mothers. In Stuart Gabel and Denise Johnston, eds, *Children of Incarcerated Parents*, 121–62. New York: Lexington.

Laub, John, Daniel Nagin, and Robert Sampson. 1998. Trajectories of change in criminal offending: Good marriages and the desistence process. *American Sociological Review* 63(2):225–38.

Lynch, James P., and William J. Sabol. 2004b. Effects of incarceration on informal social control in communities. In Mary Pattillo, David Weiman, and Bruce Western, eds., *Imprisoning America: The Social Effects of Mass Incarceration*, 135–64. New York: Russell Sage.

Mears, Daniel P., and Avinash Bhati. 2006. No community is an island: The effects of resource deprivation on urban violence in spacially and socially proximate communities. *Criminology* 44(3):509–48.

Moore, Quinn, and Heidi Shierholz. 2005. Externalities of Imprisonment: Does Maternal Incarceration Affect Child Outcomes? Paper presented at the meeting of the American Society of Criminology, Toronto.

Murray, Joseph. 2005. The effects of imprisonment on the families and children of prisoners. In Allison Liebling and Shadd Maruna, eds., *The Effects of Imprisonment*, 442–92. Cullompton, UK: Willan.

Murray, Joseph, and David Farrington. Forthcoming. Effects of parental incarceration on children. In Michael Tonry, ed., *Crime and Justice: A Review of Research*. Chicago: University of Chicago Press.

Nurse, Anne M. 2004. Returning to strangers: Newly paroled young fathers and their children. In Mary Pattillo, David Weiman, and Bruce Western, eds., *Imprisoning America: The Social Effects of Mass Incarceration*. New York: Russell Sage.

Parcel, Toby L., and Elizabeth G. Menaghan. 1993. Family social capital and children's behavior problems. *Social Psychology Quarterly* 56:120–35.

Pettit, Becky, and Bruce Western. 2004. Mass imprisonment and the life course: Race and class inequality in U.S. incarceration. *American Sociological Review* 69:151–69.

Phillips, Susan, and Barbara Bloom. 1998. In whose best interest? The impact of changing public policy on relatives caring for children with incarcerated parents. *Child Welfare* 77(5):531–41.

Phillips, Susan D., Alaatin Erkanli, Gordon P. Keeler, E. Jane Costello, and Adrian Angold. 2006. Disentangling the risks: Parent criminal justice involvement and child's exposure to family risks. *Criminology & Public Policy* 5(4):101–206.

Pogarsky, Greg, Alan J. Lizotte, and Terence P. Thornberry. 2003. The delinquency of children born to young mothers: Results from the Rochester Youth Development Study. *Criminology* 41(4):1249–86.

Pratt, Travis C., Michael G. Turner, and Alex Piquero. 2004. Parental socialization and community context: A longitudinal analysis of the structural sources of low self-control, *Journal of Research in Crime & Delinquency* 41(3):219–43.

Sabol, William J., and James P. Lynch. 2003. Assessing the longer-run effects of incarceration: Impact on families and employment. In Darnell Hawkins, Samuel Myers, Jr., and Randolph Stine, eds., *Crime Control and Social Justice: The Delicate Balance*. Westport, CT: Greenwood Press.

Sampson, Robert J., and W. Byron Groves. 1989. Community structure and crime: Testing social disorganization theory. *American Journal of Sociology* 94:774–802.

Sampson, Robert J., and John H. Laub. 1993. *Crime in the Making: Pathways and Turning Points Through Life*. Cambridge, MA: Harvard University Press.

Thomas, Adam. 2005. The Old Ball and Chain: Unlocking the Correlation between Incarceration and Marriage. Unpublished manuscript.

Thomas, James C., and Elizabeth Torrone. 2006. Incarceration as forced migration: Effects on selected community Health outcomes. *American Journal of Public Health* 96(10):1–5.

U.S. Department of Justice, Bureau of Justice Statistics. 1997. *Survey of Inmates in State and Federal Correctional Facilities*. Ann Arbor, MI: ICPSR.

Vacha, Edward F., and T. F. McLaughlin. 1992. The social structural, family, school and personal characteristics of at-risk students: Policy recommendations for school personnel. *Journal of Education* 174(3):9–25.

Venkatesh, Sudhir Alladi. 1997. The social organization of street gang activity in an urban ghetto. *American Journal of Sociology* 103:82–111.

Viboch, Marcy. 2005. *Childhood Loss and Behavioral Problems: Loosening the Links*. New York: Vera Institute of Justice.

Western, Bruce. 2006. *Punishment and Inequality in America*. New York: Russell Sage.

Western, Bruce, Leonard M. Lopoo, and Sara McLanahan. 2004. Incarceration and the bonds between parents in fragile families. In Mary Pattillo, David Weiman, and Bruce Western, eds., *Imprisoning America: The Social Effects of Mass Incarceration*. New York: Russell Sage.

Western, Bruce, Mary Patillo, and David Weiman. 2004. In Mary Pattillo, David Weiman, and Bruce Western, eds., *Imprisoning America: The Social Effects of Mass Incarceration*, 1–18. New York: Russell Sage.

Widom, Kathy Spatz. 1989. Does violence beget violence: A critical examination of the literature. *Psychological Bulletin* 106(1):3–28.

Widom, Kathy Spatz. 1994. Childhood victimization and risk for adolescent problem behaviors. In Robert D. Ketterlinus and Michael E. Lamb, eds., *Adolescent Problem Behaviors: Issues and Research*, 127–64. New York: Lawrence Earlbaum.

DISCUSSION QUESTIONS

1. Incarceration policy in the United States is having a devastating impact on poor families, on children, and especially on the marriage rate in minority communities. What public policies might you suggest to address each of these unintended consequences of incarceration?
2. "Individuals *choose* to commit crimes, and then they are punished for their offenses with incarceration. Society's punitive approach to criminal behavior is not the social problem. The *real* social problem begins and ends with bad choices on the part of troubled individuals." Respond to this statement. Say why you agree or disagree with this view.
3. What is the link between mass-incarceration and deteriorating intimate relations, including rates of sexually transmitted diseases? What is mass incarceration's connection to the rising rates of out-of-wedlock births? Are these unintended consequences of the current policy simply unavoidable? Why or why not?
4. About ten percent of African American children have a parent in prison, according to this selection. What are the greater social implications of this horrendous statistic? Is this a social problem *by itself*? Why or why not?
5. Mass-incarceration significantly reduces the pool of marriageable black men for black women, especially in America's inner-cities, substantially reducing the likelihood of getting married. Is the declining rate of marriage and the increasing increases in cohabitation among African Americans a social problem? Why or why not?

31

Mass Incarceration and the Problems of Prisoner Reentry

DEVAH PAGER

The Four Questions

1. What is the problem that Pager describes?
2. Is it a *social* problem, or is it a *personal* one? Who is harmed by this problem? Is it society, or is it the individual?
3. What are the causes of this problem? Is it society or is it the individual?
4. What can be done to alleviate this problem?

Topics Covered

Hiring discrimination

Racial stereotypes

Mass incarceration

Stigma

Racial disparities in hiring

Race and crime

Black incarceration rate

Poverty

Incarceration policy

Punishment and race

Jerome arrived at a branch of a national restaurant chain in a suburb twenty miles from Milwaukee. He immediately sensed that he was the only black person in the place. An employee hurried over to him, "Can I help you with something?" "I'm here about the job you advertised," he replied. The employee nodded reluctantly and went off to produce an application form. Jerome filled out the form, including information about his criminal background. He was given a math test and a personality test. He was then instructed to wait for the manager to speak with him. The manager came out after about ten minutes, looked over Jerome's application, and frowned when he noticed the criminal history information. Without asking any questions about the context of the conviction, the manager started to lecture: "You can't be screwing up like this at your age. A kid like you can ruin his whole life like this." Jerome began to explain that he had made a mistake and had learned his lesson, but the manager cut him off, "I'll look over your application and call if we have a position for you."

Jerome could have been any one of the hundreds of thousands of young black men released from prison each year who face bleak employment prospects as a result of their race and criminal records. In this case, Jerome happened to be working for me. He was one of four college students I had hired as "testers" for a study of employment discrimination. An articulate, attractive, hard-working young man. Jerome was assigned to apply for entry-level job openings throughout the Milwaukee metropolitan area, presenting a fictitious profile designed to represent a realistic ex-offender. Comparing the outcomes of Jerome's job search of those of three other black and white testers presenting identical qualifications

SOURCE: From Devah Pager, "Blacklisted: Hiring Discrimination in an Era of Mass Incarceration," in *Against the Wall: Poor, Black, and Male*, ed. Elijah Anderson, Philadelphia, PA: University of Pennsylvania Press, 2008, pp. 71–83. Reprinted with permission of the University of Pennsylvania Press.

with and without criminal records gives us a direct measure of the effects of race and a criminal record, and of possible interactions between the two, in shaping employment opportunities. In this essay, I consider the ways in which high rates of incarceration among African Americans may fuel contemporary stereotypes about the criminal tendencies of young black men. The evidence reviewed here suggests that the disproportionate growth of criminal justice intervention in the lives of young black men and the corresponding media coverage of this phenomenon, which presents an even more skewed representation, has likely played an important role in reinforcing deep-seated associations between race and crime, with implications for employment discrimination and broader forms of social disfranchisement.

RACIAL STEREOTYPES IN AN ERA OF MASS INCARCERATION

Over the past three decades, we have seen an unprecedented expansion of the criminal justice system, with rates of incarceration increasing more than five-fold from 1970 to 2000. Today the United States boasts the highest rate of incarceration in the world, with over two million individuals currently behind bars. The expansive reach of the criminal justice system has not affected all groups equally: African Americans have been more acutely affected by the boom in incarceration than any other group. Blacks comprise over 40 percent of the current prison population, although they are just 12 percent of the U.S. population. At any given time, roughly 12 percent of all young black men between the ages of twenty-five and twenty-nine are behind bars, compared to less than 2 percent of whites in the same age group. Roughly a third of young black men are under criminal justice supervision (Bureau of Justice Statistics 2006, 2000, Table 1.29).[1] Over the course of a lifetime, nearly one in three young black men—and well over half of young black high school dropouts—will spend some time in prison. According to these estimates, young black men are more likely to go to prison than to attend college, serve in the military, or, in the case of high school dropouts, to be in the labor market (Bureau of Justice Statistics 1997; Pettit and Western 2004). Imprisonment is no longer a rare or extreme event among our nation's most marginalized groups. Rather, it has now become a normal and anticipated marker in the transition to adulthood.

In addition to the unprecedented reach of incarceration in the lives of young black men today, these trends have troubling consequences that may extend well beyond the prison walls. There is good reason to believe that the mass incarceration of black men contributes to continuing discrimination against the group as a whole, not only those with criminal records, by reinforcing the association of criminality with African Americans that has long been a feature of racial prejudice in the U.S. African American men have long been regarded with suspicion and fear. In contrast to progressive trends in other racial attitudes that have occurred in recent decades, associations between race and crime have changed little. Survey respondents consistently rate blacks as more prone to violence than any other American racial or ethnic groups, endorsing stereotypes of aggressiveness and violence most frequently in their ratings of African Americans (Sneiderman and Piazza 1993; Smith 1991). The stereotype of blacks as criminals is deeply embedded in the collective consciousness of whites, irrespective of their level of prejudice or personal beliefs (Devine and Elliot 1995; Eberhardt et al., 2004, 7; Graham and Lowery 2004).

Although the current prevalence of racial stereotypes cannot be traced to any single source, the disproportionate growth of criminal justice intervention in the lives of young black men, compounded by skewed media coverage of this phenomenon, has likely played an important role. Experimental research shows that exposure to news coverage of a violent incident committed by a black perpetrator not only increases punitive attitudes about crime but further increases negative attitudes about blacks generally (Gilliam and Iyengar 2000; Gilliam, Iyengar, Simon, and Wright 1996; Entman 1990). The more exposure whites have to images of blacks in custody or behind bars, the stronger their expectations become regarding the race of assailants and the criminal tendencies of black strangers.

The consequences of mass incarceration may well extend far beyond the costs to the individuals behind bars, the families that are disrupted, and the communities whose residents cycle in and out.[2] The criminal justice system may itself legitimate and reinforce deeply embedded racial stereotypes, contributing to the persistent chasm in this society between black and white.

THE CREDENTIALING OF STIGMA

For each individual processed through the criminal justice system, police records, court documents, and corrections databases detail arrests, charges, convictions and terms of incarceration. Most states make these records publicly available, often through online repositories, accessible to employers, landlords, creditors, and other interested parties.[3] As increasing numbers of occupations, public services, and other social goods become off limits to ex-offenders, these records can be used as the official basis for determining eligibility or exclusion. The state serves as a credentialing institution, providing official and public certification of those among us who have been convicted of wrongdoing. The "credential" of a criminal record, like educational or professional credentials, constitutes a formal and enduring classification of social status, which can be used to regulate access and opportunity across numerous social, economic, and political domains.

In the labor market, the criminal credential has become a salient marker for employers, with increasing numbers making use of background checks to screen out undesirable applicants. The majority of employers claim that they would not knowingly hire an applicant with a criminal background. These employers show little concern about the specific information conveyed by a criminal conviction and its bearing on a particular job, but rather view this credential as an indicator of "general employability" or trustworthiness (Holzer 1996, 60).[4] Well beyond the single incident at its origin, the credential comes to stand for a broader internal disposition.

The power of the credential lies in its recognition as an official and legitimate means of evaluating and classifying individuals.[5] The negative credential of a criminal record offers formal certification of the offenders among us, and official notice of those demographic groups most commonly implicated. But credentials have effects that reach beyond their formalized domain. Particularly in cases where the certification of a particular status is largely overlapping with other status markers, such as race, gender, and age, public assumptions about who is and is not a "credential holder" may become generalized or exaggerated. Because blacks are so strongly associated with the population under correctional supervision, it becomes easy to assume that any given young black man is likely to have, or to be on his way to acquiring, a criminal record. According to legal scholar David Gole, "When the results of the criminal justice system are as racially disproportionate as they are today, the criminal stigma extends beyond the particular behaviors and individuals involved to reach all young black men, and to a lesser extent all black people. The criminal justice system contributes to a stereotyped and stigmatic view of African Americans as potential criminals" (1995, 2561). Invoking this formal category may legitimate forms of social exclusion that, based on ascriptive characteristics alone, would be more difficult to justify."[6] In this way, negative credentials make possible a new rationale for exclusion that reinforces and legitimates existing social cleavages.

To understand the workings and effects of this negative credential, we must rely on more than speculation as to when and how these official labels are invoked as the basis for enabling or denying opportunity. Because credentials are often highly correlated with other indicators of social status or stigma, especially race, gender, class, we must examine their direct and independent impact. In addition, credentials may affect certain groups differently from others, with the official marker of criminality carrying more or less stigma depending on the race of its bearer. As increasing numbers of young black men are marked by their contact with the criminal justice system, it becomes a critical priority to understand the costs and consequences of this now prevalent form of negative credential.

APPLYING FOR JOBS IN WHITE AND BLACK, WITH AND WITHOUT A CRIMINAL RECORD

This study uses an experimental audit methodology to measure the extent to which race and criminal backgrounds represent barriers to employment. The basic design of an employment audit involves sending matched pairs of individuals, called testers, to apply for real job openings in order to see whether employers respond differently to applicants on the basis of specific characteristics. The current study include four male testers, two black and two white, matched into two teams; the two black testers formed one team, and the two white testers formed a second. The testers were college students from Milwaukee who were matched on the basis of age, race, physical appearance, and general style of self-presentation. They were assigned fictitious resumes that reflected equivalent levels of education (high school degree) and work experience (steady employment across a range of entry-level jobs). Within each team, one tester was randomly assigned a "criminal record" for the first week; the pair then rotated which member presented himself as the ex-offender for each successive week of employment searches, so that each tester served in the criminal records condition for an equal number of cases.[7] By varying which member of the pair presented himself as having a criminal record, unobserved differences within the pairs of applicants were effectively controlled.

The testers participated in a common training program to become familiar with the details of their assumed profile and to ensure uniform behavior in job interviews. The training period lasted for one week, during which testers participated in mock interviews with one another and practice interviews with cooperating employers. The testers were trained to respond to common interview questions in standardized ways, and were well rehearsed for a wide range of scenarios that emerge in employment situations. Frequent communication between myself and the testers throughout each day of fieldwork allowed for regular supervision and troubleshooting in the event of unexpected occurrences.

A random sample of entry-level positions requiring no previous experience and no education beyond high school was drawn each week from the Sunday classified advertisement section of the *Milwaukee Journal Sentinel*. In addition, I drew a supplemental sample from Jobnet, a state-sponsored website for employment listings that was developed in connection with Wisconsin's W-2 Welfare-to-Work initiative.[8] I excluded from the sample those occupations with legal restrictions on ex-offenders, such as jobs in the health care industry, work with children and the elderly, jobs requiring handling firearms (e.g., security guards), and jobs in the public sector.

Each of the audit pairs was randomly assigned fifteen job openings each week. The white pair and the black pair were assigned separate sets of jobs, with the same-race testers applying to the same jobs.[9] One member of the pair applied first, with the second applying one day later; whether the ex-offender came first or second was determined randomly. A total of 350 employers were audited during the course of this study, 150 by the white pair and 200 by the black pair. The black team performed additional tests because black testers received fewer callbacks on average than whites did; in this situation, a larger sample size enables the calculation of more precise estimates of the effects under investigation.

Immediately after submitting a job application, testers filled out a six-page response form that coded relevant information. Important variables included type of occupation, metropolitan status (city/suburb), wage, size establishment, and race and sex of employer. Additionally, testers wrote detailed narratives describing the overall interaction and recording any statements on application forms or comments made by employers specifically related to race or criminal records.

The study focused only on the first stage of the employment process. Testers visited employers, filled out applications, and proceeded as far as they could during the course of one visit. If testers were asked to interview on the spot, they did so, but they did not return to the employer for a second visit. I therefore compare the results on the basis of the proportion of applications that elicited callbacks from employers. Individual voice mail boxes were set up for each tester to record employer responses. I focus on this initial stage of the employment process because it is the stage likely to be most affected by the barriers of race and a criminal record. Early on, employers have the least individualizing information about the applicant and are more likely to generalize on the basis of group-level, stereotyped characteristics. In a parallel case, a recent audit study of age discrimination found that 76 percent of the measured differential treatment occurred at this first stage of the employment process (Bendick, Jackson, and Reinoso 1994). Given that both race and a criminal record, like age, are highly salient characteristics, it is likely that as much, if not more, of the overall effects of racial criminal stigma will be detected at this stage.

A second advantage of the callback rather than a job offer as the key outcome variable is that it does not require employers to narrow their selection down to a single applicant. At the job offer stage, if presented with an ex-offender and an equally qualified non-offender, even employers with little concern over hiring ex-offenders would likely select the applicant with no criminal record, an arguably safer choice. Equating the two applicants could magnify the impact of the criminal record, as it becomes the only remaining basis for selection between the two (Heckman 1998). The callback does not present such complications. Typically employers interview multiple candidates for entry-level positions before selecting a hire. In a telephone survey following the audit, employers in this study reported interviewing an average of eight applicants for the last entry-level position filled. At the callback stage, employers need not yet choose between the ex-offender and non-offender. If the applicants appear well qualified and the employer does not view the criminal record as an automatic disqualifier, he or she can interview them both.[10]

RACIAL DISPARITIES IN HIRING OUTCOMES

Results are based on the proportion of applications submitted by each tester which elicited callbacks from employers. Three main findings appear in the audit results, presented in Figure 31.1. First, there is a large and significant effect of criminal record for white job seekers, with 34 percent of whites without criminal records receiving callbacks relative to only 17 percent of otherwise equally qualified whites with criminal records. A criminal record thus reduces the likelihood of a callback for whites by 50 percent. Second, there is some indication that the magnitude of the criminal record effect may be even larger for blacks than whites. While the interaction between race and criminal records is not statistically significant, the substantive difference is worth nothing. While the ratio of callbacks for non-offenders relative to offenders for whites was two to one (34 versus 17 percent), this

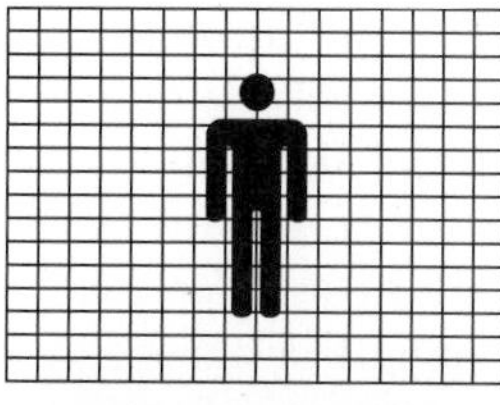

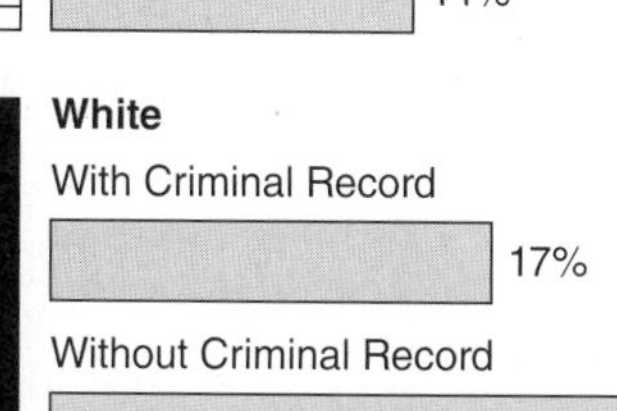

FIGURE 31.1 Effects of race and criminal background on employment. Bars represent percentage of callbacks received by each group. The effects of race and criminal record are large and statistically significant ($p < .01$). The interaction between the two is not significant in the full sample.

same ratio for blacks was close to three to one (14 versus 5 percent).[11] Finally, looking at the callback rates for black and white tester pairs side by side, the fundamental importance of race becomes vividly clear. Among those without criminal records, black applications were less than half as likely to receive a callback compared to equally qualified whites (14 versus 34 percent). This disparity implies that young black men needed to work more than twice as hard, applying to twice as many jobs, to secure the same opportunities as whites with identical qualifications. Even more striking, the powerful effects of race rival the strong stigma conveyed by a criminal record. In this study, a white applicant *with a criminal record* was just as likely to receive a callback as a black applicant without any criminal history (17 versus 14 percent).[12] Despite the fact that the white applicant revealed evidence of a felony drug conviction and reported having recently returned from a year and a half in prison, employers seemed to view this applicant as no more risky than a young black man with no history of criminal involvement. Racial disparities have been documented in many contexts, but here, comparing the two effects side by side, we are confronted with a troubling reality: in terms of one's chances of finding one job, being black in America today confers just about the same disadvantage as having a felony conviction.

In presentations of this research, I have often heard an audible gasp from the audience when I display these results. Could the effect of race really be so large? The magnitude of these effects stands in striking opposition to the prevailing wisdom that such blatant forms of discrimination have become vanishingly rare. It is tempting to think that there is something peculiar about this study, or about the time and place in which it was conducted, which offers an exaggerated view. When this study was conducted, Milwaukee was the second most segregated city in the country, implying great social distance between blacks and whites.[13] If race relations were more strained in Milwaukee than in other parts of the country, then the effects of race found there might be larger than what would be found in other urban areas. In fact, however, the magnitude of the race effect found in this study falls squarely within the range found in previous audit studies (Pager 2007a). An audit study in Washington, D.C., found that blacks were 24 percentage points less likely to receive a job offer than their white counterparts, a finding strikingly similar to the 20 percent difference between white and black non-offenders found here.[14]

Likewise, a recent field experiment by Marianne Bertrand and Sendbil Mullainathan (2003) found large effects of race among employers in Boston and Chicago. In this study, the researchers mailed resumes with racially identifiable names to employers in the two cities. Their sample was restricted to listings for sales, administrative support, clerical and customer service positions. Despite the narrower range of occupations and the higher level of qualifications presented in this study, these authors find clear evidence of racial bias. White male names triggered a callback rate of 9.19 percent, compared to 6.16 percent among black male names. The ratio of callbacks for whites to blacks (1.5), while smaller than the ratio of callbacks to white and black non-offenders from this study (2.4), strengthens our confidence that Milwaukee is not the only city in which race continues to matter.[15] A replication of our study in New York City obtained very similar results, with whites receiving callbacks at twice the rate of similarly qualified blacks (31 percent versus 15 percent), and white felons receiving callbacks at rates similar to those of blacks with no criminal history (17 percent versus 13 percent) (Pager, Western, and Bonikowski 2007).

Each of these studies reinforces the conclusion that race represents an extremely powerful barrier to job entry. The matched designs allow us to separate speculation about applicants' qualifications (supply-side influences) from the racial attributions or biases of employers (demand-side influences). While these studies remain silent on the many supply-side factors that may also contribute to the employment difficulties of young black men, they speak loud and clear about the significance of employers' racial biases or attributions in shaping the opportunities available to young black and white job seekers. Before applicants have an opportunity to demonstrate their capabilities in person, a large proportion are weeded out on the basis of a single categorical distinction.

DRIVING WHILE BLACK: ASSOCIATIONS BETWEEN RACE AND CRIME

I received a call from Andre at about two o'clock one afternoon. He was calling me from his cell phone while sitting in the back of a police car. The police had stopped him at a freeway entrance on the way to one of his assignments. Though Andre had not committed any traffic violation, the police explained they were looking for someone who matched his description: "a black man, between the ages of 21 and 25." Andre was instructed to step out of his car and asked to take a seat in the back of the police vehicle. Passersby craned their necks to catch a glimpse of the latest criminal suspect who had been apprehended. The police asked him a long series of questions and called in his information to the station to have his background thoroughly checked. In the end, the police were satisfied that Andre was not the guy they were looking for and let him go. Andre had spent over an hour in the back of that police car.

We often hear popular accounts of the problems of "driving while black," the phenomenon that blacks are pulled over arbitrarily for police checks, with the implication that in some places any black man is automatically suspect (Harris 1994).[16] Over the course of the fieldwork for this study, I witnessed some of these episodes first hand. The young men serving as testers in my field experiment were bright college kids, models of discipline and hard work; yet in the course of their daily lives, they were at times mistaken for the troublemaker types featured on the nightly news. Particularly in casual encounters, while driving or when entering a place of business, there seemed little they could do to signal their distance from the dominant stereotype of the black male.[17]

On several occasions, black testers were asked in person, before submitting their applications, whether they had a prior criminal history. For these employers, a young black man immediately aroused concern about criminal involvement, and this issue took center stage before getting to matters of education, work experience, or qualifications. None of the white testers, by contrast, was asked about his criminal history up front.[18] These experiences are consistent with Elijah Anderson's account of the suspicion with which young black men are often viewed. According to Anderson, "the anonymous black male is usually an ambiguous figure who arouses the utmost caution and is generally considered dangerous until he proves he is not" (1990, 190).[19] Overcoming this initial stereotype is one of the first challenges facing the young black male job applicant, particularly in low-wage labor markets where fewer objective indicators, such as a college degree or related work history, are available for, or relevant to, the evaluation.

The effect of race demonstrated here is especially striking by virtue of its contrast with the effect of a criminal record. Seeing the two categories side by side drives home just how much race matters in employment contexts: being black is viewed as tantamount to being a convicted felon. These effects, however, should not be seen as independent. In an era of mass incarceration, when one in three young black men will wind up in prison, black men are readily associated with criminal activity in the minds of whites. High levels of incarceration cast a shadow of criminality over the black population as whole, implicating the majority of black men who have remained crime-free....

ENDNOTES

1. To some degree, these disparities reflect differences in the level of recorded criminal activity between groups. Particularly for violent crimes such as homicide, blacks are represented in roughly equal proportions among those arrested and those imprisoned. In contrast, in the case of drug crimes, which have been a major source of prison growth since 1980, the evidence suggests that whites outnumber blacks in both consumption and distribution, but blacks are disproportionately charged

with and convicted of possession and dealing. See Sampson and Lauritsen 1997; Blumstein 1982, 1993; Tonry 1995.

2. Loïc Wacquant 2000 addresses the interdependent and reciprocal nature of racial disproportionality in punishment. In what he calls a "deadly symbiosis" between prison and ghetto, Wacquant finds a reinforcement of repression and social marginality driven by these twin institutions. In this formation, the management and containment of poor black men becomes the dominant objective, with both prison and ghetto isolating this problematic population from the distant mainstream. As increasing numbers of blacks churn from poor neighborhoods to prison and back again, the functional equivalence of these institutions—in the eyes of residents, employers and the general public—is powerfully reinforced.
3. Over 71 million criminal history records were maintained in state criminal history repositories by the end of 2003 (Bureau of Justice Statistics 2006a). As of 2005, 38 states provide public access to their criminal record repositories and 28 make some or all of this information available online (Legal Action Center 2004).
4. In answering a question about their willingness to hire an applicant with a criminal record, roughly a quarter of employers responded, "it depends." This finding suggests that at least for some employees the type of crime or the circumstances of the conviction could be significant beyond the simple fact of conviction.
5. The literature on labeling provides a parallel approach to analyzing the effects of such formal negative markers as juvenile delinquent, mentally ill, homosexual drug user, etc.: see Beaker 1963; Goffman 1963; Garfinkel 1956.
6. As psychologist John Dovidio explains, because most Americans today "consciously endorse egalitarian values, they will not discriminate directly and openly in ways that can be attributed to racism; however, because of their negative feelings they will discriminate, often unintentionally, when their behavior can be justified on the basis of some factor other than race" (2001, 835).
7. The criminal record in all cases were a drug felony, possession with intent to distribute (cocaine), and 18 months of prison time. Tester presented the information to employers by checking the box "yes" in answer to the standard application question, "Have you ever been convinced of a crime?" As additional cues, testers also reported work experience in the correctional facility, and listed their parole officer as a reference. For a more detailed discussion, see Pager 2003.
8. Employment services such as Jobnet have become a much more common method of finding employment in recent years, particularly for difficult-to-employ populations such as welfare recipients and ex-offenders. A survey conducted by Harry Holzer, Steven Raphael, and Michael Stoll (2006) found that nearly half of Milwankee employers (46 percent) use Jobnet to advertise job vacancies in their companies.
9. Comparisons of outcomes by race are thus based on between-pair comparisons. Between-pair comparisons provide less efficient estimators, but they are nevertheless unbiased, provided that there are no systematic differences between the samples of jobs assigned to each pair or between the observed characteristics of the black and white pair apart from race. For a more extensive discussion, see Pager 2003, 2007, chap. 5.
10. A more in-depth discussion of methodological concerns, including limits to generalizability, representativeness of testers, sample restrictions, and experimenter effect is presented in Pager 2007, app, 4A.
11. White not significant in the full sample this interaction becomes significant when analyzed specifically for suburban employers and among employers with whom the testers had extended personal contact: see Pager 2007a, chap. 7.
12. The sound bite that has emerged from this study is somewhat misleading. The research has been cited among politicians and the media as demonstrating that "whites with a felony conviction have a better chance at getting a job than do blacks with clean criminal histories." This isn't quite right. The difference between a callback rate of 17 percent (for white felons) and 14 percent (for black nonoffenders) is small and not statistically significant. An accurate way to summarize these findings, then, is to conclude: A white felon has about the same chance of getting a callback as a black man with no criminal background.
13. There are several ways of calculating the level of racial segregation in a metropolitan area. The most

common method is the dissimilarity index, which calculates the relative distribution of two racial groups (e.g., blacks and whites) across neighborhoods in a city. For U.S. city rankings according to this index, see http://www.censusscope.org/segregation.html.

14. The study by Bendick, Jackson, and Reinoso (1994) included an assessment of the full hiring process, from application to job offer. That the racial disparities reported here (at the first stage of the employment process) closely mirror those from more comprehensive studies provides further reassurance that this design is capturing a majority of the discrimination that takes place in the hiring process.
15. The lower callback rates overall in the Beraund and Mullainathan study (2003) may reduce the observed contrasts between resume pairs. If for example, 5 percent of employers tend to call back *all* applicants as a matter of policy the resulting contrast would be based on a very small number of employers who conduct any type of screening at the resume submission stage.
16. A recent study by the National Institute of Justice found that while black and white drivers were stopped by the police at roughly equal rates, once stopped blacks were substantially more likely to have their cars and/or bodies searched. Contrary to arguments that the use of race is warranted by higher rates of illegal activity among blacks, searches of blacks produced significantly lower yield (evidence of contraband) compared to searches of whites (3.5 versus 14.5 percent). See Bureau of Justice Statistics 2003, tables 9, 11.
17. Such predicaments have been discussed elsewhere in many contexts. Some middle-class blacks indicate that they sometimes dress up even to go grocers shopping or to the mall, as a necessary precaution for dealing with street-wary whites. See Lacy 2004; Feagin and Sikes 1994. Having the luxury of wearing sweatpants to the grocery store is not something most whites think of as among the privileges they enjoy as a function of their skin color. The testers in this study dressed in "business casual," typically a button-down shirt and slacks.
18. The level of suspicion greeting black applicants has been further documented in the work of Susan Gooden (1999). Comparing the treatment of black and white welfare recipients, Gooden found that black job applicants were required to complete a pre-application twice as often and were significantly more likely to be subjected to drug tests and criminal background checks than were their white counterparts. Interviews with black applicants in Gooden's study were shorter and less thorough. In short, blacks faced additional hurdles in the application process, while simultaneously receiving fewer opportunities to demonstrate their qualifications.
19. Farmer and Terrell (2001) begin with the assumption that the higher rates of criminal activity among African Americans provide useful information in evaluating the criminal propensities of an unknown African American individual. Their estimates suggest that such inferences alone (without other mediating information) produce a rate of error whereby, at its logical extreme, an innocent African American would be almost five times more likely to be wrongfully convicted of a violent crime than an innocent white individual and, in the case of murder, eight times more likely.

REFERENCES

Anderson, Elijah. 1990. *Streetwise Race: Class and Change in an Urban Community*. Chicago: University of Chicago Press.

Becker, Howard. 1963. *Outsiders: Studies in the Sociology of Deviance*. New York: Free Press.

Bendick, Marc. Jr., Charles Jackson, and Victor Reinoso. 1994. Measuring employment discrimination through controlled experiments. *Review of Black Political Economy* 23: 25–48.

Bertrand, Marianne, and Sendhit Mullainathan. 2003. Are Emily and Brendan More Employable Than Lakisha and Jamal. A Field Experiment on Labor Market Discrimination. National Bureau of Economic Research Working Paper 9873. Cambridge Mass, NBER.

Blumstein, Alfred. 1982. On the racial disproportionality of United States prison populations. *Journal of Criminal Law and Criminology* 73: 1259–81.

———. 1993. Racial disproportionality revisited. *University of Colorado Line Review* 64: 743–760.

Bodenhausen, Galen, V., and Mervl Lichtenstein. 1987. Social stereotypes and information processing strategies: The impact of task complexity. *Journal of Personality and Social Psychology* 52: 871–880.

Bureau of Justice Statistics. 2006a. Survey of State Criminal History Systems. NCJ 210297. Washington, D.C.: U.S. Department of Justice.

———. 2006b. Prison and Jail Inmates at Midyear 2005. Washington, D.C.: U.S. Department of Justice.

———. 2003. Characteristics of Drivers Stopped by Police. 2002. Washington, D.C.: U.S. Department of Justice.

———. 2000. Correctional Populations in the United States, 1997. NCJ 177613. Washington, D.C.: U.S. Department of justice.

———. 1997. Lifetime Likelihood of Going to State or Federal Prison. Washington, D.C.: U.S. Department of Justice.

Bushway, Shawn. 2004. Labor market effects of permitting employer access to criminal history records. *Journal of Contemporary Criminal Justice* 20: 276–91.

Bussev, Jenny and John Trasvina. 2003. Racial Preferences: The Treatment of White and African American Job Applicants by Temporary Employment Agencies in California. Berkeley, Calif.: Discrimination Research Center of the Impact Fund.

Cole, David. 1995. The paradox of race and crime: A comment on Randall Kennedy's "Politics of Distinction." *Georgetown Law Journal* 83: 2547–71.

Crocker, Jennifer, Brenda Major, and Claude Steele. 1998. Social stigma. In *Handbook of Social Psychology*. ed. Daniel Gilbert, Susan Fiske, and Gardiner Lindzey. 504–53. Boston: McGraw-Hill.

Devine, Patricia and Scott Elliot. 1995. Are racial stereotypes really fading? The Princeton trilogy revisited. *Personality and Social Psychology Bulletin* 21 (11):1139–50.

Dovidio, John F. 2001. On the nature of contemporary prejudice: The third wave. *Journal of Social Issues* 57 (4): 829–49.

Eberhardt, Jennifer I., Phillip Atiba Goff, Valerie J. Purdie, Paul G. Davies. 2004. Seeing black: Race, crime, and visual processing. *Journal of Personality and Social Psychology* 87:876–93.

Farmer, Amy and Dek Terrell. 2001. Crime versus justice: Is there a trade-off? *Journal of Law and Economics* 44 (October): 345–66.

Feagin, Joe R., and Melvin P. Sikes. 1994. *Living with Racism: The Black Middle-Class Experience*. Boston: Beacon Press.

Fiske, Susan. 1998. Stereotyping, prejudice, and discrimination. In *The Handbook of Social Psychology*, 4th ed., ed. Daniel Gilbert, Susan Fiske, and Gardner Lindzey, 357–411. Boston: McGraw-Hill.

Garfinkel, Harold. 1956. Conditions of successful degradation ceremonies. *American Journal of Sociology* 61: 420–424.

Gilliam, Franklin D., and Shanto Iyengar. 2000. Prime suspects: The influence of local television news on the viewing public. *American Journal of Political Science* 44 (3): 560–73.

Gilliam, Franklin D., and Shanto Iyengar, Adam Simon, and Oliver Wright. 1996. Crime in Black and White: The violent, scary world of local news. *Harvard International Journal of Press/Politics* 1 (6): 6–23.

Goffman, Erving. 1963. *Stigma: Notes on the Management of a Spoiled Identity*. New York: Prentice-Hall.

Gooden, Susan T. 1999. The hidden third party: Welfare recipients' experiences with employers. *Journal of Public Management & Social Policy* 5: 69–83.

Graham, Sandra and Brian S. Lowery. 2004. Priming unconscious racial stereotypes about adolescent offenders. *Law and Human Behavior* 28: 483–504.

Harris, David. 1994. Factors for reasonable suspicion: When black and poor means stopped and frisked. *Indiana Law Journal* 69: 659–93.

Heckman, James J. 1998. Characterizing Selection Bias Using Experimental Data. NBER Working Paper 6699. Cambridge, Mass.: National Bureau of Economic Research.

Holzer, Harry J. 1996. *What Employers Want: Job Prospects for Less-Educated Workers*. New York: Russell Sage.

Holzer, Harry J., Steven Raphael, and Michael Stoll. 2006. Perceived criminality, criminal background checks and the racial hiring practices of employers. *Journal of Law and Economics* 49 (2): 451–80.

Klite, Paul, Robert A. Bardwell, and Jason Salzman. 1997. Local TV news: Getting away with murder. *Harvard International Journal of Press/Politics* 2: 102–12.

Lacy, Karyn R. 2004. Black spaces, Black places: Strategic assimilation and identity construction in middle-class suburbia. *Ethnic and Racial Studies* 27 (6):908–30.

Legal Action Center. 2004. *After Prison: Roadblocks to Recentry. A Report on State Legal Barriers Facing People with Criminal Records.* ed. Paul Samuels and Debbie Mukamal. New York: Legal Action Center.

Lodder, LeAnn, Scott McFarland, and Diana White, 2003. *Racial Preferences and Suburban Employment Opportunities.* Chicago: Legal Assistance Foundation of Metropolitan Chicago.

Nunes, Ana, and Brad Seligman. 2000. *A Study of the Treatment of Female and Male Applicants by San Francisco Bay Area Auto Service Shops.* Berkeley, Calif.: Discrimination Research Center of the Impact Fund.

Oliver, Mary Beth. 1994. Portrayals of crime, race and aggression in "reality-based" police shows: A content analysis. *Journal of Broadcasting and Electronic Media* 38 (2): 179–92.

Pager, Devah. 2007a. *Marked: Race, Crime, and Finding Work in an Era of Mass Incarceration.* Chicago: University of Chicago Press.

———. 2007b. The use of field experiments for studies of employment discrimination: Contributions, critiques, and directions for the future. *Annals of the American Academy of Political and Social Sciences* 609: 104–33.

Pager, Devah. 2003. The mark of a criminal record. *American Journal of Sociology* 108 (5): 937–75.

Pager, Devah, and Bruce Western. 2005. Discrimination in low trust labor markets. Paper presented at the Annual Meetings of the American Sociological Association, Philadelphia. August.

Pager, Devah, Bruce Western, and Bart Bonikowski. 2007. Discrimination in low-wage labor markets. Working paper, Princeton University.

Pettit, Becky, and Bruce Western. 2004. Mass imprisonment and the life course: Race and class inequality in U.S. incarceration. *American Sociological Review* 69: 151–69.

Quillian, Lincoln, and Devah Pager. 2001. Black neighbors, higher crime? The role of racial stereotypes in evaluations of neighborhood crime. *American Journal of Sociology* 107 (3):717–67.

Romer, Daniel, Kathleen H. Jamieson, and Nicole J. deCouteau. 1998. The treatment of persons of color in local television news: Ethnic blame discourse or realistic group conflict? *Communication Research* 25 (3): 286–305.

Sampson, Robert J., and Janet L. Lauritsen. 1997. Racial and ethnic dispartities in crime and criminal justice in the United States. *Crime and Justice: Ethnicity, Crime and Immigration: Comparative and Cross-National Perspectives* 21: 311–74.

Smith, Tom W. 1991. *What Americans Say About Jews.* New York: American Jewish Committee.

Sneiderman, Paul M., and Thomas Piazza. 1993. *The Scar of Race.* Cambridge, Mass.: Belknap Press of Harvard University Press.

Tonry, Michael. 1995. *Malign Neglect: Race, Crime, and Punishment in America.* New York: Oxford University Press.

Turner, Margery, Michael Fix, and Raymond Struyk. 1991. *Opportunities Denied, Opportunities Diminished: Racial Discrimination in Hiring.* Washington, D.C.: Urban Institute Press.

Wacquant, Loic. 2000. Deadly symbiosis: When ghetto and prison meet the mesh. *Punishment and Society* 3 (1): 95–134.

Word, Carl O., Mark P. Zanna, and Joel Cooper. 1974. The nonverbal mediation of self-fulfilling prophecies in interracial interactions. *Journal of Experimental Social Psychology* 10: 109–20.

DISCUSSION QUESTIONS

1. "Employers actively discriminate against convicted criminals in hiring decisions, and that is their legal right. Tough luck!" What would you say to someone who holds this belief based on the information in this reading?

2. Does society need a "Don't Ask, Don't Tell" policy for employers that are interviewing convicted criminals who have served their "time" and are not under any post-conviction restrictions (i.e., those required to register as sex

offenders)? Would such a policy, mandating employers not to make inquiries about a person's prior convictions and time-served, mitigate the effects of being blacklisted in hiring decisions? Would you support such a public policy? Why or why not?

3. Is long-term stigma, in the form of publicly available criminal records on convicted individuals, a social problem? Doesn't society have a right to know who its criminals and deviants are?

4. Does the potential stigma of being criminally convicted serve a deterrent function in society for law abiding citizens? Why or why not? What are the potential unintended consequences of removing the stigma of conviction and incarceration records from public scrutiny?

5. What factors might explain the much greater magnitude of hiring discrimination for black applicants with a criminal record over that of similarly situated white applicants with a criminal record? Is this a social problem that is more detrimental to blacks than whites? Why or why not? What should be done about this aspect of the social problem?

PART IX

✵

Social Problems Related to the Family

Many social problems are related to the family, and not all of them can be included here. The four selections cover divorce, lack of child care facilities, poverty and mother-only families, and homelessness—all serious problems with implications for many individuals and for society at large.

Stephanie Coontz asks readers to be realistic about why people choose divorce and how that choice may affect children. Her point is that divorce itself is usually not the problem; rather, the nature of the marriage and how the divorce is carried out tend to create serious problems for children.

Barbara Bergmann and Suzanne Helburn view child care in the United States as a crisis situation, with rippling consequences for the future educational outcomes of children in substandard child care and even their ultimate economic productivity. The authors argue that only a federally funded program can address this problem.

William Julius Wilson takes the issue of family fragmentation among African Americans head-on, addressing structural and cultural factors that contribute to the nearly 70% of black children who are born outside of marriage. This is a subject of both immense sociological importance and great political controversy. Too often, the labels "mother-only-family" or "female-headed household" have been political *codes* for poor, underclass, disadvantaged, culturally deprived (or morally depraved), welfare-dependent, and inner-city folk. Wilson knows this political minefield all too well, and handles this issue with scientific and moral perspicacity by not *blaming the victims* for the structural causes that underlie their reality.

The last article, by the National Alliance to End Homelessness, is entitled *Ending Family Homelessness: Lessons from Communities*. This article is hopeful because it suggests that homelessness is a social problem *with* a ready solution as seen in the many examples of innovative tactics that many cities across the nation are employing to address this problem.

32

Divorce in Perspective

STEPHANIE COONTZ

The Four Questions

1. What is the problem that Coontz identifies?
2. Is it *social?* Who is hurt? Is society hurt? Is the problem caused by society? Is there widespread concern about this problem?
3. What is the cause?
4. What are Coontz's suggestions for dealing with this problem?

Topics Covered

Divorce

Family

Marriage

Socialization

Children of divorce

Conflict

We have already accepted the fact that aging Americans are increasingly unlikely to live out their lives in "traditional" nuclear families, where they can be supported and cared for entirely by a spouse or a child. We can no longer assume that a high enough proportion of kids grow up with both biological parents that society can continue to ignore the "exceptions" that were there all along. Nor can our school schedules, work policies, and even emotional expectations of family life continue to presume that every household has a husband to earn the income and a wife to take care of family needs.

The social and personal readjustments required by these changes can seem awfully daunting. Here's what a spokesman for the Institute for American Values told me during a tape-recorded debate over whether it was possible to revive male breadwinning and restore permanent marriage to its former monopoly over personal life: "The strongest point in your argument is that the toothpaste is out of the tube. There's no longer the subordinate status of women to the extent there was in earlier eras—there is simply too much freedom and money sloshing around. We may be heading into what some sociologists call a 'postmarriage society' where women will raise the children and men will not be there in any stable, institutional way. If so, we'd better build more prisons, even faster than we're building 'em now."[1]

I don't think the consequences of facing reality are quite so bleak. As my grandmother used to say, sometimes problems are opportunities in work clothes. Changes in gender roles, for example, may be hard to adjust to, yet they hold out the possibility of constructing far more honest and satisfying relationships between men and women, parents and children, than in the past. But this doesn't mean that every change is for the better, or that we don't pay a price for some of the new freedoms that have opened up. Divorce is a case in point. While divorce has rescued some adults and children from destructive marriages, it has thrown others into economic and psychological turmoil.

For the family values crusaders, this is where the discussion of how to help families begins and ends. "Let's face it," one "new consensus" proponent told me privately, "the interests of adults and children are often different, and there are too many options today for parents to pursue personal fulfillment at the expense of their children's needs. Sure there are other issues. But unless we keep the heat on about the dangers of divorce, parents will be tempted to put their own selfish concerns above the needs of their children." Fighting the "divorce culture" has to be the top priority, he argued, because "it's the one thing we can affect" by making parents realize what disastrous consequences divorce has for the future of their kids.

I have met only a tiny handful of divorced parents who didn't worry long and hard about the effects of divorce on their children (almost a third of divorced women, for example, attempted a reconciliation between the time of initial separation and the final divorce, according to data from 1987–1988).[2] And while it's true that a few pop psychologists have made irresponsible claims that divorce is just a "growth experience," I don't believe we are really a culture that "celebrates" divorce, even if well-meaning people sometimes issue overly rosy reassurances to those who have undergone this trauma.

But for the record, let me be clear. Ending a marriage is an agonizing process that can seriously wound everyone involved, especially the children. Divorce can interfere with effective parenting and deprive children of parental resources. Remarriage solves some of the economic problems associated with divorce but introduces a new set of tensions that sometimes, at least temporarily, make things even more difficult.

Surely, however, it's permissible to put the risks in perspective without being accused of glorifying single parenthood or attempting, in Barbara Dafoe Whitehead's words, "to discredit the two-parent family." And the truth is that there has been a lot of irresponsible doom-saying about "disrupted" families. In a widely distributed article in the *Atlantic Monthly,* for example, Whitehead spends eight pages explaining why kids from divorced families face almost insurmountable deficits. Then, when she's convinced the average single mom to run out immediately to find a father for her child, she lowers the boom: Children in stepfamilies turn out even worse.[3]

While it is true that children in divorced and remarried families are more likely to drop out of school, exhibit emotional distress, get in trouble with the law, and abuse drugs or alcohol than children who grow up with both biological parents, most kids, from *every* kind of family, avoid these perils. And to understand what the increased risk entails for individual families, we need to be clear about what sociologists mean when they talk about such children having more behavior problems or lower academic achievement. What they really mean to say is *not* that children in divorced families have more problems but that *more* children of divorced parents have problems.

It's an important distinction, especially if you are a divorced or divorcing parent. It doesn't mean that all kids from divorced families will have more problems. There will be outstanding kids and kids with severe problems in both groups, but there will be a slightly higher proportion of kids from never-disrupted families in the outstanding group and a slightly lower proportion of them in the group with severe problems.

As family researcher Paul Amato notes, in measures of both achievement and adjustment, "a large proportion of children in the divorced group score *higher* than the average score of children in the nondivorced group. Similarly, a large proportion of children in the non-divorced group score *lower* than the average score of children in the divorced group." Comparing the average outcomes of children in various types of families obscures the fact that there is "more variability in the adjustment of children in divorced and remarried families than in nondivorced families." So knowing there are more divorced kids who do poorly is not really helpful. The question is how many more children from divorced and never-married parents are doing poorly, and what accounts for this, since some divorced children do exceptionally well and most are within normal range.[4]

Researchers who use clinical samples, drawn from people already in therapy because their problems are severe enough that they have sought outside help, tend to come up with the highest estimates of how many children are damaged by divorce. In

1989, for example, Judith Wallerstein published a long-term study of middle-class children from divorced families, arguing that almost half experienced long-term pain, worry, and insecurity that adversely affected their love and work relationships. Her work was the basis for Whitehead's claim in the *Atlantic Monthly* that "the evidence is in: Dan Quayle was right." But this supposedly definitive study, based on a self-selected sample of only sixty couples, did not compare the children of divorced couples with those of nondivorced ones to determine whether some of their problems might have stemmed from other factors, such as work pressures, general social insecurities, or community fragmentation. Moreover, the sample was drawn from families already experiencing difficulty and referred to the clinic for therapy. Only a third of the families were judged to be functioning adequately *prior* to the divorce.[5]

Research based on more representative samples yields much lower estimates of the risks. Paul Amato and Bruce Keith recently reviewed nearly every single quantitative study that has been done on divorce. Although they found clear associations with lower levels of child well-being, these were, on average, "not large." And the more carefully controlled the studies were, the smaller were the differences reported. The "large majority" of children of divorce, wrote eleven family researchers in response to Whitehead's misuse of their data, do not experience severe or long-term problems: *Most* do not drop out of school, get arrested, abuse drugs, or suffer long-term emotional distress. "Examining a nationally representative sample of children and adolescents living in four diverse family structures," write researchers Alan Acock and David Demo, "we find few statistically significant differences across family types on measures of socioemotional adjustment and well-being."[6]

Sara McLanahan, often cited by family values crusaders for her research on the risks of divorce, points out that divorce does not account for the majority of such social problems as high school dropout rates and unwed teen motherhood. "Outlawing divorce would raise the national high school graduation rate from about 86 percent to 88 percent.... It would reduce the risk of a premarital birth among young black women from about 45 percent to 39 percent."[7]

To be sure I'm not minimizing the risks, let's take a comparatively high estimate of divorce-related problems, based on the research of Mavis Hetherington, one of the most respected authorities in the field. She argues that "20 to 25 percent of kids from divorced families have behavior problems—about twice as many as the 10 percent from nondivorced families. You can say, 'Wow, that's terrible,'" she remarks, "but it means that 75 to 80 percent of kids from divorced families *aren't* having problems, that the vast majority are doing perfectly well."[8]

The fact that twice as many children of divorce have problems as children in continuously married families should certainly be of concern to parents. But it's important to remember that the doubling of risk is not evenly distributed among all families who divorce. Some of the families who contribute to these averages have had several divorces and remarriages. A study of boys who had been involved in divorce and remarriage found that those who had experienced many transitions, such as two or more divorces and remarriages, "showed the poorest adjustment." But even here the causal relationship involved more than divorce alone. It was "antisocial mothers" who were most likely *both* to experience many marital transitions and to engage in unskilled parenting practices that in turn placed their sons at risk for maladjustment.[9]

Many of the problems seen in children of divorced parents are caused not by divorce alone but by other frequently coexisting yet analytically separate factors such as poverty, financial loss, school relocation, or a prior history of severe marital conflict. When Rand Institute researchers investigated the relation between children's test scores and residence in a female-headed family, for example, the gross scores they obtained showed a significant deficit, but the disadvantage of children in mother-headed families was reduced to "essentially zero" when they controlled for other factors. "Apparently," they concluded, "a lot of the gross difference is ... due to income, low maternal education, and other factors that frequently characterize single-parent families, rather than family structure itself."[10]

Income differences account for almost 50 percent of the disadvantage faced by children in single-parent homes. The tendency of less-educated women to have higher rates of divorce and unwed motherhood also skews the statistics. In fact, a mother's educational background has more impact on her child's welfare than her marital status. Other research suggests that the amount of television kids watch affects aggressive behavior whether or not their parents are divorced; one survey found that eating meals together was associated with a bigger advantage in school performance than was having two parents. [11]

Researchers who managed to disentangle the effects of divorce itself from the effects of a change in residence found that relocation and loss of peer support were more likely to interfere with school completion than parental separation. McLanahan's research with Gary Sandefur suggests that up to 40 percent of the increased risk of dropping out of school for children in single-parent families is attributable to moving after the divorce. A 1996 study found that the impact of family structure on schooling is "reduced substantially" when the number of school changes is controlled.[12]

Obviously, divorce often triggers financial loss, withdrawal of parental attention, and change of residence or schools. In this sense it is fair to say that divorce causes many childhood problems by creating these other conditions. But it makes more sense to adopt policies to minimize income loss or school and residence changes than to prohibit divorce across the board, for there are no hard and fast links between family structure, parental behaviors, and children's outcomes. One pair of leading researchers in the field conclude that there is "*no* clear, consistent, or convincing evidence that alterations in family structure per se are detrimental to children's development."[13]

The worst problems for children stem from parental conflict, before, during, and after divorce—*or within marriage*. In fact, children in "intact" families that are marked by high levels of conflict tend to do worse than children in divorced and never-married families. Two researchers who compared family types and child outcomes over a period of five years found that children who remained in highly conflicted marriages had more severe behavior problems than children in any other kind of family. They "were more depressed, impulsive, and hyperactive" than children from either low-conflict marriages or divorced families.[14]

In the first two years after a divorce, says Hetherington, children of divorce "look worse off than kids from intact families, even bad intact families." But after two or three years, "the kids who lived in one-parent households, with a competent mother, were doing better—with half as many behavioral problems—than the kids in the conflict-ridden homes."[15]

Furthermore, the problems found in children of divorced parents were often there months, or even years, *prior* to the parental separation. Eight to twelve years before a family breakup, researchers have found, parents who would eventually divorce were already reporting significantly more problems with their children than parents who ended up staying together. This finding suggests that "the association between divorce and poor parent–child relationships may be spurious; the low quality of the parents' marriage may be a cause of both." Alternatively, severely troubled children may help to precipitate a divorce, further distorting the averages. [16]

Some behaviors that make kids look worse off in the first few years after divorce may actually be the first steps toward recovery from damaging family patterns. For example, psychologist Richard Weissbourd cites the case of Ann, a 10-year-old girl who had become the family caretaker to cope with her father's alcoholism and her mother's long work hours. This role gave her many satisfactions and a strong sense of importance within the family, but it cut her off from friends and schoolmates. After the divorce, Ann's mother recovered from her own stress, spent more time at home, and resumed her parental role. Ann resented her "demotion" in the family and began to throw temper tantrums that landed her in a therapist's office. Yet her turmoil, far from being evidence of the destructive effects of divorce, was probably a necessary stage in the move to healthier relations with both parents and peers.[17]

One long-term study found that divorce produced extremes of *either* negative or positive behavior in children. At one end of the spectrum were children who were aggressive and insecure. These children were likely to have been exposed to a disengaged or inconsistently harsh parenting style. A significant number of these children were boys who had been temperamentally difficult early in life and whose behavior problems were made worse by family conflict or divorce.

At the other extreme were caring, competent children who were exceptionally popular, self-confident, well behaved, and academically adept. These children had the most stable peer friendships and solid relations with adults. And a high proportion—more than half—of the girls in this group came from divorced, female-headed families. Their mothers were typically warm, but not always available, and most of these girls had had to assume some caretaking responsibility for siblings, grandparents, or even their mothers at a young age. "Experiences in a one-parent, mother-headed family seemed to have a positive effect for these girls."[18]

As Mavis Hetherington sums up the research, most family members go through a period of difficulty after a divorce but recover within two to three years. Some exhibit immediate and long-term problems, while others adapt well in the early stages but have problems that emerge later. "Finally, a substantial minority of adults and children ... emerge as psychologically enhanced and exceptionally competent and fulfilled individuals."[19] It should be reassuring for divorced parents to realize that such enhancement is possible, and that we also know a lot about how to avoid the *worst* outcomes for children.

WHEN DIVORCE HAS LONG-TERM EFFECTS, AND HOW TO MINIMIZE THEM

I don't want to trivialize the consequences of divorce. Transitions of any kind are stressful for kids, mostly because they are stressful for parents and therefore disrupt parental functioning. But it is important to point out the variability in the outcomes of divorce. Divorce is only one part of a much larger group of factors affecting parental functioning and child well-being. In many cases, the conditions imperiling children existed in the family prior to the divorce and would not be solved by convincing the parents to get back together. In other cases, divorce does create new problems but parents can minimize the disruptions if they set their minds to it. For these reasons, researchers have begun to emphasize that divorce is an ongoing process beginning long before physical separation and continuing long after the divorce is finalized. It is the *process,* not the divorce itself, that "is most significant in shaping subsequent family dynamics and individual adjustment."[20]

A critical factor in children's adjustment to divorce is how effectively the custodial parent functions. Usually this means the mother. The main problem for children of divorce is when depression, anger, or economic pressures distract their mother's attention. A recent study of 200 single-parent families in Iowa found that somewhere between 20 and 25 percent of mothers became preoccupied in the aftermath of divorce, paying less attention to what their children were doing or focusing too much on negative behavior and responding to it harshly. A large part of their reactions stemmed from economic stress. But many of these distracted mothers had always been more self-centered, impatient, disorganized, and insensitive than the other mothers, traits that may have triggered their divorces in the first place. Only a small amount of dysfunctional parenting seemed to be associated with family structure alone, yet there were enough incidences for researchers to conclude that divorce does make it harder for many mothers to parent well, even when they are stable individuals who are not overwhelmed by economic stress.[21]

The main danger for children is conflict between parents during and after divorce. Few marriages disintegrate overnight; the last few months or years are often marked by severe strife. More than half of divorced couples in one national survey reported frequent fighting prior to separation. More than a third of those who fought said that the fights sometimes became physical. And children were often present during these incidents.[22]

Divorce may allow parents to back off, but sometimes it produces continuing or even escalating conflict over finances and custody. And post-divorce marital conflict, especially around issues connected with the children, is the largest single factor associated with poor adjustment in youngsters whose parents have divorced. A study of more than 1,000 divorcing families in California found that children in disputed custody cases (about 10 percent of the sample) seemed the most disturbed. They were two to four times more likely than the national average to develop emotional and behavioral problems, with boys generally displaying more symptoms than girls.[23]

Parents certainly should be educated about the potential problems associated with divorce and with raising children alone. But outlawing divorce or making it harder to get would not prevent parents from fighting or separating, and could easily prolong the kinds of conflict and disrupted parenting that raise the risks for children. While people who are simply discontented or bored with their relationship should be encouraged to try and work things out, campaigns to scare parents into staying married for the sake of the kids are simply out of touch with the real complexities and variability in people's lives. As psychologist Weissbourd writes, "divorce typically has complex costs and benefits" for children. They may be more vulnerable in some ways after the divorce and more protected in other ways.[24]

Studies show that fathers in unhappy marriages tend to treat their daughters negatively, especially when the daughters are young. These girls may benefit by getting away from this negative spillover, even if their brothers go through a hard period. Women who are dissatisfied with their marriage are at high risk of developing a drinking problem. Divorce or separation lowers such women's stress and tends to reduce their alcohol dependence. Getting sober may improve their parenting enough to counteract the negative effects of divorce on their children.[25]

As such examples reveal, open conflict is not the only process that harms children in a bad marriage. One recent study of adolescent self-image found the *lowest* self-esteem in teens of two-parent families where fathers showed little interest in their children. Such youngsters, lacking even the excuse of the father's absence to explain his lack of interest, were more likely than kids in divorced families to internalize the problem in self-blame.[26]

Given these kinds of trade-offs, it is not enough to just reiterate the risks of divorce. We also need to tell people what they can do to minimize these risks. The most important thing is to contain conflict with the former spouse and to refrain from "bad mouthing" the other parent to the children. Divorcing parents must not involve their children in the hostilities between them or demand that the child choose sides. They should not ask children to report on the other's activities, or to keep secrets about what's going on in one household.[27]

Leaders of the "new consensus" crusade are fond of saying that trying to teach people how to divorce with less trauma is like offering low-tar cigarettes to people instead of helping them quit smoking. But this is a sound bite, not a sound argument. Yes, there are clearly people who could save their marriages, or at least postpone their divorces, and should be encouraged to do so by friends, colleagues, or professionals who know their situation. But there are others whose marriages are in the long run more damaging to themselves and their children than any problems associated with divorce. In between there are many more people for whom it's a close call. Yet since "most divorced mothers are as effective as their married counterparts once the parenting boundaries are renegotiated," and since most families recover from divorce within a few years, it is neither accurate nor helpful to compare divorce to a carcinogenic substance.[28]

Is it possible for divorced parents to behave civilly? A national sample of parents who divorced in 1978 and were interviewed one, three, and five years later found that half of the couples were able to coparent effectively. The other half, unfortunately, were unable "to confine their anger to their marital differences; it infused all the relationships in the family" and made cooperative or even civil coparenting extremely rare. A more recent California study found that three to four years after separation, only a quarter of divorced parents were engaged in such "conflicted coparenting," marked by high discord and low communication. Twenty-nine percent were engaging in cooperative coparenting, characterized by high communication and low dissension, while 41 percent were engaging in a kind of parallel parenting, where there was low communication between parents but also low conflict.[29]

Time helps. In one study of couples splitting up, "strong negative feelings among women dropped from 43 to 19 percent in the two years following separation, while for men they declined from 22 to 10 percent." There are encouraging signs that more immediate results can be obtained when parents are shown how their behavior affects their children. A recent study found that simply having children fill out a questionnaire and then sending that information to their divorced parents was enough to effect "significant change in the behavior of the parents." Specialists in divorce research recommend early intervention strategies to encourage parents to think of themselves as a "binuclear" family and separate their ongoing parenting commitments from any leftover marital disagreements.[30]

We know that people can learn to manage anger, and this seems to be the key to successful coparenting. It is not necessary for parents to like each other or even to "make up." The difference between divorced parents who were and were not able to coparent effectively, writes researcher Constance Ahrons, "was that the more cooperative group *managed* their anger better. They accepted it and diverted it, and it diminished over time." Establishing boundaries between the parents' relationship with each other and their relationships with their children is critical. For instance, it helps if parents have a friend to whom they can vent about all the crazy or terrible things they think their former partner has done. This is not something that should be discussed with the children. Most parents know this in the abstract, but they often need a friend, colleague, or professional to help them prevent their feelings about the partner from spilling over into interactions with their children.[31]

What about the problem of father absence? Surveys at the beginning of the 1980s found that more than 50 percent of children living with divorced mothers had not seen their fathers within the preceding year, while only 17 percent reported visiting their fathers weekly. But more recent studies show higher levels of paternal contact. A 1988 sample found that 25 percent of previously married fathers saw their children at least once a week, and only 18 percent had not visited their children during the past year. This may mean that as divorce has become more common, fathers have begun to realize that they must work out better ways of remaining in touch with their children, while mothers may be more willing to encourage such involvement.[32]

One of the puzzling findings of much divorce research is how little impact frequent visitation with fathers has on children's adjustment after divorce. But one recent study found evidence that while divorce weakened the significance of fathers for children's overall psychological well-being, a close relationship to a father, even when contact was minimal, seemed to have a strong association with a child's happiness. Other research shows that nonresidential fathers help their children best when they continue to behave as parents, "monitoring academic progress, emphasizing moral principles, discussing problems, providing advice, and supporting the parenting decisions of the custodial mother," rather than behaving as a friendly uncle who shows up to have fun with the children for a day.[33]

People need to know this kind of information, and a truly pro-family social movement would spend much more time publicizing such findings than making sweeping pronouncements about what's good and bad for children in the abstract. Again, it's a matter of coming to terms with reality. Historically speaking, the rise of alternatives to marriage is a done deal. Right here, right now, 50 percent of children are growing up in a home with someone other than two married, biological parents. It is not a pro-child act to deny divorced parents the information they need to help them function better or to try so hard to prevent divorce that we suppress research allowing parents to weigh their options, both pro and con.

Of course we should help parents stay together where possible, but the evidence suggests that we will save more marriages by developing new family values and support systems than by exhorting people to revive traditional commitments. And the fact remains that we will never again live in a world where people are compelled to stay married "until death do us part." Some couples will not be willing to go through the hard work of renegotiating traditional gender roles and expectations. Some individuals will choose personal autonomy over family commitments. Some marriages will fail for other reasons, such as abuse, personal betrayals, or chronic conflict—and often it is in no one's interest that such marriages be saved.

That is why, shocking though it may sound to the family values think tanks, we need, as researcher William Goode suggests, to "institutionalize" divorce in the same way that marriage remains institutionalized—to surround it with clear obligations and rights, supported by law, customs, and social expectations. To institutionalize divorce is not the same as advocating it. It simply means society recognizes that divorce will continue to occur, whether we like it or not. Reducing the ambiguity, closing the loopholes, and getting rid of the idea that every divorce case is a new contest in which there are no accepted ground rules will *minimize* the temptation for individuals to use divorce to escape obligations to children. Setting up clear expectations about what is civilized behavior will cut back on the adversarial battles that bankrupt adults and escalate the bitterness to which children are exposed.[34]

As one divorce lawyer writes, "I see couples every day who never lay a hand on one another but are experts in using children as instruments of psychological torture." Such children are not served well, she argues, by a high-minded refusal to sanction divorce or a rear-guard battle to slow it down. As the president of the Family and Divorce Mediation Council of Greater New York puts it: "Blaming children's problems on a megalith called 'Divorce' is a bit like stating that cancer is caused by chemotherapy. Neither divorce nor chemotherapy is a step people hope to have to take in their lives, but each may be the healthiest option in a given situation." He suggests that mediation "can restore parents' and children's sense of well-being" better than attempts to keep people locked in unhappy marriages where pent-up frustrations eventually make postdivorce cooperation even harder to obtain.[35]

Similarly, we need to institutionalize remarriage. Experts on stepfamilies argue that clearer norms and expectations for stepparents would facilitate easier adjustments and more enduring relationships. At the legal level, we must recognize and support the commitments that stepparents make, rather than excluding them from rights and obligations to their stepchildren. In one court case, for example, a boy named Danny was raised by his stepfather from the age of one, after his mother died. The biological father did not ask for custody until Danny was seven. Although a lower court ruled that Danny should be allowed to stay with his stepparent, who had been the primary parent for six of Danny's seven years, a higher court overruled this decision, calling the stepfather "a third party" whose claims should not be allowed to interfere with the rights of the biological parent.[36]

While we must adjust our laws to validate ties between stepparents and stepchildren, we also need to develop flexible models of various ways to achieve a "good" relationship. The worst problems facing stepfamilies, experts on remarriage now believe, are produced by unrealistic fantasies about re-creating an exclusive nuclear family unit in a situation where this is impossible because the child has at least one parent who lives outside the home. The best way to succeed, researchers in the field agree, is to reject the nuclear family model and to develop a new set of expectations and behaviors.[37]

What it all comes down to is this. Today's diversity in family forms, parenting arrangements, and sex roles constitutes a tremendous sea change in family relations. We will not reverse the tide by planting our chair in the sand like King Canute and crying, "Go back! Go back!" There are things we can do to prevent the global tide of changing work situations and gender roles from eroding as many marriages as it presently does, but our primary task is to find new and firmer ground on which to relocate family life.

ENDNOTES

1. David Blankenhorn, "Can We Talk? The Marriage Strategy," *Mirabella*, March 1995, p. 91.
2. Howard Wineberg and James McCarthy, "Separation and Reconciliation in American Marriages," *Journal of Divorce and Remarriage* 20 (1993).
3. Barbara Dafoe Whitehead, "Dan Quayle Was Right," *Atlantic Monthly*, April 1993, p. 55.
4. Paul R. Amato, "Life-Span Adjustment of Children to Their Parents' Divorce," *The Future of Children* 4, no. 1 (Spring 1994), p. 147; E. Mavis Hetherington, "An Overview of the Virginia Longitudinal Study of Divorce and Remarriage with a Focus on Early Adolescence," *Journal of Family Psychology* 7, no. 1 (June 1, 1993), p. 53.
5. Judith Wallerstein and Sandra Blakeslee, *Second Chances: Men, Women and Children a Decade After Divorce* (New York: Ticknor & Fields, 1989); Frank Furstenberg, Jr., and Andrew Cherlin, *Divided Families: What Happens to Children When Parents Part* (Cambridge, Mass.: Harvard University Press, 1991), p. 68; Andrew Cherlin and Frank Furstenberg, "Divorce Doesn't Always Hurt the Kids," *Washington Post*, March 19, 1989. Judith Wallerstein and Joan Kelly, *Surviving Breakup* (New York; Basic Books, 1980). Wallerstein and Kelly suggested that there was a "sleeper effect" for young women, where problems caused by divorce were not evident until years later. But a ten-year Australian study found "no convincing evidence" for such an effect. Rosemary Dunlop and Ailsa Burns, "The Sleeper Effect—Myth or Reality?" *Journal of Marriage and the Family* 58 (May 1995), p. 375. It is possible that the young women who reported such effects to Wallerstein were engaging in an after-the-fact attempt to explain why they were having troubles.
6. Paul Amato, "Children's Adjustment to Divorce," *Journal of Marriage and the Family* 55 (1993); Paul Amato and Bruce Keith, "Parental Divorce and the Well-Being of Children: A Meta-Analysis," *Psychological Bulletin* 110 (1991); Arlene Skolnick and Stacey Rosencrantz, "The New Crusade for the Old Family," *American Prospect*, Summer 1994, p. 62; Rex Forehand, Bryan Neighbors, Danielle Devine, and Lisa Armistead, "Interparental Conflict and Parental Divorce: The Individual, Relative, and Interactive Effects on Adolescents Across Four Years," *Family Relations* 43 (1994), p. 387; Bonnie Thornton Dill, Maxine Baca Zinn, and Sandra Patton, "Feminism, Race, and the Politics of Family Values," *Report from the Institute for Philosophy and Public Policy* 13 (1993), p. 17; Alan C. Acock and David H. Demo, *Family Diversity and Well-Being* (Thousand Oaks, Calif.: Sage, 1994), p. 213; P. Lindsay Chase-Lansdale, Andrew Cherlin, and Kathleen Kiernan, "The Long-Term Effects of Parental Divorce on the Mental Health of Young Adults: A Developmental Perspective," *Child Development* 66 (1995).
7. Sara S. McLanahan, "The Two Faces of Divorce: Women's and Children's Interests," *Macro-Micro Linkages in Sociology* (Newbury Park, Calif: Sage, 1991), p. 202. She notes that these "estimates assume that all of the negative impact of family disruption is due to the disruption itself as opposed to preexisting characteristics of the parents."
8. Kathryn Robinson, "The Divorce Debate: Which Side Are You On?" *Family Therapy Networker* (May/June 1994), p. 20.
9. D. M. Capaldi and G. R. Patterson, "Relation of Parental Transitions to Boys' Adjustment Problems: I. A Linear Hypothesis. II. Mothers at Risk for Transitions and Unskilled Parenting," *Developmental Psychology* 27, no. 3 (1991), p. 489; William S. Aquilino, "The Life Course of Children Born to Unmarried Mothers: Childhood Living Arrangements and Young Adult Outcomes," *Journal of Marriage and the Family* 58 (May 1996), p. 306.
10. David Grissmer, Sheila Nataraj Kirby, Mark Berends, and Stephanie Williamson, *Student Achievement and the Changing American Family* (Santa Monica, Calif: Rand Institute on Education and Training, 1994), p. 66; Doris R. Entwisle and Karl L. Alexander, "A Parent's Economic Shadow:

Family Structure Versus Family Resources as Influences on Early School Achievement," *Journal of Marriage and the Family* 57 (May 1995), p. 399.

11. Sara McLanahan and Gary Sandefur, *Growing Up with a Single Parent: What Hurts, What Helps?* (Cambridge, Mass.: Harvard University Press, 1995), pp. 2–3; Elizabeth Kolbert, "Television Gets a Closer Look as a Factor in Real Violence," *New York Times*, December 14, 1994; Rachel Wildavsky, "What's Behind Success in School?" *Reader's Digest*, October 1994, p. 52.
12. Sameera Teja and Arnold L. Stolberg, "Peer Support, Divorce, and Children's Adjustment," *Journal of Divorce and Remarriage* 20, no. 3/4 (1993); Robert Haveman, Barbara Wolfe, and James Spaulding, "The Relation of Educational Attainment to Childhood Events and Circumstances, *Institute for Research on Poverty Discussion Paper No. 908–90*, Madison, Wisconsin, 1990, p. 28; David Demo and Alan Acock, "The Impact of Divorce on Children," in Alan Booth, ed., *Contemporary Families: Looking Forward, Looking Back* (Minneapolis: National Council on Family Relations, 1991), p. 185; Maxine Baca Zinn and Stanley D. Eitzen, *Diversity in American Families* (New York: Harper and Row, 1987), p. 317; "Frequent Moving Harmful, Study Says," *Olympian*, July 24, 1996, p. A3; Jay Teachman, Kathleen Paasch, and Karen Carver, "Social Capital and Dropping Out of School Early," *Journal of Marriage and the Family* 58 (1996), p. 782.
13. Adele Eskeles Gottfried and Allen W. Gottfried, eds., *Redefining Families: Implications for Children's Development* (New York: Plenum, 1994), p. 224.
14. Furstenberg and Cherlin, *Divided Families*, p. 70; Amato and Keith, "Parental Divorce and the Well-Being of Children," p. 40; Andrew Cherlin, "Longitudinal Studies of Effects of Divorce on Children in Great Britain and the United States," *Science*, June 7, 1991, pp. 1386–1389; Joan Kelly, "Longer-Term Adjustment in Children of Divorce," *Journal of Family Psychology* 2 (1988); Larry Lettich, "When Baby Makes Three," *Family Therapy Networker* (January/February 1993), p. 66; Forehand et al., "Interparental Conflict and Parental Divorce," p. 387; Stacy R. Markland and Eileen S. Nelson, "The Relationship Between Familial Conflict and the Identity of Young Adults," *Journal of Divorce and Remarriage* 20, no. 3/4 (1993), p. 204.
15. Hetherington quoted in Robinson, "The Divorce Debate," pp. 27–28.
16. Furstenberg and Cherlin, *Divided Families*, p. 64; Paul R. Amato and Alan Booth, "A Prospective Study of Divorce and Parent–Child Relationships," *Journal of Marriage and the Family* 58 (May 1996), pp. 356–357; Robert E. Emery and Michele Tuer, "Parenting and the Marital Relationship," in Tom Luster and Lynn Okagaki, eds., *Parenting: An Ecological Perspective* (Hillsdale, N.J.: Lawrence Erlbaum, 1993), p. 135.
17. Richard Weissbourd, "Divided Families, Whole Children," *American Prospect* (Summer 1994), p. 69.
18. E. Mavis Hetherington, "Coping with Family Transitions: Winners, Losers, and Survivors," in *Annual Progress in Child Psychiatry and Child Development* (New York: Brunner/Mazel, 1990), pp. 237–239.
19. Hetherington, "Coping with Family Transitions," p. 221.
20. Marilyn Coleman and Lawrence H. Ganong, "Family Reconfiguring Following Divorce," in Steve Duck and Julia T. Wood, eds., *Confronting Relationship Challenges*, vol. 5 (Thousand Oaks, Calif: Sage, 1995), pp. 81–85. See also: Acock and Demo, *Family Diversity and Well-Being*, p. 224; Paul R. Amato, Laura Spencer Loomis, and Alan Booth, "Parental Divorce, Marital Conflict, and Offspring Well-being During Early Adulthood," *Social Forces* 73, no. 3 (March 1995), p. 895; Nan Marie Astone and Sara S. McLanahan, "Family Structure, Parental Practice and High School Completion," *American Sociological Review* 56 (June 1991), p. 318; Forehand et al., "Interparental Conflict and Parental Divorce," p. 392.
21. Furstenberg and Cherlin, *Divided Families*, p. 71; Ronald L. Simons and Associates, *Understanding Differences Between Divorced and Intact Families: Stress, Interaction, and Child Outcome* (Thousand Oaks, Calif.: Sage, 1996), pp. 208, 210, 222. For an argument that it is almost entirely family processes rather than divorce per se that cause poor outcomes, see Teresa M. Cooney and Jane Kurz, "Mental Health Outcomes Following Recent Parental Divorce: The Case of Young Adult Offspring," *Journal of Family Issues: The Changing Circumstances of Children's Lives* 17, no. 4 (July 1996), p. 510.
22. Furstenberg and Cherlin, *Divided Families*, p. 21.
23. Janet Johnston, "Family Transitions and Children's Functioning," in Philip Cowan et al., eds., *Family, Self, and Society: Toward a New Agenda for Family*

Research (Hillsdale, N.J.: Lawrence Erlbaum, 1993); Amato, "Life-Span Adjustment of Children to Their Parents' Divorce," p. 175; James Bray and Sandra Berger, "Noncustodial Father and Paternal Grandparent Relationships in Stepfamilies," *Family Relations* 39 (1990), p. 419.

24. Weissbourd, "Divided Families, Whole Children," p. 68.
25. Philip A. Cowan, Carolyn Pape Cowan, and Patricia Kerig, "Mothers, Fathers, Sons, and Daughters: Gender Differences in Family Formation and Parenting Style," in Cowan et al., eds., *Family, Self, and Society*, p. 186; Sharon Wilsnack, Albert Klasson, and Brett Schurr, "Predicting Onset and Pernicity of Women's Problem Drinking: A Five-Year Longitudinal Analysis," *American Journal of Public Health* 81 (1991), pp. 305–318.
26. Jennifer Clark and Bonnie Barber, "Adolescents in Postdivorce and Always-Married Families: Self-Esteem and Perceptions of Fathers' Interest," *Journal of Marriage and the Family* 56 (1994), p. 609.
27. Susan Gano-Phillips and Frank D. Fincham, "Family Conflict, Divorce, and Children's Adjustment," in Mary Anne Fitzpatrick and Anita L. Vangelisti, eds., *Explaining Family Interactions* (Thousand Oaks, Calif.: Sage, 1995), p. 207.
28. Emery and Tuer, "Parenting and the Marital Relationship," pp. 138–139.
29. Constance Ahrons, *The Good Divorce: Keeping Your Family Together When Your Marriage Comes Apart* (New York: Harper Perennial, 1994), p. 6; Amato, "Life-Span Adjustment of Children to Their Parents' Divorce," p. 167.
30. Joyce A. Arditti and Michaelena Kelly, "Fathers' Perspectives of Their Co-Parental Relationships Postdivorce: Implications for Family Practice and Legal Reform," *Family Relations* 43 (January 1994), p. 65; Furstenberg and Cherlin, *Divided Families*, pp. 26–27; Kevin P. Kurkowski, Donald A. Gordon, and Jack Arbuthnot, "Children Caught in the Middle: A Brief Educational Intervention for Divorced Parents," *Journal of Divorce and Remarriage* 20, no. 3/4 (1993), p. 149; Constance Ahrons and R. B. Miller, "The Effect of Postdivorce Relationship on Paternal Involvement: A Longitudinal Analysis," *American Journal of Orthopsychiatry* 63 (1993).
31. Ahrons, *The Good Divorce*, p. 82; Emery and Tuer, "Parenting and the Marital Relationship," p. 145. See also Melinda Blau, *Families Apart: Ten Keys to Successful Co-parenting* (New York: G. P. Putnam's Sons, 1993).
32. James H. Bray and Charlene E. Depner, "Perspectives on Nonresidential Parenting," in Charles E. Depner and James H. Bray, eds., *Nonresidential Parenting: New Vistas in Family Living* (Newbury Park, Calif.: Sage, 1993), pp. 6–7.
33. Paul R. Amato, "Father-Child Relations, Mother-Child Relations, and Offspring Psychological Well-Being in Early Adulthood," *Journal of Marriage and the Family* 56 (November 1994), p. 1039; Susan Chollar, "Happy Families: Who Says They All Have to Be Alike?" *American Health* (July/August 1993); Simons and Associates, *Understanding Differences Between Divorced and Intact Families*, p. 224; Bonnie L. Barber, "Support and Advice from Married and Divorced Fathers: Linkages and Adolescent Adjustment," *Family Relations* 43 (1994), p. 433.
34. William Goode, *World Changes in Divorce Patterns* (New Haven, Conn.: Yale University Press, 1993), pp. 330, 345; Robert Emory, "Divorce Mediation: Negotiating Agreements and Renegotiating Relationships," *Family Relations AA* (1995); Cheryl Buehler and Jean Gerard, "Divorce Law in the United States: A Focus on Child Custody," *Family Relations AA* (1995).
35. "Letters to the Editor," *New York Times*, December 31, 1995.
36. Andrew Cherlin, "Remarriage as an Incomplete Institution," *American Journal of Sociology* 84 (1978); Mark A. Fine, "A Social Science Perspective on Stepfamily Law: Suggestions for Legal Reform," *Family Relations* 38 (1989); Andrew Schwebel, Mark Fine, and Maureena Renner, "A Study of Perceptions of the Stepparent Role," *Journal of Family Issues* 12 (1991); Mark A. Fine and David R. Fine, "Recent Changes in Laws Affecting Stepfamilies: Suggestions for Legal Reform," *Family Relations* 41 (1992); Andrew Cherlin and Frank Furstenberg, "Stepfamilies in the United States: A Reconsideration," *American Review of Sociology* 20 (1994), p. 378.
37. Virginia Rutter, "Lessons from Stepfamilies," *Psychology Today* (May/June 1994), pp. 66–67; Lynn White, "Growing Up with Single Parents and Stepparents: Long-Term Effects on Family Solidarity," *Journal of Marriage and the Family* 56 (1994), p. 946.

DISCUSSION QUESTIONS

1. Is the high rate of divorce "good" or "bad"? What does it "prove," in your opinion?
2. Do you think that the increasing equality between men and women influences the divorce rate? Explain your answer.
3. Why does divorce become an especially difficult problem for children?
4. When is divorce probably a good thing for children?
5. If you were in a marriage relationship with children and you decided to get a divorce, what principles should you try to follow to minimize the effects on the children?
6. According to Coontz, "the leaders of the 'new consensus' crusade are fond of saying that trying to teach people how to divorce with less trauma is like offering low-tar cigarettes to people instead of helping them quit smoking." Coontz rejects this as an unsound argument. What do you think?
7. What does William Goode mean when he suggests that we "institutionalize" divorce? What good will this do, according to Goode? Do you agree?

33

What's Wrong with Child Care in America

BARBARA BERGMANN AND SUZANNE HELBURN

The Four Questions

1. What is the problem described? What values make it a problem?
2. Is it a *social* problem? What effect, if any, does this problem have for society? Are many people affected? Is the cause social?
3. What is the cause of this problem?
4. What exactly are Bergmann and Helburn arguing the United States should do about this problem?

Topics Covered

Family

Socialization

Poverty

Gender

Child care

Government

"I'm a single mom with a one-year-old and a three-year-old and a job with a not-so-hot wage. I went down to the local

SOURCE: From Barbara Bergmann and Suzanne Helburn, *America's Child Care Problem* (New York: Palgrave, 2002), pp. 1–12. Reprinted by permission.

child care center and was quoted a price of $13,500 per year for the two kids. That would take half my pay and leave me without money for rent. So I looked around for care that's cheaper. I found a lady who already cares for three kids in her home. Really, I have no way of knowing how nice she is to the children when she is alone with them or how many hours the kids will be propped up in front of a TV set or even left to themselves. I also don't know how often her boyfriend comes around while the kids are there and how nice he is to them. She has never bothered to get a license. But her price is half the center's—still a huge part of my budget, but just barely manageable. That's what I ended up "choosing." A friend next door works in the child care center I can't afford. Her wage doesn't make it into the not-so-hot category—it's $8.50 an hour. She says the center has some good workers on the staff, but because of the pay, the center has to fill some of its slots with people pretty much off the street who have no training at all. Turnover is high; this is hard on the center's operations and on the kids, who keep getting attached to caregivers, only to lose them. She is thinking of bailing out soon herself."

—BANK LOAN OFFICER

Looking back over the twentieth century, historians may well decide that the most important transformation it brought to America was the change in the role of women and the resulting change in the way our society finances and arranges for the care and rearing of young children. Yet as we enter the twenty-first century, we haven't yet faced up to the child care needs created by women's large-scale entry into the labor market.

The American child care system, in which parents, largely unassisted, must buy the care they need in the marketplace, has not worked well. It is in the public's interest that the services children receive be of good quality, but millions of parents are unable to pay what standard-quality services currently cost, much less what they would cost if quality were improved. Parents need assistance in two ways. They need more help in meeting the cost of child care. And they need more help in assuring the safety and quality of the care their children get.

As, in the twentieth century, fertility declined, as women's educational achievement and aspirations grew, as wages and jobs available to women increased, and as single parenthood grew through the rise in divorce and out-of-wedlock births, the two-parent family with one parent at home became less and less common. Now only about a third of preschool children are cared for that way.[1] Mothers' paychecks are an important source of support for millions of families, and, in a growing proportion of cases, are an indispensable source of support. Yet the high cost of child care makes severe inroads on those paychecks, and therefore on the standard of living of families. Child care costs can take away 25 percent or more of the incomes of low-wage families who have to pay for it.[2] And millions of children are not getting the quality of care that provides the safety, nurturing, and help with development that they need.

The kind of care children get, and the effect of the cost of that care on their family's standard of living are problems that deserve national attention. Fortunately, the country is starting to turn to this issue. The high cost of child care is a major cause of low living standards in families with children and of the social pathology that such conditions often cause. Obviously, that is—or should be—a public concern. The low quality of care that many young children are receiving should also be of public concern. It affects the kind of adult population we will have in the future—it helps determine how psychologically secure, how socially mature, how economically productive the future citizens of this country will be. Equally important, the care they are getting affects the quality of the life our children are leading right now—their feelings of happiness, security, and self worth. The care children get affects parents' ability to get to work reliably, as well as the level of security they feel while at work that their children are in good hands. This in turn affects worker productivity, labor turnover, and employers' costs of production.

Though there is general agreement that the American child care problem is serious, there is little agreement on what to do about it. Conservatives say mothers (with the exception of single mothers, perhaps) should stay home with their children, and they attribute many of today's societal ills to mothers' jobholding. Libertarians would rely totally on the free market to evolve a supply of care that would be appropriate to the country's needs in terms of quality and cost, and would favor withdrawing what government subsidies and regulations are now in place. Some argue that government and employer help to families with children discriminates against the childless, while others argue that parents are aiding society by raising children and deserve society's help in doing so. Many advocates of increased help from society look to community action—by business, charities and foundations—to mobilize the resources to improve the quality and availability of care in each locality. Others hope that state and local governments will contribute more than they now do to help parents with child care, and look to solve at least part of the problem through the increased provision of free prekindergartens. Finally, there are those, the present authors among them, who believe that only a large, active, and expensive federal program, providing both finance and a national framework for quality improvement, will serve the nation's purposes adequately.

Government assistance in child care is likely to come in the form of financial help to parents and to child care providers and in an improved regulatory system, rather than in a takeover by government of what has developed as a private industry. Nevertheless, the need for help is extensive, and its cost would run in the tens of billions of dollars of additional spending each year.

Government already provides expensive help to parents with the cost of children's upbringing. It pays the full cost of primary and secondary education. Government provides and pays for a major part of college education as well. Some government help has been extended for care and education in the preschool years, but that help is grossly underfunded. Many parents of young children, under great financial stress, are eligible for help under the current regulations but receive nothing because government funds appropriated for this purpose are insufficient or because they do not know of the program. A major increase in government funding for early care and education and a far better outreach to inform parents of the programs that exist are necessary if we are to relieve parents' financial stress, improve the quality of care, and make progress in reducing child poverty.

THE AFFORDABILITY PROBLEM

Child care is a "big-ticket" item for families. Care in a center for one preschool child averages more than $5,000 a year, and in some parts of the country, it is not unusual for center care to cost more than $10,000. Noncenter care can be had for 10 percent less, and care from providers who are not registered and may evade taxes comes somewhat cheaper. Even so, for most families who have to buy care, child care costs take a big chunk of the family's budget. The problem of costs is particularly acute for families with below-average wage incomes, many of which are single-mother families. The pay from a minimum-wage job would just about cover the average fees for two preschool children in a child care center, with nothing left over for food, clothing, and shelter. But the problem of affordability is by no means restricted to poor or near-poor families. Good quality child care is expensive even relative to the income of many two-earner families....

Child care's requirement for large amounts of labor is what makes it so expensive relative to parents' resources. Yet child care is performed by people who get low pay. (Lots of labor hired at low pay still costs a lot.) Some child care workers are college graduates, but many of them have poor education and minimal or no training. Turnover is high. Reforming the industry's hiring and pay practices would contribute to better quality care but would raise costs still further.

The high cost of child care raises questions about government help to pay for it. Should our society, through government programs, help parents with the cost of child care, as it helps them with the costs of elementary school, high school, and college? Should government support child care for all families, regardless of their income, as some countries do, or should government help be restricted to the lower-income families? Should government subsidize care by nannies, or by grandmas? Should government money subsidize child care by mothers who stay home with their children?

THE QUALITY PROBLEM

The quality issue is present wherever children are cared for—whether by their mothers or by others. Children need care that keeps them safe and happy. But they can be held back if care fails to address their developmental needs. Most experts—people whose profession is the study of child care—argue that care that fails to provide stimulus to development should not be considered good quality care. For children to develop properly they need to grow mentally, physically, morally, and emotionally. Moreover, they need to be socialized to interact amicably with others.

Quality and affordability are, of course, connected problems. Child care's high cost tempts some parents to go to off-the-books providers who may save them some money but who may provide care of lower quality or doubtful safety. Most of the low-income parents who get no government help feel they have no alternative to the cheapest care they can find. More affluent parents do have more leeway, but many don't spend the extra amount it takes to provide good quality. But price is not the only problem. Parents claim to want the kind of quality that experts specify, but many parents are not able to detect mediocre child care or even bad care when they see it.

As a result, many poor-to-mediocre child care facilities flourish in America, and indeed are in the majority. A high proportion of children are cared for by solo providers—friends, relatives, nannies, and family child care providers, who are women who take children into their homes. In such care, there is virtually no oversight—by supervisors, by colleagues, by regulators, even by the parents. Some of this solo care is of good quality. But given the lack of oversight, it is difficult to find out what is going on, much less to tell good from bad care. In the absence of evidence, parents have to resort to hearsay and intuition, which do not always give accurate assessments....

THE WAY OUT OF AMERICA'S CHILD CARE PROBLEM

In this [selection], we recommend an aggressive assault on the country's child care problem, led by and mostly financed by the federal government but with most of the administration performed by state and local governments and most services given by private-sector providers. Our proposal would solve the affordability problem and would allow us to make progress on the quality problem....

We argue that care for children of families with incomes at or below the poverty line should be subsidized completely out of public funds. Such families cannot afford to spend any of their income for child care; they need all of it for food, clothing, shelter. For families with incomes above the poverty line, we propose the following standard of "affordability": No family should have to lay out for child care more than 20 percent of its income *in excess* of the poverty line. If parents and the public were to share the costs of child care on such a basis, millions of middle-class families would be helped, along with millions of lower-income families. (For example, a married couple with one three-year-old would receive substantial help if their income was less than $40,000 under such a program.) Funding should suffice to cover all families eligible for services who apply for them.

There is currently little direct provision by government of child care for children under five, and therefore a large privately-run industry has come into existence. We do not envision the possibility that a substantial move away from private sector provision is likely or necessarily desirable. Whatever the merits and demerits in the use of vouchers for children in the K-12 grades (and we consider the demerits substantial), we do consider vouchers useful in administering a subsidy program for the early care and education of children under five years of age and for before- and after-school programs. They would give parents flexibility in choosing their child care provider.

The providers of care reimbursed under the program we advocate could be for-profit or non-profit, public or private, religious or secular, home based or center based. We recommend that only licensed providers be eligible to receive reimbursement with public funds. Some caregiving for four-year-olds would take place in the prekindergarten programs that states are currently advancing, and in Head Starts. The move to full-day kindergarten for five-year-olds would provide some additional hours of care, as well as additional hours of activities aimed at getting children ready for the first grade. Children under 13 would have access to before- and after-school programs and summer programs subsidized on the same basis.

We use the rubric "affordable care of improved quality" rather than "affordable quality care" because we believe a rapid and drastic improvement in child care quality, while eminently desirable, is not a realistic goal given the current state and organization of child care. Affordability can be provided in relatively short order if the will to spend the necessary funds is there. Quality is a more difficult matter. Improving the financial help to parents will allow many to switch their children out of low-cost informal care into licensed care, where the current quality is likely to be better. But raising the quality of existing providers will not be rapid or easy. Slow and gradual progress, under policies designed to encourage such progress is, we believe, the best that can be hoped for in the near term. Making progress on the quality problem would require that the 20 or so states with low standards in the ratio of staff to children amend their requirements. States should also set training requirements for center staff members and for those running family child care homes, where children are cared for in the caretaker's home. We advocate an improved regime of inspection of licensees and less delay in the suspension of those found to be delivering service of less than minimum adequacy. An increase in the number of providers who seek and are granted accreditation and an expansion of resource and referral services would be helpful. Setting up standards that providers must meet in order to receive federal funds would also be desirable.

Efforts to improve quality could not succeed without a flow of funds to pay for the costs to providers of meeting the higher standards. We recommend that reimbursement of providers be at a level that would allow them to deliver services of a quality that is average in the United States today, with higher reimbursement rates for those who demonstrate significantly higher quality. (Providing funds to pay the cost of currently average quality would allow providers delivering quality below the current average enough funds to raise their quality. If such an improvement by the bottom half of providers were to occur, the quality average would, of course, rise accordingly.) The additional funds many providers would receive would enable them to pay higher wages, which would, over time, be expected to increase the supply of trained workers and reduce turnover among them.

The total expenditure such a program would require is, we estimate, about $50 billion a year. The federal government now spends about $15 billion on child care, and the states together spend about $4 billion. So our proposed program would require about $30 billion of new money a year. It would be unrealistic to hope that employers and voluntary philanthropic efforts in each community would suffice to fill the funding gap. Nor can we expect state and local governments, with their limited taxing power, to be able to come forth with the needed resources. Only the federal government would be able to finance a program of this magnitude and insure that children in every community in the country get the care they need.

We recognize that an expensive public program of this magnitude does not accord with the common assumption that "the era of big government is over" and with the seeming widespread acceptance of the idea that the closing of that era was a good thing. Nevertheless, certain large expenditures do from time to time get added to the federal budget. Prescription drug coverage for the elderly was advocated by both candidates in the presidential election of 2000. Additions to the military budget of funds that would be sufficient to pay for much if not all of the child care program we propose is in prospect at this writing. A much-expanded program of subsidies for child care would not be politically possible without considerable agitation for it, even in an era of budget surplus. Yet polling data indicate that there is already a basis in public opinion for considerably expanded government help with child care, particularly for lower-income working parents; in one recent poll 63 percent of respondents gave support to increased federal spending to provide child care assistance to working parents.[3]

Powerful opposition will come from those who regard the movement of mothers out of the home and into jobs as a terrible mistake. Yet most people understand that for better or worse the mothers of small children will continue to hold jobs and need child care. Whether mothers "need" to work, want to work, or find that working is the best of all the alternatives open to them, they do and will work. The practical question that faces the country is how to deal with the child care needs that result.

A system built around the principle that parents should have to pay no more than 20 percent of their income above the poverty line for child care of approved quality, if enacted and funded so that all eligibles who sought places in the system could have them, would effectively solve the "afford-ability" problem. The trend that is evident toward free public provision of full-day kindergarten and prekindergarten, which could provide care and education for children of age four and five, and the increases that have been recently made in appropriations for child care subsidies show that these are popular and politically viable programs.

A program like the one we have outlined should garner support from the public school teachers' unions and the more organized parts of the private child care industry—the for-profit companies and the religious groups that run child care centers. There is a third group, far more powerful than the first two, from which support might also be enlisted: employers. Employers stand to gain in two ways. First, a more reliable child care system would reduce absenteeism and tardiness. Second, employers would be relieved of pressure to provide subsidized child care as a fringe benefit. There is, of course, a fourth group: Parents, many of whom would be relieved of a good portion of the heavy financial pressure that paying for child care involves, and who would be relieved of the anxieties that attend sending one's child to "informal" care of doubtful reliability, quality, and safety.

ENDNOTES

1. The source of these data are the U.S. Bureau of Labor Statistics web page. We refrain from giving exact web addresses, as they are likely to change through time.
2. Lynne M. Casper, "What Does It Cost to Mind Our Preschoolers?" *Current Population Reports*, pp. 70–52. (Washington DC: U.S. Bureau of the Census, September 1995).
3. This poll was done by Princeton Survey Research Associates in January 1998.

DISCUSSION QUESTIONS

1. Does this problem rise to the level of a social problem? Do the authors make a good case? Do you agree? What core values are we considering?
2. What exactly do the authors want to change? Will this be effective? Will it lead to unintentional consequences that are serious?
3. Do you believe that child care is the responsibility of society, or must it remain the responsibility of the family? Or, perhaps, is it now the responsibility of society already?
4. What is the likelihood that the federal government will do what is suggested by the authors?
5. If you favor what the authors suggest, what single argument would you use to influence political leaders? If you do *not* favor what the authors suggest, what single argument would you use to influence political leaders?

34

The Fragmentation of the Poor Black Family: Culture vs. Structure

WILLIAM JULIUS WILSON

The Four Questions

1. What is the problem that Wilson describes?
2. Is this a *social* problem? What effect is this problem having on society? Who is affected by this problem?
3. What is the cause of this problem?
4. What can be done to address this problem?

Topics Covered

Black family

Poverty

Moynihan report

Single-parenthood

Non-marital births

Delayed entry into marriage

Structured explanations

Cultural explanation

Marriage rates

Chronic unemployment

Inner-city joblessness

In 1965, a report entitled "The Negro Family: The Case for National Action" was published by the assistant secretary of labor, Daniel Patrick Moynihan, who in 1976 would be elected the Democratic senator from New York. In what quickly became known as "the Moynihan report," the author asserted that racial inequality, combined with the breakdown of the black family, was creating a "new crisis in race relations." Moynihan combined structural and cultural arguments to analyze the deteriorating state of black families. The report created a firestorm of controversy due in part to the racial climate and popular ideology at the time it was published....

Given the volatility of race relations, liberal critics believed that Moynihan's cultural arguments amounted to blaming African Americans for their own misfortune. This criticism ignored Moynihan's careful attention to structural causes of inequality, and it created a backlash against the report that essentially shut down meaningful conversation about the role of culture in shaping racial outcomes. The Moynihan report is a particularly pertinent subject for a discussion of the black family because it not only anticipated later developments in black family fragmentation, but the controversy it generated clearly made the African American family the central focus of the structure-versus-culture debate.

In a *New York Times* obituary for Daniel Patrick Moynihan in March 2003, I was quoted as saying that the Moynihan report is an important and prophetic document.[1] I still stand by that statement. The report is important because it continues to be a reference for studies on the black family. It is prophetic because Moynihan's predictions about the fragmentation of the African American family and its connection to inner-city poverty were largely borne out.

... Dramatic statements made in the report drew press attention and were often taken out of context. For example, in his chapter entitled "The Tangle of Pathology," Moynihan boldly stated, "At the heart of the deterioration of the fabric of Negro society is the deterioration of the Negro family. It is the fundamental weakness of the Negro community at the present time" and "at the center of the tangle of pathology is the weakness of the family structure. Once or twice removed, it will be found to be the principal source of most of the aberrant, inadequate, or antisocial behavior that did not establish, but now serves to perpetuate the cycle of poverty and deprivation."[2]

Reporters and columnists organized their coverage around the attention-grabbing statements on the breakdown of the black family, and readers who had not read the actual document often had no idea that Moynihan had devoted an entire chapter to the root causes of family fragmentation, including urbanization, unemployment, poverty, and Jim Crow segregation. A *Washington Post* bylined article noted, according to "White House sources," that the Watts riot in 1964—a race riot in a suburb of Los Angeles—strengthened President Johnson's "feeling of the urgent need to restore Negro families' stability."[3] Accordingly, as Lee Rainwater and William L. Yancey observed, by the time many critics, including black critics, got around to reading the report, they "could no longer see it with fresh eyes but were instead heavily influenced by their exposure to the press coverage, particularly as this coverage tied the report to an official 'explanation' for Watts."[4]...

The logic put forth by proponents of the black-perspective explanation is interesting because it does not even acknowledge self-destructive behavior in the ghetto. This is a unique response to the dominant American belief system's emphasis on individual deficiencies, rather than the structure of opportunity, as causes of poverty and welfare. Instead of challenging the validity of the underlying assumptions of this belief system, this approach sidesteps the issue altogether by denying that social dislocations in the inner city represent any special problem. Researchers who emphasized these dislocations—such as persistent unemployment, crime, and drug use—were denounced, even when their work rejected the assumption of individual responsibility for poverty and welfare and focused instead on the structural roots of these problems.[5]...

Several trends that had earlier worried Moynihan have become much more pronounced. One-quarter of all nonwhite births were to unmarried women in 1965, the year Moynihan wrote the report on the Negro family, and by 1996 the proportion of black children born outside of marriage had reached a high of 70 percent; it then dipped slightly to 69 percent in 2005 (see Figure 34.1).[6] And in 1965 a single woman headed 25 percent of all nonwhite families; by 1996, however, the proportion of all black families headed by a single woman had

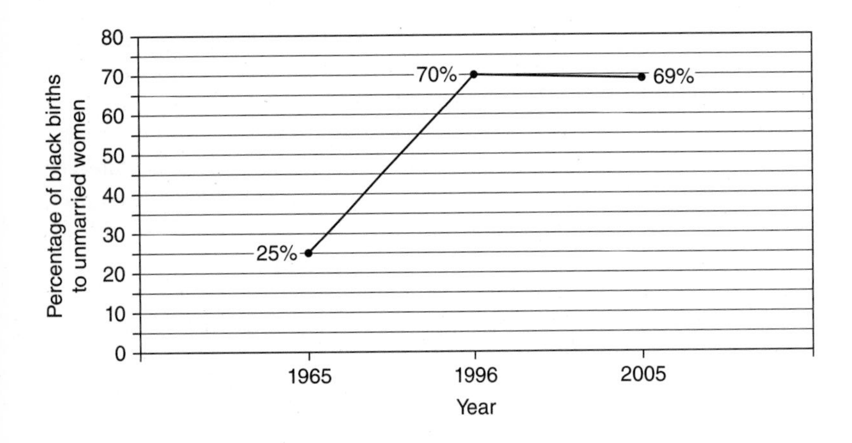

FIGURE 34.1 Nonmarital Births among Blacks*

*Statistics were not available for blacks in 1965. It is estimated that blacks comprised roughly 90 percent of those classified as "nonwhites."

SOURCE: US Department of Health, Education, and Welfare, *Trends in Illegitimacy, 1940–1965.* National Vital Statistics System, Series 21, No.15 (Washington, DC: Government Printing Office, 1974); US Department of Health and Human Services, *Nonmarital Childbearing in the United States, 1940–1999.* National Vital Statistics Reports, Vol. 48, No. 16 (Washington, DC: Government Printing Office, 2000); and US Department of Health and Human Services, *Births: Preliminary Data for 2006* (Washington, DC: Government Printing Office, 2006).

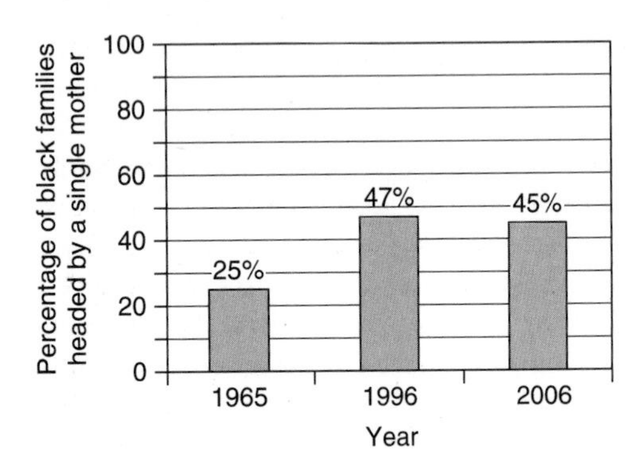

FIGURE 34.2 Black* Families Headed by a Single Mother

*Statistics were not available for blacks in 1965. It is estimated that blacks comprised roughly 90 percent of those classified as "nonwhites."

SOURCE: US Bureau of the Census, *Current Population Survey* (Washington, DC Government Printing Office, 1996), and US Bureau of the Census, *Current Population Survey* (Washington, DC Government Printing Office, 2006).

swelled to 47 percent, dropping slightly to 45 percent in 2006 (see Figure 34.2).[7]

One reason for Moynihan's concern about the decline in the rate of marriage among blacks is that children living in single-parent families in the United States, especially families in which the parents have never been married to each other, suffer from many more disadvantages than those are raised in married-parent families. A study relying on longitudinal data (data collected on a specific group over a substantial period of time) found that, in the United States, persistently poor families (defined as having family incomes below the poverty line during at least eight years in a ten-year period) tended to be headed by women, and 31 percent of all persistently poor households were headed by nonelderly *black* women.[8] This is an astounding figure, considering that African Americans account for just 12.4 percent of the entire U.S. population.[9]

As sociologist Kathryn Edin pointed out, "more children are poor today than at any time since before Lyndon Johnson's War on Poverty began three decades ago. Children living in households headed by single mothers are America's poorest demographic group. This fact is not surprising, since low-skilled single mothers who work seldom earn enough to bring their families out of poverty and most cannot get child support, medical benefits, housing subsidies, or cheap child care."[10]

In 2006, whereas the median income of married-couple families with children was $72,948, the median income of single-parent families in which the mother was divorced was $35,217. For families in which the mother had never been married, the median income was only $18,111. Likewise, whereas less than one-tenth of children in husband-wife families were living below the poverty line, nearly one-quarter of those living with divorced mothers and over half of those living with mothers who had never been married were classified as poor.[11] Recent research firmly supports the idea that some of the increase in the child poverty rate can be attributed to changes in family structure (i.e., the gradual shift from married-couple to single-parent families).[12] Finally, longitudinal studies also revealed that before passage of the welfare reform legislation in 1996, mothers who had never been married received assistance from Aid to Families with Dependent Children (AFDC) for a significantly longer period than did separated or divorced mothers.[13]

In addition to the strong connection linking single parenthood with poverty and welfare receipt, the available research indicates that children from low-income households without fathers present are more likely to be school dropouts, become teenage parents, receive lower earnings in young adulthood, be welfare recipients, and experience cognitive, emotional, and social problems. Moreover, daughters who grow up in single-parent households are more likely to establish single-parent households themselves than are those raised in married-couple households. And finally, single-parent households tend to exert less control over the behavior of their adolescent children.[14]

Although statistics that record nonmarital births provide some sense of the extent to which mothers are having children outside of marriage, they do not provide an accurate picture of how many mothers are actually parenting alone. The reason is that nonmarital-birth statistics measure marital status at the time of birth only, and do not take into account other types of co-parenting relationships, such as couples who marry after the birth of a child. For example, among women with nonmarital first births, 82 percent of whites, 62 percent of Hispanics, and 59 percent of blacks married by age forty.[15] However, according to one recent study of fragile families in seven cities, when the focus was low-income minority families, three-fourths of which were black, only 15 percent of unmarried mothers were married by the time of their child's first birthday.[16] ...

Analyzing data from our study, Martha Van Haitsma found that "network differences translate into childcare differences. Mexican women with young children are significantly more likely than their black counterparts to have regular childcare provided by a friend or relative."[17] The high proportion of two-adult Mexican American households with working fathers, particularly among immigrant Mexicans, may be an important factor in the mother's greater access to networks of child care.

Also in our study, the high percentage of black mothers who lived with young children in a single-adult household was associated with problems finding and keeping a job—what social scientists call labor force attachment.[18] If a single mother in Chicago's inner city lived in a *co-residential household*—that is, a household that included at least one other adult—and received informal child care, she significantly improved her chances of entering the labor force. Inner-city mothers who were not on welfare, lived in a co-residential household, and received informal child care had a very high (90 percent) probability of labor force activity; those who maintained sole-adult households and did not receive informal child care had a much lower (60 percent) probability of working. Of the 12 percent of inner-city welfare mothers who candidly reported that they worked at least part-time—probably in the informal economy—those who lived in a co-residential household and received informal child care were more than five times as likely to work as were those who lived in single households and did not receive informal child care.[19]

Given the sharp increase in single-parent families and out-of-wedlock births in the African American community and the research showing a relationship between these trends and economic hardships, few serious scholars would maintain that Moynihan's concerns about the changes in the black family were unjustified, even though the percentage of nonmarital births and single-mother families has increased among whites and Latinos as well.[20] What continues to be disputed is how we account for the fragmentation of the African American family. What is often overlooked is that

Moynihan attempted to synthesize structural and cultural analyses to understand the dynamics of poor black families. This relationship between structure and culture is explored in the remainder of this chapter.

THE ROLE OF STRUCTURAL FACTORS

The explanations most often heard in the public debate over the last several decades associate the increase in out-of-wedlock births and single-parent families with welfare. The general public discussions and proposals for welfare reform reflected a wide assumption that there was a direct causal link between the level or generosity of welfare benefits and the likelihood that a young woman would bear a child outside of marriage. The fact that welfare recipients received benefits for their children was assumed to provide an incentive far additional childbearing. This widespread belief led several states, as part of the 1996 welfare reform, to adopt a family cap policy that stopped increasing benefits for additional babies born. In addition, welfare policy is thought to discourage marriage because of the way income limits are used to qualify someone for benefits. In 1998 the federal poverty line was $13,133 for a single parent with two children, and it was just $3,400 more for a dual-parent household with two children ($16,530). Therefore, it is very easy for a dual-parent family to disqualify itself from welfare, even if earnings are very low.[21] Accordingly, welfare has been construed as a disincentive for marriage.

However, scientific evidence gathered in the early 1990s offered little support for these claims.[22] Research examining the association between the generosity of welfare benefits and out-of-wedlock childbearing and teen pregnancy prior to the enactment of welfare reform in 1996 indicated that benefit levels had no significant effect on the likelihood that African American girls and women would have children outside of marriage; likewise, benefit levels had either no significant effect or only a small effect on the odds that whites would have children outside of marriage. There is no evidence to suggest that welfare was a major factor in the rise of childbearing outside of marriage.

The corollary to the "welfare encourages childbirth outside of marriage" assumption is that decreases in welfare benefits also hinder nonmarital births. But this does not hold true. The states with the largest declines in AFDC benefits did not register the greatest slowdown in out-of-wedlock births.[23] The rate of out-of-wedlock teen childbearing had nearly doubled between 1975 and 1996, the year welfare reform was enacted, even though the real value of AFDC, food stamps, and Medicaid during that period had fallen, after adjustment for inflation. Indeed, while the real value of cash welfare benefits had plummeted from the mid-1970s to the mid-1990s, not only had out-of-wedlock childbearing increased, but the tendency for partners to marry following the birth of their child had decreased as well.[24]

In *The Truly Disadvantaged,* I argued that the sharp increase in black male joblessness since 1970 accounts in large measure for the rise in the rate of single-parent families, and that because jobless rates are highest in the inner-city ghetto, rates of single parenthood are also highest there. Thus, many of the structural factors that have contributed to the increase in joblessness among low-skilled black males and the corresponding sharp decline in their income discussed in the previous chapters—the decreased relative demand for low-skilled labor caused by the computer revolution, the growing internationalization of low-skilled labor, the decline of the manufacturing sector, and the growth of service industries—logically extend to a discussion of the factors that contribute to the fragmentation of black families.

Whereas almost 27.5 percent of young African American men nationwide (ages eighteen to twenty-nine) with annual earnings of over $35,000 were married in 2006, the marriage ratio decreased steadily for those earning less than that—21.7 percent for those earning between $25,000 and $35,000, 15.3 percent for those earning between $15,000 and $25,000, 7.6 percent for those earning between

$2,000 and $10,000, and only 7.5 percent for those with no reported earnings.[25]

However, research on the relationship between male employment and rates of marriage and single parenthood has yielded mixed findings. Although there is a strong association between rates of marriage and both employment status and earnings at any given point in time, national longitudinal studies suggest that these factors account for a relatively small proportion of the overall *decline* in marriage among African Americans. Harvard professor Christopher Jencks points out that the decline in the proportion of African American men who were married and living with their wives was almost as large among those who had worked throughout the previous years as among black men in general.[26] In addition, studies have shown only modest support for the hypothesis linking the sharp rise in poor, single-parent families to the declining employment status and income of young black men.[27]

However, these studies are based on national data, not data specific to inner-city neighborhoods where many experiences relate to race and poverty. How much of the decline in the black marriage rate in the inner city can be accounted for by the increasing joblessness among black males? Our study of family life and poverty in Chicago's inner city was not a longitudinal study, but we did collect retrospective (or life history) marriage and employment data that help us estimate trends over time. An analysis of respondents' retrospective data comparing the employment experiences of different age groups (cohorts) revealed that marriage rates had dropped much more sharply among jobless black fathers than among employed black fathers. But this drop applies only to the younger cohorts. Analyzing data from our survey, Mark Testa and Marilyn Krogh found that although employment had no significant effect on the likelihood that black single fathers ages thirty-two to forty-four would eventually marry, it increased that likelihood by eight times for single fathers eighteen to thirty-one years old.[28]

Joblessness among black men is a significant factor in their delayed entry into marriage and in the decreasing rates of marriage after a child has been born, and this relationship has been exacerbated by sharp increases in incarceration that in turn leads to continued joblessness. Nevertheless, much of the decline in marriages in the inner city, including marriages that occur after a child has been born, remains unexplained when only structural factors are examined.

THE ROLE OF CULTURAL FACTORS

Even though whites and Latinos have also experienced an increase in percentage of nonmarital births and single-mother families, although at a slower rate than among African Americans, social scientists continue to argue about whether unique cultural factors may account for the fragmentation of the African American family. As mentioned earlier, the controversy over the Moynihan report resulted in a persistent taboo on cultural explanations to help explain social problems in the poor black community.

If the public release of the Moynihan report was untimely in terms of the changing racial climate, it was also unfortunately published during heated debate over anthropologist Oscar Lewis's work on the culture of poverty. Although Lewis's work had been conducted among Spanish-speaking persons in Mexico, Puerto Rico, and New York City, his argument that poverty is passed from generation to generation through learned behaviors and attitudes was irresistibly attractive—or repellent—to persons interested in the plight of poor black Americans. Indeed, the Moynihan report quickly became a reference point for debates about the culture of poverty. The link between the report and Lewis's theory was made especially explicit following the publication of an article and later a book, both titled "Blaming the Victim," written by the Boston psychologist and civil rights activist William Ryan as a critique of the Moynihan report.[29] "Blaming the victim" became a slogan repeatedly used by critics of the culture-of-poverty thesis, and they made repeated reference to the Moynihan report when voicing their criticisms.

Relying on participant observation and life history data to analyze Latin American poverty, Oscar Lewis described the culture of poverty as "both an adaptation and a reaction of the poor to their marginal position in a class-stratified, highly individuated capitalistic society." However, he also noted that once the culture of poverty comes into existence, "it tends to perpetuate itself from generation to generation because of its effect on the children. By the time slum children are age six or seven they have usually absorbed the basic values and attitudes of their subculture and are not psychologically geared to take full advantage of changing conditions or increased opportunities which may occur in their lifetime."[30] Although Lewis later modified his position by placing more weight on external societal forces than on self-perpetuating cultural traits to explain the behavior of the poor, conservative social scientists embellished the idea that poverty is a product of "deeply ingrained habits" that are unlikely to change following improvements in external conditions.[31]...

... Like Oscar Lewis, Moynihan relates cultural patterns to structural factors and then discusses how these patterns come to influence other aspects of behavior. In the concluding chapter of his report, for example, Moynihan states that the situation of the black family "may indeed have begun to feed on itself." To illustrate, he notes that from 1948 to 1962, the unemployment rate among black males and the number of new AFDC cases were very highly correlated. After 1962, however, the trend reversed itself for the first time. The number of new AFDC cases continued to rise, but black male unemployment declined.[32] "With this one statistical correlation, by far the most highly publicized in the Report," states the historian Alice O'Connor, "Moynihan sealed the argument that the 'pathology' had become self-perpetuating; pathology, here measured as welfare 'dependency,' was no longer correlated with the unemployment rate: it was going up on its own."[33] Also like Oscar Lewis, Moynihan talks about the adverse effects of children being exposed to the cultural environment or, as he puts it, to the tangle of pathology in the ghetto....

The decline in the rate of marriage among inner-city black parents in the last several decades is a function not simply of deepening scarcity of jobs for low-skilled workers or of changing attitudes toward sex and marriage in society at large, but of, as Testa emphaszes, "the interaction between material and cultural constraints."[34] The important point is that "variation in the moral evaluations that different sociocultural groups attach to premarital sex, out-of-wedlock pregnancy, and nonmarital parenthood affects the importance of economic considerations in a person's decision to marry."[35] That is, discouraging economic conditions tend to reinforce any tolerance for having children without marriage or even partnering. The weaker the norms are against premarital sex, out-of-wedlock pregnancy, and nonmarital parenthood, the more economic considerations affect decisions to marry.

The data we collected in the late 1980s in our random survey of nearly 2,500 poor and nonpoor African American, Latino, and white residents in Chicago's inner-city neighborhoods shed light on this relationship.[36] Only 28 percent of the African American parents aged eighteen to forty-four were married when they were interviewed in 1987 and 1988, compared with 75 percent of the Mexican American parents, 61 percent of the white parents, and 45 percent of the Puerto Rican parents. African Americans in these neighborhoods suffered from higher levels of joblessness and higher rates of concentrated poverty (the percentage of poor families in a neighborhood), which accounted for some of the differences. But even when ethnic-group variations in work activity, poverty concentration, education, and family structure were taken into account, significant differences between inner-city blacks and the other groups, especially the Mexican Americans, remained.[37] Accordingly, it is reasonable to consider the influence of cultural variables in accounting for some of these differences.

A brief comparison between inner-city blacks and inner-city Mexican Americans (many of whom are immigrants) in terms of family perspectives provides some evidence of cultural differences. Marriage and family ties are subjects of "frequent and intense discourse" among Mexican American immigrants.[38] Mexicans come to the United States with a clear conception of a traditional family unit that features men as breadwinners. Although extramarital affairs by men are tolerated, "a pregnant, unmarried woman is a source of opprobrium, anguish, or great concern."[39] Pressure is applied by the kin of both parents for the couple to enter into marriage. Religion undoubtedly plays a role in Mexican American marital sanctions. Mexicans have been strongly influenced by Roman Catholicism, a religion that discourages divorce and out-of-wedlock pregnancies. It is therefore reasonable to assume that the Mexican cultural framing of marriage is significantly influenced by religious beliefs and other traditional conceptions of what constitutes an appropriate family unit.

The intensity of the commitment to the marital bond among Mexican immigrants will very likely decline, the longer they remain in the United States and are exposed to U.S. norms and the changing opportunity structures for men and women. Indeed, Mexican American women born in the United States are significantly more likely to experience a marital disruption (i.e., divorce or separation) compared to Mexican American women born in Mexico (40.9 percent compared to 13.1 percent respectively).[40]

Nonetheless, cultural arrangements reflect structural realities. In comparison with inner-city blacks, inner city Mexican immigrants have a stronger attachment to the labor force—they have come, after all, a very long way to find work—as well as stronger households, networks, and neighborhoods. Therefore, as long as these differences exist, attitudes toward the family and family formation among inner city blacks and Mexican immigrants will contrast noticeably.

Ethnographic data from our Chicago study revealed that the relationships between inner-city black men and women, whether in a marital or nonmarital situation, were often fractious and antagonistic. Inner-city black women routinely said they distrusted men and felt strongly that black men lacked dedication to their partners and children.[41] They argued that black males are hopeless as either husbands or fathers and that more of their time is spent on the streets than at home. As one woman, an unmarried mother of three children from a high-poverty neighborhood on the South Side, put it, "and most of the men don't have jobs ... but if things were equal it really wouldn't matter, would it? I mean OK if you're together and everything, you split, whatever ... The way it is, if they can get jobs then they go and get drunk or whatever." When asked if that's why she had not gotten married, she stated,

> I don't think I want to get married but then ... see you're supposed to stick to that one and that's a fantasy. You know, stick with one for the rest of your life. I've never met many people like that, have you? ... If they're married and have kids. Them kids come in and it seems like the men get jealous 'cause you're spending your time on them. OK they can get up and go anytime. A woman has to stick there all the time 'cause she got them kids behind their backs.

Women in the inner city tended to believe that black men got involved with women mainly to obtain sex or money, and that once these goals had been achieved, the women were usually discarded. For example, one woman from a poor neighborhood on the West Side of Chicago was asked if she still saw the father of her child. She stated, "He left before the baby was born, I was about two weeks pregnant and he said that he didn't want to he bothered and I said 'Fine—you go your way and I go mine.'"

... There was a widespread feeling among women in the inner city that black males had relationships with more than one female at a time. And since some young men left their girlfriends as soon as they became pregnant, it was not uncommon to find a black male who had fathered at least three children by three separate women. Despite the problematic state of these relationships, sex among inner-city black teenagers was widely practiced. In the ethnographic phase of our research, respondents reported that sex was an integral and expected aspect of their romantic relationships. Males especially felt peer pressure to be sexually active. They said that the members of their peer networks bragged about their sexual encounters and that they felt obligated to reveal their own sexual exploits. Little consideration was given to the implications or consequences of sexual matters for the longer-term relationship or for childbearing....

... Many of the men viewed marriage as tying them down and resulting in a loss of freedom. "Marriage. You can't have it, you can't do the things you wanna do then," stated an unemployed twenty-one-year-old unmarried father of one child from a poor neighborhood on the West Side.

The men in the inner city generally felt that it was much better for all parties to remain in a nonmarital relationship until the relationship dissolved rather than getting married and then having to get a divorce. A twenty-five-year-old unmarried West Side resident, the father of one child, expressed this view:

> Well, most black men feel now, why get married when you got six to seven womens to one guy, really. You know, 'cause there's more women out here mostly than men. 'Cause most dudes around here are killing each other like fools over drugs or all this other stuff. And if you're not that bad looking of a guy, you know, and you know a lot of women like you, why get married when you can play the field the way they want to do, you know?

A twenty-five-year-old part-time worker with a seven-year-old daughter explained why he had avoided marriage following the child's birth:

> For years I have been observing other marriages. They all have been built on the wrong foundation. The husband misuses the family or neglects the family or the wife do the same, ah, they just missing a lot of important elements. I had made a commitment to marry her really out of people pleasing. My mom wanted me to do it; her parents wanted us to do it. Taking their suggestions and opinions about the situation over my own and [I] am a grown man. These decisions is for me to make and I realized that they were going to go to expect it to last for 20, 40 years so I evaluated my feelings and came to the honest conclusion that was not right for me, right. And I made the decision that the baby, that I could be a father without necessarily living there.

Others talk about avoiding or delaying marriage for economic reasons. "It made no sense to just get married because we have a baby like other people ... do," argued an eighteen-year-old unmarried father of a two-week-old son. "If I couldn't take care of my family, why get married?"

These various responses show that marriage was "not in the forefront of the men's minds."[42] The dominant attitude among the young, single black fathers in our Chicago study was, "I'll get married in the future when I am no longer having fun and when I get a job or a better job."[43] Marriage would limit their ability to date other women or "hang out" with the boys.

... There is very little reasearch on changing norms and sanctions regarding the family in the inner city,[44] but the norms do appear to have changed. In a study of fathering in the inner-city based on a series of interviews with the same respondents over several years, Frank Furstenberg notes,

> I have no way of knowing for sure, but I think that families now exert less pressure on men to remain involved than they once did. I found no instance, for example, of families urging their children to marrry or even to live together as was common when I was studying the parents of my informants in the mid-1960s.[45]

The data from our Chicago study, however, indicate that the young men did "feel some obligation to contribute something to support their children."[46] ... As Frank Furstenberg points out,

> When ill-timed pregnancies occur in unstable partnerships to men who have few material resources for managing unplanned parenthood, they challenge, to say the least, the commitment of young fathers. Fatherhood occurs to men who

> often have a personal biography that poorly equips them to act on their intentions, even when their intentions to do for their children are strongly felt. And fatherhood [in the inner city] takes place in a culture where the gap between good intention and good performance is large and widely recognized.[47]

Black women in the inner city were more interested in marriage, but their expectations for matrimony were low and they did not hold the men they encountered in very high regard. The women feel that even if they did marry, the marriage was likely to be unsuccessful. They maintained that husbands were not as dedicated to their wives as in previous generations and that they would not be able to depend on their husbands even if they did get married.

A young welfare mother of three children from a high-poverty neighborhood on the West Side of Chicago made the following point:

> Well, to my recollection, twenty years ago I was only seven years [old] but … twenty years ago, men, if they got a woman pregnant, that if they didn't marry her, they stood by her and took care of the child. And nowadays, when a man makes you pregnant, they're goin' off and leave you and think nothing of it. And also … also, uh, twenty years ago, I find that there were more people getting married and when they got married they were staying together. I found that with a lot of couples nowadays, that when they get married they're so quick to get a divorce. I've thought about marriage myself many times, uh, but nowadays, it seems to me that when it comes to marriage, it just doesn't mean anything to people. At least the men that I talk to. And also, twenty years ago, I think families were closer [police siren]. I found now families are drifting apart, they're not as loving as they were twenty years ago. I find with a lot of families now, they're quicker to hurt you than to help you….

Finally, a twenty-seven-year-old single woman (who was childless, had four years of college, worked as a customer service representative, and lived in a high-poverty neighborhood on the South Side) talked about changes in the family structure in relation to her own personal situation. She stated that there had been a definite change in the family structure as far as the mother and father being together.

> The way things are going now you'll find more single women having kids, but not totally dependent on the guy being there. I know there's a change in friends of mine who have kids, the father isn't there with them. They're not so totally dependent on him any more. They're out there doing for themselves … You have to make it one way or another, and you can't depend on him to come through or him to be there. And a lot of them are searching for someone to be with, but if [he] comes he does and if he doesn't he doesn't. Because I always say by the time I'm thirty, if I'm not married, I know I'll still have me a child. But I wouldn't be so hung up on the idea of having somebody be there. I probably wouldn't have a child by anyone I was seeing or anything. I'd probably go through a sperm bank. I think financially I could do it, but I would need help as far as babysitting and stuff.

The ethnographic data reveal that both inner-city black males and females believe that since most marriages will eventually break up and no longer represent meaningful relationships, it is better to avoid the entanglements of wedlock altogether. For many single mothers in the inner city, remaining single makes more sense as a family formation strategy than does marriage. Single mothers who perceive the fathers of their children as unreliable or as having limited financial means will often—rationally—choose single parenthood. From the point of view of day-to-day survival, single parenthood reduces the emotional burden and shields these women from the type of exploitation that often accompanies the sharing of both living arrangements and limited resources. Men and women are extremely suspicious of each other, and their concerns range from the degree of financial commitment to fidelity. For all these reasons, women in our study often stated they did not want to get married until they were sure it was going to work out—and by "work out" they meant that they wanted a spouse who would contribute financially and emotionally to them and to their children....

Kathryn Edin and Maria Kefalas presented similar findings in a more recent study. In their book, *Promises I Can Keep: Why Poor Women Put Motherhood before Marriage,* Edin and Kefalas collected and analyzed data on low-income black, white, and Puerto Rican single mothers in Camden, New Jersey (one of America's poorest cities), and in eight poor neighborhoods in Philadelphia.[48]

Edin and Kefalas found that the low-income, young mothers they interviewed valued motherhood highly. Indeed, their identity, emotional fulfillment, personal success, and hope for the future were tied to motherhood. The thought of not being a parent or of postponing parenting until their thirties—a common practice for middle-class women—was anathema to the women in this study. However, the respondents also valued marriage and hoped to be married some day. Edin and Kefalas forcefully argue that poor women postpone marriage not because they value it lightly, but because they feel that they cannot commit to marriage until they are confident of success.

The basic problem they face, as did the women in our earlier Chicago study, is that the men to whom they have access tend not to be marriageable because of a range of problems: poor education, chronic joblessness, low earning, criminal records, spells of incarceration, drug and alcohol abuse, intimate violence, and chronic infidelity. There is a short supply of good, decent, trustworthy men in their world. The better-off men go to the better-off women. It is not surprising that the relationships these women have with the fathers of their children are plagued with physical abuse, mistrust, and infidelity. The women therefore wait until they can find a man they can trust—a man who has, over time, proven himself to be a dependable and responsible partner and father. Their dreams include financial security and having a house before getting married. And because all of these conditions are so difficult to meet, they become mothers long before they become wives, and some never marry.

Edin and Kefalas point out that unlike many affluent women, the poor women they studied do not view having a child out of wedlock as ruining their lives, because they feel that their future would be even bleaker without children. For these women, motherhood is the most important social role they believe they will ever play, and it is the surest accomplishment they can attain. Many of the women told Edin and Kefalas that they had been headed for trouble until they got pregnant and turned their lives around because of a desire to be good mothers. Many of them said that having children had been a life-altering experience and that they could not imagine living without children.

Whereas middle-class women often put off marriage and childbearing to pursue economic goals, poor women have children in the absence of better opportunities. The mothers in this study expressed confidence in their ability to provide for their children. However, because these mothers frequently fail to recognize the disadvantages that will affect their children's chances in life, this confidence is often unjustified. In this sense their cultural framing of marriage and motherhood not only shapes how they respond to poverty—it may also indirectly affect their children's odds of escaping poverty. The implication here is that the decision to forgo or delay marriage increases the odds that the family will live in persistent poverty.

In 1978, sociologist Diana Pearce coined the term "feminization of poverty" to describe the increase of female-headed households among the poor.[49] Since then, studies have repeatedly reported that married-couple families are far less likely to be in poverty than are households headed by a single mother. In one study, whereas only slightly more than 5 percent of married-couple families lived in poverty in 2004, nearly 30 percent of female-headed families were poor.[50] According to another study, "children in single-female-headed households account for more than 60% of all children in families living in poverty."[51] This effect is not unique to any one group. The poverty rate of white children in households headed by single females was almost 5 times greater than the poverty rate for those in mairried-couple families; for black children the rate was 4.5 times greater, and for Hispanic children it was 2.4 times greater. The author states, "Because of the high rates of poverty experienced by children in families headed by single females, black children in single-female-headed families account for more than 85% of all poor black children."[52] Finally, a longitudinal study that tracked a national random sample of families over several years revealed that poverty tends to be chronic for children in single-mother families where the mother was either never married or was a teenager when the child was born.[53]

It has been argued that because married-couple families frequently have two potential (and often actual) breadwinners, their chances of preventing or escaping poverty are much greater than those of families headed by single females.[54] Given the extremely high jobless rate among low-skilled black males, however, this argument—although generally true—is less applicable to poor black families in the inner city.

ENDNOTES

1. Adam Clymer, "Daniel Patrick Moynihan Is Dead, Senator from Academia Was 76," *New York Times*, March 27, 2003.
2. Daniel P. Moynihan, *The Negro Family: The Case for National Action* (Washington, DC: Office of Planning and Research, US Department of Labor, 1965), 80.
3. Lee Rainwater and William L. Yancey, *The Moynihan Report and the Politics of Controversy* (Cambridge, MA: MIT Press, 1967), 144.
4. Ibid., 154.
5. See, for example, Hare, "Challenge of a Black Scholar" and Alkalimat "Ideology of Black Social Science." This rejection even included the thoughtful argument, so clearly articulated by Kenneth Clark and Lee Rainwater in the latter 1960s, that the logical outcome of racial isolation and class subordination is that individuals are forced to adapt to the realities of the ghetto community and are therfore seriously impaired in their ability to function in any other community. See Kenneth B. Clark, *Dark Ghetto: Dilemmas of Social Power* (New York: Harper & Row, 1965); and Lee Rainwater, "Crucible of Identity: The Negro Lower-Class Family," *Daedalus* 95 (Winter 1966), 176–216.
6. US Department of Health and Human Services. *Nonmarital Childbearing in the United States, 1940–1999.* National Vital Statistics Reports, Vol. 48, No. 16 (Washington, DC: Government Printing Office, 2000); Brady E. Hamilton, Joyce A. Martin, and Stephanie J. Ventura, "Births: Preliminary Data for 2005." NCHS Health E-Stats, at www.cdc.gov/nchs/products/pubs/pubd/hestats/prelimbirths05/prelimbirths05.htm (accessed September 20, 2007).
7. US Bureau of the Census, *Household and Family Characteristics*. Current Population Survey, Series P20–495 (Washington, DC: Government Printing Office, March 1996) table 1; and US Bureau of the Census. Population Division, "Current Population Survey, 2006 Annual Social and Economic Supplement," table F1, at www.census.gov/population/socdemo/hh-fam/cps2006/tabF1.all.xls (accessed September 20, 2007).
8. Greg J. Duncan, *Years of Poverty, Years of Plenty* (Ann Arbor: Institute for Social Research, University of Michigan, 1984).
9. US Bureau of the Census, "American Community Survey" (2006), table B02001, at http://factfinder-censusgov/servlet/DatasetMainPageServlet?-program=ACS&_submenuld=datasets_2&_lang=en&_ts=
10. Kathryn Edin, "The Myths of Dependence and Self-Sufficiency: Women, Welfare, and Low-Wage Work," *Focus* 17(1995): 203–230.
11. For married couples, this information comes from US Bureau of the Census, "American Community Survey" (2006), table B19126 ("Median income") and table B17006 ("Poverty status of children"), at http://factfinder.census.gov/servlet/DatasetMainpageServlet?_program=ACS&_submenuld=datasets_2&_lang=en&_ts=; for single mother families, from

an analysis of microdata in US Bureau of the Census. "American Community Survey" (2006).

12. For a review of this research, see Adam Thomas and Isabel Sawhill, "For Richer or for Poorer: Marriage as an Antipoverty Strategy," *Journal of Policy Analysis and Management* 15 (2002): 58–59.
13. David Ellwood and Mary Jo Bane, *The Impact of AFDC on Family Structure and Living Arrangements.* Research in Labor Economics, Vol. 7 (Greenwich, CT: JA1 press, 1985); June F. O'Neill, Douglas Wolf, Laurie Bassi, and Michael Hannan, *An Analysis of Time on Welfare*, US Department of Health and Human Services Report, Contract No. HHS-10083-0048 (Washington, DC: Urban Institute, 1984).
14. Mollie A. Martin, "Family Structure and Income Inequality in Families with Children," *Demography* 48 (2006): 421–446; Paul Amato, "The Impact of Family Formation Change on Cognitive, Social and Emotional Well-Being of the Next Generation," *Furture of Children* 15 (2005): 75–96; Sara McLanahan, "Diverging Destinies: How Children Are Faring Under the Second Demographic Transition," *Demographic* 41 (2004): 607–627; Sara McLanahan and Gary Sandefur, *Growing Up With a Single Parent: What Hurts, What Helps* (Cambridge MA: Harvard University Press, 1994), Sara McLanahan and Irwin Garfinkel, "Single Mothers, the Underclass, and Social Policy," *The Annals of the American Academy of Political and Social Science* 501(1) (January 1989): 130–152; Sara McLanahan and Larry Bumpass. "Intergenerational Consequences of Family Disruption," *American Journal of Sociology* 94(1) (1988): 130–152 and Sheila Fitzgerald Krein and Andrea H. Beller, "Educational Attainment of Children from Single-Parent Families: Differences by Exposure, Gender and Race,'" *Demography* 25 (May 1988): 221–224.
15. Deborah Roempke Graefe and Daniel T. Lichter, "Marriage among Unwed Mothers: Whites, Blacks and Hispanics Compared," *Perspectives on Sexual and Reproductive Health* 34 (6) (2002): 286–292.
16. Marcia Carlson, Sara McLanahan, and Paula England, "Union Formation in Fragile Families," *Demography* 41 (2004): 237–261.
17. Martha Van Haitsma, "A Contextual Definition of the Underclass," *Focus* 12 (Spring Summer 1989): 27–31.
18. Lena Lundgren-Gaveras, "Informal Network Support, Public Welfare Support and the Labor Force Activity of Urban Low-Income Single Mothers' paper," Chicago Urban Poverty and Family Life Conference, Chicago IL, October 10–12, 1999).
19. Ibid.
20. The proportion of nonmarital births among whites and Latinos reached 24.5 and 46.4 percent, respectively, in 2005; Hamilton, Martin and Ventura, "Births."
21. Thomas and Sawhill, "For Richer or Poorer."
22. See, for example, Greg J. Duncan's testimony before the subcommittee on Human Resources of the Committee on Ways and Means Hearing on Early Childbearing. Washington, DC, July 29, 1994; and Saul D. Hoffman, Gregory J. Duncan, and Ronald B. Miney, "Marriage and Welfare Use among Young Women: Do Labor Market, Welfare and Neighborhood Factors Account for Declining Rates of Marriage among Black and White Women?" (paper, American Economics Association, New Orleans, December 1991).
23. Duncan, testimony, July 29, 1994.
24. For a thorough discussion of shifting marriage norms over time, see Andrew Cherlin. "The Deinstitutionalization of American Marriage," *Journal of Marriage and Family* 66 (2004): 848–861.
25. These figures are based on an analysis of microdata from US Bureau of the Census, "American Community Survey," (2006), at http://www.soc.jhu.edu/people/Cherlin/cherlin_jmfmarriagepaper.pdf.
26. Christopher Jencks. "Is the American Underclass Growing?" in *The Urban UnderClass*, eds. Christopher Jencks and Paul Peterson (Washington, DC: Brookings Institution, 1991), 28–102.
27. For a good review of these studies, see David Ellwood and Christopher Jencks, "The Uneven Spread of Single Parent Families in the United States. What Do We Know? Where Do We Look for Answers? in *Social Inequality*, ed. Kathryn Neckerman (New York: Sage Foundation, 2004), 8–78.
28. Mark Testa and Marilyn Krogh, "The Effects of Employment on Marriage among Males in Inner-City Chicago," in *The Decline in Marriage among African Americans: Causes, Consequences and Policy Implications*, eds. M. Belinda Tucker and Claudia Mitchell-Kernan (New York: Sage Foundation, 1995): 59–95.

29. William Ryan, *Blaming the Victim* (New York: Pantheon, 1971).

30. Oscar Lewis, "The Culture of Poverty," in *On Understanding Poverty Perspectives from the Social Sciences*, ed. Daniel Patrick Moynihan (New York: Basic Books, 1968), 188. See also Oscar Lewis, *Five Families: Mexican Case Studies in the Culture of Poverty* (New York, Basic Books, 1959); Oscar Lewis, *The Children of Sanchez* (New York, Random House, 1961); and Oscar Lewis, *La Vida: A Puerto Rican Family in the Culture of Poverty—San Juan and New York* (New York, Random House, 1966).

31. See, for example, Edward Banfield, *The Unheavenly City*, 2nd ed. (Boston: Little Brown, 1970).

32. Moynihan, *Negro Family*, 93.

33. Alice O'Connor, *Poverty and Knowledge: Social Science, Social Policy, and the Poor in Twentieth-Century U.S. History* (Princeton, NJ: Princeton University Press, 2001).

34. Mark Testa, "Male Joblessness, Nonmarital Parenthood and Marriage" (paper, Chicago Urban and Family Life Conference, Chicago, October 10–12, 1991), 16.

35. Ibid.

36. Our survey—the Urban Poverty and Family Life Study (UPFLS), conducted in 1987 and 1988—included a random sample of nearly 2,500 poor and nonpoor African American, Latino and white residents in Chicago's poor inner-city neighborhoods and is discussed in Chapter 3. Inner-city neighborhoods were defined in this study as those with poverty rates of at least 20 percent.

37. Testa, "Male Joblessness"; Testa and Krogh, "Effects of Unemployment," and Martha Van Haitsma, "A Contextual Definition of the Underclass," *Focus* 12 (Spring-Summer 1991): 27–31.

38. Richard P. Taub, "Differing Conceptions of Honor and Orientations among Low-Income African-Americans and Mexican-Americans" (paper, Chicago Urban Poverty and Family Life Conference, Chicago, IL: October 10–12, 1991), 6.

39. Ibid.

40. Julie A. Phillips and Megan M. Sweeney, "Premarital Cohabitation and Marital Disruption among White, Black and Mexican American Women," *Journal of Marriage and Family* 67 (May 2005): 296–314. See also a, R. S. Oropesa, Daniel T. Lichter, and Robert N. Anderson, "Marriage Markets and the Paradox of Mexican American Nuptiality," *Journal of Marriage and the Family* 56(4) (1994): 889–907.

41. This general distrust has also been documented by Kathryn Edin and by Christina Gibson-Davis and her colleagues. They note that women are wary of getting married in part because they are afraid that their partners will try to take more control of the household and start ordering them around. See K. Edin, "What Do Low-Income Single Mothers Say about Marriage?" *Social Problems* 47(1) (2000): 112–133 and Christina Gibson-Davis, Kathryn Edin, and Sara McLanahan, "High Hopes but Even Higher Expectations: The Retreat from Marriage among Low-Income Couples," *Journal of Marriage and Family* 67 (2005): 1301–1312. The concerns over infidelity have also been documented in Paula K. England, Kathryn Edin, and K. Linnenberg, "Love and Distrust among Unmarried Parents" (paper, National Poverty Center Conference on Marriage and Family Formation among Low-Income Couples, Washington, DC, September 4–5, 2003).

42. Taub, "Differing Conceptions," 9.

43. Robert Laseter, "Young Inner-City African American Men: Work and Family Life" (PhD dissertation, University of Chicago, 1994), 195.

44. Although marriage behavior varies by racial/ethnic and socioeconomic groups, support for the institution of marriage is relatively evenly widespread. Indeed, 70 percent of welfare recipients say they expect to marry. Low-income and minority women voice doubts about marriage to their current partners but nonetheless show a high level of support for the institution of marriage overall. See D .T. Lichter, C. D. Batson and J. D. Brown, "Welfare Reform and Marriage Promotion: The Marital Expectation and Desires of Single and Cohabitating Mothers," *Social Service Review* 78 (2004): 2–24; R. Kelly Raley, "Recent Trends and Differentials in Marriage and Cohabitation: The United States" in *The Ties That Bind: Perspectives on Marriage and Cohabitation,* ed. Linda J. Waite (New York, Aldine de Gruyter, 2000), 19–39; Jane G. Mauldon, Rebecca A. London, David J. Fein, Rhiannon Patterson, and Steven Bliss, *What Do They Think? Welfare Recipients' Attitudes Toward Marriage and Childbearing* (Cambridge, MA, Abt Associates, 2002).

45. Frank F. Furstenberg Jr., "Fathering in the Inner-City: Paternal Participation and Public Policy" (unpublished manuscript, University of Pennsylvania, 1994).
46. Laseter, "Young Inner-City African American Men," 40.
47. Furstenberg, "Fathering in the Inner-City," 29.
48. Kathryn Edin and Maria Kefalas, *Promises I Can Keep: Why Poor Women Put Motherhood before Marriage* (Berkeley: University of California Press, 2005).
49. Diana Pearce, "Feminization of Poverty, Work and Welfare," *Urban and Social Change Review* 11 (1978): 146–160.
50. Douglas J. Besharov, *Measuring Poverty after Katrina* (Washington. DC: American Enterprise Institute, 2006).
51. Eugene M. Lewit, "Children in Poverty," *Future of Children* 3 (Spring 1993): 179.
52. Ibid., 180.
53. Greg J. Duncan, "The Economic Environment of Childhood," in *Children in Poverty*, ed. Althea C. Huston (Cambridge, MA: Cambridge University Press, 1991), 23–50.
54. See, for example, Lewit, "Children in Poverty."

DISCUSSION QUESTIONS

1. Roughly 70% of black children are born to unwed mothers, and 45% of all black families are headed by single mothers, according to reported statistics in this article. Why should these statistics raise concern for sociologists, policy makers, and for society?
2. Was Daniel P. Moynihan *right* when he sounded the alarm on this emerging negative trend in his 1965 report entitled "The Negro Family"? Why or why not? Did history vindicate his initial concerns?
3. Sociologists seem to be divided on whether the cause of black family fragmentation is *structurally or culturally* driven. Which explanation appeals to you most, and why? Is it an either/or dilemma, or is it both?
4. What specific policies would you support to reverse the family fragmentation trend in America? Can anything be done on the public policy side to address this social problem? Why or why not?
5. *Poor, welfare dependent, mother-only families are a social problem!* Respond to this statement. Do you agree or disagree, and why?
6. Wouldn't any public policy that specifically seeks to reduce the number of out-of-wedlock births and the high number of "female-headed households" imply that the two-parent family is "better" than the single-mother (or single-father) family for raising children? Is the two-parent family "better"?

35

Ending Family Homelessness: Lessons from Communities

NATIONAL ALLIANCE TO END HOMELESSNESS

The Four Questions

1. What exactly is the problem? Who does this problem affect?
2. Is it a *social* problem?
3. What is the cause of this problem?
4. What can be done about the problem? What does the author suggests we do about it?

Topics Covered

Homelessness

Family

Poverty

Housing

Communities are adopting new approaches to respond to family homelessness. Resources are increasingly focused on helping families avoid homelessness and helping those that do become homeless quickly reconnect to housing in the community. Across the country, leading communities are developing, testing and sharing innovative strategies. They are collecting more reliable data on how families move in and out of homelessness, which facilitates their ability to refine interventions, evaluate outcomes, and ultimately expand successful initiatives.

The approaches being adopted are rooted in the pioneering work of the communities profiled in *Promising Strategies to End Family Homelessness.*[1] By orienting their homelessness system to focus primarily on helping families quickly access stable housing, these communities sharply reduced family homelessness. Hennepin County Minnesota, Columbus Ohio, and Westchester County New York reduced family homelessness by over 40 percent and others achieved smaller but significant results.

The recession reversed this progress. Lost jobs, inadequate benefits, and increases in foreclosures jeopardize the housing stability of millions of families. Across the country, there are reported surges in the number of families seeking eviction prevention assistance, families doubling up with extended family, and families turning to shelter because they have no where else to go.

To respond to the increased need for assistance and to help avoid sharp increases in homelessness, in 2009 Congress enacted the Homelessness Prevention and Rapid Re-Housing Program (HPRP) as part of the American Recovery and Reinvestment Act (ARRA). The program provided $1.5 billion in new resources that are allowing communities to implement the key strategies that are effective in reducing family homelessness.

With new HPRP resources and an overwhelming number of families in need, many communities are rapidly transforming their response to family homelessness to serve families better. These communities are working to maximize the effectiveness and efficiency of all of their resources to prevent families from becoming homeless and when that fails, to help families quickly return to housing.

SOURCE: From National Alliance to End Homelessness, *Ending Family Homelessness: Lessons from Communities.* Washington, DC; Author, June 2010. Reprinted by permission.

This paper provides an overview of family homelessness in the United States. It examines the key strategies that communities are implementing to respond to increased family homelessness. It also discusses the need for federal leadership.

UNDERSTANDING FAMILY HOMELESSNESS

In 2008, 84,000 families experienced homelessness on a single night, and approximately half a million people in families stayed in a homeless shelter or transitional housing program over the course of the year.[2] All indications are that the number of families experiencing homelessness is higher today (May, 2010) because of the recession. Family homelessness increased by 10 percent between July and September 2009 in seven communities that HUD monitors to identify shifts in homelessness.[3] Those communities also reported an increase in the number of families who experienced homelessness for the first time.

What Are the Characteristics of Homeless Families?

Multiple studies have found that families who experience homeless are very similar to other low-income families with a few exceptions.[4] African-American families are disproportionately represented. Parents tend to be younger, most families are headed by a single woman age 30 or under, and over 50 percent of children in homeless families are under the age of six. A high proportion of young homeless parents were in foster care as a child. Studies have found that families that become homeless are also poorer than their housed counterparts and may have fewer social supports to rely upon.

What most differentiates homeless families from other poor, housed families however is access to affordable housing. Studies have found that a housing subsidy virtually eliminates the risk of homelessness among families[5] and few families that exit a homelessness program with a housing subsidy become homeless a second time. This remains true even among those families in which the parent has significant challenges such as a mental health or substance use disorder.[6]

How Do Families Become Homeless?

A typical family homelessness episode begins with a housing crisis. A family that may already be spending a disproportionate amount of its income for housing experiences an economic jolt. This can be caused by lost employment, reduced work hours, or a rise in household expenses. Families that are suddenly unable to afford their current housing will typically work to resolve the problem on their own by moving to smaller, less expensive housing, or moving in with extended family or friends. Most will never become homeless. Some, however, will not be successful in finding a stable new housing situation, and will seek shelter.

Family conflict can also create a housing crisis that results in homelessness. Some families may seek shelter to escape domestic violence. In other cases, young parents who are no longer able to stay with their parents, friends, or extended family resort to shelter. Families whose homelessness is directly caused by violence or conflict often remain in shelter simply because they lack alternatives. Without resources to pay for housing, a shelter is their only option.

How Long do Families Remain Homeless?

The vast majority of families who become homeless exit shelter or transitional housing programs within three or four months and do not become homeless a second time. Approximately 20 percent of families have longer homeless episodes of a year or more. A small number, approximately 5 percent, have multiple homeless episodes and cycle in and out of shelter and transitional programs.[7]

KEY STRATEGIES TO ENDING FAMILY HOMELESSNESS

In the years since the Alliance profiled six communities that reduced family homelessness in *Promising Strategies to End Family Homelessness* (2006), more communities have adopted the strategies that were key to their success. The onset of the recession and new resources dedicated to ending family homelessness has facilitated the adoption of these strategies and the growth of new approaches. By monitoring successful communities, the Alliance has identified six key strategies critical to ending family homelessness.

- Prevention assistance
- Rapid Re-Housing
- Helping families pay for housing
- Targeted services
- Coordinated intake, assessment, and services
- Data for planning and program management

With new HPRP resources, and sharp increases in requests for assistance, the strategies that were successful in reducing homelessness in better economic times are being replicated, expanded, and adapted to a wide variety of communities—urban, suburban and rural—to respond to increased needs. With each community's adaptation, refinements are being made and new lessons emerge.

PREVENTION ASSISTANCE

HPRP and other homelessness prevention resources can be used to help pay rent arrears or provide rental assistance to allow families to remain stably housed. They can also be used to provide legal assistance and mediation services to help families negotiate new rental terms with their landlords. Increasingly, communities are using HPRP to help families that are doubled up or in housing they can no longer afford find new housing, allowing them to avoid shelter stays altogether.

To maximize the impact of prevention resources, programs must be well-targeted to the families most likely to become homeless. This targeting has been among the greatest challenges communities face in implementing HPRP, as the number of families facing housing-related hardship (although not necessarily homelessness) has increased enormously due to the recession. There are a range of promising strategies that communities have developed to target families at greatest risk of entering shelter. Strategies include:

- using communities' own **shelter data** to develop profiles of people at risk of homelessness;
- developing **outreach programs** that identify people at greatest risk who normally would not seek assistance; and
- providing **diversion** assistance to families who have already lost their housing and are applying for shelter.

Shelter Data

New York City's HomeBase program uses data to identify families most likely to become homeless. Local Homelessness Management Information Systems (HMIS) data is used to develop profiles of families that are at the greatest risk of homelessness. Such data ensures that the eligibility criteria for homelessness prevention programs match the families most likely to enter shelter. New York City is also using the prior addresses of families to focus prevention efforts in neighborhoods where there are high numbers of families that become homeless.

Outreach Programs

Outreach efforts identify families at heightened risk of entering shelter, often through crafting collaborations with other social service agencies serving vulnerable families. The early identification of families in unsustainable housing allows providers to immediately focus on housing search so families can move directly into new housing and avoid homelessness.

An example of a collaboration to identify families at risk of homelessness is The Children's Hospital of Philadelphia Early Head Start Program and SafeHome Philadelphia, a prevention program operated by the Philadelphia Committee to End Homelessness. When the Early Head Start encounters families that are living in untenable and hazardous housing situations, they refer them to SafeHome. SafeHome then visits the family and offers them assistance finding new housing in the community.

Some communities are targeting families in subsidized housing. While very few families with permanent rent subsidies become homeless, those that do lose their subsidy may be highly susceptible to homelessness and lengthy shelter stays. To address this, some homelessness prevention programs have crafted partnerships with local public housing authorities. Housing Families, Inc. in Massachusetts is one example. The agency receives referrals directly from two area public housing authorities and then reaches out to the families at risk of eviction to offer mediation and budget support, case management, and financial assistance to help families preserve their subsidized housing.

Diversion

Communities are also expanding the options offered to families seeking shelter. By offering financial assistance and rapid re-housing services when families apply for shelter, communities are finding that they can divert families from shelter and can instead use the resources to help them resolve the crisis that threatens their housing, or quickly relocate them into new housing in the community. Central Massachusetts Housing Alliance in Worcester, MA is operating a successful homelessness diversion program for families. In the first six months of operation, the program served 52 families. With fewer families entering the shelter system, the community placed fewer families in motels because shelters were full. Within six months, the number of families residing in motels dropped from an average of 40 families a night to 4 families a night. With an average motel stay costing $18,000 per family served, the diversion program is resulting in significant cost savings to the community.

The lessons that will emerge from the new investments in homelessness prevention and the innovative work of leading communities to use data to improve their targeting efforts and program interventions will better equip all communities to prevent homelessness and greatly advance our nation's effort to end family homelessness.

PROVIDE RAPID RE-HOUSING

The single biggest shift in serving homeless families has been the growing adoption of Rapid Re-Housing to help families quickly exit shelter and return to housing in the community. Communities such as Hennepin County, Minnesota, and Columbus, OH, have found that many families can be quickly re-housed by helping with housing search and landlord negotiation; providing rental assistance; and delivering home-based case management services. The availability of HPRP resources greatly facilitated the widespread adoption of Rapid Re-Housing.

Working with Landlords

Rapid Re-Housing providers are primarily focused on helping families quickly navigate the transition back into housing. They work aggressively to identify housing options in the community through ongoing housing search and cultivating relationships with landlords. Rapid Re-Housing providers may negotiate with landlords to reduce the rent and to persuade them to accommodate families' housing barriers. They take steps such as guaranteeing to the landlord that the landlord will not have to absorb the costs of an eviction if the placement is not successful. They make commitments to help landlords resolve any landlord-tenant disputes that may arise. The demonstrated commitment to promoting a successful housing placement, and helping landlords and tenants resolve issues that might arise, has been critical to the success of Rapid Re-Housing providers and allows providers to find housing for tenants with even the most challenging rental histories.

Help Families Become Successful Tenants

Rapid Re-Housing providers also prepare families for successful tenancy. They explore the families' housing history to identify and resolve issues that contributed to the families' housing loss. They work with families to resolve credit issues and educate them about their rights and responsibilities as tenants. Providers continue to work with the families after a housing placement to promote housing retention and link family members with community-based supports that can help them achieve their long-term goals.

Financial Assistance

Rapid Re-Housing programs typically provide financial assistance to help families pay for rent. Assistance may range from part of a security deposit to helping families access a permanent housing subsidy. Programs might also provide some financial incentives to landlords to rent to families with challenging rental histories. This may include providing several months of rent upfront or doubling the security deposit for families with very problematic rental histories.

HPRP is allowing communities to create or expand Rapid Re-Housing efforts, transforming how they respond to family homelessness. With new HPRP and other ARRA funds, The Road Home in Salt Lake City, UT, for example, is responding to a sharp, recession-related increase in family homelessness by offering Rapid Re-Housing to many more families. The program helped 200 families rapidly return to housing in the first six months of operation. The expansion of Rapid Re-Housing has allowed the program to meet the increased demand for assistance without expanding the organization's shelter capacity or turning families away. By connecting families quickly with income and employment support provided by the local TANF agency, the program is also helping to shore up the supports the families will require after they are housed.

There is a parallel movement emerging among domestic violence providers. With limited capacity to provide shelter, domestic violence providers are adopting strategies to help families quickly find safe, stable housing. The nationally recognized Volunteers of America Oregon Home Free Program has been a leader in this movement to expand options available to survivors of violence. In addition to housing search, landlord negotiation, and financial assistance, the program helps families develop safety plans, connects them to legal and immigration law services, and provides trauma-informed support and advocacy. Recognizing the innovative approach, the Centers for Disease Control and Prevention is studying the outcomes of families assisted by Home Free. Initial findings indicate that families with stable housing have better outcomes on an array of measures, including severity of Post-Traumatic Stress Disorder, depression, and quality of life.

The growing use of Rapid Re-Housing has several benefits. It significantly reduces the amount of time families reside in shelter programs, reducing the cost of community shelter programs and allowing shelter beds to serve other families in need. It also returns families more rapidly to permanent housing that can provide a firm foundation from which families can address other challenges and receive the support they require to thrive.

HELPING FAMILIES PAY FOR HOUSING

While permanent housing subsidies provide the best protection against homelessness, only a fraction of families that would benefit from a housing subsidy receives one. Families can spend years waiting for a housing subsidy to become available. As a result, most families will exit shelter having saved up some money to rent new housing in the community, but without the benefit of a permanent rental subsidy. Most of these families will not, however, become homeless again.

Temporary Rental Assistance

To help families afford housing sooner, many communities have developed their own short- and medium-term rental assistance programs. The expanding practice of giving families short- and medium-term rental assistance to help them exit shelter more rapidly has been a major factor in revolutionizing the delivery of homelessness services nationally.

Westchester County, New York, Hennepin County, Minnesota, and Columbus, Ohio are among the communities that found that providing families with shelter was often far costlier than providing them with rental assistance. Rental assistance, lasting from 1 to 18 months, allows many families to return to stable housing in the community and is a better investment for families and children than shelter. Evidence indicates that families that receive time-limited rental assistance can successfully transition into housing in the community and avoid a subsequent homeless episode. And now with HPRP funding, communities can provide up to 18 months of rental assistance to preserve housing or re-house a family that has become homeless.

Permanent Rent Subsidies

With a shortage of permanent rent subsidies to offer families experiencing homelessness, some communities are developing strategies to target the permanent rent subsidies they do have to those that may need them the most. A strategy used in New York City, prioritized homeless individuals and families with a fixed income (such as SSI or SSDI) due to a disability and child welfare-involved homeless families for a permanent rent subsidy. Because of the prioritization, the families received a permanent rent subsidy much sooner than they otherwise would have, avoiding long shelter stays and the separation of parents and children.

Over a two year period, the New York City Department of Homeless Services housed more than 1,800 families that either depend solely on a fixed income benefit or are involved with the child welfare system. To eliminate delays while permanent housing subsidies were processed, the Department of Homeless Services provided eligible families with a bridge subsidy that ended as soon as the permanent rent subsidy became available. This allowed families to immediately move out of shelter and into housing of their own.

Communities may also take a "wait and see" approach to providing permanent rent subsidies. In Salt Lake City, for example, most homeless families are being assisted with three months of rental assistance and employment support. After three months, the program re-evaluates the family's housing and employment situation. If it appears that they will not be able to continue independently, rent assistance is extended and additional services are provided for another three months. The community is prioritizing permanent rent subsidies for families who, after 18 months of HPRP-funded assistance, still appear unable to pay for housing independently.

TARGETING SERVICES

The growth of Rapid Re-Housing has greatly influenced how homeless providers deliver services to families. New fields of service provision have evolved to help families search for housing, negotiate with landlords, and maintain housing. Because Rapid Re-Housing programs generally provide only short-term assistance, services designed to help families quickly increase their income through work have taken center stage.

Mobile Services

Rapid Re-Housing has also increased the use of mobile, as opposed to facility-based, support services. Mobile or community-based services allow social workers and other service providers to deliver assistance wherever families are residing, whether the family is in a doubled-up situation and at risk of homelessness, in shelter, or in newly acquired housing in the community. Because services do not end when families leave a shelter program, this affords greater continuity in the support offered to parents and children.

Partnering with Community-Based Agencies

The diversity of families' needs has led communities to develop new partnerships with community and public social service programs to ensure families have access to the full array of quality support services they require to succeed. By focusing on helping families access community-based supports to achieve their long-term goals, homeless service providers can focus on housing.

One example of a local partnership to enhance services to families that are homeless is the STRong program in Minneapolis, Minneapolis. The program is a partnership between Reuben Lindh, Wayside House and St. Stephens Human Services to offer Rapid Re-Housing assistance coupled with child-centered support services to young homeless and at-risk parents and their young children. Clallam County, Washington offers another example of a successful local partnership, Living in Families Together (LIFT). The county has dedicated permanent supportive housing units for homeless families under the supervision of the child welfare agency, including families under court supervision. The child welfare agency provides the intensive supportive services to help families remain together. The collaboration has not only ended the homelessness of the families served, it has resulted in a reduced number of children placed into foster care.

Critical Time Intervention

Local homelessness programs are also improving the quality of services they provide by exploring evidence-based practice models. One example is Critical Time Intervention for Families, a time-limited case management model primarily targeted to homeless families that include a parent with a mental health or substance use disability. The model, first tested in Westchester County, New York, starts with very intensive support services which taper off after nine months as families stabilize and are connected with supports in the community.

COORDINATING INTAKE, ASSESSMENT, AND SERVICES

Until recently, most communities had a limited variety of housing and service interventions to respond to families experiencing homelessness. With more service and housing options available, communities are developing intake and assessment strategies to match families with the intervention that best fits their needs and makes the most efficient use of resources.

A coordinated intake process ensures that families do not have to call multiple programs to find the assistance they need. A common assessment process can ensure that families' needs are well understood before they enter a program. Finally, a referral system based on common intake and assessment ensures that families are sent to the programs that are best suited to meet their needs, not just the first program they can get into. A coordinated system can also ensure that the more service-intensive interventions (such as transitional housing and permanent supportive housing) are targeted to the families in the community that require more specialized or intensive supports.

Hamilton Family Center in San Francisco, California has developed an assessment tool to help match families to the full array of programs they offer. Dudley Apartments, a permanent supportive housing program Hamilton Family Center operates in partnership with Mercy Housing California, is reserved for families seeking assistance with the most significant barriers to housing stability, including families who have had multiple homeless episodes. The intensive services they offer onsite are designed to assist residents with mental health, substance abuse, and domestic violence issues.

Creating a Common Vision

The creation of a coordinated service system requires a significant commitment from local providers and stakeholders. A strategic plan is usually required and individual organizations must agree to work together toward a common vision. Minimizing gaps and duplication in service delivery may require individual organizations to change how they deliver services and may also require organizations to change their missions. Ongoing work is required to ensure there is consensus around how existing and new community resources will be used to assist families. Ultimately, the creation of a unified vision and a coordinated service system to achieve goals can result in more efficient use of a community's resources and a much better experience for homeless families. With a shared vision, community providers are more effective in attracting new resources and demanding greater accountability from public sector agencies in how they respond to family homelessness.

In Alameda Country, California, the crafting of the EveryOne Home Plan to End Homelessness included multiple stakeholders, including county officials, officials from the 14 cities within the County, and an array of homelessness service providers. As a result, a common vision for how homelessness could be ended in Alameda emerged. This facilitated the commitment of federal and local resources, including HPRP funds, to the key areas identified in the plan. It also helped EveryOne Home attract other ARRA resources for homelessness prevention, resulting in an overall commitment of $12 million to combat homelessness in the region. Stakeholders continue to meet regularly to promote effective implementation of their community's programs. The committees evaluate emerging data, discern what is working and what is not, and identify the program-level and system improvements needed.

USING DATA FOR PLANNING AND PROGRAM MANAGEMENT

The shifts that are underway to transform the nation's homeless service system rely on research and data that demonstrate what works. Ongoing data collection and analysis help communities refine interventions and inform how they target resources. Most importantly, data helps demonstrate what interventions are effective in reducing family homelessness. This allows communities to identify the most effective strategies to implement.

Informing Plans

Data has been instrumental in the development of local plans and in ultimately propelling communities' efforts to end homelessness. Accurate information about how many families experience homelessness over the course of the year and on any given night and the comparative costs and effectiveness of interventions, has led communities to shift their spending on homelessness and build support for new initiatives. In many communities, this information has resulted in new investments in rental assistance.

Understanding the needs of different subsets of families by analyzing data leads to more strategic targeting of service-rich interventions such as permanent supportive housing and transitional housing and informs the targeting of prevention resources to families at risk of homelessness. The data which helps refine how communities "match" families with the right intervention ensures that all homelessness resources are used to maximum effect.

Measuring Performance

Communities have utilized data to create benchmarks from which they can assess their overall progress toward the goal of ending homelessness. Progress can be measured through reductions in a community's point in time count, which provides a one day snapshot of homelessness, or reductions in the number of families seeking shelter over the course of a year. Data can capture improvements in communities' homelessness system that will likely foreshadow future reductions. Massachusetts, for example, now contrasts the rate at which families are entering the state shelter programs with the rate of exits to permanent housing. Recent increases in shelter exit with declines in new entrants into shelter indicate the state's homelessness system is poised to see overall declines in the near future. Similarly, communities are capturing the length of time families reside in programs before re-accessing housing to measure progress toward ending homelessness.

Recognizing the importance of data, HPRP and the McKinney-Vento Homeless Assistance Program require local communities to capture program and system level outcomes. Data informs policymakers, voters, philanthropists, and other stakeholders, about the effectiveness of their investments in ending homelessness and helps build support for increased investments in strategies that work. Policymakers' confidence in strategies to end homelessness has contributed to Congress awarding annual increases in the HUD McKinney-Vento Homeless Assistance Program, even as many other programs were being cut.

CONTINUING PROGRESS IN THE YEARS AHEAD

Local leaders have been relentless in their pursuit of successful program strategies that make efficient use of their limited resources. Their hard work has paid off. Continued progress, however, depends heavily on new lessons derived from local leaders at the forefront of innovation. It also depends on the leadership of the federal government and the assurance that communities will have the resources they need to prevent and end family homelessness.

In 2011, communities will be expected to implement the Homeless Emergency Assistance and Rapid Transition to Housing (HEARTH) Act, which reauthorizes the McKinney-Vento Homeless Assistance Programs at the US Department of Housing and Urban Development (HUD). This restructuring of HUD's homelessness programs has been informed by strategies communities have used to reduce homelessness. Congress can have a substantial impact on homelessness by appropriating sufficient resources to ensure that communities can effectively implement these new program strategies without pulling resources from existing successful approaches to ending homelessness.

Congress and the Administration can make the job more viable by investing heavily in housing that is affordable to very low-income families. The Housing Choice Voucher Program and the National Affordable Housing Trust Fund are critical to expanding housing opportunities for low income families. The impact will be maximized if resources are targeted to those with the greatest housing needs—households with incomes less than 15 percent of Area Median Income.

Federally funded programs that are intended to provide a safety net to vulnerable families are failing to prevent these families from experiencing homelessness. The Temporary Assistance for Needy Families (TANF) program, for example, includes among its purposes providing assistance, "so that children may be cared for in their own homes or in the homes of relatives." Despite this, families served by TANF programs experience high rates of housing instability and homelessness and recent data indicates that less than one-fifth of families entering homelessness programs receive TANF support. To make progress on family homelessness, TANF agencies must improve the program's effectiveness in ensuring low income families are afforded real opportunities to increase their self-sufficiency and avoid homelessness. Similarly, child welfare agencies and other programs serving families with high rates of homelessness and housing instability need better responses to the housing crises of the families they serve.

Congress can support initiatives to improve the outcomes of critical social safety net programs in serving vulnerable families at risk of homelessness. This may be achieved with dedicated resources to allow state and local TANF or child welfare agencies to offer intensive and comprehensive services to those families on their caseloads that have the greatest barriers to self-sufficiency, the least ability to maintain housing, and most challenges supporting their children. Congress can also stimulate innovation by supporting strategies designed to leverage both housing and support services for families with significant barriers. Programs such as the Family Unification Program (FUP) or the proposed Housing and Services for Homeless Persons Demonstration both bring together affordable housing and services for needy homeless and at-risk families.

Communities across the country have demonstrated that ending family homelessness is possible. The momentum they have created will be strengthened with the leadership and support of the federal government. Safety net programs serving vulnerable families must be improved to more effectively prevent and end homelessness and sufficient resources committed to provide low-income families with housing they can afford.

Conclusion

Approximately half a million people in families are homeless over the course of a year. Despite the economic crisis that continues to place many more families at risk, there is a clear path forward for preventing and eventually ending family homelessness. The strategies outlined in this report—prevention assistance; Rapid Re-Housing; helping families pay for housing, targeted services; coordinated intake, assessment, and services; and data for planning and program management—have been central to the success in communities that have reduced homelessness, even during this economic crisis. As these strategies continue to be adopted in communities around the country, and as they become a greater focus of federal policy and resources, more progress will be made.

ENDNOTES

1. National Alliance to End Homelessness, *Promising Strategies to End Family Homelessness*. Washington, DC: Author, 2006. Available at: http://www.endhomelessness.org/content/article/detail/999
2. U.S. Department of Housing and Urban Development, *2008 Annual Homeless Assessment Report to Congress*. Washington, DC: Office of Community Planning and Development, 2009. Available at: http://www.hudhre.info/documents/4thHomelessAssessmentReport.pdf
3. U.S. Department of Housing and Urban Development, *The Homelessness Pulse Report: Third Quarterly Report*: Washington, DC: Office of Community Planning and Development, 2010. Available at http://www.hudhre.info/documents/HomelessnessPulseProjectJan10.pdf
4. Debra J. Rog and John C. Buckner, "Homeless Families and Children," Toward Understanding Homelessness: The 2007 National Symposium on Homelessness Research. U.S. Department of Health and Human Services, Assistant Secretary for Planning and Evaluation and U.S. Department of Housing and Urban Development, Office of Policy Development and Research, 2007.
5. Abt Associates Inc., Gregory Mills, Daniel Gubits, Larry Orr, David Long, Judie Feins, Bulbul Kaul, Michelle Wood, Amy Jones & Associates, Cloudburst Consulting, The QED Group. *Effects of Housing Vouchers on Welfare Families*. Washington. DC: U.S. Department of Housing and Urban Development Office of Policy Development and Research, 2006. Available at: http://www.huduser.org/portal/publications/commdevl/hsgvouchers.html
6. Marybeth Shinn, Beth Weitzman, Daniela Stojanovic, James R. Knickman, Lucila Jimenez, Lisa Duchon, Susan James, and David H. Krantz, "Predictors of Homelessness Among Families in New York City: From Shelter Request to Housing Stability," *American Journal of Public Health*, 88(11), 1998, 1651–1657.
7. Culhane, D.P., Metraux, S., Park, J.M., Schretzman, M. & Valetne, J. (2007). "Testing a Typology of Family Homelessness Based on Patterns of Public Shelter Utilization in Four U.S. Jurisdictions: Implications for Policy and Program Planning." *Housing Policy Debate*, 18(1): 1–28.

DISCUSSION QUESTIONS

1. Given the national shortage of affordable housing, what should be the government's role in addressing this predicament? Is this a social problem that can be solved using free-market solutions? Why or why not?
2. The author suggests using child welfare and TANF (Temporary Assistance to Needy Families) funds to provide shelter for the 500,000 American families that spent at least one night in a homeless shelter in 2008. Do you support expanding current government welfare entitlements to provide affordable homes to homeless families? Why or why not?
3. Given the severe lack of affordable housing and the paucity of government funds specifically geared toward alleviating this social problem, are there specific things that individuals can do, according to the author of this selection, to reduce the likelihood of becoming homeless? What are these things?
4. The National Alliance to End Homelessness, the author of this selection, suggests that homelessness is a social problem that has a ready solution. However, the solution almost exclusively relies on the expansion of government welfare entitlements which are severely strained and underfunded. Is this really a social problem with a solution? Why or why not?

PART X

Social Problems Related to Education

Most of us know that our public schools have some serious problems. On the one hand, not everyone gets a decent education, and on the other, many people disagree over what constitutes a decent education. Perhaps we expect too much from our schools; perhaps we no longer know what we have a right to expect. Rapid social change in society is altering the educational institutions, and it is difficult to identify both the problems and ways to deal with them.

Annette Lareau and Elliot B. Weininger, in the essay *Class and the Transition to Adulthood*, make the case that the key variable in educational outcomes and scholarly achievements is social class—namely, parenting styles that are largely differentiated by socioeconomic advantage. Middle and upper-middle class parents employ, according to the authors, a "concerted cultivation" method, where they see it as their primary responsibility to encourage and enhance their children's would-be aptitudes and talents for future success. Concerted cultivation takes significant resources of money, time, personal effort, and intellect on the part of middle class parents.

James Traub believes that it is not simply schools or parents who are at fault. Children growing up in economically poor neighborhoods are not adequately prepared for the challenge of school. We cannot expect schools to excel with students who are far more influenced by forces outside the school that work against a successful education.

Jonathan Kozol identifies the ways that elementary schools reinforce class position. He has visited many schools throughout the United States, some excellent and some poor. Wealthy and middle-class children go to the best schools;

the poor are left behind. Kozol blames our priorities as a society for such inequalities, and believes that with greater resources schools in poor neighborhoods can provide much better educational opportunities to the children there. It is difficult, he concludes, for students in neighborhoods that are poor to get a decent education.

Gerald Grant writes on *Why There Are No Bad Schools in Raleigh*, outlining key policy decisions that led to an unprecedented rise in student achievement and academic standards. This book is a case-study on how a school district with the right mix of smart policies, ample resources, and *the political will* to challenge the status quo can transform itself.

David Leonhardt, a reporter for the *New York Times*, has written an excellent piece on social class, inequality, and college dropouts. Leonhardt is not a sociologist, but he certainly thinks like one, and he seems to know a great deal about why students who are not upper or middle class have a difficult time getting a college degree. Instead of giving an opportunity for members of the working and lower classes to better themselves, a good college education is increasingly becoming a way to keep people in place.

36

Social Class and Education

ANNETTE LAUREAU AND ELLIOT B. WEININGER

The Four Questions

1. What is the problem according to the authors of this selection?
2. Is it a *social* problem?
3. Who does the problem harm? Does it harm society, or does it only harm the individual?
4. What can we do about this problem?

Topics Covered

Educational aspirations

Educational inequalities

Social class and educational outcomes

Class inequalities

Concerted cultivation

Social reproduction

Parental role in educational outcomes

Class differences in educational socialization

If the educational research models are correct, then students' negotiation of the transition to higher education can be conceptualized as an individual process. Since students are on the cusp of adulthood and, in many instances, pressing for increased autonomy, this approach could be a useful one. Yet there are also signs in the literature that at earlier points in children's educational careers class differences in parent involvement have surfaced (Brantlinger 2003; Lareau 2000; Reay 1998). In addition ... even if everything goes smoothly, the process of applying to college is complex. If a student encounters trouble, for example, in securing a promised letter of recommendation, or gets a low grade in a key course, then the assistance of adults might be particularly helpful in keeping him or her on track (Deil-Amen 2007). In this section, we compare two young African American women who each aspired to a medical career. Tara Carroll grew up in a poor family and dreamed of becoming a nurse. Stacey Marshall grew up in a middle class family and hoped to become a pediatrician. Tara was raised by her grandmother when she was younger and later moved in with her mother (her father was in prison). Stacey's biological parents had raised her since birth. Both young women had a sibling who was approximately two years older. The careers they aspired to required a science background. During middle school, both had struggled with math; neither young woman was naturally gifted in the subject.

Comparing Tara's and Stacey's transitions from high school shows that the differences in parental interventions in education observed when the children were younger have persisted. For example, in keeping with the cultural logic of the accomplishment of natural growth, Tara's mother placed the responsibility for applying to college on her daughter.

> Well, I let Tara make her choices. She told me what she wanted to do, and I let her do it. I should have participated more, yes, but because of my [work] schedule, I couldn't.... I should have participated

SOURCE: From Annette Lareau and Elliot B. Weininger, "Class and the Transition to Adulthood," in *Social Class: How Does it Work?* © 2008 Russell Sage Foundation, 112 East 64th Street, New York, NY 10021. Reprinted with Permission.

> more and pushed a little more. You know, get on the Internet and find scholarships, utilize that. Get her to write and sell herself more. You have to do essays to participate for scholarships. I should have been a little bit more pushy that way.

Although she had a car, and all of the schools Tara was interested in were within a one-hour drive of their home, Tara's mother did not visit any. The visits that Tara experienced were arranged at the "health academy" in her public high school. Tara's mother also did not recall ever having seen an application. Tara filled out the forms at school, with assistance from teachers: "Every college that she applied to, all I saw was the letters, yea or nay, and that was hurtful for her—everybody kept turning her down." Her mother also lacked basic information about crucial aspects of Tara's education: "I wish I could find that letter. All I remember is that [her GPA] was above average. I don't know exactly what it was. I think it was around a B."

Tara's knowledge was limited too. As a senior, she had to ask her counselor what a "3.4" meant:

> I had went to my counselor, his name was Mr. Bradley, and he did my GPA, like combined my averages, like, add[ed] up all my grades.... I remember, it was close to a 3.5 or it was like a 3.4. [I said] something like, "Well, what's that?" He was like, "That's a B." So I knew I did good overall [in] high school.

Tara's combined SAT scores totaled 690. She applied to six or seven colleges, including a number of schools whose representatives had made visits to her high school. In addition to assistance from teachers, Tara received advice from an aunt who had graduated from college. Still, she was unable to overcome some of the challenges the applications posed:

> I only got one recommendation. I didn't know a lot of teachers in there, um, but I had one teacher I really liked. Her name was Ms. Thomas. She gave me a letter of recommendation.... I had to get one more, but I couldn't. I tried to get one from my counselors, my counselor, but he came up with an excuse, like, um [*she changes her voice*], "You know, I got a lot of people to do, and I got to take time," some ol' excuse, and he had to see if he had time. And I was supposed to get one more, and that was from my English teacher ... his name was Mr. Rogers. He was the head coordinator of my [school within a school] health academy. But he never got around to writin' it He just kept saying, "Remind me, remind me, remind me." And I never got a chance to get that letter of recommendation from him, so I only had one.

Tara applied to colleges where the median SAT scores were 100 to 300 points above her own scores. She was rejected everywhere. She went to community college "as a last resort" to pursue her hope of becoming a nurse. She enrolled in an anatomy and physiology course but soon found that she did not like the instructor. Although she went to the instructor's office hours, she still did not understand the material, did not feel comfortable asking questions, and did not know that she could have sought help from an adviser. She simply stopped going to class. Since she did not formally drop the course, she failed it and will have to take it again. She says she is "taking a break from school" but hopes to go back in the future. Put differently, although Tara had relatively high grades and a fervent desire to attend college and had energetically visited colleges and filled out numerous applications, her college search failed to yield admission to a four-year college because of her lack of informal knowledge about the higher education system and the fact that she applied to schools that were quite selective (reflecting her high aspirations). Both she and her mother found the process to be "hurtful."

Middle class parents, especially mothers, generally have active roles in the college application process. Often parents guide their children every step of the way. For example, Stacey Marshall's experience with college applications was quite different from

Tara's. From the beginning, Stacey received considerable adult support. Middle class parents, having gone to college themselves and having sought out detailed information from their social networks, generally feel comfortable providing their children with informal advice on academic matters. Stacey's mother worked in the computer industry and had a master's degree. In addition, she had long been committed to an especially active form of concerted cultivation. She reported that she had "kept tabs with the teacher" all through her daughter's primary and secondary education: "I was then and remained through high school kind of a visible parent in the school district, letting them know that I was watching what was going on."

Stacey's mother was concerned about her daughter's desire to be a premed student because she was aware that math was not Stacey's strong suit:

> She had the traumatic experience near the end of the [fifth-grade] year where she had gotten what I believe was her very first C.... Math has haunted her ever since. [*Pauses.*] Even now, there is this insecurity, this feeling that "I can't do it," which in my mind puts a damper on that desire to go into medicine, because she's got to deal with that math.

Nor, in her mother's opinion, had Stacey chosen to enroll in sufficiently challenging classes in high school: "Stacey was always a B-plus student with the ability to be A-minus, [who] never wanted to accept a challenge to go for a more challenging class. She did stick with the sciences. The counselor and I forced her into that."[1]

Indeed, Stacey's mother actively sought ways to further the development of her daughter's natural sciences skills. She mentioned, for example, that the summer before Stacey's senior year she and her husband had spent "a lot of money" (at least $1,500) to have Stacey attend a ten-day "young people in science" program suggested by a high school microbiology teacher.

After Stacey grew three inches in a single summer and was no longer competitive in gymnastics, she was persuaded to play basketball for her middle school. She became a basketball star in high school. During her senior year, she was recruited by Ivy League schools. Her combined SAT scores were 1060, and her grades were a B average. Although Stacey desperately wanted to go to Columbia and the basketball program there would have helped her gain admission, her parents told her the financial aid gap (over $15,000 per year) was too large and that therefore she could not go. (Privately, she reported still feeling "bitter" about her parents' unilateral decision.) Her mother worked hard to collect extensive information about the acceptance policies of various colleges; she guided her daughter to colleges that she thought would be a good fit for Stacey and that also reflected her daughter's desire not to be too far from home. Ultimately, Stacey was admitted to a selective state university three hours from home; as a member of the women's basketball team, she was given a full scholarship.[2]

During Stacey's freshman year, her parents continued to play the same sort of active role they had taken throughout the college application process. Her mother helped her select courses strategically: "I said, 'Look, you have got to get out of calculus. Because first semester you don't want your GPA to get too low, because then you can never dig yourself out of it.' And so she was calling me about 'what should I take?'"

Her mother's advice concerned both the immediate decision and, more importantly, the resources Stacey needed to use to make effective decisions in general:

> She emailed me first, and then she called me one night at eleven o'clock. And it was like, "Well, I can take the cinema course." I said, "*Stacey*" [*Laughs.*] I said, "Do you know anything about this? Do they even give you a description?" [She said,] "Wellllll, no, not really." But she ... was on the computer; she could see that there were seats in this class; the time was right. I said, "Sometimes you can be jumping out of the frying pan into the fire. This course may have an interesting name, but,

> one, it sounds like, yeah, you'd be watching movies, but two, you'll probably be writing about [them]. You have these other courses where you will be writing. You are telling me that you don't like to write." She said "Well, I don't know. I guess I see your point."

Stacey's mother continued:

> "In the meantime, you have an academic adviser...." I really stressed with her that *she* needed to establish communication with those people. I said, "You need to call Alesha. Call Alesha, set up an appointment, and *go to talk to her*. Talk to her about your major. And see what advice she can give to you. That is what she is getting paid for."

Indeed, in some ways, her mother appeared to be trying to teach Stacey to be less dependent:

> I think that she was used to, throughout school, to me making a call or talking to a teacher as part of my parental responsibility. A couple of times I got emails from her while she was at college with questions like, "What should I do?" or, "What do you think?" And as the year wore on, there were some things that I would say, "Look, Stacey, this is *your* responsibility. I can't go and talk to this professor. This is college. [Stacey's university] doesn't even send me your grades. [*Laughs*.] They won't tell me anything...."

Stacey Marshall clearly benefited from her mother's informal guidance. Although it is uncertain whether she will be able to successfully pursue her university's premed curriculum, it is much more likely that Stacey will graduate from college than it is that Tara will, although both are attached to the idea.

To summarize, comparing Tara Carroll's experiences with Stacey Marshall's suggests that middle class children have more informational resources at their disposal as they negotiate college than do working class and poor youth. Also, the comparison reveals a striking similarity in the overall cultural logic of child rearing that parents followed when their children were in fourth grade and when their children were in high school, even though the problems and concerns shifted considerably.

NATURAL GROWTH CONTINUES

Many students, particularly those of working class and poor origins, do not attend college. A significant number, particularly in urban communities, drop out of high school. Since Lareau's study drew on this population, it is not surprising that a number of the youth in her study struggled in high school. A number dropped out. Some graduated from high school and attempted to start, but quickly stopped, attending community college. Their mothers and fathers were extremely concerned about their educational progress. But since parents generally defined education as the school's responsibility, parents played much less of role in critically overseeing or intervening in their children's high school experiences than did middle class parents. When mothers did try to intervene, they reported that they were unable to alter the course of events. Put differently, in *Unequal Childhoods*, Lareau reported that working class and poor families spent scarce resources taking care of children and keeping them safe. Unlike the middle class parents, the working class and poor parents did not see it as their duty to continuously develop their children's talents and skills. Interviews conducted ten years later revealed continuity with these earlier patterns (see Table 36.1).

Overall, parents continued to try to keep their children safe and to protect them. Parents (many of whom were high school dropouts) continued to cede responsibility for education to the school and, in addition, to the children themselves. Unlike middle class parents, the less-privileged parents did not micro-manage core aspects of the school experience. Nor did they seem to possess potentially

TABLE 36.1 Continuities in Class Differences in Concerted Cultivation and Natural Growth

Social Class	White		African American	
Middle Class Students	Garrett Tallinger	Melanie Handlon	Alexander Williams	Stacey Marshall
	Did elaborate college search. Admitted to Ivy League schools, but chose Villanova, a small Catholic college with a good basketball program; on basketball scholarship.	Applied to colleges, but was rejected by top choices; did not want to attend the school that did offer admittance. Her mother was embarrassed by Melanie's decision not to attend college.	Did elaborate college search, with college visits; admitted to Columbia University medical school program as an undergraduate.	Did elaborate college search, with college visits; admitted to the University of Maryland. Mother helps select courses and gives college advice.
	Father guides course selection.	Attends cosmetology school.	Mother gives college advice.	
Working Class Students	Billy Yanelli	Wendy Driver	Tyrec Taylor	Jessica Irwin (bi-racial)
	Dropped out of high school; obtained a GED.	High school graduate; accepted by small Catholic college, but decided not to attend.	High school graduate; never took the SAT.	High school graduate; received solicitations from dozens of colleges owing to PSAT scores in 1300s.
	Does not know anyone attending college. Father helped with job acquisition; father intervened when drug test violation threatened job retention.	Mother extremely disappointed by Wendy's decision not to attend college. College search handled by high school; parents did not know SAT scores or that application fees were nonrefundable.	Mother wants Tyrec to become a lawyer; will help him "apply for a loan" for college. Started at community college but stopped.	Attends Temple University.
Poor students	Mark Greeley	Katie Brindle	Harold McAllister	Tara Carroll
	Dropped out of high school; wants to work "in computers." Mother pressured Mark to take GED course; he did so, but did not take certification exam. Complained of police harassment.	Dropped out of high school. Cleans houses full-time with mother. With help of eighth-grade teacher, admitted to public magnet high school. Mother nixed the most academic and prestigious school because of bad neighborhood; wanted her to stay closer to home.	Dropped out of high school; works as a waiter in a chain restaurant. Denial of opportunity to play basketball in high school a major life disappointment (coach wanted him to play football instead; Harold refused).	High school graduate; currently works taking care of disabled adults. Superficial college search with little parental input; some college visits and all applications overseen by high school personnel; applied to many colleges, but not admitted to any. Enrolled in community college; failed a key class; hopes to return.

Source: Authors' compilation

important informal knowledge, such as the academic status hierarchy of their local public high schools. In the case of Katie Brindle, a white girl from a poor background, this apparent lack of information had long-term consequences. When Katie was about to enter high school, her mother's primary concern was her daughter's safety. Her mother did not want Katie to go across town to attend a school in a different neighborhood. This more distant school was much more highly ranked academically than the one that her mother directed her daughter to attend, although it is unclear whether either she or Katie knew this. Katie recalled that she had "wanted to go to Lincoln High. But my mom said I could only take the Washington or Franklin [choices] because they were in the neighborhood."

Saying that she was "wild" and that there was nothing her mother could have done to change that, Katie reported that she "never really went to high school." Her teenage years were punctuated by truancy, fighting, alcohol, and drugs. At age seventeen, she became pregnant, dropped out of school, and gave birth to a daughter, who is now being raised mainly by Katie's sister. As with many working class and poor youth in the study, she has received substantial assistance from her family. Her mother found her a job and got up with the baby at night when Katie was recovering from the cesarean section surgery. Her sister often took care of the baby and, when Katie was "running the streets," offered to raise the child. Her mother and sister also gave her gifts of cash and food. Katie hopes to get a GED one day; in the meantime, she works with her mother cleaning houses.

An orientation to schooling similar to the Brindles' was evident in other poor and working class families. Parents showed love and concern for their children in many different ways, but since they did not define their proper role as including the management of school careers, they had limited information about the options available to their children. For example, the Greeleys, another poverty-level white family, relied on school personnel for guidance when Mark, their eldest son, was ready for high school. As one of the few whites in an inner-city middle school, Mark had had a difficult time, frequently getting into fights. His mother explained that in selecting a high school for Mark, she "went by the word of his [middle school] school counselor, who said that this would be a great school for him. It's smaller, no one has issues, no one has problems." The school, however, did not prove to be a good choice. Mark was again one of the few white students, and he again had problems. Although he later transferred to a different high school, he became increasingly disengaged and eventually dropped out.

As Mark's problems were escalating, his mother attempted to intercede. She called "the school" to find out what she could do. This effort proved futile:

> I called the school. And I asked them, what, what—you know, "My son doesn't wanna come to school—what legal—what can I do legally?" There's nothing I can do when he's seventeen—if he don't wanna go, he don't have to go. It's just that simple. They can quit when they're sixteen, but you have to have the parents' permission, and he has to have a job. But, you know, he just didn't go to school the last month, and then he was seventeen. So ... there's nothin' that I could've made him do.

Mark dropped out of high school when he was sixteen. He got a job stocking grocery shelves at night for $10 an hour. At the time of the follow-up interview, he lived at home, and his mother frequently drove him to work.

All three of Ms. Greeley's children have scored high enough on standard IQ tests to qualify for school-based gifted programs. (A teacher recommended that Mark be tested, and Ms. Greeley then had his siblings tested as well.) Nevertheless, none of the Greeley children have done well academically, although Mark's sister is likely to graduate from high school. Their mother noted ruefully that

> Pammy was one hundred forty-something [IQ score], and Rick was one hundred

> forty-something. But none of them did nothing with it. I think if situations … I don't know—I just believe that it's because of circumstances and situations and environment. They didn't flourish, you know, they didn't flourish. I mean, it might be the same anywhere. [But] I think if schools were different and they had different opportunities to go to better schools—they didn't. Unfortunately, when you're poor, you have to go to the neighborhood school, you know. They didn't have—there was no choices of going anywhere else—you have to go to the neighborhood school.

Ms. Greeley's reliance on school officials when making key decisions concerning her children's schooling, coupled with her sense of powerlessness in the face of negative events, was typical of working class and poor parents in the study, and it closely echoed the sensibility that was apparent among them some ten years earlier—namely, the supposition that responsibility for their children's education was properly relegated to educators.

Also striking is the case of Harold McAllister. An African American child from a poor family, Harold grew up in a housing project. When asked whether anyone had discussed the possibility of his pursuing his education beyond high school, Harold indicated that in ninth grade he had been in a college-oriented school:

> Like ninth grade, I was in the magnet charter, that was like the top charter—in the school they do all college-bound work. So I was cool. The magnet was in just like math and science, that's their true, like, subjects, math and science. So I was cool with that, but like my attendance wasn't that good, though.

Harold's late arrivals eventually led to his being dismissed from the school:

> So they kind of kicked me out [of] that charter and put me into another charter. The other charter, the work was easier, so I started, you know how you just start chillin' more, so if I could have stayed in [the] magnet, I would have been cool. The work was harder, but that was, like, on my pace. It gave me a challenge.

Harold received Cs and Bs at his new school, with an occasional D. His academic record was shaped in part by a sports-related setback. An avid and talented athlete, he loved basketball, but he had the build of a football player (five-eleven and 240 pounds). The same teacher coached the football and basketball teams. Worried about his knees, Harold refused to try out for the football team (which "stunk"). Despite his formidable basketball talent, Harold was not selected for the basketball team. He was convinced that the coach made that decision because he was angry about Harold's lack of desire to play football. Harold was devastated. After an appeal to the school's principal proved unsuccessful, Harold chose another school and planned to transfer, but that move never materialized:

LAREAU: And what was involved in transferring?

HAROLD: Paperwork…. Because I had, I knew like a couple of players on their basketball team. I played with them in my city league…. I've never been a problem child, you know, like my grades were average or above-average, so what was the reason for them to not let me transfer there?

LAREAU: And when did you talk to your dad about it?

HAROLD: Like, some nights after, like, I'd have a basketball game, and we'd win, like, easily with, like, all of us playing together. I'm like, hey, we can do this in high school too, though. He used to say something, but, like, I knew … he didn't really care about basketball [*laughs*], like, he never played basketball. He was boxing and, like, that's it.

LAREAU: And did you talk about it with anybody else at that time besides your dad?

HAROLD: My mom sometimes. Like, I just, I didn't, I just felt the vibe ... um, I don't know how to say it, though, um. I talked to my mom a lot, but, not a lot, but, like, sometimes, though. But, like, nothing ever happened, though.

In the end, Harold got a job busing dishes at a suburban restaurant. He worked the 4:00 PM to midnight shift. During the interview, he explained that he "just started working to get my mind off of basketball." Although he had to take four buses and the trip lasted two hours, he liked the job and soon switched to a full-time schedule. He planned to work full-time and attend high school full-time. Around this time his mother moved in with her daughter, and Harold, who said he could not "deal with all those females," took up residence with his father. His school attendance again became spotty because he frequently overslept after having gotten home very late from his job. Traditionally, Harold's mother had woken him up in time to go to school; his father did not. In the spring of his senior year, Harold dropped out. His mother was outraged:

> He had a nice time in there, he was progressing good. He was—he didn't play sports in school. He played sports outside of school. But when we moved, we moved in ... April, he had two more months of school and graduate, and Harold didn't graduate. I was mad. [*Voice rising.*] All he had to do was graduate, two months, April, May, and June.

Unlike the middle class parents in the study, Harold's mother did not create a customized and individualized school career for her son. Instead, like Ms. Greeley, Ms. McAllister depended on educators to use their professional expertise to do this. And like Ms. Greeley, Ms. McAllister felt intensely frustrated by and powerless in the face of the events leading up to her son's decision to drop out.

In all, three of the four youth from poor backgrounds dropped out of high school (Tara Carroll was the exception), and one of the four working class youth (Billy Yanelli) did so. Of those who graduated, all expressed a desire to go to college. Only one did; the others were diverted by various life events. For instance, Wendy Driver, a white working class young woman, attended a private Catholic high school after not being admitted to any of the more desirable public high schools. Her paternal grandfather paid her tuition. Although Wendy struggled academically, the school counselors helped her apply to college. She was accepted at a small Catholic college with a program for students with learning disabilities. After high school graduation, however, she told her mother and stepfather that she would not attend. She reported being worried that she could not manage the work and thought that she "would fail," especially since her girlfriends who were already in college, and who had been much better students in high school than she was, were struggling. Her parents offered other explanations. Her stepfather felt that Wendy wanted to be with her boyfriend (soon to become her husband). Her mother felt that the college, a ninety-minute drive away, seemed too far from home; Wendy had no experience being away from home. In December of what would have been her freshman year of college, she became pregnant: "It wasn't a planned pregnancy. We were being careful and everything. I was taking birth control. It just happened. I was scared of course. He was scared because he didn't know what we were going to do."

Abortion was out of the question, and although her mother "was shocked at first," she became supportive. Over the objections of Wendy's boyfriend's mother, the two married immediately, in a small ceremony. A year later, they had a large formal wedding and reception. Before she had the baby, Wendy briefly explored attending the local community college, but then abandoned the idea when she found out that she would have to take remedial non-credit courses (and be "the only white girl" in the classroom). Today she is a full-time homemaker; her husband is in the Navy. Wendy hopes someday to have a child care business in her home. In the meantime, her step-father routinely drives to her house (a three-hour trip) to bring her and the baby home for visits. She and her husband are now expecting a second (planned) child. Wendy's mother, while continuing to be disappointed by her daughter's failure to attend college, has provided substantial psychological, economic, and social support for Wendy and her family.

Tyrec Taylor, a young man from an African American working class family, attended three different high schools and came extremely close to dropping out. After a disastrous junior year that included two juvenile arrests, Tyrec's mother pleaded with his father to take out a large loan ($6,000) to cover tuition at a private school. He did, and Tyrec attended and graduated. Tyrec said that he felt he was "supposed to go to college," but he never took many college preparatory classes or the SATs. Nor did he apply to college. He completed one remedial noncredit English course at a community college. He has not returned to school since that time.

Of the eight working class and poor students in the study, only Jessica Irwin, who was biracial, enrolled in a four-year college. Jessica's parents were more highly educated than the parents of the other working class and poor students: her father had attended college, and her mother had graduated from college.[3] Nevertheless, with respect to their daughter's educational career, the Irwins followed the natural growth approach. Jessica's grade point average (94) and SAT scores (1310) generated "tons and tons of mail" from colleges that hoped to recruit her. Also, many of the colleges recruiting students categorized her as African American. Jessica's parents drove her to schools for visits, but their primary concern was how her college career would be financed. Her college search process did not have the breadth and depth of those carried out by the middle class children and their families. Although she had wanted to attend Carnegie Mellon, Jessica's parents vetoed this choice once they learned the tuition. They were unaware that the financial aid packages of private schools can be large enough to wholly or almost wholly offset the high tuition rates. Nor were they aware that a nearby elite private university offers full scholarships for twenty young people from the Irwins' home city each year. Jessica ended up attending a small public university. The full scholarship that she received pleased both her and her parents. She received high grades at college and planned to major in art.

CONCERTED CULTIVATION CONTINUES

Not all of the middle class adolescents went to college. But in every case, their parents took an extremely active role in their transition out of high school; such parental actions are rarely accounted for in the current models of educational transitions. Mothers in particular did not hesitate to intervene. For example, the mother of Garrett Tallinger, a white middle class young man, reported:

MS. TALLINGER: In talking to the guidance people, I said, "I think he's capable of honors [enrolling in the honors courses]. But obviously you-all need to decide that...." One of the things I did know was that the more competitive colleges look at what is available in a high school, and if you're not taking the most rigorous of what's available, that's a strike against you in terms of their evaluating your transcript. And so I wanted Garrett, as is true with all my kids, I want them to take the most rigorous [courses] that they're possibly capable of taking.

LAREAU: And how did you know that?

MS. TALLINGER: Well, from friends who work in admissions offices. I mean, a very good friend used to work in admissions. And she said, "It's very important." And actually then talking to some other parents that had taken their daughter on a tour of Yale and Duke. He told a funny story of visiting at Duke,

> and the question to the person guiding the admissions discussions … the question from the student was, "Well, is it better to take honors calculus and get a B or take regular calculus and get an A?" He says, "It's better to take honors calculus and get an A." [*Laughs.*]

For many middle class parents in the study, friends' experiences with the college admission process proved to be an important source of information.

Garrett did enroll in rigorous high school courses, including AP economics, AP calculus, AP English literature, and two years of both physics and history. His weighted grade point average was 4.26 (which he believes is equivalent to an unweighted average of 3.78). In guiding their son's school career, the Tallingers tapped multiple resources, including the expertise of school personnel. For example, it was Garrett's ninth grade math teacher who facilitated his entrance into the highest levels of math, including AP calculus. Nevertheless, Garrett's mother took steps promptly anytime she felt that her son's best interests were not being fully served by the school. Thus, when a scheduling problem threatened to prevent Garrett from taking college preparatory courses in both math and English, she quickly intervened:

> I did have to go fight about his schedule, because they went into this new scheduling system, and if he wanted to take the AP English and AP calculus, they were given at the same time. I was like, "C'mon, this is not—you gotta figure something out. I mean, I can't believe Garrett's the only person this is impacting. You have to figure a way for these kids to maximize their opportunities, and they have to switch things around." So that I did fight for.

Ms. Tallinger's efforts to directly manage Garrett's education closely resembled her behavior, and that of other middle class parents, some ten years earlier, when the children were in elementary school.

Garrett paid close attention to the college application process. In the follow-up interview, he described his behavior in explicitly calculative terms. He considered his first SAT scores disgraceful:

> The first time I hated my score. I got a ten-sixty, I … ten-thirty, ten-sixty? I was embarrassed; real embarrassed. So I took it again [and] got … I can't remember what I got the second time, but … 'cause I remember you can mix and match. Like you take your best verbal and your best math. My mix-and-match, I got eleven-ninety.

Even this score he felt was not a good measure of his true ability: "And I know I'm better than that. My goal was to get a twelve hundred, so, ten shy."

Choosing a college also required making calculations and strategizing. As recent articles in the *Chronicle of Higher Education* (White 2005) and books such as *College Unranked* (Thacker 2005) suggest, the college search process often involves a huge investment of time and energy for both parents and children from the middle class. Garrett considered many different colleges, at different collegiate sports division levels: Stanford, Columbia, Princeton, Memphis, Boston University, North Carolina State, Davidson, and Villanova. The gap in financial aid precluded his attending Princeton or Columbia. In the end he went to Villanova on a full basketball scholarship.

For Alexander Williams, an African American middle class young man, college visits began during sophomore year of high school. By the time he made his final choice, he had visited Duke, Columbia, Dartmouth, Brandeis, Georgetown, Haverford, Northwestern, and a few others. During the follow-up interview, he recalled being annoyed that he had had to spend his spring break in junior year visiting campuses. "I didn't want to go to visit colleges on my spring break," he said, but then quickly added, "No, it was fun." He applied early-decision to Columbia, which meant that he was committed to going if he got admitted. With an A average and SAT scores of 1350, he was reasonably confident. He explained: "I didn't apply anywhere else, and if I hadn't gotten in, I would have had about a week and a half to apply everywhere else." His father had not been enthusiastic about this strategy: "I think that my mom was pretty sure I would get in; my dad was wondering why I hadn't applied anywhere else" Put differently, compared to Tara Carroll's mother, Alexander's mother had considerable informal knowledge about the higher education system, was aware of his skills, and guided him through the process ("my mom was pretty sure I would get in"). In other cases, the middle class students visited and toured many schools but, after narrowing their preferences, applied to a small number of potential colleges.

Even after their children have begun their college careers, middle class parents often remain deeply involved. The Tallingers reported that they speak with Garrett frequently and give him advice. During the follow-up interview, his father offered this example:

> He called us the end of last semester when it was time to register for this fall's semester, and he said he was going to get up and get out of business school—he wanted to be a teacher. And it surprised me a little bit. I didn't have anything against it; I was supportive. I told him, "Do what you want, whatever makes you happy."... I said, "Garrett, you just need to realize that..."—his high school coach, who he idolizes, who's been a high school teacher now for twenty years or something... I mean, you know he makes probably high fifty thousands or sixty thousand dollars a year and will probably never make a whole lot more than that. He has a great life, loves his life. So I said, "You just need to be aware of that." He kind of went, "Oh, really?" [*Chuckles.*] I said, "Yeah." 'Cause he's just unaware of these kinds of things.... It was on a Thursday, so on Friday afternoon he called and said, "Well, I'm in the business school." So he stayed in the business school.

By providing their son with regular input during his college career, Mr. and Ms. Tallinger helped Garrett align his short-term educational choices with his long-term goals. Other middle class parents in the study saw themselves as being so central to their children's postsecondary school career that they often used the term "we" during the follow-up interviews. Ms. Handlon, for instance, in reporting with embarrassment that her daughter Melanie (a white middle class young woman) had dropped out of community college, said, "We only made it one semester."[4] This pattern of parental involvement is not unique to the young people in this study. As one college administrator notes, "The 'helicopter parent,' or hovering parent who repeatedly tries to intervene and manage his or her child's life, seems to be a growing phenomenon on many campuses" (White 2005).[5]

CONCLUSIONS

.... Our central finding is that substantial class differences exist in the role played by the family, and in particular by parents. In our ethnographic data, parental involvement is an overwhelmingly middle class phenomenon. Moreover, it takes numerous forms. Middle class parents may help their children formulate effective strategies while they are still in high school (for example, by overseeing the choice of appropriate courses), help them evaluate their individual academic strengths and weaknesses, and help them realistically assess the costs and benefits of specific courses of action (including not just whether to attend college but also, for example, what to major in once there). Thus, if, as Stephen Morgan (2005) has asserted, high school decision-makers are confronted with considerable uncertainty with respect to these crucial factors, then our data suggest that parental involvement may help them reduce this ambiguity. However, it is important to recognize that middle class parental involvement often goes beyond providing advice to include, at times, directly interceding to ensure that their children's interests (as perceived) are well served. These interventions can take the form of speaking with high school counselors about course selection, complaining to school officials when AP course schedules conflict, working closely with their children—sentence by sentence—on college essays, planning (and funding) a large number of visits to potential colleges, and other actions. Indeed, the heavy involvement of middle class parents in the complex and long-term process of preparing for and applying to college is one of our most notable findings. To varying degrees, middle class parents and their children form a collective in which concerted action on the part of each family member is carefully directed toward a shared goal over the course of a child's high school career.

Among working class and poor children, by contrast, parental involvement appears to be substantially rarer. Parents do not typically see it as part of their job to monitor the minutiae of their children's educational careers or to play a central role in making or helping their children make key decisions about their post–high school path. Thus, these parents usually do not try to help their children evaluate the potential long-term implications of their high school course selection or assess their career interests in light of their academic strengths and weaknesses. Nor did we document any conversations in working class or poor families that were comparable to the one between Garrett Tallinger and his father concerning the long-term financial payoff to a particular college program, or even to college attendance in general. Additionally, most did not have experience with the college application process themselves, and neither could they draw on the cumulative experience of a network of college-educated friends; therefore, they could not serve their children as a significant source of information. Consequently, working class and poor children are more likely to rely exclusively on teachers, counselors, and school officials for assistance when making key decisions. Moreover, direct interventions in schooling by working class or poor parents are rare; when they do occur, as with Mark Greeley's mother, they are less likely to be successful.

Working class and poor parents are no less deeply committed, however, to the well being of their children than are middle class parents. The less-privileged parents in the study helped their children frequently and in many ways. They babysat, paid car insurance, supplied transportation to and from work sites, set up job interviews, and provided other crucial emotional and financial support. Some of the working class and poor parents championed college, and some were deeply disappointed that their children did not, in the end, attend. Nevertheless, compared with middle class parents, their enthusiastic "pushing" typically was less informed and less useful in enabling their children to navigate the complexities involved in pursuing higher education....

ENDNOTES

1. Stacey's mother regretted that she did not force her daughter to take more AP and honors classes: "I kick myself for that. I think that she would have gotten in more writing and more reading. She was a kid who hated to read. She's at a college that is very challenging.... I think that she is challenged."
2. Children do not, of course, always conform to their parents' wishes. Stacey's older sister Fern, who also played basketball, chose a selective public university with an excellent basketball team that was located in a relatively rural location. Her mother was vociferously opposed to the choice because she thought, given her daughter's temperament, she would be better off in an urban setting. Fern found the basketball coach to be very difficult. (Her mother wrote a letter to school officials complaining about the treatment the coach was giving her team.) Fern transferred at the end of her freshman year to a school close to home. Not wanting to sit out a year, per NCAA rules, she transferred from a Division I school to a lower division. She liked that school much better; with a major in business, she was planning to get an MBA. On variations among siblings, see Conley (this volume).
3. Jessica Irwin's father worked in a closely supervised technical job; her mother was a babysitter. They lived in a very small apartment. In the original study, they were classified as working class on the basis of their occupational characteristics and because the study did not include a lower middle class category. The family was asked to participate in part because Jessica was the highest-achieving child in the urban classroom where the observations were conducted, and her parents exhibited the

most middle class characteristics in their family-school relationship. In the intervening years, her parents have been socially mobile. They purchased a house, and Ms. Irwin got her teaching credential. She began as a classroom aide but eventually was given her own classroom. Mr. Irwin left his job to start a business in computers; he is self-employed.

4. Melanie Handlon worked for a while in a coffee shop and then went to cosmetology school. Her case highlights the important variability within class; see Lacy and Harris (this volume) for an additional discussion of this issue.

5. According to *The Chronicle of Higher Education*, the University of Vermont has hired "parent bouncers" to help keep parents away from their children at college orientations and to allow them to make their course selections without parental input (Will, 2005).

REFERENCES

Brantlinger, Ellen. 2003. *Dividing Classes: How the Middle Class Negotiates and Rationalizes School Advantage.* New York: Routledge Falmer.

Deil-Amen, Regina. 2007. "When Aspirations Meet Reality for Low-Income Minority High School Students in Their Transition to College." Paper presented to the American Sociological Association annual meeting. New York, August 13, 2007.

Lareau, Annette. 2000. *Home Advantage: Social Class and Parental Intervention in Elementary Education*, 2nd ed. Lanham, Md.: Rowman and Littlefield.

——. 2003. *Unequal Childhoods: Class, Race, Family Life.* Berkeley, Calif.: University of California Press.

Morgan, Stephen. 2005. *On the Edge of Commitment: Educational Attainment and Race in the United States.* Stanford, Calif.: Stanford University Press.

Reay, Diane. 1998. *Class Work: Mothers' Involvement in Their Children's Primary Schooling.* London: University College London.

Thacker, Lloyd, editor. 2005. *College Unranked: Ending the College Admissions Frenzy.* Cambridge, Mass.: Harvard University Press.

White, Wendy S. 2005. "Students, Parents, Colleges: Drawing the Lines." *Chronicle of Higher Education* 52(17): B16.

DISCUSSION QUESTIONS

1. The authors of this selection describe key differences in parenting styles around education issues between working class and middle class parents. What are these differences, and what social factors account for their distinct approaches?
2. According to the authors of this selection, *concerted cultivation*, the defining characteristic of middle class families, is where "parents view it as their duty to actively foster the development of their children's potential skills and talents." Why is this form of parenting, concerted cultivation, not as common among working class and poor families?
3. What specific policies might address the disadvantages that poor and working class students face in their home environments because of likely deficits in concerted cultivation? Is this a social problem that public schools can address?
4. Educational opportunity and academic outcome seem closely tied to socioeconomic status for the participants in this study. Is it a social problem that a person's life-chances, the possibility of success or failure, are so heavily influenced by social class? Why or why not?
5. *This is a problem that begins at the level of parental involvement: schools and governments cannot transform disinterested and indifferent parents in the lower classes into mothers and fathers that take a concerted cultivation approach with their children.* Respond to this viewpoint.

37

What No School Can Do

JAMES TRAUB

The Four Questions

1. What is the problem?
2. Is the problem a *social* problem?
3. What is the cause of the problem according to Traub?
4. What does Traub suggest for dealing with this problem?

Topics Covered

School as an institution
Inequality
Human capital
Social/Cultural capital
Coleman report
Family
Neighborhood

Last fall, the New York State Education Department released the results of a fourth-grade math test and eighth-grade math and language tests. It will come as no surprise to hear that the numbers for students in New York City were dreadful. About two-thirds of city eighth graders failed the English language arts test, and about three-quarters failed in math; half the fourth graders failed the math test. The results provided fuel for those who felt that New York City schools are underfinanced, that the city uses too many uncertified teachers, that academic standards are low, that junior high schools are neglected and that the tests themselves are unfair. There's some merit in all these notions. What was not said, however, was the obvious: that the city districts that performed poorly, like those that performed well, scored almost exactly as the socioeconomic status of the children in them would have predicted. You could have predicted the fourth-grade test scores of all but one of the city's 32 districts merely by knowing the percentage of students in a given district who qualified for a free lunch. Only a few dozen of the city's 675 elementary schools scored well despite high poverty rates. In other words, good schools aren't doing that much good, and bad schools aren't doing that much harm.

This is the case in virtually every big city in the country. Though over the past 35 years we have poured billions of dollars into inner-city schools, and though we have fiddled with practically everything you could think to fiddle with, we have done almost nothing to raise the trajectory of ghetto children. This is a fact, but it is one of those painful facts that it seems better to ignore. Why say anything that could discourage the children, parents and teachers who so desperately need encouragement? Why give aid and comfort to the opponents of high standards? After all, we know that some schools succeed where others fail. We know that some reforms work better than others. In a recently published book, "Choosing Equality," Joseph P. Viteritti, a professor of public policy at New York University, describes the failure of one innovation after another before embarking on a defense of the one reform he believes in, school choice. Or

SOURCE: From James Traub, "What No School Can Do," *The New York Times Magazine*, January 16, 2000, pp. 52–55, 68–69, 81, 90–91.

perhaps, if the answer isn't choice, it's something else. Surely there must be some way of making a difference in the lives of poor children.

Indeed, you can't spend much time in the schools and not believe in the efficacy of reform. In the past few years, I have visited dozens of schools, almost all of them serving inner-city children and most of them exemplary in one way or another. I have been to charter schools where concerned parents and educators have wrested control of the classrooms from ineffective bureaucracies, and also to schools with effective leaders and schools with traditional curricula and schools with "less is more" curricula and schools with "multiple intelligences" teaching styles. The good news is that we know more about what works than we did even a decade ago; should what is left of the ideological fog surrounding education evaporate, we might even agree on a better mousetrap.

But how much better? How powerful can this one institution be in the face of the kind of disadvantages that so many ghetto children bring with them to the schoolhouse door, and return to at home? In "An American Imperative," a minutely detailed study of educational inequality published in 1995, Scott Miller, a scholar now affiliated with the College Board, wrote that while "some strategies for investing resources in disadvantaged children are substantially more productive than others ... there is little evidence that any existing strategy can close more than a fraction of the overall achievement gap" separating children with low socioeconomic status from their wealthier, largely suburban counterparts.

The idea that school, by itself, cannot cure poverty is hardly astonishing, but it is amazing how much of our political discourse is implicitly predicated on the notion that it can. The national debate over entrenched poverty virtually ended with welfare reform. We now speak of the inner-city poor as if they were a remote tribe that has not yet been exposed to the miracle of the marketplace....

An alternative explanation, of course, is that educational inequality is rooted in economic problems and social pathologies too deep to be overcome by school alone. And if that's true, of course, then there's every reason to think about the limits of school, and to think about the other institutions we might have to mobilize to solve the problem. We might even ask ourselves whether there isn't something disingenuous and self-serving in our professed faith in the omnipotence of school.

The idea that bad schools perpetuate poverty, and that good schools can go a long way toward curing it, is quite new and, like most current ideas about entrenched poverty, dates to the civil rights era. One provision of the 1964 Civil Rights Act mandated the publication of a survey "concerning the lack of availability of educational opportunities for individuals by reason of race, color, religion or national origin in public educational institutions."

The sociologist James Coleman was asked to head up the survey project. The implicit assumption behind the survey, an assumption that Coleman said at the time he shared, was that black children were failing because the segregated schools they attended were grossly underfinanced. Coleman found that black students were, in fact, highly segregated, but he also concluded that resources like school libraries and laboratories were distributed more equally than was widely believed and, far more startling, that "differences in school facilities and curriculum, which are the major variables by which attempts are made to improve schools, are so little related to achievement levels of students that, with few exceptions, their effect fails to appear even in a survey of this magnitude." Coleman found that everything schools did accounted for only 5 to 35 percent of the variation in students' academic performances, though he did find that the figures for disadvantaged students were at the higher end of the scale. He concluded that "the inequalities imposed on children by their home, neighborhood and peer environment are carried along to become the inequalities with which they confront adult life at the end of school."

This was, if true, astonishing and deeply unsettling news. Nobody believes in school the way Americans do, and no one is more tantalized by its transformative powers. School is central to the American myth of self-transcendence, whether it's

Thomas Jefferson's seedbed of the republican spirit, or the one-room schoolhouse that propelled Abe Lincoln out of frontier backwardness, or those teeming junior high schools in Brooklyn that served as the melting pot for three generations of children from Russia and Italy and Ireland. In the national myth—and of course not only in myth—school has the miraculous power of leveling inequalities even as it lifts everyone. And by the time Coleman came along we had come to expect far more of school than we ever had before, because we were firmly entrenched in a meritocratic culture in which the most important rewards were decided by school success. Coleman's argument that schools essentially passed along inequality was simply unacceptable, and largely ignored....

It is true that we know far more than we used to about what does and doesn't work. But we are only beginning to figure out how to make something work on a meaningful scale. The most intriguing development in recent years is the rise of "comprehensive," or "schoolwide," reform, terms that refer to models for change that incorporate a coherent and self-consistent vision of pedagogy, curriculum and structure—a replicable, purpose-built school. There are now two dozen such models, many of which stake their reputations on their abilities to improve the performance of inner-city children drastically. More than 9,000 schools have adopted one of the designs. The most popular model among inner-city schools, and the most successful, is a reading program called Success for All, which has expanded into a whole-school reform known as Roots and Wings. The premise of Success for All is that reading is the key to success in school, and that the best way to teach reading, or any other basic skill, is to focus relentlessly on "prevention and early intervention" rather than on remediation....

The failure of school reform to make a significant dent in educational inequality tells us something about the nature of school, and something about the circumstances that produce the inequality in the first place. School, at least as we understand it now, is not as powerful an institution as it seems. Most children do not encounter school until age 5, unless they happen to be in an unusually rigorous preschool program. Anyone who has ever reared a child knows how immense, and lasting, are the effects of those first five years. Nor is school quite as all-encompassing as it seems: academic work typically takes up only about half the time that children spend in school. And whom you hang out with, both during and after school, can matter more than what happens in the classroom. One reason so many apparently effective school interventions fade over time, according to Kenneth Wong, a professor of education policy at the University of Chicago, is that while small children are responsive to adult authority figures, adolescents pay more attention to peers.

The question of exactly what it is that inner-city culture does to children is painful and immensely contentious. The success of those Russian and Italian and Irish immigrants three-quarters of a century ago, and of many Asian and Hispanic immigrants today, makes it plain that the issue has less to do with poverty as such than with culture—with conscious values as well as unconscious behaviors. It was Kenneth Clark, who is black, who first popularized the phrase "the pathology of the ghetto," in "Dark Ghetto," published in 1965. Clark wrote bluntly about how "the stigma of racial inferiority" leads to self-destructive behavior, including violence, alcohol and drug abuse, family breakdown—every social pathology save suicide. But Clark understood this damage as emotional and psychological, not cognitive. Thus, he said that desegregation, by removing the stigma, could lead to a huge improvement in black academic performance. And he railed against what he called the "cultural deprivation" argument, insisting that there was no reason "a low economic status or absence of books in the home or 'cognitive deficit'" should "interfere with the ability of a child to learn to read."

Clark did not reckon with the cognitive harm done to children who grow up in a world without books or even stimulating games, whose natural curiosity is regularly squashed, who are isolated from the world beyond their neighborhood. A study carried out in the early 80s at the University of Kansas reached the almost unfathomable conclusion that 3-year-olds in families with professional parents used more-extensive vocabularies in daily interactions than did mothers on welfare—not to mention the children of those mothers. Here is a gap far greater than even the gulf in income that separates the middle class from the poor; it is scarcely surprising that Coleman found that the effects of home and community blotted out almost all those of school.

Social scientists use the expressions "human capital" and "social capital" to describe and quantify these effects. "Human capital" was invented by the economist Theodore Schultz in 1960 to refer to all those human capacities, developed by education, that can be used productively—the capacity to deal in abstractions, to recognize and adhere to rules, to use language at a high level. Human capital, like other forms of capital, accumulates over generations; it is a thing that parents "give" to their children through their upbringing, and that children then successfully deploy in school, allowing them to bequeath more human capital to their own children.

"Social capital" refers to the benefits of strong social bonds. Coleman defined the term to take in "the norms, the social networks, the relationships between adults and children that are of value for the children's growing up." The support of a strong community helps the child accumulate social capital in myriad ways; in the inner-city where institutions have disintegrated, and mothers often keep children locked inside out of fear for their safety, social capital hardly exists. Coleman consistently pointed out that we now expect the school to provide all the child's human and social capital—an impossibility....

I recently asked Edmund W. Gordon, the first research director of Head Start and perhaps the foremost black psychologist of his generation, what he concluded about the 35-year effort to conquer poverty through the schools. Gordon, a diminutive, bearded man of painstaking precision, said: "Back in the 60's, there was desegregation, there was the War on Poverty, there was Head Start—and most of us really thought that we were going to do it. I must say that I no longer believe that. I think that schools can be much more powerful, but I don't think they can reverse all the ill effects of a starkly disadvantaged status in society. If you go back and look at Coleman's stuff, and Bourdieu's"—Pierre Bourdieu, the French social theorist—"they're talking about human and cultural capital development. It seems to me that the logic of that argument is that schools work for people who have those varieties of capital to invest in it. And if they don't, then schooling is going to be greatly handicapped." ...

Nowadays we read the Coleman report as a "conservative" document, dashing cold water on the faith that we can diminish inequality through conscious effort, and thus end the isolation of inner-city children. But Coleman was suggesting not that we should learn to live with inequality but that we will have to look to institutions other than schools to provide human and social capital. In their review of Coleman's findings, Daniel Patrick Moynihan and Frederick Mosteller, then both professors at Harvard, made this point explicitly. They cited the psychologist Jerome Kagan to the effect that "the differences in language and number competence between lower- and middle-class children" [are] "awesome" by first grade; the critical task is thus to change "the ecology of the lower-class child in order to increase the probability that he will be more successful in attaining normative skills."

How do you alter that ecology? ... "Broadly speaking," [Christopher] Jencks and Meredith Phillips wrote in "The Black-White Test Score Gap," "there are two ways of improving 3- and 4-year-olds' cognitive skills: we can change their preschool experiences and we can change their home experiences. Changing preschools is less important but easier than changing homes."

Preschool is a device for extending the reach of school downward toward the time when the cognitive gaps are beginning to form—a means of distributing human capital to children who grow up with little to none. That few Head Start programs have actually done this may say more about the design of Head Start than about the inherent limitations of preschool. In fact, several studies of "model" programs—which typically start from birth, last all day, hire professional teachers and offer a wide range of services—show large and lasting effects, drastically reducing placement in special education and being held back and boosting academic achievement. And a long-term study of the Abecedarian Project in North Carolina, a preschool program financed largely by the federal government, shows that children who received just such early preparation were far more likely than a control group to graduate from high school and attend college and got higher grades and better scores on cognitive tests.

Last fall, I spent a day at a preschool program in Vineland, a small town in southern New Jersey where the population is for the most part Hispanic and black. Impact, which is largely state financed, has several advantages that would be hard to reproduce in East New York. The school sits among a series of low brick buildings in the middle of a campus setting; everything is clean and fresh. What's more, only a little more than half of the 239 students are poor enough to qualify for a free lunch, because working-class and some middle-class parents are delighted to have their kids spend the day in such a salubrious setting. Nevertheless, what's really striking about Impact, at least if you have been to a typical Head Start program, is how appealing the preschool classes look: the teachers are well-spoken professionals, the classes are small and richly stocked with books and art materials and computers and the kids move purposefully among activities. These were classrooms, not nursery schools. Even in the child-care center, each group of children has a teacher who is certified in early-childhood education.

Impact is not so much a preschool as a multipurpose social-service agency, offering year-round, all-day child care from birth, adult literacy programs, episodic health care, including mental health, and education and counseling for pregnant teenagers. It absorbs a larger portion of a child's life than a school does, and it makes a far more ambitious attempt to offer itself as a kind of alternative community. Impact has contracted with the Vineland public schools to take teenagers in the early stage of pregnancy and give them one year of schooling in small classes as well as intensive instruction in child-rearing, nutrition and health care. The girls I met represent the middle stage of the typical downward spiral of disadvantage: themselves often born to teenage mothers, they had already been left back once or twice, and some were on their second child, by a different boyfriend from the first.

Paula Bender, who runs Impact, says: "We're in the business of trying to break cycles. A teen mother comes in. If she had stayed in her old school, she probably would have dropped out. Her child might have had a low birth weight. She can come here and get nutrition, prenatal care and education. Then we're going to take care of her baby, and make sure that any developmental lags are diagnosed and treated. The mothers are going on to high school, and we have the children from birth."

Programs like Impact are very expensive: Bender says she is spending $8,000 to $10,000 a child, which is at least double the cost of the Head Start program down the road....

An all-day, year-round child-care and preschool program—and especially one intended specifically for the poor—smacks of a kind of Progressive Era paternalism in which mothers are expected to yield up their children to wise professionals. Back when he was just a congressman with strange ideas, Newt Gingrich once suggested that inner-city children be shipped en masse to the equivalent of Boys Town, a proposal that makes poverty sound like a crime punishable by loss of offspring. But Gingrich understood something important: ghetto children need an enveloping environment that is secure and nourishing, as the streets and often the home itself are not. And school is not enveloping enough. You can't take children away from their mothers, but you can place them in an alternative environment for much of the day (which the end of welfare, in any case, has now made indispensable). There's a strong argument for universally available after-school activities. No less important would be the restoration of the web of church, community and police-sponsored programs that once flourished in big cities. The breakdown of families means that we have to ask more of social institutions—and not just schools—than we used to.

Of course we also have to ask more of the families themselves. In Jim Sleeper's "Liberal Racism," the black economist Glenn Loury recalls vexing a symposium on race by saying: "Let's talk about who's going to read to these kids.... How come these public libraries at the corners of housing projects aren't overflowing?" The language of personal responsibility makes many liberals flinch, and Edmund Gordon, Head Start's first research director, is almost certainly right in thinking that middle-class parents are a lot more likely to respond to advice about child-rearing than are single mothers struggling to keep their heads above water. But if the cognitive gap between black and white 5-year-olds has to do with the human and social capital children do or do not acquire at home and in the community—and not, or at least not principally, with racism or unfair tests or different "cognitive styles"—then we can't do without the explicit language of values and behavior. We have to unambiguously embrace the virtues of a "middle-class parenting style." And prominent black figures have to weigh in against the antiacademic and even "oppositional" peer culture that Gordon and others say is retarding black progress.

It's not impossible to imagine a bargain in which conservatives agree to pay for comprehensive institutions like Impact or for after-school activities and liberals agree to embrace the language of personal responsibility, but we would still be asking ghetto children to thrive in an incredibly adverse environment. The idea of directly addressing that environment through jobs programs, housing, health care or the adoption of "a living wage" survives only in the fringes of political discourse. But you cannot disentangle the objective conditions of a place like East New York from the habits and values of the people who live there. The most effective solution—and the most unlikely one of all—is to move families out of the ghetto environment altogether. As Lawrence Katz, a Harvard economist, puts it, "You can't change the parents, but you can change the neighborhood." Katz points to the famous Gautreaux experiment in Chicago, in which families were given subsidies to move from high-poverty neighborhoods to the suburbs; studies have found that children in these families were far more successful academically than would have otherwise been predicted. Katz is now in the midst of a long-term study of families involved in Moving to Opportunity, a five-year-old program sponsored by the Department of Housing and Urban Development that allowed a small number of families in five cities to leave the ghetto for working-class neighborhoods. Though it is too early to look at academic results, Katz's early study found very large drops in incidents of misbehavior and sizable improvements both in children's physical health and in mothers' mental health.

A few years ago, I spent several days at the McDonough School, an elementary school in Hartford. Despite its reputation for corporate wealth, Hartford is one of the poorest cities in the country; it was there that President Clinton recently announced his extremely modest new antipoverty initiative. McDonough's population was almost 100 percent Hispanic and black and 100 percent poor. The school was decrepit, but clean and orderly cheerful and safe—an oasis of purposefulness in a neighborhood so chaotic that a few months earlier a janitor was chased off a snowblower and shot outside the school's annex. The principal, Donald Carso, was the kind of irrepressible enthusiast who made a point of greeting parents out in front of the school every morning, even in the dead of winter—a figure straight out of the "effective schools" literature. But McDonough was a failing school: three-quarters of the graduating sixth graders had scored at the remedial level on the state reading test. Carso said he had reached the limit of his powers. "In this world," he told me, making a little globe with his fingertips, "we're very successful. But we can't control all the stuff on the outside."

At bottom, the reason the kids at McDonough and practically every other elementary school in Hartford were failing, while the kids in the wealthy suburbs that began just on the other side of West Hartford were thriving, was not that the schools in Hartford were bad and the schools in the suburbs were good, but that each set of children was repeating patterns, and following trajectories, established before they arrived at school. McDonough's children lived in a world bounded and defined by poverty. One teacher told me that when she talked to her sixth graders about college, they'd say: "What's college? I don't know anyone who's been to college." They didn't know anyone who had succeeded in school, landed a good job, made it to the suburbs.

I met Carso because he had decided to testify for the plaintiff in a well-known desegregation case, Sheff v. O'Neill. He said something had to be done to spring the children free from the isolation of urban poverty; he played with the idea that the Hartford schools should simply be shut down, and the children dispersed into the surrounding suburbs. Carso said he never thought the litigation would solve the problem (it hasn't), but he hoped that it would at least force people to pay attention. "We are," he said to me almost pleadingly, "saying to our neighbors, 'Take ownership of our problem.'"

That is the issue: are we going to take ownership of the problem? The gulf between the poor and the well-off is much wider than it used to be, not only financially but also psychically. The memory of the Depression, with its "there but for the grace of God" reflex, has passed out of American politics and culture. Once it was the rich who seemed to live on an island of their own; now it's the poor. Their isolation makes them gratifyingly invisible. The drop in crime even makes the poor seem like less of a threat to the prosperous; it frees us to contemplate, along with George W. Bush, the spiritual hollowness of plenty. But here's a thought: maybe our prosperity will continue to seem hollow as long as so many 3-year-old black girls face such grim prospects.

DISCUSSION QUESTIONS

1. How important is school in young people's lives? Do you agree with Traub that the school is much less important than many people believe?
2. What might government do to help prepare children better for school? In the case of highly disadvantaged children, is it better to work through parents or directly with children?
3. Traub asks: Do we cure poverty to improve education, or do we improve education to cure poverty? What do you think? Why is this so difficult to answer?
4. Traub identifies "inner-city culture" as playing a major role in creating problems for students in school. What might be the problem here, and what creates this "inner-city culture"?
5. Traub writes, "The gulf between the poor and the well-off is much wider than it used to be, not only financially but also psychically." Traub seems to think that this is the real problem. Do you agree?

38

The Shame of the Nation

JONATHAN KOZOL

The Four Questions

1. What is the problem?
2. Is it a *social* problem?
3. What are the causes of the problem?
4. What would Kozol argue must be done?

Topics Covered

Education
Class
Poverty
Racial inequality
Public funding
Private funding
Preschool
Boutique schools
Baby Ivies
Head Start

"Dear Mr. Kozol," said the eight-year-old, "we do not have the things you have. You have Clean things. We do not have. You have a clean bathroom. We do not have that. You have Parks and we do not have Parks. You have all the thing and we do not have all the thing.... Can you help us?"

The letter, from a child named Alliyah, came in a fat envelope of 27 letters from a class of third grade children in the Bronx. Other letters that the students in Alliyah's classroom sent me registered some of the same complaints. "We don't have no gardens," and "no Music or Art," and "no fun places to play," one child said. "Is there a way to fix this Problem?" Another noted a concern one hears from many children in such overcrowded schools: "We have a gym but it is for lining up. I think it is not fair." Yet another of Alliyah's classmates asked me, with a sweet misspelling, if I knew the way to make her school into a "good" school—"like the other kings have"—and ended with the hope that I would do my best to make it possible for "all the kings" to have good schools.

The letter that affected me the most, however, had been written by a child named Elizabeth. "It is not fair that other kids have a garden and new things. But we don't have that," said Elizabeth. "I wish that this school was the most beautiful school in the whole why world."

Elizabeth had very careful, very small, and neatly formed handwriting. She had corrected other errors in her letter, squeezing in a missing letter she'd initially forgotten, erasing and rewriting a few words she had misspelled. The error she had left unaltered in the final sentence therefore captured my attention more than it might otherwise have done.

"The whole why world" stayed in my thoughts for days. When I later met Elizabeth I brought her letter with me, thinking I might see whether, in reading it aloud, she'd change the "why" to "wide" or leave it as it was. My visit to her class, however, proved to be so pleasant, and the children seemed so eager to bombard me with their questions about where I lived, and why I lived there rather than New York, and who I lived with, and how many dogs I had, and other interesting questions of that sort, that I decided not to interrupt the nice reception they had given me with questions about usages and spelling. I left "the whole why world" to float around unedited and unrevised within my mind. The letter itself soon found a resting place up on the wall above my desk.[1]

In the years before I met Elizabeth, I had visited many elementary schools in the South Bronx and in one northern district of the Bronx as well.[2] I had also made a number of visits to a high school where a stream of water flowed down one of the main stairwells on a rainy afternoon and where green fungus molds were growing in the office where the students went for counseling. A large blue barrel was positioned to collect rainwater coming through the ceiling. In one makeshift elementary school housed in a former skating rink next to a funeral parlor in another nearly all-black-and-Hispanic section of the Bronx, class size rose to 34 and more; four kindergarten classes and a sixth grade class were packed into a single room that had no windows. Airlessness was stifling in many rooms; and recess was impossible because there was no outdoor playground and no indoor gym, so the children had no place to play.

In another elementary school, which had been built to hold 1,000 children but was packed to bursting with some 1,500 boys and girls, the principal poured out his feelings to me in a room in which a plastic garbage bag had been attached somehow to cover part of the collapsing ceiling. "This," he told me, pointing to the garbage bag, then gesturing around him at the other indications of decay and disrepair one sees in ghetto schools much like it elsewhere, "would not happen to white children."

A friend of mine who was a first-year teacher in a Harlem high school told me she had 40 students in her class but only 30 chairs, so some of her students had to sit on windowsills or lean against the walls. Other high schools were so crowded they were forced to shorten schooldays and to cut back hours of instruction to accommodate a double shift of pupils. Tens of thousands of black and Hispanic students were in schools like these, in which half the student body started classes very early in the morning and departed just before or after lunch, while the other half did not begin their schoolday until noon.

Libraries, once one of the glories of the New York City system, were either nonexistent or, at best, vestigial in large numbers of the elementary schools. Art and music programs had for the most part disappeared as well. "When I began to teach in 1969," the principal of an elementary school in the South Bronx reported to me, "every school had a full-time licensed art and music teacher and librarian." During the next decade, he recalled, "I saw all of that destroyed."

School physicians were also removed from elementary schools during these years. In 1970, when substantial numbers of white children still attended New York City's schools, 400 doctors had been present to address the health needs of the children. By 1993, the number of doctors had been cut to 23, most of them part-time—a cutback that affected most acutely children in the city's poorest neighborhoods where medical provision was perennially substandard and health problems faced by children most extreme. During the 1990s, for example, the rate of pediatric asthma in the South Bronx, already one of the highest in the nation, was exacerbated when the city chose to build a medical waste incinerator in their neighborhood after a plan to build it on the East Side of Manhattan was abandoned in the face of protests from the parents of that area. Hospitalization rates for these asthmatic children in the Bronx were as much as 20 times more frequent than for children in the city's affluent communities. Teachers spoke of children who came into class with chronic wheezing and, at any moment of the day, might undergo more serious attacks, but in the schools I visited there were no doctors to attend to them.[3]

Political leaders in New York tended to point to shifting economic factors, such as a serious budget crisis in the middle 1970s, rather than to the changing racial demographics of the student population, as the explanation for these steep declines in services. But the fact of economic ups and downs from year to year, or from one decade to the next, could not convincingly explain the permanent shortchanging of the city's students, which took place routinely in good economic times and bad, with bad times seized upon politically to justify these cuts while, in the good times, losses undergone during the crisis years had never been restored.

"If you close your eyes to the changing racial composition of the schools and look only at budget actions and political events," says Noreen Connell, the director of the nonprofit Educational Priorities Panel in New York, "you're missing the assumptions that are underlying these decisions." When minority parents ask for something better for their kids, she says, "the assumption is that these are parents who can be discounted. These are kids that we don't value."[4]

The disrepair and overcrowding of these schools in the South Bronx "wouldn't happen for a moment in a white suburban school district like Scarsdale," says former New York State Commissioner of Education Thomas Sobol, who was once the superintendent of the Scarsdale schools and is now a professor of education at Teachers College in New York. "I'm aware that I could never prove that race is at the heart of this if I were called to testify before a legislative hearing. But I've felt it for so long, and seen it operating for so long, I know it's true...."[5]

During the 1990s, physical conditions in some buildings had become so dangerous that a principal at one Bronx school, which had been condemned in 1989 but nonetheless continued to be used, was forced to order that the building's windows not be cleaned because the frames were rotted and glass panes were falling in the street, while at another school the principal had to have the windows bolted shut for the same reason. These were not years of economic crisis in New York.[6] This was a period in which financial markets soared and a new generation of free-spending millionaires and billionaires was widely celebrated by the press and on TV; but none of the proceeds of this period of economic growth had found their way into the schools that served the truly poor.[7]

I had, as I have noted, visited many schools in other cities by this time; but I did not know children in those schools as closely as I'd come to know, or soon would know, so many of the children in the New York City schools. So it would be these children, and especially the ones in elementary schools in which I spent the most time in the Bronx, whose sensibilities and puzzlements and understandings would impress themselves most deeply on my own impressions in the years to come, and it would be their questions that became my questions and their accusations and their challenges, when it appeared that they were making challenges, that came to be my own.

This, then, is the accusation that Alliyah and her classmates send our way: "You have.... We do not have." Are they right or are they wrong? Is this a case of naive and simplistic juvenile exaggeration? What does a third grader know about these big-time questions about what is fair and what is not, and what is right and what is wrong? Physical appearances apart, how in any case do you begin to measure something so diffuse and vast and seemingly abstract as having more, or having less, or having not at all?

In a social order where it seems a fairly common matter to believe that what we spend to purchase almost anything we need bears some connection to the worth of what we get, a look at what we think it's in our interest to invest in children like Alliyah or Pineapple may not tell us everything we need to know about the state of educational fair play within our nation, but it surely tells us *something* about what we think these kids are worth to us in human terms and in the contributions they may someday make to our society. At the time I met Alliyah in the school-year 1997–1998, New York's Board of Education spent about $8,000 yearly on the education of a third grade child in a New York City public school. If you could have scooped Alliyah up out of the neighborhood where she was born and plunked her down within a fairly typical white suburb of New York, she would have received a public education worth about $12,000 every year. If you were to lift her up once more and set her down within one of the wealthiest white suburbs of New York, she would have received as much as $18,000 worth of public education every year and would likely have had a third grade teacher paid approximately $30,000 more than was her teacher in the Bronx.[8] ...

Even these numbers that compare the city to its suburbs cannot give an adequate impression of the inequalities imposed upon the children living in poor sections of New York. For, even within the New York City schools themselves, there are additional discrepancies in funding between schools that serve the poorest and the wealthiest communities, since teachers with the least seniority and least experience are commonly assigned to schools in the most deeply segregated neighborhoods. The median salary of teachers in Pineapple's neighborhood was less than $46,000 in 2002–2003, the lowest in the city, compared to $59,000 in one of Manhattan's recently gentrified communities, and up to $64,000 in some neighborhoods of Queens.[9]

None of this includes the additional resources given to the public schools in affluent communities where parents have the means to supplement the public funds with private funding of their own, money used to build and stock a good school library for instance, or to arrange for art and music lessons or, in many of these neighborhoods, to hire extra teachers to reduce the size of classes for their children.

This relatively new phenomenon of private money being used selectively to benefit the children only of specific public schools had not been noted widely in New York until about ten years ago when parents of the students at a public school in Greenwich Village in Manhattan raised the funds to pay a fourth grade teacher, outside of the normal budget of the school, when class size in the fourth grade otherwise was likely to increase from 26 to 32, which was the average class size in the district at the time but which, one of the parents said, "would have a devastating impact" on her son. The parents, therefore, collected $46,000—two thirds of it, remarkably, in just one night—in order to retain the extra teacher.[10]...

In principle, the parents in poor neighborhoods were free to do fund-raising too, but the proceeds they were likely to bring in differed dramatically. The PTA in one low-income immigrant community, for instance, which sponsored activities like candy sales and tried without success to win foundation grants, was able to raise less than $4,000.[11] In the same year, parents at PS. 6, a top-rated elementary school serving the Upper East Side of Manhattan, raised $200,000....

In view of the extensive coverage of this new phenomenon not only by New York City papers but by those in other cities where the same trends are observed, it is apparent that this second layer of disparities between the children of the wealthy and the children of the poor is no secret to the public any longer. Yet, even while they sometimes are officially deplored, these added forms of inequality have been accepted with apparent equanimity by those who are their beneficiaries.

"Inequality is not an intentional thing," said the leader of the PTA in one of the West Side neighborhoods where parents had been raising private funds, some of which had been obtained as charitable grants. 'You have schools that are empowered and you have schools that have no power at all.... I don't bear any guilt for knowing how to write a grant," he said, a statement that undoubtedly made sense to some but skirted the entire issue of endemic underbudgeting of public schools attended by the children of poor people who did not enjoy his money-raising skills or possible connections to grant makers.[12]

A narrowing of civic virtue to the borders of distinct and self-contained communities is now evolving in these hybrid institutions which are public schools in that they benefit from the receipt of public funds but private in the many supplementary programs that are purchased independently. Boutique schools within an otherwise impoverished system, they enable parents of the middle class and upper middle class to claim allegiance to the general idea of public schools while making sure their children do not suffer gravely for the stripped-down budgets that have done great damage to poor children like Alliyah and Pineapple.

"There are cheap children and there are expensive children," writes Marina Warner, an essayist and novelist who has written many books for children, "just as there are cheap women and expensive women."[13] When Pineapple entered PS. 65 in the

South Bronx, the government of New York State had already placed a price tag on her forehead. She and her kindergarten classmates were $8,000 babies. If we had wanted to see an $18,000 baby, we would have had to drive into the suburbs. But the governmentally administered diminishment of value in the children of the poor begins even before the age of five or six when they begin their years of formal education in the public schools. It starts during their infant years and toddler years when hundreds of thousands of children in low-income neighborhoods are locked out of the opportunity for preschool education for no reason but the accident of birth and budgetary choices of the government, while children of the privileged are often given veritable feasts of rich developmental early education.

In New York City, for example, affluent parents pay surprisingly large sums of money to enroll their youngsters in extraordinary early-education programs, typically beginning at the age of two or three, that give them social competence and rudimentary pedagogic skills unknown to children of the same age in the city's poorer neighborhoods. The most exclusive of the private preschools in New York, which are known to those who can afford them as the "Baby Ivies," cost as much as $22,000 for a full-day program. Competition for admission to these pre-K schools is so intense that "private counselors" are frequently retained, at fees as high as $300 hourly, according to The Times, to guide the parents through the application process.[14]

At the opposite extreme along the economic spectrum in New York are thousands of children who receive no preschool opportunity at all. Exactly how *many* thousands is almost impossible to know. Numbers that originate in governmental agencies in New York and other states are incomplete and imprecise and do not always differentiate with clarity between authentic pre-K programs that have educative and developmental substance and those less expensive childcare arrangements that do not. But even where states do compile numbers that refer specifically to educative preschool programs, it is difficult to know how many of the children who are served are of low income since admissions to some of the state-supported programs aren't determined by low income or they are determined by a complicated set of factors of which poverty is only one.

There is another way, however, to obtain a fairly vivid sense of what impoverished four-year-olds receive in segregated sections of our cities like the Bronx. This is by asking kids themselves while you are with them in a kindergarten class to tell you how they spent their time the year before—or, if the children get confused or are too shy to give you a clear answer, then by asking the same question to their teacher.

"How many of these children were in pre-K programs last year or the last two years?" I often ask a kindergarten teacher.

In middle- and upper-class suburbs, a familiar answer is "more than three quarters of them," "this year, almost all of them," or "virtually all...." In poor urban neighborhoods, by comparison, what I more often hear is "only a handful," "possibly as many as a fourth," "maybe about a third of them got *something* for one year...." The superintendent of the district that includes Pineapple's former school estimated in the fall of 2002 that only between a quarter and a third of children in the district had received even a single year of preschool and that less than five percent had been provided with the two years of pre-K instruction that are common in most affluent communities.[15]

Government data and the estimates of independent agencies tend to substantiate the estimates of principals and teachers. Of approximately 250,000 four-year-olds in New York State in 2001–2002, only about 25 percent, some 60,000, were believed to be enrolled in the state-funded preschool program—which is known as "Universal Pre-K" nonetheless—and typically in two-and-a-half-hour sessions rather than the more extended programs children of middle-class families usually attend. Then too, because these figures were not broken down by family income levels and because the program did not give priority to children of low income, it was difficult to know how many children in the poorest neighborhoods had been excluded from the program.[16]

Head Start, which is a federal program, is of course much better known than New York's Universal Pre-K and it has a long track-record, having been created 40 years ago by Congress at a time when social programs that expanded opportunities for children of low income were not viewed with the same skepticism that is common among many people who set public policy today. In spite of the generally high level of approval Head Start has received over the years, whether for its academic benefits or for its social benefits, or both, 40 percent of three- and four-year-olds who qualified for Head Start by their parents' income were denied this opportunity in 2001, a percentage of exclusion that has risen steeply in the subsequent four years. In some of the major cities, where the need is greatest, only a tiny fraction of low-income children in this age bracket are served. In New York City, for example, less than 13,000 four-year-olds were served by Head Start in 2001; and, in many cases, Head Start was combined with Universal Pre-K, so the children served by Head Start on its own were relatively few.[17]

There are exceptions to this pattern in some sections of the nation. In Milwaukee, for example, nearly every four-year-old is now enrolled in a preliminary kindergarten program, which amounts to a full year of all-day preschool education, prior to a second kindergarten year for five-year-olds, according to the superintendent of Milwaukee's schools.[18] In New Jersey, full-day pre-K programs have been instituted for all three- and four-year-olds in 31 low-income districts, one of the consequences of a legal action to reduce inequities of education in that state.[19] More commonly in urban neighborhoods, large numbers of children have received no preschool education[20] and they come into their kindergarten year without the minimal social skills that children need in order to participate in class activities and without even such very modest early-learning skills as knowing how to hold a pencil, identify perhaps a couple of shapes or colors, or recognize that printed pages go from left to right. A first grade teacher in Boston pointed out a child in her class who had received no preschool and, as I recall, had missed much of his kindergarten year as well, and introduced me to the boy so I could sit beside him for a while and derive my own conclusions, then confirmed my first impression when she told me in a whisper, "He's a sweetheart of a baby but knows almost absolutely nothing about anything that has to do with school!"

Two years later, in third grade, these children are introduced to what are known as "high-stakes tests," which in many urban systems now determine whether students can or cannot be promoted. Children who have been in programs like the "Baby Ivies" since the age of two have been given seven years of education by this point, nearly twice as many as the children who have been denied these opportunities; yet all are required to take, and will be measured and in many cases penalized severely by, the same examinations.

Which of these children will receive the highest scores—those who spent the years from two to four in lovely little Montessori schools and other pastel-painted settings in which tender and attentive grown-ups read to them from storybooks and introduced them for the first time to the world of numbers, and the shapes of letters, and the sizes and varieties of solid objects, and perhaps taught them to sort things into groups or to arrange them in a sequence, or to do those many other interesting things that early-childhood specialists refer to as prenumeracy skills, or the ones who spent those years at home in front of a TV or sitting by the window of a slum apartment gazing down into the street? There is something deeply hypocritical in a society that holds an inner-city child only eight years old "accountable" for her performance on a high-stakes standardized exam but does not hold the high officials of our government accountable for robbing her of what they gave their own kids six or seven years before.

There are obviously other forces that affect the early school performance of low-income children: levels of parent education, social instability, and frequently undiagnosed depression and anxiety that make it hard for many parents I have known to take an active role in backing up the efforts of their children's teachers in the public schools. Still, it is all too easy to assign the primary onus of responsibility to parents in these neighborhoods. (Where were these parents educated after all? Usually in the same low-ranking schools their children now attend.) In a nation in which fairness was respected, children of the poorest and least educated mothers would receive the most extensive and most costly preschool preparation, not the least and cheapest, because children in these families need it so much more than those whose educated parents can deliver the same benefits of early learning to them in their homes.

The "Baby Ivies" of Manhattan are not public institutions and receive no subsidies from public funds. In a number of cities, on the other hand, even this last line of squeamishness has now been crossed and public funds are being used to underwrite part of the costs of preschool education for the children of the middle class in public institutions which, however, do not offer the same services to children of the poor. Starting in spring 2001, Chicago's public schools began to operate a special track of preschool for the children of those families who were able to afford to pay an extra fee—nearly $6,000—to provide their children with a full-day program of about 11 hours, starting at the age of two if parents so desired. In a city where 87 percent of students in the public schools were black or Hispanic, the pay-for-pre-school program served primarily white children.[21]

Almost all these preschools were "in gentrified or gentrifying neighborhoods," The *Chicago Tribune* reported. "The fresh paint and new toys" in one of these programs on the North Side of Chicago were not there simply "to make preschool a happier place for the new class of toddlers" but "to keep their parents from moving to the suburbs." These and other "gold-plated academic offerings" which the city was underwriting to attract or to retain the children of the middle class had already begun to slow the "brain drain" from the public schools, The *Tribune* said. In the same year in which the pay-for-pre-K program was begun, 7,000 children from low-income families, many of whom were deemed to be "at risk," were waiting for preschool spaces that the city was unable to provide.[22]

Undemocratic practices like these, no matter how strategically compelling they may seem, have introduced a radical distorting prism to an old, if seldom honored, national ideal of universal public education that affords all children equal opportunity within the borders of a democratic entity. Blurring the line between democracy and marketplace, the private subsidy of public schools in privileged communities denounces an ideal of simple justice that is often treated nowadays as an annoying residue of tiresome egalitarian ideas, an ethical detritus that sophisticated parents are encouraged to shut out of mind as they adapt themselves to a new order of Darwinian entitlements....

ENDNOTES

1. Letters from students at P.S. 28 in the South Bronx, November 1997.
2. See Jonathan Kozol, *Savage Inequalities and Ordinary Resurrections* (New York: Crown, 1991) and Jonathan Kozol, *Ordinary Resurrections* (New York: Crown, 2000).
3. See Jonathan Kozol, *Amazing Grace* (New York: Crown, 1995).
4. Interview with Noreen Connell, May 2002.
5. Interview with Dr. Sobol, December 2003.
6. *New York Times*, February 2, 1998.
7. *Ordinary Resurrections*, cited above.
8. A Report to the Governor and the Legislature on the Educational Status of the State's Schools, Statistical Profiles of Public School Districts, New York State Education Department, April 1999 (numbers for 1997–1998). In all later citations, this document is identified as Report to the Governor and the Legislature.
9. According to the State Education Department's Report to the Governor and the Legislature, June 2004 (numbers for 2002–2003), median teacher salary was $45,500 in Bronx District 7; $59,300 in

Manhattan District 1; $64,000 in Queens Districts 25 and 26.

10. *New York Times*, September 20, 23, 24, 26, and 28, 1997.
11. *New York Times*, July 23, 1995.
12. *New York Times*, June 17, 1996.
13. *Into the Dangerous World*, by Marina Warner (London: Chatto and Windus, 1989).
14. *New York Times*, October 25, 2000. In 2003, the West Side Montessori School cost $17,884 for full-day pre-K for children ages three to five (*New York Times*, January 12, 2003), while pre-K classes for four-year-olds at New York City's Horace Mann School cost $22,500. (The Horace Mann School, Tuition Schedule 2003-2004.) Also see *New York Times*, July 31, 2002, and May 28, 2003.
15. District 7 Superintendent Myrta Rivera in conversation with author, October 2002.
16. *New York Times*, February 2, 2003; *Education Week*, April 9, 2003. The principal of Pineapple's school, P.S. 65, told me in November 2003 that only 36 children in her school had been enrolled in Universal Pre-K—less than one third of her first grade enrollment of 110 pupils.
17. Interviews with Bethany Little, director of government relations, Children's Defense Fund, November 2004 and February 2005. In New York City in 2001, Head Start served 11,000 three-year-olds, 12,600 four-year-olds, and 1,100 five-year-olds; but as many as 6,000 of the four-to-five-year-olds were in programs that combine funding from Head Start and Universal Pre-K, so as few as 8,000 four-to-five-year-olds were being served in freestanding Head Start programs at the time.
18. According to Milwaukee Superintendent of Schools William Andrekopoulos (conversation with author, June 2004), 92 percent of Milwaukee's four-year-olds are enrolled in all-day early kindergarten programs.
19. Interview with Assistant Commissioner Judith Weiss, New Jersey State Department of Education, March 2005. See also *New York Times*, November 21, 2001.
20. According to "The State of Preschool," National Institute for Early Education Research, November 22, 2004, only one in ten three- and four-year-olds in the United States were enrolled in state-supported pre-K programs in 2002.
21. *Chicago Sun-Times*, September 12 and 18, 2000, and June 10, 2002; *Chicago Tribune*, September 13 and 27 and December 15, 2000, and February 13, 2001; *New York Times*, June 15, 2001. According to Amy Hendrickson of the Chicago Teacher Center (correspondence with author, February 2005), there were 15 pay-for-pre-K programs operating in Chicago in the academic year 2004–2005.
22. *Chicago Tribune*, September 13, 2000. A spokesman for Chicago Mayor Richard Daley estimated two years later that "between 10,000 and 20,000" children "could take advantage" of pre-K programs if they were available and parents were informed of this, according to the *Chicago Tribune*, September 3, 2002. "Chicago is setting up tuition programs for middle-class children when it is unable to meet the needs of children at risk…," said the executive director of Parents United for Responsible Education. (Letter to *New York Times*, June 22, 2001.)

DISCUSSION QUESTIONS

1. Kozol argues that market forces undermine democracy in education. What does he mean? Do you agree?
2. Is the cause of the problem poverty or racism?
3. What would you try to do to improve education if you had the power?
4. Kozol describes many disparities that exist in New York schools. Who is responsible?
5. Some people believe that "just throwing money into schools will not help the problems in our educational system." Would you agree?
6. Kozol calls his book "The Shame of the Nation." Do you believe this is overstating the problem?

39

Why There Are No Bad Schools in Raleigh

GERALD GRANT

The Four Questions

1. This reading is unusual because it outlines a success story in education. What was the problem that this community was able to successfully address?
2. Was this problem a *social* one?
3. What was the initial cause of the problem? Who was affected?
4. What does Grant suggest for dealing with similar problems?

Topics Covered

Public education
Education policy
Educational outcomes
Black-white achievement gap
Class-based diversity
Education reform
Education funding

On my way to Raleigh in 2003, driving down I-81 past Frackville and Mahanoy in a Pennsylvania blizzard, I wondered if all the hype about the extraordinary success of Raleigh's public schools could be true. While reformers could tell stirring stories of high achievement in a particular urban school here or there, usually attributed to a charismatic principal, Raleigh had transformed an entire urban system in ways that dramatically raised the achievement of poor and minority students in *all* its schools. I had departed that morning from Syracuse, where only 25 percent of eighth graders passed state achievement tests in math and reading. In Raleigh, where city and suburbs had merged to form a single countywide school system that served children of all social classes and races, 91 percent passed.[1]

Astonishingly, because children in the urban core of most of America have become Ralph Ellison's "invisible" children, there had been no outcry in Syracuse, not even at one middle school where 95 percent failed the state math test. In the affluent Syracuse suburb of Fayetteville-Manlius, where only 16 percent did not pass, mass protests would have broken out if even half of the students had failed the state tests, never mind 95 percent. Enraged parents would have stormed the school headquarters demanding the ouster of the entire School Board. Similar city-suburban gaps can be found in most cities.

But the word "gap" papers over the dangerous reality that these statistics reveal—the failure to make good on an implicit bargain that America made with its poorest citizens. This bargain promised that the great income inequalities permitted in a capitalist society would he balanced by equal educational opportunities for all. The Wake County Public School System, of which Raleigh is a part, is one of the few urban school systems in America that made good on that democratic bargain.[2] Gaps in educational achievement became not only intolerable but unthinkable there. Educators didn't just talk equal educational opportunity. They delivered

SOURCES: Reprinted by permission of the publisher from "There Are No Bad Schools in Raleigh," in *Hope and Despair in the American City: Why There Are No Bad Schools in Raleigh* by Gerald Grant, pp. 91–94, 102–104, 107–108, 120–126, 201–202, Cambridge, Mass: Harvard University Press.

it to all children in the system, day after day. And they reduced the gap between rich and poor, black and white, more than any other large urban educational system in America.

In the ensuing years, in visits to more than a score of Raleigh schools, I often heard teachers and principals say, "There are no bad schools in Wake County." And they were right. Perhaps the most convincing proof of that lay not just in the stunning test scores across the whole county but in the fact that virtually all the teachers in each school I visited enrolled their own children in Wake County's public schools, most often in the same school where they taught. In Syracuse, it was rare to find teachers who sent their own children to Syracuse public schools. Most teachers with families lived in the suburbs and sent their children to affluent, predominantly white suburban schools. Like the rest of the middle class, they had abandoned the city and its schools long ago.

One of the schools I first visited in Raleigh, Bugg Elementary, lay in the southeast quadrant of the city, which was the historic black district. Although nearly a third of the Bugg children were low income and the school was still majority black—54 percent—it was a magnet school that attracted whites from across the county to its programs in art and science. In third grade 94 percent of white children and 79 percent of blacks passed the state math test. By fifth grade 100 percent of both blacks and whites passed the test. Not all of Bugg's white children were bused to school. Some walked. In a city where housing costs had soared as fast as Syracuse homes had depreciated, the modest ranch houses with neat lawns in the Bugg neighborhood were still a bargain in 2003, and the school's reputation drew white families as well as black ones to buy homes there.

Mary Page, principal of the Bugg School, spread out a large leather briefing book on her desk and gave me a mournful smile when she told me that it wasn't easy to move back to North Carolina. She had attended black segregated schools in the 1950s in Rockingham, a small town in the western part of the state where the movie house was not only segregated but had no toilets for blacks. Her life took a different course at Warren Wilson College, a Presbyterian school near Asheville, where she was one of only seven African-American students in her class. She later went to Germany with her husband, who was serving in the United States Army, and taught there, before settling down as a teacher in northern Virginia's Prince William County. She was not eager to return to North Carolina in 1997 when her husband accepted an offer to be registrar at Shaw University. Her memories of segregation in the rural south had left scars, and she had never lived in Raleigh nor heard much about the success its schools had achieved since the merger with Wake County. She arrived a year before the Wake County system announced that its goal was to have 95 percent of all K–8 students pass state exams in reading and math within five years.

"I thought it was a good goal. I believe you've got to set goals really high. If we had said 80 percent we wouldn't have gotten any more than that." But many thought it was a mistake—that Wake County could not possibly succeed and would wind up with egg on its face. Even Mary Page had doubts. Her first job was vice principal of a school across the street from a large housing project in downtown Raleigh: "To he honest, we had some children whose scores were so low and I thought, my God, how are we going to get those children there."[3] The details of how they got there, or nearly got there with 91.3 percent of all children in Raleigh and Wake County passing state tests five years later, make up one of the most inspiring stories in public education....

SETTING THE 95 PERCENT GOAL

Wake County set the goal of having 95 percent of all K–8 students achieve at or above grade level by 2003. The goal would be measured by state tests in grades 3 through 8. Karen Banks, the new director of evaluation, rejoiced at having a single, focused goal. On the day the School Board met to vote, she doubted that anyone in the room realized the galvanizing effect the goal would have. "Many people thought we were crazy," Banks said. "There were concerns that teachers would not support the goal because it was unrealistic and that we would have a backlash from parents of high-achieving children who thought they would be shortchanged." A high official in the North Carolina State Department of Education warned her that Wake was creating expectations it couldn't possibly fulfill: "Karen, why did you let them set that 95 percent goal? You all aren't going to be able to reach that high of a goal in five years."

Within the Wake County school system, however, McNeal had begun a revolution years earlier to prepare the principals and top staff for the change. That was why Banks's contribution was so important: she moved the system to data-driven thinking with her creative analyses of what worked and what did not. McNeal's own conversion had come in 1989 when he attended a workshop with Edward Deming, the organizational guru whose theories the Japanese credited for their economic miracle. McNeal arranged seminars so that his top staff could interact with Deming to learn how to analyze data to intervene more quickly and speed up change. He also made sure that principals could understand and effectively use the new data Banks's office was producing to reallocate resources within their schools McNeal did not use the word "fired," but in one year he "moved" 27 principals who were failing to act aggressively enough to develop plans to reach low-performing children.

When he was named superintendent, McNeal linked his own contract renewal and salary increments to success in reaching the 95 percent goal. "I got some calls from friends of mine in the black community," he said. "I followed two white superintendents who had received good raises whether scores went up or not. My friends were asking, 'Oh my God, is this what a black man has to do to get the top job now?'" He told them the salary plan was his own idea: "Should the Board expect me to close the academic achievement gap of students in the district? Absolutely. Can they quantify that? Absolutely. Should they expect me to recruit the best and retain the brightest? Absolutely. That's my job." McNeal understood what he was doing in setting a standard for his own performance, and felt it was one of the best things he ever did. He intended to send a message: "I wanted the whole community to own this goal and I wanted to start by showing my commitment to it."

Unlike Bush's policy, which was underfunded and left millions of children behind in high-poverty inner-city schools, Wake County has come close to its educational goal under McNeal. Not every school attained a 95 percent passing rate, but many did. While only 71 percent of third graders passed the state's math and reading tests in 1994, by 2003 more than 91 percent of all Wake students in grades 3 through 8 had done so. McNeal was named National Superintendent of the Year in 2004 by the Association of American School Administrators.

At the same time, the test score gap between black and white children shrank dramatically. When the 95 percent goal was set, only 57 percent of black children in grades 3 through 8 were passing the state math test. By 2003, 81 percent passed. White scores also rose, but the gap between whites and blacks had shrunk from 37 points to 17. The gap between Hispanic students and whites narrowed from 28 points to 11. The passing rate for poor children, defined as those with family incomes low enough to qualify for free or reduced-price lunch, rose from 55 to 80 percent in the same period.

Some have suggested that much of Wake's school achievement was a result of the economic boom Raleigh enjoyed, drawing wealthier and more education-conscious parents to its schools. But the data show that the percentage of students categorized as impoverished or minority population increased after the goal was set, due in part to large immigration of Hispanic students into the Raleigh-Durham area. Poor students in elementary schools rose from 10 to 15 percent, and minority students rose from 31 to 43 percent. The system as a whole became 40 percent minority and 27 percent African American....

Some critics have suggested that perhaps North Carolina's state exams were just not as difficult as those of New York and other northern states. But Wake County students performed well in national comparisons as well. In 1990, with less than half of Wake's seniors—most of them college-bound–taking the SAT, the county scored below the national average. Yet in 2007, with 76 percent of Wake's high school seniors taking the test (compared with a national average participation rate of 48 percent), the average score of 1057 was 40 points above the national average. Wake students scored 11 points higher than the national average on the SAT writing test that was introduced in 2006. Wake's improved scores on state tests reflect the kinds of gains in mathematical reasoning and verbal skill that national tests measure and colleges value.

How, then, did Wake do it? Was it just setting a goal and using increased testing to whip teachers and students until they met the new target? Superintendent McNeal agreed with Bridges on the principal reason. Breaking down the wall between affluent suburbs and impoverished inner cities created a "healthy balance" of rich and poor in every classroom. And in 2000, Wake was the first metropolitan school district to move away from racial balance to economic balance as the measure of a school's diversity. Any school in Wake County where more than 40 percent of pupils were poor enough to qualify for subsidized lunches was defined as being out of balance. The policy guaranteed that all schools in Wake County would have a core of middle-class students who would establish a floor of positive expectations and create student networks across class lines that would benefit poor students.[4] Through this network of friends, less privileged students would get to know parents who might help them get a job or gain admission to college or simply serve as role models. Schools with a majority of middle-class parents will not tolerate incompetent teachers, or drinking fountains that don't work, or restrooms with no toilet paper.

Wake had begun to move to a class-based definition of diversity in the late 1990s, when its board reviewed research showing the strong link between family income and achievement—namely, that the achievement of children was depressed in schools with high concentrations of poverty. It began to use socioeconomic status as one of the three key factors in its assignment policy. The other two were racial diversity and pupil achievement level. The board had also become sensitized to court rulings in the South that prohibited any racial assignment in formerly desegregated schools that had been declared free of racial discrimination. This led to the move to drop race as a factor in school assignments in Wake County. Initially, the U.S. Department of Education's Office of Civil Rights challenged Wake's policy on the grounds that the new income-based criterion was simply a proxy for race and hence discriminated against whites by giving preferential treatment to blacks. However, on review, Wake was cleared of all discrimination charges, based on its defense of the clear relevance of socioeconomic status to academic achievement among all ethnic groups.[5]...

In the jargon of sociology, the merged Wake County school system redistributed social capital by changing the networks of opportunity for poor and minority children. Merely pouring dollars into economically segregated urban schools could no more provide equal educational opportunity than spending dollars to maintain "separate but equal" racially segregated schools had done in the 1950s—although many urban schools need and deserve more dollars to help the children they have. Teachers in Wake came to believe that having a good mix of students in every school gave them the best chance of success in reaching the high goals that the countywide system had set for itself.

While McNeal agreed that balance was critical, he knew that neither a 95 percent goal nor a policy of economic diversity would create results without major changes in how schools operated. Chief among these was providing teachers with resources and giving them the freedom to create programs they were proud of while holding them accountable for results....

HOW DID WAKE COUNTY MAKE IT HAPPEN?

School reforms usually fall into one of three types: fix a broken system, disrupt the system, or replace the system. Raleigh did all three. It replaced two systems leaning toward failure with a completely new merged system. It disrupted the old ways of doing business by introducing competition, giving parents a choice by turning a third of its schools into magnet schools. And it fixed the system with new forms of teaching and management.[6]

The primary cause of Raleigh's success was having the courage and political will to merge city and suburban schools in 1976. That step alone probably accounted for more than half the test score gains. It enabled Wake to create the right balance of racial, ethnic, and economic assets in every school, so that no one school was overwhelmed by the neediest students. Social capital was redistributed fairly throughout city and county in a way that almost no urban school system outside the South had achieved. Social capital is the yeast that makes a good school rise.

The belief that "there are no bad public schools in Wake County" was widely held by parents and teachers. In urban systems where bad schools are easily identified—by low test scores, high dropout rates, concentrations of children in poverty, more crime and violence—good teachers often leave for job offers in the suburbs. As a result of this talent flight, schools serving the neediest children in the urban core are disproportionately staffed with teachers who are least qualified or poorly prepared to teach the subjects they have been assigned to teach.[7] In Raleigh, where parents put their children on waiting lists to attend inner-city schools, no teachers felt that they were in a "losing school" where the deck was stacked against them because the school had been virtually abandoned by middle-class families. Nor did they hesitate to bring their own children to school with them. Teacher absenteeism was a small fraction of what it was in Syracuse, where teachers called in sick frequently, remarking to colleagues that they needed a "mental health day."

The second important step that Wake County took was to create magnet schools offering choice among distinctive educational options. In Raleigh, it was not just token choice. Magnets had failed to achieve major desegregation in many other cities because there were only a few of them, often located in the inner city. Raleigh created a critical mass of magnet schools, particularly in the early years of merger, by turning nearly a third of the schools into schools of choice. The top leadership spent a year selling the wide variety of choices to black and white parents, convincing them that these were opportunities they couldn't afford to turn down.

All the magnets were opened in the following year and were placed in schools along the border between the city and suburbs. Neither city nor suburban children had to be bused far. The principle of two-way busing, with white children being bused into the city while black children rode to schools in the suburbs, was enshrined in Raleigh's program from the beginning. The magnet program was not just a device to achieve racial balance, however Magnet schools became laboratories to test new programs and new approaches to teaching. Programs that succeeded were exported to other schools in the system, and those that failed were later jettisoned.

Often overlooked was the opportunity for choice that the magnets offered teachers as well as parents and students. The magnet programs helped Wake recruit able teachers ... who were looking for the right niche, and gave other teachers within the system the chance to grow in new directions. In later years, magnets helped reduce the tension that afflicted some districts about the impositions of testing. Even children who were getting double the instruction in math and reading in order to catch up also got to choose a dance class or a course in film-making. And teachers were given plentiful opportunities to create such courses.

Third, instead of complaining about statewide testing, Wake County used it to set a gutsy goal and mobilize the community at a time when many doubted that 95 percent of its children could pass the state tests. This occurred four years before the No Child Left Behind federal law was passed. Studies of failed reforms have revealed that they were often poorly introduced without negotiating the prior agreements that lead teachers to buy in. Raleigh created the Wake Partnership uniting the business sector, political leadership, parents, and teachers. Thousands came to public meetings to discuss the new goal. It helped the whole community to own it, by convincing them that the schools could not do it alone. Teachers told me in interviews how galvanizing these meetings were. This wasn't a pet project of a new principal or a phony experiment in a few schools that administrators knew were close to achieving the goal already. It applied to everyone, to all schools. Teachers accepted that it made sense to break the goal into two five-year periods, getting the elementary and middle schools up to speed by 2003 and the high schools by 2008. Everyone agreed that the best way to reform high schools is to reform grade schools first. It was a stunningly clear and simple goal that allowed schools to choose different means to achieving it.[8]

Likewise, all were being held accountable—not only teachers but also principals and the top leadership. Even Superintendent McNeal agreed to have his salary pegged to the achievement levels of all students. For years, McNeal had led the system toward a more data-driven curriculum. Wake had one of the most advanced databases of any school system in the nation. Its Website—far ahead of what most school systems had at that time—provided detailed profiles of each school with in-depth analyses of its test scores and with comments posted by students and parents about safety and teachers' attitudes.

Important as data were, however, Wake did not see success on the new tests as the goal of reform but as a *means* to reform. Real reform went deeper. It involved a change from older models of management toward new cultures of leadership and teaching. Change in management styles began before the new goal was announced. Principals were taught how to use their computers to analyze the new data in ways that would improve instruction. Control of the budget was decentralized as Wake moved to school site management—that is, principals were given authority to make major shifts in their budgets to put more resources at the service of low-achieving pupils. They took advantage of seminars conducted at the central office to learn new skills.

The school district created the Wake Leadership Academy, a new training program for aspiring principals using its own cutting-edge staff and faculty at North Carolina State University. These principals-in-training went out in teams to analyze top-performing schools as well as failing ones in Raleigh and other school districts. They learned how to conduct the kind of collaborative inquiry that helped teachers develop. While classroom visits and one-on-one supervision of teachers, particularly new hires, continued to be an important part of a principal's work, Raleigh principals increasingly became managers and nurturers of teaching teams, each of which had its own teacher-leader or co-leaders. The team might be composed entirely of fourth grade teachers, or of cross-grade teams of elementary math instructors, or of high school biology teachers.

The teams were at the heart of the change in the teaching culture of Wake County. For generations teachers had worked in a system that sorted and selected students, while for the most part ignoring the dropout rate. Even the word "dropout" didn't come into widespread use until the 1960s. Up until that time, half or more of all students were expected to fail or drop out, and when they did it was the pupil's fault for not learning what teachers taught. Not much was known about what teachers were actually teaching, or how they were going about it. They worked in isolation, and only rarely engaged in joint planning or evaluation with colleagues. Experienced teachers seldom shared their practical wisdom or special techniques with novices. Only in recent decades has teaching all students to succeed become a goal. In many places it is still an empty mantra.

Wake County's adoption of the 95 percent goal in 1998 put the nail in the coffin of the old teacher belief system in Raleigh. Making it work was another thing altogether, however. Part of the task was to align what was taught with what got tested, and some teachers felt the screws were tightened too much. But most Raleigh teachers I interviewed felt that they were teaching a good curriculum and that the tests were fair. Getting most kids to pass those tests, however, was a stretch for many teachers.[9] Even prior to 1998 some Wake schools with more than a third of their students from low-income and minority families had come close to the 95 percent goal. Teachers were encouraged to visit those schools. Many found it a transforming experience—seeing is believing. Some had the courage to tell their colleagues that they had seen teachers who were doing what those colleagues were doing—only better. The creation of teacher teams moved that kind of realistic assessment to the core of conversations....

A large part of the motivation came from buying into the dream, coming to believe that a 95 percent pass rate wasn't just pie-in-the-sky, that they were actually accomplishing something that urban public schools had never done before: educating all the kids. Motivation also came from the satisfaction of being part of a community of teachers who are sharing knowledge and learning together. It was not just "implement this new program and shut up." It was about sitting down together with data printouts in hand, analyzing weaknesses, thinking about the kids, sharing best practices, coming up with ideas of what might work better, trying some of them, and seeing what works....[10]

ENDNOTES

1. These are 2003 scores. By 2006, Syracuse had fallen to 21 percent passing.
2. In this chapter and hereafter, in referring to the merged Raleigh–Wake County school system, I use Raleigh and Wake County schools interchangeably.
3. All quotations in this chapter are drawn from interviews by the author, except where noted otherwise.
4. See Richard Kahlenberg, *All Together Now Creating Middle Class Schools through Public School Choice* (Washington. DC: Brookings Institution Press, 2003), and Allan Ornstein, *Class Counts: Education, Inequality, and the Shrinking Middle Class* (New York: Rowman and Littlefield, 2007).
5. Susan Leigh Flinspach and Karen E. Banks, "Moving Beyond Race: Socioeconomic Diversity as a Race-Neutral Approach to Desegregation in the Wake County Schools," in John Charles Boger and Gary Orfield, eds., *School Resegregation: Must the South Turn Back?* (Chapel Hill: University of North Carolina Press, 2005).
6. I thank one of the anonymous reviewers of this manuscript for suggesting this typology.
7. National Commission on Teaching and America's Future. *What Matters Most: Teaching for America's Future* (New York, 1996).
8. On the value of reforming grade schools first, see E. D. Hirsch, "An Epoch-Making Report. But What About the Early Grades," *Education Week*, April 23, 2008.
9. See Ronald F. Ferguson, *Excellence with Equity: An Emerging Vision for Closing the Achievement Gap* (Cambridge: Harvard Education Press, 2007), and Richard F. Elmore, "The Limits of Change," *Harvard Education Letter*, January 2002.
10. David Armor is among those who have said one of the shortcomings of most reforms introduced in the wake of the No Child Left Behind law is that advocates assumed they could close the achievement gap without any changes in families. Wake County did not assume that. See Armor's "No Excuses: Simplistic Solution for the Achievement Gap?" *Teachers College Record*, February 12, 2004. On the value of expanding time for instruction, see Tommy M. Tomlinson ed., *Motivating Students to Learn: Overcoming Barriers to High Achievement* (Berkeley: McCutchan Publishing, 1993), and Susan H. Fuhrman, "If We're Talking About Race, Let's Talk About Education," *Education Week*, May 7, 2008.

DISCUSSION QUESTIONS

1. What does it mean to *redistribute* social capital? Why was this *redistribution* the key factor in the success of Raleigh's public schools?
2. Is the success of Raleigh's public schools a model that can be replicated in other school districts across the United States? Why or why not.
3. In order for the redistribution of social capital to occur, parents and voters must be willing to give up some control over where their children ultimately attend school. Why has this aspect of the Raleigh model been so controversial to school districts across the United States?
4. A key feature of the Raleigh model according to the author was this: "Control of the budget was decentralized ... that is, principals were given authority to make major shifts in their budgets to put more resources at the service of low-achieving pupils." Why aren't more principals given direct control over their budgets? Is this aspect of the model readily transferable? Why or why not?
5. *The Raleigh model will never work under the current fiscal environment where states are cutting per-pupil appropriations to public education, reducing teaching staff, and increasing average classroom size.* Respond to this statement.

40

The College Dropout Boom

DAVID LEONHARDT

The Four Questions

1. What is the problem?
2. Is it a *social* problem?
3. What are the causes of the problem?
4. Is there a way to alleviate the problem?

Topics Covered

Higher education

Class inequality

Economic mobility

Meritocracy

Elite colleges

One of the biggest decisions Andy Blevins has ever made, and one of the few he now regrets, never seemed like much of a decision at all. It just felt like the natural thing to do.

In the summer of 1995, he was moving boxes of soup cans, paper towels, and dog food across the floor of a supermarket warehouse, one of the biggest buildings in the area of southwest Virginia surrounding the town of Chilhowie. The heat was

SOURCE: From David Leonhardt, "The College Dropout Boom," in *Class Matters*, ed. *The New York Times*, pp. 87–104.

brutal. The job had sounded impossible when he arrived fresh off his first year of college, looking to make some summer money, still a skinny teenager with sandy blond hair and a narrow, freckled face.

But hard work done well was something he understood, even if he was the first college boy in his family. Soon he was making bonuses on top of his $6.75 an hour, more money than either of his parents made. His girlfriend was around, and so were his hometown buddies. Andy acted more outgoing with them, more relaxed. People in Chilhowie noticed that.

It was just about the perfect summer. So the thought crossed his mind: maybe it did not have to end. Maybe he would take a break from college and keep working. He had been getting Cs and Ds, and college never felt like home, anyway.

"I enjoyed working hard, getting the job done, getting a paycheck," Blevins recalled. "I just knew I didn't want to quit."

So he quit college instead, and with that, Andy Blevins joined one of the largest and fastest-growing groups of young adults in America. He became a college dropout, though nongraduate may be the more precise term.

Many people like him plan to return to get their degrees, even if few actually do. Almost one in three Americans in their mid-twenties now fall into this group, up from one in five in the late 1960s, when the Census Bureau began keeping such data. Most come from poor and working-class families.

The phenomenon has been largely overlooked in the glare of positive news about the country's gains in education. Going to college has become the norm throughout most of the United States, even in many places where college was once considered an exotic destination—places like Chilhowie, an Appalachian hamlet with a simple brick downtown. At elite universities, classrooms are filled with women, blacks, Jews, and Latinos, groups largely excluded two generations ago. The American system of higher learning seems to have become a great equalizer.

In fact, though, colleges have come to reinforce many of the advantages of birth. On campuses that enroll poorer students, graduation rates are often low. And at institutions where nearly everyone graduates—small colleges like Colgate, major state institutions like the University of Colorado, and elite private universities like Stanford—more students today come from the top of the nation's income ladder than they did two decades ago.

Only 41 percent of low-income students entering a four-year college managed to graduate within five years, the U.S. Department of Education found in a 2004 study, but 66 percent of high-income students did. That gap had grown over recent years.

"We need to recognize that the most serious domestic problem in the United States today is the widening gap between the children of the rich and the children of the poor," Lawrence H. Summers, the president of Harvard, said when announcing in 2004 that Harvard would give full scholarships to all its lowest-income students. "And education is the most powerful weapon we have to address that problem."

There is certainly much to celebrate about higher education today. Many more students from all classes are getting four-year degrees and reaping their benefits. But those broad gains mask the fact that poor and working-class students have nevertheless been falling behind; for them, not having a degree remains the norm.

That loss of ground is all the more significant because a college education matters much more now than it once did. A bachelor's degree, not a year or two of courses, tends to determine a person's place in today's globalized, computerized economy. College graduates have received steady pay increases over the past two decades, while the pay of everyone else has risen little more than the rate of inflation.

As a result, despite one of the great education explosions in modern history, economic mobility—moving from one income group to another over the course of a lifetime—has stopped rising, researchers say. Some recent studies suggest that it has declined over the last generation.

Put another way, children seem to be following the paths of their parents more than they once did. Grades and test scores, rather than privilege, determine success today, but that success is largely being passed down from one generation to the next. A nation that believes that everyone should have a fair shake finds itself with a kind of inherited meritocracy.

In this system, the students at the best colleges may be diverse—male and female and of various colors, religions, and hometowns—but they tend to share an upper-middle-class upbringing. An old joke that Harvard's idea of diversity is putting a rich kid from California in the same room as a rich kid from New York is truer today than ever; Harvard has more students from California than it did in years past and just as big a share of upper-income students.

Students like these remain in college because they can hardly imagine doing otherwise. Their parents, understanding the importance of a bachelor's degree, spent hours reading to them, researching school districts, and making it clear to them that they simply must graduate from college.

Andy Blevins says that he too knows the importance of a degree, but that he did not while growing up, and not even in his year at Radford University, sixty-six miles up the interstate from Chilhowie. Ten years after trading college for the warehouse, Blevins, who is twenty-nine, spends his days at the same supermarket company. He has worked his way up to produce buyer, earning $35,000 a year with health benefits and a 401(k) plan. He is on a path typical for someone who attended college without getting a four-year degree. Men in their early forties in this category made an average of $42,000 in 2000. Those with a four-year degree made $65,000.

Still boyish-looking but no longer rail thin, Blevins says he has many reasons to be happy. He lives with his wife, Karla, and their son, Lucas, in a small blue-and-yellow house at the end of a cul-de-sac in the middle of a stunningly picturesque Appalachian valley. He plays golf with some of the same friends who made him want to stay around Chilhowie.

But he does think about what might have been, about what he could be doing if he had the degree. As it is, he always feels as if he is on thin ice. Were he to lose his job, he says, everything could slip away with it. What kind of job could a guy without a college degree get? One night, while talking to his wife about his life, he used the word "trapped."

"Looking back, I wish I had gotten that degree," Blevins said in his soft-spoken lilt. "Four years seemed like a thousand years then. But I wish I would have just put in my four years."

THE BARRIERS

Why so many low-income students fall from the college ranks is a question without a simple answer. Many high schools do a poor job of preparing teenagers for college. Many of the colleges where lower-income students tend to enroll have limited resources and offer a narrow range of majors, leaving some students disenchanted and unwilling to continue.

Then there is the cost. Tuition bills scare some students from even applying and leave others with years of debt. To Blevins, like many other students of limited means, every week of going to classes seemed like another week of losing money—money that might have been made at a job.

"The system makes a false promise to students," said John T Casteen III, the president of the University of Virginia, himself the son of a Virginia shipyard worker.

Colleges, Casteen said, present themselves as meritocracies in which academic ability and hard work are always rewarded. In fact, he said, many working-class students face obstacles they cannot overcome on their own.

For much of his fifteen years as Virginia's president, Casteen has focused on raising money and expanding the university, the most prestigious in the state. In the meantime, students with backgrounds like his have become ever scarcer on campus. The university's genteel nickname, the Cavaliers, and its aristocratic sword-crossed coat of arms seem appropriate today. No flagship state university has a smaller proportion of low-income students than Virginia. Just 8 percent of undergraduates in 2004 came from families in the bottom half of the income distribution, down from 11 percent a decade earlier.

That change sneaked up on him, Casteen said, and he had spent a good part of the previous year trying to prevent it from becoming part of his legacy. Starting with the fall 2005 freshman class, the university will charge no tuition and require no loans for students whose parents make less than twice the poverty level, or about $37,700 a year for a family of four. The university has also increased financial aid to middle-income students.

To Casteen, these are steps to remove what he describes as "artificial barriers" to a college education placed in the way of otherwise deserving students. Doing so "is a fundamental obligation of a free culture," he said.

But the deterrents to a degree can also be homegrown. Many low-income teenagers know few people who have made it through college. A majority of the nongraduates are young men, and some come from towns where the factory work ethic, to get working as soon as possible, remains strong, even if the factories themselves are vanishing. Whatever the reasons, college just does not feel normal.

"You get there and you start to struggle," said Leanna Blevins, Andy's older sister, who did get a bachelor's degree and then went on to earn a Ph.D. at Virginia studying the college experiences of poor students. "And at home your parents are trying to be supportive and say, 'Well, if you're not happy, if it's not right for you, come back home. It's okay' And they think they're doing the right thing. But they don't know that maybe what the student needs is to hear them say, 'Stick it out just one semester. You can do it. Just stay there. Come home on the weekend, but stick it out.'"

Today, Leanna, petite and high-energy, is helping to start a new college a few hours' drive from Chilhowie for low-income students. Her brother said he had daydreamed about attending it and had talked to her about how he might return to college.

For her part, Leanna says, she has daydreamed about having a life that would seem as natural as her brother's, a life in which she would not feel like an outsider in her hometown. Once, when a high school teacher asked students to list their goals for the next decade, she wrote, "having a college degree" and "not being married."

"I think my family probably thinks I'm liberal," Leanna, who is now married, said with a laugh, "that I've just been educated too much and I'm gettin' above my raisin'."

Her brother said that he just wanted more control over his life, not a new one. At a time when many people complain of scattered lives, Andy Blevins can stand in one spot—his church parking lot, next to a graveyard—and take in much of his world. "That's my parents' house," he said one day, pointing to a sliver of roof visible over a hill. "That's my uncle's trailer. My grandfather is buried here. I'll probably be buried here."

TAKING CLASS INTO ACCOUNT

Opening up colleges to new kinds of students has generally meant one thing over the last generation: affirmative action. Intended to right the wrongs of years of exclusion, the programs have swelled the number of women, blacks, and Latinos on campuses. But affirmative action was never supposed to address broad economic inequities, just the ones that stem from specific kinds of discrimination.

That is now beginning to change. Like Virginia, a handful of other colleges are not only increasing financial aid but also promising to give weight to economic class in granting admissions. They say they want to make an effort to admit more low-income students, just as they now do for minorities and children of alumni.

"The great colleges and universities were designed to provide for mobility, to seek out talent," said Anthony W. Marx, president of Amherst College. "If we are blind to the educational disadvantages associated with need, we will simply replicate these disadvantages while appearing to make decisions based on merit."

With several populous states having already banned race-based preferences and the United States Supreme Court suggesting that it may outlaw such programs in a couple of decades, the future of affirmative action may well revolve around economics. Polls consistently show that programs based on class backgrounds have wider support than those based on race.

The explosion in the number of nongraduates has also begun to get the attention of policy makers. In 2005, New York became one of a small group of states to tie college financing more closely to graduation rates, rewarding colleges more for moving students along than for simply admitting them. Nowhere is the stratification of education more vivid than in Virginia, where Thomas Jefferson once tried, and failed, to set up the nation's first public high schools. At a modest high school in the Tidewater city of Portsmouth, not far from John Casteen's boyhood home, a guidance-office wall filled with college pennants does not include one from rarefied Virginia. The colleges whose pennants are up—Old Dominion University and others that seem in the realm of the possible—have far lower graduation rates.

Across the country, the upper middle class so dominates elite universities that high-income students, on average, actually get slightly more financial aid from colleges than low-income students do. These elite colleges are so expensive that even many high-income students receive large grants. In the early 1990s, by contrast, poorer students got 50 percent more aid on average than the wealthier ones, according to the College Board, the organization that runs the SAT entrance exams.

At the other end of the spectrum are community colleges, the two-year institutions that are intended to be feeders for four-year colleges. In nearly every one are tales of academic success against tremendous odds: a battered wife or a combat veteran or a laid-off worker on the way to a better life. But overall, community colleges tend to be places where dreams are put on hold.

Most people who enroll say they plan to get a four-year degree eventually; few actually do. Full-time jobs, commutes, and children or parents who need care often get in the way. One recent national survey found that about 75 percent of students enrolling in community colleges said they hoped to transfer to a four-year institution. But only 17 percent of those who had entered in the mid-1990s made the switch within five years, according to a separate study. The rest were out working or still studying toward the two-year degree.

"We here in Virginia do a good job of getting them in," said Glenn Dubois, chancellor of the Virginia Community College System and himself a community college graduate. "We have to get better in getting them out." ...

FEW BREAKS FOR THE NEEDY

The college admissions system often seems ruthlessly meritocratic. Yes, children of alumni still have an advantage. But many other pillars of the old system—the polite rejections of women or blacks, the spots reserved for graduates of Choate and Exeter—have crumbled.

This was the meritocracy John Casteen described when he greeted the parents of freshmen in a University of Virginia lecture hall in the late summer of 2004. Hailing from all fifty states and fifty-two foreign countries, the students were more intelligent and better prepared than he and his classmates had been, he told the parents in his quiet, deep voice. The class included seventeen students with a perfect SAT score.

If anything, children of privilege think that the system has moved so far from its old-boy history that they are now at a disadvantage when they apply, because colleges are trying to diversify their student rolls. To get into a good college, the sons and daughters of the upper middle class often talk of needing a higher SAT score than, say, an applicant who grew up on a farm, in a ghetto, or in a factory town. Some state legislators from northern Virginia's affluent suburbs have argued that this is a form of geographic discrimination and have quixotically proposed bills to outlaw it.

But the conventional wisdom is not quite right. The elite colleges have not been giving much of a break to the low-income students who apply. When William G. Bowen, a former president of Princeton, looked at admissions records recently he found that if test scores were equal a low-income student had no better chance than a high-income one of getting into a group of nineteen colleges, including Harvard, Yale, Princeton, Williams, and Virginia. Athletes, legacy applicants, and minority students all got in with lower scores on average. Poorer students did not.

The findings befuddled many administrators, who insist that admissions officers have tried to give poorer applicants a leg up. To emphasize the point, Virginia announced in the spring of 2005 that it was changing its admissions policy from "need blind"—a term long used to assure applicants that they would not be punished for seeking financial aid—to "need conscious." Administrators at Amherst and Harvard have also recently said that they would redouble their efforts to take into account the obstacles students have overcome.

"The same score reflects more ability when you come from a less fortunate background," Lawrence Summers, the president of Harvard, said. "You haven't had a chance to take the test-prep course. You went to a school that didn't do as good a job coaching you for the test. You came from a home without the same opportunities for learning."

But it is probably not a coincidence that elite colleges have not yet turned this sentiment into action. Admitting large numbers of low-income students could bring clear complications. Too many in a freshman class would probably lower the college's average SAT score, thereby damaging its ranking by *U. S. News & World Report,* a leading arbiter of academic prestige. Some colleges, like Emory University in Atlanta, have climbed fast in the rankings over precisely the same period in which their percentage of low-income students has tumbled. The math is simple: when a college goes looking for applicants with high SAT scores, it is far more likely to find them among well-off teenagers.

More spots for low-income applicants might also mean fewer for the children of alumni, who make up the fundraising base for universities. More generous financial aid policies will probably lead to higher tuition for those students who can afford the list price. Higher tuition, lower ranking, tougher admission requirements: these do not make for an easy marketing pitch to alumni clubs around the country. But Casteen and his colleagues are going ahead, saying the pendulum has swung too far in one direction.

That was the mission of John Blackburn, Virginia's easygoing admissions dean, when he rented a car and took to the road in the spring of 2005. Blackburn thought of the trip as a reprise of the drives Casteen took twenty-five years earlier, when he was the admissions dean, traveling to churches and community centers to persuade black parents that the university was finally interested in their children.

One Monday night, Blackburn came to Big Stone Gap, in a mostly poor corner of the state not far from Andy Blevins's town. A community college there was holding a college fair, and Blackburn set up a table in a hallway, draping it with the University of Virginia's blue and orange flag.

As students came by, Blackburn would explain Virginia's new admissions and financial aid policies. But he soon realized that the Virginia name might have been scaring off the very people his pitch was intended for. Most of the students who did approach the table showed little interest in the financial aid and expressed little need for it. One man walked up to Blackburn and introduced his son as an aspiring doctor. The father was an ophthalmologist. Other doctors came by too. So did some lawyers.

"You can't just raise the UVA flag," Blackburn said, packing up his materials at the end of the night, "and expect a lot of low-income kids to come out."

When the applications started arriving in his office, there seemed to be no increase in those from low-income students. So Blackburn extended the deadline two weeks for everybody, and his colleagues also helped some applicants with the maze of financial aid forms. Of 3,100 incoming freshmen, it now seems that about 180 will qualify for the new financial aid program, up from 130 who would have done so the year before. It is not a huge number, but Virginia administrators call it a start.

A BIG DECISION

On a still-dark February morning, with the winter's heaviest snowfall on the ground, Andy Blevins scraped off his Jeep and began his daily drive to the supermarket warehouse. As he passed the home of Mike Nash, his neighbor and fellow gospel singer, he noticed that the car was still in the driveway. For Nash, a school counselor and the only college graduate in the singing group, this was a snow day.

Blevins later sat down with his calendar and counted to 280: the number of days he had worked last year. Two hundred and eighty days—six days a week most of the time—without ever really knowing what the future would hold.

"I just realized I'm going to have to do something about this," he said, "because it's never going to end."

In the weeks afterward, his daydreaming about college and his conversations about it with his sister Leanna turned into serious research. He requested his transcripts from Radford and from Virginia Highlands Community College and figured out that he had about a year's worth of credits. He also talked to Leanna about how he could become an elementary school teacher. He always felt that he could relate to children, he said. The job would take up 180 days, not 280. Teachers do not usually get laid off or lose their pensions or have to take a big pay cut to find new work.

So the decision was made. Andy Blevins says he will return to Virginia Highlands, taking classes at night; the Gospel Gentlemen are no longer booking performances. After a year, he plans to take classes by video and on the Web that are offered at the community college but run by Old Dominion, a Norfolk, Virginia, university with a big group of working-class students.

"I don't like classes, but I've gotten so motivated to go back to school," Blevins said. "I don't want to, but, then again, I do."

He thinks he can get his bachelor's degree in three years. If he gets it at all, he will have defied the odds.

DISCUSSION QUESTIONS

1. Is higher education a class equalizer, or does it reinforce the advantages of birth?
2. "Grades and test scores, rather than privilege, determine success today, but that success is largely being passed down from one generation to the next." Why?
3. What encourages students to drop out of college? Does social class play a role?
4. "Overall, community colleges tend to be places where dreams are put on hold," writes Leonhardt. Do you agree? If you agree, what are the factors that discourage a four-year college degree for those arriving from community colleges? If you do not agree, what are some of the factors that actually can help students to adjust from a community college to a four-year degree?
5. "Lawrence H. Summers, when he was president of Harvard, contended: "We need to recognize that the most serious domestic problem in the United States today is the widening gap between the children of the rich and the children of the poor. And education is the most powerful weapon we have to address that problem."' Do you agree with Summer's statement?
6. "I don't like classes, but I've gotten so motivated to go back to school. I don't want to, but, then again, I do," Andrew Blevins says. "He thinks he can get his bachelor's degree in three years. If he gets it at all, he will have defied the odds," replies Leonhardt. What is going on here?

PART XI

Social Problems Related to Health Care

The *Patient Protection and Affordable Care Act (PPACA)*, signed into law by President Barack Obama on 23rd March 2010, is a significant step in the long history of political attempts to address the health care problem in the United States. The key features of the new law, much of which will take effect within the next four years, include expanded Medicaid eligibility to previously excluded groups, the end of insurance discrimination based on pre-existing conditions, setting up health insurance exchanges, and giving tax incentives so businesses can offer and/or expand health insurance coverage to their employees. There is much uncertainty, however, as to whether these unprecedented—yet necessary—changes will significantly address the rising cost of health insurance to individuals and corporations.

Katherine Swartz reminds us of the realities and risks of being uninsured in America. Swartz addresses the problems with the employer-sponsored insurance (ESI) model of delivery, and she provides case-studies of how other societies have dealt with the quandary of rising costs in their attempts to finance a system of universal access to healthcare.

Charles Dougherty looks at the problem of inequality in health care. He explores the difficulty that many of us have caring about health care problems as long as *we* are served adequately. As a result, we do not create an approach to health care similar to those of other industrialized nations, and health care for the poor continues to be a social problem.

James W. Russell introduces the differences between European systems and our own, concluding that the "majority of health care spending in Europe is publicly financed. In the U.S. the majority is privately financed." The problem every society faces is to decide how much it is willing to pay for everyone's health and how much should be the responsibility of each individual. It is not simply either/or but a matter of degree.

41

New Realities, New Risks: Uninsured in America

KATHERINE SWARTZ

The Four Questions

1. What is the key problem that Swartz identifies?
2. Why is this a *social* problem?
3. What are the underlying causes of this problem?
4. Does Swartz offer any ideas for dealing with this problem?

Topics Covered

Uninsured people

Health care system

Health care policy

Employer-sponsored insurance

Rising cost of health insurance

Adverse selection risk

Mandatory health insurance law

Health care expenditures

In 2006, 47 million Americans did not have any type of health insurance. This was 2.2 million more people than in 2005,[1] and it was the largest one-year increase in the number of uninsured since the Census Bureau started collecting insurance status data in 1979. Moreover, among people younger than sixty-five, the increase in the uninsured means that one in six nonelderly people (almost 18 percent) are now uninsured—and at risk for financial problems, receiving less care, and having poorer quality outcomes if they do need medical care....

THE SCOPE OF THE PROBLEM OF BEING UNINSURED

The extent of the uninsured problem has at least two dimensions. One relates to the significant and negative consequences for people of being without health insurance. The other concerns the number of people without any type of coverage and the primary groups of people who are uninsured.

Being Without Insurance: Financial and Health Risks

There are both financial and health-related consequences of being uninsured. Uninsured people are at risk for substantial sums of money to pay for unexpected medical care—and sometimes these bills are a financial catastrophe, causing many to declare bankruptcy. Even among people who do not declare bankruptcy, the fall-out from medical bills for emergency room visits or hospitalizations can have long-term consequences. Ironically, these outcomes include the depletion of savings and not being able to afford health insurance because the people are paying down debts that they would not have incurred had they been covered by health insurance.[2]

There is mounting evidence that not having insurance has negative effects on people's use of health care and on their health outcomes.[3] In a study of people who experienced a health shock (caused by an unintentional injury or the onset of a

SOURCE: From "Uninsured in America: New Realities, New Risks," by Katherine Swartz, in *Health At Risk: America's Ailing Health System—and How to Heal It*, Edited by Jacob S. Hacker, New York, NY: Columbia University Press, 2008, pp. 32–65.

chronic condition), uninsured individuals were less likely to obtain any medical care and were more likely not to have received any recommended follow-up care than insured individuals.[4] Moreover, the uninsured with new chronic conditions reported significantly worse short-term health changes about 3.5 months after the initial shock and those who had an unintentional injury were significantly more likely to report that they were not fully recovered and were no longer being treated compared to those with health insurance.[5] A recent study conducted by the American Cancer Society shows that uninsured people are significantly less likely to get screened for cancer, more likely to be diagnosed with an advanced stage of the disease, and less likely to survive that diagnosis than their privately insured counterparts.[6] A similar study of people with diabetes underscored the importance of health insurance for receiving needed care and effectively managing the disease—many of the uninsured diabetics interviewed for the study had serious medical complications that likely would not have occurred had they been covered by insurance.[7]

Studies of people fifty to sixty-four years of age show that those who are uninsured receive fewer clinical services, are more likely to experience health declines, and die at younger ages than insured adults in the same age group.[8] When these uninsured individuals enrolled in Medicare at age sixty-five, they had greater use of health care and had higher medical spending than previously insured adults with similar characteristics at ages fifty-nine to sixty and comparable health insurance after age sixty-five.[9]

Thus, being uninsured has significant negative effects on people's financial situation and their use of health care and health outcomes. This is increasingly worrisome as the newer, more effective treatments for cancers and neurological chronic conditions involve pharmacological options that frequently are not offered to uninsured people. Similarly, uninsured people who have traumatic injuries from car or other accidents are far less likely to receive the continuum of care needed to make timely and good recoveries. All in all, being uninsured is dangerous for one's health.

Estimating Who is Uninsured and Major Subgroups of the Uninsured

As noted earlier, 47 million people were uninsured in 2006, and 46.4 million of them were under the age of sixty-five. These estimates are based on responses by a large, nationally representative sample of people who were asked by the census in March 2007 about their health insurance coverage in the previous calendar year.[10] Because the survey does not specifically capture times when people are uninsured for part of the year, the estimate of 47 million uninsured does not fully account for all the people who were affected by the lack of insurance during 2006.[11] Since the risks of being uninsured are large even for people who are uninsured for short periods of time, the annual census data provide a lower-bound estimate of the number of people affected by not having health insurance during a year. I follow the convention of relying on the annual census data to describe the uninsured rather than surveys that follow people over longer periods of time. The longer surveys have smaller samples of people, making estimates of subgroups of the uninsured less reliable, and the dynamics of health insurance coverage complicate analyses of the characteristics of the uninsured.

The characteristics of people who are most likely to be uninsured have changed over the past three decades.... In this section, I describe the uninsured in terms of the largest subgroups among them. Three characteristics stand out as key predictors of who is uninsured: being a younger adult twenty-five to forty-four years of age, having less formal education, and having low income.

Adults Twenty-Five to Forty-Four Years of Age .... Two out of five uninsured are twenty-five to forty-four years old.... More significantly, the percents of all people in these age cohorts who are uninsured are at all-time highs: more than a quarter of all twenty-five- to thirty-four-year-olds and nearly a fifth of all thirty-five- to forty-four-year-olds were uninsured in 2006....

Less Formal Education Almost two-thirds of uninsured adults twenty-three to sixty-four years of age have not gone past high school for formal education. Having low levels of formal education is a large handicap for finding a job with employer-sponsored insurance (ESI): among adults who have not completed high school, 44 percent are uninsured; and a quarter of all adults who have high school diplomas but no further formal education are uninsured. These high fractions reflect the decline in employer willingness to offer ESI and the fact that demand for people with less education is now largely in lower-wage jobs. This is a shift from a generation ago when high school graduates could find well-paying jobs with large manufacturers. In 2006, five of the eight occupations that had the largest numbers of workers did not have high education requirements and more than a fifth of the people in each of these occupations were uninsured.

Low Income ... 50 percent of the uninsured had family incomes below $30,000 ... If we compare the incomes of the uninsured to the federal poverty level ($10,294 for an individual; $16,079 for a family of three; $20,614 for a family of four in 2006), a quarter of all the uninsured were officially in poverty in 2006. Another 29 percent had incomes between the poverty level and two times the poverty level (that is, they were near poor). Thus, 50 to 55 percent of the uninsured have low incomes (below $30,000 or below two times the poverty level) and cannot afford to purchase health insurance. In spite of the widespread (and mistaken) view that Medicaid covers all people in poverty, 34 percent of people with incomes below the poverty level were uninsured. Among all people who were near poor, 30 percent were uninsured.

Three other characteristics also describe significant but smaller subgroups among the uninsured; being middle class, being less than nineteen years old, and being foreign-born.

Middle Class Being uninsured is not just a problem for low-income people—it also is increasingly a risk for middle-class people. The threshold level of income for the middle class can be defined as the median household income (that is, the income level at which 50 percent of all households in the country have incomes below it). In 2006, the median household income was $48,201, so anyone with an income above $48,201 can be defined as being in the middle class.[12]... By this definition, 13.5 million of the 46.4 million nonelderly uninsured were middle class—29 percent of the uninsured. If we look just at working age adults (twenty-three to sixty-four years old) who are also middle class, one in ten was uninsured in 2006. This is a significant change from 1979 when just 6 percent of middle-class adults were uninsured.

Children Under Age Nineteen In 2006, 9.4 million children were uninsured—an increase of 700,000 over the previous year. The increase in the number of uninsured children appears to be due to a drop in the number of children with access to employer-sponsored coverage–either because the parent lost access to employer coverage or because the employer raised the out-of-pocket cost of dependent coverage. In 2006, 63.4 percent of all children under age nineteen were covered by private insurance (employer-sponsored or direct purchased) and 24.6 percent had public coverage (Medicaid or SCHIP, the State Children's Health Insurance Program). The fraction of all children who were uninsured was 12.1 percent, which was a slight increase over the last few years. Still, the fraction is significantly lower than it was in 1979 when 17 percent of all children were uninsured.[13]

Between the late 1980s and 2005, the number of uninsured children and the fraction of all children who were uninsured declined steadily—primarily because eligibility criteria for Medicaid were expanded to include older children at higher income levels and in 1997, the State Children's Health Insurance Program (SCHIP) was created.[14] States now have the option of covering children under Medicaid if their family incomes are as much as 185 percent of the poverty level. Uninsured children are eligible for SCHIP if their families have incomes that are too high for Medicaid but below a level set by each state. As of 2007, forty-one states and the District of Columbia had SCHIP income eligibility caps at or above 200 percent of the poverty level. SCHIP funding has included monies for outreach efforts, which have been credited with enrolling large numbers of children eligible for SCHIP and Medicaid....

Foreign-Born Status Just over a fifth of the uninsured in 2006—ten million people—were not born in the United States and were not citizens. Another 2.3 million uninsured were foreign-born and were naturalized citizens. Not quite half (46.6 percent) of the foreign-born population who were not yet citizens was uninsured. This is in contrast to 15 percent of Americans born in the United States and 19.8 percent of naturalized citizens who were uninsured.

People who report themselves as foreign-born but not citizens on the census data are almost all here legally. Illegal immigrants are very unlikely to answer census survey questions. Thus, the foreign-born who are not citizens include people who have not yet lived here long enough to apply for citizenship and people who may expect to return to their country of origin some time in the future. Foreign-born residents who have been in the United States for longer periods of time are less likely to be uninsured than people who immigrated within the past five years.

A majority of foreign-born noncitizens are younger adults with low levels of formal education, who earn low wages—and do not have ESI at their jobs. But even middle-class and well-educated foreign-born noncitizens are more likely to be uninsured than their native-born counterparts. Although almost 40 percent of noncitizen residents have middle-class incomes, more than a quarter of them are uninsured....

Underinsured People In addition to the major groups of people without any health insurance, there are a substantial number of people who are underinsured—that is, although they have health insurance, it is not adequate in terms of protecting them against catastrophic medical expenses. People can be underinsured for a variety of reasons, including that they have out-of-pocket medical expenses that exceed 10 percent of income (or 5 percent of income for a low-income person, with income below 200 percent of the poverty level), or deductibles that are greater than 5 percent of income. A recent analysis of people with health insurance policies estimates that sixteen million adults nineteen to sixty-four years of age were underinsured in 2003.[15] As more people have health policies with annual deductibles upwards of $5,000, the number of underinsured is sure to increase. The underinsured are sheltered from some medical care costs, but since they are not protected against catastrophic costs they face the same financial and health risks as the uninsured.

PRINCIPAL REASONS PEOPLE ARE UNINSURED

Decline in Employer-Sponsored Health Insurance—and a Signal Shift in Employer Attitudes About ESI

The large increase in the number of uninsured between 2005 and 2006 reflects a now almost decade-old decline in the percentage of people with ESI. In 2000, 68.3 percent of the population younger than sixty-five years of age had employer-sponsored coverage: in 2006, the fraction was 62.9 percent.[16] During this same time period, there was a steady decline in the fraction of firms that sponsor health insurance (from 69 percent in 2000 to 61 percent in 2006).[17] In practice, the shrinking of ESI coverage is greater than these statistics suggest. Many firms that offer ESI to "regular" employees are increasingly using workers hired through contract houses (often known as contract workers) and temp agencies, and other self-employed people who work on specific tasks for long periods of time. When companies hire people in these ways, the workers are not technically employees and are not included in the fringe benefit plans the firms offer.[18] Younger adults are particularly likely to be employed as contract workers, which helps explain their significant representation among the uninsured....

The decline in the proportion of firms offering ESI and the increased use of contract workers are warning signs that it is unlikely that ESI will continue to cover a majority of Americans. These signs indicate the extent to which companies are moving to limit their financial risk of increases in health care costs over which they do not have direct control. For employers, the fastest rising labor cost has been health insurance. Since 1996, premiums for ESI (both actual policies purchased from insurers and premium equivalent costs for self-insured plans) have grown every year; the increases reflect the doubling of health care spending since 1996. Between 2001 and 2007, premiums for firms with more than three employees increased 78 percent—outpacing general inflation, which rose 17 percent.[19]

These premium increases occurred in spite of most employers shifting more out-of-pocket costs onto the workers in the form of higher deductibles and co-payments, and implementing more restrictions on pharmaceuticals and mental health benefits. These cost-shifting moves helped reduce the rate at which premiums rose but they also shifted more medical care costs onto people who became sick. In addition, although the average share of premiums paid by employees (16% for single policies and 28% for family policies) have remained constant since 1999, there is some evidence to suggest that workers paying the highest shares are those with lower incomes.[20] A major survey of ESI premiums recently found that workers in firms with a high proportion of low-wage workers (35 percent or more of workers earn less than $21,000 a year) pay a higher share of family policy premiums than do workers in firms with lower proportions of low-wage workers.[21] Although the rate of increase in premiums between 2006 and 2007 (6.1 percent) was the smallest since 1999, it was still double the general rate of inflation.

Companies' willingness to sponsor group coverage is increasingly strained by the fact that health insurance costs have been rising faster than inflation and labor productivity. In a world of increased global competition, firms are moving to reduce their financial exposure to costs over which they have little control—and health insurance is high on that list. Policy analysts have long focused on small firms because a much smaller fraction of them offer ESI to their employees compared with large firms; for example, only 36 percent of firms with less than ten employees offer ESI while 95 percent of firms with more than one hundred employees do.[22] But now large firms are raising questions about how much they should be contributing toward health insurance. The CEO of General Motors, Rick Wagoner, has suggested that most of the costs of people with catastrophic medical expenses should be shifted to a federal government program, much like the costs of a natural disaster have become a federal responsibility. This shift in large companies' attitudes toward the cost of ESI, which comes on top of the increased use of temporary and contract workers in order to avoid including them in fringe benefits, deserves attention. It indicates that ESI as we have known it is not something to be counted on in the future.

High Premiums for Individual (Nongroup) Insurance—Countering the Risk of Adverse Selection

People who do not have access to ESI have only one choice for purchasing health insurance: the individual insurance market in their state—the market in which insurers sell policies covering individuals (and their dependents) rather than policies covering groups of people. Individual insurance is far more expensive than ESI because insurers face the risk that a disproportionate number of people who want to purchase individual policies are at higher risk of having high medical costs than the general population. This risk is known as adverse selection. As a result, premiums for individual policies can cost between $100 and $400 per month depending on a person's age and which state he or she lives in, and the less expensive policies come with a deductible of $1,000 or $2,000; premiums for family policies typically cost more than $700 per month and have a deductible of $5,000 or more.

In spite of the fact that many younger adults do not have ESI and are good candidates to purchase individual coverage, the rise in health care costs has driven up both premiums and the risk of adverse selection in the individual market. There is a catch-22 vicious cycle about this. Increasingly, the people who purchase individual insurance are forty-five and older—ages when health care spending tends to increase. It is not uncommon for those who are older or have medical conditions to face premiums in the individual market of $12,000 or more per year or to be offered policies that do not include care related to their conditions. Those who are younger and healthy also generally face premiums that are high relative to what they think their health care costs are likely to be because insurers expect that adverse selection is occurring also among the younger adults. The result is that individual policies are unattractive and unaffordable to younger adults. Even those who are earning middle-class incomes may decide that any "normal" medical care they might use would cost less than the premiums they would face in the individual market.

Unaffordability of Health Insurance for Lower Income People

In addition to individual market policies being generally quite expensive, almost half of the uninsured (22.8 million) are people with annual incomes below $30,000. Among these low-income uninsured are 4.4 million children, many of whom may be eligible for Medicaid and SCHIP. Some of the 18.4 million uninsured adults with incomes below $30,000 may have access to ESI but the evidence is that only a small fraction of people (around 3 percent) who are offered ESI turn it down and are uninsured.[23] But most likely, the reason most of these lower income people are uninsured is that they do not have access to ESI and simply cannot afford to purchase insurance in the nongroup market. Premiums upwards of $8,000 per year for family policies on incomes of $30,000 or less would account for more than a quarter of the family's income—a fraction that most of us would consider totally unaffordable.

HOW CAN WE REDUCE THE NUMBER OF UNINSURED?

Near-Term Route

Expand Public Programs to Cover All Poor and Near Poor As noted earlier, a quarter of the uninsured—11.5 million people—had incomes below the poverty level in 2006. They represent a third of all poor people; a major share of the poor who do have health coverage are covered by Medicaid and a small share have ESI. Most of the uninsured in poverty are children whose parents have not enrolled them in Medicaid (for which they are eligible) and adults who are not eligible for Medicaid and do not have ESI. Uninsured people with incomes below the poverty level simply cannot afford to purchase health insurance on their own. They will be insured only if a public program is created to cover them or if Medicaid eligibility is expanded to include them.

Another 13.6 million of the uninsured (29 percent) are near poor, with incomes between one and two times the poverty level. Most people with incomes below two times the poverty level live in families with no more than three people, so $32,000 is approximately the top income for most of the near poor. Near poor uninsured adults generally have hourly wages not much above the minimum wage ($10 per hour is equivalent to $20,000 per year if a person works a forty-hour week for fifty weeks). People with incomes below $32,000 cannot afford to buy family coverage in the individual market: and anyone who has had a health problem in the past or who is between forty-five and sixty-four years of age is highly likely to be rejected for coverage or to face individual premiums of more than $1,000 per month. Thus, the only way they will gain coverage is if there is a public health coverage program or a highly subsidized program that involves private insurance plans. For example, a highly subsidized program like SCHIP that involves private insurance plans might be attractive to near-poor people even if they had to pay a nominal premium to enroll....

Use Tax Credits to Subsidize People Who Do Not Have Access to ESI and Buy Policies in the Individual Insurance Market People who do not have access to ESI where they work and are not self-employed are not eligible for a tax subsidy of their premium payments. This is in direct contrast to the tax code subsidization of health insurance for people who have ESI or are self-employed. People whose employers offer ESI do not pay payroll or income taxes on the share of premiums that employers pay.[24] Similarly, people who are self-employed may deduct the full amount of the premium they pay for health insurance from their income (so long as the premium does not exceed the income) before determining their income tax. This means that the self-employed are able to buy health insurance with pretax income (that is, income that is not subject to income or payroll taxes). Thus, there is a large tax advantage in receiving part of compensation in the form of employer contributions to health insurance premiums or being self-employed and purchasing health insurance.

The tax code treatment of employer payments for health insurance reduced federal revenues by more than $200 billion in fiscal year 2007.[25] (As a point of comparison, the revenues foregone by the tax code subsidy for ESI amount to more than the federal government's share of Medicaid expenses in fiscal year 2007.) Moreover, the tax code subsidization of ESI is more valuable to higher income people because at higher incomes, the marginal income tax rate is higher: not having to pay that tax on the employer share of ESI provides a large subsidy. When the tax subsidy is taken into account, the Tax Policy Center of the Urban Institute and the Brookings institution estimates that the after-tax premium is on average equal to 6.8 percent of income for all tax units that have health coverage—but it is less than this for all tax units with incomes above $100,000.[26] Anyone with an income between $200,000 and $500,000, for example, is estimated to have an after-tax premium of only 2.7 percent of income, while someone with an income between $30,000 and $40,000 has an after-tax premium equal to 13.9 percent of income.

We could extend this tax advantage to anyone who does not have access to ESI and is not self-employed but earns enough to pay income taxes. The simplest form of this strategy consists of tax credits, where a person's taxes are reduced by some amount of money if the person purchases health insurance. One version of tax credits would have individuals receive a $1,000 credit and families a $2,500 credit for purchasing coverage. In his 2007 State of the Union speech. President Bush went much farther—he proposed abolishing the current tax code treatment of ESI and replacing it with a tax credit system of $7,500 for individuals and $15,000 for families.[27] The advantage of tax credits over permitting people to deduct the premium from their income is that tax credits do not encourage people to buy overly generous policies: tax credits are the same for people regardless of which policy they buy....

Reduce the Risk of Adverse Selection As noted earlier, a major reason premiums are high in the small-group and individual insurance markets is that there is a potential for adverse selection in those markets. Since the risk of adverse selection is what causes insurers to charge high rates in the small-group and individual markets, strategies are needed to reduce the risk to insurers of having a disproportionate number of very-high-cost enrollees.

One such strategy involves three policies that would work best if they were jointly implemented: merge the small-group and individual markets; create a government-sponsored reinsurance program to spread the costs of people with the very highest medical costs among all taxpayers; and require everyone to purchase a minimum level of health insurance.

Merge Small-Group and Individual Markets In most states, the number of people with small-group insurance is substantially larger than the number with individual coverage.[28] This is consistent with insurers' concerns that the individual market is a high-risk market where young, healthy people do not purchase coverage. The result is that premiums are particularly high in the individual market, providing a further disincentive for young and healthy people to buy policies....

By itself, however, a policy of merging the individual and small-group markets may reduce premiums by only 10 to 15 percent for people who previously had to buy policies in the individual market.[29] Given what we know about how price sensitive people are when it comes to buying health insurance, a decline of 10 to 15 percent may not cause a significant number of uninsured individuals to purchase policies. Thus, a merger of these two markets needs to be one part of a multipart strategy to reduce adverse selection.

A Government-Sponsored Reinsurance Program for the Small-Group and Individual Markets A more direct way of addressing insurers' concern with adverse selection is to create a government-sponsored reinsurance program for the small-group and individual insurance markets.[30] Reinsurance is essentially insurance for insurers—either for the possibility that some individuals will have very expensive medical care or that a group of individuals will together have medical expenses above some threshold. The former type—known as excess-of-loss reinsurance—is particularly suited to addressing insurers' concern with adverse selection because it provides protection against the possibility that some individuals may have very high costs. A government-sponsored reinsurance program would pick up a large share (anywhere from 50 to 95 percent) of the costs for people who have annual medical spending that qualifies for the reinsurance. The reinsurance could cover people with medical costs in the top 1 or 2 percent of the population—or it could pick up a large share of individuals' annual costs in a limited slice (or corridor) of health care expenses, perhaps between $30,000 and $100,000, for example. Note that reinsurance does not cover all of the expenses above the threshold for when it kicks in: the original insurance plan is still responsible for a share of the costs. That share can be 50 or 10 or 5 percent, and it can have different values for different corridors of the distribution of health care spending. Because the original insurer retains responsibility for a share of a person's costs, it has a strong incentive to manage the care of the people who are very sick, thereby reducing unnecessary care and expenses.

Reinsurance would immediately bring down insurers' risk of having a disproportionate number of very-high-cost people. With less risk, the insurers can then lower the premiums. The Healthy New York program (restricted to people who have incomes below 250 percent of the poverty level or who work in small firms and earn less than $35,000) has premiums that are about half of what they are in the standard individual market in New York .[31] The premiums are much lower than might be expected on the basis of knowing that people in the top 1 percent of the medical expenditure distribution account for about 30 percent of all medical spending in the United States. The larger reduction in Healthy New York premiums seems to be because the risk to insurers of having a large number of people with very high costs is reduced by the reinsurance.

The biggest hurdle in considering a reinsurance program is its costs, especially if it were to cover almost all of the costs of people whose medical expenses put them in the top 1 or 2 percent of the population. The program costs could be reduced, however, if the reinsurance were structured to cover a particular slice of the costs of higher cost people or if the original insurer had responsibility for more than 5 percent of the costs. In Healthy New York, for example, the reinsurance covers 90 percent of the annual costs of people whose expenses are between $5,000 and $75,000; expenses below $5,000 and above $75,000 are fully the responsibility of the HMO. There are many ways to structure a reinsurance program and the costs of various alternative structures could be estimated in order to choose one that balances a need to reduce premiums and be cost-effective. It also would be easier to administer a reinsurance program for a merged small group and individual market than for the two markets separately.

Require People to Buy Health Insurance The third policy to reduce adverse selection is a requirement that everyone buy at least a minimum level of health insurance. (People with low incomes would be subsidized, as noted above.) The only way to reduce the threat of adverse selection for a market as a whole is to require that people purchase coverage. This is why insisting that people have at least a minimum level of health insurance is gaining attention among policymakers and the public—it is the surest way to bring younger, healthy people into the pool of people covered by policies sold in the small-group and individual insurance markets....

These near-term policies are useful for helping people who currently do not have health insurance. But they do not address the increasing threat that employers will limit their financial contributions to ESI and perhaps their role even in sponsoring group coverage to employees. I turn next to a second course of actions and policies that do that.

Long-Term Restructuring of Financing and Organizing Health Insurance

... The shift in companies' attitudes about their role in offering ESI offers an opportunity for rethinking how we might achieve universal coverage. In particular, we need to plan for major changes in how health insurance is financed and how people could obtain coverage.

Restructure Financing of Health Insurance Two principles should be put forth at the outset of considering how we might restructure the financing of health insurance in the United States. First, employers gain great benefits from having both a healthy workforce and a healthy pool of potential workers. Because of this, employers and companies should continue to pay part of health insurance costs in the United States. Second, everyone should be in the new system so the financing will take account of the fact that lower income people cannot pay as much as higher income people. How the full cost of health insurance is divided up among individuals and companies, and among individuals by income, then depends on what we know about incentives embedded in various types of taxes and the distributional effects of different ways we might structure such taxes and premium payments.

Other countries have wrestled with how to structure the financing of their health insurance and they offer examples for the United States. In particular, several European Union countries have revised their health insurance systems within the last decade as part of their efforts to slow the growth in health care spending and in response to the EU's requirement that health insurance be portable across country borders. Most of the countries rely on a mix of individual and employer premiums or taxes to pay for the insurance, and many countries have a mix of private and public health insurance plan options. The Netherlands and Germany offer two examples.

The Netherlands altered its health insurance system just two years ago (January 2006). All adults pay a nominal premium (about 1050 euros in 2006, or about $1,500 using late 2007 exchange rates between the dollar and euro) that depends on which of thirty-three private health plans the adult chooses.[32] In addition to the premium, employers pay a payroll tax of 6.5 percent on employees' income up to 30,105 euros in 2006 (about $43,000)—or a maximum of about 2,000 euros ($2,900) per year.[33] Self-employed people and retirees pay 4.4 percent of their income. Low-income people can apply for a subsidy, which is dependent on a person's income; about 30 percent (5 million people) of the population receive such subsidies.

Germany requires that all people with annual incomes below about 47,250 euros ($67,570 using late 2007 exchange rates) participate in a public insurance system; about 10 percent of the population is exempt from the social system. The social system has a basic benefits package for which employees pay 7.5 percent of their salary (up to 47,250 euros) and employers pay 6.6 percent of their workers' salaries (up to 47,250 euros) for insurance. Children and nonworking spouses are covered "free of charge"; there are no distinctions between individual and family policies. However, in the private insurance system where people can purchase supplementary coverage (chiefly to cover private hospital rooms and perhaps greater choice of health care providers), family members are charged separately. Private insurance premiums are set to reflect the expected risk of individuals—and the insurance companies can turn down people for coverage (unlike the social insurance system).

The United States could restructure its financing of health insurance along lines similar to those used by the Netherlands, Germany, and others in the EU and the OECD. We could create a mix of financing sources that includes individual premiums and income-related taxes. The individual premiums could be relatively modest and would depend on which health plan was chosen. The income-related taxes could apply to all of a person's income or there could be a cap on the income which is taxable (as with the Social Security tax). Since we also want companies to contribute to the health insurance financing, companies could be required to pay a payroll tax that would be dedicated to health insurance. Such a financing structure also could be used to eliminate separate programs for the poor. In Europe, lower income people are offered the same choices at least for basic coverage as higher income people. Their payments are subsidized by higher income people and companies, but they are not shifted into a separate program like Medicaid.

Reorganize How We Choose Health Plans With employers less likely to sponsor health insurance, we will need an alternative arrangement for how we choose health insurance. Recall that in this new world of reorganizing how we finance health insurance and choose among health plans, everyone would be covered and the only difference in what anyone would pay would be the nominal per-person premium—which would depend on the health plan chosen. How we choose health insurance plans could still involve ESI: many employers may be willing to sponsor insurance options as a way of attracting and retaining employees. But for people who do not have employers that sponsor group policies, the pooling mechanisms described earlier would be the most efficient way of offering health plan options from which people could choose. The pooling mechanism agency would set standards for a minimum basic benefits package that every policy must cover and require insurers to accept all applicants. In exchange for requiring insurers to accept all applicants, the pooling agency also would establish an adjustment to the premium based on a person's "risk" of needing higher cost medical care so the health plans would not have an incentive to provide less than optimal care for high-cost people. The risk adjustment would be funded out of revenues raised by taxing individuals and companies. The Dutch have such a risk-adjustment system and risk-adjustment models have been used in conjunction with Medicare and Medicaid for more than a decade....

Slow the Growth in Health Care Spending For the past decade, health care spending per capita and health insurance premiums have been rising faster than both general price inflation and productivity growth. The gulf between the growth in insurance costs and productivity is why many employers have opted out of sponsoring group health benefits and caused those that do offer ESI to increase employee cost sharing so workers may have more incentive to hold down health care spending. Although there may be benefits to the increased spending on health care,[34] it also is clear that health care cannot continue to absorb an ever-larger share of families' budgets or federal and state budgets. There are opportunity costs to spending on health—such spending is crowding out other priorities such as education, repairing and expanding infrastructure, and national security.

By the same logic, the United States will not be able to afford to provide everyone with a basic level of health insurance unless the growth in health care spending is brought under control. Two types of health care spending are driving the overall growth in spending. One type includes expensive but not super-high-cost diagnostic, surgical, and pharmaceutical services that are no longer considered high risk so more people are taking advantage of them, especially because insurance covers their costs. Hip replacements and chemotherapies for some cancers are examples of this driver of health care spending. The other type consists of very new and very expensive diagnostic and treatment options for cancers, chronic conditions, and acute traumas. New radiological scanning devices and new forms of radiation surgery such as the

Cyberknife® are examples of the technical innovations that are capable of saving some people's lives. However, they are enormously expensive—and the money spent on such machines, as well as the teams of people needed to manage them, is money that cannot be used for other purposes.

To control the growth in health care spending, there will need to be limits on what medical services are included under basic health insurance coverage. As a society, we may want basic insurance to cover the first type of expensive services, especially since they improve the quality of life for large numbers of people. But the second type of health care spending raises issues of how far we want to go with basic insurance coverage. Cost-sharing mechanisms, such as deductibles and co-insurance or co-payments, can help make people think about the costs of health care but they have almost no effect on the treatment choices of people who are very sick and have expenses above $10,000 deductibles. Slowing the growth in health care spending ultimately will not be possible unless the basic benefits package excludes treatments that are enormously expensive and not cost-effective. A new and independent federal agency is needed to determine the cost-effectiveness of new diagnostic tests and medical therapies, and only those deemed cost-effective should be covered by the basic health plans. Limiting what basic health insurance pays for is unavoidable if we are to control health care spending.

As long as we do not have everyone covered by a basic level of health insurance, it is relatively easy for the health care system to face very few limits on what is covered by insurance. The result is what we have seen for the past four decades: an increasing fraction of the population without health insurance while those with insurance have both access to excellent health care and relatively low out-of-pocket expenses for such care. A new financing and organizational structure that ensures that everyone has basic health insurance would enable us to slow the growth in health care spending....

ENDNOTES

1. Author's analysis of the *March Current Population Survey*, released by the U.S. Census Bureau. August 28, 2007.
2. Kaiser Commission on Medicaid and the Uninsured, *In Their Own Words: The Uninsured Talk About living Without Health Insurance*. September 2000, www.kff.org/uninsured/loader.cfm?url=/commonspot/security/getfile.cfm&PageID=13470.
3. See especially J. Hadley, "Sicker and Poorer: The Consequences of Being Uninsured," *Medical Care Research and Review* 60. no. 2 Supplement (2003): 35–755; and Institute of Medicine, *Care Without Coverage: Too Little, Too Late* (Washington, DC: National Academy Press, 2002).
4. J. Hadley, "Insurance Coverage, Medical Care Use, and Short-term Health Changes Following an Unintentional Injury or the Onset of a Chronic Condition," *JAMA* 297, no. 10 (2007): 1073–84.
5. Ibid.
6. E. Ward, M. Halpern, N. Schrag, V. Cokkinides, C. DeSantis, P. Bandi, R. Siegel, A. Stewart, and A. Jemal, "Association of Insurance with Cancer Care Utilization and Outcomes," *CA: A Cancer Journal for Clinicians* 58, no. 1(2008) 9–31.
7. K. Pollitz, E. Bangit, K. Lucia, M. Kofman, K. Montgomery, and H. Whelan, *Falling Through the Cracks: Stories of How Health Insurance Can Fail People with Diabetes*, American Diabetes Association and Georgetown University Health Policy Institute, February 8, 2005, http://web.diabetes.org/Advocacy/healthresearchreport0505.pdf.
8. J. M. McWilliams, A. M. Zaslavsky, E. Meara, and J. Z. Ayanian, "Impact of Medicare Coverage on Basic Clinical Services for Previously Uninsured Adults," *Journal of American Medical Association* 290 (2003): 757–64; J. Z. Ayanian, J. S. Weissman, E. C. Schneider, J. A. Ginsburg, and A. M. Zaslavsky, "Unmet Health Needs of Uninsured Adults in the United States," *JAMA* 284 (2000); 2061–69; D. W. Baker, J. J. Sudano, J. M. Albert, E. A. Borawski, and A. Dor, "A Lack of Health Insurance and Decline in Overall Health in Late Middle Age," *New England Journal of Medicine* 345 (2001):

1106–12; J. M. McWilliams, A. M. Zaslavsky, E. Meara, and J. Z. Ayanian, "Health Insurance Coverage and Mortality among the Near-Elderly," *Health Affairs* 23, no. 4 (2004): 223–33.

9. J. M. McWilliams, E. Meara, A. M. Zaslavsky, and J. Z. Ayanian, "Use of Health Services by Previously Uninsured Medicare Beneficiaries," *New England Journal of Medicine* 357 (July 12, 2007): 143–53.

10. The survey is the Current Population Survey (CPS), which is conducted every month and surveys more than fifty thousand households across the country. The CPS is the source for monthly estimates of the unemployed numbers. Every month (except December), additional questions are asked about specific topics, and since 1980, questions about health insurance have been included in the March supplemental survey. The questions related to health insurance coverage ask about specific types of health insurance (e.g., employer-sponsored coverage, Medicare, Medicaid, self-purchased policies), and only people who answer "no" to every question are considered uninsured. This means that someone who may have been uninsured for part of the previous year but had employer-sponsored coverage in the other months is not counted as uninsured.

11. People can be uninsured for quite different lengths of time. Some people lose health insurance when they lose a job and then gain it again it they get a job with coverage—this could take only a month or several months. Other people may be uninsured for years because they are in jobs that do not have group health insurance as a benefit and they cannot afford to purchase their own individual policies. Studies of spells without health insurance show that about half of them last no more than six or seven months—but another quarter of all spells go on for a year or more; see K. Swartz, J. Marcotte, and T. D. McBride, "Spells without Health Insurance: The Distribution of Durations when Left-Censored Spells are Included," *Inquiry* 30, no. 1(2003): 77–83. Several studies using surveys that followed people for two or four years show that between 30 and 38 percent of the population younger than sixty-five were uninsured or at least a month, and 9 to 12 percent were always uninsured; see P. F. Short, D. R. Graefe, and C. Schoen, "Churn, Churn, Churn: How Instability of Health Insurance Shapes America's Uninsured Problem," *Issue Brief* (New York: Commonwealth Fund. November 2003); P. F. Short and D. R. Graefe, "Battery-Powered Health Insurance? Stability in Coverage of the Uninsured," *Health Affairs* 22. No. 6 (2003): 244–55; K. Kilen, S. Glied, and D. Ferry, "Entrances and Exits: Health Insurance Churning, 1998–2000." Issue Brief (New York: Commonwealth Fund. September 23. 2005). These numbers bracket the estimate that almost 18 percent of the population under sixty-five were uninsured in 2006 and illustrate how the dynamics of health insurance coverage affect the estimates of how many people are uninsured. If we restrict our focus-to people who are long-term uninsured, we estimate a smaller number of uninsured; and if we include anyone who has gone at least a month without coverage over a year's time, we estimate a larger number.

12. Households are comprised of single individuals as well as married-couple families and related families. As a result, 56 percent of all individuals in 2006 were middle class—they were either individuals or in families with incomes greater than the median household income.

13. In 1979, 10.8 million children were uninsured; they accounted for 17 percent of all children and 39 percent of the uninsured population under age sixty-five.

14. The Omnibus Budget Reconciliation Act of 1987 expanded Medicaid eligibility such that now states must cover all children ages six to eighteen with family incomes below the poverty level, and all children under age six with family incomes below 133 percent of the poverty level. States have the option of covering children with incomes up to 185 percent of the poverty level.

15. C. Schoen, M. M. Doty, S. R. Collins, and A. L. Holmgren, "Insured But Not Protected: How Many Adults Are Underinsured?" *Health Affairs* Web Exclusive W5 (June 14. 2005): 289–302. http://content.healthaffairs.org/cgi/reprint/hlthaff.w5.289v1. See also earlier estimates of the underinsured population: P. J. Farley, "Who Are the Underinsured?" *Milbank Quarterly* 63. no. 3 (1985): 476–503, and P. F. Short and J. S. Banthin, "New Estimates of the Underinsured Younger than 65," *JAMA* 274. no. 16 (1995): 1302–1306.

16. The percentage of the nonelderly population covered by ESI held steady in the mid-1990s (at 64.6 percent on average) and then increased in the

second half of the 1990s to a high of 68.4 percent in 2000 (P. Fronstin, "Sources of Health Insurance and Characteristics of the Uninsured: Analysis of the March 2007 Current Population Survey," *Issue Brief* No. 310 [Washington. DC: Employee Benefit Research Institute, October 2007]). The decline since 2000 is the longest since the CPS began to ask questions about health insurance coverage in 1980.

17. G. Claxton, I. Gil, B. Finder, B. DiJulio, S. Hawkins, J. Pickreign, H. Whitmore, and J. Gabel, *Employer Health Benefits*: 2006 *Annual Survey* (Menlo Park. CA: Henry J. Kaiser Family Foundation: Chicago. IL: Health Research and Educational Trust, 2006).
18. K. Swartz, *Reinsuring Health: Why More Middle-Class People Are Uninsured and What Government Can Do* (New York: Russell Sage Foundation. 2006).
19. C. Claxton, J. Gabel, B. Dijulio, J. Pickreign, H. Whitmore, B. Finder, P. Jacobs, and S. Hawkins, "Health Benefits in 2007: Premium Increases Fall to an Eight-Year Low, While Offer Rates and Enrollment Remain Steady," *Health Affairs* 26. no. 5 (2007): 1407–16.
20. Ibid.
21. Ibid.
22. Agency for Healthcare Research and Quality, Center for Financing, Access and Cost Trends, Medical Expenditure Panel Survey, 2003, Insurance Component, Table 1.A.2, http://info.ahrq.gov.
23. J. Hass and K. Swartz, "The Relative Importance of Worker, Firm, and Market Characteristics for Racial/Ethnic Disparities in Employer-Sponsored Health Insurance," *Inquiry* 44. no. 3 (2007): 280–302; P. J. Cunningham, *Choosing To Be Uninsured: Determinants and Consequences of the Decision to decline Employer-Sponsored Health Insurance* (Washington. DC. Center for Studying Health System Change. 1999); P. F. Cooper and B. S. Schone, "More Offers, Fewer Takers for Employment-Based Health Insurance: 1987 and 1996," *Health Affairs* 16. no. 6 (1997): 142–49.
24. The payroll taxes are primarily for Social Security and Medicare, which together equal 7.65 percent of earnings (up to $97,500 in 2007 for Social Security but with no maximum for Medicare). Employers also do not pay these payroll taxes on what they pay for health insurance premiums (or premium equivalents if the employers self-insure their workers' health care costs). In addition, workers do not pay income taxes on the amount of money that the employer pays for the premium; this particularly benefits higher income workers who face higher marginal tax rates.
25. L. Burman, "Taking a Check Up on the Nation's Health Care Tax Policy: A Prognosis," Statement before the U.S. Senate Committee on Finance, March 8, 2006.
26. Tax Policy Center, The Urban Institute and Brookings Institution, Table 107-0054: Current Law Tax Benefits for Health Insurance, Distribution of Subsidies by Cash Income Class, Nondependent Tax Units with Head or Spouse Under 65, February 6. 2007, http://www.taxpolicycenter.org/numbers/displayatab.cfm?DocID=1457.
27. See http://www.whitehouse.gov/stateoftheunion/2007/initiatives/healthcare.html. For an analysis of the proposed standard deduction for health insurance, see L. Burman, J. Furman, G. Leiserson, and R. Williams, *The President's Proposed Standard Deduction for Health Insurance: An Evaluation*, Urban Institute and Brookings Institution Tax Policy Center, February 14, 2007, http://www.taxpolicycenter.org/UploadedPDF/411423_Presidents_Standard_Deduction.pdf. Many people believe the Bush administration's proposal would encourage employers to stop offering ESI. Workers would then face higher premiums than they pay now because the advantages of group purchasing by employers, especially large employers, would no longer exist.
28. For example, in Massachusetts, before the health insurance reform legislation was passed in 2006, about 900,000 people were covered by small-group policies and only about 45,000 people were covered by individual market policies. Massachusetts has a labor market that provides strong incentives for small firms to offer ESI so the relative share of people covered by the two markets is probably more extreme there than in other states.
29. The commission in Massachusetts charged with evaluating the likely effects of merging the small-group and individual markets estimated that premiums for people who had been covered by small-group policies would go up between 1

and 1.5 percent and premiums for people who had been purchasing coverage in the individual market would be reduced by 14 percent.

30. Swartz, *Reinsuring Health.*
31. All of the HMOs in the state must participate in Healthy New York (HNY)—just as they must participate in the state's standard individual (self-pay) insurance market. The benefits package for HNY is leaner than that in the individual market, and people in HNY are subject to somewhat higher cost-sharing expenses. The reinsurance funds from the state's Tobacco Settlement Funds.
32. If a person chooses a plan with a deductible, the nominal premium will be lower. The highest deductible in 2006 was 500 euros (or about $650 at the exchange rate in 2006) per person. See the Royal Netherlands Government, Ministry of Health, Welfare and Sport, "The New Care System in the Netherlands," May 2006: http://www.minvws.nl/en/folders/z/2006/the-new-health-insurance-system-in-three-languages.asp.
33. Technically, the individuals pay the 6.5 percent tax on their income up to 30,150 euros, but employers are required to reimburse this amount to their employees.
34. D. M. Cutler and M. McClellan, "Is Technological Change Worth It?" *Health Affairs* 20, no. 5 (2001): 12–29.

DISCUSSION QUESTIONS

1. According to Katherine Swartz, there are roughly 47 million uninsured people in the United States. What are some of the key predictors of uninsured Americans?
2. Illegal immigrants constitute a significant portion of the uninsured in the United States, according to the author of this selection. What obligation does society have to uninsured individuals who are not citizens, especially those numbered as illegal immigrants?
3. As healthcare premiums increase, the number of employer-sponsored insurance plans is decreasing as "companies are moving to limit their financial risk of increases in health care costs over which they do not have direct control," this according to Katherine Swartz. What can be done about the problem of increasing healthcare premiums and the resulting shift in financial risk to workers *from* employers?
4. The Netherlands and Germany present a model of healthcare delivery that seems to be working much better than our system. What specific aspects of their healthcare delivery systems do you find noteworthy? What aspects of their healthcare delivery model would not work in the United States, and why?
5. *In light of evidence of rising increases in health care premiums and medical costs, the United States desperately needs a "public option," namely a single payer system modeled on the Canadian or British design.* Respond to this statement.

42

Protection of the Least Well-Off

CHARLES J. DOUGHERTY

The Four Questions

1. What problem does Dougherty identify? What values are violated?
2. Is this a *social* problem?
3. What causes this problem?
4. What does Dougherty suggest for addressing this problem?

Topics Covered

Health care
Moral responsibility
Rights
Inequality
Poverty
Racism
Social Darwinism
Role of government

FOUNDATIONS

... [A]n affirmative response to protect others from vulnerability may be the primary moral mechanism that leads individuals from their own narrow concerns into special moral relationships with others. Protection of vulnerability may be the moral motive that moves duty from focusing on the particular self toward concern for concrete others. It may be the same motive that drives morality toward universality toward concern for anyone's vulnerability. If so, then protection of the most vulnerable, those least well-off, should be an especially powerful moral value.

SPECIAL RIGHTS

This same point can be captured in the language of dignity and rights. Human dignity provides the grounds for a claim of universal and equal human rights. No person has more or less human dignity, an estimate of worth that is incalculably great in every case. Hence dignity is the basis at once for the special value placed on each individual and for the moral equality of all persons.

From this perspective, all human rights are the same. In addition to equal human rights, however, special rights are and ought to be recognized in special situations. Special rights, by law and by custom, are introduced into circumstances of marked inequality as a practical means of restoring or preserving equality. Children have rights that parents do not enjoy, students have rights that teachers do not have, and prisoners have rights that their jailers do not have. Children, students, and prisoners have these rights because members of these groups are so easily exploited in relations with parents, teachers, and jailers. The point is not that individuals in these groups have different standing as persons. They do not have more human rights. Rather, circumstances of unequal authority, knowledge, and power require special rights on the part of the more vulnerable party to ensure equality. Because of the general correlativity of rights and duties—one person's right creates another person's duty—parents, teachers, and jailers have special obligations in their relationships that correspond to the special rights of children, students, and prisoners.

This same point holds for patients. Doctors and other health care professionals have special obligations to patients, primarily a fiduciary responsibility to put the interests of their patients first. Corresponding to that duty are special rights on the part of patients. Hospital patients have special rights, for another example, rights not held by hospital administrators or by the doctors and nurses who work in hospitals.[1] Once again, these are not human rights in the most basic sense, but equalizing special rights designed to restore the balance inherent to universal human rights in characteristically unbalanced situations.

This point can be generalized and applied to any group that suffers a significant disadvantage in its relationship with others. Individuals have rights against society because of the graphic inequality between the individual and the group; this is especially clear in legal contexts. In criminal cases, for example, charges are expressed as the *United States versus Jane Smith* or the *People of the Commonwealth of Pennsylvania against Richard Jones*. Similarly, one can speak of minorities as having certain rights against the majority, rights designed to ensure protection in their vulnerable situation. Human rights protect our general vulnerabilities; special rights protect special vulnerabilities. The moral drive to protect the most vulnerable should thus be expressed as acknowledgment of their special rights.

HEALTH DISPARITIES

Some of the most marked vulnerabilities of people arise from the inequalities created by health status.[2] Some persons are born free of genetically inherited illness; others' genes doom or disable them from the start. Some are born into conditions supportive of healthy growth and development; others are thwarted by their surroundings. Some live and work in relatively safe conditions; others are subject to high rates of accident and injury. As a result of these differences, many Americans enjoy long, healthy lives; others have shorter life spans and greater morbidity.

The facts of these inequalities alone would be challenging enough to a moral framework committed to the equality inherent in human dignity, but other contemporary realities in the United States make the situation more challenging yet. Disproportionate suffering from high mortality and morbidity rates are linked systematically to other inequalities on the American scene.[3] Race, ethnicity, general socioeconomic status, and level of education are highly correlated with states of health. As a general rule, membership in a minority race or ethnic group, low socioeconomic standing, and fewer years of formal education create a strong likelihood of increased morbidity and shortened life span.

Not all the specific linkages involved are known; some that are known to some degree are not known clearly. Much remains to be learned, for example, about the power of genetic inheritance and the racial and ethnic factors at stake in that inheritance. It is not inconceivable that some of the racial and ethnic disparities in average birth weight or rates of hypertension may have a genetic basis or component. Yet some linkages are obvious even to the casual observer. It is very clear that minority racial and ethnic status overlaps in many cases with low socioeconomic status and fewer years of formal education. This means that whatever genetic factors may be involved in racial or ethnic health deficits, their effects are exaggerated by poverty and lack of education.

There are multiple links between poverty and illness, causation running both ways.[4] People living in poverty have less healthy diets, live in more polluted and dangerous surroundings (including areas with high levels of crime and street violence), have higher rates of many of the lifestyles and behaviors that trigger disease, and face substantial barriers in accessing timely and comprehensive health care. Lack of education can mean less knowledge about the prevention and causes of disease and less sophistication about the appropriate use of the health care system. Even where the relevant knowledge is available, poverty conspires against the ability to defer gratification. It encourages living for the moment because the future promises so little. This disposition militates against prevention and the development of positive health habits.

Many of these disadvantaging conditions can be traced more or less to racism and ethnic prejudice. Generations of slavery of African-Americans, for example, places this group at a unique disadvantage. Despite efforts of many to overcome racism, skin color remains a powerful trigger for prejudice throughout the United States. Continued prejudice based on ethnicity or linguistic and cultural differences certainly shapes the social conditions that affect the health status of many minority groups. There can be little doubt, for example, that bias against minorities accounts in part for the high rates of unemployment typical of minority communities.[5] High unemployment itself contributes to diminished health status through the other conditions it produces: poverty, drug and alcohol abuse, street crime, domestic violence, family disintegration, homelessness, and hopelessness.[6] It may be overly simple but not inaccurate to conclude that present inequalities in health status among racial and ethnic minorities are directly related to the relative poverty of their socioeconomic circumstances, which is itself linked to racial and ethnic prejudice in American life. Even admitting a role for genetic chance and a range of personal and social responsibilities, one clear line of causation of health-related inequality is rooted in a morally shameful dimension of America's past and present social realities.

Others fall into categories of exceptional vulnerability in health terms for reasons that are less clear. There has been an appalling increase in homelessness over the last several decades.[7] Since the mid-1960s, the proportion of public spending for children has declined while that for the elderly has increased.[8] Now the elderly are among the least poor of the age cohorts in the United States, whereas children are among the poorest. The frail elderly and those of any age who must rely on home assistance or institutional care continue to suffer for want of a rational long-term care system. Those with chronic mental illnesses are also among the most vulnerable. These Americans, along with members of minority racial and ethnic groups suffering from high rates of morbidity and mortality, are candidates for special rights designed to counterbalance the inequalities that they experience.

RELIGIOUS VIEWS

Put in terms of rights and duties, special rights for the least well-off create obligations on the part of relatively healthy, well-off Americans to design a system that takes the former's needs into consideration. The value of human dignity demands it.

Special concern for those who are least well-off can also be understood as a value in itself. Historically, Judeo-Christian societies have pointed with pride to special institutions and personal sacrifices made on behalf of the least well-off. There is more than one injunction in the Bible to the effect that a society is judged by how it cares for those least able to care for themselves.[9] Christianity has multiple historical links to the least well-off. It was born in the modest circumstances of its founder, became noteworthy (in part) because of the miraculous cures of some of the least well-off, and spread rapidly in its early years among the poor and among slaves.[10]

In the American context, many religiously based institutions in health care, education, and the social services focus their missions on special service to the poor and the otherwise marginalized. From this religious perspective, the vulnerabilities of poverty, sickness, and ignorance create a special spiritual opportunity, the potential for a link to the divine through service to the least well-off. Those providing care to these populations discharge a religious mission, one valued in itself.

This same point can be made in secular terms as well. It is a matter of moral common sense that fairness demands special advocacy for those least able to make their own voices heard. Wealthy, influential, and well-educated Americans will make their interests clear in any large social change, especially in debates about reform of the health care system. Middle class Americans, despite periodic feelings that "the system" is unresponsive to them, vote in large numbers and therefore have the attention of politicians who must seek re-election. Many of the least well-off, by contrast, are unsophisticated about politics and do not vote in large numbers. If their interests are to be defended effectively, others who are moved by a sense of fairness must champion their cause....

BARRIERS TO JUSTICE

There are serious inequities in the health care arena in the United States and strong religious and secular reasons for wanting to correct these inequities, for wanting to improve the situation of the least well-off. This conviction, however, is not persuasive to many Americans for several reasons. The most subterranean of these reasons is the persistence of racism and ethnic prejudice. Without putting it in so many words, perhaps even denying it were it made explicit, many Americans simply resist any efforts thought to be directed to the improvement of the lot of racial and ethnic minorities.[11] This view is obviously incompatible with the core of the Judeo-Christian tradition and the progressive ideals of the Enlightenment as well as with the specific theories of justice examined here. Nonetheless, racism is one of the abiding moral challenges in American life in general and in health care reform in particular. Little progress can be made in combatting it unless its influence is named, uncomfortable as that may be.

A second view, more respectable and therefore more frequently aired in contemporary political debates, although itself also deeply incompatible with the best in American religious and secular traditions, is a crude social Darwinism committed to a "dog-eat-dog" interpretation of society. This view is often rooted in a rugged individualism that denies the moral relevance of most social bonds.[12] Some versions recognize the significance of family and group connections but deny the relevance of moral obligations outside the immediacy of these intimate circles. This view is linked frequently with a heroic conception of success: Hard work inevitably brings success; poverty is the inevitable consequence of laziness and lack of self-discipline. On this account, there is no justification for sympathy for the plight of the least well-off because those living in disadvantaged conditions deserve the circumstances in which they live. Their poverty is an expression of their own lack of effort. The poor need only exert the requisite will power and concerted effort to pull themselves out of their disadvantaged circumstances. Failure to do so is a continual moral indictment, not of society, but of the least well-off themselves.[13]

A third cultural barrier to the conclusion that action on behalf of the least well-off is morally imperative is the most widespread. It may not be consistent with the dominant religious tradition, but it is consistent with some major themes in American secular traditions. In this view, the goal of greater equality for those living at the margins of society is admirable. The stark inequities in the United States should be softened or erased. The problem lies in appeal to government to achieve these goals. Whenever the government acts, it always entails unacceptable increases in taxation and regulation. Worse yet, government action to improve the worst conditions has proven itself to be ineffective, counterproductive, or both.[14]

In this view, the only acceptable weapons to combat the circumstances of the least well-off are voluntary efforts, education, and charity. Organized efforts on the part of government to provide services, to redistribute income, or to eradicate poverty have not succeeded and cannot do so. In the best cases, government efforts serve only to increase spending and swell bureaucracies. In the worst cases, the effect of government is to make its intended beneficiaries less self-reliant and therefore worse off than before government intervention. What is needed most is will power and self-reliance; what government creates is dependency.[15]

REMOVING BARRIERS

These are powerful barriers, but rejoinders can be made to each. Racial and ethnic prejudice is despicable and must be exposed for what it is. Although progress has not been nearly as rapid as justice demands, the United States has made important strides in this arena in the last several generations. The Civil Rights Movement ended racial barriers in the law and helped to make explicit public expressions of racism unacceptable culturally. The numbers of members of minority races and ethnic groups in positions of leadership and influence have grown throughout the United States. Genuine health care reform offers one of the few opportunities on the political horizon to make additional progress in this area in the near term. Creation of a system of universal access to high-quality health care for all Americans would be a clear and systemic rejection of racism. It would also be an important and symbolically powerful act of resistance to the slide toward "two Americas": an affluent, optimistic majority and an impoverished, alienated minority.[16]

With respect to the "dog-eat-dog" worldview, it is worth remembering the significant public subsidies that have supported the development and the delivery of health care since the end of World War II.[17] Substantial amounts of public money were spent on the expansion of the hospital sector and on efforts to increase the numbers of doctors. Massive subsidies from the tax system were used (and are still being used) to spread health insurance through the ranks of the middle classes. A great deal of money has been invested (and is still invested) in America's own federally run and tax-supported national health insurance program, Medicare. The rugged individualism assumed by the dog-eat-dog worldview is simply not applicable to health care. Furthermore, it is deeply incompatible with the basic orientation of caring that motivates the health care enterprise at its core. Health care is predicated on the dependence of people on one another. Nothing is more obvious in the face of serious or prolonged illness.

Finally, the lesson drawn by those who espouse the goal of improving the lot of the least well-off but reject the use of government as a means to that end, the lesson that government has shown itself inept, is simply overly broad. Certainly, some government programs have failed. Certainly, taxation and the spread of bureaucracy are substantive political issues that require debate and many prudential judgments; but the allegation that the federal government is simply incapable of creating a health care system that reduces or eliminates some of the most stark health status inequalities in the United States at an acceptable cost is belied by international experience. Nations around the world have accomplished this goal on their own terms.

No nation in the world is quite like the United States in every respect. Of course, this can be said equally about every other nation in the world. Nevertheless, many nations in the world share large portions of our national experience. Other nations have deep roots in the Judeo-Christian or the Enlightenment tradition or both. Others are immigrant nations or have marked racial and ethnic diversity. Others share our democratic politics and postindustrial economy. Others share our scientific-medical culture.

In virtually all the major trading partners of the United States who share some or all of these social realities, the government has erased the worst health care inequalities by creating universal health insurance coverage and by making special efforts on behalf of those who are least well-off in society. Not all of these international efforts succeed, nor are any of the successes transportable to the United States without adjustment for our American national experience and circumstances. Yet the fact remains that significant progress can be made in these areas because significant progress has been made elsewhere under reasonably similar conditions. Can health care reform be impossible for the United States, when it has been done with tolerable success by Canada, Japan, Germany, Great Britain, South Korea, Australia, France, Sweden, New Zealand, and others?

The view that the federal government is incompetent in the health care arena is also belied by domestic experience with the Medicare program. Since the mid-1960s, the United States has maintained a single national medical norm, a basic level of health care, for all elderly Americans regardless of medical conditions, socioeconomic circumstances, or state of residence. This achievement has strong political support, including the protection of the powerful American Association of Retired Persons (AARP). By and large, Americans regard Medicare as their own program, one that serves them and their families. There is little credible evidence that this age-linked national insurance program has created socially destructive dependencies or made people less concerned for their health because their care is insured. Certainly there is room for greater efficiency and cost-consciousness in Medicare. Experiments with managed care, new forms of cost-sharing (especially for the wealthy elderly), and other cost-cutting measures are appropriate. There is no widespread public sentiment, however, to "get the government out" of Medicare, privatize it, or send it to the states. For all its flaws, the federal government has succeeded with Medicare, succeeded in creating and conducting a comprehensive national program of health coverage despite having focused on the most expensive phase of the life span.

STRATEGIES FOR JUSTICE

What exactly could the United States do to eliminate, or at least to mitigate, some of the worst health care conditions and to improve the lot of the least well-off Americans? Three general strategies suggest themselves. First, a universal health insurance system must be devised that includes a basic benefit package comprehensive enough to attract and maintain the support of the middle class so that the poor and the middle classes would rely on the same package. Even if some Americans who are very affluent would buy more coverage, the political power and influence of the middle class would assure that the basic benefit package is comprehensive and that the services covered by the benefit package are of high quality. In short, health care reform should tie the fate of the poor to that of the middle class.

Second, whatever health system evolves out of the present changes, greater investment must be made to address public health needs and to fund health initiatives in poorer communities. It would not be sufficient to achieve universal coverage without also responding to the fact that some areas of the nation and some groups would enter universal coverage with significant health status deficits and with significant insufficiencies in their health care delivery infrastructures.

Finally, to address these deficits, mechanisms would have to exist in a reformed health care system to draw health care professionals and resources toward underserved areas in both rural regions and the inner cities, perhaps by creating enhanced reimbursements for providers in those areas. Efforts like the National Health Service Corps, which provides scholarships to medical students in exchange for future service in underserved areas, could be expanded.[18] Financial arrangements are needed to underwrite the ability of institutions in these communities to purchase needed medical equipment and to upgrade physical plants. This is especially imperative for academic health centers, which provide a great deal of uncompensated care in America's largest cities. Because of this financial drain and the costs of their educational programs, academic health centers are unwelcome partners in emerging health care networks. They are therefore at a significant disadvantage in a competitive environment.[19] As a general strategy, more must be done to support institutions and professionals who serve the underserved. Commitment to improving the lot of the least well-off means adopting affirmative action measures within a general commitment to the human equality dictated by respect for human dignity.

It is obvious that present changes in the health care marketplace will not and cannot achieve these goals. Competition and cost-cutting will not bring universal coverage or health care affirmative action. In fact, they are very likely to make the situation worse. Markets alone cannot serve the interests of the least well-off. Despite American reluctance to admit the point, the logic of markets, worldwide health care experience, and the success of Medicare indicate both the necessity and possibility of government action in this arena.

THE SICK AND DYING

There is one additional category of vulnerable Americans whose plight deserves special emphasis as the health care system evolves. Ironically, these are the sick and the dying, Americans who have been the traditional focus of the health care system.

This could change, and to their disadvantage. In the present climate, it is not uncommon for critics from all political leanings to condemn the entire health care system as a "sick care system."[20] These critics demand greater emphasis on prevention and wellness programs. Certainly, they have a point. Prevention and wellness are indeed understressed in the present system, and more should be done in this arena. Yet it is worth remembering that the primary clients for prevention and wellness programs are healthy people, those who are already well-off in health care terms.

At the same time, one of the most important and growing influences in the new health care marketplace is capitation. The main financial incentive at the heart of capitation provides a very strong temptation to enroll healthy, underutilizing populations and to avoid or disenroll sick, overutilizing populations. Healthy people pay their monthly fees and make no claims on the capitated network's resources. The sick and dying, by contrast, are loss leaders from a financial perspective; they drain resources from the network. Greater stress on prevention and the financial incentives of capitation therefore threaten the traditional commitment of the system to the care of those most in need, the sick and the dying.

There is much to be said in favor of both prevention and capitation. It is certainly true that American health care could do far more to prevent illness and to keep Americans well. All things considered, it is better to prevent illness than to perfect means of dealing with it when it occurs, but not all illness is preventable in every case, and dying is not preventable in any case. Regardless of preventive efforts, there will always be sick and dying patients in need of assistance from the health care system. In the rush to develop more wellness programs and to keep people out of hospitals, this important fact must not be forgotten. Regrettable though it may be, most people are not interested in health and health care when they are healthy. It is when they are sick, and especially when they are dying, that people need health care and the services of dedicated professionals and institutions.

There are many benefits to be expected from capitation, not the least of which is the incentive in networks to keep enrollees healthy, an incentive suggested by the term "health maintenance organization." Efforts today to keep enrollees free of disease tomorrow can serve the financial interests of a capitated plan. Capitation can also be expected to increase economic efficiency. Care for the most vulnerable, however, means that capitated networks must be carefully regulated to avoid abuses: disenrollment of those who need them most, use of preexisting illness exclusions, and adoption of marketing techniques that are designed to build a network profile of healthy underutilizing enrollees. Ways will have to be developed to curb these abuses, from simple legal prohibition of disenrollment and preexisting illness exclusions to risk-adjusted reimbursement and incentives or mandates to enroll high-risk populations.

In the years ahead there will be considerable experimentation with preventive programs and ways to control the negative side of capitation. Throughout this experimentation, the moral goal must be clear. A special value and a special obligation are associated with care of the most vulnerable, the least well-off. In health care, the primary focus must remain on care of the sick and the dying. They have a special right to health care, and society has a special duty to care for them.

ENDNOTES

1. See, generally, Charles J. Dougherty and Sandra L. Dougherty, "Moral Reconstruction in the Hospital: A Legal and Philosophical Perspective on Patient Rights," *Creighton Law Review* 14, no. 4, supplement (1980–1981): 1409–34.
2. Charles Dougherty, "Equality and Inequality in American Health Care," *Freedom and Equality* (Washington, D.C.: Federation of State Humanities Councils, 1992), pp. 6–17; and Lu Ann Aday, *At Risk in America* (San Francisco: Jossey-Bass, 1993).

3. Charles Dougherty, *American Health Care* (New York: Oxford University Press, 1988), pp. 3–19.
4. Ibid., p. 5.
5. *A Matter of Fact* 19 (July–December 1993) (Ann Arbor, Mich.: Pierian Press, 1994), 108–109.
6. J. Morris, D. Cook, and A. Shaper, "Loss of Employment and Mortality," *British Medical Journal* 308, no. 6937 (1994): 1135–39; D. Ezzy, "Unemployment and Mental Health: A Critical Review," *Social Science and Medicine* 37, no. 1 (1993): 41–52; and S. Wilson and G. Walker, "Unemployment and Health: A Review," *Public Health* 107, no. 3 (1993): 153–62.
7. *A Matter of Fact* 17 (July–December 1992) (Ann Arbor, Mich.: Pierian Press, 1993): 188.
8. Daniel Callahan, "Reforming the Health Care System for Children and the Elderly to Balance Cure and Care," *Academic Medicine* 67, no. 4 (1992): 219–22.
9. Generally, see Robert Murray, *The Cosmic Covenant: Biblical Themes of Justice, Peace, and the Integrity of Creation* (London: Sheed and Ward, 1992).
10. Paul Maier, *First Christians: Pentecost and the Spread of Christianity* (New York: Harper and Row, 1976).
11. See, e.g., Joseph A. Califano, "The Challenge to the Health Care System: Can the Third Biggest Business Take Care of the Medically Indigent? A Personal Perspective," in *Health Care for the Poor and Elderly: Meeting the Challenge*, ed. Duncan Yaggy (Durham, N.C.: Duke University Press, 1984), pp. 45–57.
12. Charles Dougherty, "The Excess of Individualism," *Health Progress* 73, no. 1 (January 1992): 22–28.
13. Majorie Hope and James Young, *The Faces of Homelessness* (Lexington, MA: Lexington Books), pp. 197–201.
14. On expanding government bureaucracies, see Robert Higgs, *Crisis and Leviathan: Critical Episodes in the Growth of American Government* (New York: Oxford University Press, 1987); and James Q. Wilson, *Bureaucracy: What Government Agencies Do and Why They Do It* (New York: Basic Books, 1989).
15. On the problem of dependency, see June Axinn and Mark Stern, *Dependency and Poverty: Old Problems in a New World* (Lexington, MA: Lexington Books, 1988); and Tamar Ann Mehuron, editor, *Points of Light: New Approaches to Ending Welfare Dependency* (Washington, D.C.: Ethics and Public Policy Center, 1991).
16. Fred R. Harris and Roger W. Wilkins, eds., *Quiet Riots: Race and Poverty in the United States* (New York: Pantheon Books, 1988); and Bill E. Lawson, *The Underclass Question* (Philadelphia: Temple University Press, 1992).
17. Paul Starr, *The Social Transformation of American Medicine* (New York: Basic Books, 1982), pp. 335–78.
18. V. Stone, J. Brown, and V. Sidel, "Decreasing the Field Strength of the National Health Service Corps: Will Access to Care Suffer?" *Journal of Health Care for the Poor and Underserved* 2, no. 3 (1991): 347–58.
19. J. Jonas, S. Etzel, and B. Barzansky, "Educational Programs in U.S. Medical Schools, 1993–1994," *JAMA* 272, no. 9 (1994): 694–701; and John K. Iglehart, "Health Care Reform and Graduate Medical Education," *The New England Journal of Medicine* 330, no. 16 (1994): 1167–71.
20. See, e.g., E. Bevis, "Alliance for Destiny: Education and Practice," *Nursing Management* 24, no. 4 (1993): 56–61.

DISCUSSION QUESTIONS

1. What do you believe a democratic society owes the individual in terms of medical care?
2. Why is this selection titled "Protection of the Least Well-Off"?
3. How does economic and racial inequality play a role in health care?
4. The author clearly favors "the creation of a system of universal access to high-quality health care for all Americans." What does this mean? Do you agree? What is the downside?

43

Social Policy in Health Care: Europe and the United States

JAMES W. RUSSELL

The Four Questions

1. What is the problem?
2. Is it a *social* problem?
3. What are the causes of the problem?
4. What does Russell recommend be done?

Topics Covered

Health care

Health insurance

Private insurance

Government and health care

Europe and the United States

"It is vital for people to be protected from having to choose between financial ruin and loss of health."

—WORLD HEALTH ORGANIZATION[1]

Each year eighteen thousand people die in the United States because of lack of health insurance coverage.[2] With forty-five million citizens lacking health insurance coverage and not counting many millions more whose coverage is inadequate, it is undeniable that the American system is seriously deficient compared to its European counterparts, where all citizens have full coverage. The problem goes further. The United States has the world's most expensive health care system (see Table 43.1). Americans pay an average annual cost of $5,724 per person for their health care, well over twice as high as the average Western European cost of $2,471.

If there is any area of social policy in which Americans are likely to be aware that their system has flaws, it is in health care. Americans generally acknowledge that their high and rising cost of health care is a problem and they are less satisfied with their health system than on average are Western Europeans. Only 40 percent of Americans are satisfied with their health care system compared to 57 percent of Western Europeans (Table 43.1).

But Americans are less likely to lay the cause on the privatized nature of their health system, assuming that medical care is expensive by its very nature. The majority of health care spending in Europe is publicly financed. In the United States the majority is privately financed (Table 43.1). The major actors in the American health system—insurance companies, pharmaceutical corporations, and physicians—derive the world's highest health-related profits and incomes, largely because they are immune to government regulation or control as they are in Europe. American physicians are the world's highest paid. On average they receive 5.5 times the income of average workers. In contrast, German physicians, the highest paid in Europe, receive only 3.4 times the average income of workers. In other European countries the income disproportions are less: Switzerland–2.1, France–1.9, Sweden–1.5, United Kingdom–1.4.[3]

SOURCE: Social Policy in Health Care: Europe and the United States by James W. Russell. From *Double Standard: Social Policy in Europe and the United States,* © 2006 Rowman & Littlefield.

TABLE 43.1 Health Care System Financing and Cost

	Percent Public Financing	Per Capita Cost	Percent Public Satisfied With Health Care System
Andorra	70.5	$1,908	—
Austria	69.9	$2,220	83
Belgium	71.2	$2,515	77
Denmark	82.9	$2,583	76
Finland	75.7	$1,943	74
France	76.0	$2,736	78
Germany	78.5	$2,817	50
Greece	52.9	$1,814	19
Ireland	75.2	$2,367	48
Italy	75.6	$2,166	26
Luxembourg	85.4	$3,066	72
Malta	71.8	$ 965	—
Monaco	79.6	$4,258	—
Netherlands	65.6	$2,564	73
Norway	83.5	$3,409	—
Portugal	70.5	$1,702	24
San Marino	79.2	$3,094	—
Spain	71.3	$1,640	48
Sweden	85.3	$2,512	59
Switzerland	57.9	$3,446	—
United Kingdom	83.4	$2,160	56–57
European average	*74.4*	*$2,471*	
United States	44.9	$5,274	40

SOURCE: World Health Organization. *The World Health* Report 2005 (Geneva: World Health Organization, 2005), Annex Tables 5, 6; Organization for Economic Cooperation and Development (OECD). *Health Data 2005* (Paris: OECD, 2005).

TABLE 43.2 Health Care System Quality and Outcomes

	WHO ranking	Infant Mortality	Life Expectancy
France	1	4	80
Italy	2	5	81
San Marino	3	—	81
Andorra	4	—	81
Malta	5	6	79
Spain	7	4	80
Austria	9	4	79
Norway	11	3	79
Portugal	12	5	77
Monaco	13	—	81
Greece	14	5	79
Luxembourg	16	5	79
Netherlands	17	5	79
United Kingdom	18	5	79
Ireland	19	6	78
Switzerland	20	5	81
Belgium	21	8	79
Sweden	23	4	81
Germany	25	4	79
Finland	31	3	79
Denmark	34	8	77
European Average		*4.9*	*79.4*
United States	37	7	77

SOURCE: World Health Organization. *The World Health Report 2000* (Geneva: World Health Organization, 2000), Annex Table 1 and *The World Health Report 2005*, Annex Tables 1, 2b.

Americans acknowledge that the growing number of uninsured persons is a problem. But they are less likely to be aware that their expensive system ranks poorly in quality as well as coverage. Two issues are involved in determining the quality of health care: the quality of a health care system and the average health conditions of a citizenry. The World Health Organization (WHO) ranked in order the quality of the health care systems of 191 countries. All twenty-one Western European systems ranked higher than the United States (Table 43.2).

American health conditions are inferior to those in Europe in terms of infant mortality and longevity, the two most common measures used in international comparisons. Of one thousand babies born live, seven die within their first year of life in the United States compared to an average 4.9 in Western Europe. The average American life span of 77 years is 2.9 years less than the average Western European life span of 79.4.

DEVELOPMENT OF NATIONAL HEALTH INSURANCE

Modern health systems began during the last part of the nineteenth century. Before then there was little professional health coverage for whole populations and no national plans to provide such coverage. Families cared for, and pooled their financial resources to provide for, sick members. Religious and other charities might be available to help the poor. In Europe and the United States, early labor unions set up sickness funds for members.

German Chancellor Bismarck established the prototype of national health insurance in 1883, in large part to counter the public support the socialist labor unions were gaining from the popularity of their sickness funds. Mandatory contributions from employers and employees funded the Bismarckian system. By the 1920s a number of European countries, including Belgium, Norway, and Britain, had followed the German example and set up similar models. None of the systems, though, were extensive enough to cover their entire national population nor were their benefits comprehensive enough to cover all situations. In 1930 social health insurance still covered less than half of the working populations of Europe.[4]

The Soviet Union, beginning in 1917, and post–World War II Eastern European communist countries were the first to establish comprehensive health systems that covered entire populations. Those accomplishments stimulated Western European leaders to establish their own comprehensive systems. In the same way that Chancellor Bismarck wanted to counter the influence of socialist labor unions, post-war Western European leaders sought to counter the influence and appeal of communist parties that were significant in a number of major countries, including Italy and France.

In 1946, following recommendations from the 1942 Beveridge Report, Britain passed the National Health Service Act. Sweden passed similar legislation the same year. By establishing centralized, government-owned health systems, the British and Swedish approaches differed from the Bismarckian social insurance approach. Both represented the first noncommunist approaches to universal health care coverage.

The state thus was the major actor in the development of the Western European health systems by either mandating establishment of social insurance funds tied to workplaces or establishing national health services. Social insurance offered mandated coverage to employed workers; national health services to all citizens. In time the social insurance systems would develop supplementary government-financed plans to extend coverage to unemployed and other sectors of the population ineligible for workplace-related insurance.

Following passage of the 1935 Social Security Act in the United States, several bills were introduced but never passed in Congress to extend Social Security provisions to health care for the entire population. American labor union leaders supported these attempts to establish national health insurance. But by the 1950s they adapted to the apparent reality that it was unlikely to be accomplished. Instead, health insurance became one of the issues for which they bargained with management. It became one of the advantages of union membership, since unionized workplaces were more likely to have health insurance as a fringe benefit. Health insurance development in the United States thus partially followed the Bismarckian model in that for the most part it was tied to employment as a fringe benefit. But unlike in Europe the United States government never established or mandated coverage for the workers or the population as a whole, leaving such coverage up to the outcome of private market-driven activities, labor union collective bargaining, and employer largesse.

If the European health systems were established by government action, the American system evolved out of patchwork private market opportunities around health insecurities and residual government actions that protected particular populations such as veterans, the elderly, and the poor.

COMPARATIVE HEALTH SYSTEMS

There are three health system models in Europe and the United States. They exist on a continuum between the most market driven and the most government controlled. At the two ends of the continuum are privately organized-for-profit health care that exists in the United States and national health systems—sometimes called the Beveridge model—that exist in the Scandinavian countries, the United Kingdom, Italy, Spain, and Portugal. In between but closer to the government pole, are the social insurance systems that exist in Germany, France, and the Benelux countries.

In the private health care system of the United States, health care is treated as a commodity that is sold to consumers on market principles. The four major health care actors—physicians, insurance companies, hospitals, and providers of medical goods, such as pharmaceutical companies—operate the private businesses. State involvement and regulation is minimized. Individuals purchase health care services as commodities according to how much they can afford. Those who are ineligible for or who cannot afford health insurance go without.

Most Americans who have health insurance have it as a fringe benefit in their employment. Having health benefits tied to employment is seen to American conservatives as a way to encourage work. Instead of being a right, health insurance is a reward reserved for those who work. It is treated as the commodity that is available only to those who perform adequately in the marketplace.

The obvious problem of tying health insurance to the workplace is that it will not cover the unemployed. There is no national requirement in the United States, unlike in Europe, that employers actually include health insurance as a fringe benefit. This has become readily apparent, as a number of employers have begun to cease offering insurance altogether as it has become increasingly more expensive or they've begun to require workers to pay out of pocket greater portions of the costs. In a number of cases, employers have begun to cease paying the insurance costs of retirees.

Unlike European social insurance models in which the state mandates that employers and employees contribute to a fund, there is no such state mandate in the United States. Employee health insurance exists only if employers voluntarily offer it to their workers, or workers extract it as a fringe benefit from them. Therein lies the reason why so many working Americans lack health insurance.

In the social insurance model (Germany, France, the Benelux countries), there are a multiplicity of different health funds attached to unions, cooperatives, charities, and private companies. Employees and employers share payments into different such social insurance funds. Social insurance models have obtained universal coverage by virtue of two features. First, they are mandatory. Employers must offer social insurance plans and employees must enroll in them. Second, for those who are unemployed or otherwise ineligible for workplace-related coverage, governments have compulsory tax-financed back-up plans.

National health services—in the Scandinavian countries, the United Kingdom, Italy, Spain, and Portugal—organize health care as a publicly controlled service much like the provision of public education. Individuals pay for health care services through their taxes and are eligible to receive them by virtue of being residents or citizens of the country, thereby ensuring universal coverage. Governments own hospitals and may directly employ physicians on salaries.

None of the country health systems, though, are pure representations of any of the models. All have complex combined features. All have state and private funding. Doctors on government salary coexist with doctors in private practice. Private insurance plans exist alongside social insurance. This has led analysts to conclude that while there are pure models of different types of health systems, in practice all countries combine features from the different models.[5]

The United States, which has the most privately organized system, also has government-financed Medicare insurance for the elderly and Medicaid insurance for the poor. All of the European countries with social insurance or national health systems also allow supplemental private insurance to be sold. If the United States uses government-based insurance to partially compensate for the failure of the private market to find it profitable to cover all citizens, European countries allow private insurance companies to make businesses out of providing coverage for gaps in national coverage. From the point of view of citizens, purchasing supplementary private insurance is a way of topping off their basic plans. The difference is that whereas the private market on its own cannot cover citizens who lack purchasing power, government plans could become more comprehensive if that became a budgeted priority depending upon political conditions.

Mixed systems in Europe in which private insurance is allowed a role are a compromise between state-administered provision of basic needs and the market. They ensure universal coverage while allowing freedom of choice of coverage according to the different purchasing powers of consumers. Mixed provision systems, though, inevitably compromise egalitarian access, since they allow more choices for basic necessities to those with higher incomes.

Switzerland provides an unusual example of a European country that has achieved universal coverage with mainly private insurance. But for that to be achieved, the Swiss government had to implement regulatory and supplementary actions that violate pure free market principles. It requires its citizens to purchase health insurance in the same way that states of the United States require their motorists to purchase automobile insurance. Government regulations then require that private insurance plans pool their funds so that risks are shared equally. That avoids companies' selling policies only to healthy citizens. Swiss national or canton governments then provide coverage for citizens who are unable to purchase private insurance.

REFORM POLICY

Two questions underlie the politics of health policy: what type of primary model—national health service, social insurance, or private—a country will have and what type of secondary features it will adopt as reforms. Reforms in themselves are neither good nor bad. While often being presented publicly as purely pragmatic responses to technical problems, reforms usually carry consequences for underlying social policy goals.

In Europe, the primary issue for health care reform is the extent to which privatization and market features will expand within the otherwise state-controlled or regulated health care landscape. It is common to argue in the United States that the generous European welfare state is no longer affordable in an era of globalized competition. That argument, though, is difficult to sustain for health care policy when Europe has achieved universal higher quality coverage at less than half the cost of the American privatized system. Those who would reform in the direction of the American privatized model seek either to reduce state spending or to increase opportunities for private health businesses.

The underlying question is where the line will be between socialized coverage that is available for all citizens and privatized coverage that is available for only those who can afford it. Private insurance companies and their ideological allies would like to expand the latter. However, the inevitable consequences of adopting more features of privatized health care systems in Europe will be increased health inequality.

In the United States, the two glaring problems are increasing coverage and containing costs. With forty-five million—15 percent of the population—without health care coverage, the immediate necessity is to devise a national plan that will expand coverage to the entire population. It is not just a question, though, of expanding token coverage—insurance that covers only a few health problems or in which there are substantial costs borne out of pocket by consumers. It must be coverage in which people are, in the words of the World

Health Organization, "protected from having to choose between financial ruin and loss of health."[6] At the same time, the out-of-control costs of American health care caused primarily by private profiteering must be reined in.

To obtain those goals, the United States could adopt features of either the European social insurance or national health system models. Most of the American health system is closest to the social insurance model since it is workplace based. The most obvious reform to expand coverage based on that model would be to make it mandatory as it is in Europe. All American workplaces would be required to have health plans. To contain the costs of those plans, the government would have to regulate and negotiate the permitted charges of insurance companies, physicians, and other providers. Socializing insurance risk pools could contain costs further.

There are a number of health care reform advocates in the United States who would prefer a Canadian-style single payer system in which the government establishes a tax-funded health insurance agency that covers all citizens. Such a centralized body would then be able to negotiate fees with physicians and other private providers. At the same time the single-payer system falls short of a European national health system model, since it allows private physicians and other providers.[7]

Kaiser Permanente, a nonprofit insurance company, provides a model that is close to a European national health system approach for reforming health care in the United States. Henry J. Kaiser, an eccentric capitalist who sought a cost-effective way to provide health care for his workers, founded it in 1945. Kaiser wanted to make profit off of the work of his workers, not their health care needs. Healthy workers were more profitable workers.

What made Kaiser Permanente unusual in the American health care landscape was that from its beginning it emphasized preventive care and placed its doctors on salary. This made it like a European national health service and controversial in the American context, causing it to be denounced as socialistic. But it was able to more efficiently deliver health care services at lower costs than its traditional market-driven counterparts.

Kaiser physicians received lower incomes than those in private practice. But the tradeoff for them was that they worked fewer hours and could concentrate on providing health care without being distracted by problems of running a business, as are physicians in private practice.

In the early decades of Kaiser Permanente, there were marked differences in the experience of having it rather than traditional health insurance. It cost less. It was centralized, usually in a Kaiser hospital. It was prepaid. Kaiser members simply showed their identification cards and were treated. In traditional private insurance plans, members would pay up front and then be reimbursed, requiring a mountain of forms to keep track of. In traditional plans, physicians often charged higher fees than what the insurance would pay, causing unpredictable and aggravating out-of-pocket expenses for patients. In Kaiser there were no extra charges.[8]

Kaiser's emphasis on not-for-profit preventive health care has produced a potential model for health care reform. Kaiser members are generally more satisfied with their plans than are members of for-profit insurance plans. As a result of its preventive practices, including monitoring and controlling blood pressure and cholesterol, Kaiser members in Northern California have a 30 percent lower death rate from heart disease than the general population.[9]

In the end any reform that will actually make a difference in terms of resolving the serious problems of the American health care system will have to expand the government sector. The most effective way but least likely of political success would be to establish a system in which citizens would pay a new dedicated tax, as Medicare is paid, to finance government health insurance instead of them or their employers continuing to pay for increasingly more expensive private insurance as they do now. Less effective but more politically possible, would be to expand the coverage of existing government programs such as Medicare. Either way, any reform will have to rein in current private profiteering.

ENDNOTES

1. World Health Organization, *World Health Report 2000* (Geneva: World Health Organization, 2000), 24.
2. Institute of Medicine of the National Academies, *Insuring America's Health* (Washington, DC: National Academies Press, 2004), 8.
3. Uwe E. Reinhardt, Peter S. Hussey, and Gerard F. Anderson, "Cross-National Comparisons of Health Systems Using OECD Data, 1999," *Health Affairs*, vol. 21, no. 3 (2002): 175.
4. Cited in Richard Freeman, *The Politics of Health in Europe* (Manchester: Manchester University Press, 2000), 26.
5. See Freeman, *The Politics of Health in Europe*, 5; and Robert H. Blank and Viola Burau, *Comparative Health Policy* (Hampshire, UK: Palgrave Macmillan, 2004), 22.
6. World Health Organization, *World Health Report 2000*, 24.
7. In June 2005, the Canadian Supreme Court ruled that the Quebec part of the national tax-based single-payer health care system was in violation of the province's bill of rights because it forbid private health insurance. Critics charged that the single-payer system was plagued by shortages of services. The purpose of the Quebec case was to allow private insurance companies to sell policies to cover the shortages. Defenders of the Canadian system correctly argue that introduction of private insurance will inevitably lead to health inequality between those who can and cannot afford the policies. Such shortages as do exist could be resolved by greater public funding without introducing health inequality.
8. Originally, private insurance companies simply passed the extra charges of physicians and other providers on to the consumers. In line with national efforts to rein in health care costs, private health insurance companies in the 1990s began to control physician and provider fees more closely. They directly contracted physicians into their networks. In return for receiving the business of the insurance plan's members, the physicians were required to charge only negotiated set fees.
9. Steve Lohr, "Is Kaiser the Future of American Health Care?," *New York Times*, October 31, 2004, section 3, 1. In line with its emphasis on prevention, Kaiser Permanente referred to itself as a health maintenance organization or HMO. Ironically, for-profit health insurance companies began to develop their own versions of HMOs in the 1980s in order to lower health costs. Their versions differed, though, in that they were being used to maintain or increase profits rather than deliver cost-effective preventive health care. In the eyes of the public, the term HMO, which originally had a progressive connotation, became increasingly identified with corporate greed.

DISCUSSION QUESTIONS

1. European societies seem to have a different view of health care than the United States. Can you identify the basic differences in philosophy?
2. Why is it that the United States does not develop national health service similar to European models? Is it a different philosophy, is it fear, is it politics, is it because we have a different history, or … ?
3. Russell clearly sees the importance of reform in the United States toward a European model. Do you agree? If so, what are the negative consequences? If you do not agree, what are the negative consequences?

PART XII

Social Problems Related to Political Institutions, Violence, and Terrorism

Part XII examines seven social problems that deal threats to democratic society. Charles Reich introduces the issue of power in society by describing the corporation as the "invisible government," showing its antidemocratic tendencies and its concerns with profit rather than people. He points out corporate ties to government and describes what he calls the "real constitution of the land."

Michael Parenti, in a polemic against military empire, explicates why annual spending of roughly $800 billion on the armed forces leads to a quagmire of social, economic, political, and moral problems for the United States. The question that Parenti poses is one that reaches across the aisle of political rancor of *left vs. right* ideologies: Can the United States afford such sustained levels of unprecedented military expenditures as the world's only superpower? And if "yes," what are the unintended social and economic costs to American society and the global community?

Andrew Bacevich, a professor and retired officer in the United States Army, writes about war as a social problem. Bacevich worries that our current approach to terrorism is leading to a "War Without Exits" strategy that ultimately threatens democracy itself.

Mark Juergensmeyer gives an excellent introduction to the relationship of religion and violence. He examines how religion may sometimes increase violence and sometimes legitimate it. His central point is that in order to respond successfully to religion-based terrorism, we must first understand why groups resort to violence in the first place.

Abby Ferber and Michael Kimmel ask us to examine radical militia groups in our society. This article shows the gendered and racial character of an increasingly popular movement that draws support largely from young and middle-aged white men.

David Altheide is concerned with the "politics of fear" created by media and government that influences us to give up our civil rights and control over our own lives.

The final selection is by Adam Jones from his book *Genocide: A Comprehensive Introduction.* In this essay, Jones deals with genocide denial as a problem with both social and moral dimensions. Why do people deny genocides, and what are the broader social, legal, and moral implications of these denials for the victims of genocide, the survivors of genocidal conflicts, and for the societies that have to mend after such ghastly exercises in violence?

44

The Corporation as Invisible Government

CHARLES A. REICH

The Four Questions

1. What is the problem that Reich identifies?
2. Why is this a *social* problem? Who gets hurt? Is society hurt? Is this an important concern to many people?
3. What causes this problem?
4. What are Reich's suggestions for improvement?

Topics Covered

Free market
Government
Corporation
Welfare state
Managerial system
Elites

The invisible government hides behind two myths: the myth of the "free market," and the myth of "big government." In fact, the most important changes in America have been the disappearance of the free market and the ineffectuality of public government. Yet public government is all that we see and hear about. We are not told that the growth of public government was a response—a secondary phenomenon. The primary trend has been the growth of private, corporate government. Public government has been repeatedly called upon to protect individuals and society from harm caused by private government, including the Depression of the 1930s. As private corporations and their operations became national and international in scope, state and local governments proved unable to regulate activities beyond their borders.

Private economic government is a far more important factor in the lives of individuals than public government. Private government controls people by controlling their ability to make a living. In order to get a job, have a career, escape the abyss of being rejected or discarded, people will accept the dictates of corporate and institutional employers, even when these dictates go far beyond anything that public government could constitutionally impose. Employers can and do demand a degree of subservience and conformity that public government could never require. Economic punishment is a more effective weapon than the punishment inflicted by law. Dismissal can be a more efficient means of destroying people than the death penalty. Public government is limited in what it can do to individuals by the provisions of the Constitution; private government is subject to no such limitations.

Prior to World War II, the presence of private economic power was a major issue in American life. William Jennings Bryan's populism, Theodore Roosevelt's trustbusting, Woodrow Wilson's New Freedom, and Franklin D. Roosevelt's New Deal were all responses to private economic power. Public government, labor unions, and small business were all viewed as a counterforce. Then, after World War II, the whole subject of private government vanished from public discourse. During the past forty years, private power grew far beyond the size that had previously caused such concern,

SOURCE: From Charles A. Reich, *Opposing the System*. Crown Publishers, Inc., 1995. Reprinted by permission of Gerard McCauley Agency.

but it remained out of sight. Voters forgot why public government existed and saw it, rather than private government, as the cause of problems facing the individual. We refused to recognize that a basic change had taken place in American society—that access to wealth and position had come under tight control, and that the manager, rather than the public official, held primary authority over people's lives. Charles Lindblom writes:

> It has been a curious feature of democratic thought that it has not faced up to the private corporation as a peculiar organization in an ostensible democracy. Enormously large, rich in resources, the big corporations, we have seen, command more resources than do most government units. They can also, over a broad range, insist that government meet their demands, even if these demands run counter to those of citizens expressed through their polyarchal controls. Moreover, they do not disqualify themselves from playing the partisan role of a citizen—for the corporation is legally a person. And they exercise unusual veto powers. They are on all these counts disproportionately powerful, we have seen. The large private corporation fits oddly into democratic theory and vision. Indeed, it does not fit.[1] ...

Every form of legal control over the corporation has failed. Control by the stockholders—the supposed owners of corporations—was lost to management. Ownership was thus separated from control, with management free to pursue its own course without genuine supervision by the "owners." The device of nonvoting stock was utilized to keep most stockholders powerless while a small inside group—or sometimes just a single individual—reserved voting rights for themselves. A second kind of legal control—the antitrust laws—also failed. Passed by Congress at the end of the nineteenth century to prevent monopolies, anticompetitive mergers and other restraints of trade, the antitrust laws have remained on the books but have simply not been enforced. The Department of Justice under both Republican and Democratic administrations has looked the other way while ever more gigantic mergers have taken place, and the courts have "interpreted" the antitrust laws to permit many of the practices which the authors of the laws sought to prevent.

A third attempted legal limit on corporate power was regulatory legislation, some dating back to the Progressive Era, some a product of the New Deal. Today, much of this regulation has been repealed or rejected. Another restraint on corporate power, the labor union, has been so weakened that it no longer serves as the counterforce it was expected to be during the New Deal. As corporate operations have become international in scope, even sovereign states have been unable to exert effective control over multinationals, which in some cases are virtually sovereign states themselves. Today, the *Fortune* 500 dominate American life. By the year 2010, will it be the *Fortune* 50 or the *Fortune* 5?

So long as corporations, no matter what their size, are seen as remaining in the area of business, it is possible to imagine that the other areas of society continue to function as before. But if corporations begin to exercise functions that are governmental, then a structural change takes place that alters the premise of constitutional government.

In 1993, Lauren Allen was a sales associate in the sporting goods department of the new Wal-Mart in Johnstown, New York. The mother of a two-year-old son and a four-year-old daughter, she had separated from her husband. Several months after starting work, she began dating Samuel Johnson, another sales associate in the same department who was single at the time, seeing him after work for bowling or a meal, often in the company of other coworkers. They fell in love. Then Wal-Mart discovered their relationship. The corporation's handbook for employees states: "Wal-Mart strongly believes in and supports the 'family unit.' A dating relationship between a married associate and another associate, other than his or her own spouse, is not consistent-with this belief and is prohibited." Ms. Allen was summoned to the store manager's office and asked if she was dating

Mr. Johnson. When she acknowledged the relationship, both were dismissed from their jobs. Neither was able to obtain comparable employment elsewhere. Ms. Allen found work folding hospital sheets and towels at a dry cleaner; Mr. Johnson worked as a stock clerk in a lumberyard. The young couple, aged twenty-three and twenty, live together. "I don't think it was right, what they did," Ms. Allen said. "I felt it was my personal life. It wasn't interfering with my job.... I didn't think it was right that they could dictate to me in that way."[2]

The rules that corporations make and enforce are not adopted democratically, nor are they enforced with the fairness required by due process of law. The workplace is authoritarian and hierarchical. It is ruled from the top down. Because the workplace is an economic necessity for most Americans, they have little choice but to submit to its discipline. But this creates a major conflict with democracy. When corporations make and enforce their own laws, they supersede the democratic process and threaten the core values of representative government.

We are told that democracy is the best form of government, and we have fought wars and spent trillions of dollars in a seemingly successful effort to preserve democracy here and abroad. Yet the democratic model is losing out to the authoritarian model in our daily lives here at home. Following the corporate lead, virtually all of our institutions, from schools and colleges to Little League, are based on the top-down model. If we spend most of our time working under authority, how can we say we are a democratic society? When and where do we practice democracy?

No matter how much governmental power corporations exercise over their employees, corporations are exempt from the Bill of Rights. They do not have to allow their employees freedom of speech, or due process of law, except where statutes explicitly so require. The reason for this exemption is that the Bill of Rights applies only to actions by the government and its agencies, and no matter how "governmental" the actions of corporations may in fact be, the courts continue to treat corporations as if they were truly private when it comes to the Bill of Rights. Corporations are glad to accept their exempt status, since rights are obstacles to economic efficiency, and corporations do not welcome legislation that requires them to observe employees' rights.

The result of this corporate exemption from the Bill of Rights is that employees lose many of their rights as citizens when inside the workplace. Moreover, workplace rules may often extend to nonworkplace times and places, so that the employee's space for full citizenship can shrink even further. An employee who exercises his or her free speech rights outside the workplace is not protected against adverse employer reaction, even though this speech is protected by the Bill of Rights against governmental retaliation. This makes the Bill of Rights a hollow guarantee for many employees, since they may risk jobs or promotions by exercising their full rights as citizens. And the real impact of this diminution of citizenship is felt not only by the comparatively few employees who get fired for exercising their rights but by the great mass of employees who are afraid to risk their employers' disapproval. These employees allow their economic dependency to become docility; they are afraid to be free. The corporations which dominate these employees have successfully leveraged economic power into political power that undercuts the Constitution.

As the size and power of business enterprises increased, this great change caused the size and nature of government to increase in response. So-called big government is thus a creation of corporate power, not its antithesis. The Constitution set up a national government of carefully limited and specifically enumerated powers, a government that would not and could not interfere with a large zone of economic and personal freedom for individuals. But toward the end of the nineteenth century the growth of corporations created problems which the government was soon called upon to solve. Small businesses and farmers were squeezed or forced out of business. Consumers found themselves confronting prices fixed by monopolies. Workers experienced disadvantage in bargaining with their corporate employers. Workers and consumers found that corporations had little concern for their health and safety. The economy was subjected to wild swings from boom to bust, causing periods of frightening insecurity, mass unemployment, and misery. Increasingly, the people turned to government for help in dealing with these and other problems which even the most competent of individuals, farmers, and small businesses could not cope with by themselves....

Today, we have forgotten the reason for the growth of government, because we deny and repress the fact of corporate governmental power. Today's rhetorical attacks on "big government" for interfering with business have largely succeeded in obscuring the fact that it is big business, not big government, that primarily regulates the lives of ordinary Americans. Until "big government" intervened, corporations were free to exploit child labor. Until "big government" protected workers' bargaining, corporations could pay less than a living wage. Until "big government" provided unemployment compensation and Social Security benefits, the worker cast off by a corporate employer had nowhere to turn. When corporate management of the economy resulted in the loss of jobs by millions of willing workers during the Great Depression, it was "big government" that came to the rescue.

The use of the term "welfare state" is a way of disguising the fact that the real issue is economic insecurity and poverty, attributable to the narrowing of economic opportunity and the loss of economic independence that American workers have experienced as a result of corporate dominance of the labor market. The "welfare state" was needed because of corporate-caused deprivation. On their own, individuals below the top cannot command enough for their only asset—their labor—to pay for their lifetime needs, including education, security, and the support of families.

Similar in principle is the role of government in undertaking to support public services and public goods, which individuals cannot produce on their own and for which corporations do not take responsibility—from national defense and police to national parks, airports, highways, and schools. Schools, for example, benefit individuals but also render an essential service to corporations by providing an educated workforce. Highways are another service to economic activity as well as to the public.

Once government had been expanded to enable it to regulate, to offer positive assistance, and to extend its influence into the economic life of the nation, then a step unforeseen by the New Deal thinkers occurred: All of these new powers were redirected to favor corporate power and disfavor labor, small farmers and businesses, and the individual. Regulatory agencies created to restrain corporate power were filled with appointees friendly to big business. The social programs intended to help the needy were steadily cut back and hedged with restrictions while aid to corporations became more openhanded. Trillions were spent on the cold war defense buildup; much of this money went to enrich giant corporate defense contractors such as Boeing, General Dynamics, and General Electric. The tax laws became an elaborate source of favors and giveaways as, under the rubric of "tax expenditures" and "tax incentives," tax breaks and tax subsidies for big business were written into the Internal Revenue code. Oil leases, timber sales, and other subsidized distributions of public resources flowed to corporations where once (as under the Homestead Act) public lands had been a source of support for individuals.

A crucial illustration of how governmental powers intended to strengthen democracy were transformed into vital supports for corporate power is the story of how radio and television licenses were distributed. All of the radio and television channels are publicly owned and remain so. The stations and networks that utilize these channels are licensees of the federal government. Originally it was intended that broadcast licenses be allocated by the Federal Communications Commission to a broad variety of community groups and interests representative of American pluralism and obligated to use their broadcasting privileges to further "the public interest." What has happened instead is that corporate giants, especially those already in control of other forms of mass communication such as newspapers, gathered the lion's share of the licenses, often controlling multiple stations. In addition, the major networks and stations have been given over exclusively to commercial use, so that they are dominated by large corporate advertisers. Despite the fact that these broadcast licenses are worth many millions of dollars to the licensees, the licensees neither share their profits with the public (as owners) nor perform any but the most token public services. As great as is this giveaway of wealth, the giveaway of power is even more significant. It means that corporate America completely dominates the *public* airways, and nothing that is disapproved of by corporate America is likely to be heard by the public....

As government became more clearly allied with business, the legal system lost whatever neutrality it possessed and became a pliable tool of power. The wealth that government distributes and the advantages that government confers must be based upon some form of legal authorization. But when law is employed to justify this process, the ideal of "equal justice under law" disappears and respect for the law is lost. While an unemployed worker receives less than enough compensation to live on, a rancher obtains government-subsidized grazing privileges of great value, a lumber company gets to use the national forests for very substantial profit, and the recipient of a defense contract obtains a privilege worth millions of dollars. Most legally authorized favors and advantages are the result of successful lobbying or other uses of power and influence. Groups like the poor, which lack power and influence, end up with the smallest share of government bounty. Obtaining government wealth thus becomes an insiders' game. Those on the outside are the losers.

The law is full of justifications of the disparate treatment accorded the many classes of citizens and businesses. But these explanations cannot hide favors based upon power and influence. The neutrality, fairness, and dignity of law is destroyed by its involvement in this game of government favors. Lawlessness becomes difficult to distinguish from the advantageous use of law. The person who acquires wealth with the help of a gun may see himself *[sic]* as no different from the person who acquires wealth with the help of a tax break or favorable agency ruling, except that the latter is not subject to punishment and usually walks off with a far greater gain. Why should the dispossessed respect the authority of law? The outsider's crime may seem morally no different from the insider's coup.

The increasing merger of governmental and corporate interests has been further solidified by the emergence of a trained group of managers who moved easily and interchangeably from the corporate to the public sector and back again. The existence of this managerial elite allows both sectors to make important decisions on the basis of shared knowledge and shared assumptions without any conspiracy or overt coordination. The most basic choices can be taken out of the hands of the people and made in undemocratic fashion without any visible signs of how this is accomplished. Even members of the elite do not necessarily recognize the coordinated decision-making process of which they are a part.

The managerial elite is formed on the basis of shared knowledge and assumptions. The shared knowledge is acquired in school, college, professional school, and later on at work. Like other significant features of the System, this shared knowledge has no name, is never directly taught, is not written down, and prefers invisibility. The starting point of the elite's shared knowledge is that every member of the elite discovered the road to success, which the vast majority never find. For example, the road to success involves passing tests, from elementary school to graduate or professional school, and then in the lower rungs of the public or corporate bureaucratic ladder. The many years of test-passing give rise to a knowledge of what is required to satisfy those who create and administer the tests, and what will bring about rejection. Our present managers may be the most thoroughly tested group of individuals who ever lived on the planet. What are they tested for? The answer is: skills that advance the interests of the System. But these skills are not necessarily the same as the substantive skills of the doctor, the auto worker, the farmer. The people who run our society may not be the best at the usual categories of human achievement, but they are the best at rising to the top of the System....

The elite live in a different country than the rest of Americans. It is not possible to understand the System and its actions without understanding this fact. The elite sees its own ascendancy as just, and cannot understand the anger below. Yet the rules for success used by the elite are often very different from the rules observed by ordinary people. This leads the elite to believe that those below "cannot be told" the real reasons for decisions that are made. The question becomes what should the people be told, not what are the facts. The perplexity of the voter who tries first one party and then the other, winding up always with the same elite, shows how democracy has given way to the rule of the System's managers.

Shared knowledge leads to shared assumptions, which are even more crucial than knowledge in making it possible for the elite managers to work together without "conspiracy." These invisible shared assumptions are the real Constitution, the real fundamental law, which guides the System:

First. Impersonal economic "forces" produce better decisions and choices than can be made by even the most thoughtful individuals or groups attempting to weigh the competing interests of society. Often called "market forces," they are venerated and deferred to even when the "market" is impaired, dominated, or nonexistent.

Second. "Economic growth" is a measure of the well-being of the society as a whole. Such growth benefits all the elements and individuals in society. There is no need for "Gross Domestic Product" to be offset by a compilation of "Gross Domestic Cost." "Growth" can be counted on to create new jobs and wider individual opportunity.

Third. All of the important social values affected by the economic system are measurable and quantifiable. Intangible or "soft" interests such as trust, loyalty, community, natural beauty, or sacredness are separate from the economic process and may be "willed" by individuals of strong enough character and a high enough sense of personal responsibility. For example, care of children is a matter of "will," not economics.

Fourth. The treatment of people in their role as employees does not affect the rest of their lives. For example, an authoritarian workplace does not diminish or impair a worker's ability to practice democracy outside the workplace. Summary layoffs and terminations after years of service do not affect workers' capacity for loyalty, or for commitment to loved ones. Coercion on the job does not lead to an increase of coercive behavior off the job.

Fifth. The market may be counted on to supply human and social needs even though a massive and unrelenting effort is made to influence consumer preferences, no opposing view of needs is allowed to reach the public by way of the media, and the consumer has received no information about society's need for balance, diversity, or long-term investment. No alternative method of planning for the future is required.

Sixth. People are rational actors whose behavior can be motivated, controlled, channeled, and deterred by threats and promises of economic gain and loss, by repression and punishment, by constant competition against others, and by orders issued by those in superior positions. Economic self-interest is expected to dominate human behavior, but people can be expected to forgo self-interest in their personal lives in favor of heightened responsibility to others and to "the community."

Seventh. The ultimate product of society is the best possible System, not the best possible human beings.

Given these shared assumptions, the corporate elite and the governmental elite together make the most important decisions affecting society, often overriding what voters choose at the polls. What remains for presidential candidates to debate or for Congress to vote on is most often trivial and irrelevant, incapable of overturning the more basic decisions that are made in nondemocratic fashion. It is as if the Constitution had been amended to include the following provision:

> *Managerial Review*
>
> Policies and choices made by voters as part of the electoral process shall be subject to review by a Board of Managers

> representing the governmental and corporate sectors. The Board may, in its absolute discretion, confirm or reverse decisions by the electorate or substitute other policies for those chosen by the electorate.

... The decision that most directly affects the lives of ordinary Americans concerns the structure of work and the workforce. The workforce has been reshaped into an ever steeper pyramid, with fewer and fewer people enjoying a greater and greater share of wealth and privilege at the top; more and more workers thrown into a "lower tier" of reduced pay, benefits, and security; and a permanent surplus of unemployed at the bottom ready to take any job that becomes available and thus making the working population easily replaceable and therefore unable to demand higher wages and better terms and conditions. It is this decision that is destroying the middle class, driving more and more people into desperation and poverty, separating the population into antagonistic classes of extreme wealth and powerless subservience, and creating the dissatisfaction that drives the demand for change.

Everyone's life has been changed by this decision. Only forty years ago the shape of the workforce was very different, with a broad and affluent middle class as the most important stabilizing feature of American life. Now, workers who "play by the rules" are steadily losing ground, and frustration, fear, and anger has stolen their dreams. Opportunity is greatly diminished by this decision, security is lost, antagonism is engendered, community is destroyed. And yet Americans have passively accepted this prime example of unwanted social change, believing it to be not a decision at all, but the result of impersonal economic forces that no one can or should control.

What this explanation overlooks is that the present structure of the workforce is not the result of competition and the free play of economic forces, but the product of the absence of competition. So long as corporate power was offset by strong labor unions and by a government representing all classes in society, an upwardly mobile workforce existed. Only when corporate power reached the point of ability to control the workforce without interference by unions, government, or competition from other corporations for workers, did the situation change. The change took place because of the growth of uncontested power. Global competition does not dictate the extreme inequality of our hierarchical workforce. As is well known, the other industrial countries with which we compete, such as Japan and Germany, have far less disparity between the pay of top executives and workers. Sharing the pie more equally among executives and workers would not diminish the profitability of the enterprise as a whole....

A second major decision that affects the lives of ordinary Americans is how we will spend our money—on toys for children or schools for children, on entertainment or public libraries, on expensive automobiles or preserving natural beauty. In a free economy, this is a choice that is supposed to be made by the people. When we find ourselves asking, as we now do, "What has happened to the money?"—for schools, for universities, for child care, for parks, for other elements of any good society—we are revealing that we have lost this sovereign power of choice. If we had consciously made a decision not to spend our resources on such goods as education, there would be no mystery even if we had made an unwise choice. The mystery arises when we cannot find the money for the goods we say we want....

Thus the two great decisions that make the most difference to ordinary Americans were not made by "big government" or by the voters but by unelected managers who are well isolated against the effects of their own decisions. Why do we passively accept this situation when we find the results so painful and unsatisfactory? The answer is that we have been led to believe that these decisions are being made not by managers but by "the free market." Despite overwhelming evidence that the free market is a myth, it is universally spoken of as if it were the central truth about our economy. It is not.

The great irony of the free-market myth is that it is perpetuated by the very same corporations which have done everything possible to eliminate the free market. From the very beginning, the corporation has had one primary goal—control. The purpose of large-size, massive accumulations of capital and the organization of production by many workers is usually given as "efficiency"—and the large corporation has indeed proved more efficient at many tasks than individuals or small firms. Thus it is difficult to imagine anything smaller than a large corporation running a steel mill or producing automobiles or operating an airline. "The efficiency of large scale" is an axiom of economics. But what underlies efficiency is *control.* The larger the corporation, the better it can control the entire process of acquiring raw materials, manufacturing a product, and distributing and marketing the product. The efficiency of Henry Ford's assembly line lay in its control of the activities of thousands of workers, suppliers, and distributors. Hence from the very beginning the effect of the large corporation was to undermine those two cherished eighteenth-century institutions which lay at the foundation of America—the free market and democracy.

Corporations wanted a market, but not a free market. They wanted the power to fix the terms and conditions of employment on a take-it-or-leave-it basis and to fix the cost of materials and the price of finished products in the same way. Therefore, the natural tendency of the corporate system was toward *combination* and *concentration.* Corporations combined *vertically* to control each step of the production process, from acquiring raw materials to consumer sales. Independent suppliers or distributors might retain a measure of freedom to bargain, but where they became subsidiaries of the manufacturer that freedom gave way to control. Corporations also combined *horizontally* to control every area of the country; for example, having identical retail outlets from Maine to Florida and from New York to California. Horizontal combination means that whenever a consumer tries to buy a product, she *[sic]* finds the identical choice, not a range of choices....

The key to both the ability of the System to make the most important economic decisions by command and its ability to conceal this fact by using the free-market myth lies in the System's control of communications, knowledge, and opinion. No challenge to the free-market myth is heard on television or radio; the economics profession solidly perpetuates the myth; and meanwhile the decisions that are actually made are then pressed on the American people from every direction. Charles Lindblom describes a "command economy" as one that embraces the *"full range of methods of control.... It seeks to control the mind, as far as possible, by controlling all forms of communication...."* (italics in original).[3] We too live under a system that seeks to control the mind by making invisible what is really happening and covering the truth with myth.

The managerial system has proven immensely successful for business. It is far more efficient, at least in the short run, than the slower democratic model. In theory, managerialism might also provide the best possible government for a high-tech society in the twenty-first century. If top executives produced a healthy and happy society as the bottom line of their endeavors, they might be worth the multimillion-dollar salaries some are now paid. But for a managerial system to succeed, it would have to be subjected to a larger framework of control. The managers would have to be held responsible for serving the best interests of society as a whole. They would have to preserve our basic values of equality, individual rights, and ultimate respect for the people. In contrast, our present managerial system is not controlled by any outside structure. It does not even purport to serve the best interests of society, and it escapes responsibility for the disastrous conditions it creates.

Driven by the goal of economic efficiency, harnessing the forces of competition and human desire, utilizing the latest advances in organization and technology, the managerial system has proved to be a dynamic and expansive competitor to the older forms of constitutional democracy and the free market. The fact that the managerial system is *in competition* with these older forms, rather than simply coexisting with them, has not been adequately recognized. Instead, it has always been tacitly assumed that the managerial system would confine itself to business activities and not threaten the larger contours of democracy, the Constitution, and the free market. This assumption, that management would remain confined within the existing structure of society, has proven false. Instead, management has demonstrated a powerful drive for expansion, leading it to seek and gain control over both government and the free market, until the managerial system has reached a point where it pervasively influences everything else in society. But the very characteristics which give the managerial system its power make it unsuitable for the governance of an entire society. Applied to society as a whole, the strengths of managerialism become weaknesses, and its drive for efficiency becomes a power to destroy.

In its drive for economic efficiency, the managerial system is single-minded: It has but one goal, the maximization of profit. By contrast, a society must have multiple goals, must promote a thriving diversity, must seek a balance among many interests. The single-mindedness of a management seeking purely economic gain rejects all other social and human needs. For example, the managerial system is highly selective; it wants only those workers it can utilize with maximum efficiency and casts aside other workers who are deemed surplus. But a society as a whole has to be inclusive, and it must offer a place to everyone. It cannot reject people because they are inefficient. A managerial system seeks flexibility; it is ready to discard people the moment they are not needed. On the other hand, most people require security and stability; human beings are not as adaptable as machine parts and cannot build lives upon a foundation of perpetual uncertainty. A managerial system is authoritarian and hierarchical; it does not recognize the values of equality, democratic participation, or organic community. It treats people as interchangeable machine parts, whereas a true community regards every individual as unique. A managerial system has no concern with human needs. It will make and sell whatever brings a profit, and refuse to make whatever does not bring a profit, irrespective of what people need to become well-functioning individuals, families, and communities.

Rights are inefficient. Fair procedures are inefficient. Constitutional limitations are inefficient. Democratic dialogue is inefficient. Self-government is inefficient. Nonconformity is inefficient. Dignity is inefficient. The managerial system is impatient with all of these. If the cheap labor of an authoritarian country can lower the cost of a product, management prefers that choice to the work of individuals endowed with rights and dignity.

These limitations of a managerial system prevent it from being a sustainable model for an entire society. But they have not prevented it from taking over control of society and its institutions. This is one of the great weaknesses of present-day public philosophy: It assumes, without any basis, that such a takeover will not occur despite the fact that managerialism has amply demonstrated the power to cross any and all boundaries. Instead, corporate wealth and influence have allowed management to conscript government. The state is supposed to prevail over all other forms in society because the state has a monopoly of force, but corporate wealth has been able to prevail over force. Nothing is more crucial to classical economic thought than the separation of government from the business sector, yet the merger of public and private is a fact of life. Economic power can be leveraged to exert political power. Corporate wealth is leveraged to control expression in the media. Corporate power over livelihood becomes control over the private lives of workers. Corporate dominance of entertainment is leveraged to control American culture. Corporate donations to universities becomes control over scholarship and thought.

The managerial system presents virtually every possible contrast with both the conservative and the liberal vision of a pluralistic, self-governing republic. A great intellectual failure of both conservatives and liberals has been their inability or unwillingness to recognize this fundamental conflict. And since the conflict itself is unobserved, neither conservatives nor liberals have proposed any method of resolving or presiding over the conflict, whether by means of structure, philosophy, a return to older traditions, or some new social invention.

The invisibility and denial which obscure the unprecedented power concentrated in economic government also enable it to escape the responsibility that should accompany such power. Today all of the talk about responsibility is directed at the most powerless people in our society. The powerful call for responsibility by the dispossessed but are silent about the responsibilities of power. They deny that responsibility inescapably accompanies power. A managerial system might serve society well if it were also a responsible system. What is wrong with the present System is not necessarily that it is managerial, but that it is not accountable to the people it serves, and instead serves itself.

The temptations under the present System for those in power, whether in the public or corporate sectors, to serve themselves, particularly in a financial way, are overwhelming. The safeguards are virtually nonexistent. The merger of corporate and governmental sectors brings out and combines the irresponsible side of each. The government becomes infected with the profit motive; the corporate sector uses governmental power without constitutional restraint. Both seek to minimize responsibilities, the corporate sector because responsibilities are costs which reduce profits, the government because responsibilities are bureaucratic failures for which no public official wants to accept blame. Both resort to lies and deception and whenever possible avoid candor with the public. Both seek to shift blame to someone else. Both reflect the absence of a presiding authority; corporations are no longer responsible to their owners, and government is no longer responsible to the sovereign people....

ENDNOTES

1. Charles E. Lindblom, *Politics and Markets* (New York: Basic Books, 1977), p. 356.
2. Jacques Steinberg, "At Wal-Mart, Workers Who Dated Lose Jobs," *New York Times*, July 14, 1993, p. A16.
3. Lindblom, *Politics and Markets*, pp. 238–39.

DISCUSSION QUESTIONS

1. Does it matter for democracy if the individual is dominated by the corporation rather than the government?
2. "Economic punishment is a more effective weapon than the punishment inflicted by law." Do you agree with this statement?
3. Do you believe that if you work for a corporation, you must be willing to give up your rights, or do you think that you still have rights? What rights should you be able to retain when you work for someone else?
4. "Today we have forgotten the reason for the growth of government." Do you believe that it is to the advantage or disadvantage of the corporation that people distrust government?
5. React to what the author describes as the shared assumptions of those who govern the "invisible government": market forces take precedence over careful decision-making; economic growth is the measure of social well-being; social values are measurable; profit and gain are people's basic motivation; the System is more important than people.
6. Do you believe Reich is overstating his case? Are corporations the master of our lives?

45

The U.S. Global Military Empire

MICHAEL PARENTI

The Four Questions

1. What is the problem Parenti outlines in this article?
2. Is this a *social* problem? Is this a problem of cultural values and politics? Does society see this as a problem?
3. Who is being harmed by this problem?
4. What is to be done about this problem? Does the author imply any ways of dealing with it?

Topics Covered

Global superpower

U.S. military empire

Government military expenditures

Cost of foreign wars

Hidden costs of U.S. military empire

The United States is said to be a democracy, but it is also the world's only superpower, with a global military empire of a magnitude never before seen in history. What purpose does this empire serve?

A GLOBAL KILL CAPACITY

The U.S. military has a nuclear overkill capacity of more than 8,000 long-range missiles and 22,000 tactical ones, along with ground and air forces ready to strike anywhere and a fleet larger in total tonnage and firepower than all the other navies of the world combined. With only 5 percent of the earth's population, the United States devotes more to military expenditures than all the other industrialized nations put together. Over the last half century, U.S. leaders deployed thousands of nuclear weapons and hundreds of thousands of military personnel to over 350 major bases and hundreds of minor installations spanning the globe. This massive deployment supposedly was needed to contain a Soviet Union bent on world domination—although evidence indicates that the Soviets were never the threat they were made out to be by our Cold War policymakers.[1]

Despite the overthrow of the USSR and other Eastern European Communist nations in 1990–92, U.S. military allocations continued at budget-busting stratospheric levels, and U.S. overseas military strength remained deployed in much the same pattern as before, with its Cold War arsenal of long-range nuclear missiles aimed mostly at the former Soviet Union, an enemy that no longer exists. In recent years the list of sites targeted by U.S. nuclear weapons actually grew by 20 percent, including targets in Russia, Belarus, Ukraine, Kazakhstan, China, Iran, Iraq, and North Korea.[2]

Along with direct yearly military appropriations, which rose to about $500 billion by 2006, there are the indirect costs of war and empire: veterans benefits, including health care and disability costs; federal debt payments due to military spending, over $150 billion each year; covert military and intelligence operations; the 70 percent of federal research and development funds that goes to the military; space weapon programs; military aid to

SOURCE: From Parenti, Michael, *Democracy for the Few*, 8th edition, Belmont, CA: Cengage/Wadsworth, 2008, pp. 77–91.

other countries; "supplementary appropriations" for specific wars, as in Iraq (over $100 billion in 2006); and defense expenses picked up by nonmilitary agencies including the Energy Department's nuclear weapons programs, which consumes more than half of that department's budget. Taken together, actual military spending for fiscal year 2006 came to almost $800 billion. The United States is also the world's largest arms merchant, with some $37 billion in weapons exports to other nations in 2004. It costs our taxpayers billions of dollars a year to subsidize these sales, but the profits go entirely to the corporate arms dealers.[3]

The federal budget is composed of *discretionary spending* (the monies that the Congress allocates each year) and *mandatory spending* (the monies that must be allotted in compliance with already existing authorizations, such as payments on the national debt or Social Security). In the discretionary budget, more money is spent on the military than on all domestic programs combined. Under Bush Jr. the arms budget increased 41 percent in four years. Meanwhile the Department of Defense (DOD), also called "the Pentagon" after its enormous five-sided headquarters, proposed to spend $2.3 trillion from 2006 to 2011.[4]

The Bush Jr. administration made an accelerated effort to develop the Strategic Defense Initiative, or "Star Wars." First proposed in the mid-1980s, Star Wars is a ground- and space-based missile program that supposedly would intercept and destroy all incoming ballistic warheads launched by other nations. Over the years, the military has spent about $120 billion unsuccessfully trying to create this "space shield." Despite the lack of progress, Bush Jr. increased the Star Wars budget by 20 percent for 2007. If the Star Wars project ever does prove successful, it will make the nuclear arsenals of other nations obsolete and deprive them of any deterrence against U.S. nuclear missiles. This in turn will encourage them to spend more to update their own long-range attack system.[5] Also, Star Wars is in violation of the Outer Space Treaty, signed by ninety-one nations, including the United States, which bans weapons of mass destruction in space.

A chief of the U.S. Space Command enthused: "We're going to fight from space and we're going to fight into space. We will engage terrestrial targets someday—ships, airplanes, land targets—from space."[6] The professed goal of the U.S. Space Command is to dominate "the space dimension of military operations to protect U.S. interests *and investments*."[7]

Bush Jr. also asked Congress to fund a new nuclear weapon, nicknamed the "bunker buster," that would penetrate hardened underground bunkers while capable of yielding an explosive force seventy times the size of the Hiroshima bomb.[8] The military already can beam powerful electromagnetic or pulsed radio-frequency radiation transmissions back to earth, seriously impairing the mental capacity of whole populations, causing severe physiological disruption or perceptual disorientation for an extended period, according to the Air Force.[9]

PENTAGON PROFITS, WASTE, AND THEFT

The DOD's procurement program is rife with fraud and profiteering. Its own auditors admit the military cannot account for one-fourth of what it spends, over $100 billion a year. Such sums do not just evaporate; they find their way into somebody's pockets. President Bush Jr.'s secretary of defense, Donald Rumsfeld, admitted that "according to some estimates we cannot track $2.3 trillion in transactions." When Bush Jr. called "for more than $48 billion in new defense spending," this caused retired Vice Admiral Jack Shanahan to comment, "How do we know we need $48 billion since we don't know what we're spending and what we're buying."[10]

If the Pentagon's missing funds were returned to state and local governments on a pro rata basis, they all would be able to pay off their debts, vastly improve their educational and health services, provide housing for the homeless, and still have funds left over for other things—all just on what the military "misplaces."

There's more to this story: It was reported that the Pentagon was storing $41 billion in excess supplies gathering dust or rusting away. The U.S. Army allocated $1.5 billion to develop a heavy-lift helicopter, even though it already had heavy-lift helicopters and the Navy was building an almost identical one. Congress voted for C-130 cargo planes that the Air Force did not want—because they were so dysfunctional—and extra B-2 bombers that the Pentagon never requested. The Air Force started to develop an F/A-22 fighter plane in 1986 that has cost $29 billion and was still not combat ready twenty years later in 2005. The Pentagon approved a plan to spend $16 billion to lease one hundred jetliner refueling tankers from Boeing, which cost more than buying the planes. The tankers were built by Boeing in part with Pentagon funds.[11]

The Government Accountability Office (GAO), watchdog agency for Congress, reported that the Pentagon had no sure way of knowing how $200 billion was spent waging war in Iraq and Afghanistan. The GAO identified instances in which costs were off by 30 percent or more. Multi-billion-dollar Pentagon contracts were plagued by "inadequate planning and inadequate oversight," according to the GAO controller-general. A third or more of the government property that Halliburton and its subsidiaries were paid to manage in Iraq could not be located by auditors. Contractors were repeatedly paid for work never performed. Halliburton grossly overcharged the Pentagon for fuel supplies, construction, meals for troops, and other services, while delivering substandard equipment and contaminated water to troops and civilians at U.S. bases on Iraq.[12] Two members of Congress concluded that Halliburton was systematically overcharging on hundreds of requisitions every day, with an enormous cumulative cost to the taxpayer of billions of dollars. Millions in cash were found stuffed in footlockers and filing cabinets. "The general feeling," concluded one Army contracting officer, "is that the contractor [Halliburton] is out of control."[13]

Defense contractors have been known to make out duplicate bills to different military agencies, getting paid twice for the same service. Tests have been rigged and data falsified to make weapons appear more effective than they actually are. Military acquisition officials have negotiated contracts with defense companies while at the same time negotiating for jobs with those very same firms. Many top defense contractors have been under criminal investigation, but most fraud goes unpunished. The public purse is pilfered on small items too. The military paid $511 for light bulbs that cost ninety cents and $640 for toilet seats that cost $12. And after paying Boeing Aircraft $5,096 for two pairs of pliers, the tough Pentagon procurers renegotiated the price down to $1,496—a real bargain.[14]

Billions are spent on military pensions that go mostly to upper-income senior officers. Vast sums have been expended at military bases for golf courses, polo fields, restaurants, and officers' clubs, replete with gold-plated chandeliers, oak paneling, and marble fixtures. There is a Pentagon-leased luxury hotel outside Disney World in Florida that requires an annual federal subsidy of $27 million. Two golf courses at Andrews Air Force Base in Maryland were not enough; so a third one costing $5 million was built. And in the midst of intense budget cutting of human services, Congress allocated $1 billion for seven luxury aircraft to service the Pentagon's top commanders. Meanwhile, of the 15,000 disabled troops returned from Iraq by 2004, many went for months without receiving pay and medical benefits to which they were entitled.[15]

For the corporate contractors military spending is wonderful, for the following reasons:

- There are almost no risks. Unlike automobile manufacturers who must worry about selling the cars they produce, the weapons dealer has a guaranteed contract.
- Almost all contracts are awarded without competitive bidding at whatever price a corporation sets. Many large military contracts have cost overruns of 100 to 700 percent. To cite a notorious example, the C-5A transport plane had a $4 billion cost overrun (and its wings kept falling off).
- The Pentagon directly subsidizes corporate defense contractors with free research, and

development, public lands, building, renovations, and yearly cash subsidies totaling in the billions.[16]

- Defense spending does not compete with the consumer market and is virtually limitless. There are always more advanced weapons of destruction to develop and obsolete weaponry to replace.

In recent years the DOD has been privatizing various functions that used to be performed by military personnel. Kitchen duty, laundry, fuel supplies, military prison construction, heavy equipment maintenance, and certain security assignments are now contracted out to private companies that perform these tasks, often with little oversight and for outrageously padded prices.[17]

Military spending is much preferred by the business community to other forms of government expenditure. Public monies invested in occupational safety, environmental protection, drug rehabilitation, or public schools provide for human needs and create jobs and buying power. But such programs expand the *nonprofit public sector*, bringing no direct returns to business, if anything, shifting demand away from the private market. In contrast, a weapons contract injects huge amounts of public funds directly into the *private corporate sector* at a rate of profit that is generally two or three times higher than what other investments yield.

U.S. leaders say that military spending creates jobs. So do pornography and prostitution, but there might be more worthwhile ways of creating employment. In any case, civilian spending generates more jobs than military spending; $1 billion (1990 value) of military procurement creates an average of 25,000 jobs, but the same amount would create 36,000 jobs if spent on housing, 41,000 jobs in education, and 47,000 in health care.[18]

To put military spending in perspective, consider the following: the $800 million Congress saved in 1997 by cutting Supplementary Security Income for 150,000 disabled children amounts to less than one-third the cost of building and maintaining one B-2 bomber.[19] The $5.5 trillion spent just for nuclear weapons over the last half-century exceeded the combined federal spending on education, social services, job programs, the environment, general sciences, energy production, law enforcement, and community and regional development during that same period.[20]

To keep America on its arms-spending binge, corporate lobbyists and the Department of Defense itself spend millions of dollars on exhibitions, films, publications, and a flood of press releases to boost various weapons systems. The DOD finances military-related research projects at major universities and propagates the military viewpoint at hundreds of conferences and in thousands of brochures, articles, and books written by "independent scholars" in the pay of the Pentagon.[21]

HARMING OUR OWN

The U.S. military inflicts numerous hidden costs upon the economy, the environment, and human life. The armed services use millions of acres of land at home and abroad in bombing runs and maneuvers, causing long-lasting damage to vegetation, wildlife, and public health. Military target ranges on the Puerto Rican island of Vieques, in South Korea, and even within the United States are heavily contaminated with petroleum products, uranium, and other carcinogenic heavy metals.

The military uses millions of tons of ozone-destroying materials. It contaminates the air, soil, and groundwater with depleted uranium, plutonium, tritium, and other toxic wastes, while amassing vast stockpiles of lethal chemical and biological agents. There are some 20,000 radioactive and toxic chemical sites on military bases and nuclear weapons plants and laboratories across the United States. Many of these have repeatedly released radioactive and poisonous wastes into the air and waterways, including millions of gallons dumped illegally into makeshift evaporation pits and seepage basins, causing a contamination that will require significant cleanup costs. In fact, the government now admits that most of these sites will never be cleaned up and will need "permanent stewardship" for generations to come.[22]

Populations at home and abroad have been sickened by nuclear bomb tests. After decades of denial, the government is conceding that American workers who helped make nuclear weapons were exposed to radiation and chemicals that produced cancer and early death. The Department of Energy admits that it would cost astronomical sums and take decades to clean up the contamination generated by nuclear arms production and testing. Instead, it spent $40 billion on nine new plants capable of designing additional nuclear weapons.[23]

During the 1950s the U.S. Army conducted germ warfare experiments in American cities, causing numerous civilian illnesses and deaths. The U.S. Coast Guard, responsible for policing our waterways, has dumped more than 100,000 used batteries containing lead, mercury, and other chemicals into our rivers and lakes. The U.S. military is a major polluter, using vast amounts of ozone-depleting materials and generating 500,000 tons of toxins yearly. The Pentagon admitted to Congress that some 17,500 military sites violate federal environmental laws.[24]

The military is also a danger to its own ranks. Every year hundreds of enlisted personnel are killed in vehicular accidents, firing exercises, practice flights, maneuvers, and other readiness preparations. During World War II, the Navy tested the effects of poison gas using sailors as guinea pigs. As many as 60,000 took part in the experiments, many suffering long-term disabilities. Tens of thousands of veterans have been sickened or have died from exposure to atomic testing during the 1950s or from toxic herbicides used in the Vietnam War. And more than 200,000 Gulf War veterans may have been exposed to depleted uranium or other highly hazardous materials, including anthrax inoculations that are suspected of causing serious illness. In 1994, Senator John Rockefeller (D-W.V.) issued a report revealing that for at least fifty years the Department of Defense used hundreds of thousands of military personnel in human experiments involving intentional exposure to such dangerous substances as mustard and nerve gas, ionizing radiation, and psychochemicals. The Pentagon eventually declassified reports showing that U.S. soldiers and sailors had been secretly exposed to toxic chemical agents that caused serious ailments.[25]

The Department of Defense also is one of the biggest and cruelest users of animal experimentation. Animals "are burned, shot, bled, irradiated, dosed with biological, nuclear, and chemical weapons, assaulted with cannonades of noise, exposed to deadly viruses," and then studied as they suffer lingering deaths.[26]

NOTES

1. Tom Gervasi, *The Myth of Soviet Military Supremacy* (Harper & Row, 1986); Fred Kaplan, *Dubious Specter: A Skeptical Look at the Soviet Nuclear Threat* (Institute for Policy Studies, 1980).
2. *Defense Monitor* (newsletter of the Center for Defense Information, Washington, D.C.), September/October 2000.
3. David McGowan, *Derailing Democracy* (Common Courage Press, 2000), 95; and *Defense Monitor*, September/October 2005.
4. *New York Times*, 6 December and 27 December 2005.
5. *New York Times*, 20 June 2000; and "U.S. Missile Defense Project May Spur China Nuke Buildup," *Oakland Tribune*, 10 August 2000.
6. Quoted in Karl Grossman, "U.S. Violates World Law to Militarize Space," *Earth Island Journal*, Winter/Spring 1999.
7. U.S. Space Command, *Vision for 2020*, quoted in McGowan, *Derailing Democracy*, 196 (italics added).
8. See Alliance for Nuclear Accountability, http://www.ananuclear.org/rnep.html#update.
9. Gar Smith and Clare Zickuhr, "Project HAARP: The Military's Plan to Alter the Ionosphere." *Earth Island Journal*, Fall 1994.
10. For Rumsfeld and Shanahan quotes, see, respectively, "The War on Waste," CBS News Report, 29 January 2002; and *Defense Monitor*, May 2000.

11. "An Arms Race with Ourselves," Business Leaders for Sensible Priorities, New York, n.d. www.businessleaders.org; and *New York Times*, 24 March 2005; Michael Sherer, "Buy First, Fly Later," *Mother Jones*, January/February 2005; *San Francisco Chronicle*, 24 May 2003.
12. *New York Times*, 16 June and 27 November 2004, 29 December 2005, and 25 January 2006.
13. Reps. Henry Waxman and John Dingell, "Whistleblowers Sound the Alarm on Halliburton in Iraq," *Multinational Monitor*, March 2004; and Army contracting officer quoted in Associated Press report, 31 October 2004.
14. Report by Russel Mokhiber, *Multinational Monitor*, October 2004; *The Pentagon Follies*, Council for a Livable World and Taxpayers for Common Sense (Washington, D.C., 1996).
15. See *The Pentagon Follies; New York Times*, 6 August 1990, 1 April 1996 and 8 October 1999; *Los Angeles Times*, 18 February 2005.
16. Michael Sniffen, "No Open Bidding for Most Pentagon Contracts," *Grand Theft Pentagon* (Common Courage, 2005). Associated Press report, 30 September 2004; for an overview see Jeffrey St. Clair, *Grand Theft Pentagon* (Common Courage, 2005).
17. P. W. Singer, *Corporate Warriors* (Cornell University Press, 2003); Nelson Schwartz, "The Pentagon's Private Army," *Fortune*, 3 March 2003.
18. "Why We Overfeed the Sacred Cow," *Defense Monitor*, no. 2, 1996.
19. Robert Scheer, "Our Rained-Out Bomber," *The Nation*, 22 September 1997.
20. *The U.S. Nuclear Weapons Cost Study Project*, Brookings Institution, Washington, D.C., 2000.
21. Bob Feldman, "War on the Earth," *Dollars and Sense*, March/April 2003.
22. "Feds Say Nuke Sites Will Never Be 'Clean,'" *Oakland Tribune*, 8 August 2000.
23. Seth Shulman, *The Threat at Home: Confronting the Toxic Legacy of the U.S. Military* (Beacon Press, 1992); *New York Times*, 29 January 2000.
24. *Washington Post*, 9 June 1980; *New York Times*, 29 November 1988 and 2 September 1998; and Tyrone Savage, "The Pentagon Assaults the Environment," *Nonviolent Activist*, July/August 2000.
25. For all these various cases, see Department of Defense, *Worldwide U.S. Military Active Duty Military Personnel Casualties*, Directorate for Information Operations and Reports M07, n.d.; *San Francisco Chronicle*, 12 June 1991; Michael Uhl and Tod Ensign, *G. I. Guinea Pigs* (Putnam, 1980); Sam Smith, "Research and Experiments on the Home Front," *Justice Xpress*, Summer 2002; *Citizen Soldier* newsletter, January 2003.
26. Phil Maggitti, "Prisoners of War: The Abuse of Animals in Military Research," *Animal Agenda* 14, no. 3, 1994.

DISCUSSION QUESTIONS

1. According to the author of this selection, in 2006 the United States spent almost $800 billion in military related expenditures. Is this a social problem? Why or why not?
2. What are some possible unintended consequences (for society) of the U.S.'s current level of military spending?
3. *Social problems affect people, so when the author claims that "The Department of Defense also is one of the biggest and cruelest users of animal experimentation ..." this is not a social problem.* Respond to this statement.
4. If you were to write a one-page letter to your Congressman or Congresswoman advising on the current level of military spending in the United States, what would your letter say? What guidance would you proffer to your representative?
5. The United States spent $5.5 trillion, according to the author, on nuclear weapons over the past half century in its arms-race with the former Soviet Union and other potential nation-state threats. Was this an effective deterrent strategy in your opinion? Why or why not?

46

War without Exits

ANDREW BACEVICH

The Four Questions

1. What is the problem for Bacevich in this selection?
2. Is this a *social* problem? Who does this problem adversely affect?
3. What is the principal cause of this problem?
4. Is this a problem with a solution? What does the author say about addressing this conundrum?

Topics Covered

Superpower and military costs

"Long war" commitments

Foreign military involvement

Political policy

Economic and social costs of "Long war"

Deficit spending on wars

For the United States, the passing of the Cold War yielded neither a "peace dividend" nor anything remotely resembling peace. Instead, what was hailed as a historic victory gave way almost immediately to renewed unrest and conflict. By the time the East-West standoff that some historians had termed the "Long Peace" ended in 1991, the United States had already embarked upon a decade of unprecedented interventionism.[1] In the years that followed, Americans became inured to reports of U.S. forces going into action—fighting in Panama and the Persian Gulf, occupying Bosnia and Haiti, lambasting Kosovo, Afghanistan, and Sudan from the air. Yet all of these turned out to be mere preliminaries. In 2001 came the main event, an open-ended global war on terror, soon known in some quarters as the "Long War."[2]

Viewed in retrospect, indications that the Long Peace began almost immediately to give way to conditions antithetical to peace seem blindingly obvious. Prior to 9/11, however, the implications of developments like the 1993 bombing of the World Trade Center or the failure of the U.S. military mission to Somalia that same year were difficult to discern. After all, these small events left unaltered what many took to be the defining reality of the contemporary era: the preeminence of the United States, which seemed beyond challenge.

During the 1990s, at the urging of politicians and pundits, Americans became accustomed to thinking of their country as "the indispensable nation." Indispensability carried with it both responsibilities and prerogatives.

The chief responsibility was to preside over a grand project of political-economic convergence and integration commonly referred to as globalization. In point of fact, however, globalization served as a euphemism for soft, or informal, empire. The collapse of the Soviet Union appeared to offer an opportunity to expand and perpetuate that empire, creating something akin to a global Pax Americana.

The indispensable nation's chief prerogative, self-assigned, was to establish and enforce the norms governing the post–Cold War international order. Even in the best of circumstances, imperial

SOURCE: Edited selection from "Introduction: Without Exits" from *The Limits of Power: The End of American Exceptionalism* by Andrew J. Bacevich.

policing is a demanding task, requiring not only considerable acumen but also an abundance of determination. The preferred American approach was to rely, whenever possible, on suasion. Yet if pressed, Washington did not hesitate to use force, as its numerous military adventures during the 1990s demonstrated.

Whatever means were employed, the management of empire assumed the existence of bountiful reserves of power—economic, political, cultural, but above all military. In the immediate aftermath of the Cold War, few questioned that assumption.[3] The status of the United States as "sole superpower" appeared unassailable. Its dominance was unquestioned and unambiguous. This was not hypernationalistic chest-thumping; it was the conventional wisdom.

Recalling how Washington saw the post–Cold War world and America's place in (or atop) it helps us understand why policy makers failed to anticipate, deter, or deflect the terrorist attacks of September 11, 2001. A political elite preoccupied with the governance of empire paid little attention to protecting the United States itself. In practical terms, prior to 9/11 the mission of homeland defense was unassigned.

The institution nominally referred to as the Department of Defense didn't actually do defense; it specialized in power projection. In 2001, the Pentagon was prepared for any number of contingencies in the Balkans or Northeast Asia or the Persian Gulf. It was just not prepared to address threats to the nation's eastern seaboard. Well-trained and equipped U.S. forces stood ready to defend Seoul or Riyadh; Manhattan was left to fend for itself.

Odd as they may seem, these priorities reflected a core principle of national security policy: When it came to defending vital American interests, asserting control over the imperial periphery took precedence over guarding the nation's own perimeter.

After 9/11, the Bush administration affirmed this core principle. Although it cobbled together a new agency to attend to "homeland security," the administration also redoubled its efforts to shore up the Pax Americana and charged the Department of Defense with focusing on this task. This meant using any means necessary—suasion where possible, force as required—to bring the Islamic world into conformity with prescribed American norms. Rather than soft and consensual, the approach to imperial governance became harder and more coercive.

So, for the United States after 9/11, war became a seemingly permanent condition. President George W. Bush and members of his administration outlined a campaign against terror that they suggested might last decades, if not longer. On the national political scene, few questioned that prospect. In the Pentagon, senior military officers spoke in terms of "generational war," lasting up to a century.[4] Just two weeks after 9/11, Secretary of Defense Donald Rumsfeld was already instructing Americans to "forget about 'exit strategies'; we're looking at a sustained engagement that carries no deadlines."[5]

By and large, Americans were slow to grasp the implications of a global war with no exits and no deadlines. To earlier generations, place names like Iraq and Afghanistan had been synonymous with European rashness—the sort of obscure and unwelcoming jurisdictions to which overly ambitious kings and slightly mad adventurers might repair to squabble. For the present generation, it has already become part of the natural order of things that GIs should be exerting themselves at great cost to pacify such far-off domains. For the average American tuning in to the nightly news, reports of U.S. casualties incurred in distant lands now seem hardly more out of the ordinary than reports of partisan shenanigans on Capitol Hill or brush fires raging out of control in Southern California.

How exactly did the end of the Long Peace so quickly yield the Long War? Seeing themselves as a peaceful people, Americans remain wedded to the conviction that the conflicts in which they find themselves embroiled are not of their own making. The global war on terror is no exception. Certain at our own benign intentions, we reflexively assign responsibility for war to others, typically malignant Hitler-like figures inexplicably bent on denying us the peace that is our fondest wish….

The impulses that have landed us in a war of no exits and no deadline come from within. Foreign policy has, for decades, provided an outward manifestation of American domestic ambitions, urges, and fears. In our own time, it has increasingly become an expression of domestic dysfunction—an attempt to manage or defer coming to terms with contradictions besetting the American way of life. Those contradictions have found their ultimate expression in the perpetual state of war afflicting the United States today.

Gauging their implications requires that we acknowledge their source: They reflect the accumulated detritus of freedom, the by-products of our frantic pursuit of life, liberty, and happiness.

Freedom is the altar at which Americans worship, whatever their nominal religious persuasion. "No one sings odes to liberty as the final end of life with greater fervor than Americans," the theologian Reinhold Niebuhr once observed.[6] Yet even as they celebrate freedom, Americans exempt the object of their veneration from critical examination. In our public discourse, freedom is not so much a word or even a value as an incantation, its very mention enough to stifle doubt and terminate all debate.

...This heedless worship of freedom has been a mixed blessing. In our pursuit of freedom, we have accrued obligations and piled up debts that we are increasingly hard-pressed to meet. Especially since the 1960s, freedom itself has undercut the nation's ability to fulfill its commitments. We teeter on the edge of insolvency, desperately trying to balance accounts by relying on our presumably invincible armed forces. Yet there, too, having exaggerated our military might, we court bankruptcy.

The United States today finds itself threatened by three interlocking crises. The first of these crises is economic and cultural, the second political, and the third military. All three share this characteristic: They are of our own making.... Writing decades ago, Reinhold Niebuhr anticipated that predicament with uncanny accuracy and astonishing prescience. As such, perhaps more than any other figure in our recent history, he may help us discern a way out.

As pastor, teacher, activist, theologian, and prolific author, Niebuhr was a towering presence in American intellectual life from the 1930s through the 1960s. Even today, he deserves recognition as the most clear-eyed of American prophets. Niebuhr speaks to us from the past, offering truths of enormous relevance to the present. As prophet, he warned that what he called "our dreams of managing history"—born of a peculiar combination of arrogance and narcissism—posed a potentially mortal threat to the United States.[7] Today, we ignore that warning at our peril.

Niebuhr entertained few illusions about the nature of man, the possibilities of politics, or the pliability of history. Global economic crisis, total war, genocide, totalitarianism, and nuclear arsenals capable of destroying civilization itself—he viewed all of these with an unblinking eye that allowed no room for hypocrisy, hokum, or self-deception. Realism and humility formed the core of his worldview, each infused with a deeply felt Christian sensibility.

Realism in this sense implies an obligation to see the world as it actually is, not as we might like it to be. The enemy of realism is hubris, which in Niebuhr's day, and in our own, finds expression in an outsized confidence in the efficacy of American power as an instrument to reshape the global order.

Humility imposes an obligation of a different sort. It summons Americans to see themselves without blinders. The enemy of humility is sanctimony, which gives rise to the conviction that American values and beliefs are universal and that the nation itself serves providentially assigned purposes. This conviction finds expression in a determination to remake the world in what we imagine to be America's image.

In our own day, realism and humility have proven in short supply. What Niebuhr wrote after World War II proved truer still in the immediate aftermath of the Cold War: Good fortune and a position of apparent preeminence placed the United States "under the most grievous temptations to self-adulation."[8] Americans have given themselves over to those temptations. Hubris and sanctimony have become the paramount expressions of American statecraft. After 9/11, they combined to produce the Bush administration's war of no exits and no deadlines.

President Bush has likened today's war against what he calls "Islamofascism" to America's war with Nazi Germany—a great struggle waged on behalf of liberty.... Yet that commitment, however well intentioned, begs several larger questions: As actually expressed and experienced, what is freedom today? What is its content? What costs does the exercise of freedom impose? Who pays?

These are fundamental questions, which cannot be dismissed with a rhetorical wave of the hand. Great wartime presidents of the past—one thinks especially of Abraham Lincoln speaking at Gettysburg—have not hesitated to confront such questions directly....

Freedom is not static, nor is it necessarily benign. In practice, freedom constantly evolves and in doing so generates new requirements and abolishes old constraints. The common understanding of freedom that prevailed in December 1941 when the United States entered the war against Imperial Japan and Nazi Germany has long since become obsolete. In some respects, this must be cause for celebration. In others, it might be cause for regret.

The changes have been both qualitative and quantitative. In many respects, Americans are freer today than ever before, with more citizens than ever before enjoying unencumbered access to the promise of American life. Yet especially since the 1960s, the reinterpretation of freedom has had a transformative impact on our society and culture. That transformation has produced a paradoxical legacy. As individuals, our appetites and expectations have grown exponentially. Niebuhr once wrote disapprovingly of Americans, their "culture soft and vulgar, equating joy with happiness and happiness with comfort."[9] Were he alive today, Niebuhr might amend that judgment, with Americans increasingly equating comfort with self-indulgence.

The collective capacity of our domestic political economy to satisfy those appetites has not kept pace with demand. As a result, sustaining our pursuit of life, liberty, and happiness at home requires increasingly that Americans look beyond our borders. Whether the issue at hand is oil, credit, or the availability of cheap consumer goods, we expect the world to accommodate the American way of life.

The resulting sense of entitlement has great implications for foreign policy. Simply put, as the American appetite for freedom has grown, so too has our penchant for empire. The connection between these two tendencies is a causal one. In an earlier age, Americans saw empire as the antithesis of freedom. Today, as illustrated above all by the Bush administration's efforts to dominate the energy-rich Persian Gulf, empire has seemingly become a prerequisite of freedom.

There is a further paradox: The actual exercise of American freedom is no longer conducive to generating the power required to establish and maintain an imperial order. If anything, the reverse is true: Centered on consumption and individual autonomy, the exercise of freedom is contributing to the gradual erosion of our national power. At precisely the moment when the ability to wield power—especially military power—has become the sine qua non for preserving American freedom, our reserves of power are being depleted.

One sees this, for example, in the way that heightened claims of individual autonomy have eviscerated the concept of citizenship. Yesterday's civic obligations have become today's civic options. What once rated as duties—rallying to the country's defense at times of great emergency, for example—are now matters of choice. As individuals, Americans never cease to expect more. As members of a community, especially as members of a national community, they choose to contribute less.

Meanwhile, American political leaders—especially at the national level—have proven unable (or unwilling) to address the disparity between how much we want and what we can afford to pay. Successive administrations, abetted by Congress, have deepened a looming crisis of debt and dependency through unbridled spending. As Vice President Dick Cheney, a self-described conservative, announced when told that cutting taxes might be at odds with invading Iraq, "Deficits don't matter."[10] Politicians of both parties certainly act as if they don't.

Expectations that the world beyond our borders should accommodate the American way of life are hardly new. Since 9/11, however, our demands have become more insistent. In that regard, the neoconservative writer Robert Kagan is surely correct in observing that "America did not change on September 11. It only became more itself."[11] In the aftermath of the attacks on the World Trade Center and the Pentagon, Washington's resolve that nothing interfere with the individual American's pursuit of life, liberty, and happiness only hardened. That resolve found expression in the Bush administration's with-us-or-against-us rhetoric, in its disdain for the United Nations and traditional American allies, in its contempt for international law, and above all in its embrace of preventive war.

When President Bush declared in his second inaugural that the "survival of liberty in our land increasingly depends on the success of liberty in other lands," he was in effect claiming for the United States as freedom's chief agent the prerogative of waging war when and where it sees fit, those wars by definition being fought on freedom's behalf. In this sense, the Long War genuinely qualifies as a war to preserve the American way of life (centered on a specific conception of liberty) and simultaneously as a war to extend the American imperium (centered on dreams of a world remade in America's image), the former widely assumed to require the latter.

Yet, as events have made plain, the United States is ill-prepared to wage a global war of no exits and no deadlines. The sole superpower lacks the resources—economic, political, and military—to support a large-scale, protracted conflict without, at the very least, inflicting severe economic and political damage on itself. American power has limits and is inadequate to the ambitions to which hubris and sanctimony have given rise.

Here is the central paradox of our time: While the defense of American freedom seems to demand that U.S. troops fight in places like Iraq and Afghanistan, the exercise of that freedom at home undermines the nation's capacity to fight. A grand bazaar provides an inadequate basis upon which to erect a vast empire.

Meanwhile, a stubborn insistence on staying the course militarily ends up jeopardizing freedom at home. With Americans, even in wartime, refusing to curb their appetites, the Long War aggravates the economic contradictions that continue to produce debt and dependency. Moreover, a state of perpetual national security emergency aggravates the disorders afflicting our political system, allowing the executive branch to accrue ever more authority at the expense of the Congress and disfiguring the Constitution. In this sense, the Long War is both self-defeating and irrational.

Niebuhr once wrote, "One of the most pathetic aspects of human history is that every civilization expresses itself most pretentiously, compounds its partial and universal values most convincingly, and claims immortality for its finite existence at the very moment when the decay which leads to death has already begun."[12] Future generations of historians may well cite Niebuhr's dictum as a concise explanation of the folly that propelled the United States into its Long War.

In an immediate sense, it is the soldier who bears the burden of such folly. U.S. troops in battle dress and body armor, whom Americans profess to admire and support, pay the price for the nation's collective refusal to confront our domestic dysfunction. In many ways, the condition of the military today offers the most urgent expression of that dysfunction. Seven years into its confrontation with radical Islam, the United States finds itself with too much war for too few warriors—and with no prospect of producing the additional soldier needed to close the gap. In effect, Americans now confront a looming military crisis to go along with the economic and political crises that they have labored so earnestly to ignore.

The Iraq War deserves our attention as the clearest manifestation of these three crises, demonstrating the extent to which they are inextricably linked and mutually reinforcing. That war was always unnecessary. Except in the eyes of the deluded and the disingenuous, it has long since become a fool's errand. Of perhaps even greater significance, it is both counterproductive and unsustainable.

Yet ironically Iraq may yet prove to be the source of our salvation. For the United States, the ongoing war makes plain the imperative of putting America's house in order. Iraq has revealed the futility of counting on military power to sustain our habits of profligacy. The day of reckoning approaches. Expending the lives of more American soldiers in hopes of deferring that day is profoundly wrong. History will not judge kindly a people who find nothing amiss in the prospect of endless armed conflict so long as they themselves are spared the effects. Nor will it view with favor an electorate that delivers political power into the hands at leaders unable to envision any alternative to perpetual war.

Rather than insisting that the world accommodate the United States, Americans need to reassert control over their own destiny, ending their condition of dependency and abandoning their imperial delusions. Of perhaps even greater difficulty, the combination of economic, political, and military crisis summons Americans to reexamine exactly what freedom entails. Soldiers cannot accomplish these tasks, nor should we expect politicians to do so. The onus of responsibility falls squarely on citizens.

ENDNOTES

1. John Lewis Gaddis, *The Long Peace: Inquiries into the History of the Cold War* (New York, 1989).
2. As a description of the post-9/11 conflict, the phrase is attributed to General John Abizaid, commander of U.S. Central Command from 2003 to 2007. Bradley Graham and Josh White, "Abizaid Credited with Popularizing the Term 'Long War,'" *Washington Post*, February 3, 2006.
3. Those who did question the assumption of permanent American supremacy tended to be foreigners, which makes them easier to ignore. See, for example, Emmanuel Todd, *After the Empire: The Breakdown of the American Order* (New York, 2002).
4. Bill Gertz, "General Foresees 'Generational War' against Terrorism," *Washington Times*, December 13, 2006.
5. Donald Rumsfeld, "A New Kind of War," *New York Times*, September 27, 2001.
6. Reinhold Niebuhr, *The Irony of American History* (New York, 1952), p. 91.
7. Ibid., p.3.
8. D. B. Robertson, *Love and Justice: Selections from the Shorter Writings of Reinhold Niebuhr* (Cleveland, Ohio, 1957), p. 97.
9. Reinhold Niebuhr, *The World Crisis and American Responsibility* (New York, 1958), p. 125.
10. Quoted in Ron Suskind, *The Price of Loyalty: George W. Bush, the White House, and the Education of Paul O'Neill* (New York, 2004), p. 291.
11. Robert Kagan, "Power and Weakness," *Policy Review*, June–July 2002.
12. Reinhold Niebuhr, *Beyond Tragedy* (New York, 1937), p. 39.

DISCUSSION QUESTIONS

1. What does it mean that the United States is pursuing a policy of "global war with no exits and no deadline"? Why is this problem one of *social* concern?
2. What are the unintended consequences to society and for the economy of wars without exits?
3. What does the author mean when he says, "Freedom is the altar at which Americans worship, whatever their nominal religious persuasion"? How does our abiding belief in freedom contribute to the irrationality of wars without exits?

4. Do you agree with Bacevich that the United States faces "three interlocking crises" of economics, politics, and culture? Why or why not? What is the principal crisis in each of these arenas?
5. Provide an explication in your own words of the central paradox of our time, according to Bacevich. Do you agree with this paradox? Why or why not?

47

The Global Rise of Religious Violence

MARK JUERGENSMEYER

The Four Questions

1. What exactly is the problem described?
2. Is it a *social* problem?
3. What is the cause of this problem, according to Juergensmeyer?
4. Does Juergensmeyer describe what can be done about this situation?

Topics Covered

Religion

Violence

Terrorism

Culture

When plastic explosives attached to a Hamas suicide bomber ripped through the gentrified Ben Yehuda shopping mall in Jerusalem in December 2001, the blast damaged not only lives and property but also the confidence with which most people view the world. As in prior acts of terrorism on this same popular mall, the news images of the bloodied victims projected from the scene portrayed the double arches of a McDonald's restaurant in the background, their cheerful familiarity appearing oddly out of place with the surrounding carnage. Many who viewed these pictures saw symbols of their own ordinary lives assaulted and vicariously felt the anxiety—the terror—of those who experienced it firsthand. After all, the wounded could have included anyone who has ever visited a McDonald's—which is to say virtually anyone in the developed world. In this sense, the blast was an attack not only on Israel but also on normal life as most people know it.

This loss of innocence was keenly felt by Americans as they watched in horror at the televised images of the September 11, 2001 assaults on New York City's World Trade Center. But even in the years immediately prior to the destruction of the World Trade Center, Americans had been targets of a diverse series of terrorist attacks: ethnic shootings in California and Illinois in 1999; the attack on American embassies in Africa in 1998; abortion clinic bombings in Alabama and Georgia in 1997; the bomb blast at the Olympics in Atlanta and the destruction of a U.S. military housing complex in Dhahran, Saudi Arabia, in 1996; the tragic destruction of the federal building at Oklahoma City in 1995; and the 1993 explosion at the World Trade Center, which was an eerie forecast of the terror to come scarcely eight years later. These incidents and a host of violent episodes associated with American religious extremists—including the Christian militia, the Christian Identity movement, and Christian anti-abortion activists—have brought Americans into the same uneasy position occupied by many in the rest of the world. Increasingly, global society must confront religious violence on a routine basis.

The French, for example, have dealt with subway bombs planted by Algerian Islamic activists, the British with exploding trucks and buses ignited by Irish Catholic nationalists, and the Japanese with nerve gas placed in Tokyo subways by members of a Hindu-Buddhist sect. In India residents of Delhi have experienced car bombings by both Sikh and Kashmiri separatists, in Sri Lanka whole sections of the city of Colombo have been destroyed both by Tamils and by Sinhalese militants, Egyptians have been forced to live with militant Islamic attacks in coffeehouses and riverboats, Indonesians witnessed the bombing of Bali nightclubs by al Qaeda-related activists, Algerians have lost entire villages to attacks perpetrated allegedly by supporters of the Islamic Salvation Front, and Israelis and Palestinians have confronted the deadly deeds of both Jewish and Muslim extremists. For many Middle Easterners, terrorist attacks have become a way of life.

In addition to their contemporaneity, all these instances share two striking characteristics. First, they have been violent—even vicious—in a manner calculated to be terrifying. And, second, they have been motivated by religion.

THE MEANING OF RELIGIOUS TERRORISM

The ferocity of religious violence was brought home to me in the midst of the conflict in Northern Ireland when I received the news that a car bomb had exploded in a Belfast neighborhood I had visited the day before. The following day firebombs ripped through several pubs and stores, apparently in protest against the fragile peace agreement signed earlier in 1998. It was a disturbing repetition of another close call, one that I had experienced in Israel. A suicide bombing claimed by the militant wing of the Palestinian Muslim political movement, Hamas, tore apart a bus near Hebrew University the day after I had visited the university on, I believe, the very same bus. The pictures of the mangled bodies on the Jerusalem street and the images of Belfast's bombed-out pub, therefore, had a direct and immediate impact on my view of the world.

What I realized then is the same thing that all of us perceive on some level when we view pictures of terrorist events: on a different day, at a different time, perhaps in a different bus, one of the bodies torn to shreds by any of these terrorist acts could have been ours. What came to mind as I heard the news of the Belfast and Jerusalem bombings, however, was not so much a feeling of relief for my safety as a sense of betrayal—that the personal security and order that is usually a basic assumption of public life cannot in fact be taken for granted in a world where terrorist acts exist.

That, I take it, is largely the point: terrorism is meant to terrify. The word comes from the Latin *terrere*, "to cause to tremble," and came into common usage in the political sense, as an assault on civil order, during the Reign of Terror in the French Revolution at the close of the eighteenth century. Hence the public response to the violence—the trembling that terrorism effects—is part of the meaning of the term. It is appropriate, then, that the definition of a terrorist act is provided by us, the witnesses—the ones terrified—and not by the party committing the act. It is we—or more often our public agents, the news media—who affix the label on acts of violence that makes them terrorism. These are public acts of destruction, committed without a clear military objective, that arouse a widespread sense of fear.

This fear often turns to anger when we discover the other characteristic that frequently attends these acts of public violence: their justification by religion. Most people feel that religion should provide tranquility and peace, not terror. Yet in many of these cases religion has supplied not only the ideology but also the motivation and the organizational structure for the perpetrators. It is true that some terrorist acts are committed by public officials invoking a sort of "state terrorism" in order to subjugate the populace. The pogroms of Stalin, the government-supported death squads in El Salvador, the genocidal killings of the Khmer Rouge in Cambodia, ethnic cleansing in Bosnia and Kosovo, and government-spurred violence of the Hutus and Tutsis in Central Africa all come to mind. The United States has rightfully been accused of terrorism in the atrocities committed during the Vietnam War, and there is some basis for considering the nuclear bombings of Hiroshima and Nagasaki as terrorist acts.

But the term "terrorism" has more frequently been associated with violence committed by disenfranchised groups desperately attempting to gain a shred of power or influence. Although these groups cannot kill on the scale that governments with all their military power can, their sheer numbers, their intense dedication, and their dangerous unpredictability have given them influence vastly out of proportion with their meager military resources. Some of these groups have been inspired by purely secular causes. They have been motivated by leftist ideologies, as in the cases of the Shining Path and the Tupac Amaru in Peru, and the Red Army in Japan; and they have been propelled by a desire for ethnic or regional separatism, as in the cases of Basque militants in Spain and the Kurdish nationalists in the Middle East.

But more often it has been religion—sometimes in combination with these other factors, sometimes as the primary motivation—that has incited terrorist acts. The common perception that there has been a rise in religious violence around the world in the last decades of the twentieth century has been borne out by those who keep records of such things. In 1980 the U.S. State Department roster of international terrorist groups listed scarcely a single religious organization. Almost twenty years later, at the end of the twentieth century, over half were religious.[1] They were Jewish, Muslim, and Buddhist. If one added to this list other violent religious groups around the world, including the many Christian militia and other paramilitary organizations found domestically in the United States, the number of religious terrorist groups would be considerable. According to the RAND–St. Andrews Chronology of International Terrorism, the proportion of religious groups in the late 1990s increased from sixteen of forty-nine terrorist groups to twenty-six of the fifty-six groups listed the following year.[2] For this reason U.S. government officials frequently proclaim terrorism in the name of religion and ethnicity, as one of them put it, "the most important security challenge we face in the wake of the Cold War."[3]

Throughout this study we will be looking at this odd attraction of religion and violence. Although some observers try to explain away religion's recent ties to violence as an aberration, a result of political ideology, or the characteristic of a mutant form of religion—fundamentalism—these are not my views. Rather, I look for explanations in the current forces of geopolitics and in a strain of violence that may be found at the deepest levels of religious imagination.

Within the histories of religious traditions—from biblical wars to crusading ventures and great acts of martyrdom—violence has lurked as a shadowy presence. It has colored religion's darker, more mysterious symbols. Images of death have never been far from the heart of religion's power to stir the imagination. One of the haunting questions asked by some of the great scholars of religion—including Émile Durkheim, Marcel Mauss, and Sigmund Freud—is why this is the case. Why does religion seem to need violence, and violence religion, and why is a divine mandate for destruction accepted with such certainty by some believers?

These are questions that have taken on a sense of urgency in recent years, when religious violence has reappeared in a form often calculated to terrify on a massive scale. These contemporary acts of violence are often justified by the historical precedent of religion's violent past. Yet the forces that combine to produce religious violence are particular to each moment of history. For this reason, I will focus on case studies of religious violence both within their own cultural contexts and within the framework of global social and political changes that are distinctive to our time.

This is a [selection] about religious terrorism. It is about public acts of violence at the turn of the century for which religion has provided the motivation, the justification, the organization, and the world view. In this [selection], I have tried to get inside the mindset of those who perpetrated and supported such acts. My goal is to understand why these acts were often associated with religious causes and why they have occurred with such frequency at this juncture in history. Although it is not my purpose to be sympathetic to people who have done terrible things, I do want to understand them and their world views well enough to know how they and their supporters can morally justify what they have done.

What puzzles me is not why bad things are done by bad people, but rather why bad things are done by people who otherwise appear to be good—in cases of religious terrorism, by pious people dedicated to a moral vision of the world. Considering the high-sounding rhetoric with which their purposes are often stated, it is perhaps all the more tragic that the acts of violence meant to achieve them have caused suffering and disruption in many lives—not only those who were injured by the acts, but also those who witnessed them, even from a distance.

Because I want to understand the cultural contexts that produce these acts of violence, my focus is on the ideas and the communities of support that lie behind the acts rather than on the "terrorists" who commit them. In fact, for the purposes of this study, the word "terrorist" is problematic. For one thing, the term makes no clear distinction between the organizers of an attack, those who carry it out, and the many who support it both directly and indirectly. Are they all terrorists, or just some of them—and if the latter, which ones? Another problem with the word is that it can be taken to single out a certain limited species of people called "terrorists" who are committed to violent acts. The implication is that such terrorists are hell-bent to commit terrorism for whatever reason—sometimes choosing religion, sometimes another ideology, to justify their mischief. This logic concludes that terrorism exists because terrorists exist, and if we just got rid of them, the world would be a more pleasant place.

Although such a solution is enticing, the fact is that the line is very thin between "terrorists" and their "non-terrorist" supporters. It is also not clear that there is such a thing as a "terrorist" before someone conspires to perpetrate a terrorist act. Although every society contains sociopaths and others who sadistically enjoy killing, it is seldom such persons who are involved in the deliberate public events that we associate with terrorism, and few studies of terrorism focus exclusively on personality. The studies of the psychology of terrorism deal largely with social psychology; that is, they are concerned with the way people respond to certain group situations that make violent public acts possible.[4] I know of no study that suggests that people are terrorist by nature. Although some activists involved in religious terrorism have been troubled by mental problems, others are people who appear to be normal and socially well adjusted, but who are caught up in extraordinary communities and share extreme world views.

Most of the people involved in acts of religious terrorism are not unlike Dr. Baruch Goldstein, who killed over thirty Muslims as they were praying at the Tomb of the Patriarchs in Hebron. Goldstein was a medical doctor who grew up in a middle-class community in Brooklyn and received his professional training at Albert Einstein College of Medicine in the Bronx. His commitment to an extreme form of Zionism brought him to Israel and the Kiryat Arba settlement, and although he was politically active for many years—he was Rabbi Meir Kahane's campaign manager when he ran for the Israeli parliament—Goldstein did not appear to be an irrational or vicious person. Prior to the attack at Hebron, his most publicized political act had been a letter to the editor of the *New York Times*.[5] If Goldstein had deep and perverse personality flaws that eventually surfaced and made him a terrorist, we do not know about them. The evidence about him is to the contrary: it indicates that, like his counterparts in Hamas, he was an otherwise decent man who became overwhelmed by a great sense of dedication to a religious vision shared by many in the community of which he was a part. He became convinced that this vision and community were profoundly assaulted, and this compelled him to a desperate and tragic act. He was certainly single-minded about his religious concerns—even obsessed over them—but to label Goldstein a terrorist prior to the horrible act he committed implies that he was a terrorist by nature and that his religiosity was simply a charade. The evidence does not indicate either to be the case.

For this reason I use the term "terrorist" sparingly. When I do use it, I employ it in the same sense as the word "murderer": it applies to specific persons only after they have been found guilty of committing such a crime, or planning to commit one. Even then I am somewhat cautious about using the term, since a violent act is "terrorism" technically only in the eyes of the courts, more publicly in the eyes of the media, and ultimately only in the eyes of the beholder. The old saying "One person's terrorist is another person's freedom-fighter" has some truth to it. The designation of terrorism is a subjective judgment about the legitimacy of certain violent acts as much as it is a descriptive statement about them.

When I interviewed militant religious activists and their supporters, I found that they seldom used the term "terrorist" to describe what their groups had done. Several told me that their groups should be labeled militant rather than terrorist. A Lutheran pastor who was convicted of bombing abortion clinics was not a terrorist, he told me, since he did not enjoy violence for its own sake. He employed violence only for a purpose, and for that reason he described these events as "defensive actions" on behalf of the "unborn."[6] Activists on both sides of the struggle in Belfast described themselves as "paramilitaries." A leader in India's Sikh separatist movement said that he preferred the term "militant" and told me that "terrorist" had replaced the term "witch" as an excuse to persecute those whom one dislikes.[7] One of the men associated with the al Qaeda network essentially agreed with the Sikh leader, telling me that the word "terrorist" was so "messy" it could not be used without a lot of qualifications.[8] The same point of view was expressed by the political leader of the Hamas movement with whom I talked in Gaza. He described his movement's suicide attacks as "operations."[9] Like many activists who used violence, he likened his group to an army that was planning defensive maneuvers and using violence strategically as necessary acts. Never did he use the word "terrorist" or "terrorism."

This is not just a semantic issue. Whether or not one uses "terrorist" to describe violent acts depends on whether one thinks that the acts are warranted. To a large extent the use of the term depends on one's world view: if the world is perceived as peaceful, violent acts appear as terrorism. If the world is thought to be at war, violent acts may be regarded as legitimate. They may be seen as preemptive strikes, as defensive tactics in an ongoing battle, or as symbols indicating to the world that it is indeed in a state of grave and ultimate conflict....

SEEING INSIDE CULTURES OF VIOLENCE

Terrorism is seldom a lone act. When Dr. Baruch Goldstein entered the Tomb of the Patriarchs carrying an automatic weapon, he came with the tacit approval of many of his fellow Jewish settlers in the nearby community of Kiryat Arba. When the five Aum Shinrikyo scientists boarded subway trains in Tokyo headed towards the Kasumigaseki terminal and unleashed their deadly containers of sarin gas, they were acting according to the instructions of their organization's leaders. When Rev. Paul Hill stepped from a sidewalk in Pensacola, Florida, and shot Dr. John Britton and his security escort as they prepared to enter their clinic, he was cheered by a certain circle of militant Christian anti-abortion activists around the United States. When Mohammad Atta and other members of the al Qaeda network boarded commercial airlines in Boston and Newark which minutes later plunged into the twin towers of the World Trade Center, eventually causing them to crumble into dust, they came as part of a well-orchestrated plan that involved dozens of co-conspirators and thousands of sympathizers in the United States, Europe, Afghanistan, Saudi Arabia, and elsewhere throughout the world.

As these instances show, it takes a community of support and, in many cases, a large organizational network for an act of terrorism to succeed. It also requires an enormous amount of moral presumption for the perpetrators of these acts to justify the destruction of property on a massive scale or to condone a brutal attack on another life, especially the life of someone one scarcely knows and against whom one bears no personal enmity. And it requires a great deal of internal conviction, social acknowledgment, and the stamp of approval from a legitimizing ideology or authority one respects. Because of the moral, ideological, and organizational support necessary for such acts, most of them come as collective decisions —such as the conspiracy that led to the release of nerve gas in the Tokyo subways and the Hamas organization's carefully devised bombings.

Even those acts that appear to be solo ventures conducted by rogue activists often have networks of support and ideologies of validation behind them, whether or not these networks and ideologies are immediately apparent. Behind Yitzhak Rabin's assassin, Yigal Amir, for instance, was a large movement of Messianic Zionism in Israel and abroad. Behind convicted bomber Timothy McVeigh and Buford Furrow, the alleged attacker of a Jewish day-care center, was a subculture of militant Christian groups that extends throughout the United States. Behind America's Unabomber, Theodore Kaczynski, was the strident student activist culture of the late 1970s, in which one could easily become infected by the feeling that "terrible things" were going on.[10] The 1993 bombing of the World Trade Center was initially thought to be the work of only a small group of individuals linked to a blind Egyptian sheik; only later was it found to have wider connectons to the world-wide al Qaeda network of Islamic activism associated with Osama bin Laden. In all of these cases the activists thought that their acts were supported not only by other people but by a widely shared perception that the world was already violent: it was enmeshed in great struggles that gave their own violent actions moral meaning.

This is a significant feature of these cultures: the perception that their communities are already under attack—are being violated—and that their acts are therefore simply responses to the violence they have experienced. In some cases this perception is one to which sensitive people outside the movement can readily relate—the feeling of oppression held by Palestinian Muslims, for example, is one that many throughout the world consider to be an understandable though regrettable response to a situation of political control. In other instances, such as the imagined oppression of America's Christian militia or Japan's Aum Shinrikyo movement, the members' fears of black helicopters hovering over their homes at night or the allegations of collusion of international governments to deprive individuals of their freedoms are regarded by most people outside the movements as paranoid delusions. Still other cases—such as those involving Sikh militants in India, Jewish settlers on the West Bank, Muslim politicians in Algeria, Catholic and Protestant militants in Northern Ireland, and anti-abortion activists in the United States—are highly controversial. There are sober and sensitive people to argue each side. In many cases, such as in the terrorist acts perpetrated by the al Qaeda activists associated with Osama bin Laden, specific political grievances are magnified into grand spiritual condemnations.

Whether or not outsiders regard these perceptions of oppression as legitimate, they are certainly considered valid by those within the communities. It is these shared perceptions that constitute the cultures of violence that have flourished throughout the world—in neighborhoods of Jewish nationalists from Kiryat Arba to Brooklyn where the struggle to defend the Jewish nation is part of daily existence, in mountain towns in Idaho and Montana where religious and individual freedoms are thought to be imperiled by an enormous governmental conspiracy, and in pious Muslim communities around the world where Islam is felt to be at war with the surrounding

*"responses to the violence they have experienced

secular forces of modern society. Although geographically dispersed, these cultures in some cases are fairly small: one should bear in mind that the culture of violence characterized by Hamas, for example, does not implicate all Palestinians, all Muslims, or even all Palestinian Muslims....

In order to respond to religious terrorism in a way that is effective and does not produce more terrorism in response, I believe it is necessary to understand why such acts occur. Behind this practical purpose in writing this book, however, is an attempt to understand the role that violence has always played in the religious imagination and how terror could be conceived in the mind of God.

These two purposes are connected. One of my conclusions is that this historical moment of global transformation has provided an occasion for religion—with all its images and ideas—to be reasserted as a public force. Lurking in the background of much of religion's unrest and the occasion for its political revival, I believe, is the virtually global devaluation of secular authority and the need for alternative ideologies of public order. It may be one of the ironies of history graphically displayed in incidents of terrorism, that the answers to the questions of why the contemporary world still needs religion and of why it has suffered such public acts of violence, are surprisingly the same.

ENDNOTES

1. "Global Terror," *Los Angeles Times*, August 8, 1998, A16.
2. Bruce Hoffman, *Inside Terrorism* (New York: Columbia University Press, 1998), 91.
3. Warren Christopher, "Fighting Terrorism: Challenges for Peacemakers," address to the Washington Institute for Near East Policy, May 21, 1996. Reprinted in Warren Christopher, *In the Stream of History: Shaping Foreign Policy for a New Era* (Stanford, CA: Stanford University Press, 1998), 446.
4. See, for example, the essays from a conference on the psychology of terrorism held at the Woodrow Wilson International Center for Scholars, in Walter Reich, ed., *Origins of Terrorism: Psychologies, Ideologies, Theologies, States of Mind* (New York: Cambridge University Press, 1990).
5. Baruch Goldstein, letter to the editor, *New York Times*, June 30, 1981.
6. Interview with Rev. Michael Bray, Reformation Lutheran Church, Bowie, Maryland, April 25, 1996.
7. Interview with Sohan Singh, leader of the Sohan Singh Panthic Committee, Mohalli, Punjab, August 3, 1996.
8. Interview with Mahmud Abouhalima, convicted coconspirator in the World Trade Center bombing case, federal penitentiary, Lompoc, California, September 30, 1997.
9. Interview with Abdul Aziz Rantisi, cofounder and political leader of Hamas, Khan Yunis, Gaza, March 1, 1998.
10. Lance W. Small, an assistant professor of mathematics at the University of California, Berkeley, at the time Kaczynski taught there, quoted in David Johnston and Janny Scott, "The Tortured Genius of Theodore Kaczynski," *New York Times*, May 26, 1996, Al. According to the authors, Kaczynski's brother David thought that Kaczynski was unaffected by any particular political movement at the time.

DISCUSSION QUESTIONS

1. This is a very challenging selection, in part because it is difficult to pin down exactly what the problem is, much less the cause. It is about violence and it is about religion. Religion

sometimes encourages violence; violence sometimes shapes religion. Do you agree? Discuss both aspects.

2. Perhaps religion does not really encourage violence; perhaps its role is to justify violence, violence that has other causes. What do you think?
3. What is "terror," according to Juergensmeyer? Do you agree with him? Definition is very important: If it is carefully done it must clearly distinguish what events are included in its category and what are not. Can you do this in defining *terror?*
4. "One person's terrorist is another person's freedom-fighter." Do you agree? What happens in trying to define *terrorism* if you agree with this? Is definition possible?
5. What is the one common characteristic that all religious terrorists have?
6. In your opinion, is it necessary to see things from the point of view of terrorists to understand the root cause of violence by them?

48

White Male Militia

ABBY L. FERBER AND MICHAEL S. KIMMEL

The Four Questions

1. What is the problem? Is it really a problem?
2. Is it a *social* problem? Are large numbers of people hurt? Is society hurt? Are the causes of the problem social? Is it recognized as a problem by many people?
3. What are the causes of the problem?
4. How can this problem be alleviated?

Topics Covered

Rural America

Ideology

Militia movement

Christian identity theology

Social movements

Downward mobility

Racism

Manhood

THE RURAL CONTEXT

The economic restructuring of the global economy has had a dramatic effect on rural areas throughout the industrial world.[1] The Reagan Revolution in general meant corporate downsizing, declining real wages, changing technology, increasing the gap between the wealthy and everyone else, uncertainty in the stock market, new waves of Latino and Asian immigrants to the United States, and a steady

SOURCE: From Abby L. Ferber and Michael S. Kimmel, *Home-Grown Hate: Gender and Organized Racism*, 1997. Reprinted by permission.

decline in manufacturing jobs, which were replaced by lower-paying jobs in the service sector. Increased capital mobility and the elimination of tariff barriers have also weakened the bargaining power of labor and left the average American worker feeling vulnerable and betrayed.[2] Between 1980 and 1985 alone, 11 million American workers lost their jobs through plant closings and layoffs. Of those who found new jobs, over half experienced downward mobility.[3] For rural Americans, economic uncertainty was compounded by threats to traditional Western industries like logging, mining, ranching, and farming, where consolidation has also proceeded rapidly and markedly.[4] Squeezed between corporate capital (agribusiness) and the federal government (regulations, environmentalism, and the like), farmers felt themselves to be the "victim[s] of the global restructuring of the rural world."[5]

Since 1980, nearly three quarters of a million medium and small-sized family farms have been lost. During the 1980s farm crisis, "farmers faced the worst financial stress since the Great Depression."[6] For the affected farmers, it dashed the American Dream of upward mobility and replaced it with the stark reality of downward mobility. As Osha Davidson notes: "Many of the new rural poor had not only shared American cultural goals—they had achieved them for a time. They had been *in* the middle class, *of* the middle class. They had tasted the good life and then had fallen from it." Davidson also notes the irony—crucial to our analysis—that "the victims of this blight, the inhabitants of the new rural ghettos, have always been the most blindly patriotic of Americans, the keepers of the American dream."[7] The state of crisis continues today, as "family farms, which use little hired labor and whose households are sustained through farming alone, are being edged out."[8] Many are now speaking of a "new farm crisis," but it is more accurate to say that the crisis of the 1980s never truly ended.[9]

Moreover, for many the continuing farm crisis is a gender crisis, a crisis of masculinity, as we argue below. Many white, rural, American men feel under siege and vulnerable, unsure of their manhood. They are furious and are looking for someone to blame.

Some direct their rage inwards, even to the point of suicidal thoughts and actions. Others direct this seething rage outwards. "Many debt ridden farm families will become more suspicious of government, as their self-worth, their sense of belonging, their hope for the future deteriorates," predicted Oklahoma psychologist Glen Wallace in 1989. "The farms are gone," writes Joel Dyer, "yet the farmers remain. They've been transformed into a wildfire of rage, fueled by the grief of their loss and blown by the winds of conspiracy and hate-filled rhetoric."[10]

It is not only many rural men that have faced wrenching economic transformations, however; many urban men, too, have tasted the good life and fallen from it. "It is hardly surprising, then, that American men—lacking confidence in the government and the economy, troubled by the changing relations between the sexes, uncertain of their identity or their future—began to *dream*, to fantasize about the powers and features of another kind of man who could retake and reorder the world. And the hero of all these dreams was the paramilitary warrior."[11] The militia movement is one embodiment of these dreams, one that is strikingly rural both in the population it draws from and in the location of the majority of its activities.

THE MILITIA MOVEMENT

The militia movement cannot be easily defined; there is no central organization or leadership. The movement is comprised of loosely connected paramilitary organizations who "perceive a global conspiracy in which key political and economic events are manipulated by a small group of elite insiders."[12] The movement consists of numerous unrelated groups who form private armies, share distrust in the government, and have armed themselves to fight back. According to the Militia Watchdog, an Internet organization that tracks the militia movement, the militias originated with the Posse Comitatus and the patriot movement from the 1970s and 1980s. (The Southern Poverty Law Center [SPLC], which tracks right-wing extremist groups, lists the militias as a subset of the patriot movement.) Like their predecessors, militia members believe that the U.S. government has become totalitarian and seeks to disarm its citizenry and create a "One World Government."

Militia members believe that traditional political reform is useless and that they must resist our laws and attack the government. They believe that armed confrontation is inevitable. While not unified in any traditional sense, the movement is nonetheless tied together through the Internet, where groups and individuals share stories and tips. At places like survivalist expos and gun shows, they sell literature, recruit new members, and purchase arms and survivalist gear.[13] Some groups sell their wares via mail-order catalogs, and many meet on the weekends to train in guerilla-warfare tactics. Militia organizations frequent *Soldier of Fortune* conventions, subscribe to the magazine, and draw members from *Soldier of Fortune* enthusiasts. The Militia of Montana has had booths at *Soldier of Fortune* expositions, where they peddle T-shirts that read "Angry White Guy" and bumper stickers proclaiming "I Love My Country, but I Hate My Government."[14]

The Militia of Montana (with the deliciously unironically gendered acronym "MOM") may serve as the prototype American militia. Founded by former Aryan Nations member John Trochman with his brother and nephew in the aftermath of Ruby Ridge, MOM was the first significant militia organization and the largest national distributor of militia propaganda. At MOM meetings and through mail-order distribution, members sell their own manuals as well as a wide variety of books and videos including, *A Call to Arms, Battle Preparations Now, The Pestilence (AIDS), America in Crisis, The Illuminati Today, Booby Traps,* and *Big Sister Is Watching You* (discussed below). Numerous manuals encourage and train readers in kidnapping, murder, and explosives, urging acts of terrorism.[15]

Estimates of the militia movement's size and appeal vary. Since the first militias began appearing in the early 1990s, their numbers have expanded to include anywhere from 50,000 to 100,000 members in at least forty states.[16] In 1996, the number of militias and similar patriot groups hit an all-time high of 858, with militia units or organizers in every state.[17] That number declined to 435 in 1998, of which 171 were classified as militias, and in 2001, only 158 patriot groups remained active, with 73 classified as militias. This decline is attributable to a variety of factors. According to Mark Potok of the SPLC:

> They have gone home, disillusioned and tired of waiting for the revolution that never seems to come. They have been scared off, frightened by the arrests of thousands of comrades for engaging in illegal "common-law" court tactics, weapons violations and even terrorist plots. And they have, in great numbers, left the relatively non-racist Patriot world for the harder-line groups that now make up most of the radical right."[18]

In many respects the militia movement is dead, although a variety of militia and patriot groups still exists. It represented one brief, strong manifestation of the radical right, drawing from the organization and ideology of the white-supremacist movement in conjunction with other influences. Many who once participated in the movement have since moved into other radical-right organizations. There are few sharp lines dividing the various subgroups of the radical right. People flow between groups and have overlapping memberships and allegiances.[19] At its peak, one observer wrote, it was the "convergence of various streams of fanatical rightwing beliefs that seems to be sweeping the militia movement along. Overlapping right wing social movements with militant factions appear to be coalescing within the militias."[20] This demonstrates the important point that far-right groups are intricately interconnected and share a basic antigovernment, anti-Semitic, racist, and sexist/patriarchal ideology. Equally, the extent of involvement in the movement varies; some men simply correspond through e-mail and read militia literature; others attend training sessions, stockpile food supplies and weapons, and resist paying taxes. The most dangerous form small, secret cells of two to ten people who plan sabotage and terrorism. The cells have been linked to several terrorist acts, including the Oklahoma City bombing, the derailment of an Amtrak train in Arizona, and multiple bomb plots targeting "the Southern Poverty Law Center, offices of the Anti-Defamation League, federal buildings, abortion clinics and sites in the gay community."[21]

Militias provide training in weapons use, target practice, intelligence gathering, encryption and decryption, field radio operation, navigation, unarmed defense, the manufacture of explosives, and demolition, among other things.[22] Much of the instruction is provided by Vietnam veterans, Gulf War veterans, and active military and law enforcement officers.

Social Composition of the Militia

Who are the militia members? While no formal survey of the militias has been undertaken, several demographic characteristics can nonetheless be discerned. Numerous researchers have documented the rural nature of the militia movement; its roots are strongest in the intermountain Montana and Idaho panhandle.[23] Potok notes that the militia movement is "almost entirely rurally based."[24] Historian Carol McNichol Stock situates the militia movement within the historical tradition of rural radicalism on the left and right, rooted in the values of producerism and vigilantism. Stock explores what she labels the ideology of rural producer radicalism: "the desire to own small property, to produce crops and foodstuffs, to control local affairs, to be served but never coerced by a representative government, and to have traditional ways of life and labor respected ... is the stuff of one of the oldest dreams in the United States."[25]

The militias are also Christian, and the movement is strongest in states with high concentrations of fundamentalist Christians. Many have embraced Christian Identity theology, which gained a foothold on the far right in the early 1980s. About half of the militia members in South Carolina, for example, are also followers of Christian identity.[26] The Christian Identity focus on racism and anti-Semitism provides the theological underpinnings of the shift from a more "traditional agrarian protest" to the paramilitarism of the militias. While Christian Identity is surely a fringe movement, easily distinguished from mainline Protestantism, this move was nevertheless "encouraged by the evangelism—secular and sacred—of the New Right as well."[27] It is from the Christian Identity movement that the far right gets its theological claims that Adam is the ancestor of the Caucasian race, while nonwhites are pre-Adamic "mud people" without souls, and Jews are the children of Satan.[28] According to this doctrine, Jesus was not Jewish but northern European; his Second Coming, heralded by the Apocalypse, is at hand, and followers can bring the Apocalypse closer. It is the birthright of Anglo-Saxons to establish God's Kingdom on earth; America and Britain's "birthright is to be the wealthiest, most powerful nations on earth...able, by divine right, to dominate and colonize the world."[29]

That militia members are white and male is also evident but equally analytically significant. According to several researchers, militia members tend to be middle-aged, in their late thirties to fifties, while the active terrorists tend to be somewhat younger—in their twenties.[30] Many hate crimes are committed by teenagers; in their early twenties, they "graduate" to militias and other radical-right organizations.[31] Like other groups of ethnic nationalists, the militias and their followers consist of two generations of dispossessed and displaced lower-middle-class men—small farmers, shopkeepers, craftsmen, skilled workers. Some are men who have worked all their adult lives, hoping to pass on the family farm to their sons and retire comfortably. They believed that if they worked hard, their legacy would be assured, but they leave their sons little but a legacy of foreclosures, economic insecurity, and debt. Tom Metzger, head of White Aryan Resistance (WAR), estimates that while 10 percent of his followers are skinheads, most are "businessmen and artisans."[32] The sons of these farmers and shopkeepers expected to—and felt entitled to—inherit their fathers' legacy.[33] As Tim McVeigh, from Lockport, New York, wrote in a letter to the editor in his hometown paper just a few years before he blew up the federal building in Oklahoma City, "the American dream of the middle class has all but disappeared, substituted with people struggling just to buy next week's groceries."[34] And when it became evident that they would not inherit that legacy, some of them became murderously angry —at a system that had emasculated their fathers and threatened their manhood.

Of course, the militias are not composed entirely of men. Lori Linzer, a militia researcher at the Anti-Defamation League, found that while there are small numbers of women involved in the movement, they are most likely to become involved with Internet discussions and Web sites and less likely to be active in paramilitary training and other militia activities. Though many women have played active roles in the movement, it remains "vastly, mainly, white Christian men."[35]

Many militia and white-supremacist groups have sought to establish refuge in rural communities, where they could practice military tactics, stockpile food and weapons, hone their survivalist skills, and become self-sufficient in preparation for Armageddon, the final race war, or whatever cataclysm they envision. For example, while preparing for the year 2000 (Y2K), some groups set up "covenant communities." These were self-sufficient and heavily armed rural settlements of white people, who feared that "when the computers crash, government checks to minorities in the inner cities will stop. Then starving Hispanics and blacks will flood into the rural parts of America, armed to the teeth and willing to stop at nothing in order to wrench food from the tables of white Christians."[36]

In addition, rural areas are seen by far-right extremist leaders as a strong base for possible recruiting. Accurately reading the signs of rural decline and downward mobility, these leaders "see an opportunity to increase their political base by recruiting economically troubled farmers into their ranks."[37] While "the spread of far-right groups over the last decade has not been limited to rural areas alone," writes Davidson, "the social and economic unraveling of rural communities—especially in the midwest—has provided far-right groups with new audiences for their messages of hate."[38] For many farmers facing foreclosures, far-right promises to help them save their land have been appealing, offering farmers various schemes and legal maneuvers to help prevent foreclosures, blaming the farmers' troubles on Jewish bankers and the One World Government. In stark contrast to the government indifference encountered by rural Americans, a range of far-right groups, most recently the militias, have seemingly provided support, community, and answers.

In that sense, the militias are simply following in the footsteps of the Ku Klux Klan, the Posse Comitatus, and other far-right patriot groups that recruited members in rural America throughout the 1980s. In fact, rural America has an especially entrenched history of racism and an equally long tradition of collective local action and vigilante justice. There remains a widespread notion "that Jews, African-Americans, and other minority-group members 'do not entirely belong,'" which may, in part, "be responsible for rural people's easy acceptance of the far right's agenda of hate."[39] "The far right didn't create bigotry in the Midwest; it didn't need to," Davidson concludes. "It merely had to tap into the existing undercurrent of prejudice once this had been inflamed by widespread economic failure and social discontent."[40]

Many militia members are also military veterans. Several leaders served in Vietnam and were shocked at the national disgust that greeted them as they returned home after that debacle.[41] Some veterans believed they were sold out by the government caving in to effeminate cowardly protesters; they can no longer trust the government to fight for what is right. Louis Beam, for example, served eighteen months in Vietnam before returning to start his own paramilitary organization, which was broken up in the 1980s by lawsuits. He now advocates "leaderless resistance," the formation of underground terrorist cells.

Bo Gritz, a former Green Beret in Vietnam, returned to Southeast Asia several times in clandestine missions to search for prisoners of war (POWs) and was the real-life basis for the film *Rambo*. He used his military heroism to increase his credibility among potential recruits; one brochure described him as "this country's most decorated Vietnam veteran" who "killed some 400 Communists in his illustrious military career."[42] In 1993, Gritz began a traveling training program, Specially Prepared Individuals for Key Events (SPIKE), a rigorous survival course in paramilitary techniques. Gritz and a colleague, Jack McLamb, a retired police officer, created their own community, Almost Heaven, in Idaho. Gritz captures the military element of the militias. They believed they were entitled to be hailed as heroes, as had earlier generations of American veterans, not to be scorned as outcasts. Now he symbolizes "true" warrior-style masculinity, the reward for men who join the militia.

What characterizes these scions of small-town rural America—both the fathers and the sons—is not only their ideological vision of producerism threatened by economic transformation, their sense of small-town democratic community, an inclusive community that was based on the exclusion of broad segments of the population; it is also their sense of entitlement to economic, social, political, and even military power. To cast the middle-class, straight, white man as the hegemonic holder of power in America would be to miss fully the daily experience of these straight white men. They believe themselves to be *entitled* to power—by a combination of historical legacy, religious fiat, biological destiny, and moral legitimacy. But they believe they do not have power. That power, in their view, has been both surrendered by white men—their fathers—and stolen from them by a federal government controlled and staffed by legions of the newly enfranchised minorities, women, and immigrants, all in service to the omnipotent Jews who control international economic and political life. "Heaven help the God-fearing, law-abiding Caucasian middle class," explained Charlton Heston to a recent Christian Coalition convention, especially:

> Protestant or even worse evangelical Christian, Midwest or Southern or even worse rural, apparently straight or even worse admittedly [heterosexual], gun-owning or even worse NRA card–carrying average working stiff, or even worst of all, male working stiff. Because not only don't you count, you're a downright obstacle to social progress.[43]

Downwardly mobile rural white men—those who lost the family farms and those who expected to take them over—are squeezed between the omnivorous jaws of capital concentration and a federal bureaucracy which is at best indifferent to their plight and at worst facilitates their further demise.

Militia Ideology

Militia ideology reflects this squeeze yet cannot fully confront its causes. Rooted in heartland conservatism, the militias have no difficulty blaming the federal government for their ills but are loathe to blame capitalism. After all, they are strong defenders in capitalist economics of the self-made man, and many have served in the armed forces defending the capitalist system that ensures individual freedom.[44] As a result, a certain rhetorical maneuvering must displace the analysis of capitalism onto another force that distorts and disfigures the pure capitalist impulse. This rhetorical maneuvering is the deployment of racism, sexism, homophobia, and anti-Semitism into a rhetoric of emasculating "others" against whom the militias' fantasies of the restoration of American masculinity are played out.

Central to the militia ideology is its antistatist position. It is big government, not big capital, that is eroding Americans' constitutional rights. International economic arrangements such as NAFTA or GATT are understood as politically disenfrachising white American workers. Recent governmental initiatives such as the Brady Bill, which requires a waiting period before the purchase of handguns and that ban certain assault rifles, are seen as compromising the Constitutional right to bear arms and are perceived as a threat to white men's ability to protect and defend their families. Gun control is seen as a further attempt by the government to emasculate white men. The 1993 FBI/ATF assault on the Branch Davidian compound in Waco, Texas, and the 1992 standoff and shootout with white separatist Randy Weaver at Ruby Ridge, Idaho, which resulted in the death of Weaver's wife and son, have further exacerbated their distrust in the federal government.[45] Restrictions on the right to bear arms are perceived as just further steps in the government's attempt to disarm and eventually control all citizens, leading inevitably to a UN invasion and the New World Order.

Militia publications are replete with stories of government conspiracies, and many believe the U.S. government is working with international forces to control U.S. citizens. For example, some argue that black helicopters are spying on citizens, monitoring devices are being implanted in newborns, Hong Kong police forces are being trained in Montana to disarm U.S. citizens, and markings on the back of road signs are secret codes to direct invading UN forces.[46] In response, militias have established "common law courts"—self-appointed groups that usurp the authority of the law, staging their own trials and issuing their own legal documents.

In many respects, the militias' ideology reflects the ideologies of other fringe groups on the far right, from whom they typically recruit and with which they overlap. While the militias may not be as overtly racist and anti-Semitic as some other white-supremacist groups, many researchers have documented the extensive links between the two.[47] For example, from white-supremacist groups they embrace the theory of the international Jewish conspiracy for world control. From Christian Identity groups they take their idiosyncratic reading of the Bible, which holds that Jews are descendants of Satan (through Cain), that people of color are "pre-Adamic mud people," and that Aryans are the true people of God. Militia member Rodney Skurdal used Christian Identity theology to justify his refusal to pay taxes, arguing that "[if] we the white race are God's chosen people … and our Lord God stated that 'the earth is mine,' why are we paying taxes on His 'land'?"[48]

And from all sides the militias take racism, homophobia, nativism, sexism, and anti-Semitism. Like antistatism, these discourses provide an explanation for the feelings of entitlement thwarted, fixing the blame squarely on "others" whom the state must now serve at the expense of white men. What is central to our analysis here is that the unifying theme of all these discourses, which have traditionally formed the rhetorical package Richard Hofstadter has labeled "paranoid politics," is *gender*. Specifically, it is by framing state policies as emasculating and problematizing the masculinity of these various "others" that rural white militia members seek to restore their own masculinity.

In this way, militias can claim a long historical lineage. Since the early nineteenth century, American manhood has pivoted around the status of breadwinner—the independent, self-made man who supports his family by his own labor. The breadwinner was economically independent, king of his own castle, embedded in a political community of like-minded and equally free men. When this self-made masculinity has been threatened, one of American men's responses has been to exclude others from staking their claim to manhood. Like the Sons of Liberty who threw off the British yoke of tyranny in 1776, these contemporary Sons of Liberty see "R-2," the Second American Revolution, as restorative—to retrieve and refound traditional masculinity on the exclusion of others. The entire rhetorical apparatus that serves this purpose is saturated with gendered readings—of the problematized masculinity of the "others," of the emasculating policies of the state, and of the rightful masculine entitlement of white men.

That such ardent patriots as militia members are so passionately antigovernment might strike the observer as contradictory. After all, are these not the same men who served their country in Vietnam or in the Gulf War? Are these not the same men who believe so passionately in the American Dream? Were they not the backbone of the Reagan Revolution? Indeed they are and were. Militia members face the difficult theoretical task of maintaining their faith in America and capitalism and simultaneously providing an analysis of an indifferent state, at best, or an actively interventionist one, at worst, coupled with a contemporary articulation of corporate capitalist logic that leaves them, often literally, out in the cold.

It is through a decidedly gendered and sexualized rhetoric of masculinity that this contradiction between loving America and hating its government, loving capitalism and hating its corporate iterations, is resolved. First, like others on the far right, militia members believe that the state has been captured by evil, even Satanic forces; the original virtue of the American political regime has been deeply and irretrievably corrupted. Environmental regulations, state policies dictated by urban and Northern interests, the Internal Revenue Service—in their view, these are the outcomes of a state now utterly controlled by feminists, environmentalists, blacks, and Jews.[49]

According to this logic, the welfare state has been captured by feminists, so that now, like all feminists and feminist institutions, it serves to emasculate white manhood. Several call for the invalidation of the thirteenth and fourteenth Amendments, which eliminated slavery and provided equal protection.[50] One leader, John Trochman, argues that women must relinquish the right to vote and to own property.[51] One book sold by MOM well illustrates these themes. In *Big Sister Is Watching You: Hillary Clinton and the White House Feminists Who Now Control America—And Tell the President What to Do,* Texe Marrs argues that Hillary Clinton and her feminist coconspirators were controlling the country and threatening American's rights and our national sovereignty. Marrs describes "Hillary's Hellcats" and "Gore's Whores"—a "motley collection [including] lesbians, sex perverts, child molester advocates, Christian haters, and the most doctrinaire of communists."[52] These women—such as Jocelyn Elders, Janet Reno, Maya Angelou, Donna Shalala, Laura D'Andrea Tyson, Roberta Actenberg, and Ruth Bader Ginsburg—are said to be members of the "conspiratorial Council on Foreign Relations and the elitist Trilateral Commission, [they] attend the annual conclave of the notorious Bilderbergers [and are] hirelings of the left-wing, radical foundations designed to promote the New World Order."[53] These "feminist vultures ... the most militant of the militant ... femiNazis ... control a heartless police establishment more efficient than Stalin's."

Marrs claims that "Big Sister" intends nothing short of a New World Order, to be accomplished through a "10 Part Plan" which includes:

> the replacement of Christianity with feminist, new-age spirituality.... History will be rewritten, discarding our true heroes.... Homosexuality will be made noble, and the male-female relationship undesirable.... Patriotism will be smashed, while multiculturalism shall be exalted, and the masses will come to despise white, male dominated society as a throwback to the failed age of militarism and conflict. The masses shall be taught to revile nationalism, patriotism, and family ... abortion and infanticide ... encouraged.... Women will dominate in all walks of life—in law, medicine, literature, religion, economics, entertainment, education, and especially in politics.[54]

In this vision, feminism, multiculturalism, homosexuality, and Christian-bashing are all tied together, part and parcel of the New World Order. On the other hand, Christianity, traditional history, heterosexuality, male domination, white racial superiority and power, individualism, meritocracy, and the value of individual hard work describe the True, Right America which is at risk and must be protected. These facets are all intertwined, so that multicultural textbooks, women in government, and legalized abortion can individually be taken as signs of the impending New World Order.

Marrs's book displays tremendous anxiety over changing gender roles. Working within the difference-versus-equality framework, race and gender equality are unthinkable and necessarily a threat to white men. Increased opportunities for women can only lead to the oppression of men. Marrs proclaims: "In the New Order, woman is finally on top. She is not a mere equal. *She is Goddess.*"[55] Hillary, he writes, "represents the ascendance and preeminence of Woman ... *Woman in Control.*"[56] However, real power is never truly granted women, who must continue to be thought of as passive and incapable of possessing real power. Marrs argues: "Hillary Rodham Clinton is Big Brother in drag. She is in effect, a politically correct, transvestite Big Brother."[57] Not really a woman after all, she comes to represent the confusion of gender boundaries and the demasculinization of men, symbolizing a future where men are not allowed to be real men.

This text suggests several themes of interest to us here. The notion that the state has been taken over means that it no longer acts in the interests of "true" American men. The state is an engine of gender inversion, feminizing men, while feminism masculinizes women. Feminist women, it turns out, are more masculine than men are. Not only does this call the masculinity of white men into question, but it uses gender as the rhetorical vehicle for criticizing "other" men. Typically, problematizing the masculinity of these others takes two forms simultaneously: other men are both "too masculine" and "not masculine enough." We call this the "Goldilocks Paradox," after the fairy-tale heroine who found chairs too big or too small, porridge too hot or too cold, never "just right." So, too, the "others" are seen as too masculine—violent rapacious beasts with no self-control—or not masculine enough—weak, helpless, effete, incapable of supporting a family.

Hence, in the logic of militias and other white-supremacist organizations, gay men are both promiscuously carnal and sexually voracious, and effete fops who do to men what men should only have done to them by women. Black men are both violent hypersexual beasts, possessed of an "irresponsible sexuality," seeking white women to rape, and less than fully manly, "weak, stupid, lazy."[58] In *The Turner Diaries,* the apocalyptic novel that served as the blueprint for the Oklahoma City bombing and is widely read and peddled by militias, author William Pierce depicts a nightmarish world where white women and girls are constantly threatened and raped by "gangs of Black thugs."[59] Blacks are primal nature—untamed, cannibalistic, uncontrolled, but also stupid and lazy—and whites are the driving force of civilization. "America and all civilized society are the exclusive products of White man's mind and muscle," is how *The Thunderbolt* put it.[60] Whites are the "instruments of God," proclaims *The Turner Diaries.* "[T]he White race is the Master race of the earth... the Master Builders, the Master Minds, and the Master warriors of civilization." What can a black man do but "clumsily shuffle off, scratching his wooley head, to search for shoebrush and mop."[61]

Most interesting is the portrait of the Jew. On the one hand, the Jew is a greedy, cunning, conniving, omnivorous predator; on the other, the Jew is small, beady-eyed, and incapable of masculine virtue. By asserting the hypermasculine power of the Jew, the far right can support capitalism as a system while decrying the actions of capitalists and their corporations. According to militia logic, it is not the capitalist corporations that have turned the government against them, but the international cartel of Jewish bankers and financiers, media moguls, and intellectuals who have already taken over the U.S. state and turned it into ZOG (Zionist Occupied Government). The United States is called the "Jewnited States," and Jews are blamed for orchestrating the demise of the once-proud Aryan man....

Embedded in this anti-Semitic slander is a critique of white American manhood as soft, feminized, weakened—indeed, emasculated. Article after article decries "the whimpering collapse of the blond male," as if White men have surrendered to the plot.[62] According to *The Turner Diaries,* American men have lost the right to be free; slavery "is the just and proper state for a people who have grown soft."[63] Yet it is here that the militias simultaneously offer White men an analysis of their present situation and a political strategy for retrieving their manhood. As *National Vanguard* puts it:

> As Northern males have continued to become more wimpish, the result of the media-created image of the "new male"—more pacifist, less authoritarian, more "sensitive," less competitive, more androgynous, less possessive—the controlled media, the homosexual lobby and the feminist movement have cheered ... the number of effeminate males has increased greatly ... legions of sissies and weaklings, of flabby, limp-wristed, non-aggressive, non-physical, indecisive, slack-jawed, fearful males who, while still heterosexual in theory and practice, have not even a

> vestige of the old macho spirit, so deprecated today, left in them.[64]

It is through the militias that American manhood can be restored and revived—a manhood in which individual white men control the fruits of their own labor and are not subject to the emasculation of Jewish-owned finance capital, a black- and feminist-controlled welfare state. It is a militarized manhood of the heroic John Rambo—a manhood that celebrates their God-sanctioned right to band together in armed militias if any one—or any governmental agency—tries to take it away from them. If the state and capital emasculate them, and if the masculinity of the "others" is problematic, then only real White men can rescue this American Eden from a feminized, multicultural, androgynous melting pot. "The world is in trouble now only because the White man is divided, confused, and misled," we read in *New Order*. "Once he is united, inspired by a great ideal and led by real men, his world will again become livable, safe, and happy."[65] The militias seek to reclaim their manhood gloriously, violently. On the recorded message of the Militia of Michigan, one could hear the following telling narrative:

> Stand firm, stand strong, Militia Men! America has much need of you today. Be vigilant now, as never before. Evil is trying to steal our country away. Perhaps tomorrow or in a thousand years you will receive the rewards you are due. Our flag will fly, our spirit will soar, and it will happen because of you. History will record many of your names, stories will tell of where you've been.[66]

In the ideology of the white-supremacist movement and their organized militia allies, it is racism that will again enable white men to reclaim their manhood. The amorphous groups of white supremacists, skinheads, and neo-Nazis may be the symbolic shock troops of this movement, but the rural militias are their well-organized and highly regimented infantry.

ENDNOTES

1. Bonanno et al., 1994; Jobes, 1997: 331.
2. Gouveia and Rousseau, 1995.
3. See Weis, 1993.
4. Jobes, 1997.
5. Dyer, 1997: 61.
6. Lobao and Meyer, 1995: 6.
7. Davidson, 1996: 118, 9.
8. Labao and Meyer, 1995: 61; see also Hanson, 1996.
9. Bell, 1999.
10. Dyer, 1997.
11. Gibson, 1994: 11.
12. Junas, 1995a: 227.
13. Rand, 1996.
14. Lamy, 1996: 26.
15. Stern, 1996a: 78.
16. Potok, interview, July 21, 1999.
17. Southern Poverty Law Center, 1999.
18. Potok, interview, December 23, 2002.
19. Berlet and Lyons, 1995.
20. Ibid., 24.
21. Coalition for Human Dignity, 1995; Southern Poverty Law Center, 1999: 23.
22. Southern Poverty Law Center, 1999: 20.
23. Corcoran, 1997; Dyer, 1997; Stern, 1996a; Stock, 1996.
24. Potok, interview, July 21, 1999.
25. Stock, 1996: 16.
26. Potok, interview, July 21, 1999.
27. Stock, 1996: 173.
28. Aho, 1990; Anti-Defamation League, 1988; Barkun, 1997.
29. Zeskind, 1999: 19.
30. See Aho, 1990; Linzer, 1999.
31. See O'Matz, 1996.
32. Serrano, 1990.

33. Junas, 1995a.
34. In Dyer, 1997: 63.
35. Potok, interview, July 21, 1999.
36. Southern Poverty Law Center, 1999, 13.
37. Young, 1990: 15.
38. Davidson, 1996: 109.
39. See Snipp, 1996: 127, 122.
40. Davidson, 1996: 120.
41. Gibson, 1994: 10.
42. Mozzochi and Rhinegaard, 1991: 4.
43. Citizens Project, 1998/1999: 3.
44. See Kimmel, 1996.
45. Dees, 1996; Southern Poverty Law Center, 1997; Stern, 1996a.
46. Southern Poverty Law Center, 1997.
47. Crawford and Burghart, 1997: 190.
48. Stern, 1996a: 89.
49. See Dyer, 1997.
50. Stern, 1996a: 82.
51. Crawford and Burghart, 1997; Stern, 1996a: 69.
52. Marrs, 1993: 11.
53. Ibid., 13.
54. Ibid., 22–23.
55. Ibid., 28.
56. Ibid., 31.
57. Ibid., 28.
58. NS Mobilizer, cited in Ferber, 1998a: 81.
59. Pierce, 1978: 58.
60. Cited in Ferber, 1998: 76.
61. Pierce, 1978: 71; in *New Order*, cited in Ferber, 1998a: 91.
62. In Ferber, 1998a: 127.
63. Pierce, 1978: 33.
64. Cited in Ferber, 1998a: 136.
65. In Ferber, 1998a: 139.
66. Pierce, 1978: 207.

REFERENCES

Aho, James A. 1990. *The Politics of Righteousness: Idaho Christian Patriotism*, Seattle: University of Washington Press.

Anti-Defamation League. 2000. *Hate Groups in America: A Record of Bigotry and Violence*. New York: ADL.

Barkun, Michael. 1997. "Racist Apocalypse: Millennialism on the Far Right." *The Year 2000: Essays on the End*, ed. Charles B. Strozier and Michael Flynn. New York: New York University Press.

Bell, M. M. 1999. "The Social Construction of Farm Crises." Presented at the Annual meetings of the Rural Sociological Society, Aug. 5, Chicago.

Berlet, Chip. 2002, and Matthew Lyons. 1995. "Militia Nation." *The Progressive*, June (22–25).

Bonanno, A., L. Busch, W. Friedland, L. Gouveia, and E. Mingione, eds. 1994. *From Columbus to Conagra: The Globalization of Agriculture and Food*. Lawrence, KS: University Press of Kansas.

Corcoran, James. [1990] 1997. *Bitter Harvest: The Birth of Paramilitary Terrorism in the Heartland*. New York: Penguin Books.

Crawford, Robert, and Devin Burghart. 1997. "Guns and Gavels: Common Law Courts, Militias and White Supremacy." *The Second Revolution: States Rights, Sovereignty, and Power of the County*, ed. Eric Ward. Seattle: Peanut Butter Publishing.

Davidian, Blanche. 1996. "The Performance of Patriotism: Infiltration and Identity at the End of the World." *Theatre* 27:1 (7–28).

Dees, Morris. 1996. *Gathering Storm: America's Militia Threat*. New York: Harper Perennial.

Dyer, Joel. [1997] 1998. *Harvest of Rage: Why Oklahoma City is only the Beginning*. Boulder: Westview Press.

Ferber, Abby L. 1998a. *White Man Falling: Race, Gender and White Supremacy*. Lanham, MD: Rowman and Littlefield.

Gibson, James William. 1994. *Warrior Dreams: Violence and Manhood in Post-Vietnam America*. New York: Hill and Wang.

Gouveia, Lourdes, and Mark O. Rousseau. 1995. "Talk is Cheap: The Value of Language in the World Economy—Illustrations from the United States and Quebec." *Sociological Inquiry* 65:2 (156–180).

Hanson, Victor D. 1996. *Fields without Dreams: Defending the Agrarian Idea.* New York: Free Press.

Jobes, Patrick C. 1997. "Gender Competition and the Preservation of Community in the Allocation of Administrative Positions in Small Rural Towns in Montana: A Research Note." *Rural Sociology* 62:3 (315–334).

Junas, Daniel. 1995a. "The Rise of Citizen Militias: Angry White Guys with Guns." *Eyes Right: Challenging the Right Wing Backlash*, ed. C. Berlet. Boston: South End Press.

Kimmel, Michael S. 1996. *Manhood in America: A Cultural History.* New York: The Free Press.

Lamy, Philip. 1996. *Millennium Rage: White Supremacists, and the Doomsday Prophecy.* New York: Plenum Press.

Linzer, Lori. 1999. Anti-Defamation League researcher. Telephone interview, July 16.

Marrs, Texe. 1993. *Big Sister is Watching You: Hillary Clinton and the White House Feminists Who now Control America—And Tell the President What to Do.* Austin, TX: Living Truth Publishers.

O'Matz, Megan. 1996. "More Hate Crimes Blamed on Juveniles." *Morning Call*, October 23.

Pierce, William. 1978. *The Turner Diaries.* Hillsboro, VA: National Vanguard Books.

Potok, Mark. 1999. Southern Poverty Law Center researcher. Telephone interviews, July 16, 19, and 21 and December 23, 2002.

Rand, Kristen. 1996. *Gun Shows in America.* Washington, DC: Violence Policy Center.

Serrano, Richard. 1990. "Civil Suit Seeks to Bring Down Metzger Empire." *Los Angeles Times*, February 18.

Snipp, C. Matthew. 1996. "Understanding Race and Ethnicity in Rural America." *Rural Sociology* 61:1 (125–142).

Southern Poverty Law Center. 1999. *Intelligence Report* Spring. Montgomery: SPLC.

———. 1997. *False Patriots: The Threat of Antigovernment Extremists.* Montgomery: SPLC.

Stern, Kenneth S. 1996a. *A Force upon the Plain: The American Militia Movement and the Politics of Hate.* New York: Simon & Schuster.

Weis, Lois. 1993. "White Male Working Class Youth: An Exploration of Relative Privilege and Loss." *Beyond Silenced Voices: Class, Race and Gender in United States Schools*, eds. L. Weis and M. Fine. Albany, NY: SUNY Press.

Young, Thomas J. 1990. "Violent Hate Groups in Rural America." *International Journal of Offender Therapy and Comparative Criminology* 34:1 April (15–21).

DISCUSSION QUESTIONS

1. What are some of the forces that are changing rural America? Who is hurt by the changes?
2. Do you believe the authors are describing a serious problem?
3. It is often argued that people who join militia groups and sometimes become violent are simply "crazy." This is not how the authors describe the situation. What do the authors believe attract people to these groups?
4. The authors try to bring in gender issues throughout their analysis. Is there truly a question of "masculinity" in these groups? Does the issue of masculinity help us understand, or does it confuse, the situation?

49

Terrorism and the Politics of Fear

DAVID ALTHEIDE

The Four Questions

1. What is the problem?
2. Is it a *social* problem?
3. What is its cause?
4. Does Altheide suggest ways to alleviate the problem?

Topics Covered

Politics of fear
Social construction
Fear
Social control
Socialization
Mass media
Popular culture
Politicians of fear
Propaganda

The politics of fear is paradoxical. The complexities are illustrated in various crime control efforts as well as military interventions discussed in this book. On the one hand, the policies, programs, and changes that occur are perceived as beneficial in the short run because they keep us safe, solve problems, and prosecute—and kill—those who threaten us. On the other hand, public perceptions change over time as more people come to regard such policies as reckless, destructive, and serving the interest of the manipulators. Recall that excesses and egregious civil rights violations by the FBI and the CIA were made public; these agencies were reined in by congressional action. However, the collective memory seems to last about as long as the next crisis, when entertainment-oriented news media fan the flames of "emergency" and shut out the soothing language of context and perspective. The problem, then, is that we are all increasingly implicated as being manipulators. More of us enjoy the alleged safety and security that is credited to the formal agents of social control with whom we have entrusted more of our lives. Part of the challenge, then, is to recognize how publics are cultivated through the mass media to accept the *ethic of control:* problems can and should be solved by more invasive control. After a brief overview of how citizens become involved in reducing their own citizenship rights, I will suggest some ways to offset, if not overcome, the pervasive politics of fear.

DEFINING THE POLITICS OF FEAR

... [T]he politics of fear refers to decision makers' promotion and use of audience beliefs and assumptions about danger, risk, and fear in order to achieve certain goals. The politics of fear should correspond well with the amount of formal social control in any society. The source of fear may be an authority, God, or an internal or external enemy. Tracking

SOURCE: From David Altheide, *Terrorism and the Politics of Fear*, Rowman & Littlefield, 2006, pp. 207–221. Used by permission.

the expanded control efforts over time can illustrate how the politics of fear has evolved in any social order. Moreover, behind most efforts to enact more control will be a series of events and accounts about "what should be done." Changes in public language and in the discourse of fear will also accompany social control changes. However, as emphasized here, once such changes are enacted, they symbolically enshrine the politics of fear even when public perceptions about the specific source of that fear process may diminish.

The politics of fear is exercised during times of conflict, but it accumulates and gradually informs policy and everyday-life behavior, even if there are occasional bouts of resistance. The politics of fear does not imply that citizens are constantly afraid of, say, a certain enemy day in and day out. The object of fear might change, but fear of threats to one's security is fairly constant. The context of control promotes this, as do numerous messages about menaces that justify general social control measures....

Any response to the politics of fear must first recognize that it is a social construction that is linked to cultural meanings produced by the mass media. I have stressed that individual politicians are not to "blame" and should not be given credit for a country like the United States to wage war. There are numerous checks and balances in the U.S. system, and Congress, journalists, and the mass media, as well as public opinion, can—and ultimately will—stop reigns of terror that emerge from the politics of fear. It is not a "power grab" per se that enables conspirators like the members of the Project for a New American Century to gain control of the reins of government and dominate foreign and domestic policy. This takes a lot of work, a lot of cooperation from many people of goodwill who, ironically, are just trying to do the right thing, to protect their families and country from harm. It is what these social actors take into account about their past, how they draw on manipulated media images to understand their situation, and in turn what meanings they project into the future.

Fear promotes fear. Fear limits our intellectual and moral capacities, it turns us against others, it changes our behavior and perspective, and it makes us vulnerable to those who would control us in order to promote their own agendas. The politics of fear simply translates these "concerns" into preventive action; claims are made that the "bad situation" can be fixed through more control. This is true regardless of whether the hot issue is crime, illegal drugs, immigration, or international conflict. In most cases, the control is focused on regulating individuals rather than on broader social issues (e.g., poverty and oppressive foreign relations) that have contributed to the problem. More recently however, the work linking fear to the politics of fear has become far more sophisticated; the recent "war on terrorism," for example, rests on important changes that have occurred in our culture and social institutions and owes less to cunning individuals who simply ride these cultural changes.

A truism in social science is that all social products reflect the process that made them. Fear is a product, and the politics of fear is part of the process. It is a process that includes the mass media because in the modern world we know very little beyond our immediate experience that is not mass mediated. I have argued in this book that the mass media, popular culture, and the process of media logic are the key to our strengths and weaknesses. We are an entertainment-oriented society, and virtually everything that is meaningful to us and taken for granted is part of a "program" that repeats, resonates, and reproduces our lives. Today, propaganda abounds; we just call it by different names. In Hitler's day, there was not a lot of propaganda, so he and one of his top henchmen, Josef Goebbels, created unique blends of glamorized falsehoods, refining the delivery of emotional symbols and slogans across radio, movies, and newspapers and building on a historical context of "Germany against the world." In our day, things are different; all major mass media are governed by propaganda, often in the guise of advertising and marketing, which rely on simplification, distortion, and emotional appeals to increase the "bottom line," the cultural culmination of profitability and success (Ewen 1976, 1999; Jackall 1994, 1999). Anything that brings in the market—that is, the people and their dollars—is permissible. The current generation—the age of my students—is the most marketed generation in history. Every aspect of their lives is fair game to commercial manipulators. This is critical because the most effective form of social control is when it is taken for granted as part of the normal course of things and is not even recognized as "control" but just as "what everyone knows." Social scientists refer to this as "normalization," and this occurs through a "socialization process" whereby individual members of society essentially acquire the expectations, assumptions, and patterns of everyday life. Soon, the way things are is the way things should be. This is the process that makes meaningful social change so difficult. This is also the process of the politics of fear.

Most of my students were born into the politics of fear and know nothing else. The pervasive surveillance that regulates more of our lives is part of their taken-for-granted baseline of experience (Marx 1988; Staples 2000). To some extent, this is true of each generation, but this one is different. For example, my generation (born in the mid-1940s) was taught about the dreaded superpower that challenged us: the Soviet Union, Russia, the communists, or the "commies." And the "other side" learned the same thing about us. We learned that there were many aspects of our lives that had to be regulated in order to "protect us," including foreign travel. There was a lot of concern about "internal" enemies and infiltration by the "commies." Senator Joseph McCarthy, eventually brought down by independent journalists like Murrow and others who dared to face him, raised havoc among American legislators, cultural creators, intellectuals, and activists. The Cold War chilled the culture. But unlike the present politics of fear, the technology of control did not penetrate our bodies. That is different today. Many of my students have had urinalysis and other drug screenings. They were born at a time of pervasive control, and for the most part they have normalized it; most see nothing wrong with being asked "to pee in a cup if you've got nothing to hide." They do not see the control. This is part of the politics of fear. Social routines and activities change to reflect the ethic of control: problems can and should be solved by more invasive control. This is especially the case when control is justified to contain threats to personal safety and national security. Few of my students are aware of a prior time when individual rights as citizens would not permit such bodily transgressions, where there were clear limits to how far surveillance could go.

Crime and war are two sides of the politics of fear that have drastically changed our culture and paved the way to widespread acceptance of the latest justification known as the "war on terrorism." The politics of fear becomes part of culture, and it changes through the cultural process. One way to understand how this works is to monitor popular culture and the mass media (Surette 1998). While the politics of fear operates alongside cultural products, the cultural products demonstrate the results. We can examine some of these changes about crime, punishment, and the shifting focus from protecting the individual to protecting the state and the interests of the mass audience. This has important implications for citizenship.

Consider television programs about crime. Numerous scholars have noted how mass-media scenarios, narratives, and rhetoric have shifted since the 1960s from more concern with rights of the accused to the rights of the prosecution (Cavender 2004; Ericson 1995; Garland 2001; Surette 1998). Typically, popular-culture rhetoric does not celebrate both simultaneously. Moreover, the nature of rhetoric and cultural narratives entails treating individuals and the state in opposition; both cannot be promoted at the same time; rather, when one is promoted, the other is often disparaged and delegitimated or treated as morally contemptible. Many readers will recognize this in portrayals of the "rights of the accused" that are often presented in crime dramas. Also referred to as the Miranda rights (named after a famous legal case, *Miranda v. Arizona),* these are commonly depicted when an arresting officer informs an arrested person that they "have the right to remain silent... have the right to an attorney," and so on. Typically, the arresting officer will make a derogatory comment about the suspect, such as "you've heard it before ... you probably know it better than we do." The message, as Surette (1998) argues, is that the courts and legal protections like the Bill of Rights are hampering law enforcement and are helping criminals, terrorists, and other merchants of fear....

We become accustomed to more control, and it is gradually taken for granted. It becomes part of our cognitive and emotional baseline of experience, even how we structure our living conditions (Ellin 1997). It seems normal when it is expressed and somewhat different when it is challenged. The language of social control agencies pervades cultural experience. Each new step is a feature of the politics of fear and the cultural context of our age. And propaganda plays a large part in these efforts.

The politics of fear is a feature of the ecology of communication, which refers to the structure, organization, and accessibility of information technology, various forums, media, and channels of information (Altheide 1995). It provides a conceptualization and perspective that joins information technology and communication (media) formats with the time and place of activities. Routine activities and perspectives about everyday life reflect political decisions that have been made to increase social control activities that were justified to combat sources of fear. Political decisions have cumulative effects on social life as they "backwash," or flow over time, from their originating event and debates to seep into other aspects of everyday life. Crime control policies of the 1970s, for example, still inform everyday routines by police agencies and other social institutions. Numerous efforts to prevent crime that are taken for granted include "stop and frisk," "no-knock searches," "preventive detention," "presumptive arrest," and "police DUI checkpoints."

Technological changes promote the politics of fear as well. Fundamentally, the face of fear is expanded surveillance; it is ubiquitous and penetrating, ranging from satellite cameras to monitoring weather, troop movements, and terror suspects to invasive drug testing (e.g., urinalysis) and increasingly DNA surveillance to detect health risks. Surveillance is more pervasive because of technology that is less obtrusive and that can do more things for lower costs. More "miniaturization" of microchips, improved optics, and better wireless communication contribute to more communication devices like computers, cameras (including closed-circuit television), cell phones, personal digital assistants (PDAs or "Palm Pilots"), and iPods, but all these promote surveillance. Surveillance is virtually everywhere: work, home, school, stores and malls, sports stadiums, highways, airports, and even restrooms (Marx 1988; Staples 2000). For example, most cell phones come with a GPS locator that enables anyone with appropriate communication gear to find where you are at any time. Cameras abound in public, and they are becoming less expensive.

Controlling our borders is also about controlling us. I refer not just to the occasional capture of the "bad guy" but rather to how our expectations about everyday life become muted to numerous transgressions of basic civil rights as citizens. Recall that virtually no control movement put into effect is justified explicitly as "we want to control and regulate all citizens so that we can have more power over them." Rather, the case tends to be put in very apologetic if not painful terms, such as "unfortunately, we have to give up a few rights for our own protection" or, as more broadly stated after the 9/11 attacks, "the world changed that day," meaning that everything could justifiably be viewed as different from then on. This included control and surveillance.

Previous chapters referred to the massive changes in civil rights that occurred as a result of the USA Patriot Act and attendant legislation that accompanied the creation of the Department of Homeland Security. One of the big items was to increase surveillance along the U.S. borders, mainly in order to prevent terrorists from entering the United States. Tens of thousands of new jobs have been added to the Border Patrol, and several hundred million dollars have been spent for more technologically enhanced security along the borders. Very few—if any—bona fide terrorists have been captured as a result of this infusion of dollars and control along the border (although its proponents always argue the "negative," which is basically that these expanded efforts have deterred numerous attacks). (Yet, as the embarrassingly slow response by federal agencies to Hurricane Katrina's devastation of New Orleans in 2005 shows, there has been little attention paid to basic infrastructure repair and maintenance, emergency medical response, and systematic evacuation procedures.) But the security has had consequences. First, numerous foreign visitors, as well as U.S. citizens reentering the United States, have been checked and reminded again about the power of others over their bodies. Second, while drug arrests were common at the Mexican border for decades, the expanded surveillance approach did help nab people with criminal records. Thus, for all practical purposes, this portion of investment in Homeland Security—justified to keep us safe from terrorists—has provided us with more criminals....

The discourse of fear underlies modern propaganda. It comes from claims makers' construction of certain atypical events (e.g., the abduction of a child) as typical, common, and likely to happen to "you." These events are presented as symptomatic of all social life. These repeated propaganda messages are presented through mass-media entertainment formats to draw on audiences' emotions of fear on the one hand while providing a refreshed perspective for framing and interpreting subsequent events as further examples of the need for more control on the other. Thus, propaganda involving symbols of fear and threat contributes to how situations are defined and shaped by the expanding symbolic fear machinery. Public expectations about order accompany the new symbolic frameworks, but this leads to more examples of disorder, which in turn call for yet tighter controls to protect the moral foundations from the dark forces of fear. So strong are these symbolic parameters that anyone who questions the process or challenges the assumptions is likely to be the most visible and easily targeted threat to order.

The politics of fear quickly transforms many people into politicians of fear. We begin to self-monitor our language, behavior, and perspective. I have in mind the way in which everyday-life activities are monitored for compatibility with prevailing language, discourse, and assumptions underlying the politics of fear before they are carried out. One important consequence is that social actors become aware of this threat and begin to monitor their conduct through what Marx (1988) has referred to as "auto surveillance." This may be done by simply refraining from certain activities (e.g., not renting pornographic videos because someone will find out or writing letters to the editor of a repressive newspaper) or not going to certain places or participating in activities that challenge official rulings and programs (e.g., protest marches and demonstrations). A Canadian commented on the implications of this expanding surveillance gaze:

> "At some point, you start asking yourself, as you do in societies that aren't free, should I do this particular thing or not?" says Radwanski. "Not because it might be illegal or wrong, but because of how it might look to watchers of the state." (*Toronto Star,* May 12, 2003, p. A01)

Altering language is one of the most important ways that people display the politics of fear. The use of disclaimers or amending the meaning of our words prior to uttering them is increasingly common (Hewitt and Stokes 1975). For example, someone who opposes a foreign war might say to another whose views differ, "I support our troops, but I am not sure that this war was justified," or, "I am very patriotic and concerned about being attacked, but we need to plan our military action more carefully." Disclaimers enable us to maintain membership while skirting the edge of an issue that fundamentally challenges the very foundation and meaning of that membership. It is a covering device to protect us from outrage and scorn and, above all, from having our own legitimacy questioned by family, friends, peers, and fellows with whom we speak. Such caution is widespread in a world run on fear.

Citizenship is affected by the politics of fear. Successful politicians of fear obliterate the sanctity of citizenship, and they do this one case at a time. Bush administration officials engaged in numerous civil rights violations by arresting people and holding them without charge, denying access to attorneys, and conspiring with foreign governments to torture persons suspected of terrorism (Herbert 2005; Jehl and Johnston 2005).

> The Bush administration's secret program to transfer suspected terrorists to foreign countries for interrogation has been carried out by the Central Intelligence Agency under broad authority that has allowed it to act without case-by-case approval from the White House or the State or Justice Departments, according to current and former government officials.
>
> The unusually expansive authority for the C.I.A. to operate independently was provided by the White House under a still-classified directive signed by President Bush within days of the Sept. 11, 2001, attacks at the World Trade Center and the Pentagon, the officials said.
>
> The process, known as rendition, has been central in the government's efforts to disrupt terrorism, but has been bitterly criticized by human rights groups on grounds that the practice has violated the Bush administration's public pledge to provide safeguards against torture. (Jehl and Johnston 2005)

They count on individual cases "blowing over" and not getting much attention, weakening the opposition, and, above all, silencing any news organizations that insist on publicizing such illegal conduct. These illegal acts are ensconced in a rhetoric of patriotism and moral justification, all wrapped up in the slogan "war on terrorism." Individual rights, we learn now (and learned in Hitler's Germany), are secondary to collective security; this means that citizenship becomes a matter of convenience, not a right that can be situationally lifted for a campaign of fear....

It would be immoral to not offer some general guidelines for a way out. I cannot condone the politics of fear that have dominated much of the modern world. I take a firm position on this and offer the following modest comments. The politics of fear is relevant for social life because it influences our activities, meanings, routines, and perspectives. It is difficult to undo the policies and procedures that expand control and fear, partly because, as I noted previously, it becomes taken for granted by the next generation. These effects can be reduced through critical thinking and awareness of the social changes and the implications of blanket adjustments in security and policy. First, this requires good investigation and clear language about the context, nature, and consequences of certain changes. Too often journalists and others who have the public ear operate as though all these changes are unproblematic and helpful. They are seldom reviewed, discussed, and examined critically, especially months or years following the event or events that helped launch them. Second, very little of any consequence occurs in our society without popular culture. I expect that we will see more investigative reports, movies, and television programs that dramatize the injustice and oppression that result from this expansive control. Third, the courts and aggressive attorneys—probably younger ones—should pursue both individual and class-action suits on behalf of those whose basic civil rights have been violated by specific policies and legislation. A fourth suggestion that I offer is partly informed by my doubts that the third point mentioned will be very successful in higher courts in the United States since it is our own government that is culpable and has been very reluctant to enforce court rulings against itself. The suggestion is that we continue to pursue international tribunals for redress against the illegal actions of the United States. Fifth, politicians and other decision makers should be held accountable for their actions. This may entail symbolic protests against their speaking engagements, writing letters, or simply stating their egregious sins in appropriate forums. Sixth, this is the time for educators—parents, teachers, religious leaders, and good neighbors—to show courage and speak truth to power. We should inform our students and citizens about war programming and the deadly role it plays in darkening our future. We must tell the young people about another way, about the implications of social control and bad decisions. Seventh, scholars and researchers of all persuasions should attend once again to the subtle forms of propaganda, deviance, and resistance. The relevance of these for all aspects of social life, symbolic communication, and moral conduct should be explored and promoted when morally justifiable. The foundation of this moral reasoning, in my opinion, must be citizenship and civil rights. In our endeavor, let us not become what we're trying to undo, let us not forget how moral absolutism and entertainment got us to this point. Above all, we must continue to tell our students and whoever will listen to be aware of the propaganda project but to not be afraid.

REFERENCES

Altheide, David L. 1995. *An Ecology of Communication: Cultural Formats of Control.* Hawthorne, NY: Aldine de Gruyter.

Cavender, Gray. 2004. "Media and Crime Policy: A Reconsideration of David Garland's *The Culture of Control.*" *Punishment and Society* 6:335–348.

Ellin, Nan. 1997. *Architecture of Fear.* New York: Princeton Architectural Press.

Ericson, Richard V. ed. 1995. *Crime and the Media.* Brookfield, VT.: Dartmouth University Press.

Ewen, Stuart. 1999. *All Consuming Images: The Politics of Style in Contemporary Culture.* New York: Basic Books.

Garland, David. 2001. *The Culture of Control: Crime and Social Order in Contemporary Society.* Chicago: University of Chicago Press.

Herbert, Bob. 2005. "Our Friends, the Torturers." *New York Times* February 18, www.nytimes.com/2005/02/18/opinion/18herbert.html?ex=1109394000&en=a84651749cc4fld5&ei=5070.

Jackall, Robert ed., 1994. *Propaganda.* New York: New York University Press.

Jackall, Robert, and Janice M. Hirota. 1999. *Experts with Symbols: Advertising, Public Relations, and the Ethics of Advocacy.* Chicago: University of Chicago Press.

Jehl, Douglas, and David Johnston. 2005. "Rule Change Lets C.I.A. Freely Send Suspects Abroad to Jails." *New York Times*, March 6, 2005, www.nytimes.com/2005/03/06/politics/06intel.html?ex=1110776400&en=e36cc36fc5ef2f81&ei=5070.

Marx, Gary T. 1988. *Undercover: Police Surveillance in America.* Berkeley: University of California Press.

Staples, William G., 2000. *Everyday Surveillance: Vigilance and Visibility in Postmodern Life.* Lanham, MD.: Rowman & Littlefield.

Surette, Ray. 1998. *Media, Crime and Criminal Justice: Images and Realities.* Belmont, CA: West/Wadsworth.

DISCUSSION QUESTIONS

1. Is the "war against terrorism" an example of the politics of fear?
2. Altheide writes: "Any response to the politics of fear must first recognize that it is a social construction that is linked to cultural meanings produced by the mass media." What does he mean by this? Is he correct? Is this important?
3. "Remember 9/11. The world changed that day." What does this statement mean? What are its implications? How far are you willing to go in order to protect society?
4. Altheide's "social problem" is very important for him. Some people would argue that this problem is not important. What do you think?

50

Genocide: Memory, Forgetting, and Denial

ADAM JONES

The Four Questions

1. What is the problem that Jones describes?
2. Is this a *social* problem? Is this a problem concerning morality and values of individuals?
3. Who is harmed by this problem? Are there any victims?
4. What can or should be done about this problem? What does the author infer?

Topics Covered

Genocide

Genocide denial

Discourse of genocide denial

Violence

Denial and free speech

Denial is viewed increasingly as a final stage of genocide, and an indispensable one from the viewpoint of the *génocidaires*. "The perpetrators of genocide dig up the mass graves, burn the bodies, try to cover up the evidence and intimidate the witnesses. They deny that they committed any crimes, and often blame what happened on the victims."[1] As Richard Hovannisian has written:

> Following the physical destruction of a people and their material culture, memory is all that is left and is targeted as the last victim. Complete annihilation of people requires the banishment of recollection and the suffocation of remembrance. Falsification, deception, and half-truths reduce what was to what may have been or perhaps what was not at all.[2]

The phenomenon of genocide denial is overwhelmingly associated with the Jewish Holocaust. Since this resurged in the public consciousness in the early 1960s, a diverse and interlinked network of Holocaust deniers has arisen. In Europe, a centuries-old tradition of anti-semitism underlies their activities, which overlap with neo-Nazi violence against Jews and their property. In North America, the neo-Nazi element is also strong. In both "wings" of the denialist movement, however, academic figures—such as Arthur Butz in the US, David Irving in Great Britain, and Robert Faurisson in France—have also sought to provide a veneer of respectability for the enterprise.

We will consider specific denial strategies below, but before we do, it is important to stress that the Jewish Holocaust is not officially denied by any state or national elite (though denial is common intellectual currency in the Arab and Muslim world).[3] Thus, in the West at least, deniers of the Jewish catastrophe remain relatively marginal figures, with little access to the mainstream.

However, the broader phenomenon of genocide denial is far more deeply entrenched, often representing a societal consensus rather than a fringe position. Individual and collective narcissism play a pivotal role in buttressing denial; in many contexts, a denialist stance heads off cognitive dissonance between one's preferred view of self and country,

SOURCE: From Adam Jones, *Genocide: A Comprehensive Introduction*, New York, NY: Routledge: 2006, pp. 345–361.

and the grimmer reality. There is also usually an element of material self-interest. Denial can pay well, since it fortifies the status quo and serves powerful and prosperous constituencies, both political and corporate. Positive rewards are combined with sanctions. Failure to deny (that is, a determination to acknowledge) may result in the loss or denial of employment; decreased social standing; dismissal as a "kook" and a "radical"; and so on.

Among the most common discourses of genocide denial are the following:

"***Hardly anybody died.***" Reports of atrocities and mass killings are depicted as exaggerated and self-serving. (The fact that some reports *are* distorted and self-interested lends credibility to this strategy.) Photographic and video evidence is dismissed as fake or staged. Gaps in physical evidence are exploited, particularly an absence of corpses. Where are the bodies of the Jews killed by the Nazis? (Incinerated, conveniently for the deniers.) Where are the bodies of the supposed thousands of Kosovars killed by Serbs in 1999? (Buried on military and police bases, or dumped in rivers and down mineshafts, as it turned out.) When the genocides lie far in the past, obfuscation is easier. Genocides of indigenous peoples are especially subject to this form of denial. In many cases, the groups in question suffered near-total extermination, leaving few descendants and advocates to press the case for truth.

"***It was self-defense.***" Murdered civilians—especially adult males—are depicted as "rebels," "brigands," and "terrorists." The state and its allies are justified in eliminating them, though unfortunate "excesses" may occur. Deniers of the Armenian genocide, for example, play up the presence of armed elements and resistance among the Armenian population—even clearly defensive resistance. Likewise, deniers of Nazi genocide against Jews turn cartwheels to demonstrate "that *Weltjudentum* (world Jewry) had declared war on Germany in 1933, and the Nazis, as the ruling party of the nation, had simply reacted to the threat."[4] Jews were variously depicted as predatory capitalists, decadent cosmopolitans, and architects of global communism. The organizers of the third "classic" genocide of the twentieth century, in Rwanda, alleged that the assault on Tutsis was a legitimate response to armed invasion by Tutsi rebels based in Uganda, and the supposed machinations of a Tutsi "fifth column" in Rwanda itself.

A substrategy of this discourse is the claim that "***the violence was mutual.***" Where genocides occur in a context of civil or international war, they can be depicted as part of generalized warfare, perhaps featuring atrocities on all sides. This strategy is standard among deniers of genocides by Turks, Japanese,[5] Serbs, Hutus, and West Pakistanis—to name just a few. In Australia, Keith Windschuttle has used killings of whites by Aboriginals to denounce "The Myths of Frontier Massacres in Australian History."[6] (See also "*We* are the real victims," below.) Sometimes the deniers seem actually oblivious to the content of their claims, reflecting deeply embedded stereotypes and pervasive ignorance, rather than malicious intent. As I write these words, a CNN International reporter has just referred to the world standing by and "watch[ing] Hutus and Tutsis kill each other" during the Rwanda genocide of 1994.[7]

"***The deaths weren't intentional.***" The difficulties of demonstrating and documenting genocidal intent are exploited to deny that genocide occurred. The utility of this strategy is enhanced where a longer causal chain underlies massive mortality. Thus, when diverse factors combine to cause death, or when supposedly "natural" elements such as disease and famine account for many or most deaths, this denialist discourse is especially appealing. It underpins most denials of indigenous genocides, for example. Deniers of the Armenian and Jewish holocausts also contend that most deaths occurred from privations and afflictions that were inevitable, if regrettable, in a wartime context—in any case, not genocidal.

"***There was no central direction.***" Frequently, states and their agents establish a degree of deniability by employing freelance agents such as paramilitaries (as in Bosnia-Herzegovina and Darfur), criminal elements (e.g., the *chétés* in the Armenian genocide), or members of the targeted groups themselves (Jewish *kapos* in the Nazi death camps; Mayan peasants conscripted for genocide against Mayan populations of the Guatemalan highlands).

State attempts to eliminate evidence may mean that documentation of central direction, as of genocidal intent, is scarce. Many deniers of the Jewish Holocaust emphasize the lack of a clear order from Hitler or his top associates to exterminate European Jews. Armenian genocide denial similarly centers on the supposed freelance status of those who carried out whatever atrocities are admitted to have occurred.

"***There weren't that many people to begin with.***" Where demographic data provide support for claims of genocide, denialists will gravitate towards the lowest available figures for the targeted population, or invent new ones. The effect is to cast doubt on mortality statistics by downplaying the victims' demographic weight at the outbreak of genocide. This strategy is especially common in denials of genocide against indigenous peoples, as well as the Ottoman genocide against Armenians.

"***It wasn't/isn't 'genocide,' because...***" Here, the ambiguities of the UN Genocide Convention are exploited, and combined with the denial strategies already cited. Atrocious events do not qualify as "genocide" ... because the victims were not members of one of the Convention's specified groups; because their deaths were unintended; because they were legitimate targets; because "only" specific sectors of the target group were killed; because "war is hell"; and so on.

"***We would never do that.***" Collective pathological narcissism occludes the acknowledgment, or even the conscious consideration, of genocide. When the state and its citizens consider themselves pure, peaceful, democratic, and law-abiding, responsibility for atrocity may be literally unthinkable. In Turkey, notes Taner Akçam, anyone "dar[ing] to speak about the Armenian Genocide ... is aggressively attacked as a traitor, singled out for public condemnation and may even be put in prison."[8] In Australia, "the very mention of an Australian genocide is ... appalling and galling and must be put aside," according to Colin Tatz. "A curious national belief is that simply being Australian, whether by birth or naturalisation, is sufficient inoculation against deviation from moral and righteous behaviour."[9] Comedian Rob Corddry parodied this mindset in the context of US abuses and atrocities at Abu Ghraib prison near Baghdad. "There's no question what took place in that prison was horrible," Corddry said on *The Daily Show with Jon Stewart*. "But the Arab world has to realize that the US shouldn't be judged on the actions of a ... well, we shouldn't be judged on actions. It's our principles that matter, our inspiring, abstract notions. Remember: just because torturing prisoners is something we did, doesn't mean it's something we *would* do."[10]

"***We are the real victims.***" For deniers, the best defense is often a strong offense. With its "Day of Fallen Diplomats," Turkey uses Armenian terrorist attacks against Turkish diplomatic staff to pre-empt attention to the Turkish genocide against Armenians. In the case of Germany and the Nazi holocaust, there is a point at which a victim mentality concentrating on German suffering leads to the horrors that Germans inflicted, on Jews and others, being downgraded or denied. In the Balkans, a discourse of genocide was first deployed by Serb intellectuals promoting a nationalist–xenophobic project; the only "genocide" admitted was that against Serbs, whether by Croatians in the Second World War (which indeed occurred), or in Kosovo at the hands of the Albanian majority (which was a paranoid fantasy). Notably, this stress on victimhood provided powerful fuel for unleashing the genocides in the first place.[11]

DENIAL AND FREE SPEECH

What are the acceptable limits of denialist discourse in a free society? Should *all* denial be suppressed? Should it be permitted in the interest of preserving a liberal public sphere?

In recent years, many countries in the West have grappled with these questions. Varied approaches have been adopted, ranging from monitoring denialist discourse, to punitive measures including fines, imprisonment, and deportation. At the permissive end of the spectrum lies the United States. There, notorious deniers of the Jewish Holocaust, as well as neo-Nazi and Ku Klux

Klan-style organizations, operate mostly unimpeded, albeit sometimes surveilled and infiltrated by government agents. A much harder line has been enforced in France and Canada. In France, Holocaust denier Robert Faurisson was stripped of his university teaching position and hauled before a court for denying that the Nazi gas chambers had existed. Eventually, in July 1981, the Paris Court of Appeals assessed "personal damages" against Faurisson, based on the likelihood "that his words would arouse in his very large audience feelings of contempt, of hatred and of violence towards the Jews in France."[12]

In Canada, Alberta teacher Jim Keegstra "for twelve years... indoctrinated his students with Jewish conspiracy explanations of history... biased statements principally about Jews, but also about Catholics, Blacks, and others."[13] His passage through the Canadian justice system was labyrinthine. In 1982, Keegstra was dismissed from his job and, in 1984, charged with promoting racial hatred. In 1985, he was convicted, and sentenced to five months in jail and a $5,000 fine. The decision was overturned by the Alberta Court of Appeal, however, citing Canada's Charter of Rights and Freedoms; but Canada's Supreme Court subsequently ruled (narrowly) that hate speech was *not* constitutionally protected. In 1992, Keegstra was retried in Alberta and convicted. Once again, the conviction was overturned on appeal, this time on procedural grounds. At the time of writing, it was possible the case would be heard a second time before the Supreme Court of Canada.[14]

Undoubtedly the most famous trial involving a genocide denier is the libel case brought in 2000 by David Irving, an amateur historian of some early repute who nonetheless cast doubt and aspersions on the genocide of the Jews. Deborah Lipstadt accused Irving of genocide denial in her book *Denying the Holocaust*, referring to him as a "discredited" scholar and "one of the most dangerous spokespersons for Holocaust denial."[15] She also pointed to his links with neo-fascist figures and movements. Irving exploited Britain's loose libel laws to file a suit for defamation. The resulting trial became a *cause célèbre*, with prominent historians taking the stand to outline Irving's evasions and obfuscations of the historical evidence, as well as the character of his personal associations. The final, 350-page judgment by Judge Charles Gray cited Irving for nineteen specific misrepresentations, and contended that they were deliberate distortions to advance a denialist agenda. Irving's suit was dismissed, leaving him with a two-million-pound bill for legal costs—although he was subject to no legal sanction per se.

The spectrum of policies towards deniers, from permissive to prosecutory, is mirrored by the debate among genocide scholars. Those who call for punitive measures against deniers stress the link between denial and genocide, including future genocides, as well as the personal suffering that denial inflicts on a genocide's survivors and their descendants. This argument is made eloquently by Roger Smith, Eric Markusen, and Robert Jay Lifton, who hold that:

> denial of genocide [is] an egregious offense that warrants being regarded as a form of contribution to genocidal violence. Denial contributes to genocide in at least two ways. First of all, genocide does not end with its last human victim; denial continues the process, but if denial points to the past and the present, it also has implications for the future. By absolving the perpetrators of past genocides from responsibility for their actions and by obscuring the reality of genocide as a widely practiced form of state policy in the modern world, denial may increase the risk of future outbreaks of genocidal killing.

They especially condemn the actions of some professional scholars in bolstering various denial projects:

> Where scholars deny genocide, in the face of decisive evidence that it has occurred, they contribute to a false consciousness that can have the most dire reverberations. Their message, in effect, is: murderers did not really murder; victims were not really killed; mass murder requires no confrontation, no reflection, but should be ignored, glossed over. In this way scholars

> lend their considerable authority to the acceptance of this ultimate human crime. More than that, they encourage—indeed invite—a repetition of that crime from virtually any source in the immediate or distant future. By closing their minds to truth such scholars contribute to the deadly psychohistorical dynamic in which unopposed genocide begets new genocide.[16]

The opposing view does not dispute the corruption of scholarship that genocide denial represents. However, it rejects the authority of the state to punish "speech crimes"; it stresses the arbitrariness that governs *which* genocide denials are prohibited; and it calls for proactive engagement and public denunciation in place of censorship and prosecution. A leading exponent of such views is the political scholar and commentator Noam Chomsky, whose most bitter controversy revolves around a defense of the right of Robert Faurisson to air his denialist views. In an essay titled "Some Elementary Comments on the Rights of Freedom of Expression," published (without his prior knowledge) as a Foreword to Faurisson's *Mémoire en défense*, Chomsky depicted calls to ban Faurisson from teaching, even to physically attack him, as in keeping with authoritarian tradition:

> Such attitudes are not uncommon. They are typical, for example, of American Communists and no doubt their counterparts elsewhere. Among people who have learned something from the 18th century (say, Voltaire) it is a truism, hardly deserving discussion, that the defense of the right of free expression is not restricted to ideas one approves of, and that it is precisely in the case of ideas found most offensive that these rights must be most vigorously defended. Advocacy of the right to express ideas that are generally approved is, quite obviously, a matter of no significance .... Even if Faurisson were to be a rabid anti-Semite and fanatic pro-Nazi ... this would have no bearing whatsoever on the legitimacy of the defense of his civil rights. On the contrary, it would make it all the more imperative to defend them.[17]

Each of these perspectives brings important ideas to the table. To expand on Smith *et al's* reasoning: in most societies, some speech is subject to legal sanction—libelous, threatening, and obscene speech, for instance. It can reasonably be asked whether genocide denial does not do greater harm to society, and pose a greater threat, than personal libels or dirty words. Is not genocide denials libel against an entire people? And is the threat it poses not extreme, given that denial may sow the seeds of future genocides?

The case is a powerful one, and yet I find myself generally in agreement with Chomsky. Free speech *only* has meaning at the margins. Banning marginal discourses undermines liberal freedoms. Moreover, only a handful of deniers—principally those assailing the Jewish and Armenian holocausts—have attracted controversy for their views. The president (François Mitterrand) of the same French state that prosecuted Robert Faurisson not only actively supported Rwanda's *génocidaires*—before, during, and after the 1994 catastrophe—but when asked later about the genocide, responded: "The genocide or the genocides? I don't know what one should say!" As Gèrard Prunier notes, "this public accolade for the so-called 'theory of the double genocide' [i.e., by Tutsis against France's Hutu allies, as well as by Hutus against Tutsis] was an absolute shame."[18] It advanced a key thesis of genocide deniers: that the violence was mutual or defensive in nature. But Mitterrand's words were widely ignored; he was certainly in no danger of being arraigned before a tribunal. Likewise, the Canadian state that prosecuted Jim Keegstra was the same that shamefully "dodged implementation of the Genocide Convention" by "quietly redefining the crime in the country's domestic enforcement statute so as to omit any mention of policies and actions in which Canada was and is engaged," specifically its genocide against native people.[19] *Quis custodiet custodiens*—who will judge the judges?

One wonders, as well, whether the names and mendacious views of people such as Irving, Faurisson, and Keegstra would be remotely as prominent, if prosecutions and other measures had not been mounted against them.[20] (Indeed, it makes me queasy to print them here.) These individuals, and the initiatives they sponsor, are best confronted with a combination of monitoring, marginalization, and effective public refutation. Such refutation can be accomplished by visible and vocal denunciation, informed by conscientious reportage and scholarship, as well as through proactive campaigns in schools and the mass media.

While genocide denial in the public sphere may be destructive, for genocide scholars and students its consequences may actually be productive. Professional deniers have spurred scholarship in areas that otherwise might not have attracted it.[21] Moreover, not all "denial" is malevolent. Whether a genocide framework should be applied in a given case is often a matter of lively *and legitimate* debate. In recent decades, the character and content of mass killing campaigns in Bosnia and Kosovo, Darfur, Biafra (Nigeria), East Timor, Guatemala, and Vietnam have been intensively analyzed and hotly disputed. I believe this is to be encouraged, even though I find some of the views expressed to be disturbing and disheartening. Keeping denial of *all* genocides out of the realm of crime and punishment may be the price we pay for this vigorous exchange.[22]

ENDNOTES

1. Gregory H. Stanton, "Eight Stages of Genocide," http://www.genocidewatch.org/8stages.htm.
2. Richard G. Hovannisian, "Denial of the Armenian Genocide in Comparison with Holocaust Denial," in Hovannisian, ed., *Remembrance and Denial: The Case of the Armenian Genocide* (Detroit, MI: Wayne State University Press, 1999), p. 202.
3. For a summary, with many examples, see Anti-Defamation League, "Holocaust Denial in the Middle East: The Latest Anti-Israel Propaganda Theme," http://www.adl.org/holocaust/denial_ME/Holocaust_Denial_Mid_East_prt.pdf.
4. Michael Shermer and Alex Grobman, *Denying History: Who Says the Holocaust Never Happened and Why Do They Say It?* (Berkeley, CA: University of California Press, 2002), p.40.
5. Japanese denial of its genocidal atrocities against the Chinese and other Asian peoples during the Second World War is one of the most notorious and best-studied cases. "Japan has always presented itself as the victim of the war and has consistently ignored and repressed any attempts to focus on its aggression and war crimes.... The Japanese government and society have conducted an intensive and successful repression of any information about the war in which Japan is not presented as a peace-loving nation or in which anything negative about its history is mentioned." Elazar Barkan, *The Guilt of Nations: Restitution and Negotiating Historical Injustices* (New York: W.W. Norton, 2000), pp. 50, 60. Japanese governments and education authorities have persistently depicted the Japanese invasion and occupation of its "Greater Asian Co-prosperity Sphere" as a defensive response to a campaign of US economic warfare and political isolation; downplayed the horrors of events such as the Nanjing Massacre of 1937–38 (see Chapter 2); and played up Japanese suffering at Allied hands, especially in the area-bombing raids and atomic attacks. For an overview, see Gavan McCormack, "Reflections on Modern Japanese History in the Context of the Concept of Genocide," ch. 16 in Robert Gellately and Ben Kiernan, eds., *The Specter of Genocide: Mass Murder in Historical Perspective* (Cambridge: Cambridge University Press, 2003), pp. 265–88.
6. For a summary and debunking, see Ben Kiernan, "Cover-up and Denial of Genocide: Australia, the USA, East Timor, and the Aborigines," *Critical Asian Studies*, 34: 2 (2002), pp. 180–82.
7. CNN International broadcast, December 31, 2004.
8. Taner Akçam, *From Empire to Republic: Turkish Nationalism and the Armenian Genocide* (London: Zed Books, 2004), P. 209.
9. Colin Tatz, *With Intent to Destroy: Reflecting on Genocide* (London: Verso, 2003), p. 137.
10. Corddry quoted in Alan Shapiro, "American Treatment of Iraqi and Afghan Prisoners:

An Introduction," TeachableMoment.org, http://www.teachablemoment.org/high/prisoners.html.

11. A grimly ironic variation on this theme is the exceptionalist approach adopted by some victims of genocide or their descendants, claiming an "exclusivity of suffering" at the expense of other victim groups. Hence, for many years a tacit understanding prevailed among politically powerful sectors of Turkish and Israeli society to marginalize the Armenian genocide by proclaiming the uniqueness and incommensurability of the Jewish Holocaust. Thea Halo has contended, in turn, that too many scholars, activists, and scholar-activists have focused so intensively on Armenian suffering during and after the First World War that they have effectively denied comparable atrocities inflicted by the Ottomans upon other Christian populations of their realm, notably Assytians and Pontic Greeks. Halo writes: "The expropriation of an evil so egregious and monumental, strips the other nameless victims of that same monumental evil, of their rightful place in history, as if they never existed, thereby assuring that their Genocide is complete." Halo, "The Exclusivity of Suffering: When Tribal Concerns Take Precedence over Historical Accuracy," unpublished research paper, 2004.
12. Paris court judgment cited in Shermer and Grobmans, *Denying History*, pp. 10–11.
13. David Bercuson and DouglasWertheimer, quoted in Luke McNamara, "Criminalising Racial Hatred: Learning from the Canadian Experience," *Australian Journal of Human Rights*, 1: 1 (1994). Available online at http://www.austlii.edu.au/au/other/ahric/ajhr/V1N1/ajhr 1113.html.
14. See McNamara, "Criminalising Racial Hatred." The case of Ernst Zündel, a German-born denier of the Jewish Holocaust, has also attracted polemics in recent years. Zündel became "a political hot potato to immigration officials in Canada and the United States." Moving from Canada to the US when the Canadian government denied his application for citizenship, Zündel was then deported back to Canada, and then on to Germany, where he sits in prison at the time of writing. See CBC News Online, "Indepth: Ernst Zundel," September 30, 2004, http://www.cbc.ca/news/background/zundel/; "Holocaust Denier Behind Bars in Germany," CTV.ca, March 2, 2005.
15. Richard J. Evans, *Lying About Hitler: History, Holocaust, and the David Irving Trial* (New York: Basic Books, 2001), p.6. Evans was one of the historians who testified at the trial; his book provides an excellent overview of the proceedings.
16. Roger W. Smith, Eric Markusen, and Robert Jay Lifton, "Professional Ethics and the Denial of the Armenian Genocide," in Hovannisian, ed., *Remembrance and Denial*, pp. 287, 289.
17. Noam Chomsky, "Some Elementary Comments on the Rights of Freedom of Expression," http://www.zmag.org/chomsky/articles/8010-free-expression.html. Much controversy attached to Chomsky's comment in this essay that "As far as I can determine, he [Faurisson] is a relatively apolitical liberal of some sort."
18. Gèrard Prunier, *The Rwanda Crisis: History of a Genocide* (New York, NY: Columbia University Press, 1997), p. 339.
19. "Thus, the [UN] convention's prohibitions of policies causing serious bodily or mental harm to members of a target group and/or effecting the forcible transfer of their children were from the first moment expunged from Canada's 'legal understanding.' In 1985, the Canadian statute was further 'revised' to delete measures intended to prevent births within a target group from the list of proscribed policies and activities.... At least one Canadian court, moreover, has recently entered a decree making it a criminal offense for anyone to employ the term ['genocide'] in any other way." Ward Churchill, "Genocide by Any Other Name: North American Indian Residential Schools in Context," in Adam Jones, ed., *Genocide, War Crimes and the West: History and Complicity* (London: Zed Books, 2004), pp. 83–84.
20. After the David Irving decision, historian Andrew Roberts claimed the judgment against Irving was at best a partial victory, since "the free publicity that this trial has generated for him and his views has been worth far more than could ever have been bought for the amount of the costs." Quoted in Evans, *Lying About Hitler*, p. 235.
21. According to Colin Tatz, "for all the company they keep, and for all their outpourings, these deniers assist rather than hinder genocide and Holocaust research," in part by "prompt[ing] studies by men and women of eminence...who would otherwise

not have written on genocide." Tatz, *With Intent to Destroy*, pp. 139–140.

22. For a discussion of responsible versus malicious denial of a genocide framework for Cambodia, see Ben Kiernan, "Bringing the Khmer Rouge to Justice," *Human Rights Review*, 1: 3 (April–June 2000), pp. 92–108.

DISCUSSION QUESTIONS

1. The author makes the argument that *denial* is the "final stage of genocide, and an indispensable one..." Why is denial an integral part of all genocidal acts on the part of perpetrators and their defenders?
2. Provide a current example of some of Jones' "most common discourses of genocide denial" that you've come across in the media (internet, print, radio, or television).
3. As noted by the author, several countries around the world have passed laws criminalizing deniers of genocide. What do you think about these laws? Is genocide denial really an "egregious offense" worthy of punishment? Why or why not?
4. When does "free speech" become a social problem?
5. What can society do to prevent genocide and its denial?

PART XIII

Social Problems Related to Population and the Environment

These five selections turn our attention to world problems. For some people, these are the most important problems of all. We live on planet Earth, and what each society does affects all the others because of population and environmental problems. Because both are world problems, it is very difficult to solve them overnight. They are compounded by the fact that there is great disagreement. Usually, each society's people place the blame on other societies and overlook their own contribution to these problems. In this context, what truly is "progress"? The consequences of the problems described here are tremendously important for everyone in the world.

Anthony Giddens, in a selection from his book entitled *The Politics of Climate Change*, presents an argument for curbing our profligate consumption of fossil fuels: oil, gas, and coal. He takes on the important issue of peak oil, and raises profound questions on the social implications of continuing along the same path of over-consuming non-renewable resources.

Environmental activist, Bill McKibben, argues that we are living through "A Very Special Moment" in the history of planet earth–one that has greater importance than any moment that has gone before. This moment presents a social problem, one that we might be able to overcome and one that will become very costly if we ignore it.

John Powell and his colleagues consider the roles of race and social class in the Hurricane Katrina disaster. Katrina exposed a long history of racial and class chasms—especially in the residentially segregated black and poor wards of New Orleans. Not surprisingly, these communities have suffered most from this disaster.

Laura Wagner offers a first-person account of one of the worst earthquakes in recent decades that devastated the Caribbean nation of Haiti. Wagner makes a poignant statement that there is no such thing as a "natural disaster" because it is the economically deprived and socially vulnerable members of society that always suffer more acutely from these events. But Wagner is also eminently hopeful about the capacity of the Haitian people to rebuild their communities and restore their lives in the long run.

Finally, Laurence Kotlikoff and Scott Burns examine the consequences of population dependency brought about by large numbers of elderly in the world, with fewer and fewer individuals to support them. This, of course, threatens Social Security, retirement resources, and family stability in the United States. However, the population is becoming older all over the world, and many societies have even greater problems than these.

51

Climate Change: Running Out, Running Down

ANTHONY GIDDENS

The Four Questions

1. What does Giddens suggest as the central problem in this piece?
2. Is this problem a *social* one? Who does this problem affect?
3. What is the cause of this problem?
4. What can be done to address this problem? How can individuals address this problem? How can society as a whole address this problem?

Topics Covered

Climate change
Fossil fuels
Pollution
Peak oil
Political policy

Oil, gas and coal, the three dominant energy sources in the world, are all fossil fuels, producing greenhouse gases on the large scale. Reducing our dependence on them, or (most notably in the case of coal) making them far cleaner environmentally than they are at the moment, is imperative for mitigating climate change. The technologies required both to reduce our vulnerability to energy shortfalls and to reduce carbon emissions are one and the same; they include wind, wave and solar energy, hydroelectricity and thermal power. Lifestyle change is likely to be of key importance in both spheres, particularly when directed at curbing profligate habits of energy use.

The industrial revolution in its country of origin, Britain, was fuelled by coal—or, more accurately, by the scientific and technological discoveries which turned coal into a dynamic energy source. The changeover from burning wood—previously the prime energy source—was not easy, since it meant a transformation of habits. By the mid-seventeenth century wood was running out as a source of fuel; but many initially detested the sooty coal that came to replace it, and which, in the end, actually helped create a whole new way of life based on cities and machine production.

The turn to coal ushered in the world we now inhabit, in which the energy of the individual citizen or worker is of trivial importance compared to that produced from inanimate resources. As Richard Heinberg has observed in relation to the US:

> If we were to add together the power of all the fuel-fed machines that we rely on to light and heat our homes, transport us, and otherwise keep us in the style to which we have become accustomed, and then compare that total with the amount of power that can be generated by the human body, we would find that each American has the equivalent of over 150 'energy slaves' working for us twenty-four hours a day.[1]

Oil has never replaced coal, but it began to mount a challenge to coal's dominance from the turn of the twentieth century onwards. For a while,

SOURCE: Excerpt from Anthony Giddens, *The Politics of Climate Change*, pp. 35–48. Polity Press, 2009. Reprinted by Permission.

in the early part of that century, the US was the biggest oil producer in the world and for a long period was largely self-sufficient in oil. During much of that time the US was an anti-imperial power, with quite a different philosophy from the dominant imperial formation, the British Empire; for instance, the US opposed the Franco-British intervention in Suez in 1957, partly on strategic grounds, but also on moral ones. Of course, these roles were later reversed, as the US came to see the Middle East as more and more vital to its interests. Yet it is worth restating the obvious—the history of oil is the history of imperialism, in one guise or another....

OPEC, the Organization of the Petroleum Exporting Countries, was set up by the producing nations to act as a counter-balance to the influence of the oil corporations. It was followed over the years by the widespread and progressive take-over of oil assets by state-owned companies in those nations. OPEC was founded in 1960 and for some while there were no major shocks affecting energy prices or world supply. However, the leaders of OPEC were outraged by the support given by the US and other Western countries to Israel in the Arab-Israeli war of 1973. Oil exports to the US, Britain and some other states were blocked, while OPEC raised the price of oil by 70 per cent, precipitating economic recession in the industrial countries.

I mention these well-known episodes because they bring home how close the connections at some points between international politics and energy security are (and will continue to be); and also because they serve as a reminder that whether or not the oil will flow does not depend upon the assessment of resources alone, but on how those resources intersect with geopolitics.

French emissions of greenhouse gases are markedly lower today than they might otherwise have been because, following the oil crisis provoked by OPEC's actions, France took the decision to become more independent of world energy markets and invested heavily in nuclear power. Japan also took note and introduced policies to regulate energy use and promote energy conservation. Today it is among the most energy-efficient of the industrial countries and is in the vanguard of clean energy technology, for instance in the car industry. Its emissions are relatively high, however, because of its dependence upon coal for electricity production. Sweden instituted a range of energy-saving policies and started to reduce its oil dependency, a process that is still continuing. Far more waste is currently recycled in Japan and Sweden than in most other industrial countries. Having no indigenous resources of its own in the 1970s, Denmark took fright and initiated measures to transfer parts of its electricity production to renewable energy sources, particularly windpower. At the same time, Brazil made the decision to invest in biofuels and now has a higher proportion of motor transport running off them than any other country, although the environmental benefits are dubious because of the deforestation involved.

The US was also obliged to react. Its responses included considering plans to invade Saudi Arabia, but also, more realistically, introducing measures to conserve energy, in the shape of the Energy Policy Conservation Act.[2] It was a significant intervention, because it showed that the wasteful energy habits of American consumers could be curbed if the impetus was strong enough. The aim of one section of the Act was to double the energy efficiency of new cars within 10 years. The target wasn't reached, but major improvements were nevertheless achieved. However, as the sense of crisis receded, fuel consumption rose again, soon to become lower per mile travelled than it had been before.

PEAK OIL

The debate about the limits of the world's fossil fuel resources is of great consequence for climate change policy. In 1956 the American geologist Marion King Hubbert made the now famous prediction that indigenous oil production in the US would peak in 1970—a prediction that was widely rejected early on, but which turned out to be valid, even though the actual level of oil production was still going up in 1970. Peak oil calculations depend upon assessments of what in the oil industry is known as the "ultimate reserves" a given country or oilfield has. It does not refer to how much oil exists, but to how much can ever be extracted—usually a much smaller amount.[3]

The controversies surrounding peak oil are as intense as those concerned with global warming, and the two debates in fact closely resemble one another. There are those who believe that there is plenty of oil and gas to go round. They do not accept that we should be worried about future sources of supply. In their view there are sufficient resources to last for a long while, even given the rising levels of economic growth of the large developing countries and even given the growing world population. David Howell and Carol Nakhle, for example, argue that there is enough of the "known, relatively easy-to-extract stuff" to last for at least another 40 years. More reserves, they continue, are certain to be found. Under the melting ice of the Arctic, "billions of tonnes of oil and billions of cubic metres of gas lie waiting." New oilfields are available for exploration in Alaska, off the coast of Africa and offshore in Brazil. Even in the much-explored Middle East, a possible further cornucopia awaits.[4]

Such authors are the functional equivalents of the climate change sceptics—they are saying, "Crisis, what crisis?" Mainstream opinion is less sanguine, or at least has become so over the past few years, and is represented by the bulk of industry analysts and the official publications of the major oil countries. It holds that there may be enough oil (and even more gas) to continue to expand levels of production for some while. However, no one knows, almost by definition, how much there is in as yet unexplored fields or what the difficulties of recovering it may be. The international Energy Agency (IEA), set up to monitor oil production after the 1970s oil embargo, predicted in 2007 that there will be no peak m oil production before 2030.[5]

Others believe that the world is rapidly approaching peak oil and that the adjustments that will have to be made by the industrial and industrializing countries, perhaps in the quite near future, are of epic proportions. As one prominent writer expresses it, we are likely to confront "the kind of dramatic, earth-shattering crisis that periodically threatens the very survival of civilization. More specifically, it is an energy crisis brought about by the conflict between the rising global demand for energy and our growing inability to increase energy production".[6] These words come from the investment analyst Stephen Leeb, who in the early 2000s predicted that world oil prices would reach $100 a barrel, a claim regarded by many at the time as ridiculous. Before 2008, Leeb was one of very few individuals talking of the possibility of oil prices reaching $200 a barrel or more. By the middle of that year—prior to the financial crisis—talk of such a possibility became commonplace; at the same time, it was publicly endorsed by the investment bank Goldman Sachs. Oil prices had risen to $147 a barrel by July 2008, but by December, as recession started to bite, they fell back to $40 a barrel.

Leeb is one among a clutch of writers who hold that orthodox claims that world oil supplies will not peak for another 40 or 50 years are fundamentally mistaken.[7] The disagreements between those who write about oil production centre upon two main issues—how much recoverable oil there is in existing fields, and what the chances are of large new oil deposits being found. As in the climate change discussion, it will make a great

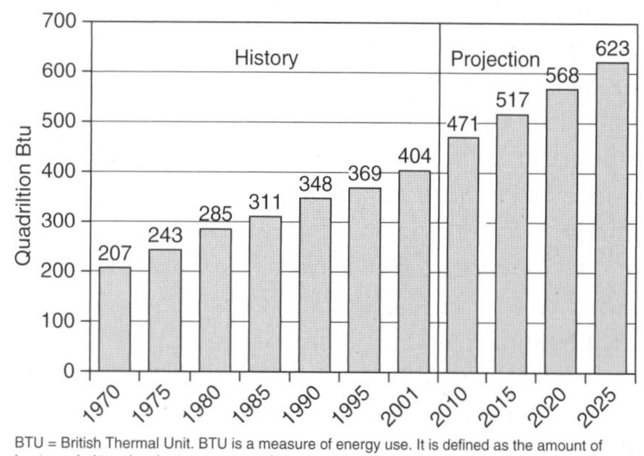

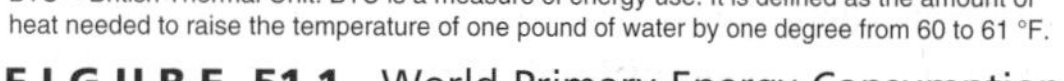
BTU = British Thermal Unit. BTU is a measure of energy use. It is defined as the amount of heat needed to raise the temperature of one pound of water by one degree from 60 to 61 °F.

FIGURE 51.1 World Primary Energy Consumption

SOURCE: Energy Information Administration.

difference to humanity's future who is right, or more nearly right.

The amount of new oil discovered each year has been declining for some while. According to David Strahan, discovery was at its highest point as long ago as 1965: "These days, for every barrel we discover, we now consume at least three".[8] Most of the world's biggest oilfields were identified long before that date. Of the 50 highest-producing oil countries, 18 have now passed their peak, even according to conservative estimates. If one includes the smaller producers, more than 60 oil-producing countries have done so too. Their production losses have so far been offset by growth in other areas and by improvements in extraction and processing technology.

Such authors reject the idea that big new oil and gas fields will be opened up under the Arctic or anywhere else, pointing to the extraction difficulties that will be involved. They argue that production in areas of the world outside the OPEC nations and Russia has remained static for years, in spite of successful finds in a range of countries. Russia's output growth of oil, although currently on the increase, looks likely to founder. The world will probably continue to have to look to OPEC and to the Middle East with all its tensions and problems.

In 2008, at the time at which oil prices reached their high point, the IEA produced a new report on global oil output.[9] The survey asserted once again that peak oil production is not imminent, but it did find that output from some of the world's largest oilfields is declining faster than earlier estimates had suggested. A significant increase in investment will be needed just to maintain the current level of production. The IEA calculates that the oil-producing countries and the oil companies will have to invest a total of about $360 billion a year until 2030 in order to replace falling oil production and to be able to meet the likely demands of the developing countries. Investment decisions by OPEC will be critical. The IEA expects oil produced from non-conventional sources, such as Canada's oil sands, to play a decisive role in keeping up levels of output. Yet the methods of extraction used in such cases are heavily polluting in terms of greenhouse gas emissions.

Supposing the theorists of peak oil are right, can natural gas step into the place of oil to some degree? After all, gas produces lower emissions than either oil or coal, and can be used for at least some of the purposes to which oil is put—for instance, cars can be converted to run on compressed natural gas without too much difficulty. It is often said that world supplies of gas far outstrip those of oil; some say there is enough to last the world for some 70–80 years from now, even given growing demand. David Victor and colleagues have suggested that there will be a worldwide move towards this energy resource.[10] By 2050, they argue, gas could supplant oil to become the most important energy source in the world. According to them, there is enough gas available to last for a century at today's rates of consumption. Yet, as in almost every aspect of energy security, controversy exists here too. There is a large distance indeed between the most optimistic estimate of recoverable gas reserves (20,000 trillion cubic feet) and the lowest (8,000 trillion cubic feet).

According to Julian Darley, the most difficult energy problem facing the United States—and by extension other major industrial countries too—in the near future, will in fact not be to do with oil but with gas.[11] In the early 2000s, as happened earlier with oil, for the first time more natural gas was consumed than was discovered—"it is the dread hand of depletion writing on the wall."[12] Natural gas has a vital role in the production of fertilizers—it has been called the single most important ingredient in the diets of human beings. At the moment there is no known replacement for it in this role. There are more than six billion people in the world, and that number will probably rise to nine billion later this century. In spite of the damage they can do to the soil, one cannot see how, without fertilizers, the world could possibly feed such numbers.

It is normally assumed that, in contrast to oil and gas, one thing we can be sure of is that the world has vast supplies of coal at its disposal. However, some are now saying that world coal supplies might be more limited than has hitherto been supposed.[13] There may have been large-scale over-reporting of coal reserves. Energy Watch, a German energy consultancy, has looked at the reserves listed by coal-producing countries and found that they have stayed the same even though those countries continued to mine extensively. For example, although China has mined 20 per cent of its coal since 1992, its listed reserves remain unchanged. Countries that have revised their figures have done so in a sharply downward direction, suggesting that improved techniques of assessment have produced more sober estimates than those made previously. Energy Watch has calculated that coal supplies may peak far earlier than is conventionally thought, perhaps as soon as 2025. Majority opinion, to repeat, is against such a conclusion. Indeed, one of the main worries about the world possibly running out of oil is that there could be an upswing in the use of coal.

SWEATING THE ASSETS

Electricity generation is a major source of energy consumption and of the generation of greenhouse gases. To see what has happened in this area, we have to look first of all at the institutions and practices set up at the time when energy was cheap, because the thinking behind them now looks remarkably shallow. In the period following the Second World War, energy was the *locus classicus* of the state planning that was everywhere in vogue. Coal-mining was widely nationalized, while miners in numerous countries enjoyed an almost mythic status, partly because of the dangers of their jobs, but also because of the centrality of their work to the economy.

Partly as a hangover from the war years, security of energy supply was a core concern, to which government control was the response. The widespread turn to nuclear power in the 1950s and 1960s was also guided everywhere by the state. Many believed that this source would eventually provide energy in abundance; instead, it proved obstinately expensive and, in the public mind at least, hazardous. Apart from in one or two countries—as mentioned earlier, particularly France—the nuclear option was largely suspended. In many countries, nuclear power stations built decades ago are still in use, although they are now approaching the end of their lives....

Since the early 2000s, what Helm calls a new energy paradigm has emerged. It is (or was) marked by rises in the price of oil and gas well beyond what seasoned observers had thought possible. But it also involves a return to the protection of national energy supplies, modernization of plant, investment for the future, a consciousness of the finitude of oil and gas resources, recognition of the key importance of foreign policy to energy security—and an awareness of the need to integrate energy policy with the struggle to limit climate change. Political considerations have come once more to intrude deeply into energy markets because of their concentration in the hands of states which use them as instruments of domestic and foreign policy.

The US is heavily dependent upon Middle Eastern oil producers, as are Europe and Japan, although all are now scrambling to diversify their sources of supply. Russia's attempt to return as a great power is based upon its fossil fuel resources and the high prices they currently command. China's very rapid rate of economic growth has led the country to take far more of an international role than it had done previously, as it makes its presence felt in the Middle East, Africa and Latin America in pursuit of oil, gas and minerals.

The EU countries import almost half their gas from Russia and a substantial proportion of their oil too. The implications on both sides for climate change policy are considerable. Russia is maximizing its income from oil and gas without the modernization that could have come about had the country permitted the introduction of outside investment and encouraged effective management. Gazprom, Russia's largest state-owned company, is notoriously inefficient and poorly managed. Domestic and industrial consumers in Russia get their energy at heavily subsidized prices, a policy that is changing only slowly, but which does nothing to promote energy conservation.

The Russian leadership has developed a confrontational approach to the EU, with consequences that also spill over into the area of climate change. It has firmly rejected EU approaches to find a meeting point: "We intend to retain state control over the gas transport system and over Gazprom. We will not split Gazprom up. And the European Commission should not have any illusions. In the gas sector, they will have to deal with the state."[14]

Russia has found it quite easy to do individual deals with EU member-states and thereby undercut European unity. A notable example is the Nord Stream pipeline project, which brings together Gazprom and two of Germany's biggest energy companies. Although practical and diplomatic problems mean that it may never be built, the very idea of the project runs contrary to the spirit of European solidarity. Since energy security and responding to climate change are so closely linked, an EU that cannot speak with one voice on the first could find its capabilities to make progress with the second compromised—a serious matter, since, in terms of concerns to combat global warming, it aspires to be the world leader.

A report produced for the meeting of the European Council in March 2008 pointed to some of the security questions that could be posed by the melting of the Arctic ice.[15] The mineral wealth believed to lie beneath the Arctic might become a flash-point for new tensions. In 2007, a group of Russian scientists planted a flag on the Arctic seabed, implying that much of it is Russian territory, a case that Russia is now vigorously pursuing in international courts. Few prescriptions of international law apply to the Arctic. Most of the seabed is uncharted. The US, Norway and Denmark are all lodging ownership claims. The US is the only significant country not to have ratified the UN Convention on the Law of the Sea, which provides at least one forum for discussion for competing claims to the Arctic. However, it has largely abided by its conventions, and pressures for the country formally to ratify grew after the Russian action. In 2008, President George W. Bush recommended ratification to the US Senate.

THE STRUGGLE FOR RESOURCES

Only the US is ahead of China in terms of oil and gas consumption. In 2007 China accounted for about 40 per cent of the worldwide growth in demand for oil. Its level of demand will rise by about 6 per cent a year over the next decade if its rates of growth are sustained and its energy policies remain the same. In casting around for oil, China is pursuing an aggressive foreign policy—following, it could be said, in the footsteps of Britain and the US. China does not work through oil corporations as Western countries tend to do, but its objectives are much the same. It essentially buys oilfields in different countries for its own use, setting the terms of sale locally. Countries where China has done such deals range from Venezuela to Indonesia, Oman, Yemen—and Sudan.

It has also made substantial inroads into the Middle East, to the chagrin of the US. Saudi Arabia has become the largest oil supplier to China, and the Chinese have been allowed to explore for gas within the country. China has forged a close relationship with Iran and is importing increasing amounts of gas and oil from that state. American oil companies are prohibited from doing business with Iran as a result of an Act of Congress, so cannot get a look in. In the meantime, along with Russia, China blocks the imposition of sanctions on Iran, which most other nations in the international community support in their attempt to stop the country from acquiring nuclear weapons and the rocket systems able to deliver them. "Both sides behave as if an oil shortage is looming, and that it's 'us or them.'"[16]

India has not yet adopted such a high foreign policy profile as China in respect of oil, but it will need much more fuel as its economy advances and as consumer tastes change. In China, there has been a steep rise in car use over the past decade, with no heed at all being paid to environmental considerations. Much the same is set to happen in India. The Tata Nano car was unveiled in that country in January 2008. Costing 100,000 rupees (£1,300), it is by far the cheapest new car in the world; millions of Indians, even those on a relatively modest income, will, for the first time, be able to buy a car. The Tata Nano has a 33bhp petrol engine, which, because of being so small, is reasonably fuel-efficient. Yet the sheer numbers likely to appear on the roads will certainly result in large-scale environmental consequences.

Outside India, there are also plans to market the car in Latin America, South-East Asia and Africa. The chief scientist of the IPCC, Rajendra Pachauri, himself an Indian, says he is "having nightmares about it." The response from the industrialist Ratan Tata sums up perfectly some of the dilemmas surrounding both energy security and climate change: "We need to think of our masses. Should they be denied the right to an individual form of transport? "[17]

Commentators on energy security have started to speak of "Chindia" to refer to the combined impact of Chinese and Indian economic growth on world oil and energy markets. For most of the time since the Second World War the rise in demand for oil has been only about half that of overall economic growth. Since 2000, however, that proportion has increased to 65 per cent.[18] At the moment, measured on a per capita basis, Chindia consumes one-seventh of the total for the industrial countries. If Chindia joins the high-income group of countries within the next 20 years, as seems likely, growth in worldwide energy demand will dramatically accelerate.

In the areas of both climate change and energy security, the main divergence between the more sanguine and the apocalyptic writers is time—how much time remains before large changes will have to be made to the ways in which we live. Even if the effects of climate change are progressive rather than abrupt, and will mainly affect subsequent generations rather than ourselves, the lesson should still be to prepare early, and to start now. Exactly the same is true of energy security, even if those who say that oil and gas have several decades to run turn out to be right....

ENDNOTES

1. Richard Heinberg, *The Party's Over* (Gabriola Island: New Society Publishers, 2003), p. 31.
2. Paul Middleton, *A Brief Guide to the End of Oil* (London: Robinson, 2007), ch. 3.
3. David Strahan, *The Last Oil Shock* (London: Murray, 2007), p. 40.
4. David Howell and Carole Nakhle, *Out of the Energy Labyrinth* (London: Tauris, 2007), pp. 88–92.
5. International Energy Agency, *World Energy Outlook* 2007 (Paris: OECD/IEA, 2007).
6. Stephen Leeb, *The Coming Economic Collapse* (New York: Warner, 2007), p. 1.
7. See, for example, Kenneth Deffeyes, *Hubbert's Peak: The Impending World Oil Shortage* (Princeton: Princeton University Press, 2001); Paul Roberts, *The End of Oil* (London: Bloomsbury, 2004); Michael T. Klare, *Resource Wars* (New York: Holt, 2002); Matthew R. Simmons, *Twilight in the Desert* (New York: Wiley, 2005); Strahan, *The Last Oil Shock.* Strahan's title unconsciously echoes that of Fred Pearce on global warming, quoted in chapter 1: *The Last Generation.*
8. Strahan, *The Last Oil Shock,* p. 60.
9. Carola Hoyos and Javier Blas, 'Investment is the Key to Meeting Oil Needs', *Financial Times*, 29 October 2008.
10. David Victor et al., *Natural Gas and Geopolitics* (Cambridge: Cambridge University Press, 2006).
11. Julian Darley, *High Noon for Natural Gas* (White River Junction: Chelsea Green, 2004).
12. Ibid. p. 5.
13. David Strahan, 'Lump Sums', *Guardian,* 5 March 2008.
14. Speech of then-President Vladimir Putin, quoted in Edward Lucas, *The New Cold War* (London: Bloomsbury, 2008), p. 212.
15. Adele Airoldi, *The European Union and the Arctic* (Copenhagen: Nordic Council of Ministers, 2008).
16. Strahan, *The Last Oil Shock*, p. 180.
17. Reported in Times Online, 11 January 2008.
18. Leeb, *The Coming Economic Collapse*, p. 77.

DISCUSSION QUESTIONS

1. The over-consumption of oil, gas, and coal is a social problem and a major contributor to greenhouse gas. The author suggests that "Lifestyle change is likely to be of key importance … at curbing profligate habits of energy use." What lifestyle changes have you made to curb your over-consumption of fossil fuels?
2. What is "peak oil"? Is it a social problem, or is it a potential solution to our over-consumption and dependence on foreign oil? How will peak oil affect our reliance on coal as an energy source, and what are other unintended consequences of peak oil? Explain your response.
3. How do China and India figure into the current global "struggle for resources"? Why are these countries important to any discussion on curbing greenhouse gases, peak oil, and global environmental degradation?
4. Why does Tata's Nano represent a troubling trend for the global environment, according to the author? Do you agree (or disagree) with his alarm? Why?
5. Is climate change a social problem that can be solved one-nation-at-a-time? Or is it a problem that demands an unprecedented (and unlikely) level of global cooperation? Explain your response.

52

A Very Special Moment

BILL MCKIBBEN

The Four Questions

1. What exactly is the problem?
2. What causes the problem, according to the author?
3. Is this a *social* problem? Or is it something beyond the social world? Perhaps it is biological or environmental (a physical problem)?
4. What can be done? McKibben's selection describes many possible options.

Topics Covered

Progress
Overpopulation
Developing world
Environment
Consumerism
Technology
Future of the planet

We may live in a special time. We may live in the strangest, most thoroughly different moment since human beings took up farming, 10,000 years ago, and time more or less commenced. Since then time has flowed in one direction—toward *more,* which we have taken to be progress. At first the momentum was gradual, almost imperceptible, checked by wars and the Dark Ages and plagues

SOURCE: Excerpt from Bill McKibben, "A Special Moment in History: The Future of Population," in *Atlantic Monthly* 281, No. 5, May 1998, pp. 55–76. Reprinted by permission.

and taboos; but in recent centuries it has accelerated, the curve of every graph steepening like the Himalayas rising from the Asian steppe. We have climbed quite high. Of course, fifty years ago one could have said the same thing, and fifty years before that, and fifty years before *that*. But in each case it would have been premature. We've increased the population fourfold in that 150 years; the amount of food we grow has gone up faster still; the size of our economy has quite simply exploded.

But now—now may be the special time. So special that in the Western world we might each of us consider, among many other things, having only one child—that is, reproducing at a rate as low as that at which human beings have ever voluntarily reproduced. Is this really necessary? Are we finally running up against some limits? ...

If we live at a special time, the single most special thing about it may be that we are now apparently degrading the most basic functions of the planet. It's not that we've never altered our surroundings before. Like the beavers at work in my back yard, we have rearranged things wherever we've lived. We've leveled the spots where we built our homes, cleared forests for our fields, often fouled nearby waters with our waste. That's just life. But this is different. In the past ten or twenty or thirty years our impact has grown so much that we're changing even those places we don't inhabit—changing the way the weather works, changing the plants and animals that live at the poles or deep in the jungle. This is total. Of all the remarkable and unexpected things we've ever done as a species, this may be the biggest. Our new storms and new oceans and new glaciers and new springtimes—these are the eighth and ninth and tenth and eleventh wonders of the modern world, and we have lots more where those came from.

We have gotten very large and very powerful, and for the foreseeable future we're stuck with the results. The glaciers won't grow back again anytime soon; the oceans won't drop. We've already done deep and systemic damage. To use a human analogy, we've already said the angry and unforgivable words that will haunt our marriage till its end. And yet we can't simply walk out the door. There's no place to go. We have to salvage what we can of our relationship with the earth, to keep things from getting any worse than they have to be.

If we can bring our various emissions quickly and sharply under control, we *can* limit the damage, reduce dramatically the chance of horrible surprises, preserve more of the biology we were born into. But do not underestimate the task. The UN's Intergovernmental Panel on Climate Change projects that an immediate 60 percent reduction in fossil-fuel use is necessary just to stabilize climate at the current level of disruption. Nature may still meet us halfway, but halfway is a long way from where we are now. What's more, we can't delay. If we wait a few decades to get started, we may as well not even begin. It's not like poverty, a concern that's always there for civilizations to address. This is a timed test, like the SAT: two or three decades, and we lay our pencils down. It's *the* test for our generations, and population is a part of the answer.

WHEN we think about overpopulation, we usually think first of the developing world, because that's where 90 percent of new human beings will be added during this final doubling....

We fool ourselves when we think of Third World population growth as producing an imbalance, as Amartya Sen points out. The white world simply went through its population boom a century earlier (when Dickens was writing similar descriptions of London). If UN calculations are correct and Asians and Africans will make up just under 80 percent of humanity by 2050, they will simply have returned, in Sen's words, "to being proportionately almost exactly as numerous as they were before the European industrial revolution."

And of course Asians and Africans, and Latin Americans, are much "smaller" human beings: the balloons that float above their heads are tiny in comparison with ours. Everyone has heard the statistics time and again, usually as part of an attempt to induce guilt. But hear them one more time, with an open mind, and try to think strategically about how we will stave off the dangers to this planet.

Pretend it's not a moral problem, just a mathematical one.

- An American uses seventy times as much energy as a Bangladeshi, fifty times as much as a Malagasi, twenty times as much as a Costa Rican.
- Since we live longer, the effect of each of us is further multiplied. In a year an American uses 300 times as much energy as a Malian; over a lifetime he will use 500 times as much.
- Even if all such effects as the clearing of forests and the burning of grasslands are factored in and attributed to poor people, those who live in the poor world are typically responsible for the annual release of a tenth of a ton of carbon each, whereas the average is 3.5 tons for residents of the "consumer" nations of Western Europe, North America, and Japan. The richest tenth of Americans—the people most likely to be reading this magazine—annually emit eleven tons of carbon apiece.
- During the next decade India and China will each add to the planet about ten times as many people as the United States will—but the stress on the natural world caused by new Americans may exceed that from new Indians and Chinese combined. The 57.5 million Northerners added to our population during this decade will add more greenhouse gases to the atmosphere than the roughly 900 million added Southerners.

These statistics are not eternal. Though inequality between North and South has steadily increased, the economies of the poor nations are now growing faster than those of the West. Sometime early in the next century China will pass the United States as the nation releasing the most carbon dioxide into the atmosphere, though of course it will be nowhere near the West on a per capita basis.

For the moment, then (and it is the moment that counts), we can call the United States the most populous nation on earth, and the one with the highest rate of growth. Though the U.S. population increases by only about three million people a year, through births and immigration together, each of those three million new Americans will consume on average forty or fifty times as much as a person born in the Third World. My daughter, four at this writing, has already used more stuff and added more waste to the environment than most of the world's residents do in a lifetime. In my thirty-seven years I have probably outdone small Indian villages.

Population growth in Rwanda, in Sudan, in El Salvador, in the slums of Lagos, in the highland hamlets of Chile, can devastate *those places*. Growing too fast may mean that they run short of cropland to feed themselves, of firewood to cook their food, of school desks and hospital beds. But population growth in those places doesn't devastate *the planet*. In contrast, we easily absorb the modest annual increases in our population. America seems only a little more crowded with each passing decade in terms of our daily lives. You can still find a parking spot. But the earth simply can't absorb what we are adding to its air and water.

So if it is we in the rich world, at least as much as they in the poor world, who need to bring this alteration of the earth under control, the question becomes how. Many people who are sure that controlling population is the answer overseas are equally sure that the answer is different here. If those people are politicians and engineers, they're probably in favor of our living more efficiently—of designing new cars that go much farther on a gallon of gas, or that don't use gas at all. If they're vegetarians, they probably support living more simply—riding bikes or buses instead of driving cars.

Both groups are utterly correct. I've spent much of my career writing about the need for cleverer technologies and humbler aspirations. Environmental damage can be expressed as the product of Population × Affluence × Technology. Surely the easiest solution would be to live more simply and more efficiently, and not worry too much about the number of people.

But I've come to believe that those changes in technology and in lifestyle are not going to occur easily and speedily. They'll be begun but not finished in the few decades that really matter. Remember that the pollution we're talking about is not precisely pollution but rather the inevitable result when things go the way we think they should: new filters on exhaust pipes won't do anything about that CO_2. We're stuck with making real changes in how we live. We're stuck with dramatically reducing the amount of fossil fuel we use. And since modern Westerners are practically machines for burning fossil fuel, since virtually everything we do involves burning coal and gas and oil, since we're wedded to petroleum, it's going to be a messy breakup....

Changing the ways in which we live has to be a fundamental part of dealing with the new environmental crises, if only because it is impossible to imagine a world of 10 billion people consuming at our level. But as we calculate what must happen over the next few decades to stanch the flow of $C0_2$, we shouldn't expect that a conversion to simpler ways of life will by itself do the trick. One would think offhand that compared with changing the number of children we bear, changing consumption patterns would be a breeze. Fertility, after all, seems biological—hard-wired into us in deep Darwinian ways. But I would guess that it is easier to change fertility than lifestyle.

PERHAPS our salvation lies in the other part of the equation—in the new technologies and efficiencies that could make even our wasteful lives benign, and table the issue of our population. We are, for instance, converting our economy from its old industrial base to a new model based on service and information. Surely that should save some energy, should reduce the clouds of carbon dioxide. Writing software seems no more likely to damage the atmosphere than writing poetry.

Forget for a moment the hard-ware requirements of that new economy—for instance, the production of a six-inch silicon wafer may require nearly 3,000 gallons of water. But do keep in mind that a hospital or an insurance company or a basketball team requires a substantial physical base. Even the highest-tech office is built with steel and cement, pipes and wires. People working in services will buy all sorts of things—more software, sure, but also more sport utility vehicles. As the Department of Energy economist Arthur Rypinski says, "The information age has arrived, but even so people still get hot in the summer and cold in the winter. And even in the information age it tends to get dark at night."

Yes, when it gets dark, you could turn on a compact fluorescent bulb, saving three fourths of the energy of a regular incandescent. Indeed, the average American household, pushed and prodded by utilities and environmentalists, has installed one compact fluorescent bulb in recent years; unfortunately, over the same period it has also added seven regular bulbs. Millions of halogen torchere lamps have been sold in recent years, mainly because they cost $15.99 at Kmart. They also suck up electricity: those halogen lamps alone have wiped out all the gains achieved by compact fluorescent bulbs. Since 1983 our energy use per capita has been increasing by almost one percent annually, despite all the technological advances of those years.

As with our homes, so with our industries. Mobil Oil regularly buys ads in leading newspapers to tell "its side" of the environmental story. As the company pointed out recently, from 1979 to 1993 "energy consumption per unit of gross domestic product" dropped 19 percent across the Western nations. This sounds good—it's better than one percent a year. But of course the GDP grew more than two percent annually. So total energy use, and total clouds of CO_2, continued to increase.

It's not just that we use more energy. There are also more of us all the time, even in the United States. If the population is growing by about one percent a year, then we have to keep increasing our technological efficiency by that much each year—and hold steady our standard of living—just to run in place. The President's Council on Sustainable Development, in a little-read report issued in the winter of 1996, concluded that "efficiency in the use of all resources would have to increase by more than fifty percent over the next four or five decades just to keep pace with population growth." Three million new Americans annually means many more cars, houses, refrigerators. Even if everyone consumes only what he consumed the year before, each year's tally of births and immigrants will swell American consumption by one percent.

We demand that engineers and scientists swim against that tide. And the tide will turn into a wave if the rest of the world tries to live as we do. It's true that the average resident of Shanghai or Bombay will not consume as lavishly as the typical San Diegan or Bostonian anytime soon, but he will make big gains, pumping that much more carbon dioxide into the atmosphere and requiring that we cut our own production even more sharply if we are to stabilize the world's climate.

The United Nations issued its omnibus report on sustainable development in 1987. An international panel chaired by Gro Harlem Brundtland, the Prime Minister of Norway, concluded that the economies of the developing countries needed to grow five to ten times as large as they were, in order to meet the needs of the poor world. And that growth won't be mainly in software. As Arthur Rypinski points out, "Where the economy is growing really rapidly, energy use is too." In Thailand, in Tijuana, in Taiwan, every 10 percent increase in economic output requires 10 percent more fuel. "In the Far East," Rypinski says, "the transition is from walking and bullocks to cars. People start out with electric lights and move on to lots of other stuff. Refrigerators are one of those things that are really popular everywhere. Practically no one, with the possible exception of people in the high Arctic, doesn't want a refrigerator. As people get wealthier, they tend to like space heating and cooling, depending on the climate."

In other words, in doing the math about how we're going to get out of this fix, we'd better factor in some unstoppable momentum from people on the rest of the planet who want the very basics of what we call a decent life. Even if we airlift solar collectors into China and India, as we should, those nations will still burn more and more coal and oil. "What you can do with energy conservation in those situations is sort of at the margin," Rypinski says. "They're not interested in fifteen-thousand-dollar clean cars versus five-thousand-dollar dirty cars. It was hard enough to get Americans to invest in efficiency; there's no feasible amount of largesse we can provide to the rest of the world to bring it about."

The numbers are so daunting that they're almost unimaginable. Say, just for argument's sake, that we decided to cut world fossil-fuel use by 60 percent—the amount that the UN panel says would stabilize world climate. And then say that we shared the remaining fossil fuel equally. Each human being would get to produce 1.69 metric tons of carbon dioxide annually—which would allow you to drive an average American car nine miles a day. By the time the population increased to 8.5 billion, in about 2025, you'd be down to six miles a day. If you carpooled, you'd have about three pounds of $C0_2$ left in your daily ration—enough to run a highly efficient refrigerator. Forget your computer, your TV, your stereo, your stove, your dishwasher, your water heater, your microwave, your water pump, your clock. Forget your light bulbs, compact fluorescent or not.

I'm not trying to say that conservation, efficiency, and new technology won't help. They will—but the help will be slow and expensive. The tremendous momentum of growth will work against it. Say that someone invented a new furnace tomorrow that used half as much oil as old furnaces. How many years would it be before a substantial number of American homes had the new device? And what if it cost more? And if oil stays cheaper per gallon than bottled water? Changing basic fuels—to hydrogen, say—would be even more expensive. It's not like running out of white wine and switching to red....

There are no silver bullets to take care of a problem like this. Electric cars won't by themselves save us, though they would help. We simply won't live efficiently enough soon enough to solve the problem. Vegetarianism won't cure our ills, though it would help. We simply won't live simply enough soon enough to solve the problem.

Reducing the birth rate won't end all our troubles either. That, too, is no silver bullet. But it would help. There's no more practical decision than how many children to have. (And no more mystical decision, either.)

The bottom-line argument goes like this: The next fifty years are a special time. They will decide how strong and healthy the planet will be for centuries to come. Between now and 2050 we'll see the zenith, or very nearly, of human population. With luck we'll never see any greater production of carbon dioxide or toxic chemicals. We'll never see more species extinction or soil erosion. Greenpeace recently announced a campaign to phase out fossil fuels entirely by mid-century, which sounds utterly quixotic but could—if everything went just right—happen.

So it's the task of those of us alive right now to deal with this special phase, to squeeze us through these next fifty years. That's not fair—any more than it was fair that earlier generations had to deal with the Second World War or the Civil War or the Revolution or the Depression or slavery. It's just reality. We need in these fifty years to be working simultaneously on all parts of the equation—on our ways of life, on our technologies, and on our population.

As Gregg Easterbrook pointed out in his book *A Moment on the Earth* (1995), if the planet does manage to reduce its fertility, "the period in which human numbers threaten the biosphere on a general scale will turn out to have been much, much more brief" than periods of natural threats like the Ice Ages. True enough. But the period in question happens to be our time. That's what makes this moment special, and what makes this moment hard.

DISCUSSION QUESTIONS

1. Would you argue that the problem described here is caused by society or by some other agent?
2. What single approach to this problem would you argue for? What can we really do?
3. Some people believe that in the long run there is really no serious problem since humankind has the ability to create whatever technology is needed to keep up with the world's needs. Do you agree?
4. In what way is social inequality the cause of the problem? Is it the rich? Is it the poor? Is it both?
5. What is "progress"? How would McKibben define "progress"?
6. Are you concerned? If not, why not? If you are, is there anything you can do?

53

The Crisis and Opportunity of Katrina

JOHN A. POWELL, HASAN KWAME JEFFRIES, DANIEL W. NEWHART, AND ERIC STIENS

The Four Questions

1. What is the problem that the authors describe?
2. Is this a *social* problem? Who does this problem disproportionally affect?
3. What is the ultimate cause of this problem?
4. What can be done to lessen the impact of this problem?

Topics Covered

Natural disasters

Social inequality

Racism

Segregation

Poverty

Community

> "You simply get chills every time you see these poor individuals ... so many of these people ... are so poor and they are so black, and this is going to raise lots of questions for people who are watching this story unfold."
>
> —WOLF BLITZER ON CNN, SEPTEMBER 1, 2005[1]

Immediately after Hurricane Katrina ravaged New Orleans and the Gulf Coast, journalists and laypeople struggled to find the words to express their outrage over the situation. In a very real way, the devastation wrought by the storm challenged normative perspectives on race and class in this country. The disturbing images of poor African Americans struggling to survive in an abandoned city and the inadequate response of the government forced uncomfortable thoughts into the national consciousness. Suddenly, race and class mattered, and mattered more than most people were prepared to acknowledge. When a national dialogue began, however, it was clear that existing vocabulary was incapable of explaining what everyone was seeing. Like Wolf Blitzer, many people were left stumbling over the links between race and class, and left trying to figure out why Katrina's destructive force disproportionately impacted African-American and poor communities (Berube and Katz 2005). Katrina raised many questions for everyone who watched the story unfold. How we answer these questions is vitally important.

Soon after the levees broke, politicians and pundits tried feverishly to ease our discontent. They assured us that nature is colorblind, and that the government response, although clearly inadequate, was not a result of racial animus. We were told that class and poverty, rather than race, were the keys to understanding the crisis. Conservatives even went so far as to drape poverty in the rhetoric of welfare-as-dependency, arguing that government assistance had created a culture of victimization. Progressives, for their part, talked about the absence of an adequate safety net to deal with persistent poverty. Yet, questions about *why* African Americans are more likely than whites to be poor, and why poor African Americans are more likely to

SOURCE: From John A. Powell, Hasan K. Jeffries, Daniel W. Newhart, and Eric Stiens, "Towards a Transformative View of Race: The Crisis and Opportunity of Katrina." In *There Is No Such Thing as a Natural Disaster: Race, Class, and Hurricane Katrina*, eds. Chester Hartman and Gregory D. Squires. New York, NY: Routledge, Taylor & Francis Group. Reprinted by Permission.

live in areas of concentrated poverty, are questions that were neither asked nor answered (Muhammed et al. 2004). The mainstream media did make some effort to engage the issue of race, but the resulting discussions either suffered from a reliance on racial stereotypes or failed to move beyond race-based human interest stories. There was little critical discussion of how historical patterns of segregation contributed to the racial layout of the city, and how structures worked together to produce racial disparities and economic inequality.[2]

For far too long, Americans have been trapped in an individualistic mode of thinking about race and racism that requires there to be a racist actor in order for there to be a racist action, and that separates race and class into distinct categories. As the full extent of the damage to New Orleans became clear, the nation as a whole struggled to make sense of the situation by filtering visual images and sound bites through the dominant individualistic framework. Consequently, people asked: Is President Bush a racist or simply incompetent?[3] Were so many poor African Americans affected by the storm because they were poor or because they were black, or was it because of their culture? Would the response to Katrina have been different had New Orleans been mostly white? How could so many things have gone wrong in a country that prides itself on responsibility and opportunity?

Unfortunately, thinking about racism narrowly as a product of individual intent is not particularly helpful. Not only does it tend to be divisive because the conversations that follow often center on assigning blame and finding culpability—rather than on change—but it also diverts attention from the function of structures and institutions in perpetuating disparities, while simultaneously locating racism in the mind of individuals. Furthermore, it limits thinking about the negative effects of structures on non-whites, even though there is strong evidence that suggests that were it not for racialized structures most Americans would be substantially better off (Barlow 2003). Thus, broadening how we think and talk about race is critically important for making sense of today's world. Doing so also raises critical questions about the shrinking middle class, our anemic investment in public space, the meaning of merit in a purported meritocracy, and the promises and failures of our democratic experiment—all of which concern every American. Once we are able to discuss race and racism in these broad terms, we will be able to construct a response not only to the damage wrought by Katrina, but also to that which occurs across the country every day.

Katrina was a nightmare for the Gulf Coast, but it also created a unique opportunity to rebuild the Gulf Coast region in a way that transforms preexisting, long-lasting structural arrangements. More specifically, it created an opportunity for reexamining the connections between race and class, and deciphering precisely how race has been inscribed spatially into our metropolitan areas. In short, it has provided a rare chance to discuss the links between race, equity, justice, and democracy.

Race, as a transformative tool, can and should be applied to more than just the rebuilding effort in New Orleans. Racialized poverty, segregation, and the decaying infrastructure of our central cities are common problems plaguing urban areas nationwide. Used properly, race allows us to examine how institutional failings affect everyone, and enables us to re-imagine a society where democracy and democratic ideals are not constricted and undermined by structural arrangements....

CLASS AND RACE: PATTERNS OF DAMAGE, PATTERNS OF RESPONSE

... The endless parade of forlorn black faces and desperate and angry black voices emanating from the city made it obvious even to the most skeptical observer that race was a part of the unfolding story. Less obvious, though, was whether or not racism was involved. Consequently, when the mainstream media finally broached the subject, it did not seek to discover how racism contributed to this catastrophe, but only whether it was a factor. Of course, the racism that the media searched for once it began looking was limited only to individuals....

Those who rejected racism as a contributing factor to the disaster, as well as those who knew it was involved, somehow focused so much of their attention on identifying or dismissing the racist behavior of individuals, including that of the President, that the overall discourse on the role of race and racism lacked substance. For the most part, there was no discussion of the myriad ways race informed the social, economic, and political factors that converged long before Katrina made landfall to make New Orleans ripe for a disaster that would hit the city's black residents the hardest. Generally, just about everyone failed to discuss local patterns of residential segregation. They ignored the fact that grossly disproportionate numbers of African Americans lived in neighborhoods that were below sea level. Some pointed out that African Americans comprised 98% of the Lower Ninth Ward, but said little if anything about how this came to be (Greater New Orleans Community Data Center 2005a). Similarly, some noted the sickeningly high poverty rate among the city's black residents, but said nothing about how racialized poverty contributed to the crisis. Neither the concentration of subsidized housing, nor the lack of car ownership among poor blacks—which made it impossible for many African Americans to flee New Orleans because the city's middle-class-oriented evacuation plan was predicated on people leaving in their own vehicles—were mentioned. Racialized disinvestment in schools, public health, and other critical institutions in the core city, which impacts the suburbs as well, has existed for decades, but unlike the wind and the water, it garnered little attention. We do not believe that anyone intended to strand poor blacks in New Orleans. But it was predictable, given that we tend to regard poor people differently than we do others.

The inability of Americans, both white and black, conservative and progressive, to analyze the Katrina disaster in a way that would have rendered visible the central role of structural racism in the disaster resulted from the narrow way in which we tend to understand racism. The normative conceptualization of racism is that it is a deliberately harmful discriminatory act perpetrated by people who possess outmoded racial beliefs. It is the aberrant behavior of white supremacists and is easily identified by the discriminatory intent of perpetrators. Furthermore, it is static. It is an offense committed by a particular person at a specific moment in time. Racism happens, and then it ends.

Racism, however, is as much a product of systems and institutions as it is a manifestation of individual behavior. Indeed, structural arrangements produce and reproduce racial outcomes and can reinforce racial attitudes. The normative conceptualization of racism as primarily a psychological event, therefore, is inadequate and incomplete. Above all else, it fails to consider the ways institutions work together to perpetuate racial disparities (powell 2006). It is hardly surprising, then, that in the discourse on the Katrina crisis, almost no one acknowledged that the connections between public and private institutions over space and time limited the resources and opportunities of the city's black residents and conspired to place them directly in harm's way....

A NEW LENS, A NEW VOCABULARY, A NEW AVENUE FOR CHANGE

In New Orleans, the geography of race led directly to a disproportionate storm impact (Muro and Sohmer 2005). People of color were more likely to be living in areas of lower elevation, and therefore were at greater risk of being affected by flooding. Indeed, areas with less opportunity and high concentrations of poor people of color, such as the Ninth Ward, experienced the most damage (powell, Reece, and Gambhir 2005). Consequently, a purely class analysis is incapable of fully explaining what happened. The people who lived in these areas were poor, but they were also African-American, and like many other African-American urban dwellers, they had been concentrated in a depressed, low-opportunity area.

There is no doubt that the United States is facing a crisis of racialized poverty. Indeed, the way race plays out through our use of space and how that impacts access to opportunities for people of color is *the* civil rights struggle for the twenty-first century. Crises are tragic and horrible, but also offer tremendous opportunities for growth and change. Throughout the Gulf Coast, especially in New Orleans, we have an amazing opportunity to rebuild collectively. We can, if we choose to, rebuild the political and economic structures of that city, as well as the physical ones, with justice and equity in mind. We can, in fact, use a racially aware response to transform the region into a place where justice thrives, and New Orleans in particular into a metropolitan area that avoids many of the pitfalls that have caused other metropolitan areas to stagnate. An equitable metropolis is a thriving metropolis.

This process must be both coordinated and democratic. Without public participation, reconstruction will benefit those who already have a disproportionate voice in this country: primarily rich, wealthy, and white men. We must think about voting rights for displaced citizens to ensure that the most affected communities play a central role in reconstruction. A large part of this process will be developing a communications and technology infrastructure to allow the public access to information and the decision-making process.

First, we must eliminate concentrated poverty. Prior to Katrina, New Orleans had one of the highest rates of concentrated poverty in the country, second highest among the nation's 50 largest cities (Berube and Katz 2005). Most of these areas need to be rebuilt, but they must not be rebuilt as static replicas of what they were, nor can this opportunity be used as an excuse for displacing residents. There is tension here, to be sure, between the right of displaced residents to return and creating affordable housing that is not concentrated in a few sections of the city. This tension, however, can be mitigated in large part by involving those most affected by the storm in the planning process.

There are multiple proposals on the table for how this can be accomplished including: housing voucher programs, expanding the Low Income Housing Tax Credit, inclusionary zoning, and models based on previously successful programs such as the Gautreaux experiment in Chicago and HUD's Moving to Opportunity program (Polikoff 2006; Rosenbaum and DeLuca 2000). It is not within the scope of this chapter to critique the merits of each of these. However, it is clear that racial integration as well as economic integration needs to be a part of any effort to build mixed-income communities. The very scale of the rebuilding that needs to occur can be an advantage, as well as a challenge, because it offers an opportunity for a re-envisioning of what an integrated and livable city might look like.

Also, a racially and economically just framework has to focus on access to opportunity. It is clear that this needs to occur in New Orleans and the Gulf Coast area, but it also must include the thousands of displaced residents who may or may not return voluntarily. We run the risk of simply shifting the black urban poor from one city to another, and from one opportunity-deprived community to another. The planning process must include provisions for connecting the "Gulf Coast Diaspora" to affordable housing, job training, economic opportunities, quality education, transportation, and heath care. We must treat the citizens of New Orleans as a democracy should—that is, with equal opportunity for all.

We must focus on providing access to opportunity explicitly, both during the reconstruction process as well as afterwards. The reconstruction of the Gulf Coast will be labor-intensive and require tens of thousands of people working in tandem in order to be successful. Current and former residents of the Gulf Coast should be hired first for these jobs, and local laborers should be paid a living wage. We must put job-training programs into place so that residents are able to take advantage of the opportunities available during the redevelopment. Citizens need to have meaningful oversight of the billions of dollars that will be brought in by private development corporations to guarantee that development occurs in a way that benefits their communities, and not simply the shareholders of these companies.

Transit problems were a principal reason that the violent impact of Katrina was so disproportionately shouldered by poor African Americans. We must not allow this to occur in the future, and state and local authorities should implement immediate plans for the evacuation of residents who lack access to personal transportation. Transit also remains a key factor in connecting people with parts of metropolitan areas where opportunity flourishes, job growth is occurring, and high-quality schools exist. The expansion of public transit has to be a priority going forward.

Access to educational opportunities significantly affects well-being later in life. Not only must the public schools of New Orleans be repaired, but planners must think proactively about the linkages between residential integration and school integration. In 2003–2004, for example, 46.9% of public schools in Orleans Parish were in the "Academically Unacceptable" category, while only 5.7% of schools across Louisiana were in this category (Greater New Orleans Community Data Center 2005b). Planners must begin to consider how residential segregation, which leads to school segregation, is affecting these test scores, causing all of our children to suffer now, which in turn shuts them out from opportunity for their futures.

Lastly, health and environmental concerns are going to remain a part of life in the Gulf Coast area for decades. Officials need to take all precautions to ensure the safety of workers involved in the cleanup and redevelopment process. They need to mandate uniform standards for cleanup so that some communities do not disproportionately shoulder the burden of exposure to toxins. Most importantly, there needs to be a long-term monitoring and a grievance system established to ensure the health and safety of Katrina survivors, one that provides affordable access to healthcare if health problems arise in the future. The scale of redevelopment that is needed offers an opportunity to construct the city and surrounding areas with the next hundred years in mind. There have been sizable advances in "green buildings" over the past two decades, as well as in more environmentally-friendly methods of construction and waste disposal. Redevelopment plans need to include environmental planning as an explicit part of the process.

Throughout this process we must be proactively attentive to the ways in which all of these aspects of opportunity—housing, education, job training, employment, health care and transportation—interact with one another structurally. Adopting a regional approach to planning, therefore, is essential. Segregation, fragmentation, and concentrated poverty create barriers to opportunity for people of color and undermine the vitality and competitiveness of the entire region. An approach to rebuild in a just way must look at regions as a whole unit and create ways to more equitably distribute resources and opportunity throughout the region. It is not a coincidence that some of the poorest parts of New Orleans are also the places where the African-American population is very high (Berube and Katz 2005; Bullard 2005). It is important to consider how segregated space interacts with race and poverty, economic health, and democratic norms. The resource disparity between cities and suburbs hurts not only inner-city residents and those who live in areas that have become isolated, but also encourages a dysfunctional, fragmented system. This system encourages destructive competition such as sprawl, inefficient duplication, and divestment in infrastructure and people. The health of the city and older suburbs is linked to the health of the entire region.

Finally, we must keep the discourse of race and racism alive and inclusive, rather than subterranean and divisive. This will take some strategizing, given the current inadequacy of public discourse. We should support national and local media campaigns, community initiatives, grassroots organizations, interdenominational efforts, and political maneuvering to transform our understanding of race and class. We will need to pay particular attention not only to the needs of poor and middle class blacks and other non-whites, but to those of poor and middle-class whites as well. Globalization and devolution place the vulnerable on precarious footing that will require us to work together to recreate more equitable life opportunities in New Orleans and throughout the nation.

CONCLUSION

Given the hesitancy of the United States to confront or discuss race, even after a disaster like Hurricane Katrina brought race to our attention, it is time for a new way of speaking about race and racism. As we stumbled over words to describe the pictures that appeared in our newspapers and on our television sets, we discovered that we did not have an adequate frame for articulating what was going on, not only in New Orleans but also in this country every day. Katrina demonstrated that race and class are still salient in the United States, but discussing and understanding *how* they matter is an important part of envisioning a racially just and democratic society.

What we point towards ... is an idea that racial disparities are recreated by the interactions between structures and institutions. Access to opportunity remains extremely limited for the urban poor who are disproportionately African-American because of the way we have structured our cities, taxation policies, and transportation systems. Space in this country is highly racialized—our central cities and inner-ring suburbs are declining rapidly while our outer-ring suburbs and exurbs grow at a tremendous pace. This, combined with the increasing fragmentation of our governance structures and the willingness of municipalities to wall themselves off from metropolitan areas, has had the effect of redirecting money and resources away from urban communities, which are stagnating economically because of a shrinking tax base, lack of jobs, and declining public schools, and which are being poisoned by antiquated factories, landfills, and other polluting industries....

In the case of Katrina, those in impoverished, segregated, urban neighborhoods were uprooted from their homes as the waters rose but had no chance to escape because policymakers assumed they had the means to escape.

We all know that segregated, poor neighborhoods exist; yet we hesitate to talk about them, if we talk about them at all. We discuss the social problems that arise from having spaces cut off from opportunity, yet we refuse to include race in the discussion. Instead, we embrace any and all alternative explanations for this phenomenon. Too often, we revert to a colorblind stand, talking about poverty in the United States as if it can be divorced from race and is not one of the factors that *creates* racial disparities....

We must create structures in which new possibilities and new interests are free for all to participate in. Doing this will set us on the right course for creating a true democracy, one in which citizenship is a guarantee of access to opportunity, rather than an entrenched caste system....

A new New Orleans has an opportunity to be a model for the entire world. As a historically black city, it can now be at the forefront of addressing some of the patterns of metropolitan development that we argue are central to the way racism still functions in this country. We must evolve our thinking beyond the narrow view of individual acts and actions to a vision of race that accounts for the ways they interact with structures. By doing this, we can make race a useful tool capable of bolstering democracy and ushering in meaningful transformative change that benefits the nation as a whole.

REFERENCES

Alesina, Alberto and Edward Glaeser. 2004. *Fighting Poverty in the US and Europe: A World of Difference*. Oxford: Oxford University Press.

Barlow, Andrew. 2003. *Between Fear and Hope: Globalization and Race in the United States*. New York: Rowman & Littlefield.

Berube, Alan and Bruce Katz. 2005. *Katrina's Window: Confronting Concentrated Poverty Across America* [online]. Washington, D.C.: Brookings Institution [cited January 15, 2006]. http://www.brookings.edu/metro/pubs/20051012_Concentratedpoverty.pdf.

Blitzer, Wolf. 2005. Aftermath of Hurricane Katrina; New Orleans Mayor Pleads for Help; Race and Class Affecting the Crises? [online]. CNN.com. September 1 [cited January 15, 2006]. http://transcripts.cnn.com/TRANSCRIPTS/0509/01/sitroom.02.html.

Bonilla-Silva, Eduardo. 2003. *Racism without Racists: Color-Blind Racism and the Persistence of Racial Inequality in the United States*. Lanham, MD: Rowman and Littlefield.

Bullard, Robert. 2005. Katrina and the Second Disaster: A Twenty-Point Plan to Destroy Black New

Orleans [online]. Atlanta: Clark Atlanta University. December 23, 2005 [cited January 15, 2006]. http://www.ejre.cau.edu/Katrinaupdate.html.

Bullard, Robert and Beverly Wright. 2005. *Legacy of Unfairness: Why Some Americans get Left Behind* [online]. Atlanta: Clark Atlanta University. September 29, 2005 [cited January 15, 2006]. http://www.ejrc.cau.edu/Exec%20Summary%20Legacy.html.

Clyne, Meghan. 2005. President Bush Is "Our Bull Connor," Harlem's Rep. Charles Rangel Claims [online]. *New York Sun*. September 23, 2005 [cited January 15, 2006]. http://www.nysun.corn/article/20495.

Fagan, Amy. 2005. NAACP fights Bush Social Security Plan [online]. *The Washington Times*. April 11, 2005 [cited January 17, 2006]. http://washingtontimes.com/national/20050411-103458-1817r.htm.

Foner, Philip and David Roediger. 1989. *Our Own Time: A History of American Labor and the Working Day*. New York: Greenwood Press.

Ford, Richard. 1995. The Boundaries of Race: Political Geography in Legal Analysis. 107 *Harvard Law Review* 449, 451.

Frye, Marilyn. 1983. *The Politics of Reality: Essays in Feminist Theory*. Freedom, CA: The Crossing Press.

Greater New Orleans Community Data Center. 2005a. Lower Ninth Ward Neighborhood: People & Household Characteristics [online]. Greater New Orleans Community Data Center. March 23. 2005 [cited January 15, 2006]. http://www.gnocdc.org/orleans/8/22/people.html.

——2005b. Orleans Parish: Schools and Testing [online]. July 12, 2005 [cited January 15, 2006]. http://www.gnocdc.org/orleans/education.html.

Guinier, Leni and Gerald Torres. 2002. *The Miner's Canary: Enlisting Race, Resisting Power, Transforming Democracy*. Cambridge, MA: Harvard University Press.

Katznelson, Ira. 2006. *When Affirmative Action Was White: An Untold History of Racial Inequality in Twentieth-Century America*. New York: W.W. Norton & Company.

Klinenberg, Eric. 2002. *Heat Wave: A Social Autopsy of Disaster*. Chicago: University of Chicago Press.

Laguerre, Michel. 1999. *Minoritized Space: An Inquiry Into the Spatial Order of Things*. Berkeley, CA: Institute of Governmental Studies Press.

Mahoney, Martha R. 2003. Class and Status in American Law: Race, Interest, and the Anti-Transformation Cases. *Southern California Law Review*, May.

Manza, Jeff. 2000. Race and the Underdevelopment of the American Welfare State. *Theory and Society* 29:819–832.

Massey, Douglas and Nancy Denton. 1993. *American Apartheid: Segregation and the Making of the Underclass*, Cambridge, MA: Harvard University Press.

Mazza, Patrick and Eden Fodor. 2000. Taking Its Toll: The Hidden Casts of Sprawl in Washington State [online]. Climate Solutions [cited February 8, 2006]. http://www.climatesolutions.org/pubs/pdfs/sprawl.pdf.

Muhammed, Dedrick, Attieno Davis, Meizhu Lui, and Betsy Leondar-Wright. 2004. The State of the Dream 2004: Enduring Disparities in Black and White [online]. United for a Fair Economy [cited February 8, 2006]. http://www.faireconomy.org/press/2004/StateoftheDream2004.pdf.

Muro, Mark and Rebecca Sohmer. 2005. New Orleans after the Storm: Lessons from the Past, a Plan for the Future [online]. Washington, D.C.: Brookings Institution [cited January 15, 2006]. http://www.brookings.edu/metro/pubs/20051012_New Orleans.pdf.

Orr, Larry, Judith D. Feins, Robin Jacob, Eric Beecroft, Lawrence F. Katz, Jeffrey B. Liebman, and Jeffrey R. Kling. 2003. Moving to Opportunity Interim Impacts Evaluation [online]. U.S. Department of Housing and Urban Development Office of Policy Development and Research [cited February 8, 2006]. http://www.huduser.org/Publications/pdf/MTOFullReport.pdf.

Polikoff, Alex. 2006. *Waiting for Gautreaux*. Chicago, IL: Northwestern University Press.

powell, john a. 2004. *Regional Equity, Race, and the Challenge to Long Island* [online]. The Kirwan Institute for the Study of Race and Ethnicity [cited February, 8 2004]. http://www.kirwaninstitute.org/multimedia/presentations/LongIslandERASEThursdayMay6.ppt.

—— 2006, Rebuilding Communities and Addressing Structural Racism [tentative title]. *Clearinghouse Review*, May–June.

powell, john a, Reece Jason and Gambhir, Samir. 2005. *New Orleans Opportunity Mapping—An Analytical Tool to Aid Redevelopment,* Columbus, OH: The Kirwan Institute for the Study of Race and Ethnicity.

Roediger, David. 1999. *The Wages of Whiteness: Race and the Making of the American Working Class*. New York, NY: Verso.

Rosenbaum, James and DeLuca, Stefanie. 2000. Is Housing Mobility the Key to Welfare Reform? Lessons from Chicago's Gautreaux Program [online]. Washington, D.C.: Brookings Institution [cited February 8, 2005]. http://www.brookings.edu/dybdocroot/es/urban/rosenbaum.pdf.

Rosner, Jay. 2001a. Expert report submitted on behalf of Intervening Defendants (Student Intervenors), *Grutter v. Bollinger*, 137 F. Supp. 2d 821. E.D. Mich. March 27, 2001. No. 97-75928.

Rosner, Jay. 2001b. Expert Reports on Behalf of Student Intervenors: Disparate Outcomes by Design: University Admissions Tests. 12 *La Raza Law Journal* 377, 377–386.

Rusk, David. 1999. *Inside Game/Outside Game: Winning Strategies for Saving Urban America.* Washington, D.C.: Brookings Institution.

Schill, Michael and Wachter, Susan. 1995. The Spatial Bias of federal Housing Law and Policy: Concentrated Poverty in Urban America. 143 *University of Pennsylvania Law Review* 1285, 1286–1290.

The Center for Social Inclusion. 2005. Thinking Change: Race, Framing and the Public Conversation on Diversity. What Social Science Tells Advocates About Winning Support for Racial Justice Policies [online]. New York [cited February 8, 2006]. http://www.kirwaninstitute.org/projects/DAP%20report%20ThinkingChange.pdf.

Yancey, George. 2003. *Who Is White? Latinos, Asians, and the New Black/Non-Black Divide.* Boulder, CO: Lynne Rienner Publishers.

Young, Iris Marion. 1990. *Justice and the Politics of Difference.* Princeton, NJ: Princeton University Press.

—— 2001. Equality of Whom? Social Groups and Judgments of Injustice. *Journal of Political Philosophy*, Vol. 9. Issue 1, March: 1–18.

—— 2003. Responsibility and Structural Injustice. Prepared for Princeton University political theory workshop. December 4.

ENDNOTES

1. Transcript from CNN's "The situation room": http://transcripts.cnn.com/TRANSCRIPTS/0509/01/sitroom.O2.html.
2. This is in some ways reminiscent of when President Bush stated that the Social Security program should be changed because it discriminates against blacks. "African-American males die sooner than other males do, which means the system is inherently unfair to a certain group of people. And that needs to be fixed," Mr. Bush said at a Social Security event Jan. 11, 2005. His rationale was that African Americans do not live as long as white people in the U.S., so they receive fewer benefits. Yet, there was no inquiry into why this disparity in life expectancy exists and what might be done to address that disparity.
3. Kanye West's comments, Sep. 2, 2005—find full story at http://www.washingtonpost/com/wp-dyn/content/article/2005/09/03/AR2005090300l65.html.

DISCUSSION QUESTIONS

1. How did the geography of race and socioeconomic deprivation come to disproportionally affect African Americans in Hurricane Katrina? Moreover, what does the experience of Katrina tell us about who is most affected by "Natural Disasters"?
2. The authors argue that Hurricane Katrina unveiled the problem of racism and its long–running influence on the history of the city of New Orleans. What was racism's connection to this "Natural Disaster"? Do you agree with the authors that race and racism were the main features of this tragedy? Why or why not?
3. What are some ways the authors suggest that society might better protect its most vulnerable

and least well off citizens from "Natural Disasters" like Katrina?

4. What can the most economically vulnerable citizens do to protect themselves against situations like Katrina? (Or, are they wholly reliant on the government for assistance in such predicaments?)
5. "Natural Disasters" do not exist because these acts invariably affect the most economically and socially vulnerable members of society: Their lives are most seriously impacted by such tragedies. Respond to this statement.

54

Haiti: A Survivor's Story

LAURA WAGNER

1. What is the problem that Wagner describes in this selection?
2. Is this a *social* problem? Who does this problem affect?
3. What is the cause of this problem?
4. What can society do about this problem? Are there any suggestions that would alleviate the way this problem affects its sufferers?

Topics Covered

Global poverty

Natural disasters and economic vulnerabilities

Community resilience

Hope

I was sitting barefoot on my bed, catching up on ethnographic field notes, when the earthquake hit. As a child of the San Francisco area, I was underwhelmed at first. "An earthquake. This is unexpected," I thought. But then the shaking grew stronger. I had never felt such a loss of control, not only of my body but also of my surroundings, as though the world that contained me were being crumpled.

I braced myself in a doorway between the hallway and the kitchen, trying to hold on to the frame, and then a cloud of darkness and cement dust swallowed everything as the house collapsed. I was surprised to die in this way, but not afraid. And then I was surprised not to be dead after all. I was trapped, neither lying down nor sitting, with my left arm crushed between the planks of the shattered doorway and my legs pinned under the collapsed roof. Somewhere, outside, I heard people screaming, praying, and singing. It was reassuring. It meant the world hadn't ended.

I want you to know that before the earthquake, things in Haiti were normal. Outside Haiti, people hear only the worst—tales that are cherry-picked, tales that are exaggerated, tales that are lies. I want you to understand that there was poverty and oppression and injustice in Port-au-Prince, but there was also banality. There were teenage girls who sang along hilariously with the love ballads of Marco Antonio Solis, despite not speaking

SOURCE: From Laura Wagner, "Haiti: A Survivor's Story." This article first appeared in Salon.com, at http://www.salon.com. An online version remains in the Salon archives. Reprinted by permission.

Spanish. There were men who searched in vain for odd jobs by day and told never ending *Bouki* and *Ti Malis* stories and riddles as the sun went down and rain began to fall on the banana leaves. There were young women who painted their toenails rose for church every Sunday, and stern middle-aged women who wouldn't let me leave the house without admonishing me to iron my skirt and comb my hair. There were young students who washed their uniforms and white socks every evening by hand, rhythmically working the detergent into a noisy foam. There were great water trucks that passed through the streets several times a day, inexplicably playing a squealing, mechanical version of the theme from *Titanic*, which we all learned to ignore the same way we tuned out the overzealous and confused roosters that crowed at 3 a.m. There were families who finished each day no further ahead than they had begun it and then, at night, sat on the floor and intently followed the Mexican *telenovelas* dubbed into French—their eyes trained on fantastic visions of alternate worlds in which roles become reversed and the righteous are rewarded, dreaming ahead into a future that might, against all odds, hold promise.

I need to tell you these things, not just so that you know, but also so that I don't forget.

I think I was under the rubble for about two hours. Buried somewhere in what had been the kitchen, a mobile phone had been left to charge, and now it kept ringing. The ring tone was sentimental, the chorus of a pop love song. There was something sticky and warm on my shirt. I thought it was *sòs pwa*, a Haitian bean soup eaten over rice, which we'd had for lunch. I thought it was funny that *sòs pwa* was leaking out of the overturned refrigerator and all over me. I thought, "When I get out, I will have to tell Melise about this." Melise was the woman who lived and worked in the house. I spent a large part of every day with her and her family—gossiping and joking, polishing the furniture with vegetable oil, cooking over charcoal, and eating pounded breadfruit with our hands. She said my hands were soft. Her palms were so hard and calloused from a lifetime of household work that she could lift a hot pot with her bare hands. She called me her third daughter. I thought Melise would laugh to see me drenched in her *sòs pwa* from the bottom hem of my shirt up through my bra. It took me some time to figure out that what I thought was *sòs pwa* was actually my blood. I wrung it out of my shirt with my free right hand. I couldn't tell where it was coming from.

Melise did not make it out of the house. She died, we assume, at the moment of collapse. According to others, who told me later, she cried out, "*Letènel,* oh *letènel*!" and that was all. (The word is Creole for the French "*l'Éternel,*" a cry out to God.) She had been folding laundry on the second floor—the floor that crumbled onto the first floor, where I was pinned, thinking wildly of *sòs pwa.* Melise worked and lived in that house for 15 years. She dreamed of one day having her own home and being free. She talked about it all the time. She died in the wreckage of a place she did not consider her home.

I want to write everything down—those mundane remembrances of how life was before—because I am afraid that as time passes, people will become fossilized, that their lives and identities will begin to be knowable only through the facts of their deaths. My field notes are buried in that collapsed house. Those notes are an artifact, a record of a lost time, stories about people when they were just people—living, ordinary people who told dirty jokes, talked one-on-one to God, blamed a fart on the cat, and made their way through a life that was grinding but not without joy or humor, or normality. I don't want my friends to be canonized.

I had been in Port-au-Prince for six months, conducting research on household workers and human rights. As a young American woman not affiliated with any of the large organizations that dominate the Haitian landscape, I was overwhelmed every day by the fierce generosity of Haitians. People who had little were eager to share their food, their homes, their time, their lives. Now I'm cobbling together this narrative—these nonconsecutive remembrances—in surreal and far-removed settings: first a hospital bed in South Miami, then a Cinnabon-scented airport terminal, now a large public university during basketball season. I can't do anything for those same people who gave of themselves so naturally and unflinchingly. My friends, who for months insisted on sharing whatever food they had made, even if I had already eaten, promising me "just a little rice" but invariably giving more. My friends, who walked me to the taptap (taxi) stop nearly every day.

Now that the first journalistic burst has ended, now that the celebrity telethons have wrapped, the stories you hear are of "looters" and "criminals" set loose on a post-apocalyptic wasteland. This is the same story that has always been told about Haiti, for more than 200 years, since the slaves had the temerity to not want to be slaves anymore. This is the same trope of savagery that has been used to strip Haiti and Haitians of legitimacy since the Revolution. But at the moment of the quake, even as the city and, for all we knew, the government collapsed, Haitian society did not fall into Hobbesian anarchy. This stands in contradiction both to what is being shown on the news right now and to everything we assume about societies in moments of breakdown.

In the aftermath of the earthquake, there was great personal kindness and sacrifice in the midst of natural and institutional chaos and rupture. My friend Frenel, who worked cleaning and maintaining the house, appeared within minutes to look for survivors. He created a passage through the still-falling debris using only a flashlight and a small hammer—the kind you would use to nail a picture to a wall. Completely trapped, the nerves in my left arm damaged, I could not help him save me. He told me calmly, "Pray, Lolo, you must pray," as he broke up the cement and pulled it out, piece by piece, to free me. Once I was out, he gave me the sandals off his own feet. As I write this, I am still wearing them. At the United Nations compound, where Frenel ultimately guided and left me, everyone sat together on the cracked asphalt, bleeding and dazed, holding hands and praying as the aftershocks came. A little boy who had arrived alone trembled on my lap. Another family huddled under the metallic emergency blanket with us. Their child looked at me warily—a foreigner, covered in blood and dusted white with cement powder. His grandmother told him, "*Ou mèt chita. Li malad, menm jan avek nou.*" (You can sir. She's sick, too, just like us.)

Social scientists who study catastrophes say there are no natural disasters. In every calamity, it is inevitably the poor who suffer more, die more, and will continue to suffer and die after the cameras turn their gaze elsewhere. Do not be deceived by claims that everyone was affected equally—fault lines are social as well as geological. After all, I am here, with my white skin and my U.S. citizenship, listening to birds outside the window in the gray-brown of a North Carolina winter, while the people who welcomed me into their lives are still in Port-au-Prince, within the wreckage, several of them still not accounted for.

As I sat waiting to be flown out, trying to convince myself that I was just another injured person using up scant food and resources, a non-Haitian man whom I presumed worked for the U.N. approached me.

"Can you do me a favor?" he asked. "Could you write something down?"

I nodded, and he handed me a pen and paper.

"Tear the paper in half, and on the first half write 'unidentified local female' in block letters. Then on the second piece of paper write the same thing."

I looked up. There were bodies loaded into the back of a pickup truck. The woman's floral-print dress was showing and her feet were hanging out. There were not enough sheets and blankets for the living patients, never mind enough to adequately wrap the dead. The U.N. guy looked at me and sort of smiled as I numbly tore the paper and wrote.

"After all, you need something to do. All the bars are closed," he said.

I stared at she bodies on the truck, and I hated him. I did not know which, if any, of my friends had survived. I imagined the people I love—Marlène, one of my best friends; Damilove, the mother of my god-daughter—wrapped up in some scrap of cloth with their feet hanging out and some asshole tagging them with a half-piece of scrap paper that says they are anonymous, without history, unknown.

I am telling you two things that seem contradictory: that people in Haiti are suffering horribly, and that Haitians are not sufferers in some preordained way. What I mean is that suffering is not some intrinsic aspect of Haitian existence, it is not something to get used to. The dead were once human beings with complex lives, and those in agony were not always victims.

In Haiti I was treated with incredible warmth and generosity by people who have been criminalized, condemned, dehumanized, and abstractly pitied. They helped me in small, significant ways for the six months I was there, and in extraordinary ways in the hours after the quake. Now I cannot help them. I cannot do anything useful for them from here, except to employ the only strategy that was available to us all when we were buried in collapsed houses, listening to the frantic stirrings of life aboveground: to shout and shout until someone responds.

DISCUSSION QUESTIONS

1. What is the main point of Laura Wager's first-person account of the Haitian earthquake and its aftermath?
2. How do you understand the author's sentiment when she writes that "Suffering is not some intrinsic aspect of Haitian existence, it is not something to get used to"?
3. The author is clear about the tremendous affliction that the earthquake caused and the immense challenges that lay ahead for the Haitian people. Yet, she is *hopeful* in her belief in the strength of the island's people. What role does "hope" play in your assessment of social problems? Is it impractical to say that hope plays an indelible role in this process?

55

The Elderly: A Demographic Tidal Wave

LAURENCE J. KOTLIKOFF AND SCOTT BURNS

The Four Questions

1. Is there a problem here? What exactly is it?
2. Is it a *social* problem? What are the possible effects on large numbers of people and on society?
3. What are some of the factors causing this problem?
4. Do the authors give some way to alleviate the problem?

Topics Covered

Baby boomers

Life expectancy

Birth rates

Dependency ratio

Replacement rate

Elderly

Age distribution

SOURCE: From Laurence J. Kotlikoff and Scott Burns, *The Coming Generational Storm: What You Need to Know about America's Economic Future*, pp. 1–29.

Old age isn't so bad when you consider the alternative.

—MAURICE CHEVALIER

THE DEMOGRAPHIC TIDAL WAVE

Like it or not, ready or not, everyone reading this [selection] will experience the greatest demographic change in human history. In less than a century, the United States will move from being "forever young" to being "forever old." The largest part of the change will happen in the next thirty years as the baby boomers retire. The most dramatic changes will be experienced outside the United States, throughout the entire industrialized world.

You can get a visceral idea of what we are facing by considering an extreme: the rising population of people at least 100 years old. By mid-century the U.S. centenarian population will exceed 600,000.[1] That's ten times the number of centenarians around today.

Housing them will require a city slightly larger than the current population of Washington, D.C. (567,000), slightly smaller than the current population of San Francisco (735,000), nearly four times the current size of Anaheim, California (165,000), and nearly equal to the combined populations of Abilene, Texas (110,000), Akron, Ohio (222,000), Albany, New York (105,000), Allentown, Pennsylvania (113,000), and Amarillo, Texas (105,000). Indeed, if you check the long list of cities in America with populations of at least 100,000, only 18 are large enough to accommodate the advancing legion of centenarians.[2]

Why is this happening?

We can only remind you of the old proverb: "Be careful what you wish for. You may get it."

Imagine that you were alive in 1900. Life expectancy at birth was 47 years. The median age of all Americans (half younger, half older) was only 22.9 years. Only 4.1 percent of the population was 65 or older. Life was a constant battle. It was a struggle to be born. It was a struggle to survive infancy, let alone survive childhood. It was still another struggle to survive adulthood. It was common for a father and mother to survive one or more of their children. A husband could lose both his wife and a newborn child during childbirth. The only certainty was that survival and longevity exacted a major toll in grief.

Presented with the magical Monkey's Paw—the one with the power to grant three irreversible wishes—your first wish would have been obvious: *Let life be longer.*

And your wish would have come true.

Today, life expectancy at birth is about 76 years, a gain of 29 years. Life expectancy at 65 is now 17 years, up from 12 years in 1900. Better still, gains in life expectancy at 65 seem to be accelerating.

As a consequence, the population age 65 and over had reached 12.4 percent by 2000, nearly double the 6.8 percent of population under 5 years of age. In the course of 100 years, children had gone from outnumbering the elderly by three to one, to being rarely seen or heard (Table 55.1).

By the 1950s and 1960s, the number of kids under age 5 still exceeded the number of oldsters. Kids were about 11 percent of the population in those years, while oldsters were less than 9 percent.[3] Fertility rates, however, were hitting their baby boom peaks—levels not seen since the 1920s. With the fertility rate surging toward four children per woman, a rate that would double the population in a generation, some started to worry about too much of a good thing.[4] From 76 million in 1900, our population had doubled to 151 million in 1950. Our population looked poised to double again by the millennium. In fact, it came close: 286 million. It would have exceeded 300 million without the baby bust

TABLE 55.1 Trading Places: Youngsters Decline, While Oldsters Rise

Decade	Population under Age 5	Population Age 65 and over (%)
1900	12.1%	4.1%
2000	6.8	12.4

SOURCE: www/infoplease.com/ipa/a0110384.html.

that started in the 1970s. Today, Social Security's actuaries project we'll hit 300 million in 2006.[5]

Long lives are a great gift, but they can make a very crowded world. They can also make a very hungry world. While there was little worry that America would starve, there had already been warnings that unlimited population growth could mean hunger and starvation in China, India, and Africa.

So you made a second wish on the Monkey's Paw: *Let us all have smaller families.*

And you got your wish.

From birthrates well over 2.1 children per couple, the long-term replacement rate for population, birthrates plummeted. In some countries, birthrates fell so far that many nations in Europe will experience population declines early in this century. In the United States, the decline in births was significant, but we're still hovering near the replacement rate.

Taken by itself, the change in birthrates isn't cataclysmic. Basically, it works to accentuate the effects of the first wish, for longevity. Until the population reaches a steady state, a transition that will be measured in generations, there will be an increase in the number of old people relative to the number of young people.

But this calculation doesn't consider the baby boomers. The proverbial "pig in the python" generation that has dominated American concerns since birth, they came of age as your wish for smaller families was coming true. They had smaller families than their parents. Soon they will be starting to retire. The bumper crop of boomers born in 1946 will be reaching the age at which most people start taking Social Security, 62, in 2008. That's just four years and one presidential election away.

Unfortunately, the number of children coming of age and joining the workforce won't be nearly as large. Basically, all the forces that can enlarge the retired elderly population are in overdrive. The forces that would expand the younger (and working) population paying Social Security and Medicare taxes are in reverse. The result is a kind of perfect demographic storm.

As we said earlier, we'll see the bulk of this change over the next thirty years, but it will continue quite a bit longer. The best way to understand the magnitude of the change is to visualize it in the form of graphs that divide the population in five-year bands from those under 5 years of age all the way to 80 and over (Figure 55.1).

In 1900, the age distribution of our population was similar to what characterized all past human history. It was a pyramid—widest at the bottom and narrowing with each successive five-year interval. Only 4 percent of the population was 65 or older. There were no five-year bands beyond "65 or older." Centenarians were rare. Retirement was short. There were plenty of adult children to sustain the elderly.

By 2000, the age distribution was a very different shape. The pyramid is gone. Today the profile looks more like a house with a very tall roof. The Census Bureau has no bands beyond "80 and older." But if it did, the very top would have a narrow lightning rod—the "100 and over" population—reaching for the heavens. Instead of being the largest group at the base of the pyramid, children under 5 are about the same in number as the other groups all the way up to those in their mid-30s. The 65 and older population is now 12.4 percent of the total.

By 2030, the age distribution has a different shape again. This time it is more like a barrel. It goes almost straight up, with only minor shrinkage to the 60–64 age group. The steady shrinkage from death that defined the traditional pyramid now appears to *begin* at age 65. The population age 65 and over will have grown to 19.4 percent of the population, a huge increase in thirty years. This figure, by the way, is the *intermediate projection* used by Social Security. Other projections, including some by Social Security, have higher figures. Basically the portion of the population age 65 and over will nearly double over the next 30 years.

If the percentage of people age 65 and over nearly doubles in the next thirty years, another part of the population will have to shrink proportionately. And that's the rub: the shrinkage will be in the working-age population, the people who pay employment taxes. Back in 1950 (when we were still worried about runaway population growth), the number of workers per Social Security beneficiary was 16.5. By 2000, the ratio had fallen to 3.4. In the process, most workers started paying more in

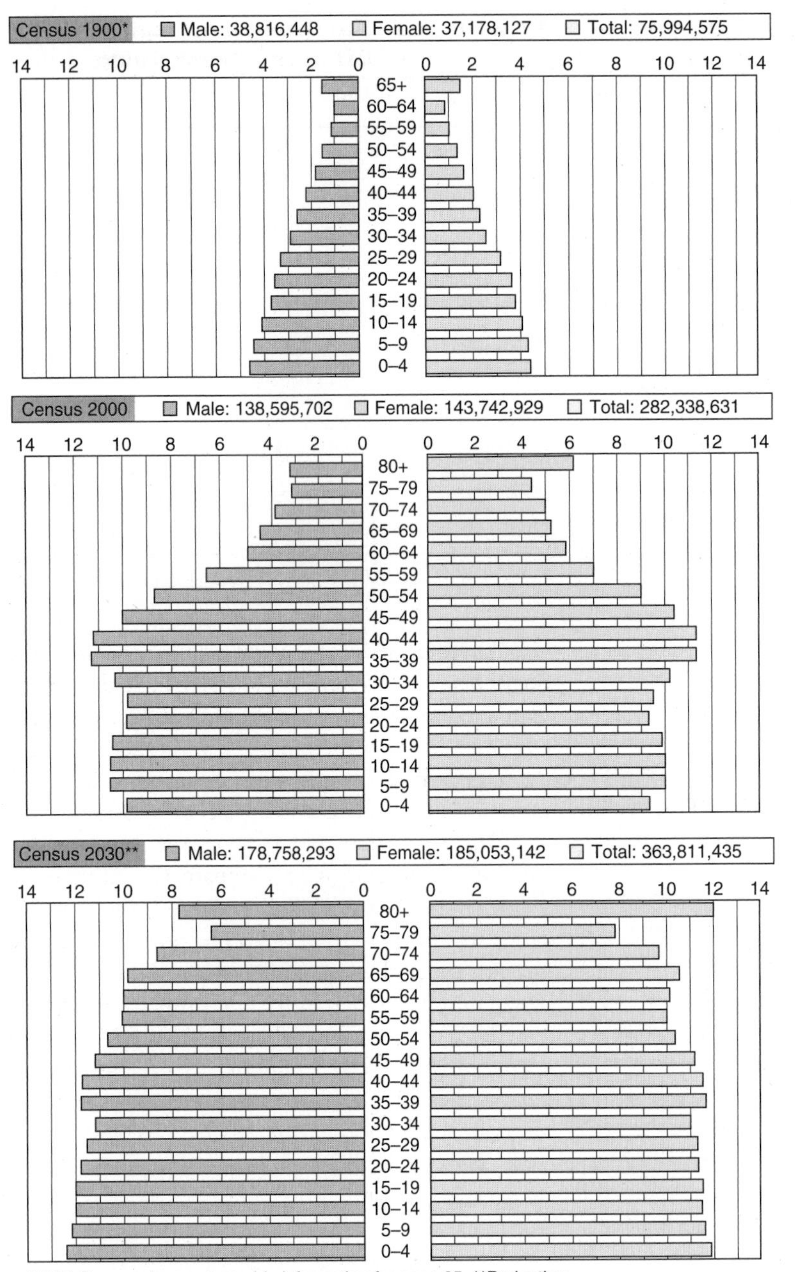

FIGURE 55.1 Population Profiles: U.S. Population by Age and Sex, in Millions.

SOURCE: U.S. Census Bureau.

employment taxes than they pay in income taxes. The employment tax rose fivefold.[6] The wages subject to the Social Security tax rose as well, rising from $3,000 in 1950 to $87,000 in 2003.

Between now and 2030, we'll have the last big surge: the retirement of the boomers. At the end, we'll be close to having only two covered workers per beneficiary. Instead of having sixteen workers chip in to support one senior citizen, we'll have only two. That's a gigantic promise-killing change for Social Security. In only eighty years, the intrinsic cost of supporting retirees will have increased eightfold. In the thirty years to 2030, the intrinsic cost of supporting retirees will rise 70 percent.

Many have seen the coming wave. It is not news. It has been the subject of books, articles, and academic studies for decades. One of the most popular books on the subject, Ken Dychtwald's *Age Wave: The Challenges and Opportunities of an Aging America,* was published in 1989.

Unfortunately, when it comes to action, we're paralyzed. It's as though, having seen the perverse results of the first two wishes on the Monkey's Paw, we're afraid to make the third wish. In fact, the problems we face won't go away. Inaction will make the problem worse, not better. This is a permanent problem, not a temporary one.

GETTING OLD AND STAYING OLD

The aging of America isn't a temporary event. We won't be getting older this year or this decade, and then turning back and getting younger. We are well into a change that is permanent, irreversible, and very long term.

Where we had 35.5 million people age 65 and older in 2000, we'll have 69.4 million in 2030. During those thirty years, the *dependency ratio*—the ratio of those 65 and older to those 20 to 64—will rise from 21.1 percent to 35.5 percent. That's a major increase.

Don't look around for evidence. You won't see it. We're in a quiet period. While the number of senior citizens rises each year, growth in the number of possible workers has been keeping pace since 1985. It will continue to keep up until the boomers start to retire in 2008, just one presidential election in the future.

The dependency ratio—which has hovered around 20 percent since 1985, ranging from a low of 20.6 percent projected for 2005 to a high of 21.6 percent in 1995—will start a major rise around 2015 when it hits 23.8 percent (Table 55.2). By 2030 it will hit 35.5 percent.

And it won't stop there.

TABLE 55.2 The Dependency Ratio Takes Off ... and Keeps Going

Year	Ratio
1985	20.1%
1995	21.6
2005	20.6
2010	21.2
2015	23.8
2020	27.5
2025	31.9
2030	35.5
2040	36.8
2060	39.2
2080	43.2

SOURCE: www.ssa.gov/OACT/nVTR02/V_demographic.html.

The intermediate population projections from the Social Security Administration show the elderly population continuing to grow through 2080, rising from 69.4 million in 2030 to 96.5 million in 2080. During the same period the aged dependency ratio will continue to climb, reaching 43.2 percent by 2080.

The most dramatic way to see how rapidly the nation is aging is to compare the number of seniors—those age 65 and over—to the number of young people. In 2000 there were 82 million people under the age of 20 in the United States. Their numbers dwarfed the 35.5 million seniors.

By 2030, however, there will be 88.6 million young people and 69.4 million seniors, approaching parity. In 2080, only fifty years later, the number of seniors, 96.5 million, will finally exceed the number of young people, 95.8 million....

WHERE HAVE ALL THE CHILDREN GONE?

As years go, 1957 was notable, if primitive.

Remote control for television sets had yet to be invented. This meant the 47.2 million sets in 39.5

million American households had their channels changed *by hand,* if you can imagine that. The sets were also relatively small and offered pictures only in black and white.

Elvis Presley dominated popular music with songs like "Love Me," "Too Much," "All Shook Up," and "Let Me Be Your Teddy Bear." Jack Kerouac's *On the Road* was published, introducing us to drugs and karma at the same time. Velcro was patented, General Foods introduced Tang, Ford introduced the Edsel, and a domestic first-class postage stamp cost only 3 cents.

The Soviet Union, flexing its technological muscles, successfully tested its first intercontinental ballistic missile, which grabbed our attention. Then they put the first satellite in space, *Sputnik,* which *really* grabbed our attention.

Oliver Hardy of Laurel and Hardy fame died, but Lyle Lovett was born. Lyle's birth was not much noticed at the time because 1957 was the year we produced the most babies ever born in America in a single year. We did it at the rate of one baby every seven seconds, closing the year with 4.3 million newborns, from a total population of only 172 million.

You can get an idea of what a staggering feat that was by doing a modern comparison. In spite of having 100 million more Americans today, we produced fewer than 4 million newborns a year through the late 1990s.

The difference is the birthrate. Often measured by total fertility rate—the number of expected lifetime births per woman—this figure also tells us if we can expect a growing or shrinking population in the future. If the total fertility rate is 2.1, population will stabilize after a number of decades. This is the birthrate sought by ZPG (Zero Population Growth) and other organizations interested in population control.

In 1957 our total fertility rate hit a record, 3.68, a figure that approaches the fertility rates of underdeveloped nations. The rate (not to mention the 4.3 million babies) frightened those concerned with population growth and set in motion Malthusian fears of global starvation. In fact, neither the number of babies born nor the total fertility rate has been that high since.

After 1957 the total fertility rate declined. It hit 2.48 in 1970 and dropped to 2.27 in 1971. And it was well below replacement rate through most of the 1970s and early 1980s. It bottomed at 1.74 in 1976. The baby bust of the seventies and the Birth Dearth of the eighties replaced the baby boom of the fifties (see Figure 55.2).

If longer life expectancies and lower birthrates can radically reshape the age distribution of a population, causing major changes in the number of retirees depending on younger workers, the juxtaposition of a baby boom with a period of baby bust works to accentuate the change still further.

Today, we teeter around the replacement rate, surrounded by ominous portents of lower birthrates in the future. As we'll show shortly, a combination of higher education, later marriages, and delays in the birth of the first child may reduce future total fertility rates below the 2.1 replacement rate. If that happens, our future will have the stresses currently being faced in Europe, Japan, and China.

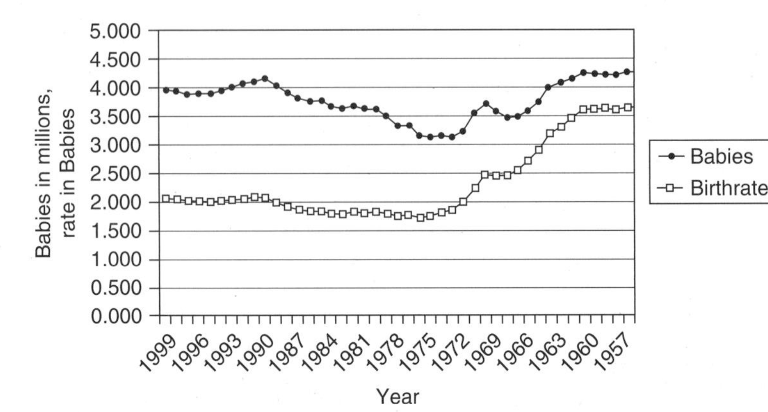

FIGURE 55.2 From Boom to Near Bust.

SOURCE: Population Reference Bureau a-FERTJJSFerti/1.x/s.

However, the operative word here is *may*. Whatever problems we face due to the rising number of elderly people, the only certainty is that we can't expect much sympathy from the rest of the world. Elsewhere, the problems are far worse. If we don't adapt and cope, there will be no one to bail us out.

You can understand this if you divide the nations of the world in two dimensions: life expectancy and birthrate. Let's make life expectancy the vertical axis and birthrate the horizontal axis, creating four quadrants (Figure 55.3). Setting the cross-hairs at a life expectancy of 60 and a birthrate of 2.1 we find two quadrants of disaster, one of major upheaval, and one of possible Panglossian balance. In the United States, we're on the tattered edge of Panglossian balance—and still face a future with gigantic generational imbalances.

Like Sisyphus, most of the world plugs on in a traditional Malthusian Hell. Life expectancies are low because public health efforts are limited, incomes are low, and starvation is a daily possibility. Birthrates are high to cope with the losses.

The starkest picture of the future comes from Postmodern Malthusian Hell, dominated by the states of the former Soviet Union. There, *both* life expectancies and birthrates have plummeted. The total fertility rate has fallen to 1.3, and life expectancies for men dropped to 57 years. As a consequence, the population of Russia could fall from its current 148 million to only 58 million by 2040—less than forty years off. This is a decline of 60 percent. That's what happened during the plague years of fourteenth-century Europe. The only difference is that Russia, with forty years, will have time to bury its dead. Cities hit by the plague lost so many people in such a short time that burying the dead became impossible.[7] …

EMPTY NESTS

Robert Schroeder died an enviable death.

Widowed at 82, he spent the last two years of his life in active contact with his children, grandchildren, and great grandchildren. All lived in the same city, less than an hour from him, and all loved his company. He worked part time, drove his car, enjoyed the security of living independently in a continuing care community, and embraced every day as a new opportunity.

On the morning of a party to celebrate his eighty-fourth birthday, he had a mild stroke and inhaled some material into his lungs, starting an

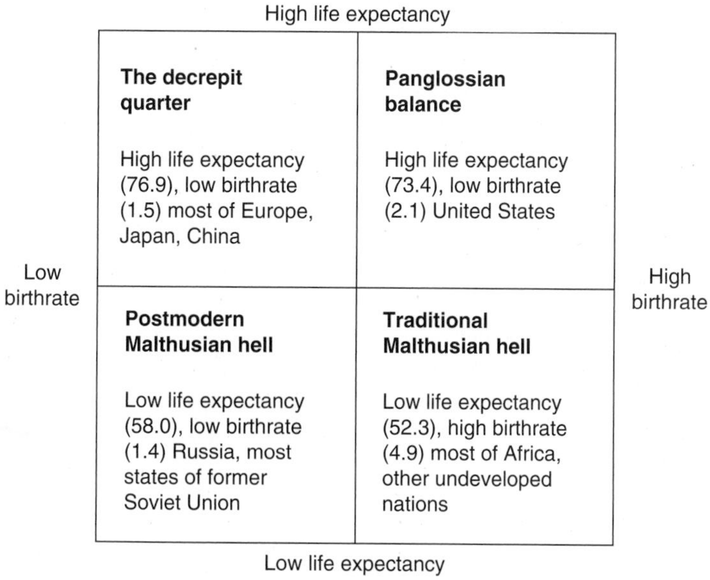

High life expectancy

Low birthrate	High birthrate
The decrepit quarter High life expectancy (76.9), low birthrate (1.5) most of Europe, Japan, China	**Panglossian balance** High life expectancy (73.4), low birthrate (2.1) United States
Postmodern Malthusian hell Low life expectancy (58.0), low birthrate (1.4) Russia, most states of former Soviet Union	**Traditional Malthusian hell** Low life expectancy (52.3), high birthrate (4.9) most of Africa, other undeveloped nations

Low life expectancy

FIGURE 55.3 The Four Quadrants of Demography.

infection. Within four days the infection was overwhelming his body. No amount of antibiotic could stop it. Three days after that, conscious to the very last, he looked silently at each of the eleven family members gathered around him. Each was touching him. And then he was gone.

George Blasius followed him only a few months later. After George married his third wife, they moved to Sarasota. When she had a stroke and had to stay in a nursing home, he visited every day, rolling the oxygen tank he needed to cope with his emphysema. Finally, he fell in the night and was hospitalized for exhaustion. With a son from his first marriage, a stepson and three sons from his second marriage, and a near-son from his third marriage, he was not without support or help. "The boys" (who ranged in age from 38 to 62) were spread around the country—Maine, New Jersey, Texas, Colorado, and California.

They flew in, one after another. The ones with air miles contributed tickets, and they rotated their visits. In the final months they pooled enough money so that one of the brothers could stay in Florida. On Father's Day three boys went to lunch with him. The next day two boys and a grandson helped him in a physical therapy session. He died that night.

Several months later, to escape the conflict over where he should be buried, the boys gathered again. This time they took a yacht volunteered by a family friend into the Gulf of Mexico. They shared straight shots of George's beloved Canadian Club and anointed the water. Then, one by one, they shared the task of dropping his ashes into the glassy swells of the gulf.

These are two of our stories. If you don't have a story about the death of a parent, you are either very young or very lucky. Like it or not, these stories are destiny: each of us will have a story about how our mother and father died.

The hard part is that the stories are changing. As marriages shorten, as mothers have fewer children, and as we become geographically dispersed around the country, the care and support elderly people could hope for—no, expect—from their children is shrinking. Sometimes it simply isn't there. They are old and alone.

We've worried about social connection for a long time. David Reisman did it in *The Lonely Crowd* in 1950; Philip Slater did it in *The Pursuit of Loneliness* in 1970. Most recently, Robert Putnam did it in *Bowling Alone* (2000). These concerns were rooted in observation of broad social change. They could always be dismissed as dour speculation or the inevitable social nagging of the intellectual community.

Today our worry has hard demographic roots. Table 55.3 examines the marital status of women in later life. While two-thirds of women in their 50s are married, the odds drop precipitously when they are 70 and older. Then, only one woman in three is still married. With higher divorce rates among younger women, the proportion is likely to grow in the future.

Most men escape this fate. But they do it the hard way, by dying.

After spouses, the next measure of connection is children. In traditional societies, adults had children to care and provide for them in their old age. Today, we have fewer children, they are geographically mobile, and their sense of obligation may have been reduced by divorce.

You can get some idea about the odds against family care by considering the findings of a study. In "How Much Care Do the Aged Receive from Their Children?" the lives of 5,000 elderly people were examined.[8] In addition to considering whether they were single or married, the study examined the number of children, the number of daughters, how many lived less than an hour away, and the amount

TABLE 55.3 Women Alone

Marital Status	50–59	60–69	70 and over
Never married	5.0	3.7	4.2
Currently divorced	16.6	11.5	5.6
Currently widowed	8.2	21.6	54.5
Total unattached	29.8	36.8	64.3
Total married	67.4	61.3	35.0

SOURCE: Rose M. Kreider and Jason M. Fields, "Number, Timing, and Duration of Marriages and Divorces: 1996," *Current Population Reports*, February 2002.

of time their children spent with them on a weekly basis. The researchers learned:

- That 22.4 percent of the elderly have no children
- That another 19.8 percent had only one child
- That 40.5 percent have no daughters
- That most of the single elderly live by themselves
- That 10 percent of those with children had no children within an hour's distance
- That over 40 percent of the "vulnerable" elderly lived by themselves
- That less than 20 percent of the elderly live with their children
- That institutionalized elderly have less contact with children, not more
- That transfers of money from child to parent (or vice versa) were rare, regardless of income

It's not a pretty picture—and that's the way things were over ten years ago. As Phyllis Diller once joked, "Always be nice to your children because they are the ones who will choose your rest home."[9]

What will familial care and support look like in 2030? That's a matter of speculation. One optimistic speculation is that the relatively high number of siblings that baby boomers have may offset their smaller number of children. Others have speculated that definitions of "family" or "kin" might include the stepchildren as well as natural children. The counter-argument is that siblings, however willing, may not be capable of providing care and support because they will be relatively old themselves. How many newspaper stories have you read about a 60- or 70-year-old daughter being exhausted by caring for her 90-year-old mother? Similarly, while the number of people we are connected to may expand through two or more marriages, multiple marriages also work to expand the number of aging people children and stepchildren may be called on to help. Call it a wash.

Kenneth W. Wachter, a researcher at the University of California at Berkeley, examined the issue using computer simulations of family patterns.[10] The results offered some hope: the increasing number of stepchildren and step-grandchildren will offset most of the anticipated decline of close biological kin. Numbers, however, aren't the same as actual support. Unless the kinship ties with stepchildren become stronger in the future, they won't be a substitute for biological kin.

What we are heading toward is a nation in which familial and institutional caring will be strained and reduced at the same time. The natural caring of family is strained and reduced by the same demographic change that will put the nonfamily substitutes—Social Security and Medicare—under extreme financial pressure.

BIG, BLUE, AND WRINKLED ALL OVER

America is getting older. But it isn't alone.

The entire planet is aging, much of it faster than we are. Most of the developed world is aging faster for two reasons. First, Americans aren't contenders in the Life Expectancy Olympics, so we're not increasing our elderly population as fast as other countries. Second, our birthrate, while flagging, isn't so low that our population will shrink over the next fifty years.

In most of Europe, it will. The problems we face are mild compared to those of many other nations.

You won't see much about this on TV. There, the cardinal rule is, "If it bleeds, it leads." You won't read much about this in most newspapers either. Stories about population, life expectancy and birthrates tend to get put in the Sunday "think piece" bin.

In fact, virtually nothing in the daily news will change how we live and what we do more than the global population shift now under way. No story is more newsworthy. We are heading toward a planet that is big, blue, and wrinkled all over.

How old? How wrinkled?

Very. At the Second World Assembly on Ageing in Madrid in 2002, those attending heard astounding figures. By 2050 the number of older persons *in the world* would exceed the number of young people *for the first time in history*. The number of children is expected to grow by only 140 million, but the number of people in their 60s is expected to grow by 600 million. The number of people in their 70s will grow by 448 million. The number of people in their 80s will grow by 253 million. And the number of people at least age 90 will grow by 56 million. While children worldwide outnumber older people by three to one today, the ratio will be one to one by 2050. This is a gigantic change.

Another way to see the shift is to examine the changes in median age between 2000 and 2050. As you can see from Table 55.4, we're going from a world in its 20s to a world in its 30s—and that includes the entire less developed world. The developed world will age much more. By 2050 the median age in Japan will be 53. The entire developed world will be in its 40s.

The world of 2050 is how the developed nations of the world could be described in 1998. Over the next fifty years, nations like Mexico, Peru, Brazil, and India will catch up. Currently Mexico has nearly eight children (those under 15) for every elderly person (those 65 and over). In India, the ratio is seven; in Brazil, six; and in Peru, eight. After centuries where few people were old and nurturing children was the primary social concern, children will become a small minority around the globe. The primary social concern will be caring for the elderly.

Meanwhile, the developed nations will continue to age. In Europe, there will be 2.1 old people *per child* by 2025, increasing to 2.6 by 2050. Europe will definitely be "the old country."

This will not be a phase. We will be older forever.

Two powerful forces, rising life expectancy and declining birthrates, drive the aging of the planet.

While the population of the United States will still be rising in 2050, the population of the *entire developed world* will be slightly lower than it is today (Table 55.5). Some areas and countries will experience sharp declines. The population of Europe (as defined in the UN. report), for instance, will shrink by 124 million, an amount greater than the current population of France and Italy combined.[11] The only way both countries will avoid an increasingly quaint (and abandoned) countryside is to sell their homes and towns to wealthy foreigners seeking a second or third home.

The population of Russia will fall by 41.2 million, or 28 percent—and some consider that estimate wildly optimistic. Lacking a Tuscany or Provence, Russia could develop whole areas like the "Buffalo

TABLE 55.4 The Advancing Median Age in Years

Country or Area	2000	2025	2050	Change 2000/2050
World	26.5	32.0	36.2	9.7
United States	35.5	39.3	40.7	5.2
China	30.0	39.0	43.8	13.8
More developed regions	37.4	44.1	46.4	9.0
Europe	37.7	45.4	49.5	11.8
Japan	41.2	50.0	53.1	11.9

SOURCE: United Nations, *World Population Ageing: 1950-2050* (New York: United Nations, 2002).

TABLE 55.5 The Aging World and Coming Population Decline

Country or Area	2000	2025	2050	%60+ in 2050
United States	283.2	346.8	397.1	26.9
More developed regions	1,191.4	1,218.8	1,181.1	33.5
Europe	727.3	683.5	603.3	36.6
France	59.2	62.7	61.8	32.7
Germany	82	78.9	70.8	38.1
Italy	57.5	52.4	43	42.3
Russia	145.5	125.7	104.3	37.2
Japan	127.1	123.8	109.2	42.3
China	1,275.1	1,470.8	1,462.1	29.9

SOURCE: United Nations, *World Population Ageing: 1950-2050* (New York: United Nations, 2002).
NOTE: Population figures are in millions.

Commons" that some American environmentalists dream of for the mountain states.[12] They would like to see large areas of New Mexico, Colorado, Wyoming, and Montana return to vast, unfenced grazing areas unpopulated by humans. The difference is that Russia will be an ecological disaster area.

Japan will likely be the oldest nation in the world, with 42 percent of its population 60 or older. Fifteen percent will be 80 and over. The centenarian population will be approaching 1 million—even though Japan's population will be down to 109 million. In a single century Japan will have gone from a nation with 4.6 children for every old person to a nation with 3.4 old people per child.

One of the most dramatic changes will be in the most populous nation in the world, China. While its population will continue to grow in the early part of this century, it will be declining by 2050. It will also be getting older fast, with the median age rising from 30 in 2000 to 39.0 in 2025 and 43.8 in 2050. By 2025, China will have 287.6 million people who are at least 60 years old, a number that exceeds the entire current population of the United States. By 2050, China will have 437 million people who are at least 60 years old. Ranked as a subnation, Old China will be larger than the total population of any nation in the world, except India.

In addition, China's population will be profoundly unbalanced because males will continue to outnumber females at all ages through the early 60s. Referred to as the "missing girls problem," the imbalance has its roots in a long-standing cultural preference for male children, a preference that was exacerbated by a government edict limiting the number of children a woman could have. Today, males age 15 to 59 outnumber females in the same age range by 25.4 million. By 2025 the mismatch will reach 30.7 million, 31.3 million by 2050.

Examining the gender gap closely, one demographer noted that "by 2020, for example, the surplus of China's males in their 20s will likely exceed the entire female population of the island of Taiwan!" The demographer calculates that about one in six Chinese men will either have to find a bride outside China or remain unmarried.

Will we see population growth anywhere?

Perhaps. While most of us read or hear the news of HIV prevalence in Africa and envision a complete population collapse, the United Nations researchers have projected that the current devastation will recede and that high birthrates will lead to major population growth there. A United Nations report from 2000 says, "Even in Botswana, where HIV prevalence is 36 percent or in Swaziland and Zimbabwe, where it is above 25 percent, the population is projected to increase significantly between 2000 and 2050: by 37 percent in Botswana, 148 percent in Swaziland and 86 percent in Zimbabwe."[13]

Similarly, less developed nations not burdened with major HIV incidence are likely to continue their high birthrates. Basically, while the less developed world will continue to be a Malthusian nightmare of high birthrates, the developed world will be plunging into a new demography: inescapable old age.

Is it the end of the world?

No. History and life are full of surprises. But the raw numbers tell us two important things. The first is that human demography, driven by simple changes in life expectancy and childbearing, is about to trump the power of economic growth. The second is that the shift in the United States may be major, but it's minor compared to what the rest of the developed world will experience.

If we get into trouble, there will be no rescue. We're on our own.

ENDNOTES

1. John W. Rowe and Robert L. Kahn, *Successful Aging* (New York: Dell), 1999, p. 6. Other sources have smaller projections. The United Nations, which regularly projects population and demographic figures for every nation and area of the world, estimates 473,400 centenarians in

the United States by 2050 and 3,218,900 in the world.

2. U.S. Department of Commerce, *Statistical Abstract of the United States,* (Washington, D.C.: U.S. Government Printing Office, 1997), table 46.
3. "Population Distribution by Age, Race, Nativity, and Sex Ratio," www.infoplease.com/ipa/A0110384.html.
4. Population Reference Bureau, "U.S. Fertility Trends: Boom and Bust and Leveling Off," www.prb.org.
5. Old Age, Survivors and Disability Insurance, "Assumptions and Methods Underlying Actuarial Estimates," 2002, www.ssa.gov/OACT/TR/TR02/V_demographic.html.
6. The combined employer/employee tax rate rose from 3 percent in 1950 to 15.3 percent in 2000.
7. Barbara W. Tuchman, *A Distant Mirror: The Calamitous Fourteenth Century* (New York: Knopf, 1978), p. 97.
8. Laurence J. Kotlikoff and John N. Morris, "How Much Care Do the Aged Receive from Their Children? A Bimodal Picture of Contact and Assistance," in *The Inquires in the Economics of Aging* (Cambridge, Mass.: National Bureau of Economic Research, 1989). ed. David A. Wise.
9. http://www.brainyquote.eom/quotes/quotes/p/phyllisdill36115.html.
10. Kenneth W. Wachter, "2030s Seniors: Kin and Step-Kin" (working paper, Berkeley University of California, April 1995), paper www.demog.berkeley.edu/'~wachter/WorkingPapers/kinpaper.pdf
11. United Nations Report, "World Population Aging: 1950-2050," 2002.
12. Richard S. Wheeler, *The Buffalo Commons* (New York: Forge, 1998).
13. A more recent report took a much darker view and chopped future population estimates in Africa.

DISCUSSION QUESTIONS

1. Kotlikoff and Burns frame the demographic tidal wave as a social problem. Why is this a social problem? Do you believe it is a social problem? Who will be hurt?
2. What accounts for the falling fertility rates of women in the Western, post-industrialized world, and what are some possible consequences of this trend? Does this, itself, create a problem?
3. What unique problems face the United States arising from longevity, according to the authors?
4. Do you see reasons why the authors might be wrong in their predictions?
5. What role, if any, can governments play in addressing this social problem?
6. The real problem identified concerning population is the population explosion in the world. Do the authors agree that this is a serious problem?

Index